Construction Craft Laborer

Level Two

Trainee Guide
Third Edition

PEARSON

Boston Columbus Indianapolis New York San Francisco Amsterdam
Cape Town Dubai London Madrid Milan Munich Paris Montreal Toronto
Delhi Mexico City São Paulo Sydney Hong Kong Seoul Singapore Taipei Tokyo

NCCER
President: Don Whyte
Director of Product Development: Daniele Dixon
Construction Craft Laborer Project Manager: Jamie Carroll
Senior Production Manager: Tim Davis
Quality Assurance Coordinator: Debie Hicks

Pearson
Director, Global Employability Solutions: Jonell Sanchez
Head of Global Certifications/Associations: Andrew Taylor
Associate Director: Sasha Jiwani
Editorial Assistant: Collin Lamothe
Program Manager: Alexandrina B. Wolf
Operations Supervisor: Deidra M. Skahill
Art Director: Diane Ernsberger
Digital Studio Project Managers: Heather Darby, Tanika Henderson
Directors of Marketing: David Gesell, Margaret Waples
Field Marketer: Brian Hoehl

Desktop Publishing Coordinator: James McKay
Permissions Specialist: Adrienne Payne
Production Specialist: Adrienne Payne
Editor: Karyn Payne

Composition: NCCER
Text Fonts: Palatino and Univers
Printer/Binder: LSC Communications
Cover Printer: LSC Communications

Credits and acknowledgments for content borrowed from other sources and reproduced, with permission, in this textbook appear at the end of each module.

Copyright © 2015, 2009, 1999 by NCCER, Alachua, FL 32615, and published by Pearson Education, Inc., New York, NY 10013. All rights reserved. Printed in the United States of America. This publication is protected by Copyright and permission should be obtained from NCCER prior to any prohibited reproduction, storage in a retrieval system, or transmission in any form or by any means, electronic, mechanical, photocopying, recording, or likewise. For information regarding permission(s), write to: NCCER Product Development, 13614 Progress Blvd., Alachua, FL 32615.

Perfect Bound   ISBN-13: 978-0-13-413096-5
                ISBN-10:   0-13-413096-0

PEARSON

Preface

To the Trainee

Being a construction craft laborer (CCL) means working in a diverse industry where you'll be needed for a variety of exciting positions. Opportunities in the construction field abound—just look around. From high-rise buildings to tunnel excavations, you'll see that construction craft laborers are needed everywhere.

As a CCL, you will have the opportunity to work in a wide array of jobs ranging from highway infrastructure projects to assisting carpenters and other craftworkers on construction sites. Construction craft laborers must be able to operate many types of tools, instruments, and equipment as they work on job sites that may change daily. From building bridges to demolition, CCLs will always be in high demand in the ever-expanding construction industry.

According to the Bureau of Labor Statistics, the predicted job growth in the field of construction craft laborer is higher than average. With the proper education and on-the-job-learning, you'll be well equipped to earn a living as a CCL in the booming construction industry.

New with *Construction Craft Laborer Level Two*

This third edition of *Construction Craft Laborer Level Two* features a new instructional design that organizes the material into a layout that mirrors the learning objectives. The new format engages trainees and enhances the learning experience by presenting concepts in a clear, concise manner. For example, trade terms are defined at the beginning of each section in which they appear and each section concludes with a brief section review.

The images and diagrams have been updated to exemplify the most current practices in the industry, with special emphasis on safety. In addition, *Elevated Work* (28207) was replaced with *Working From Elevations* (75122-13). To address green construction practices, this edition of CCL includes the third edition of *Your Role in the Green Environment*, which has been updated to LEED v4.

We invite you to visit the NCCER website at **www.nccer.org** for information on the latest product releases and training, as well as online versions of the *Cornerstone* magazine and Pearson's NCCER product catalog.

Your feedback is welcome. You may email your comments to **curriculum@nccer.org** or send general comments and inquiries to **info@nccer.org**.

NCCER Standardized Curricula

NCCER is a not-for-profit 501(c)(3) education foundation established in 1996 by the world's largest and most progressive construction companies and national construction associations. It was founded to address the severe workforce shortage facing the industry and to develop a standardized training process and curricula. Today, NCCER is supported by hundreds of leading construction and maintenance companies, manufacturers, and national associations. The NCCER Standardized Curricula was developed by NCCER in partnership with Pearson, the world's largest educational publisher.

Some features of the NCCER Standardized Curricula are as follows:

- An industry-proven record of success
- Curricula developed by the industry for the industry
- National standardization providing portability of learned job skills and educational credits
- Compliance with the Office of Apprenticeship requirements for related classroom training (*CFR 29:29*)
- Well-illustrated, up-to-date, and practical information

NCCER also maintains the NCCER Registry, which provides transcripts, certificates, and wallet cards to individuals who have successfully completed a level of training within a craft in NCCER's Curricula. *Training programs must be delivered by an NCCER Accredited Training Sponsor in order to receive these credentials.*

Special Features

In an effort to provide a comprehensive, user-friendly training resource, we have incorporated many different features for your use. Whether you are a visual or hands-on learner, this book will provide you with the proper tools to get started as a contruction craft laborer.

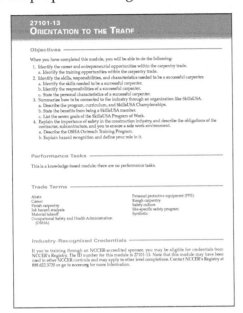

Introduction

This page is found at the beginning of each module and lists the Objectives, Performance Tasks, and Trade Terms for that module. The Objectives list the skills and knowledge you will need in order to complete the module successfully. The Performance Tasks give you an opportunity to apply your knowledge to real-world tasks. The list of Trade Terms identifies important terms you will need to know by the end of the module.

Special Features

Features present technical tips and professional practices from the construction industry. These features often include real-life scenarios similar to those you might encounter on the job site.

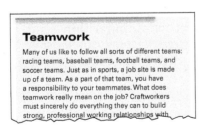

Color Illustrations and Photographs

Full-color illustrations and photographs are used throughout each module to provide vivid detail. These figures highlight important concepts from the text and provide clarity for complex instructions. Each figure reference is denoted in the text in *italics* for easy reference.

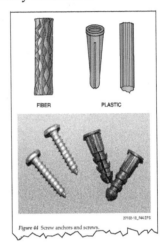

Figure 44 Screw anchors and screws.

Notes, Cautions, and Warnings

Safety features are set off from the main text in highlighted boxes and are organized into three categories based on the potential danger of the issue being addressed. Notes simply provide additional information on the topic area. Cautions alert you of a danger that does not present potential injury but may cause damage to equipment. Warnings stress a potentially dangerous situation that may cause injury to you or a co-worker.

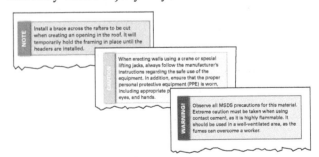

Going Green

Going Green looks at ways to preserve the environment, save energy, and make good choices regarding the health of the planet. Through the introduction of new construction practices and products, you will see how the "greening of the world" has already taken root.

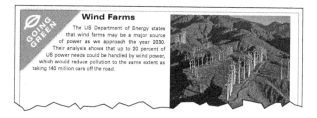

Did You Know?

The Did You Know? features offer hints, tips, and other helpful bits of information from the trade.

> **Did You Know?**
> **Balloon Framing**
> Balloon framing is frequently used in hurricane-prone areas for gable ends. In fact, this type of framing may be required by local building codes.

Step-by-Step Instructions

Step-by-step instructions are used throughout to guide you through technical procedures and tasks from start to finish. These steps show you not only how to perform a task but also how to do it safely and efficiently.

Step 1 Select the proper-size toggle bolt and drill bit for the job.

Step 2 Check the toggle bolt for damaged or dirty threads or a malfunctioning wing mechanism.

Step 3 Drill a hole completely through the surface to which the part is to be fastened.

Step 4 Insert the toggle bolt through the opening in the item to be fastened.

Step 5 Screw the wings onto the end of the toggle bolt, ensuring that the flat side of the wing is facing the bolt head.

Trade Terms

Each module presents a list of Trade Terms that are discussed within the text and defined in the Glossary at the end of the module. These terms are denoted in the text with **bold, blue type** upon their first occurrence. To make searches for key information easier, a comprehensive Glossary of Trade Terms from all modules is located at the back of this book.

> Joists must be doubled where extra loads require additional support. When a partition runs parallel to the joists, a double joist is placed underneath. Joists must also be doubled around openings in the floor frame for stairways, chimneys, etc., to reinforce the rough opening in the floor. These additional joists used at such openings are called trimmer joists. They support the headers that carry short joists called tail joists. Double joists should be spread where necessary to accommodate plumbing.
> In residential construction, floors traditionally

Section Review

The Section Review features helpful additional resources and review questions related to the objectives in each section of the module.

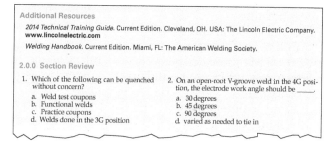

Review Questions

Review Questions reinforce the knowledge you have gained and are a useful tool for measuring what you have learned.

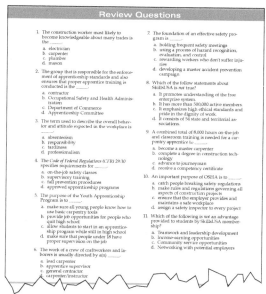

NCCER Standardized Curricula

NCCER's training programs comprise more than 80 construction, maintenance, pipeline, and utility areas and include skills assessments, safety training, and management education.

Boilermaking
Cabinetmaking
Carpentry
Concrete Finishing
Construction Craft Laborer
Construction Technology
Core Curriculum:
    Introductory Craft Skills
Drywall
Electrical
Electronic Systems Technician
Heating, Ventilating, and
    Air Conditioning
Heavy Equipment Operations
Highway/Heavy Construction
Hydroblasting
Industrial Coating and Lining
    Application Specialist
Industrial Maintenance
    Electrical and Instrumentation
    Technician
Industrial Maintenance
    Mechanic
Instrumentation
Insulating
Ironworking
Masonry
Millwright
Mobile Crane Operations
Painting
Painting, Industrial
Pipefitting
Pipelayer
Plumbing
Reinforcing Ironwork
Rigging
Scaffolding
Sheet Metal
Signal Person
Site Layout
Sprinkler Fitting
Tower Crane Operator
Welding

Maritime
Maritime Industry Fundamentals
Maritime Pipefitting
Maritime Structural Fitter

Green/Sustainable Construction
Building Auditor
Fundamentals of Weatherization
Introduction to Weatherization
Sustainable Construction
    Supervisor
Weatherization Crew Chief
Weatherization Technician
Your Role in the Green
    Environment

Energy
Alternative Energy
Introduction to the Power
    Industry
Introduction to Solar
    Photovoltaics
Introduction to Wind Energy
Power Industry Fundamentals
Power Generation Maintenance
    Electrician
Power Generation I&C
    Maintenance Technician
Power Generation Maintenance
    Mechanic
Power Line Worker
Power Line Worker: Distribution
Power Line Worker: Substation
Power Line Worker:
    Transmission
Solar Photovoltaic Systems
    Installer
Wind Turbine Maintenance
    Technician

Pipeline
Control Center Operations,
    Liquid
Corrosion Control
Electrical and Instrumentation
Field Operations, Liquid
Field Operations, Gas
Maintenance
Mechanical

Safety
Field Safety
Safety Orientation
Safety Technology

Management
Fundamentals of Crew
    Leadership
Project Management
Project Supervision

Supplemental Titles
Applied Construction Math
Tools for Success

Spanish Translations
Acabado de concreto: nivel uno
Aislamiento: nivel uno
Albañilería: nivel uno
Andamios
Carpintería:
    Formas para carpintería, nivel
    tres
Currículo básico: habilidades
    introductorias del oficio
Electricidad: nivel uno
Herrería: nivel uno
Herrería de refuerzo: nivel uno
Instalación de rociadores: nivel
    uno
Instalación de tuberías: nivel uno
Instrumentación: nivel uno, nivel
    dos, nivel tres, nivel cuatro
Orientación de seguridad
Paneles de yeso: nivel uno
Seguridad de campo

Acknowledgments

This curriculum was revised as a result of the farsightedness and leadership of the following sponsors:

Industrial Management & Training Institute
Alfred State College
SpawGlass Holding, LP
Maintenance & Construction Technology
Construction Industry Training Council of Washington

This curriculum would not exist were it not for the dedication and unselfish energy of those volunteers who served on the Authoring Team. A sincere thanks is extended to the following:

Mark Bonda
Thomas Murphy
Mark Worrell
John Stronkowski
Dave Perrin

NCCER Partners

American Fire Sprinkler Association
Associated Builders and Contractors, Inc.
Associated General Contractors of America
Association for Career and Technical Education
Association for Skilled and Technical Sciences
Construction Industry Institute
Construction Users Roundtable
Construction Workforce Development Center
Design Build Institute of America
GSSC – Gulf States Shipbuilders Consortium
ISN
Manufacturing Institute
Mason Contractors Association of America
Merit Contractors Association of Canada
NACE International
National Association of Minority Contractors
National Association of Women in Construction
National Insulation Association
National Technical Honor Society
NAWIC Education Foundation
North American Crane Bureau
North American Technician Excellence
Pearson
Pearson Qualifications International
Prov
SkillsUSA®
Steel Erectors Association of America
U.S. Army Corps of Engineers
University of Florida, M. E. Rinker School of Building Construction
Women Construction Owners & Executives, USA

Contents

Module One
Reinforcing Concrete

Explains the selection and uses of different types of reinforcing materials. Describes requirements for bending, cutting, splicing, and tying reinforcing steel and the placement of steel in footings and foundations, walls, columns, and beams and girders. (Module ID 27304-14; 15 Hours)

Module Two
Vertical Formwork

Covers the applications and construction methods for types of forming and form hardware systems for walls, columns, and stairs, as well as slip and climbing forms. Provides an overview of the assembly, erection, and stripping of gang forms. (Module ID 27308-14; 22.5 Hours)

Module Three
Horizontal Formwork

Describes elevated decks and formwork systems and methods used in their construction. Covers joist, pan, beam and slab, flat slab, composite slab, and specialty form systems and provides instructions for the use of flying decks, as well as shoring and reshoring systems. (Module ID 27309-14; 15 Hours)

Module Four
Heavy Equipment, Forklift, and Crane Safety

Covers the safety hazards and precautions necessary when working near heavy equipment. It also covers the general safety requirements for the use of forklifts and cranes. (Module ID 75123-13; 5 Hours)

Module Five
Steel Erection

Covers common safety precautions related to steel-erection work, including controlled decking zones, hazardous materials and equipment precautions, tool safety, and appropriate personal protective equipment. (Module ID 75110-13; 2.5 Hours)

Module Six
Electrical Safety

Describes the basic precautions necessary to avoid electrical shock, arc, and blast hazards. It also describes the lockout/tagout procedure. (Module ID 75121-13; 5 Hours)

Module Seven
Introduction to Construction Equipment

Introduces construction equipment, including the aerial lift, skid steer loader, electric power generator, compressor, compactor, and forklift. An overview of general safety, operation, and maintenance procedures is provided. (Module ID 27406-14; 7.5 Hours)

Module Eight
Rough Terrain Forklifts

Covers the uses of forklifts on construction sites. Includes instructions for lifting, transporting, and placing various types of loads, as well as safety, operation, and maintenance procedures. (Module ID 22206-13; 22.5 Hours)

Module Nine
Oxyfuel Cutting

Explains the safety requirements for oxyfuel cutting. Identifies oxyfuel cutting equipment and setup requirements. Explains how to light, adjust, and shut down oxyfuel equipment. Trainees will perform cutting techniques that include straight line, piercing, bevels, washing, and gouging. (Module ID 29102-15; 17.5 hours)

Module Ten
Elevated Masonry

Describes how to work safely and efficiently on elevated structures. Explains how to maintain a safe work environment, ensure protection from falls, how to brace walls from outside forces, and how to identify common types of elevated walls. Stresses safety around equipment such as cranes and hoists. (Module ID 28301-14; 15 Hours)

Module Eleven
Working from Elevations

Explains the use of fall-protection equipment. It also covers safety precautions related to elevated work surfaces, including ladders, scaffolding, and aerial lifts. (Module ID 75122-13; 5 Hours)

Module Twelve
Your Role in the Green Environment

Geared to entry-level craftworkers, *Your Role in the Green Environment* provides pertinent information concerning the green environment, construction practices, and building rating systems. This edition has been updated to reflect LEED v4 with emphasis on standards for building design and construction. The updated content features contemporary issues such as net zero buildings and an expanded focus on issues relevant to international construction.

Your Role in the Green Environment, LEED v4, Third Edition has been approved by GBCI for 15 hours of general continuing education to support LEED professionals (Module ID 70101-15; 15 Hours)

Glossary

Index

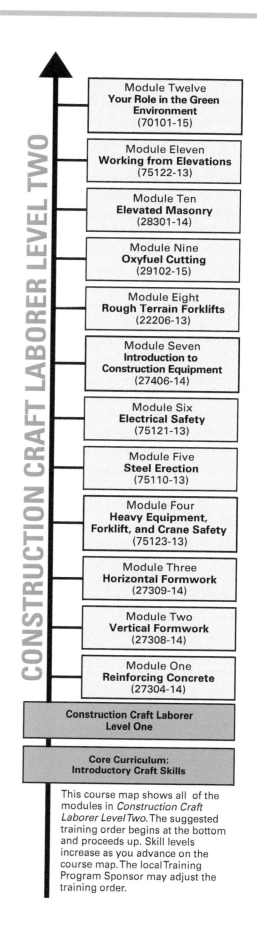

This course map shows all of the modules in *Construction Craft Laborer Level Two*. The suggested training order begins at the bottom and proceeds up. Skill levels increase as you advance on the course map. The local Training Program Sponsor may adjust the training order.

27304-14

Reinforcing Concrete

Overview

In any structure built with concrete, the foundation and any concrete walls and floors will be reinforced. Failure to use the correct amount and type of reinforcing material may result in a catastrophic failure of the structure. The majority of reinforcing material is steel bars that are inserted into the form before the concrete is placed. It is critical to use the correct reinforcing material and supports, place them in accordance with the project drawings, and use the correct methods to tie the reinforcing material. While this process may seem simple on the surface, it requires training and experience in order to do it correctly.

Module One

Trainees with successful module completions may be eligible for credentialing through the NCCER Registry. To learn more, go to **www.nccer.org** or contact us at **1.888.622.3720**. Our website has information on the latest product releases and training, as well as online versions of our *Cornerstone* magazine and Pearson's product catalog.

Your feedback is welcome. You may email your comments to **curriculum@nccer.org,** send general comments and inquiries to **info@nccer.org**, or fill in the User Update form at the back of this module.

This information is general in nature and intended for training purposes only. Actual performance of activities described in this manual requires compliance with all applicable operating, service, maintenance, and safety procedures under the direction of qualified personnel. References in this manual to patented or proprietary devices do not constitute a recommendation of their use.

Copyright © 2015 by NCCER, Alachua, FL 32615, and published by Pearson Education, Inc., New York, NY 10013. All rights reserved. Printed in the United States of America. This publication is protected by Copyright, and permission should be obtained from NCCER prior to any prohibited reproduction, storage in a retrieval system, or transmission in any form or by any means, electronic, mechanical, photocopying, recording, or likewise. To obtain permission(s) to use material from this work, please submit a written request to NCCER Product Development, 13614 Progress Blvd., Alachua, FL 32615.

From *Construction Craft Laborer, Level Two, Trainee Guide*, Third Edition. NCCER.
Copyright © 2015 by NCCER. Published by Pearson Education. All rights reserved.

27304-14
REINFORCING CONCRETE

Objectives

When you have completed this module, you will be able to do the following:

1. List applications of reinforced concrete.
   a. Describe how forces are resisted in concrete through the use of reinforcing bars.
   b. List applications for reinforced structural concrete.
   c. Discuss how posttensioned concrete is created.
2. Describe the general requirements for working with reinforcing steel, including tools, equipment, and fabricating methods.
   a. List general safety precautions when working with reinforcing steel.
   b. Describe the general characteristics of reinforcing steel.
   c. Discuss how reinforcing steel is fabricated.
   d. Explain the purpose of bar supports.
   e. Explain how welded-wire fabric reinforcement is used to reinforce concrete.
3. Describe methods by which reinforcing bars may be bent and cut in the field.
   a. Describe how to cut rebar.
   b. Describe how to bend rebar.
4. Explain the methods for placing reinforcing steel.
   a. Discuss the proper method for tying and splicing reinforcing steel.
   b. Explain the proper procedure for placing reinforcing steel.

Performance Tasks

Under the supervision of your instructor, you should be able to do the following:

1. Use appropriate tools to cut and bend reinforcing bars.
2. Demonstrate five types of ties for reinforcing bars.
3. Demonstrate proper lap splicing of reinforcing bars using wire ties.
4. Demonstrate the proper placement, spacing, tying, and support for reinforcing bars.

Trade Terms

Abutment	Dowel	Schedule
Band	Far face	Simple beam
Bar list	Flat slab	Single-curtain wall
Beam	Girder	Sleeve
Bent	Hickey bar	Span
Bundle of bars	Hook	Staggered splices
Caissons	Inserts	Stirrups
Column	Lapped splice	Strips
Column spirals	Near face	Support bars
Column ties	Pile cap	Temperature bars
Concrete cover	Pitch	Template
Contact splice	Placing drawings	Tie
Continuous beam	Rebar horses	Tie wire
CRSI	Reinforced concrete	Weephole
Double-curtain wall	Retaining wall	

Industry-Recognized Credentials

If you're training through an NCCER-accredited sponsor, you may be eligible for credentials from NCCER's Registry. The ID number for this module is 27304-14. Note that this module may have been used in other NCCER curricula and may apply to other level completions. Contact NCCER's Registry at 888.622.3720 or go to **www.nccer.org** for more information.

Code Note

Codes vary among jurisdictions. Because of the variations in code, consult the applicable code whenever regulations are in question. Referring to an incorrect set of codes can cause as much trouble as failing to reference codes altogether. Obtain, review, and familiarize yourself with your local adopted code.

Contents

Topics to be presented in this module include:

1.0.0 Reinforced Steel Applications 1
   1.1.0 Resistance of Forces by Reinforcing Bars 2
   1.2.0 Use of Reinforced Structural Concrete 4
      1.2.1 Buildings 5
      1.2.2 Bridges 5
   1.3.0 Posttensioned Concrete 6
2.0.0 Working with Reinforcing Steel 9
   2.1.0 General Safety Precautions 10
   2.2.0 Reinforcing Bars 10
   2.3.0 Fabrication 12
      2.3.1 Tools 13
      2.3.2 Fabricated Bars 13
      2.3.3 Bar Lists 13
      2.3.4 Bar-List Information 15
      2.3.5 Sample Bar List 16
      2.3.6 Special Details of Fabrication 16
      2.3.7 Tolerances in Fabrication 17
      2.3.8 Bundling and Tagging 17
   2.4.0 Bar Supports 19
      2.4.1 Steel-Wire Bar Supports 19
      2.4.2 Precast Concrete Blocks 22
      2.4.3 Other Types of Bar Supports 22
      2.4.4 Identification of Bar Supports 24
      2.4.5 Miscellaneous Accessories 24
   2.5.0 Welded-Wire Fabric Reinforcement 24
      2.5.1 Plain-Wire Reinforcement 25
      2.5.2 Deformed Welded-Wire Fabric Reinforcement 26
3.0.0 Bending and Cutting Reinforcing Steel 29
   3.1.0 Cutting Rebar 29
   3.2.0 Bending Rebar 29
4.0.0 Placing Reinforcing Steel 32
   4.1.0 Tying and Splicing Reinforcing Steel 32
      4.1.1 Tying Reinforcing Steel 33
      4.1.2 Splicing Reinforcing Steel 34
   4.2.0 Placing Reinforcing Steel 35
      4.2.1 Placing Bars in Footings and Foundations 35
      4.2.2 Placing Bars in Walls 37
      4.2.3 Placing Bars in Columns 39
      4.2.4 Placing Bars in Beams and Girders 42

Figures and Tables

Figure 1 Rebar placement 3
Figure 2 Stirrup placement 3
Figure 3 Continuous beam 3
Figure 4 Cantilevered beam and beam cage 4
Figure 5 Cantilevered retaining wall 4
Figure 6 Spread footing 4
Figure 7 Stairway 5
Figure 8 Column ties 5
Figure 9 Joist floor system 5
Figure 10 Precast wall unit 6
Figure 11 Beam bridge 6
Figure 12 Rigid-frame bridge 6
Figure 13 Box-girder bridge 7
Figure 14 Examples of tendon profiles 7
Figure 15 Tendons being tensioned 7
Figure 16 An unbonded posttensioned system 8
Figure 17 Reinforcing bar identification 12
Figure 18 Typical bar bends 14
Figure 19 Standard hook details for reinforcing bars 15
Figure 20 Sample bar list 18
Figure 21 Additional sample bar list 20
Figure 22 Examples of rebar labels 21
Figure 23 Typical wire bar supports 23
Figure 24 Typical precast concrete bar supports 24
Figure 25 Typical all-plastic supports 25
Figure 26 Standee support 25
Figure 27 Miscellaneous accessories 25
Figure 28 Typical plain-wire, welded-wire fabric reinforcement
in roll form 26
Figure 29 Manual rebar cutter 29
Figure 30 Rebar being sheared to length by a portable power shears 30
Figure 31 Hickey bar and rebar bender 30
Figure 32 Fabricated hickey bars 30
Figure 33 Bending jig 30
Figure 34 Bending reinforcing bars 30
Figure 35 Multiple bending of closed column ties on a power
stirrup bender 31
Figure 36 Types of ties 33
Figure 37 Nail-head tie 33
Figure 38 Percentage of ties 34
Figure 39 Pliers used for tying rebar 34
Figure 40 Contact splice 34
Figure 41 Spaced-lap splice 35
Figure 42 Square footing 36
Figure 43 Corner bars 36

Figure 44 Pile cap 37
Figure 45 Slab-bolster support 38
Figure 46 Block support 39
Figure 47 Nail support 39
Figure 48 Spacers and spreaders 40
Figure 49 Corner details 41
Figure 50 Dowel template 41
Figure 51 Guide boards 42
Figure 52 Dowel cage 42
Figure 53 Standing column ties 43
Figure 54 Closed column ties 44
Figure 55 Rebar horse 45
Figure 56 Spirals 45
Figure 57 Beam placing sequence 45
Figure 58 Closed stirrup 47
Figure 59 Capped stirrup 47
Figure 60 One-way reinforced slab 47
Figure 61 Two-way reinforced slab 48

Table 1	ASTM Standard Metric and Inch-Pound Reinforcing Bars............11
Table 2	Reinforcing-Bar Steel Types 13
Table 3	Reinforcing-Bar Grades... 13
Table 4	Standard Hook Specifications 16
Table 5	Maximum Prefabricated Radii 19
Table 6	Plain Wire Sizes (*ASTM A82*) 26
Table 7	Common Styles of Plain Welded-Wire Fabric 27
Table 8	Deformed Wire Sizes (*ASTM A496*) 27

Section One

1.0.0 Reinforced Steel Applications

Objective

List applications of reinforced concrete.
a. Describe how forces are resisted in concrete through the use of reinforcing bars.
b. List applications for reinforced structural concrete.
c. Discuss how posttensioned concrete is created.

Trade Terms

Abutment: The supporting substructure at each end of a bridge.

Beam: A horizontal structural member.

Bent: A self-supporting frame having at least two legs and placed at right angles to the length of the structure it supports, such as the columns and cap supporting the spans of a bridge.

Caissons: Piers usually extending through water or soft soil to solid earth or rock; also refers to cast-in-place, drilled-hole piles.

Column: A post or vertical structural member supporting a floor beam, girder, or other horizontal member and carrying a primarily vertical load.

Column spirals: Columns in which the vertical bars are enclosed within a spiral that functions like a column tie.

Column ties: Bars that are bent into square, rectangular, U-shaped, circular, or other shapes for the purpose of holding vertical column bars laterally in place and that prevent buckling of the vertical bars under compression load.

Concrete cover: The distance from the face of the concrete to the reinforcing steel; also referred to as fireproofing, clearance, or concrete protection.

Continuous beam: A beam that extends over three or more supports (including end supports).

Girder: The principal beam supporting other beams.

Reinforced concrete: Concrete that has been placed around some type of steel reinforcement material. After the concrete cures, the reinforcement provides greater tensile and shear strength for the concrete. Almost all concrete is reinforced in some manner.

Retaining wall: A wall that has been reinforced to hold or retain soil, water, grain, coal, or sand.

Simple beam: A beam supported at each end (two points) and not continuous.

Span: The horizontal distance between supports of a member such as a beam, girder, slab, or joist; also, the distance between the piers or abutments of a bridge.

Stirrups: Reinforcing bars used in beams for shear reinforcement; typically bent into a U shape or box shape and placed perpendicular to the longitudinal steel.

Concrete is arguably one of the most important construction materials. Properly installed and reinforced, it can safely serve as the supporting structures for large buildings, bridges, roads, and dams. Without proper reinforcement, however, concrete structures are accidents waiting to happen. Skilled and knowledgeable craftworkers are required to select, place, and tie steel reinforcing bar (rebar) and welded-wire fabric reinforcement in concrete formwork for foundations, walls, floors, beams, columns, and pilings.

Concrete has good compressive strength, but it is relatively weak in tension or if subjected to lateral or shear forces. Many kinds of proprietary reinforcement have been used for concrete in the past. Today, steel is generally the material used. This is because it has nearly the same temperature expansion and contraction rate as concrete. In addition, modern reinforcement conforms to American Society for Testing and Materials (ASTM) International standards that govern both its form and the types of steel used. As an alternative to steel reinforcement, fibers made from steel, fiberglass, or plastic such as nylon are sometimes added to concrete mixtures to provide reinforcement.

As discussed in the module *Properties of Concrete*, concrete is a mixture of cement, fine and coarse aggregates, water, and possibly one or more admixtures. By varying the proportions of the mixture, concrete with different compressive strengths can be obtained. These strengths typically vary from about 2,000 to 6,000 pounds per square inch (psi). Concrete is also available in higher strengths. Concrete usually sets firm in a matter of hours and typically attains design strength in about 28 days. Reinforced concrete is a combination of concrete and steel. Reinforced concrete combines the compression resistance of concrete with the tension and shear resistance of steel.

The adhesion of concrete to the surface of steel reinforcing bars (rebar) and/or welded-wire reinforcement and the resistance provided by the bar deformation, or lugs, keep the bars from slipping through the concrete. This adhesion is called the concrete bond; it makes the two materials act as one.

One of the primary purposes of reinforcing steel is to control cracking caused by tension and shear loads on the concrete. Cracking can also be caused by expansion and contraction of the concrete due to temperature changes and concrete shrinkage. Concrete has a high compressive strength but a low tensile strength, so shrinking of concrete as it cures, along with bending or shear forces, can cause cracking. To avoid cracking, concrete is reinforced with steel, which has a very high tensile strength, so that in combination the final product can resist forces from any direction. Although some cracking is inevitable, the use of reinforcing steel results in small cracks rather than large ones. Unreinforced concrete is likely to develop large cracks, which can lead to concrete failure.

In some instances, reinforcing steel is used strictly to control cracking in a slab, and does not provide any structural support. Welded-wire fabric reinforcement is often used for this purpose, but #3 rebar (in-lb) on 12" centers may be used instead (see the section *Reinforcing Bars* for information on rebar sizes). In that application, the rebar is referred to as temperature steel.

Only the correct amount and type of reinforcement, placed in the correct locations, can serve the intended purpose. Concrete reinforcement must always be selected and placed according to the engineer's drawings and specifications.

1.1.0 Resistance of Forces by Reinforcing Bars

Reinforcing bars are most effectively used in the following applications:

- *Simple beams (slabs, joists, and girders)* – *Figure 1* shows that the top half of the beam is in compression and the bottom half is in tension, so the steel is placed in the lower half, far enough from the bottom to achieve the proper amount of concrete cover (discussed in the section *Working with Reinforcing Steel*). To resist diagonal tension, stirrups are placed vertically across the beam. The stirrups are spaced more closely near the support and farther apart near the middle of the span to offset the force of shear, as shown in *Figure 2*.
- *Continuous beams* – These are beams that deflect downward between supports and have an upward thrust over the supports. Steel is required at the bottom between supports and at the top over supports, as shown in *Figure 3*.
- *Overhang, interior support, and cantilevered beams* – Tension bars must be placed in the top of the overhang and cantilever and carried back into the main span or support. As shown in *Figure 4*, a reinforcing steel framework, known as a beam cage, may be constructed to provide support for the concrete.

> **NOTE**
> The deflection shown in Figures 1, 3, and 4 has been exaggerated for illustration purposes.

- *Cantilevered retaining walls* – Main bars are required on the side toward the earth. See *Figure 5*.

Reinforcing Bars

Reinforcing steel is one means of providing concrete with shear and tension strength. Here, bundles of reinforcing bar are being readied for a concrete slab.

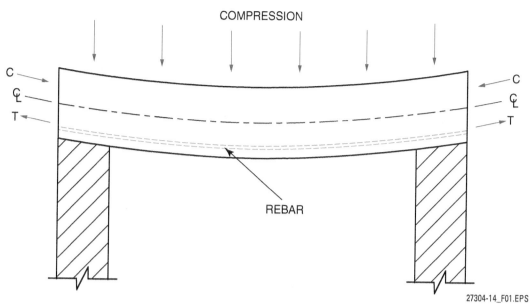

Figure 1 Rebar placement.

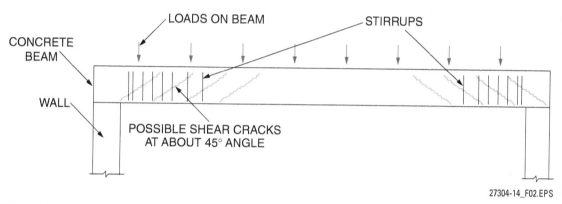

Figure 2 Stirrup placement.

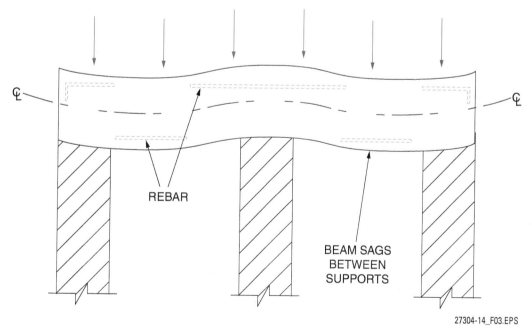

Figure 3 Continuous beam.

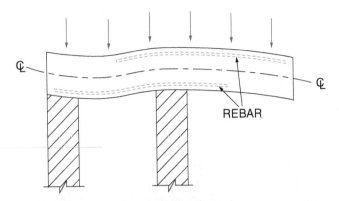

- *Continuous footings* – These carry column loads at two or more points. Straight bars are placed near the top of the slab between the columns. Truss bars are placed under the column ends. Bottom crossbars prevent curling of the concrete.
- *Spread footings* – Bars are placed in two directions (at right angles to each other) and located a prescribed distance from the bottom of the footing, as shown in *Figure 6*. (Note that the deflection is exaggerated.)
- *Inside corners* – Reinforcing bars extend past the corner from each direction and are hooked for anchorage, if necessary.
- *Stairs and landings* – Bars continue across the tension point and are bent into the stair and landing slabs. See *Figure 7*.
- *Columns* – Reinforcing steel for compression forces is most commonly used in columns. If concrete alone were used, the column height would be very limited. Reinforcing bars are about 20 times stronger than an equivalent area of concrete, so they are used to carry part of the column load. Vertical column bars are in compression and will buckle if not restrained. Column ties (*Figure 8*) or column spirals act to prevent buckling.

1.2.0 Use of Reinforced Structural Concrete

Reinforced structural concrete has various applications in multistory building frames and floors, walls, shell roofs, folded plates, bridges, and prestressed or precast elements of all types. The architectural expression of form combined with functional design can be readily achieved with reinforced structural concrete. Architects, engineers, and contractors recognize that there are inherent economic and production values in the use of reinforced structural concrete, as evidenced by the many structures in which it is used.

Figure 4 Cantilevered beam and beam cage.

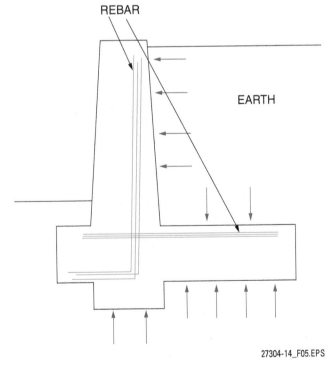

Figure 5 Cantilevered retaining wall.

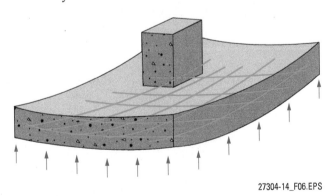

Figure 6 Spread footing.

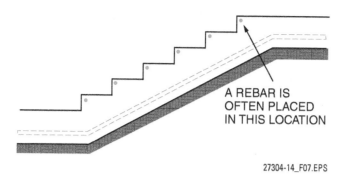

Figure 7 Stairway.

1.2.1 Buildings

Building frames and floors of reinforced concrete are constructed using the joist floor system depicted in *Figure 9*. The illustration shows an exterior rectangular and an interior round column. The reinforced concrete joist floor shown in the figure consists of a series of ribs (small beams) with a top slab, which are all cast at one time with the supporting beams.

Special reinforced concrete construction includes arches, shells, and domes. A concrete arch is curved in one direction only, and the arch thickness is variable, being thinner at the center and thicker toward the supports. A barrel-shell roof is also curved in one direction, but the spans are supported between rigid frames. A dome roof is a slab of double curvature.

Curtain walls are usually prefabricated off site, then delivered for placement. Curtain walls

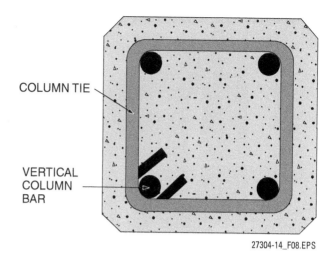

Figure 8 Column ties.

consist of one or more layers of vertical and horizontal reinforcement in a wall supported by the structural steel or concrete frame of the building, independent of the wall below (*Figure 10*).

1.2.2 Bridges

Some bridge construction units resemble those used in building construction and serve similar purposes. These units include footings, piers, caissons, walls, beams, and slabs.

A beam bridge (*Figure 11*) is commonly used for short spans. If a single span is used, the end supports are called abutments or end bents. When

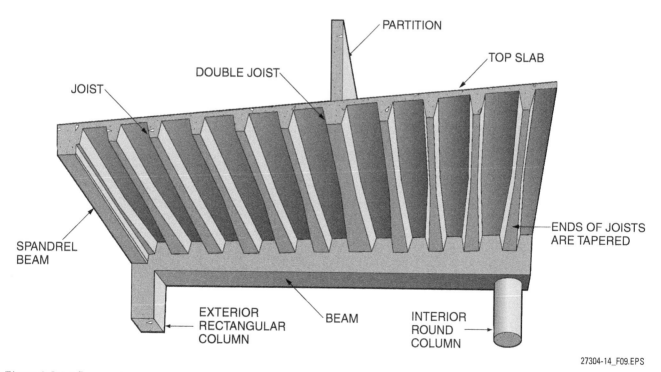

Figure 9 Joist floor system.

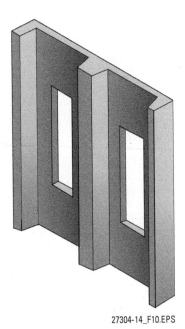

Figure 10 Precast wall unit.

supported by abutments and piers in a manner similar to that of the beam bridge.

1.3.0 Posttensioned Concrete

When a load is applied to a conventional concrete slab such as an elevated parking garage deck, the concrete will tend to sag and may develop cracks. The use of rebar in the slab tends to combat this problem, but is not enough to prevent cracking under heavy load.

The solution is posttensioning, in which a steel tendon is placed in the form with the ends protruding from it. The tendon is placed in accordance with a posttensioning profile drawing developed by a qualified engineer. *Figure 14* is an example of posttensioning profiles for posttensioned beams. The multispan-beam profile diagram shows how the tendons go across the top of a column. A tendon consists of a bar or strand, along with its associated anchoring hardware and sheath. The strand is typically made from steel wires twisted together. Once the concrete has hardened around the tendon, one end of the tendon is anchored off, and the other end is tensioned using a special hydraulic jack (*Figure 15*). When the desired tension has been reached, the other anchor is secured.

There are two types of posttensioning: bonded and unbonded. In bonded posttensioning, a steel

two or more spans are used, the intermediate supports are called piers or intermediate bents.

A rigid-frame bridge (*Figure 12*) generally consists of footings, walls, and a deck slab cast as a unit, which forms a U-shaped element.

The arch bridge has many variations, one of which is the multiple-span, open-spandrel bridge configuration. The box-girder bridge (*Figure 13*) is

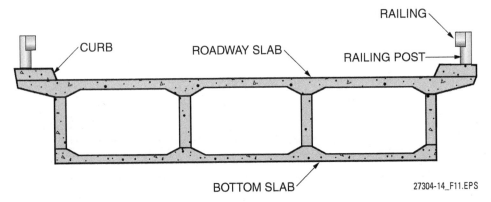

Figure 11 Beam bridge.

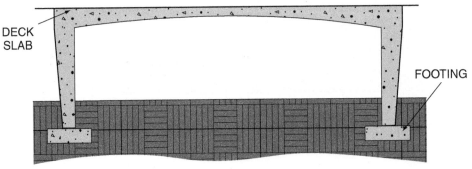

Figure 12 Rigid-frame bridge.

Figure 13 Box-girder bridge.

or plastic duct is inserted in the form. After the concrete is placed in the forms and has hardened, the strand is threaded through the duct. A tensioning force is applied, and then the duct is filled with grout.

In an unbonded system, the strand is covered with a corrosion-inhibiting grease and encased in a waterproof plastic sheath. The entire assembly is placed into the form before the concrete is placed. In some cases, the tendon sheaths are tied to the rebar for support while the concrete is placed (*Figure 16*).

The twisted-wire strand is used in large structures. The threaded bar is more common in smaller structures. A bearing plate and nut are used to anchor the bar.

Figure 15 Tendons being tensioned.

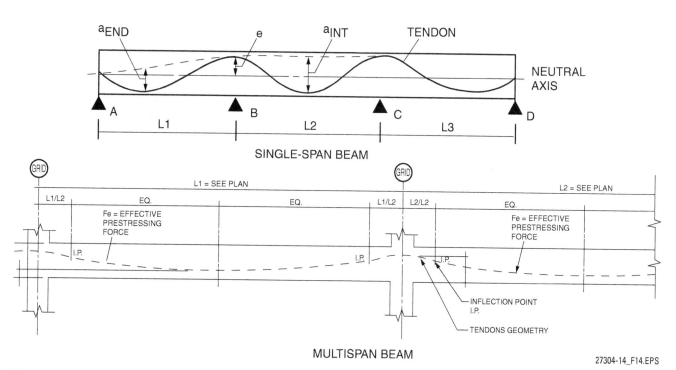

Figure 14 Examples of tendon profiles.

27304-14 **Reinforcing Concrete** Module One 7

Figure 16 An unbonded posttensioned system.

1.0.0 Section Review

1. When resisting compression forces, steel reinforcing bars are stronger than an equivalent area of concrete by about _____.
    a. two times
    b. five times
    c. 10 times
    d. 20 times

2. The concrete bridge type commonly used for short spans is the _____.
    a. box-girder bridge
    b. beam bridge
    c. arch bridge
    d. rigid-frame bridge

3. In a bonded posttensioning system, the duct is filled with grout.
    a. True
    b. False

SECTION TWO

2.0.0 WORKING WITH REINFORCING STEEL

Objective

Describe the general requirements for working with reinforcing steel, including tools, equipment, and fabricating methods.
  a. List general safety precautions when working with reinforcing steel.
  b. Describe the general characteristics of reinforcing steel.
  c. Discuss how reinforcing steel is fabricated.
  d. Explain the purpose of bar supports.
  e. Explain how welded-wire fabric reinforcement is used to reinforce concrete.

Trade Terms

Bar list: A bill of materials for a job site that shows all bar quantities, sizes, lengths, grades, placement areas, and bending dimensions to be used.

Bundle of bars: A bundle consisting of one size, length, or mark (bent) of bar, with the following exceptions: very small quantities may be bundled together for convenience, and groups of varying bar lengths or marks that will be placed adjacent to one another may be bundled together.

CRSI: Concrete Reinforcing Steel Institute.

Dowel: A bar connecting two separately cast sections of concrete. A bar extending from one concrete section into another is said to be doweled into the adjoining section.

Hickey bar: A hand tool with a side-opening jaw used in developing leverage for making in-place bends on bars or pipes.

Hook: A 180-degree (semicircular) or 90-degree turn at the free end of a bar to provide anchorage in concrete. For stirrups and column ties only, turns of either 90 degrees or 135 degrees are used.

Inserts: Devices that are positioned in concrete to receive a bolt or screw to support shelf angles, machinery, etc.

Pitch: The center-to-center spacing between the turns of a spiral.

Placing drawings: Detailed drawings that give the bar size, location, spacing, and all other information required to place the reinforcing steel.

Sleeve: A tube that encloses a bar, dowel, anchor bolt, or similar item.

Tie: A reinforcing bar bent into a box shape and used to hold longitudinal bars together in columns and beams. Also known as stirrup ties.

Tie wire: Wire (generally #16, #15, or #14 gauge) used to secure rebar intersections for the purpose of holding them in place until concreting is completed.

Craftworkers are employed by a contractor to place reinforcing bars in the concrete forms as indicated on the rebar placing drawings. One of the first jobs an apprentice will likely encounter will be the unloading, storing, and handling of reinforcing bars.

Bars must be placed carefully and accurately in order to conform to the requirements of the placing drawings. The engineer's structural drawings often have instructions for certain work to conform to a standard typical detail or note. Placing drawings must clearly indicate such items as top and bottom bars, bars hooked around other bars, and on which side or face of the member the bar is to be located. Bars must be fitted around sleeves, inserts, holes, and other openings as shown on the plans. The craftworker, supervisor, contractor, and job inspector all have the responsibility to see that bars are correctly placed.

So that the strength of any concrete member is not adversely affected, bars must be placed and held in position as shown on the placing drawings. If, for example, the top bars are lowered or the bottom bars are raised ½" more than specified in a 6"-deep slab, the load-carrying capacity could be reduced by 20 percent. A sufficient number of supports should be used in order to place the bars in their specified locations.

Considerable care is necessary if top bars are to be held in the proper position. These bars often interfere with other bars at right angles to the top bars, or with other items (often called embedments) buried in the slab, such as ducts and wiring conduits. The designer usually takes care of these potential problems, but sometimes the top bars cannot be placed to the ½" tolerance and must be relocated.

Bars may receive heavy abuse on the job. In order to ensure that they stay in position, they should be securely wired and held against movement. The supervisor, architect, or other authority inspects and checks the steel for proper placement before the concrete is placed. In all cases, no variation may be made from the placing drawings without proper authorization.

The structural engineer determines the amount of concrete cover for each part of the job. This de-

27304-14 Reinforcing Concrete

cision is based on various considerations, such as the building codes, fire hazards, exposure to weather, and the possibility of corrosion. Where not specified, however, the minimum standards established by the American Concrete Institute (ACI) should be used as guidelines.

ACI 318-95 (Building Code Requirements for Structural Concrete) section 7.7.1 includes the following guidelines for concrete cover:

- 3" at sides where concrete is cast against earth and on bottoms of footings or other principal structural members where concrete is deposited on the ground
- 2" for bars larger than #5 where concrete surfaces would be exposed to the weather after removal of forms or would be in contact with the ground; 1½" for #5 bars and smaller
- 1½" over spirals and ties in columns
- 1½" to the nearest bars on the top, bottom, and sides of beams and girders
- ¾" for #11 and smaller bars on the top, bottom, and sides of joists, and on the top and bottom of slabs where concrete surfaces are not directly exposed to the ground or weather; 1½" for #14 and #18 bars
- ¾" from the faces of all walls not directly exposed to the ground or weather for #11 and smaller bars; 1½" for #14 and #18 bars

Always refer to the structural notes for clearance and coverage requirements.

2.1.0 General Safety Precautions

Developing an overall safety consciousness is very important in any phase of construction work. The following general safety precautions apply to the placing and tying of reinforcing bars:

- Wear proper PPE, including hard hats, boots, and leather gloves.
- Do not wear loose or ragged clothing.
- Ensure your footing is always solid.
- Wear approved fall protection equipment.
- Keep the work area clean.
- Never drop any material to a lower level.
- Block piles of reinforcing steel to prevent sideways movement.
- Wear goggles with the proper shade of lens when cutting with a torch.
- Know and observe the proper hand signals for hoisting equipment. Only qualified riggers and signal persons can perform these tasks.
- Never ride on any material being hoisted.
- Lift bars or bundles properly to avoid strain.
- Always know the location of co-workers.
- Pull any projecting nails.
- When carrying bars with a partner, make sure the load is balanced. Each person should be positioned about one-quarter the length of the rebar from each end. Lift and set all bars in unison.
- Never land hoisted bundles or drop carried bars on formwork.
- Be alert for concrete buggies or any hoisted material that is swinging.
- Do not hoist bundles by the #9-gauge wire wrappings that tie bars together. Use proper slings and chokers. Double chain slings are recommended; nylon slings are not allowed for this application.
- Adequately guy and support reinforcing steel for vertical structures such as piers and columns to prevent collapse.
- Brake and secure rolled-out wire reinforcement to prevent dangerous recoiling action.
- Report all unsafe conditions.
- Always use American National Standards Institute (ANSI)-approved rebar caps on exposed ends of bars.
- Bend down loose ends of tie wires with pliers after each tie. If this is not done, the wire may puncture gloves, boots, or skin.
- Wear gloves when handling rebar. Make sure the gloves are heat resistant; rebar can get very hot lying in the sun.

2.2.0 Reinforcing Bars

Reinforcing bars, often called rebar, are available in several grades. These grades vary in yield strength, ultimate strength, percentage of elongation, bend-test requirements, and chemical composition. In addition, reinforcing bars can be coated with different compounds, such as epoxy, for use in concrete where corrosion could be a problem. To obtain uniformity throughout the United States, ASTM International has established standard specifications for these bars. These grades will appear on bar-bundle tags, in color coding, in rolled-on markings on the bars, and/or on bills of materials. The specifications are as follows:

- *ASTM A615, Standard Specification for Deformed and Plain Carbon-Steel Bars for Concrete Reinforcement*
- *ASTM A996, Standard Specification for Rail-Steel and Axle-Steel Deformed Bars for Concrete Reinforcement* (this standard replaces *A616* and *A617*)

- *ASTM A706, Standard Specification for Low-Alloy Steel Deformed Bars and Plain Bars for Concrete Reinforcement*

The standard configuration for reinforcing bars is the deformed bar. Different patterns may be impressed on the bars, depending on which mill manufactured them, but all are rolled to conform to ASTM specifications. The deformation improves the bond between the concrete and the bar, and prevents the bar from moving in the concrete.

Plain bars are smooth and round without deformations on them and are used for special purposes, such as for dowels at expansion joints where the bars must slide in a sleeve, and for expansion and contraction joints in highway pavement.

Deformed bars are designated by a number in 11 standard sizes (metric or inch-pound), as shown in *Table 1*. The number denotes the approximate diameter of the bar in eighths of an inch or in millimeters. For example, a #5 bar has an approximate diameter of ⅝". The nominal dimension of a deformed bar (nominal does not include the deformation) is equivalent to that of a plain bar having the same weight per foot.

As shown in *Figure 17*, bar identification is accomplished by ASTM specifications, which require that each bar manufacturer roll the following information onto the bar:

- Letter or symbol to indicate the manufacturer's mill
- Number corresponding to the size number of the bar (*Table 1*)
- Symbol or marking to indicate the type of steel (*Table 2*)
- Marking to designate the grade

The grade represents the minimum yield (tension strength) measured in kilopounds per square inch (ksi) that the type of steel used will withstand before it permanently stretches (elongates) and will not return to its original length (*Table 3*). Today, Grade 420 is the most commonly used rebar. Bars are normally supplied from the mill bundled in 60' lengths.

Bar fabrication is accomplished for straight bars by cutting them to specified lengths from the 60' stock. Bent bars are cut to length the same as straight bars, and then they are assigned to a bending machine that is best suited for the type of bend and size of the bar. *Table 1* provides size and weight information for various bars so that proper handling and bending equipment can be selected.

Uncoated reinforcing steel that has not been contaminated by oil, grease, or preservatives will normally rust when stored, even for short lengths of time under cover. A number of studies, some conducted over 70 years ago, have shown that rust and tight mill scale actually improve the bond between the steel and the concrete. Other studies have shown that normal handling (moving, bending, etc.) of extremely rusted reinforcement steel prepares it sufficiently for proper bonding with concrete without additional effort to remove the rust.

To reduce congestion in cast-in-place construction when multiple hooked-end bars converge from several directions, *ASTM A970* permits the use of headed reinforcing bars (HRBs). The head is a rectangular or round steel plate and can be welded, threaded, or forged on the end of the rebar.

Table 1 ASTM Standard Metric and Inch-Pound Reinforcing Bars

Bar Size		Nominal Characteristics*					
		Diameter		Cross-Sectional Area		Weight	
Metric	[in-lb]	mm	[in]	mm	[in]	kg/m	[lbs/ft]
#10	[#3]	9.5	[0.375]	71	[0.11]	0.560	[0.376]
#13	[#4]	12.7	[0.500]	129	[0.20]	0.944	[0.668]
#16	[#5]	15.9	[0.625]	199	[0.31]	1.552	[1.043]
#19	[#6]	19.1	[0.750]	284	[0.44]	2.235	[1.502]
#22	[#7]	22.2	[0.875]	387	[0.60]	3.042	[2.044]
#25	[#8]	25.4	[1.000]	510	[0.79]	3.973	[2.670]
#29	[#9]	28.7	[1.128]	645	[1.00]	5.060	[3.400]
#32	[#10]	32.3	[1.270]	819	[1.27]	6.404	[4.303]
#36	[#11]	35.8	[1.410]	1006	[1.56]	7.907	[5.313]
#43	[#14]	43.0	[1.693]	1452	[2.25]	11.38	[7.65]
#57	[#18]	57.3	[2.257]	2581	[4.00]	20.24	[13.60]

*The equivalent nominal characteristics of inch-pound bars are the values enclosed within the brackets.

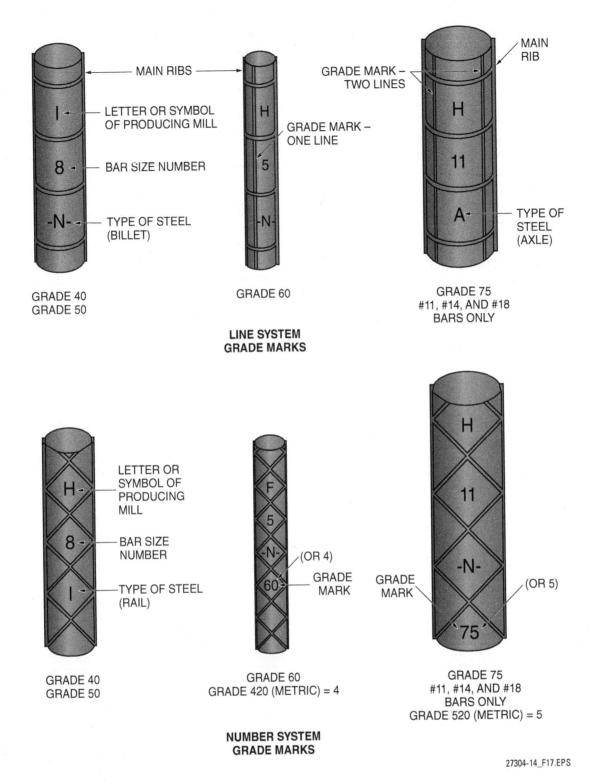

Figure 17 Reinforcing bar identification.

ASTM A955M provides information on stainless steel reinforcing bars (SSRBs). These bars, which have been used for some time, are used in highly corrosive environments or when nonmagnetic reinforcement must be used. Specification A955M defines these rebars in metric units, and the requirements parallel those in ASTM A615.

2.3.0 Fabrication

In a discussion of reinforcing steel, the terms *fabricating* and *manufacturing* should not be confused. The manufacturing of reinforcing steel produces a deformed bar from a specific type and grade of steel that is rolled to a stock straight length. The fabrication of reinforcing steel is the process of

Table 2 Reinforcing-Bar Steel Types

Symbol/Marking	Type of Steel
A	Axle (ASTM A617)
S or N	Billet (ASTM A615)
I or IR	Rail (ASTM A616)
W	Low-alloy (ASTM 706) (for welded lap, butt joints, etc.)

Table 3 Reinforcing-Bar Grades

Grade	Identification	Minimum Yield Strength
40 and 50	None	40,000 to 50,000 psi (40 to 50 ksi)
60	One line or the number 60	60,000 psi (60 ksi)
70	Two lines or the number 70	70,000 psi (70 ksi)
420	The number 4	60,000 psi (60 ksi)
520	The number 5	75,000 psi (75 ksi)

cutting and bending reinforcing bars to suit the particular needs of the job as required by the placing drawings. Most fabrication of reinforcing bars for commercial structures is done in shops, but a certain amount of it is done in the field.

2.3.1 Tools

The tools and equipment used by craftworkers working with reinforcing steel include the following:

- Hard hat
- ANSI-approved footwear
- 2" leather belt
- Tie-wire reel (needs to be good quality; don't skimp on the cost)
- Tool pouch
- Side-cutting pliers (wire cutters)
- Tape measure
- Keel holder (for holding soapstone, chalk, or crayon to mark bars)
- Leather-palm gloves (heat and puncture resistant)
- Safety glasses
- Level
- Plumb bob

- Bolt cutters (for cutting rebar)
- Sledgehammer (for aligning rebar)
- Hickey bar for bending rebar (#5 and less)

For the job of cutting welded-wire fabric reinforcement or small-size wires, side-cutting pliers or a bolt cutter are used. Rods up to ⅝" in diameter are cut with some type of hand-operated machine or cutting torch. Larger rebars can be cut with a portable band saw, chop saw, or cutting torch.

2.3.2 Fabricated Bars

The American Concrete Institute (ACI) and the Concrete Reinforcing Steel Institute (CRSI) have standardized the most common types of bar bends and assigned each a number or a letter preceding a number. These designations are used by engineers, fabricators, inspectors, and ironworkers. Information concerning bar bends is usually found on placing drawings and bar lists.

Figure 18 illustrates the typical bar bends standardized by ACI. The circled letter and/or number adjacent to each bar-bend detail indicates its type.

Unless otherwise noted, all hooks are formed in accordance with the recommended sizes for 180-degree hooks, as specified in the latest edition of *ACI 315, Details and Detailing of Concrete Reinforcement*. The dimensions and geometry of standard hooks are shown in *Figure 19* and *Table 4*.

It is important not to omit any of the letters from a standard-bend type. Doing so will result in a different shape of bar. For example, a Type 4 bar without the A or G dimension would be a straight-truss bar.

Each dimension of the standardized bar bends has been assigned a letter. These dimensions are read out-to-out, that is, from the outside diameter of the bars.

2.3.3 Bar Lists

As indicated by its name, a bar list is a list of reinforcing bars contained in a shipment. It is prepared by the fabricator and will include a comprehensive list of reinforcing bars for an entire shipment or a separate list of bars for each truckload within a shipment. The fabricators receive their information from the construction drawings.

MMFX$_2$® Rebar

MMFX$_2$® rebar is a corrosion-resistant rebar that provides five times more corrosion resistance and twice the strength of standard rebar. While stainless steel rebar provides the corrosion resistance, it is also very expensive when compared to MMFX$_2$® rebar. MMFX$_2$® rebar is used in a variety of commercial, industrial, waterway, and transportation projects.

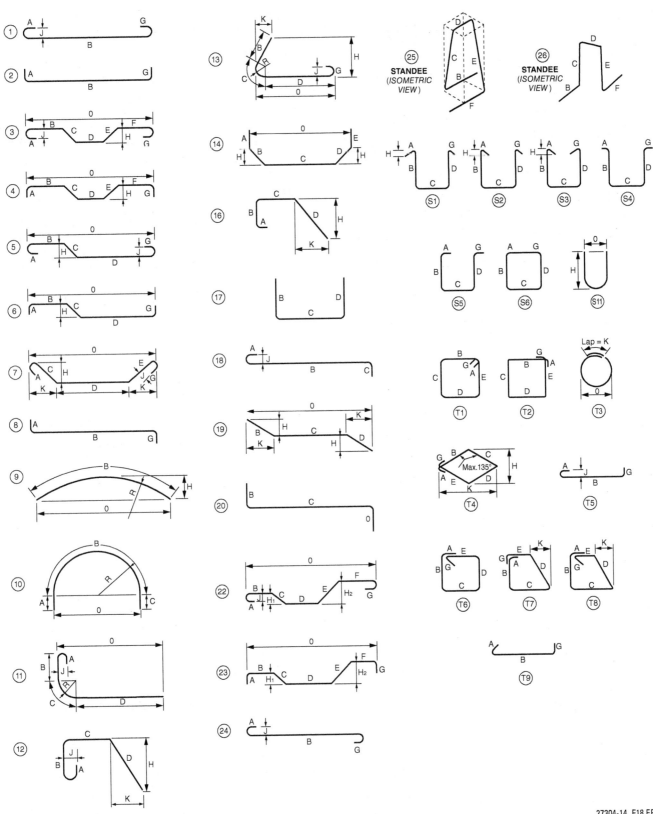

Figure 18 Typical bar bends.

RECOMMENDED END HOOKS, ALL GRADES OF STEEL

BAR SIZE	D		180° HOOKS		90° HOOKS
			A OR G	J	A OR G
#10 (#3)	60	(2¼")	125 (5")	80 (3")	150 (6")
#13 (#4)	80	(3")	150 (6")	105 (4")	200 (8")
#16 (#5)	95	(3¾")	175 (7")	130 (5")	250 (10")
#19 (#6)	115	(4½")	200 (8")	155 (6")	300 (12")
#22 (#7)	135	(5¼")	250 (10")	180 (7")	375 (14")
#25 (#8)	155	(6")	275 (11")	205 (8")	425 (16")
#29 (#9)	240	(9½")	375 (15")	300 (11¾")	475 (19")
#32 (#10)	275	(10¾")	425 (17")	335 (13¼")	550 (22")
#36 (#11)	305	(12")	475 (19")	375 (14¾")	600 (24")
#43 (#14)	485	(18¼")	675 (27")	550 (21¾")	775 (31")
#57 (#18)	610	(24")	925 (36")	725 (28½")	1050 (41")

NOTE: All metric dimensions are in millimeters (mm). Numbers shown in parentheses indicate inch-pound sizes and dimensions.

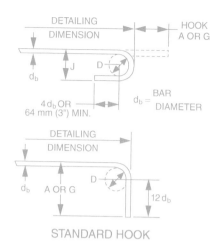

STANDARD HOOK

STIRRUP HOOKS (TIE BENDS SIMILAR)

BAR SIZE	D		90°	135°	
			A OR G	A OR G	H*
#10 (#3)	40	(1½")	105 (4")	105 (4")	65 (2½")
#13 (#4)	50	(2")	115 (4½")	115 (4½")	80 (3")
#16 (#5)	65	(2½")	155 (6")	140 (5½")	95 (3¾")
#19 (#6)	115	(4½")	305 (12")	205 (7¾")	115 (4½")
#22 (#7)	135	(5¼")	355 (14")	230 (9")	135 (5¼")
#25 (#8)	155	(6")	410 (16")	270 (10¼")	155 (6")

NOTE: All metric dimensions are in millimeters (mm). Numbers shown in parentheses indicate inch-pound sizes and dimensions.
*H dimension is approximate.

SEISMIC STIRRUP/TIE

BAR SIZE	D		135° SEISMIC HOOK	
			A OR G	H*
#10 (#3)	40	(1")	110 (4¼")	80 (3")
#13 (#4)	50	(2")	115 (4½")	80 (3")
#16 (#5)	65	(2½")	140 (5½")	95 (3¾")
#19 (#6)	115	(4½")	205 (7¾")	115 (4½")
#22 (#7)	135	(5¼")	230 (9")	135 (5¼")
#25 (#8)	155	(6")	270 (10¼")	155 (6")

NOTE: All metric dimensions are in millimeters (mm). Numbers shown in parentheses indicate inch-pound sizes and dimensions. *H dimension is approximate.

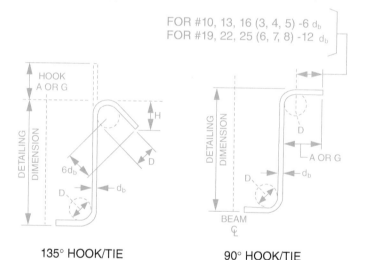
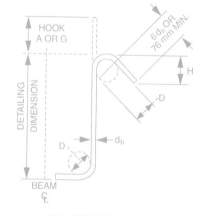

Figure 19 Standard hook details for reinforcing bars.

A bar list is useful to everyone involved in placing and fabricating reinforcing steel. The fabricator uses it for bending, tagging, shipping, and invoicing. The field crew uses it for checking quantities when unloading a shipment, sorting bars, delivering the correct bars to the correct area of the job, and for the actual placing of the bars in the forms.

2.3.4 Bar-List Information

Fabricators generally use their own forms for bar lists. Sometimes straight bars and bent bars will be placed on separate sheets, or they may be placed on the same sheet. However, certain information will be common to all bar lists. The bars are first classified as one of four types, three of which refer to bent bars. These types are as follows:

Table 4 Standard Hook Specifications

Material (ASTM Specification Number)	Bend Diameters ACI Standard[1]	Bend Diameters ACI Stirrup/Tie[2]	ASTM[3]	Bar Size	Grades
A615/615M	6d[4]	4d	3.5d	10, 13, 16 (3, 4, 5)	300, 420 (40, 60)
	6d	6d	5d	19 (6)	300, 420, 520 (40, 60, 75)
	6d	6d	5d	22, 25 (7, 8)	420, 520 (60, 75)
	8.5d	N/A	7d	29, 32, 36 (9, 10, 11)	420, 520 (60, 75)
	10.7d	N/A	9d(90°)	43, 57 (14, 18)	420, 520 (60, 75)
A706/A706M	6d	4d	3d	10, 13, 16 (3, 4, 5)	420 (60)
	6d	6d	4d	19, 22, 25 (6, 7, 8)	420 (60)
	8.5d	N/A	6d	29, 32, 36 (9, 10, 11)	420 (60)
	10.7d	N/A	8d	43, 57 (14, 18)	420 (60)

[1] ACI standard bends 90° and 180°. Finished bend diameter affected by spring-back for bars 29 to 57 (9 to 18).
[2] ACI stirrup/tie bends 90° and 135° for bars 10 to 25 (3 to 8) only.
[3] ASTM bend tests 180° unless otherwise specified.
[4] d = the nominal diameter of the bar.

- *Straight bars* – This category consists of standard straight reinforcing bars.
- *Heavy bending bars* – This category consists of metric bar sizes #13 through #57 (in-lb sizes #4 through #18) that are bent at no more than six points in one plane. Single-radius bending is also grouped in this category.

> **NOTE**
> In the following discussion and throughout this module, metric bar sizes are listed first and in-lb bar sizes are noted in parentheses.

- *Light bending bars* – This category includes all #10 (#3) bars, all stirrups and ties, and all bars #13 through #57 (#4 through #18) bent at more than six points in one plane. It also includes all single-plane radius bending with more than one radius in any bar, or a combination of radius or other type of bending in one plane.
- *Special bending bars* – This category includes all bending to special tolerances, all radius bending in more than one plane, all multiple-plane bending with one or more radius bends, and all bending for precast units.

The bars are then grouped by sizes and lengths under each classification. The largest size and the longest length of each size will be listed first. Each bar list usually contains the following information:

- Name of project
- Customer name
- Job location
- Part of job
- Placing-drawing reference number
- Grades of steel
- Order number

2.3.5 Sample Bar List

Figure 20 illustrates a sample bar list that might be prepared by a fabricator and sent with the shipment of reinforcing bars to the job site. Notice that the top of the sheet contains all the necessary information about the project, customer, etc. Also, notice the divisions labeled Straight Bars, Light Bending, and Heavy Bending.

Each bar is described by the information found under the column headings. Most of the headings are self-explanatory, listing the quantity, bar size, and length of the bars. The mark refers to the location of the bent bars as found on the placing drawing. The bend type refers to one of the standard circled letter and/or number bar bends shown in *Figure 18*. Placement and dimensions of the bends are identified in columns A through H, J, and O as required, and as shown in *Figure 18*.

Bar lists can also include a safety data sheet that lists hazardous ingredients, physical data, fire and explosion information, and reactivity information for the metals used in the rebar.

2.3.6 Special Details of Fabrication

Special details of fabrication include the following:

- *Hooks* – Hooks refer to those parts of a bar with dimensions designated A or G by ACI standards. (Refer to *Figure 18* for examples of these designations.) All other dimensions are properly called bends. There are three types of hooks in general use, each described by the size of the angle the hook encompasses— a 90-degree hook, a 135-degree hook, and a 180-degree hook (refer to *Figure 19*).

- *Spirals* – Spirals are made of smooth bar stock or of wire shaped like a coil spring. They have spacers attached on opposite sides that keep the turns of the spiral at equal distances. Center-to-center spacing of the turns of a spiral is called pitch. Spirals are usually shipped collapsed and must be opened or expanded at the job site prior to assembly. The spacers should be arranged to provide equal pitch around the spiral column.
- *Radial bending* – Curved reinforcing bars are used for tanks, bins, culverts, domes, tunnels, and other curved structures. These bars are either fabricated in a shop or sprung into shape in the field, depending on the maximum radius of the bend. *Table 5* shows the maximum radii that are generally shop fabricated. Any bars of a larger radius will usually be shipped straight.

> **WARNING!** Place rebar according to the structural drawings and specifications. Do not place according to field decisions or architectural drawings.

2.3.7 Tolerances in Fabrication

Tolerances are necessary in normal fabricating operations. The engineers typically allow for tolerances in their designs. The usual tolerances are as follows:

- *Straight bars* – ±1" (25.4 mm) in length.
- *Hooked bars* – #7 or smaller, the overall length may deviate by ±½" (12.7 mm); #8 or larger, the tolerance is ±1" (25.4 mm).
- *Truss bars* – #7 and smaller, the overall length may deviate by ±½" (12.7 mm); #8 bars and larger, the tolerance is ±1" (25.4 mm); for all bars, the H dimension may vary only 0" to ½" (0 mm to 12.7 mm).
- *Spirals* – ±½" (12.7 mm) tolerance.
- *Column ties* – ±½" (12.7 mm) tolerance.

Even though fabricators usually supply the job site with the proper number and types of bars, some field fabrication is generally necessary. This fabrication may range from straightening a bent bar or dowel to replacing a lost or badly damaged truss bar.

2.3.8 Bundling and Tagging

Straight bars are bundled and tagged. Other bars are bent first, then bundled and tagged. Unassembled spirals are bundled immediately after fabrication and tagged. Assembled spirals are tagged individually and shipped collapsed, if so designed.

Each bundle of bars should contain bars of one size, length, and mark. *Mark* is the term used to designate the part of the structure for which the bars are intended. Bundles are generally secured by wraps of #9-gauge wire spaced 10' to 15' apart, with a minimum of two ties per bundle. Each bundle should be tagged. The tag should be made of a durable material such as metal or rope fiber. Metal tags should be embossed; fiber tags should be marked with waterproof ink. In general practice, the tag will show the purchaser by name, address, and order number. The bundle of bars should be identified by the number of pieces, size and length of straight bars, mark number for bent bars, and grade of steel for both bent and straight bars. Tags for bent bars generally give information on bending dimensions.

Concrete Bridge

This cast-in-place, 1,345' concrete frame is made up of eight spans, including six interior spans of 177' each. It is supported by 7' octagonal columns set atop 8'-diameter piles.

				ORDER NO.	JC1147
PROJECT	HOCKEY ARENA			ORG. NO.	WQ31
CUSTOMER	M & G CONSTRUCTION			SHEET	1 OF 3
LOCATION	MIAMI, FLORIDA			DATE	5/10/01
MAT'L FOR	1ST FLOOR BEAMS & COLUMNS			MADE BY JAC	CHECKED BY ABC

ITEM	QTY.	BAR SIZE	FINAL LENGTH	MARK	BEND TYPE	A	B	C	D	E	F	G	H	J	O
1				STRAIGHT BARS											
2	4	#7	22-0												
3	4		17-6												
4															
5	2	#5	28-3												
6	2		17-6												
7				HEAVY BENDING											
8	2	#9	36-0	1B 901	3		10-0	2-3	12-4	2-3	9-2		1-7		
9	2		35-7	1B 902	3		10-0	2-3	12-4	2-3	9-2		1-7		
10															
11	2	#8	23-8	1B 801	1	1-1	22-7								
12															
13	2	#7	25-2	1B 703	3	10	2-3	2-8½	9-10	2-8½	6-10		1-11		
14															
15	2	#6	26-2	1B 601	3	8	5-7	2-7	8-6	2-7	5-7	8	1-10		23-4
16															
17				LIGHT BENDING											
18	22	#4	5-6	U401	S2	4½	1-11	11	1-11			4½			
19	34		5-2	U402	S1	4½	1-9	11	1-9			4½			
20															
21	26	#3	6-4	U301	S2	4	2-6	8	2-6			4			
22	24		6-0	U302	T1	4	2-0	8	2-0	8		4			
23															
24	12	#2	6-11	U201	T1	3½	1-4	1-5	1-9	1-5		3½			
25	20		3-11	U202	T1	3½	10	10	10	10		3½			

Figure 20 Sample bar list.

Table 5 Maximum Prefabricated Radii

Bar Size	Maximum Prefabricated Radius
#3	10'
#4	10'
#5	15'
#6	40'
#7	40'
#8	60'
#9	90'
#10	110'
#11	110'
#14	180'
#18	300'

WARNING! Do not rig to the #9-gauge wire bundle wrap; rig to the rebar instead.

Many rebar fabricators use computer-based applications that automatically generate bar lists and print labels for rebar bundles. *Figure 21* shows another example of a bar list, and *Figure 22* shows two labels made from the bar list. Note that the labels are directly related to two of the entries on the bar list.

WARNING! OSHA standard 29 *CFR* 1926.701(b) requires that all protruding reinforcing steel, onto and into which workers could fall, shall be guarded to eliminate the hazard of impalement.

27304-14_SA03.EPS

2.4.0 Bar Supports

Bar supports, sometimes called accessories, are used to support, hold, and space reinforcing bars and mats or wire reinforcement before and during concrete placement. Bar supports are made from steel, concrete, or plastic. When used with coated reinforcement steel, the supports should be coated with the same material or made of concrete or plastic to prevent corrosion. A sufficient number of supports and the correct size of the supports must be used to prevent the reinforcement from shifting out of position or deforming when the concrete is placed.

NOTE: For the rebar to function as designed, bar-support sizing is critical for bar placement and positioning as intended by the engineer.

2.4.1 Steel-Wire Bar Supports

Steel-wire bar supports are divided into five classes based on how well the support will prevent rust spots or similar blemishes from forming on the surface of the concrete. These classes are as follows:

- *Class A, bright basic* – Class A offers no protection against rusting. Therefore, it is used in situations where surface blemishes can be tolerated.
- *Class B, pregalvanized* – Class B offers minimal protection against rusting. It is used where nominal protection for a short amount of time is required.
- *Class C, plastic protected* – Class C offers moderate protection against rusting. It is used in situations where the surface of the concrete will be subjected to moderate exposure or where sandblasting or light grinding of the concrete surface is required.
- *Class D, stainless protected* – Class D offers more protection against rusting than Class C and is used in the same situations.
- *Class E, special stainless* – Class E is used where the concrete surface will be exposed to moderately severe conditions or where heavy grinding or sandblasting of the concrete surface is required.

Steel Reinforcement Protection

Steel reinforcement of any kind must be covered by enough concrete to be adequately protected; otherwise, the steel will rust, causing damage to the concrete. Some examples of minimum concrete coverage are:

- *Footings* – 3"
- *Concrete surface exposed to weather* – 2" for bars larger than #5, 1½" for bars #5 and smaller
- *Slabs, walls, joists* – ¾"
- *Beams and girders* – 1½"

<table>
<tr><td colspan="6">Client: SAMPLE CLIENT 1 Project: SAMPLE IMPERIAL PROJECT
Address: New York, Elm Street, 67 Type:
Code: 2001-1 Number: 1 Module: FACTORY BLDG.
Description: FIRST FLOOR FRAMING PLAN</td></tr>
<tr><th>Mark</th><th>Total</th><th>Size</th><th>Long</th><th>Wt. lb</th><th>Wt. lb/Page = 951,05</th></tr>
<tr><td></td><td></td><td></td><td></td><td></td><td>FIRST FLOOR</td></tr>
<tr><td>5K9</td><td>4</td><td>#5</td><td>6'4½"</td><td>26,60</td><td>0..> 6'4½"</td></tr>
<tr><td>4K10</td><td>5</td><td>#5</td><td>7'8"</td><td>39,98</td><td>2..> 10" 6' 10"</td></tr>
<tr><td>4K12</td><td>3</td><td>#9</td><td>12'4½"</td><td>126,01</td><td>2..> 1'-7" 9'2½"</td></tr>
<tr><td>4K13</td><td>3</td><td>#9</td><td>6'10"</td><td>69,70</td><td>2A..> 1'-7" 5'-3"</td></tr>
<tr><td>3K8</td><td>2</td><td>#3</td><td>10'10"</td><td>8,15</td><td>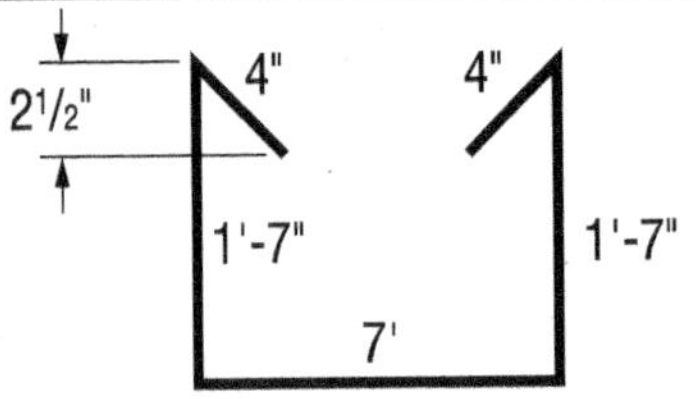
2½"
4"
4"
1'-7"
1'-7"
7'</td></tr>
<tr><td>3K2</td><td>6</td><td>#3</td><td>6'3"</td><td>14,10</td><td>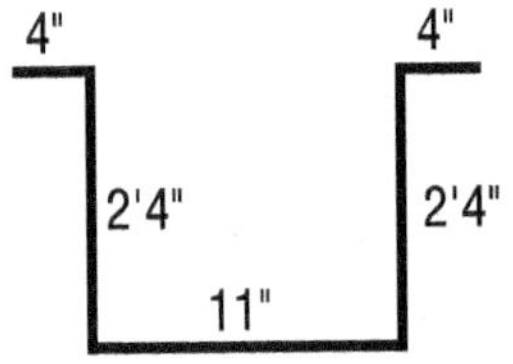
4"
4"
2'4"
2'4"
11"</td></tr>
<tr><td>3K3</td><td>6</td><td>#3</td><td>1'7"</td><td>3,57</td><td>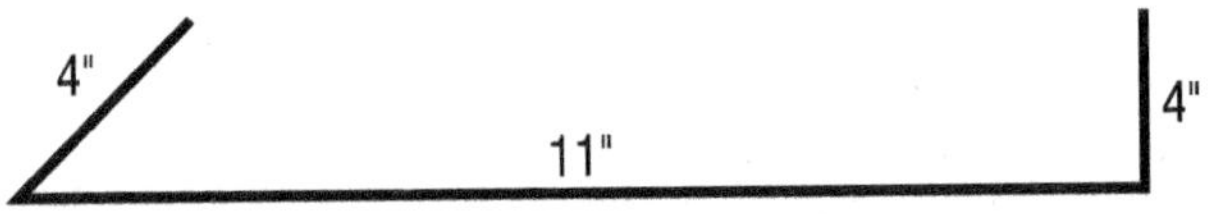
4"
11"
4"</td></tr>
<tr><td>8K6</td><td>4</td><td>#8</td><td>6'10"</td><td>72,98</td><td>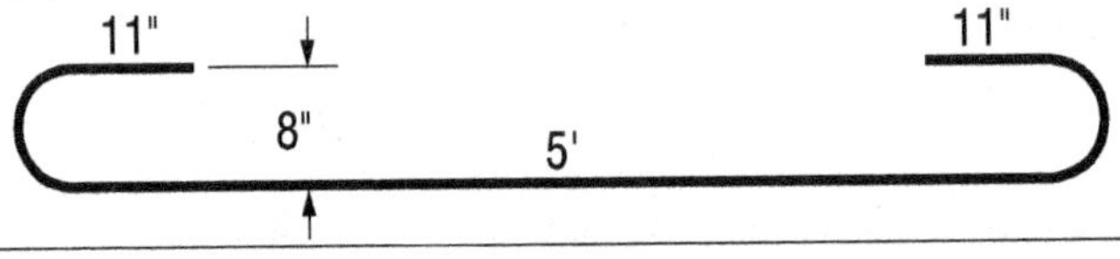
11"
11"
8"
5'</td></tr>
<tr><td>9K1</td><td>3</td><td>#9</td><td>6'7"</td><td>67,15</td><td>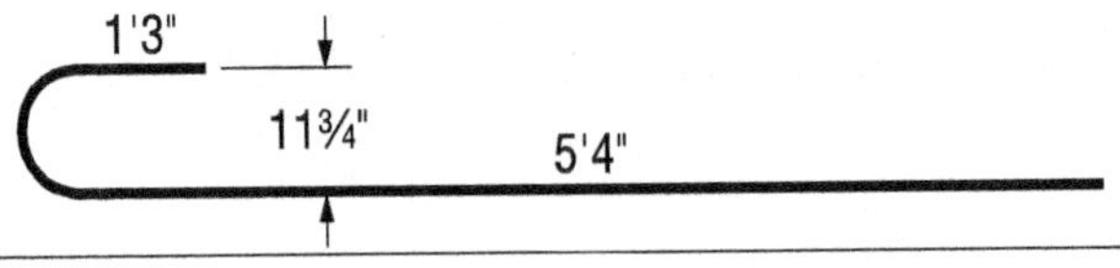
1'3"
11¾"
5'4"</td></tr>
<tr><td>10K4</td><td>3</td><td>#10</td><td>40'6"</td><td>522,81</td><td>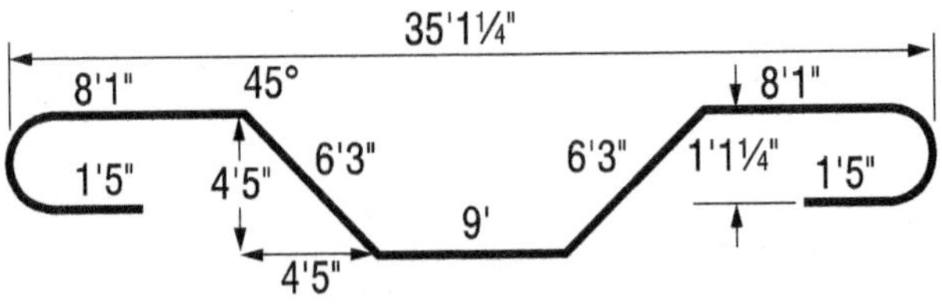
35'1¼"
8'1"
45°
8'1"
1'5"
4'5"
6'3"
6'3"
1'1¼"
1'5"
9'
4'5"</td></tr>
</table>

3 = 35,82 5 = 66,58 8 = 72,98 9 = 262,86 10 = 522,81 TOTAL = 951,05

27304-14_F21.EPS

Figure 21 Additional sample bar list.

Client: SAMPLE CLIENT 1   Project: SAMPLE IMPERIAL PROJECT

Address: New York, Elm Street, 67

Code: 2001-1   Number: 1   Module: FACTORY BLDG.

Description: FIRST FLOOR FRAMING PLAN   Type:

Location: FIRST FLOOR   Machine: Label: 3

Row: 6   Mark: 3K2   Bars: 6   #3   Long: 6'3"   Wt: 14,10

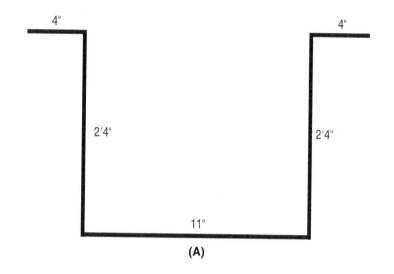

(A)

Client: SAMPLE CLIENT 1   Project: SAMPLE IMPERIAL PROJECT

Address: New York, Elm Street, 67

Code: 2001-1   Number: 1   Module: FACTORY BLDG.

Description: FIRST FLOOR FRAMING PLAN   Type:

Location: FIRST FLOOR   Machine: Label: 7

Row: 10   Mark: 10K4   Bars: 3   #10   Long: 40'6"   Wt: 522,81

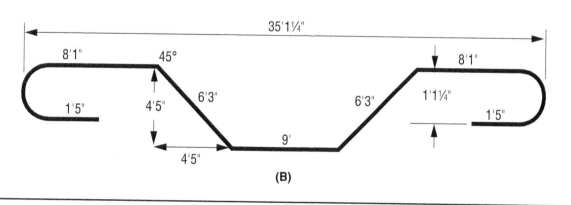

(B)

Figure 22 Examples of rebar labels.

These classes of steel-wire bar supports can be used in many different configurations. *Figure 23* shows various wire bar-support configurations. Some of the more common supports are as follows:

- *Slab bolster (SB)* – Used under the rebars placed in slabs. The top wire is corrugated, providing firm tying of bars with any spacing. The legs are on 5" centers. Slab bolsters are stocked in heights of ¾" to 2", in lengths of 5' and 10'.
- *Slab bolster upper (SBU)* – Provides support for rebars in slabs on cork or fill. This is essentially a slab bolster with two wires welded lengthwise along the base. SBUs are available in the same sizes as slab bolsters, but usually in Class A only, except on special order.
- *Slab bolster plate (SBP)* – Similar in design to slab bolster supports except that steel plates, instead of wires, are welded lengthwise along the base. The applications and sizes are the same as those for the slab bolster uppers. They are also called sand plates.
- *Beam bolster (BB)* – Supports rebars placed in beams and girders regardless of the spacing. The legs are spaced 2½" apart. They are available in 5' lengths and may be cut to the particular beam widths needed on the job. BBs are available in different heights.
- *Beam bolster upper (BBU)* – Provides support between layers of bars. Two wires are welded lengthwise along the base of a beam bolster. The heights and lengths are the same as those for the beam bolster. BBUs are available in Class A only, except on special order.
- *Bar chair (BC)* – Supports miscellaneous reinforcing steel and may, in certain applications, be used as a substitute for slab bolsters. BCs are available in different heights.
- *Joist chair (JC)* – Supports rebars placed in concrete ribs or joists of metal-pan or cinder-fill concrete slabs. JCs are available in different heights and 4", 5", or 6" widths.
- *High chair (HC)* – Formed from steel wire. It has upturned legs. HCs are available in different heights.
- *Continuous high chair (CHC)* – Supports top slab bars and truss bars at beams, girders, and columns. Available heights are the same as those for the individual high chairs.
- *Continuous high chair upper (CHCU)* – A continuous high chair with two wires welded lengthwise along the base. This supports the upper layers of reinforcing steel and may also be used on fill or cork. CHCUs are available in 5' lengths in the heights available for the individual high chairs. Except on special order, only Class A is available.
- *Joist chair upper (JCU)* – Supports reinforcing steel in pan-joist or waffle-dome work. The supporting bar is either #4 reinforcing bar or ½" plain round bar. Standard width is 14". JCUs are available in heights up to 3½" in Class A only with upturned or end-bearing legs.

2.4.2 Precast Concrete Blocks

Precast concrete blocks, also known as dobies, may also be used as bar supports, especially in footings where the bars usually clear the bottom by 3".

Blocks should be placed about 5' off center each way under the footing mat. At least four blocks should be used per mat. There are a number of precast concrete bar supports in use, as illustrated in *Figure 24*. The most common are the plain- and wired-block types. Any blocks used should have the same compressive strength as the concrete being placed.

2.4.3 Other Types of Bar Supports

A number of all-plastic supports are used in place of steel or concrete when corrosion must be avoided or appearance is a factor. Some of these plastic supports are shown in *Figure 25*.

A steel standee (*Figure 26*) may also be used as a bar support. The standee is a U-shaped support having two legs bent at 90-degree angles in opposite directions. It is used as a high chair and placed on a lower mat of bars to support an upper mat of bars.

Reinforcing-Bar Sizes

There are 11 standard sizes of manufactured reinforcing bars. These are identified by a number expressed in inch-pound units ranging from #3 to #18, or in metric units with sizes ranging from #10 to #57. In either system, the size number represents the approximate diameter of the bar. In the inch-pound system, the size number represents eighths of an inch; in the metric system it represents millimeters (mm). For example, a #5 bar has an approximate diameter of ⅝". When expressed in metric system units, the corresponding #16 bar has an approximate size of 16 mm (about ⅝").

BAR SUPPORT ILLUSTRATION	BAR PLASTIC ILLUSTRATION PLASTIC CAPPED OR DIPPED	SYMBOL	TYPE OF SUPPORT	TYPICAL SIZES
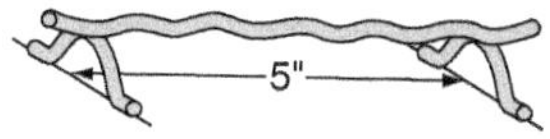 5"	CAPPED 5" 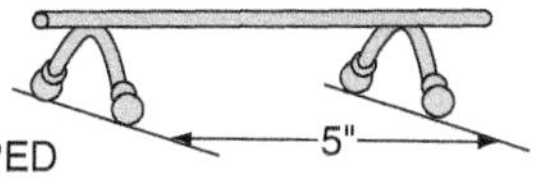	SB	Slab Bolster	3/4, 1, 1 1/2, and 2 inch heights in 5 ft and 10 ft lengths
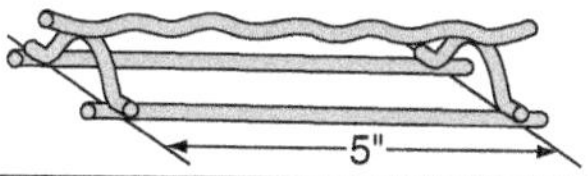 5"		SBU* SBP ↓	Slab Bolster Upper or Slab Bolster Plate	Same as SB
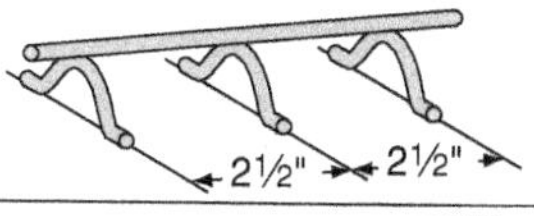 2 1/2" 2 1/2"	CAPPED 2 1/2" 2 1/2" 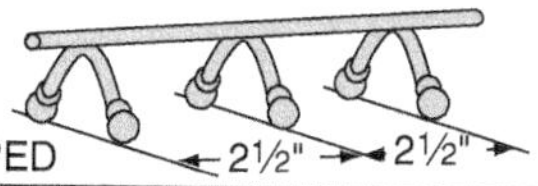	BB	Beam Bolster	1, 1 1/2, 2 to 5 inch heights in increments of 1/4 inch in lengths of 5 ft
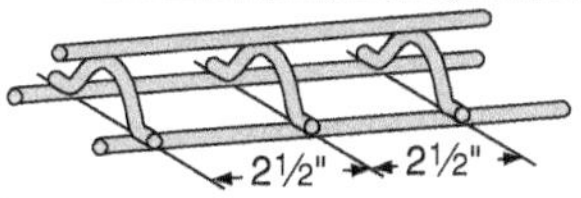 2 1/2" 2 1/2"		BBU*	Beam Bolster Upper	Same as BB
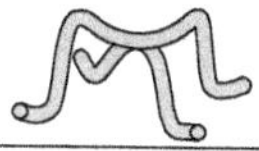	DIPPED 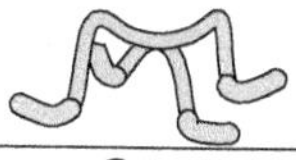	BC	Individual Bar Chair	3/4, 1, 1 1/2, and 1 3/4 inch heights
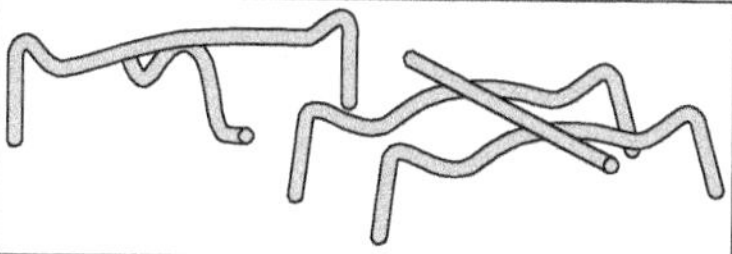 DIPPED	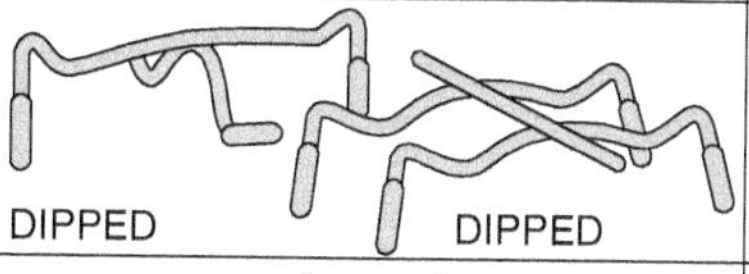 DIPPED DIPPED	JC	Joist Chair	4, 5, and 6 inch widths and 3/4, 1, and 1 1/2 inch heights
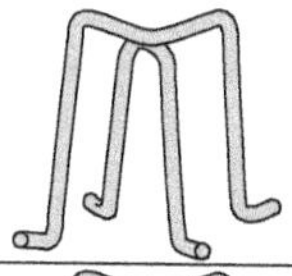	CAPPED 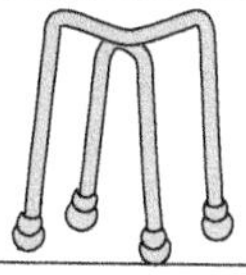	HC	Individual High Chair	2 to 15 inch heights in increments of 1/4 inch
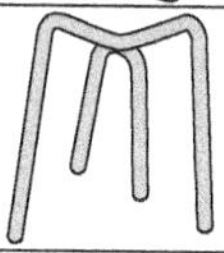		HCM*	High Chair for Metal Deck	2 to 15 inch heights in increments of 1/4 inch
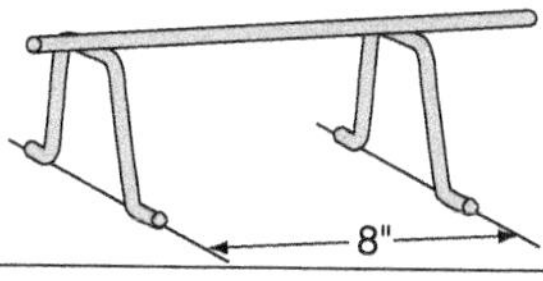 8"	CAPPED 8" 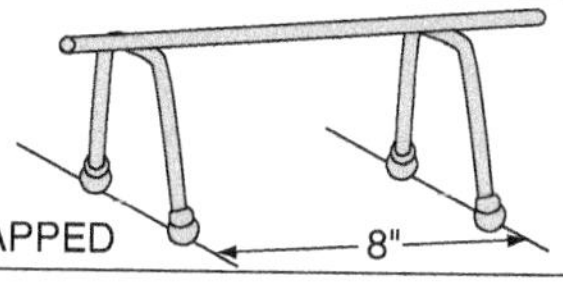	CHC	Continuous High Chair	Same as HC in 5 ft and 10 ft lengths
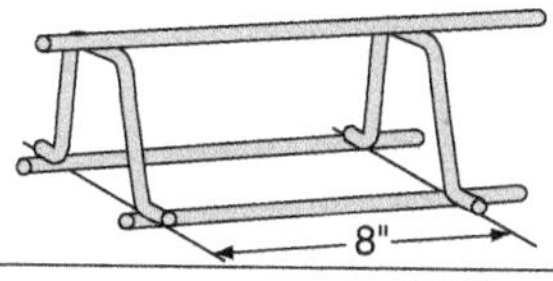 8"		CHCU*	Continuous High Chair Upper	Same as CHC
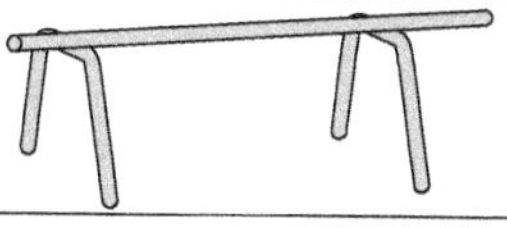		CHCM*	Continuous High Chair for Metal Deck	Up to 5 inch heights in increments of 1/4 inch
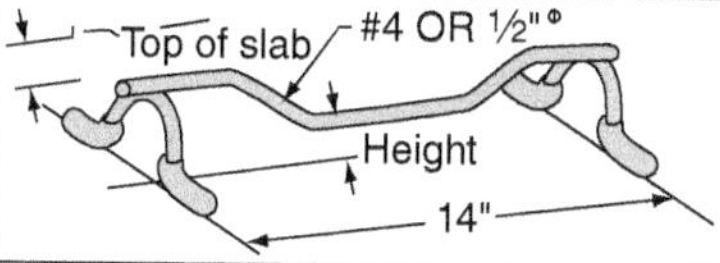 Top of slab #4 OR 1/2"⌀ Height 14"	Top of slab #4 OR 1/2"⌀ Height 14" DIPPED 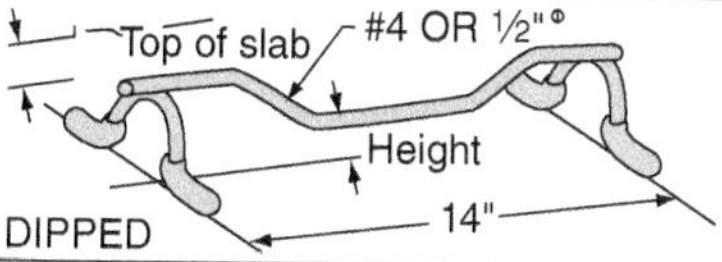	JCU**	Joist Chair Upper	14-inch span and heights range from 1 inch to 3 1/2 inches in 1/4 inch increments
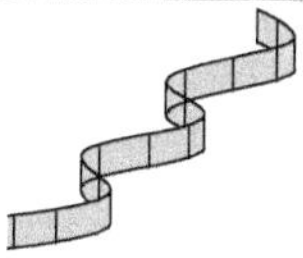		CS	Continuous Support	1 1/2 to 12 inch in increments of 1/4 inch in lengths of 6 to 8 ft

Note: 1 inch = 25.4 mm
* Usually available in Class A only, except on special order.
** Usually available in Class A only, with upturned or end bearing legs.
↓ Slab Bolster Plate (SBP) has steel plates instead of wires welded lengthwise along the base.

Figure 23 Typical wire bar supports.

BAR SUPPORT ILLUSTRATION	SYMBOL	TYPE OF SUPPORT**	TYPICAL SIZES	DESCRIPTION
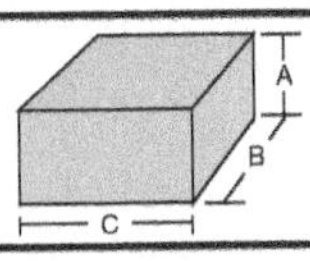	PB	Plain Block	A - ¾" to 6" B - 2" to 6" C - 2" to 48"	Used when placing rebar off grade and formwork. When C dimension exceeds 16", a piece of rebar should be cast inside block.
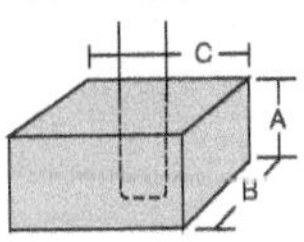	WB	Wired Block	A - ¾" to 4" B - 2" to 3" C - 2" to 3"	Generally 16-gauge tie wire is cast in block, commonly used against vertical forms or in positions necessary to secure the block by tying to the rebar.
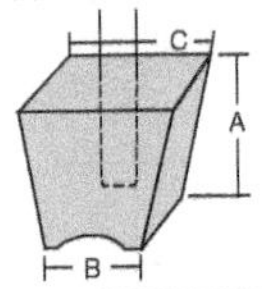	TWB	Tapered Wired Block	A - ¾" to 3" B - ¾" to 2½" C - 1¼" to 3"	Generally 16-gauge tie wire is cast in block, commonly used where minimal form contact is desired.
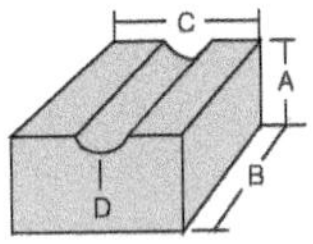	CB	Combination Block	A - 2" to 4" B - 2" to 4" C - 2" to 4" D - fits #3 to #5 bar	Commonly used on horizontal work.
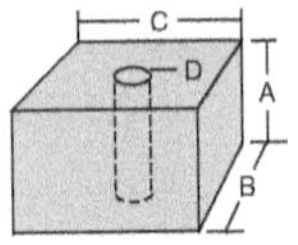	DB	Dowel Block	A - 3" B - 3" to 5" C - 3" to 5" D - hole to accommodate a #4 bar	Used to support top mat from dowel placed in hole. Block can also be used to support bottom mat.
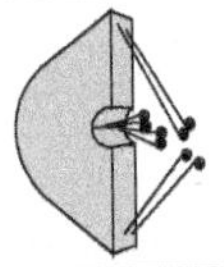	DSSS	Side Spacer – Wired	Concrete cover, 2" to 6"	Used to align the rebar cage in a drilled shaft.* Commonly 16-gauge tie wires are cast in spacer. Items for 5" to 6" cover have 9-gauge tie wires at top and bottom of spacer.
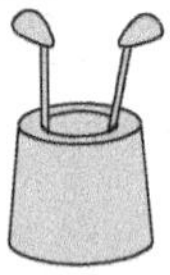	DSBB	Bottom Bolster – Wired	Concrete cover, 3" to 6"	Used to keep the rebar cage off of the floor of the drilled shaft.* Item for 6" cover is actually 8" in height with a 2" shaft cast in the top of the bolster to hold the vertical bar.
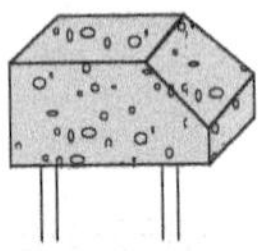	DSWS	Side Spacer for Drilled Shaft Applications	Concrete cover, 3" to 6"	Generally used to align rebar in a drilled shaft. Commonly manufactured with two sets of 16-gauge annealed wires, assuring proper clearance from the shaft wall surface.

* Also known as pier, caisson, or cast-in-drilled hole.
** Blocks should be the same compressive strength as the concrete being placed.

Figure 24 Typical precast concrete bar supports.

2.4.4 Identification of Bar Supports

Specifications, drawings, and details identify bar supports by listing their nominal height, length, type, and class of protection. For example, 6 × 5-CHC-A would signify a Class A continuous high chair having a height of 6" and a length of 5'.

2.4.5 Miscellaneous Accessories

Figure 27 shows a standard roll of tie wire and a wire tie (pigtail) used to secure lengths of reinforcing steel to each other or to various supports.

2.5.0 Welded-Wire Fabric Reinforcement

When reinforcement is required for concrete pavement, parking lots, driveways, or floor slabs, welded-wire fabric reinforcement (shown as WWF or WWR on construction drawings) can be used instead of individual rebars. Welded-wire fabric reinforcement consists of longitudinal and transverse steel wires electrically welded together to form a square or rectangular mesh or mat. Depending on the wire diameter, which can range up to ¾" or more, welded-wire fabric reinforcement is available in roll form or in 8' × 16' flat

BAR SUPPORT ILLUSTRATION	SYMBOL	TYPE OF SUPPORT	TYPICAL SIZES	DESCRIPTION
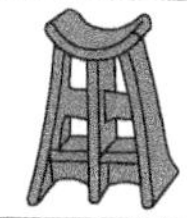	BS	Bottom Spacer	Heights ¾" to 6"	Generally for horizontal work. Not recommended for ground or exposed aggregate finish.
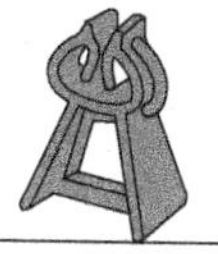	BS-CL	Bottom Spacer	Heights ¾" to 2"	Generally for horizontal work; provides bar clamping action. Not recommended for ground or exposed aggregate finish.
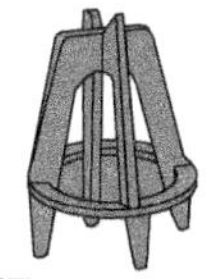	HC	High Chair	Heights ¾" to 5"	For use on slabs or panels.
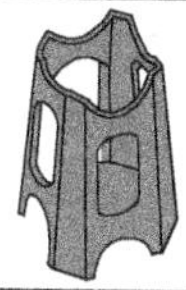	HC-V	High Chair, Variable	Heights 2½" to 6¼"	For horizontal and vertical work. Provides for different heights.
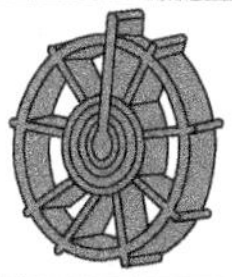	WS	Wheel Spacer	Concrete Cover ⅜" to 3"	Generally for vertical work. Bar clamping action and minimum contact with forms. Applicable for concrete-reinforcing steel.
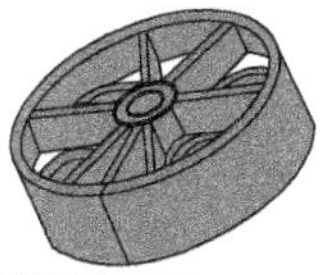	DSWS	Side Spacer for Drilled-Shaft Applications	Concrete Cover 2½" to 6"	Generally used to align rebar in a drilled shaft.* Two-piece wheel that closes and locks onto the stirrup or spiral, assuring proper clearance from the shaft wall surface.
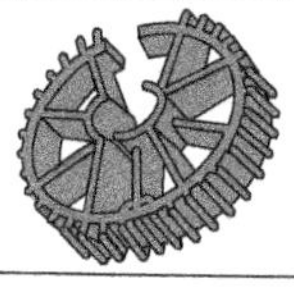	VLWS	Locking Wheel Spacer for All Vertical Applications	Concrete Cover ¾" to 6"	Generally used in both drilled shaft and vertical applications where excessive loading occurs. Surface spines provide minimal contact while maintaining required tolerance.

* Also known as pier, caisson, or cast-in-drilled hole.

Figure 25 Typical all-plastic supports.

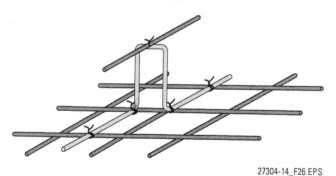

Figure 26 Standee support.

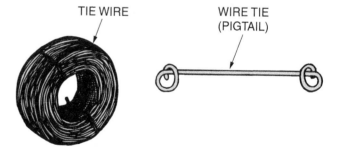

Figure 27 Miscellaneous accessories.

mats. *Figure 28* illustrates some standard sizes of plain-wire welded-wire fabric reinforcement in roll form. Bar supports can be used to space and secure welded-wire fabric reinforcement as well as rebar.

2.5.1 Plain-Wire Reinforcement

Plain wire for welded-wire fabric reinforcement is produced in accordance with *ASTM A82* and is designated by a size number consisting of a W (or MW for metric) followed by the nominal cross-sectional area of the wire in hundredths of a square inch (or square millimeters for metric).

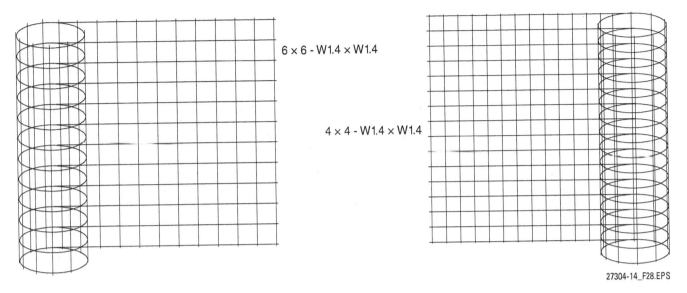

Figure 28 Typical plain-wire, welded-wire fabric reinforcement in roll form.

The wire is manufactured in sizes from W-0.5 to W-45 (MW-3 to MW-290 for metric), as shown in *Table 6*.

Welded-wire fabric reinforcement is designated by a style code such as 4 × 6 – W-10 × W-6 which, in this case, indicates the longitudinal wires are size W-10 on 4" centers and the transverse wires are W-6 on 6" centers. Any style where the transverse wire provides the minimum steel size necessary for fabrication and handling is called one-way reinforcement. Where adequate reinforcement is provided in both transverse and longitudinal directions, the reinforcement is called two-way reinforcement.

Common styles of plain welded-wire fabric reinforcement are shown in *Table 7*. The table also shows the old way of designating welded-wire fabric reinforcement using American Wire Gauge (AWG) wire gauge sizes.

2.5.2 Deformed Welded-Wire Fabric Reinforcement

Deformed wires, manufactured to *ASTM A496*, are also used in making welded-wire fabric reinforcement. The deformations along the wires are intended to improve bonding of the concrete for better crack control. The wires are manufactured and designated in a range from D-1 to D-45 (MD-6 to MD-290 for metric), as shown in *Table 8*. Deformed welded-wire fabric reinforcement is produced to *ASTM A497* and is designated the same as plain welded-wire fabric reinforcement except that the W or MW is replaced by a D or MD. Using the example in the previous paragraph, 4 × 6 – W-10 × W-6 would be 4 × 6 – D-10 × D-6 if deformed wire were used in place of plain wire.

Table 6 Plain Wire Sizes (*ASTM A82*)

Size	Diameter (in)	Area (in^2)	Weight (lb/ft)
W-0.5	0.080	0.005	0.017
W-1	0.113	0.010	0.034
W-1.4	0.135	0.014	0.048
W-1.5	0.138	0.015	0.051
W-2	0.159	0.020	0.068
W-2.5	0.178	0.025	0.085
W-2.9	0.192	0.029	0.096
W-3	0.195	0.030	0.102
W-3.5	0.211	0.035	0.119
W-4	0.225	0.040	0.136
W-4.5	0.240	0.045	0.153
W-5	0.252	0.050	0.170
W-5.5	0.264	0.055	0.187
W-6	0.276	0.060	0.204
W-7	0.298	0.070	0.238
W-8	0.319	0.080	0.272
W-10	0.356	0.100	0.340
W-12	0.390	0.120	0.408
W-14	0.422	0.140	0.476
W-16	0.451	0.160	0.544
W-18	0.478	0.180	0.612
W-20	0.504	0.200	0.680
W-22	0.529	0.220	0.748
W-24	0.553	0.240	0.816
W-26	0.575	0.260	0.884
W-28	0.597	0.280	0.952
W-30	0.618	0.300	1.020
W-31	0.628	0.310	1.054
W-45	0.757	0.450	1.531

Table 7 Common Styles of Plain Welded-Wire Fabric

Style Designation		Longitudinal or Transverse Steel Area (in²/ft)	Approximate Total Weight (lbs/100 ft²)
Current Designation (by W-number)	Previous Designation (by steel wire gauge)		
Rolls			
6 × 6 × W-1.4 × W-1.4	6 × 6 × 10 × 10	0.028	19
6 × 6 × W-2.0 × W-2.0	6 × 6 × 8 × 8*	0.040	27
6 × 6 × W-2.9 × W-2.9	6 × 6 × 6 × 6	0.058	39
6 × 6 × W-4.0 × W-4.0	6 × 6 × 4 × 4	0.080	54
4 × 4 × W-1.4 × W-1.4	4 × 4 × 10 × 10	0.042	29
4 × 4 × W-2.0 × W-2.0	4 × 4 × 8 × 8*	0.060	41
4 × 4 × W-2.9 × W-2.9	4 × 4 × 6 × 6	0.087	59
4 × 4 × W-4.0 × W-4.0	4 × 4 × 4 × 4	0.120	82
Sheets			
6 × 6 × W-2.9 × W-2.9	6 × 6 × 6 × 6	0.058	39
6 × 6 × W-4.0 × W-4.0	6 × 6 × 4 × 4	0.080	54
6 × 6 × W-5.5 × W-5.5	6 × 6 × 2 × 2†	0.110	75
4 × 6 × W-4.0 × W-4.0	4 × 4 × 4 × 4	0.120	82

*Exact W-number size for eight gauge is W-2.1
†Exact W-number size for two gauge is W-5.4

Table 8 Deformed Wire Sizes (*ASTM A496*)

Size	Diameter (in)	Area (in²)	Weight (lbs/ft²)	Size	Diameter (in)	Area (in²)	Weight (lbs/ft²)
D-1	0.113	0.01	0.034	D-17	0.465	0.17	0.578
D-2	0.159	0.02	0.068	D-18	0.478	0.18	0.612
D-3	0.195	0.03	0.102	D-19	0.491	0.19	0.646
D-4	0.225	0.04	0.136	D-20	0.504	0.20	0.680
D-5	0.252	0.05	0.170	D-21	0.517	0.21	0.714
D-6	0.276	0.06	0.204	D-22	0.529	0.22	0.748
D-7	0.298	0.07	0.238	D-23	0.541	0.23	0.782
D-8	0.319	0.08	0.272	D-24	0.553	0.24	0.816
D-9	0.338	0.09	0.306	D-25	0.564	0.25	0.850
D-10	0.356	0.10	0.340	D-26	0.575	0.26	0.884
D-11	0.374	0.11	0.374	D-27	0.586	0.27	0.918
D-12	0.390	0.12	0.408	D-28	0.597	0.28	0.952
D-13	0.406	0.13	0.442	D-29	0.608	0.29	0.986
D-14	0.422	0.14	0.476	D-30	0.618	0.30	1.020
D-15	0.437	0.15	0.510	D-31	0.628	0.31	1.054
D-16	0.451	0.16	0.544	D-45	0.757	0.45	1.531

Synthetic-Fiber Reinforcement

Among the concerns associated with welded-wire fabric reinforcement are that it is labor intensive to install and often gets bent out of position by workers walking on it. In recent years, there has been a trend to replace welded-wire fabric reinforcement with fibers made of synthetic material in slabs-on-grade and elevated decks. These fibers are added to the concrete mix by the ready-mix supplier. The fiber additives have proven effective in reducing shrinkage and controlling cracking. In some applications, steel fibers are added to the concrete mix, alone or in combination with synthetic fibers. The use of steel fibers is said to help improve the structural strength and impact resistance of concrete.

Additional Resources

29 CFR 1926, *Safety and Health Regulations for Construction*, latest edition. Washington, D.C.: Occupational Safety and Health Administration.

ACI 315, *Details and Detailing of Concrete Reinforcement*, Latest Edition. Farmington Hills, MI: American Concrete Institute.

ACI 318-95, *Building Code Requirements for Structural Concrete*, Latest Edition. Farmington Hills, MI: American Concrete Institute.

ASTM A615, *Standard Specification for Deformed and Plain Carbon-Steel Bars for Concrete Reinforcement*, Latest Edition. West Conshohocken, PA: ASTM International.

ASTM A706, *Standard Specification for Low-Alloy Steel Deformed Bars and Plain Bars for Concrete Reinforcement*, Latest Edition. West Conshohocken, PA: ASTM International.

ASTM A996, *Standard Specification for Rail-Steel and Axle-Steel Deformed Bars for Concrete Reinforcement*, Latest Edition. West Conshohocken, PA: ASTM International.

2.0.0 Section Review

1. ACI specifies that reinforcing spirals and ties in columns must have a concrete cover of _____.
   a. 3"
   b. 2½"
   c. 2"
   d. 1½"

2. After each tie is made, the loose ends of tie wires must be _____.
   a. bent down
   b. covered with tape
   c. cut off
   d. wrapped around the rebar

3. The standard configuration for reinforcing bar is the _____.
   a. perforated bar
   b. deformed bar
   c. ridged bar
   d. plain bar

4. The U-shaped bar support that is used as a high chair on a lower mat of bars to support an upper mat of bars is called a _____.
   a. tie wire
   b. spreader
   c. bulkhead
   d. standee

5. When specifying welded-wire fabric reinforcement, such as 4 × 6 – D-10 × D-6, the letter D indicates _____.
   a. deformed wire
   b. plain wire
   c. driven rebar
   d. wire depth

Section Three

3.0.0 Bending and Cutting Reinforcing Steel

Objective

Describe methods by which reinforcing bars may be bent and cut in the field.
a. Describe how to cut rebar.
b. Describe how to bend rebar.

Performance Task

Use appropriate tools to cut and bend reinforcing bars.

The two main field tasks associated with rebar are cutting and bending the material to the correct specifications. Although rebar is commonly fabricated off site, some field fabrication may also be required.

3.1.0 Cutting Rebar

Rebar can be cut to close tolerances using a portable band saw or abrasive (chop) saw. These methods are suitable for those situations that require a bar to be cut to close specifications.

Manual rebar cutters (*Figure 29*) may also be used to cut rebar and are available in various sizes. Rods up to ½" in diameter may be cut quickly and easily using manual rebar cutters.

Manual or portable power shears (*Figure 30*) are also used to cut rebar, especially portable power shears on those job sites where a great deal of cutting must be done. These machines are available in various sizes and require special care during use. Only personnel who have been properly trained should use power equipment.

3.2.0 Bending Rebar

Reinforcing bars should be bent cold. Rebar may be bent by hand using a tool called a hickey bar or another type of rebar bender (see *Figure 31*). Hickey bars are not recommended for field bending of rebar larger than #5. *Figure 32* illustrates two types of hickey bars: straight and offset. For additional information on bending rebar in the field, refer to the latest publication of the CRSI.

Jigs may also be set up on a table. *Figure 33* shows a simple bending jig made of angle iron welded to a flat plate. The plate is then bolted securely to a table or other stable surface. Jigs may be used singly or in combination to fabricate column ties or bars requiring special bends.

Fabrication shops use power benders (*Figures 34* and *35*). There are portable models available that may be found on some job sites. Power benders are usually able to bend any size rebar to any desired shape. No preheating is necessary when bending the larger sizes of bars with a power bender. Only workers trained in their use should use a power bender.

Figure 29 Manual rebar cutter.

Rebar Cutter

A portable rebar cutter like the one shown here can be used to cut rebar up to and through size #6 (¾").

CUTTER

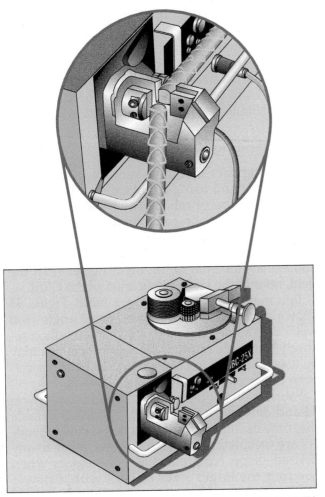

Figure 30 Rebar being sheared to length by a portable power shears.

HICKEY BAR **REBAR BENDER**

Figure 31 Hickey bar and rebar bender.

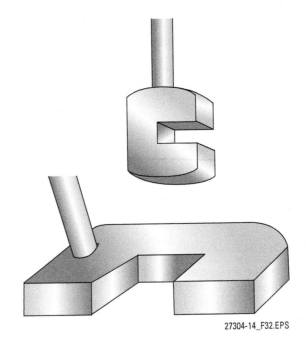

Figure 32 Fabricated hickey bars.

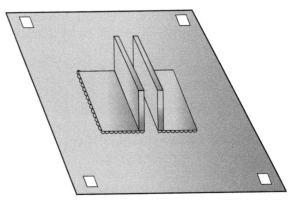

Figure 33 Bending jig.

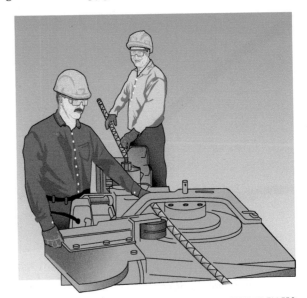

Figure 34 Bending reinforcing bars.

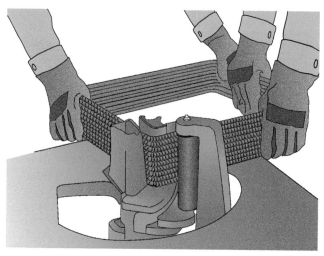

Figure 35 Multiple bending of closed column ties on a power stirrup bender.

3.0.0 Section Review

1. Manual rebar cutters are used to cut rebar up to 1 inch in diameter.
   a. True
   b. False

2. When rebar must be bent in the field, rebar smaller than #5 bar can be bent using a manual tool called a _____.
   a. hockey stick
   b. monkey bar
   c. hickey bar
   d. rod wrapper

Section Four

4.0.0 Placing Reinforcing Steel

Objective

Explain the methods for placing reinforcing steel.
a. Discuss the proper method for tying and splicing reinforcing steel.
b. Explain the proper procedure for placing reinforcing steel.

Performance Tasks

Demonstrate five types of ties for reinforcing bars.

Demonstrate proper lap splicing of reinforcing bars using wire ties.

Demonstrate the proper placement, spacing, tying, and support for reinforcing bars.

Trade Terms

Band: Reinforcing steel in columns that is wrapped around the vertical bars to counteract compression forces.

Contact splice: A means of connecting reinforcing bars by lapping in direct contact.

Double-curtain wall: A concrete wall that contains a layer of reinforcement at each face.

Far face: The face farthest from the viewer (as of a wall); may be the outside or inside face, depending on whether one is inside looking out or outside looking in.

Flat slab: A concrete slab reinforced in two or more directions, with drop panels but generally without beams, and with or without column capitals.

Lapped splice: The joining of two reinforcing bars by lapping them side by side, or the length of overlap of two bars; similarly, the side and end overlap of sheets or rolls of welded-wire fabric.

Near face: The face nearest the viewer, which may be inside or outside, depending on whether one is inside looking out or outside looking in.

Pile cap: A structural member placed on the tops of piles and used to distribute loads from the structure to the piles.

Rebar horses: Wood or metal supports that are used in groups of two or more to hold main reinforcing in a convenient position for placing ties while prefabricating column, beam, or pile cages.

Schedule: A table on placing drawings that lists the size, shape, and number of bars each way, and the mark number of the bars if they are bent.

Single-curtain wall: A concrete wall that contains a single layer of vertical or horizontal reinforcing bars in the center of the wall.

Staggered splices: Splices in bars that are not made at the same point.

Strips: Bands of reinforcing bars in flat-slab or flat-plate construction. The column strip is a quarter-panel wide on each side of the column center line and runs from column to column. The middle strip is half a panel in width, filling in between column strips, and runs parallel to the column strips.

Support bars: Bars that rest upon individual high chairs or bar chairs to support top bars in slabs or joists, respectively. They are usually #4 bars and may replace a like number of temperature bars in slabs when properly lap spliced; also used longitudinally in beams to provide support for the tops of stirrups. Also called raiser bars.

Temperature bars: Bars distributed throughout the concrete to minimize cracks due to temperature changes and concrete shrinkage.

Template: A device used to locate and hold dowels, to lay out bolt holes and inserts, etc.

Weephole: A drainage opening in a wall.

When reinforcing steel is properly cut and bent to specifications, it then must be placed into the forms and tied down to prevent the steel from moving during concrete placement. In some cases, shorter pieces of rebar must be spliced together to extend longer distances. This section provides information on the proper tying, splicing, and placing of reinforcing steel.

4.1.0 Tying and Splicing Reinforcing Steel

The wire used for tying rebar is usually 16-gauge black, soft-annealed wire. Galvanized wire is also available. Some applications may require the use of a heavier-gauge wire such as #15 or #14. The lower the number, the thicker the wire. The most common types of ties are shown in *Figure 36*. Each tie has a particular application.

The snap tie is the simplest and most basic of all ties. Several other ties end in a snap tie. This type of tie is normally used in flat, horizontal work to prevent the reinforcing bars from moving during concrete placement. The snap tie is made by wrapping the wire diagonally once around the two crossing bars. This should be done so that the ends of the wire end up on top to facilitate easy twisting. The ends are then twisted together with a pair of pliers until they are very tight against the bars.

The wrap-and-snap tie, or wall tie, is normally used when tying rebars placed in walls to keep the horizontal bars from shifting during concrete placement. The tie is made by wrapping the wire 1½ times around the vertical bar, then diagonally around the horizontal bar, ending in a snap tie.

The saddle tie is often used to hold the hooked ends of bars in position when tying footing or other mats. It is also used to secure column ties to vertical bars. The tie is made by passing the wire halfway around one of the bars (either vertically or horizontally) on each side of the crossing bar. The wires are brought squarely around the crossing bar, then up and around the first bar, where they are twisted.

The wrap-and-saddle tie is used to secure column ties to vertical bars when there might be a strain on the ties. It is sometimes used to secure heavy mats that are to be lifted by crane. The tie is made like the saddle tie with one exception: the wire is first wrapped 1½ times around the first bar.

The figure-eight tie is sometimes used in walls instead of the wrap-and-snap tie, but because it takes longer to tie, it is not usually recommended.

The nail-head tie (*Figure 37*) is used when nails are employed to hold wall bars away from forms. The tie is made by wrapping the wire once around the nail head, crossing the wire, and then wrapping it around the outside bar of the wall mat. The bar is drawn tight against the nail head by twisting the ends of the wire.

Reinforcing bars are tied to prevent their movement during normal construction processes or concrete placement. Tying adds nothing to the strength of the finished structure. Therefore, it is not necessary to tie rebars at every intersection. In most cases, tying 25, 33, or 50 percent of the intersections is sufficient. However, the perimeter of the mat must be 100 percent tied. When tying bars in mats that are being assembled in place, the size of the reinforcing bars should dictate the percentage of ties. It is an accepted practice to stagger the intersections being tied. This practice gives the mat added rigidity (*Figure 38*).

For pre-assembled mats, enough intersections must be tied to make the mats sufficiently rigid for handling by crane. A general rule of thumb for tying pre-assembled mats is to tie the perimeter 100 percent and the interior 25 percent to 50 percent unless the job specifications require otherwise.

4.1.1 Tying Reinforcing Steel

Since most tying is done in flat, horizontal formwork, learning to tie stiff-legged instead of in a squatting position will eliminate many aches and pains. After finishing a tie, if the wire from the coil is left bent, the next tie will be easier to make and less likely to cause an eye injury.

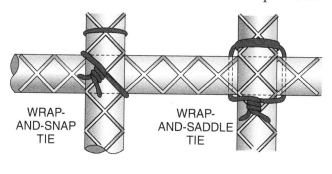
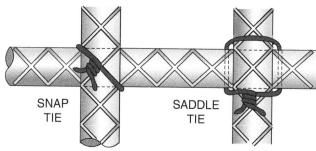
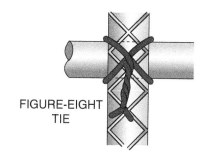

Figure 36 Types of ties.

Figure 37 Nail-head tie.

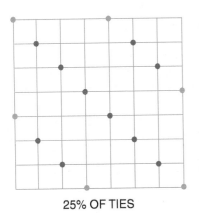
25% OF TIES

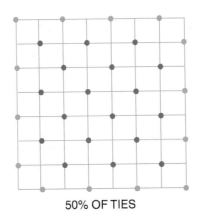
50% OF TIES

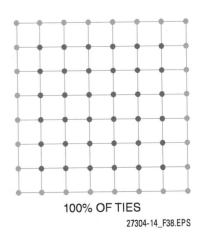
100% OF TIES

Figure 38 Percentage of ties.

To be used effectively, pliers must become an extension of your hand (*Figure 39*). Practice makes perfect in this regard. Develop a rhythm to your work. Wasted motion increases fatigue. In any trade, the true masters are those who make their work look effortless through no waste of motion.

4.1.2 Splicing Reinforcing Steel

Because in most situations it is impossible to provide full-length bars that run continuously throughout a structure, splicing reinforcing bars is common. The placing drawings will show the location and type of splice to use. Sometimes several methods of splicing will be listed by the placing drawings, and the best method may be chosen from among them. No splice should be used, however, without consulting the proper authority.

There are three basic types of splices used in reinforcing steel work. These are lapped splices, welded splices, and mechanical-coupling splices.

As suggested by the name, a lapped splice joins two pieces of reinforcing steel by placing them side by side. In general, three variables affect the length of the lap. These are:

- *Strength of the concrete* – The stronger the concrete, the shorter the splices.
- *Grade of reinforcing steel* – The higher the grade, the shorter the splices.
- *Size of the bars* – The larger the bar, the longer the splices.

These are general statements that have been included to provide an understanding of why certain lengths for certain splices are specified. In actual practice, the amount of overlap is determined by a variety of design considerations as governed by *ACI 318*, which contains tables that provide detailed requirements for lap splice lengths. The placing drawings will usually provide all the necessary information concerning the lengths of the splices. A rule of thumb is to provide a lap equal to 30 times the bar diameter or 12", whichever is greater.

There are two kinds of lapped splices. The contact splice shown in *Figure 40* is made by placing the bars next to each other so that they touch (based on the latest ACI specification).

The spaced-lap splice shown in *Figure 41* is made by placing the bars a certain distance from one another without any actual contact. The maximum center-to-center spacing of the bars must not exceed 6".

Figure 39 Pliers used for tying rebar.

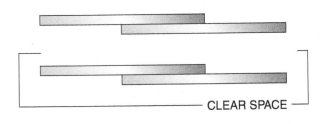

MIN. FOR BEAMS = REBAR DIAMETER OR 1" WHICHEVER IS LARGER

MIN. FOR COLUMNS = 1.5 x DIAMETER OR 1½", WHICHEVER IS LARGER

Figure 40 Contact splice.

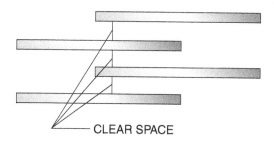

FOR BEAMS = d OR 1"
WHICHEVER IS LARGER

FOR COLUMNS = 1½d
OR 1½", WHICHEVER IS LARGER

Figure 41 Spaced-lap splice.

Of the two, the contact splice is preferred over the spaced-lap splice because it provides more strength to the finished concrete.

ACI prohibits lap splices of #14 and #18 bars except when these bars are spliced to smaller-sized dowels at footings.

There are basically two types of mechanical splices: coupling splices and end-bearing splices. They are differentiated by the types of forces they resist.

- *Coupling splices* – These splices resist both compression and tension forces. In other words, they are able to withstand forces that try to crush them and pull them apart. One such coupling splice is the forged-steel sleeve. This splice is made by heating a special sleeve until it becomes workable. The ends of the bars are then placed into it, and a portable hydraulic forge crimps the sleeve, forcing the inside of the sleeve against the deformations of the bars. As the sleeve cools, the shrinkage forces cause a tighter bond. Threaded couplers are also available.
- *End-bearing splices* – A bolted-clamp assembly is available that is used primarily in splicing vertical rebar for columns. The splice is made first by bolting a clamp to the bottom bar. The upper bar is then placed and plumbed, and the upper bolts of the clamp are tightened. Reducers are available that enable a bar of one size to be spliced to a bar of the next smaller size.

4.2.0 Placing Reinforcing Steel

This section provides an overview of the mechanics of placing reinforcing steel.

4.2.1 Placing Bars in Footings and Foundations

A footing is the part of a foundation that rests upon the earth. Footings are placed under columns, piers, and walls to provide the necessary support for these structural members.

The function of a footing is to transmit the concentrated load of a structure to the ground at a pressure that is safe and will not cause settling of the structure. Reinforcing steel is placed within the concrete footing to counteract the forces that tend to bend or break the concrete.

Footings vary greatly in terms of size and shape, depending on the type of load they must carry and the condition of the ground. Generally, the softer the ground, the larger the footing. Location and placement of bar is critical to achieve engineered design strength.

The foundation placing drawings show the individual footings. These drawings usually con-

---

Rebar Location and Coverage Is Critical

Location and placement of reinforcing bars is critical to ensure that the concrete reaches its design strength and is able to support the intended loads. The following ACI guidelines for rebar placement should be followed unless specific dimensions are provided by the architect or engineer:

- 3" at sides where concrete is cast against earth and on bottoms of footings or other principal structural members where concrete is deposited on the ground
- 2" for bars larger than #5 where concrete surfaces would be exposed to the weather after removal of forms or would be in contact with the ground; 1½" for #5 bars and smaller
- 1½" over spirals and ties in columns
- 1½" to the nearest bars on the top, bottom, and sides of beams and girders
- ¾" for #11 and smaller bars on the top, bottom, and sides of joists, and on the top and bottom of slabs where concrete surfaces are not directly exposed to the ground or weather; 1½" for #14 and #18 bars
- ¾" from the faces of all walls not directly exposed to the ground or weather for #11 and smaller bars; 1½" for #14 and #18 bars

Always refer to the structural notes for clearance and coverage requirements.

tain a schedule that lists the size, length, number of bars each way, and mark number of the bars if they are bent. The spacing of the bars may also be provided; if it is not, the bars should be spaced evenly within the footing, based upon the number of bars indicated by the placing drawings. See *Figure 42*.

Very often, bars are assembled into mats prior to installation. All the bars to be placed in one direction are laid out evenly across sawhorses. Then the bars that are located in the opposite direction are laid across the first bars and evenly spaced. The bars are then tied securely at every second or third intersection within the interior of the mat and are usually tied at all intersections around the perimeter to facilitate easier rigging of the mat. Snap ties are normally used for this purpose. If the mat is not to be placed immediately, it should be tagged with the footing number and stored off the ground in a place where it can be easily delivered to the installation site.

The footing mat is usually placed on precast concrete blocks of the proper height. Successive mats are supported in a number of ways depending on the particular needs of the job.

The reinforcing steel for continuous wall footings is normally assembled in place in the trench. Usually, two or more vertical bars are crossed by horizontal bars at spacings indicated by the placing drawings. Dowels, if required, must project above the footing the distance called for by the specifications to allow for the proper vertical splice.

Where two wall footings meet, the horizontal bars may either extend a specified distance into the intersecting footing or be designed with corner bars, depending on the method specified. See *Figure 43*.

A pile cap is a structural member placed on the tops of piles. It is used to distribute loads from the structure to the piles. Placing reinforcing steel in pile caps is similar to placing reinforcing steel in footings. The bottom mat rests on individual high chairs placed directly on the piles. Bar placement, however, may not always be at right angles, as shown in *Figure 44*.

Instead of individual column footings or continuous wall footings, a single slab of concrete may be used to support all the units of a structure. The thickness of this slab varies greatly, depending on the soil conditions and loads to be supported. In general, the thickness of the slab is 1' to 5', but it may be as much as 15'.

A mat of reinforcing steel with bars running in two directions is normally used in slabs. The bottom mat generally rests about 3" above subgrade on concrete blocks placed 5' apart in each direction. The top of the mat, or the top layer of bars, is usually about 2" or 3" below the surface of the concrete.

The placing drawings provide all the necessary information concerning the slabs, including the number of pieces, sizes, length, and spacing of the bars. Successive layers of bars are supported by standees or other suitable bar supports.

It is not difficult to determine the spacing of bars within a footing. The first step is to establish the center line of the footing. If an odd number of rebar is to be placed within the footing in one direction, one rebar will sit directly on the center line. If an even number of rebar is to be placed

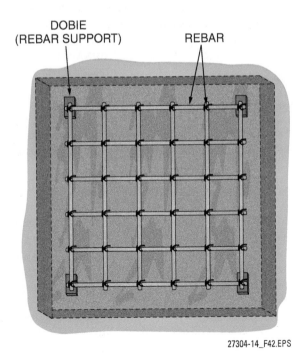

Figure 42 Square footing.

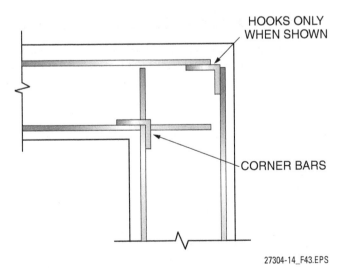

Figure 43 Corner bars.

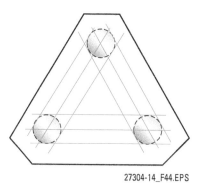

Figure 44 Pile cap.

within the footing, one rebar will lie on either side of the center line. Proceed as follows:

Step 1 Convert all dimensions to inches.

Step 2 Subtract 1 from the number of rebar to be placed in one direction (that is, from those rebar that will lie north-south or east-west).

Step 3 Divide the number of inches by the number of rebar to be placed in one direction minus 1. A remainder indicates the number of inches from each end that a bar will lie.

For example, suppose a mat that is 15' square is called for by the placing drawings. The mat, according to the information found on the placing drawings, is to be made from 22 #5 reinforcing bars, with each bar having a length of 14'-6".

First, the center line should be established and marked. This is true even if the mat is being pre-assembled on sawhorses.

Next, find out where the first bar will be placed in relation to the center line. Since the mat is to be assembled from a total number of 22 bars, the mat will be made of 11 bars running north-south and 11 bars running east-west. Eleven is an odd number; therefore, the first bar will be placed directly on the center line.

Fifteen feet converts to 180". Subtract the center bar from the number of bars to be placed in one direction. This yields 10. Divide 180" by 10; the result is 18". This is the spacing of the bars.

Place the first bar directly on the center line, the next bar 18" away from it (on center), and the next 18" from the previous, etc. Do the same for the bars running in the opposite direction, and tie the bars securely together at as many intersections as needed.

Consider another example. Suppose the same mat is to be made using 24 #5 reinforcing bars instead of 22. First, establish the center line. The mat will be made of 12 bars placed north-south and 12 bars placed east-west. This means that a bar will be placed on each side of the center line.

In this case, 180" is to be divided by 12 minus 1, or 11. This yields 16" with a remainder of 4". This means that two bars will straddle the center line and be 16" apart. (Each of the first two bars will be 8" from the center line.) Place the remaining bars on 16" centers from each other for the rods running in both directions. Tie the mat securely. There will be 4" from the last bar running north-south to the end of the bars running east-west.

4.2.2 *Placing Bars in Walls*

For the purpose of placing reinforcing steel, each wall has two faces, labeled simply the far face and the near face, which is a matter of orientation depending on whether the viewer is inside or outside.

Mechanical Rebar Splices

An example of a bolted-clamp assembly used for splicing vertical rebars for larger columns is shown in this figure.

The two basic types of reinforced concrete walls are single-curtain wall and the double-curtain wall. The single-curtain wall is a type of reinforced concrete wall in which a single layer of reinforcing steel is placed in the center between the faces. A double-curtain wall is a reinforced concrete wall in which a layer of reinforcing is placed at each face. A retaining wall, which may be either a single-curtain or double-curtain wall, is used to hold earth or fill in place.

When constructing a reinforced concrete wall, the formwork is usually erected for one face and braced from the outside so that the reinforcing steel may be placed within it. The first step is usually to fasten the vertical bars of the outside face to the dowels projecting upward from the footing. Then, one horizontal bar is typically wired to the vertical bars to keep them plumb. A wrap-and-snap tie is generally used, but a figure-eight tie can be used if more stability is required. The remaining vertical and horizontal bars are then tied at every third intersection, using no fewer than three ties per bar. If the wall being tied is a double-curtain wall, the rebar is tied to the bars of the outside face, and the inside face is set in the same manner. These mats of reinforcing steel can also be constructed on a level surface (prefabricated) and then placed and stabilized in the form.

Reinforcing-steel wall mats must be supported and/or spaced away from the formwork at the top to maintain the proper cover. There are three ways to do this. The first way is to use short lengths of slab bolsters (as shown in *Figure 45*), beam bolsters, or individual bar chairs. These are stapled to the forms, and the mat is wired to them. Class C, D, or E chairs should be used in areas of exposed concrete to avoid rust.

A second way is to use precast concrete blocks with embedded tie wires. See *Figure 46*. This is the same method used when placing bars in footings. This method is typically used for below-grade applications.

Another method to support wall mats is to drive nails into the forms, leaving them exposed for the required amount of cover, and then tie the

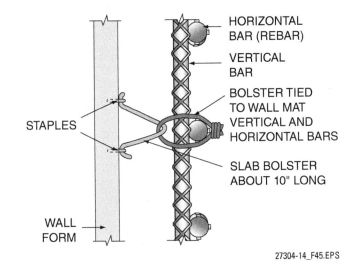

Figure 45 Slab-bolster support.

Automatic Rebar-Tying Tool

This pneumatic tool is designed to wrap, twist, and cut wire in one operation. According to the manufacturer, a person using this tool can tie rebar five times faster than a person manually tying the rebar. Another claimed advantage is the elimination of the repetitive wrist-twisting action that can lead to the condition known as carpal tunnel syndrome. The machine is fed by a spool of wire. Each spool can make more than 100 ties. An extension bar is available so that ties can be made without bending over.

wall mat to the nails using the nail-head tie (*Figure 47*). This method typically sees limited use; workers can be scratched or cut by an exposed nail and the potential rusting of the nails can be a concern.

In double-curtain walls, the mats must be spaced a specified distance from each other. This information is provided on the placing drawings. One way to space the mats is to use spreader bars. These bars are usually made from #3 rebar and bent into a U or Z shape. They are essentially standees meant to be used in walls instead of slabs.

Prefabricated wire spreaders are also available that both support and spread the wall mats. They are used especially for those situations in which the concrete will be exposed. *Figure 48* illustrates a combination of all the supports and spacers discussed.

When two walls meet at a corner, the bars of the outside faces may be bent 90 degrees or may extend for a specified distance into the intersecting wall and be spliced to a corner bar. The bars of the inside faces usually extend into the intersecting wall but may be provided with hooks. As in all cases, the placing drawings provide all the necessary information. *Figure 49* illustrates the various methods in use for both single-curtain walls and double-curtain walls.

Reinforcing steel is placed in cantilevered retaining walls in much the same way as it is placed in supported walls, except for the following:

- Bottom-footing reinforcing bars may be bent upward to be used as dowels for the back face of the wall.
- Vertical reinforcing bars are two or three different heights with perhaps every second or third bar extending to the top of the wall. Others terminate at specified cutoff points.
- A weephole is usually supplied for drainage. Care must be taken not to alter the position of the pipe when tying reinforcing steel.

4.2.3 Placing Bars in Columns

A column is a vertical member used to support a floor beam, girder, or other structural member. The main load to which a column is subjected is compression. To counteract the compression force,

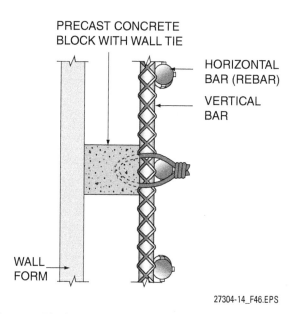

Figure 46 Block support.

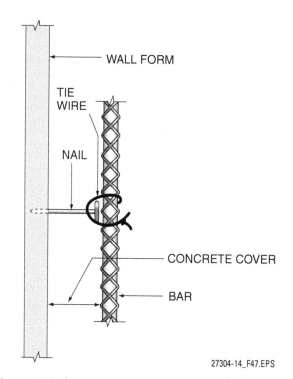

Figure 47 Nail support.

Fabricating Rebar Mats

When it is necessary to fabricate similar rebar mats, you can increase your productivity by making a template. Do this by nailing clean 2 × 4s to the tops of four sawhorses. Then lay out and mark the spacing for the north-south and east-west rebars on the 2 × 4s. Following this, drive nails on both sides of all rebar layout marks. These nails will serve as guides for rod placement and subsequent mat layout and fabrication. This method not only increases your productivity, it also helps maintain accurate spacing of the rebar while they are being placed and tied into position.

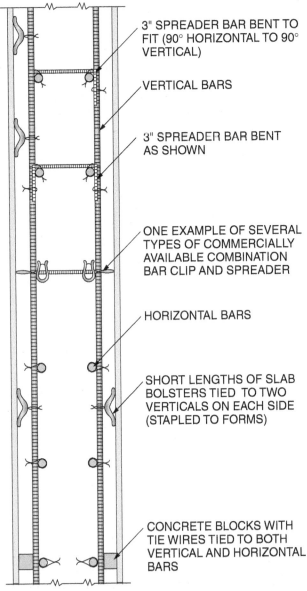

Figure 48 Spacers and spreaders.

the reinforcing steel used in columns is wrapped in **bands** around the vertical bars. This configuration of reinforcing steel, combined with the good resistance concrete exhibits against compression, makes a properly designed and tied column a very strong part of a structure.

Columns that rest on footings are connected to those footings by dowels. A dowel is a steel reinforcing bar that connects two separately cast sections of concrete. Care must be taken to place dowels in the exact places identified in the placing drawings. Dowels must project a specified distance above the footing and be placed accurately with respect to the location of the reinforcing steel to be placed within the column. Omitting or misplacing dowels can cause serious problems.

A common practice in dowel placement is to use a **template** (*Figure 50*). A template is made of boards with holes drilled in them to serve as guides for the dowels. Guide boards having a tie or a piece of spiral affixed to them at the location where the dowels are to be placed can also be used (*Figure 51*).

If the dowels have hooks, additional stability can be obtained by tying the hooked end to the footing mat of reinforcing steel.

Dowels should not be pushed into wet concrete. It is very difficult to maintain proper dowel alignment this way, and corrections after the concrete has set are difficult and costly. Whenever possible, templates should be used to obtain the correct positioning.

When placing dowels that are to be butt-spliced to the column vertical bars, extra care is needed because the splices must be staggered. The dowel cage must be constructed with **staggered splices** in mind, and the placing drawings must be studied very carefully.

Be alert for dowels required for grade beams that join the sides of footings or for dowels that extend into foundation walls or other structural units to be built later.

Another method of placing dowels is to construct a dowel cage, as shown in *Figure 52*. A dowel cage is made by assembling the dowels having 90-degree hooks in such a manner that the hooks provide a stable means of support to the footing mat. The correct placement of the dowels above the footing must be ensured.

Rebar for columns can also be assembled as rebar cages. The bands wrapped around the vertical bars are called column ties. They must be placed outside the vertical bars. These ties can be square, rectangular, U-shaped, circular, or any other shape designated by the engineers. Standards have been developed concerning the placing of column ties. *Figure 53* shows the standard placing of column ties in columns containing an even number of vertical bars. The information in *Figure 53* applies to those cages that are either pre-assembled or erected in place on freestanding, butt-spliced vertical bars.

Figure 54 shows the standard placing of column ties for lap-spliced pre-assembled cages only. The dotted lines indicated in *Figure 54* for the 6-, 8-, and 10-bar columns show how the column ties are to be tied if the distance between the centers of the bars is over 6".

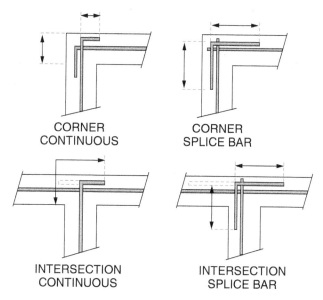

Figure 49 Corner details.

Each pattern consists of an outside closed tie with pairs of U-shaped bars that are lap-spliced and hooked at each end.

The usual procedure for tying pre-assembled reinforcing column steel is to first lay out all the vertical rebar for one side across supports, called rebar horses, as shown in *Figure 55*. Then the vertical column ties are laid out on the vertical rebars and spaced according to the placing drawings. It is generally acceptable to alternate the position of the hooks of the column ties when placing them in sets. The column ties are then wired to the vertical rebar using a saddle tie or a wrap-and-saddle tie for heavy bars. The remaining vertical rebar are put in place and wired to the column ties. When tying columns, it is extremely important to make sure all vertical rebar are lined up and even on the bottom end so that all rebar of the finished column cage will sit firmly on the footing. On 4- and 6-bar columns, every tie should be wired to every vertical rebar at every intersection to achieve good stability. On columns of 6 bars or more, tie 100 percent of the column-corner bars using saddle ties.

Large square or rectangular units may require diagonal wire bracing for greater stability. The bracing should be twisted with pliers until a sufficient amount of tension is obtained.

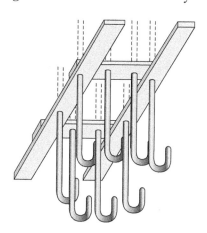

Figure 50 Dowel template.

Cleanouts

When constructing wall or column forms, it is good practice to construct one or more 6" cleanouts at the bottom of the form. The cleanouts provide easy access to the interior of the wall or column forms and allow debris to be removed before concrete is placed. Cleanouts are typically located at the corners and intersections of the form. Always ensure the cleanouts are closed and properly braced before concrete is placed in the forms.

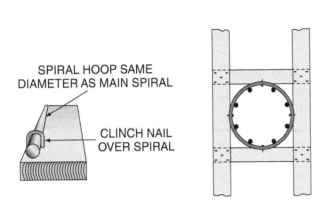
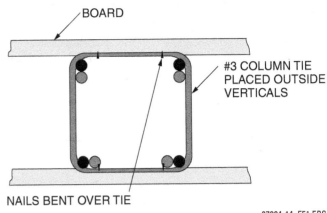

Figure 51 Guide boards.

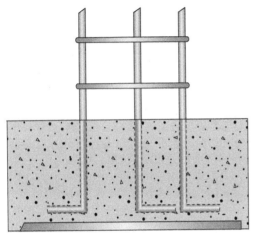

Figure 52 Dowel cage.

Spirals are made of plain rebar or wire shaped like a coil spring. They have spacers attached on opposite sides that keep the turns of the spiral at equal distances. In terms of strength, a column using spirals is generally stronger than a column using square or rectangular column ties.

Spirals are usually shipped collapsed and must be opened or knocked down at the job site prior to assembly. To pre-assemble spirals, two vertical bars are placed inside the spiral and supported at either end by rebar horses. See *Figure 56*. After two bars are placed, the remaining number of vertical rebar specified in the placing drawings are inserted and spaced according to the specifications. The bars are then wired to the spiral to achieve the necessary stability and rigidity.

Column forms should be supported at three or four points, if possible. The supports should be placed as near to the bottom, midsection, and top of the column as possible. Column forms may be supported with nails driven into the inside of the forms in much the same way nails are used to support wall reinforcing. However, precast concrete blocks with embedded tie wires are more commonly used.

If precast blocks are used, they should be wired to the column at intersections of the column ties and vertical bars. This will prevent the blocks from spinning when the column is lowered into place within the form.

4.2.4 Placing Bars in Beams and Girders

There is no standard sequence that can be listed concerning the placing of bars in beams and girders due to the individual differences in beam height and bar arrangement in any given structure. This section is intended to provide a general overview of bar placement methods and sequences. As in all steel reinforcing work, the placing drawings will dictate the best sequence. They must be studied carefully so that an efficient plan of placing may be developed.

A beam is a horizontal structural member used to carry loads from a floor to columns, walls, and girders. A girder is a principal beam. The main difference between the two is that beams support other parts of a structure, while girders support beams. This difference becomes important when placing reinforcing steel in beams and girders because it affects the sequence of placing rebar.

In general, girders will be lower than the beams they support, so all bottom horizontal bars, truss bars, and stirrups must be placed in girders before any reinforcement is placed in beams. The actual placing procedures, however, are similar for both beams and girders. The placing of reinforcing steel in beams and girders begins after column vertical rebar are positioned and concrete has been placed to the bottom of the lowest beam or girder. Reinforcing steel is generally placed from bottom to top; that is, first in girders, then in beams, and finally in slabs and joists.

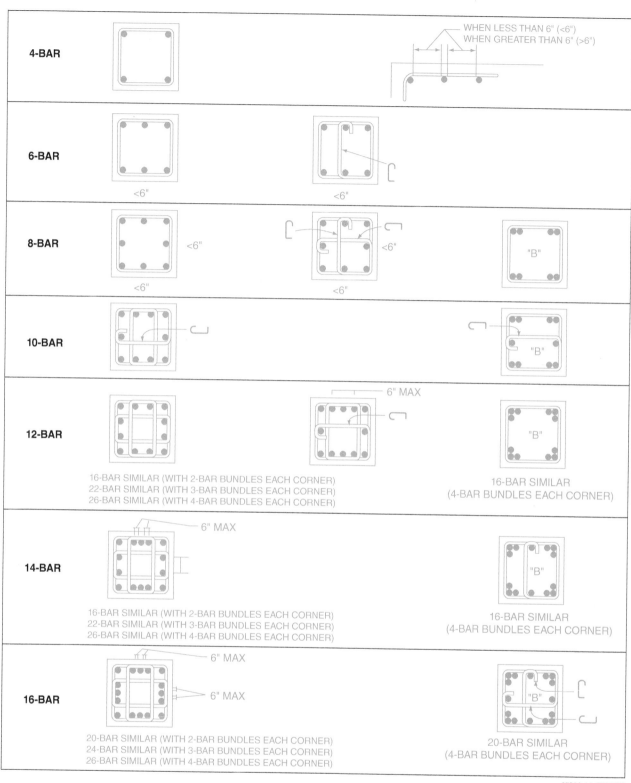

Figure 53 Standing column ties.

Placing Dowels

For dowels used in a continuous wall footing, the horizontal leg of the dowel is placed perpendicular to the length of the footing. Column or pier dowels placed in square, rectangular, or circular footings are placed such that the horizontal leg radiates from the center of the dowel configuration.

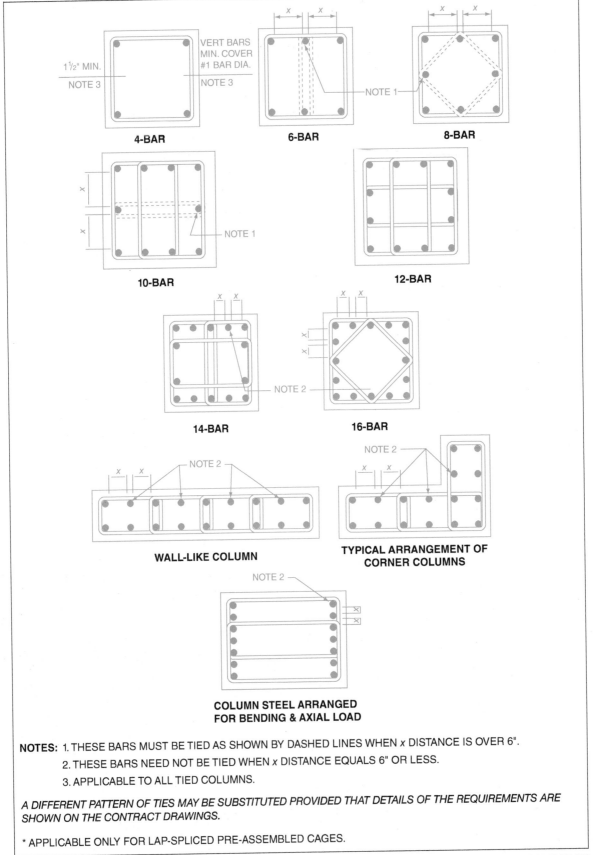

Figure 54 Closed column ties.

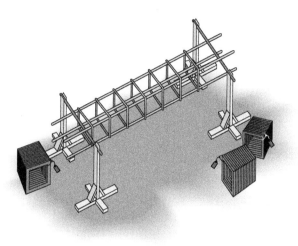

Figure 55 Rebar horse.

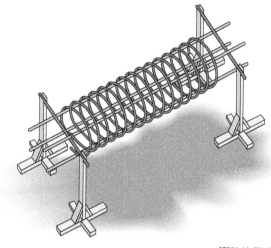

Figure 56 Spirals.

This sequence usually minimizes the need to manipulate rebar under one another at the points of intersection.

Figure 57 illustrates the general procedure for placing reinforcing steel in beams within the formwork. To place reinforcing steel in beams, proceed as follows:

Step 1 The beam bolsters are placed in the forms on centers not to exceed 5'.

Step 2 The stirrups and stirrup support bars are then placed in the forms. The forms should be marked with the proper spacing as found in the placing drawings.

Step 3 The straight bottom bars are placed next. In order to prevent these bars from moving during the concrete pour, they are often wired to the beam bolsters.

Step 4 If more than one layer of bars is required, the upper beam bolsters or bar separators are placed in the forms.

Step 5 The truss bars are placed last. A truss bar is a bar that has been bent in such a way that it serves as both top and bottom reinforcement. If truss bars are used in beams that require reinforcing placed in two layers, the truss bars should be placed directly over those in the lower layer, not in the spaces between the bars.

The reinforcing steel for beams and girders may also be pre-assembled. The sequence is usually as follows:

Step 1 Two straight bars are used as templates and marked with keel (marking crayon) or soapstone using the spacings found on the placing drawings.

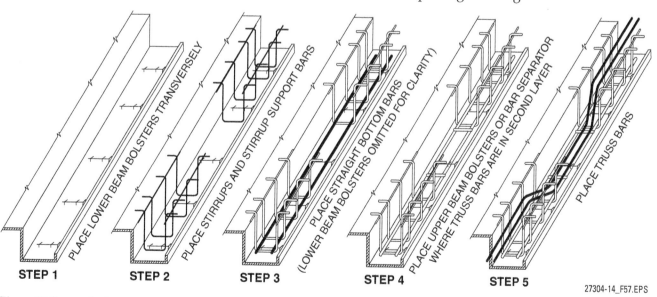

Figure 57 Beam placing sequence.

27304-14 Reinforcing Concrete   Module One 45

Step 2 The template bars are placed on rebar horses.

Step 3 Stirrups are placed on the marks and tied in place using saddle ties.

Step 4 A side bar is tied to the stirrups.

Step 5 At this point, the beam may be taken off the rebar horses and placed flat on the deck or ground. Diagonal wire braces may be tied to it to provide added rigidity.

Step 6 The bars may then be placed into the form and the bottom layers of reinforcing may be added.

Occasionally, the placing drawings will require closed ties instead of open, U-shaped stirrups. See *Figure 58*. Closed ties may be one piece with hooks in a corner or two pieces, known as cap ties, as shown in *Figure 59*.

The easier tie to use is, of course, the cap tie, because the bottom piece can be placed in the same way as a stirrup. After all other bars have been placed, the top piece of the cap tie may be placed and wired to the lower piece. If the drawings call for one-piece closed ties, these ties must be used. They can sometimes be slipped into place over the lengthwise bars. Usually, however, they must be sprung open enough so that they can be worked around the bars running lengthwise. This tends to twist the reinforcing steel out of shape and is also time consuming.

A joist is a small beam that is placed parallel between the main beams in floor construction. When joists are joined at the top to make a continuous structure, this structure is called a joist slab. Joist slabs are used in situations where high loads are anticipated.

Placing reinforcing steel in joists begins when all the beam reinforcement has been placed. In general, the sequence for placing reinforcing steel in joists is as follows:

Step 1 Place joist chairs in the form, beginning 1" from the edge of each support. If possible, joist chairs should be spaced as close as 5', or as specified.

Step 2 Place the straight bottom bars on the joist chairs. If necessary, thread the bars between the beam stirrups and under the top beam bar. The bottom bars must extend into the supports at each end according to the specifications found in the placing drawings.

Step 3 Place the truss bars on the joist chairs next to the straight bottom bars. The bent-up ends of the truss bars must cross over all top bars in the beam and extend into the adjoining section by the amount required by the placing drawings. If no truss bars are required, two bottom straight bars are generally used. Likewise, one or two top bars are generally extended into the adjoining section. The extended ends of the truss bars or straight bars are supported on individual chairs placed on support bars. They may also rest on upper-joist chairs. The placing drawings will indicate the proper method of support.

Step 4 Place the distribution ribs next in one or two lines. These are also called continuous header joists or bridging ribs. They

Reinforcing-Steel Framework

This is an example of pre-assembled reinforcing-steel framework for a column that will eventually be raised into place. For ease of construction, rebar is often installed horizontally and then raised and placed into forms.

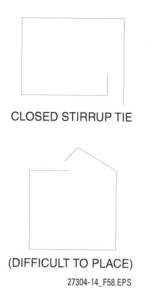

Figure 58 Closed stirrup.

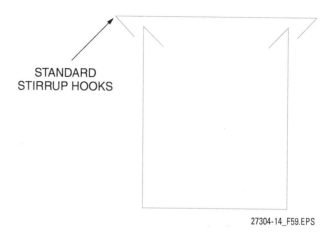

Figure 59 Capped stirrup.

extend the full length of the joist bay, are as deep as the joists, and are placed at right angles to the main joists. They provide lateral bracing for long spans of joists. Generally, there will be one rib placed at midspan for spans ranging from 18' to 24'. Bars in these ribs are usually shown on the floor joist plans.

Step 5 Place the temperature bars last. Their purpose is to minimize cracks due to changes in temperature and the normal shrinkage of concrete. Temperature bars are either #3 bars or welded-wire fabric reinforcement. If welded-wire fabric reinforcement is used, it should be unrolled so that it arches upward and then bends straight. Welded-wire fabric reinforcement is laid across the joists.

According to the direction of the main reinforcing run, slabs can be classified into two types. As its name indicates, a one-way slab contains reinforcement that runs in one direction between supports. A two-way slab contains reinforcement that runs in two directions between supports.

Reinforcement in one-way slabs consists of alternating straight bars and truss bars, or straight top and bottom bars running in one direction (*Figure 60*). Temperature bars are placed at right angles to the main reinforcing bars. Reinforcement is placed in one-way slabs in the same direction as the slab distributes the load applied to it.

The general procedure for placing and tying reinforcing steel in one-way slabs with straight bars and truss bars is as follows:

Step 1 Place slab bolsters so that they will lie at right angles to the main reinforcing. The placing drawings provide the proper spacing.

Step 2 Place main reinforcing according to the placing drawings. Each group of steel is tied in place as it is set.

Step 3 Place high chairs at right angles to the main reinforcing.

Step 4 Place temperature bars at right angles to the main reinforcing.

Step 5 Place truss bars on high chairs. Ensure the truss bars are parallel with the main reinforcing.

Reinforcement is placed in a two-way slab in the same direction as the slab distributes the load applied to it. See *Figure 61*. The reinforcement usually consists of straight and truss bars or straight top and bottom bars arranged in strips called column strips and middle strips.

The width of each strip is generally one-half the distance between the centers of the columns. Column strips usually receive more reinforcement than middle strips. A proper placing se-

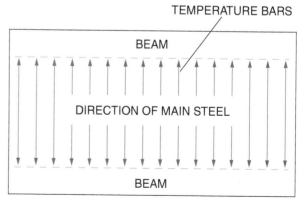

Figure 60 One-way reinforced slab.

quence must be followed to avoid threading the bars. Study the placing drawings very closely.

A general sequence for placing reinforcing steel in a two-way flat slab with straight and truss bars is as follows:

Step 1 Place continuous lines of slab bolsters in an east-west direction. Proper spacing is found on the placing drawings.

Step 2 Place the required lengths of slab bolsters in the east-west column strips at right angles to the slab.

Step 3 Place bottom straight bars running north-south in column and middle strips.

Step 4 Place bottom straight bars running east-west in column strips.

Step 5 Place three rows of #4 support bars on high chairs in an east-west direction at the head of each column. Tie the middle support bar to the column verticals.

Step 6 Place truss bars running north-south in column strips.

Step 7 Place straight top bars. These bars are usually placed within the bend-down point of the truss bars running east-west.

Step 8 Place truss bars running east-west in column strips. Usually, the east-west truss bars that rest upon the north-south, straight, middle-strip, and bottom bars are tilted sideways so that they rest upon the top bars running north-south.

Step 9 Place three more rows of #4 support bars on high chairs in north-south and east-west column strips. Two rows should be placed at all slab edges.

Step 10 Place truss bars running north-south in the middle strip.

Reinforcing bars must be fabricated to conform to the designs drawn by the engineers for the particular structure. Most of the fabrication is done in shops, but a certain amount must always be done in the field. Fabrication of reinforcing bars includes the cutting, bending, and splicing of bars in accordance with certain tolerances and codes established by the American Concrete Institute.

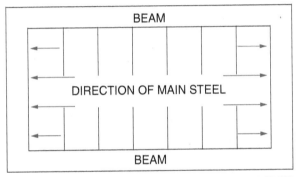

Figure 61 Two-way reinforced slab.

4.0.0 Section Review

1. The wire used for tying rebar is usually _____.
   a. 12 gauge
   b. 15 gauge
   c. 16 gauge
   d. 20 gauge

2. The part of the foundation that transmits the concentrated load of a structure to the ground is the _____.
   a. wall
   b. sill plate
   c. footing
   d. double top plate

Summary

Concrete used in structures must be properly reinforced with steel bars or welded-wire fabric reinforcement. Workers who place and tie rebar must be able to identify the various rebar types and sizes, as well as the various types of supports used to suspend rebar in the concrete.

In addition to describing concrete reinforcing materials and accessories, this module presented general information on cutting, bending, placing, splicing, and tying reinforcing steel for reinforced concrete members. Even though there are general methods of placing reinforcing steel in various types of structural components, the placing drawings provide the exact specifications and locations governing the reinforcing steel. The most efficient system of bar placement should be determined from the placing drawings.

Review Questions

1. Unreinforced concrete typically can have a compressive strength of 2,000 to _____.
   a. 4,000 psi
   b. 5,000 psi
   c. 6,000 psi
   d. 7,000 psi

2. The adhesion of concrete to reinforcing steel is known as _____.
   a. chemical transformation
   b. the concrete bond
   c. the curing process
   d. sublimation

3. Column ties or column spirals are used with vertical reinforcement to prevent _____.
   a. buckling
   b. slump
   c. cracking
   d. compression

4. The supports at either end of a beam bridge are called abutments or _____.
   a. terminal piers
   b. anchor blocks
   c. end bents
   d. cantilevers

5. Strands of a posttensioning tendon are typically made from _____.
   a. copper
   b. aluminum
   c. steel
   d. brass

6. Placement of reinforcing bars may sometimes conflict with the position of other items buried in the slab (such as wiring conduits) that are referred to as _____.
   a. embedments
   b. submerged facilities
   c. mechanicals
   d. obstacles

7. Bundles of reinforcing bars should be hoisted using _____.
   a. nylon slings
   b. the bundle's wire wrapping
   c. heavy manila rope
   d. double chain slings

8. Expansion and contraction joints in highway construction can use short lengths of plain rebar called _____.
   a. pegs
   b. transfer pins
   c. dowels
   d. slip units

9. Rebar is manufactured to a standard length of _____.
   a. 45 feet
   b. 60 feet
   c. 75 feet
   d. 90 feet

10. Stainless steel reinforcing bars are typically used in _____.
    a. cantilevered structures
    b. high-temperature environments
    c. highly corrosive environments
    d. submerged structures

11. Each dimension of a standardized bar bend is designated by a _____.
    a. letter
    b. number
    c. letter/number combination
    d. symbol

12. The fabrication tolerance for straight reinforcing bars is a variation in length of _____.
    a. ½ inch
    b. ¾ inch
    c. 1 inch
    d. 1½ inches

13. Precast concrete blocks, which can be used as bar supports in footings, are also known as _____.
    a. darbies
    b. dillies
    c. derbies
    d. dobies

14. Plastic bar supports are used in situations where _____.
    a. #3 or smaller rebar is specified
    b. corrosion must be avoided
    c. cost is a factor
    d. ambient temperatures will be less than 80°F

15. Welded-wire fabric reinforcement is available in flat mats measuring _____.
    a. 2' × 4'
    b. 3' × 6'
    c. 4' × 4'
    d. 8' × 16'

16. Rebar with a diameter of ½" is designated as _____.
    a. #3
    b. #4
    c. #5
    d. #6

17. When a great deal of rebar cutting must be done on the job site, the cutting method of choice is _____.
    a. an oxyacetylene torch
    b. bolt cutters
    c. power shears
    d. a chop saw

18. The simplest type of rebar tie is the _____.
    a. snap tie
    b. figure-eight tie
    c. saddle tie
    d. wrap-and-snap tie

19. The recommended posture for tying horizontal, flat reinforcement is _____.
    a. squatting
    b. sitting
    c. kneeling
    d. standing stiff-legged

20. When making a spaced-lap splice, the center-to-center spacing of the bars must not exceed _____.
    a. 2"
    b. 4"
    c. 6"
    d. 8"

Trade Terms Quiz

Fill in the blank with the correct term that you learned from your study of this module.

1. The quantities, lengths, sizes, and grades of all bar materials to be used are shown on the _____.

2. Bars of a single length, size, or mark (bent) are often fastened together in a unit known as a(n) _____.

3. _____ are columns consisting of vertical reinforcing bars surrounded by a spiral that functions as a column tie.

4. Information on steel-reinforced concrete construction is available from _____.

5. When discussing a wall or other vertical concrete structure, the _____ is the one most distant from the person viewing it.

6. A(n) _____ is a hand tool used to provide leverage when manually bending bars or pipes.

7. _____ are anchoring devices embedded in concrete to receive bolts or screws that will attach shelf angles or other items, such as machinery.

8. The method of joining two reinforcing bars by placing them side by side and fastening them together is known as a(n) _____.

9. _____ is the common term for concrete that has been placed around metal bars or other reinforcing materials that improve its tensile and shear strength.

10. A noncontinuous beam that is supported at its two ends is referred to as a(n) _____.

11. The horizontal distance that a beam or girder stretches between supports is its _____.

12. Also called raiser bars, _____ rest upon individual high chairs to support the top bars in a slab.

13. _____ is twisted to fasten rebar together and hold it in the desired location until concrete is placed.

14. A(n) _____ is used to locate and hold dowels when laying out bolt holes and inserts in concrete construction.

15. In flat-slab or flat-plate construction, _____ are bands of reinforcing bars.

16. _____ are used in concrete to minimize shrinkage and cracks due to temperature changes.

17. A reinforced wall used to hold back soil or materials such as coal or sand is referred to as a(n) _____.

18. A(n) _____ is a concrete slab with drop panels that is reinforced in two directions.

19. Placed atop a pile, a(n) _____ is used to distribute loads from a structure to the supporting pile.

20. Two adjacent separately cast sections of concrete are connected using metal bars called _____.

21. The distance from the topmost layer of reinforcing steel to the face of the concrete is called by various names, but _____ is most common.

22. _____ are cast-in-place, drilled hole piles. The term also refers to piers extending down to solid earth or rock through a layer of soft soil or water.

23. The support structure located at each end of a bridge is a(n) _____.

24. A(n) _____ is a frame with two or more legs that is placed perpendicular (at a right angle) to the length of the structure it supports.

25. A method of reinforcing-bar connection that involves lapping the bars in direct contact is called a(n) _____.

26. When discussing a wall or other vertical concrete structure, the _____ is the one closest to the person viewing it.

27. In a spiral, the _____ is the center-to-center spacing between the turns.

27304-14 Reinforcing Concrete

28. Bars, dowels, or anchor bolts may be enclosed by tubes known as _____.

29. Bars that are spliced at different points are said to be joined with _____.

30. Moisture is drained from the interior of a wall by passing through a(n) _____.

31. A(n) _____ is a construction document that lists, in table form, similar items.

32. The principal beam that supports other beams is called a(n) _____.

33. _____ are support devices, usually used in pairs, to hold reinforcement at a comfortable level for fastening while prefabricating columns.

34. A beam that extends over at least three supports (including the end supports) is classified as a(n) _____.

35. Also known as column ties, _____ are reinforcing steel wrapped around the vertical bars serving as column reinforcement.

36. _____ are steel bars wrapped around the vertical reinforcement of a column to prevent buckling under compression load.

37. A concrete wall with a layer of reinforcement for each face is a(n) _____.

38. Information on reinforcing-bar size, location, spacing and other needed data is contained in the _____.

39. _____ are U-shaped or box-shaped reinforcing bars placed perpendicular to the longitudinal bars in beams to improve shear strength.

40. _____ are semicircular or right-angle bends on the free end of a reinforcing bar to anchor it in concrete.

41. A horizontal structural member is classified as a(n) _____.

42. A(n) _____ is a structural member carrying a primarily vertical load as it supports a beam or other horizontal member.

43. A reinforcing bar that is bent into a box shape, a(n) _____ holds together the longitudinal reinforcing bars in a beam or column.

44. In a(n) _____, horizontal or vertical reinforcing bars are arranged in a single layer positioned in the center of the wall.

Trade Terms

Abutment	Column ties	Girder	Rebar horses	Stirrups
Band	Concrete cover	Hickey bar	Reinforced concrete	Strips
Bar list	Contact splice	Hook	Retaining wall	Support bars
Beam	Continuous beam	Inserts	Schedule	Temperature bars
Bent	CRSI	Lapped splice	Simple beam	Template
Bundle of bars	Double-curtain wall	Near face	Single-curtain wall	Tie
Caissons	Dowel	Pile cap	Sleeve	Tie wire
Column	Far face	Pitch	Span	Weephole
Column spirals	Flat slab	Placing drawings	Staggered splices	

Trade Terms Introduced in This Module

Abutment: The supporting substructure at each end of a bridge.

Band: Reinforcing steel in columns that is wrapped around the vertical bars to counteract compression forces.

Bar list: A bill of materials for a job site that shows all bar quantities, sizes, lengths, grades, placement areas, and bending dimensions to be used.

Beam: A horizontal structural member.

Bent: A self-supporting frame having at least two legs and placed at right angles to the length of the structure it supports, such as the columns and cap supporting the spans of a bridge.

Bundle of bars: A bundle consisting of one size, length, or mark (bent) of bar, with the following exceptions: very small quantities may be bundled together for convenience, and groups of varying bar lengths or marks that will be placed adjacent to one another may be bundled together.

Caissons: Piers usually extending through water or soft soil to solid earth or rock; also refers to cast-in-place, drilled-hole piles.

Column: A post or vertical structural member supporting a floor beam, girder, or other horizontal member and carrying a primarily vertical load.

Column spirals: Columns in which the vertical bars are enclosed within a spiral that functions like a column tie.

Column ties: Bars that are bent into square, rectangular, U-shaped, circular, or other shapes for the purpose of holding vertical column bars laterally in place and that prevent buckling of the vertical bars under compression load.

Concrete cover: The distance from the face of the concrete to the reinforcing steel; also referred to as fireproofing, clearance, or concrete protection.

Contact splice: A means of connecting reinforcing bars by lapping in direct contact.

Continuous beam: A beam that extends over three or more supports (including end supports).

CRSI: Concrete Reinforcing Steel Institute.

Double-curtain wall: A concrete wall that contains a layer of reinforcement at each face.

Dowel: A bar connecting two separately cast sections of concrete. A bar extending from one concrete section into another is said to be doweled into the adjoining section.

Far face: The face farthest from the viewer (as of a wall); may be the outside or inside face, depending on whether one is inside looking out or outside looking in.

Flat slab: A concrete slab reinforced in two or more directions, with drop panels but generally without beams, and with or without column capitals.

Girder: The principal beam supporting other beams.

Hickey bar: A hand tool with a side-opening jaw used in developing leverage for making in-place bends on bars or pipes.

Hook: A 180-degree (semicircular) or 90-degree turn at the free end of a bar to provide anchorage in concrete. For stirrups and column ties only, turns of either 90 degrees or 135 degrees are used.

Inserts: Devices that are positioned in concrete to receive a bolt or screw to support shelf angles, machinery, etc.

Lapped splice: The joining of two reinforcing bars by lapping them side by side, or the length of overlap of two bars; similarly, the side and end overlap of sheets or rolls of welded-wire fabric reinforcement.

Near face: The face nearest the viewer, which may be inside or outside, depending on whether one is inside looking out or outside looking in.

Pile cap: A structural member placed on the tops of piles and used to distribute loads from the structure to the piles.

Pitch: The center-to-center spacing between the turns of a spiral.

Placing drawings: Detailed drawings that give the bar size, location, spacing, and all other information required to place the reinforcing steel.

Rebar horses: Wood or metal supports that are used in groups of two or more to hold main reinforcing in a convenient position for placing ties while prefabricating column, beam, or pile cages.

27304-14 Reinforcing Concrete

Module One 53

Reinforced concrete: Concrete that has been placed around some type of steel reinforcement material. After the concrete cures, the reinforcement provides greater tensile and shear strength for the concrete. Almost all concrete is reinforced in some manner.

Retaining wall: A wall that has been reinforced to hold or retain soil, water, grain, coal, or sand.

Schedule: A table on placing drawings that lists the size, shape, and number of bars each way, and the mark number of the bars if they are bent.

Simple beam: A beam supported at each end (two points) and not continuous.

Single-curtain wall: A concrete wall that contains a single layer of vertical or horizontal reinforcing bars in the center of the wall.

Sleeve: A tube that encloses a bar, dowel, anchor bolt, or similar item.

Span: The horizontal distance between supports of a member such as a beam, girder, slab, or joist; also, the distance between the piers or abutments of a bridge.

Staggered splices: Splices in bars that are not made at the same point.

Stirrups: Reinforcing bars used in beams for shear reinforcement; typically bent into a U shape or box shape and placed perpendicular to the longitudinal steel.

Strips: Bands of reinforcing bars in flat-slab or flat-plate construction. The column strip is a quarter-panel wide on each side of the column center line and runs from column to column. The middle strip is half a panel in width, filling in between column strips, and runs parallel to the column strips.

Support bars: Bars that rest upon individual high chairs or bar chairs to support top bars in slabs or joists, respectively. They are usually #4 bars and may replace a like number of temperature bars in slabs when properly lap spliced; also used longitudinally in beams to provide support for the tops of stirrups. Also called raiser bars.

Temperature bars: Bars distributed throughout the concrete to minimize cracks due to temperature changes and concrete shrinkage.

Template: A device used to locate and hold dowels, to lay out bolt holes and inserts, etc.

Tie: A reinforcing bar bent into a box shape and used to hold longitudinal bars together in columns and beams. Also known as stirrup ties.

Tie wire: Wire (generally #16, #15, or #14 gauge) used to secure rebar intersections for the purpose of holding them in place until concreting is completed.

Weephole: A drainage opening in a wall.

Additional Resources

This module presents thorough resources for task training. The following resource material is suggested for further study.

29 CFR 1926, Safety and Health Regulations for Construction, Latest Edition. Washington, D.C.: Occupational Safety and Health Administration.
ACI 315, Details and Detailing of Concrete Reinforcement, Latest Edition. Farmington Hills, MI: American Concrete Institute.
ACI 318-95, Building Code Requirements for Structural Concrete, Latest Edition. Farmington Hills, MI: American Concrete Institute.
ASTM A615, Standard Specification for Deformed and Plain Carbon-Steel Bars for Concrete Reinforcement, Latest Edition. West Conshohocken, PA: ASTM International.
ASTM A706, Standard Specification for Low-Alloy Steel Deformed Bars and Plain Bars for Concrete Reinforcement, Latest Edition. West Conshohocken, PA: ASTM International.
ASTM A996, Standard Specification for Rail-Steel and Axle-Steel Deformed Bars for Concrete Reinforcement, Latest Edition. West Conshohocken, PA: ASTM International.
Manual of Standard Practice, Latest Edition. Concrete Reinforcing Steel Institute (CRSI).
Placing Reinforcing Bars. 2005. Concrete Reinforcing Steel Institute (CRSI).

Figure Credits

Courtesy of Portland Cement Association, CO01, Figure 13, Figures 15–16, SA02
www.rebar.net, Figure 21, Figure 22
Courtesy of JET Tools, Figure 29
Benner-Nawman, Inc., SA04
BN Products – USA, Figure 31

Klein Tools, Inc., Figure 39
Bar Splice Products, Inc., SA05
MAX USA Corp., SA06
Haskell, Figure 50b
Brice Building Company, SA07

Section Review Answers

Answer	Section Reference	Objective Reference
Section One		
1. d	1.1.0	1a
2. b	1.2.2	1b
3. True	1.3.0	1c
Section Two		
1. d	2.0.0	2
2. a	2.1.0	2a
3. b	2.2.0	2b
4. d	2.4.3	2d
5. a	2.5.2	2e
Section Three		
1. False	3.1.0	3a
2. c	3.2.0	3b
Section Four		
1. c	4.1.0	4a
2. c	4.2.1	4b

NCCER CURRICULA — USER UPDATE

NCCER makes every effort to keep its textbooks up-to-date and free of technical errors. We appreciate your help in this process. If you find an error, a typographical mistake, or an inaccuracy in NCCER's curricula, please fill out this form (or a photocopy), or complete the online form at **www.nccer.org/olf**. Be sure to include the exact module ID number, page number, a detailed description, and your recommended correction. Your input will be brought to the attention of the Authoring Team. Thank you for your assistance.

Instructors – If you have an idea for improving this textbook, or have found that additional materials were necessary to teach this module effectively, please let us know so that we may present your suggestions to the Authoring Team.

NCCER Product Development and Revision
13614 Progress Blvd., Alachua, FL 32615

Email: curriculum@nccer.org
Online: www.nccer.org/olf

❏ Trainee Guide ❏ Lesson Plans ❏ Exam ❏ PowerPoints Other _____

Craft / Level: _____ Copyright Date: _____

Module ID Number / Title: _____

Section Number(s): _____

Description: _____

_____

_____

_____

Recommended Correction: _____

_____

_____

_____

Your Name: _____

Address: _____

_____

Email: _____ Phone: _____

27308-14

Vertical Formwork

Overview

Forms are used in the construction of everything from retaining walls to bridges to high-rise buildings. Vertical forms are used in building walls, columns, and stairs. Slip forms and climbing forms are used in building piers, silos, elevator and stair cores for buildings, and even entire building structures. The ability to work with concrete formwork is essential in commercial and industrial carpentry work. A residential carpenter may occasionally work with forms, but for the commercial and industrial carpenter, working with forms is a way of life. A commercial or industrial carpenter is involved in erecting and bracing the forms, and in stripping, cleaning, and storing the forms after the concrete has hardened.

Module Two

Trainees with successful module completions may be eligible for credentialing through the NCCER Registry. To learn more, go to **www.nccer.org** or contact us at **1.888.622.3720**. Our website has information on the latest product releases and training, as well as online versions of our *Cornerstone* magazine and Pearson's product catalog.

Your feedback is welcome. You may email your comments to **curriculum@nccer.org**, send general comments and inquiries to **info@nccer.org**, or fill in the User Update form at the back of this module.

This information is general in nature and intended for training purposes only. Actual performance of activities described in this manual requires compliance with all applicable operating, service, maintenance, and safety procedures under the direction of qualified personnel. References in this manual to patented or proprietary devices do not constitute a recommendation of their use.

Copyright © 2015 by NCCER, Alachua, FL 32615, and published by Pearson Education, Inc., New York, NY 10013. All rights reserved. Printed in the United States of America. This publication is protected by Copyright, and permission should be obtained from NCCER prior to any prohibited reproduction, storage in a retrieval system, or transmission in any form or by any means, electronic, mechanical, photocopying, recording, or likewise. To obtain permission(s) to use material from this work, please submit a written request to NCCER Product Development, 13614 Progress Blvd., Alachua, FL 32615.

From *Construction Craft Laborer, Level Two, Trainee Guide*, Third Edition. NCCER.
Copyright © 2015 by NCCER. Published by Pearson Education. All rights reserved.

27308-14
VERTICAL FORMWORK

Objectives

When you have completed this module, you will be able to do the following:

1. Identify the basic types of concrete wall forms.
    a. Explain the importance of formwork planning.
    b. List the parts and accessories of concrete wall forms.
    c. Describe applications of panel form systems.
    d. Describe applications of gang forms.
2. Describe applications for patented wall-form systems.
    a. List applications for curved forms.
    b. Describe how to frame wall openings.
3. Explain how to properly assemble and set forms.
    a. Explain how to assemble forms.
    b. Explain how to set forms.
4. Identify the types of column forms.
    a. List applications for fiber and steel column forms.
    b. List applications for job-built column forms.
5. List applications of vertical slipforming and describe each.
    a. Identify slip-form components.
    b. Describe applications of climbing forms.
6. Describe how to construct stair forms.
7. List various vertical architectural and specialty forms, and describe applications for each.
    a. Describe how smooth finishes are created.
    b. Describe how textured surfaces are created.
    c. Explain the use of insulating concrete forms (ICFs).

Performance Tasks

Under the supervision of your instructor, you should be able to successfully complete two of the following three tasks:

1. Erect, plumb, and brace an instructor-selected wall form.
2. Erect, plumb, and brace an instructor-selected column form.
3. Erect, plumb, and brace a stair form.

Trade Terms

Architectural concrete
Batten
Brace
Bracing collar
Buck
Bulkhead
Climbing form
Fiberglass-reinforced plastic (FRP)

Flight
Form liner
Insulating concrete form (ICF)
Landing
Lifting eye
Rustication line
Sheathing
Shiplap

Slip form
Spreader
Spud wrench
Strongback
Waler
Yoke assembly

Industry-Recognized Credentials

If you're training through an NCCER-accredited sponsor, you may be eligible for credentials from NCCER's Registry. The ID number for this module is 27308-14. Note that this module may have been used in other NCCER curricula and may apply to other level completions. Contact NCCER's Registry at 888.622.3720 or go to **www.nccer.org** for more information.

Code Note

Codes vary among jurisdictions. Because of the variations in code, consult the applicable code whenever regulations are in question. Referring to an incorrect set of codes can cause as much trouble as failing to reference codes altogether. Obtain, review, and familiarize yourself with your local adopted code.

Contents

Topics to be presented in this module include:

1.0.0 Concrete Walls Forms 1
   1.1.0 Planning Formwork 1
   1.2.0 Wall Forms, Parts, and Accessories 2
      1.2.1 Parts and Accessories 3
   1.3.0 Panel Form Systems 4
   1.4.0 Gang Forms 7
2.0.0 Patented Wall-Form Systems 10
   2.1.0 Curved Wall Forms 10
   2.2.0 Framing Wall Openings 13
3.0.0 Assembling and Setting Forms 17
   3.1.0 Assembling Forms 17
   3.2.0 Setting Forms 17
4.0.0 Column Forms 20
   4.1.0 Fiber and Steel Column Forms 20
      4.1.1 Fiber Forms 20
      4.1.2 Steel Column Forms 20
   4.2.0 Job-Built Column Forms 24
5.0.0 Vertical Slipforming 27
   5.1.0 Slip-Form Components 27
   5.2.0 Climbing Forms 29
6.0.0 Stair Forms 31
7.0.0 Vertical Architectural and Specialty Forms 37
   7.1.0 Creating Smooth Finishes 37
   7.2.0 Creating Textured Surfaces 38
   7.3.0 Insulating Concrete Forms 38

Figures

Figure 1   Examples of concrete forms used in commercial construction work 1
Figure 2   Examples of basic hardware used in form assembly 3
Figure 3   Plywood form with walers 4
Figure 4   Strongbacks on a manufactured form 4
Figure 5   Wall ties and spreaders 5
Figure 6   Adjustable-turnbuckle form aligner 6
Figure 7   Examples of form panels 6
Figure 8   Gang forms 7
Figure 9   Gang-form assembly 7
Figure 10 Steel panel system components 11
Figure 11 Steel-framed wood panel system components 12
Figure 12 Gang wall form 13
Figure 13 Heavy-duty wall-form system 14
Figure 14 Light-, medium-, and heavy-duty hardware 15

Figures (continued)

Figure 15 Example of curved wall forms 15
Figure 16 Radius form template 15
Figure 17 Cutting a curved form plate from a plywood panel 15
Figure 18 End view of plywood panel showing how to
            cut saw kerfs to bend panels 15
Figure 19 Wall-form openings 16
Figure 20 Pins for positioning new gang-form sections 18
Figure 21 Materials commonly used for column forms 21
Figure 22 Square column form constructed using wall-form panels 22
Figure 23 One-piece reinforced-fiberglass round column form 23
Figure 24 Fiber column forms 23
Figure 25 All-steel square column form 23
Figure 26 Heavy-duty form components 24
Figure 27 Job-built column form 25
Figure 28 Typical spacing of round-column form clamps 25
Figure 29 Column form clamps 26
Figure 30 Slip-form construction 27
Figure 31 Building-core construction 27
Figure 32 Slip-form assembly. 28
Figure 33 Climbing form 30
Figure 34 Shaft form 30
Figure 35 Shaft-form corner section 30
Figure 36 Basic stair form 31
Figure 37 Basic stair layout 32
Figure 38 Laying out treads and risers on form sides 33
Figure 39 Open-sided suspended stairs and form 33
Figure 40 Form for suspended stairs between walls 34
Figure 41 Earth-supported stairs and form 34
Figure 42 Form for earth-supported stairs between walls 35
Figure 43 Cantilevered stairs and form 35
Figure 44 Freestanding stairs and form 36
Figure 45 Textured surface 38
Figure 46 Insulating concrete forms (ICFs) 38
Figure 47 ICFs used for commercial structures 38

SECTION ONE

1.0.0 CONCRETE WALL FORMS

Objective

Identify the basic types of concrete wall forms.
a. Explain the importance of formwork planning.
b. List the parts and accessories of concrete wall forms.
c. Describe applications of panel form systems.
d. Describe applications of gang forms.

Trade Terms

Architectural concrete: Concrete that serves as the architectural finish material.

Brace: A diagonal supporting member used to reinforce a form against the weight of the concrete.

Insulating concrete form (ICF): Concrete form system in which concrete is cast between two expanded polystyrene (EPS) foam panels.

Sheathing: Plywood, planks, or sheet metal that make up the surface of a form.

Slip form: A form that is moved continuously while the concrete is being placed.

Spreader: A wood or metal device used to hold the sides of a form apart.

Strongback: An upright supporting member attached to the back of a form to stiffen or reinforce it, especially around door and window openings.

Waler: Horizontal wood or metal members installed on the outside of form walls to strengthen the sheathing and stiffen the walls.

At one time, concrete forms were built in place on the job site and torn down after a single use. Salvage was limited to individual boards or timbers. However, increasing labor and material costs and the need for mass production in heavy commercial construction have brought about many changes in formwork.

Manufacturers' patented forms, prefabricated forms, and reusable form panels have become standard construction equipment. Panels can be ganged into large units for efficient wall forming (see the section *Gang Forms*). Tying, fastening, bracing, and supporting accessories continue to increase in number and variety. New materials have been applied to form construction by the patented-form industry, and new uses for conventional materials have been found. Plastics, fiberglass, steel, aluminum, and rubber, both as raw materials and in patented prefabricated shapes, have simplified the forming of concrete to meet contemporary architectural demands.

The most common vertical forms encountered on a job site are wall forms and column forms. In this module, larger form systems commonly encountered in heavy commercial construction will be covered. Other types of vertical forms, including slip forms, stair forms, and insulating concrete forms (ICFs), will also be covered. *Figure 1* shows examples of forms used in commercial construction work.

1.1.0 Planning Formwork

Formwork represents one-third to one-half of the total installed cost of concrete. Therefore, contractors consider the following factors:

- The most cost-efficient forming materials that will safely handle the load
- Reusability of the forms

Figure 1 Examples of concrete forms used in commercial construction work.

- Effective planning for use of forms
- Crew makeup
- Amount of labor needed to assemble, strip, and clean the forms

Other factors that enter into the planning process are the availability and capacity of cranes, the space available to assemble and move formwork, and the numbers and types of penetrations for pipe, conduit, and openings.

Contractors will generally select the most economical formwork for a particular job. However, if a contractor expects to use the same kinds of forms on other projects, or several times on the same project, it is often more economical to purchase more durable forms. The more times a form can be reused, the lower the cost per project.

It's not just a matter of finding the least expensive form. The formwork must be used efficiently. For example, if formwork is rented for $10,000 a month and is used once during the month, the cost per concrete structure is $10,000. If the formwork is used 10 times during the month, the cost per structure is only $1,000.

Contractors try to plan so that crews are available to place forms, place concrete, and strip forms in the shortest amount of time possible. A contractor loses money every time a job has to stop because equipment is not on hand and workers have to stand around and wait. Delays can be especially costly when the forms are rented or when the forms are needed for another project.

> **NOTE**
> Formwork manufacturers design their products with safety in mind and test them to make certain that they perform as intended within appropriate safety allowances. For your safety, and to ensure the proper installation and use of their formwork, manufacturers provide application guides detailing the component parts of their formwork, assembly instructions, and technical and safety information. All persons who are involved directly or indirectly with the use of a particular form system should always make themselves familiar with the contents of the appropriate application guide before attempting to install formwork.

The amount of formwork needed can be determined by the amount of wall to be formed, the time available in the schedule, and the curing time required for the concrete before the formwork can be moved.

Every form system is different. It takes time and training for work crews to become familiar with a particular form system well enough to set it up, strip it, and tear it down efficiently. For that reason, a contractor is likely to develop expertise with one particular system and to use it whenever possible.

Another aspect of planning is allocation of work areas at the site. Space is needed to assemble, clean, and store formwork. Most large projects will have a shop area where the forms can be assembled. Moving the forms requires special planning. If the forms are to be moved with a crane, there has to be a clear area in which to operate. This is a major concern. It would be costly to have to dismantle the crane in order to move the forms from one part of the site to another.

1.2.0 Wall Forms, Parts, and Accessories

A wall form is a retainer or mold that is constructed to provide the desired shape, support, and finish to a concrete wall. Wall forms, or any of the types of forms discussed in this module, all have one thing in common: manufacturers design formwork to produce the lightest form that will support the weight of concrete, along with any craftworkers (such as carpenters and rebar installers) who might need to work on the form.

There are a variety of wall-form systems. Simple job-built wall forms up to 4' high can be made of lumber and plywood secured with duplex nails and separated by wall ties. Other systems use manufactured form panels and attaching hardware, with lumber or metal walers and strongbacks. Some manufacturers offer proprietary systems in which all the formwork components are available from the same source. In many instances, horizontal and vertical supports are built into the form panels, so separate walers and strongbacks are unnecessary.

Formwork is designed for simple construction and to support the loads of the concrete and craftworkers. Formwork can be assembled with metal clamps or special wedge pin connections that are secure, yet make the form easy to assemble and disassemble (*Figure 2*). Handles, or a wood strip that has been nailed or screwed to the form, provide a means of grabbing the form side so that it can be pulled away from the concrete. This minimizes or eliminates the use of pry bars or hammers in prying the form loose. It also reduces damage. Some systems provide lifting eyes and other hardware that make it easy to move the form sections.

The weight of plastic concrete puts enormous pressure on formwork. To avoid formwork failure or stretching, it is important to use forms that are designed and built correctly and have the proper bracing for the job at hand.

Safety should always be first in your mind. Occupational Safety and Health Administration (OSHA) 1926 Subpart Q, *Concrete and Masonry Construction*, provides information regarding safety regulations. Specifically, 29 *Code of Federal Regulations* (*CFR*) 1926.703 covers the safety regulations related to cast-in-place concrete and formwork. To ensure safety, wall forms must be constructed exactly as designed, following a safe erection procedure. Forms must be checked before each use. If they are damaged in any way, they must be repaired. Parts that are missing must be replaced. If damaged formwork is used, it could lead to weakness in the system and cause form failure. For safety purposes, it is also important to know the required rate of concrete placement, the height of the placement, and the amount of time between pours. Above all, contractors should be concerned with the safety of their crews.

1.2.1 Parts and Accessories

The following describes various wall-form parts and accessories:

- *Walers* – Formwork components used to keep the forms aligned and through which the ties are fastened. Walers can either run vertically or horizontally, depending on the design of the formwork. However, they typically run horizontally. Walers are also sometimes called ribs, stringers, or rangers. See *Figure 3*. Walers are placed directly against the form.
- *Strongbacks* – Vertical supports used to align, straighten, support, and level the walers. Along with the ties, strongbacks support and align the walers. Only straight members should be used as strongbacks. See *Figure 4*. Strongbacks are typically placed against the walers, but other manufacturer designs may be used.
- *Ties and spreaders* – A tie is a wire or rod used to keep the form sides pulled together. Spreaders keep the form walls properly spaced before the concrete is placed. Most manufactured ties are equipped with a spreader. This spreader may be a washer held in place by a deformation on the tie, or it may be formed by a stiff tie and compatible connecting hardware (*Figure 5*).
- *Stakes and braces* – Needed to anchor the form and keep it in proper alignment. Stakes are made of steel or wood in various lengths to meet job conditions. Braces may be made of 2 × 4 or 2 × 6 lumber, but a number of form man-

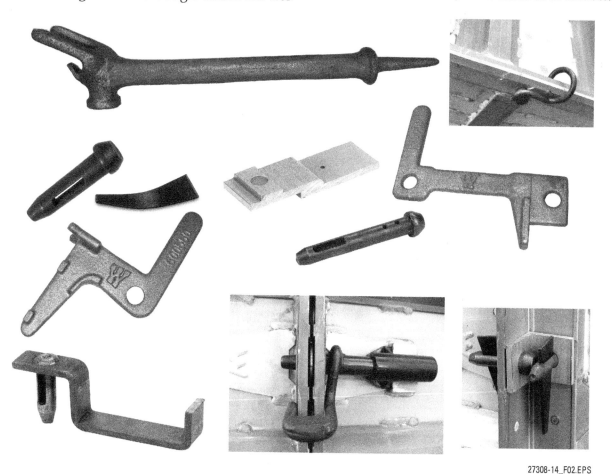

Figure 2 Examples of basic hardware used in form assembly.

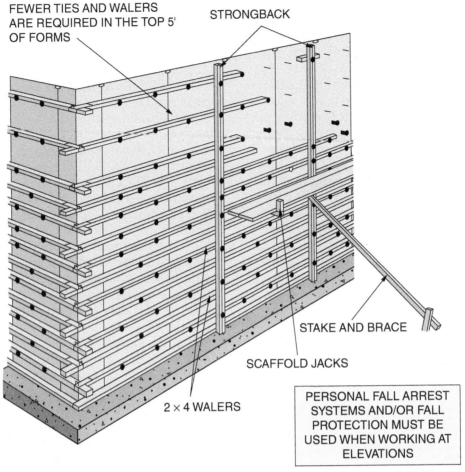

Figure 3 Plywood form with walers.

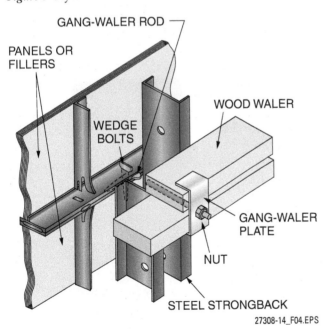

Figure 4 Strongbacks on a manufactured form.

ufacturers make braces with adjustable turnbuckles to help in aligning the forms (*Figure 6*). Many contractors use pipe braces.
- *Plates* – Serve to align the bottom of the form and provide a means for fastening the form to the footing. Using plates for alignment helps the job go faster. Plates are not used on all wall forms. In many cases, they are used on only one side of the wall form.

1.3.0 Panel Form Systems

Patented manufactured panels are available for building all kinds of wall forms. Although these panel form systems vary widely in detail, there are basically five types of systems:

- *Unframed plywood panels* – Unframed plywood panels are sometimes backed by steel braces. Locking and tying hardware are the essential parts of this form system. The lock that holds the ties is frequently part of the waler support.

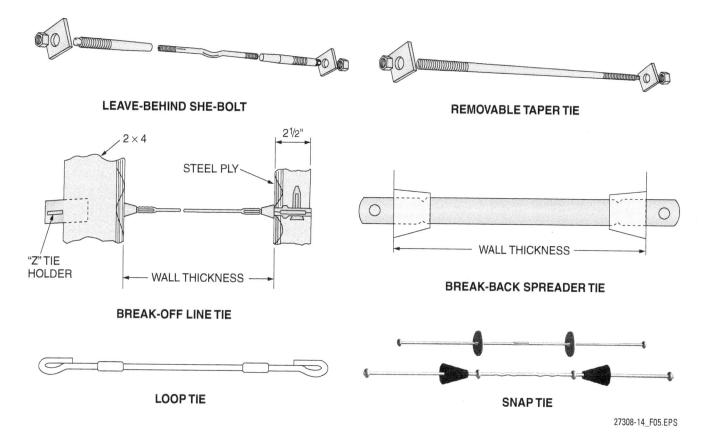

Figure 5 Wall ties and spreaders.

- *All-metal panels* – All-metal panels (*Figure 7A*) are made up of metal plates supported by a metal frame. These panels are available in steel as well as aluminum. All-aluminum forms consist of an aluminum framework and an aluminum face sheet, which is usually ⅛" thick. They generally are available in 3' × 8' sections. They are lightweight and easy to handle. Because of the cost of aluminum, however, they are very expensive and more subject to theft. Aluminum can also react chemically with concrete, causing the concrete to adhere to the form and making it unsuitable for architectural concrete.

> **NOTE**
> When installing form panels, use the largest-size panel that is available for the project you are working on.

Because they are heavier than aluminum forms, all-steel forms are available in smaller modules, usually 2' × 4'. Although steel forms are less efficient than aluminum because of the smaller size, they are very durable and can be used repeatedly if properly maintained and stored. Steel forms can be used for architectural concrete if a liner is used.

- *Plywood panels and metal frames* – These forms may be ganged or hand-set. They consist of ½" or ⅝" plywood panels recessed into a steel or aluminum framework (*Figure 7B*). They may not require bracing and are used in all types of construction.
- *Heavy steel-framed panels* – These panels are made to handle the greater pressures of cast-in-place concrete (*Figure 7C*). They are built with lumber, plywood, or synthetic sheathing. Because they are designed for heavy loads, they have hardware built into the panels for crane handling and are often ganged together for longer runs.
- *Plastic panels* – Some manufacturers are now making form panels of heavy-gauge plastic. These panels are light and easy to handle. The horizontal and vertical support members are built into the panel as it is formed.

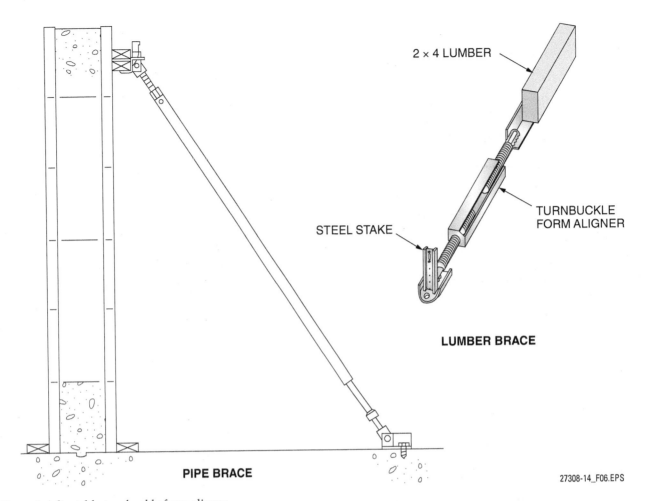

Figure 6 Adjustable-turnbuckle form aligner.

(A) ALL-METAL PANEL   (B) STEEL-FRAMED PLYWOOD PANEL   (C) HEAVY-DUTY PANEL

Figure 7 Examples of form panels.

1.4.0 Gang Forms

Steel and plywood panels and all-steel panels are designed so they can be connected into large panels called gang forms or gangs (*Figure 8*).

Using special ties and connecting hardware supplied by the manufacturer, large panels can be built on the ground, where it is easier to work. The final gang form may be 40' to 50' wide. These gang panel forms are raised using cranes or other lifting equipment.

In addition to walls and columns, gang forms are commonly used to build retaining walls, sound barriers, and bridge abutments. Gang forms are designed with extra strength to withstand the stresses of being lifted and moved. They are considered to be easier to strip than hand-set forms because a crane applies the lifting force to the form.

> **WARNING!**
> A crane should never be used to break a gang form free from the concrete because it could endanger workers. The form should be pried loose from the top while the crane is supporting it.

A large portion of the cost of concrete-in-place is the concrete formwork. Any system that works satisfactorily in achieving the end results in a safe and economical way is advantageous. The advantages of gang forms are:

- *Less time in erecting the form* – Gang forms can be built on the ground (*Figure 9*), then raised by crane to their position.
- *Less time in stripping* – Stripping is quicker because the large forms are stripped as a unit.
- *Reuse* – If the forms are assembled and used correctly, then stripped and cleaned with reasonable care, most systems can be reused. Some form panels can be used hundreds of times before they need to be replaced.

Figure 8 Gang forms.

Figure 9 Gang-form assembly.

Wall Form Construction in Process

This is an example of a hand-set wall form using plywood panels installed in manufactured frames, along with wooden walers, strongbacks, and braces. The braces are secured to temporary concrete blocks called deadmen, which were poured for that purpose. On the other side of the form, you can see that the plywood panels have been sprayed with form-release agent. The rebar mat has been placed in preparation for construction of the other half of the form on the footing.

Gang Forms

The heavier the form material, the larger the crane required to move it. Some contractors use aluminum form systems that can be moved with a small crane. This can save considerable expense. Steel frames are much heavier, but are also able to withstand greater lateral pressure. Some guidelines to ensure the safe lifting of gang forms are as follows:

- Attach the crane rigging only to lifting brackets made for use with the form system.
- Ensure that the rigging does not overload any one lifting bracket.
- Use a lifting (spreader) beam for all lifts, especially those involving gang forms that have several lifting brackets.
- When the gang form is being lifted, make sure it hangs plumb and straight.
- Use a minimum of two tag lines to safely control the movement of the gang form while it is being lifted.
- Ensure the gang form is adequately braced and securely fastened in position before releasing the lifting mechanism.
- Be sure to insert a tie-off connector in the concrete for the last form section in a gang. Never tie off to the form itself when breaking the form loose.

Additional Resources

OSHA 29 *CFR* 1926, *Safety and Health Regulations for Construction,* Latest Edition. Washington, D.C.: Occupational Safety and Health Administration.

1.0.0 Section Review

1. In earlier times, forms were built on site, then torn apart with little material salvaged.
    a. True
    b. False

2. Of the total installed cost of concrete, formwork can represent up to _____.
    a. 25 percent
    b. 33 percent
    c. 50 percent
    d. 66 percent

3. The vertical supporting element used to reinforce formwork is the _____.
    a. waler
    b. strongback
    c. form backer
    d. strongtie

4. When a gang form is being lifted, its movement should be controlled by using a minimum of two _____.
    a. guide braces
    b. belaying ropes
    c. hoisting chains
    d. tag lines

Section Two

2.0.0 Patented Wall-Form Systems

Objective

Describe applications for patented wall-form systems.
  a. List applications for curved forms.
  b. Describe how to frame wall openings.

Trade Terms

Buck: A frame placed inside a concrete form to provide an opening for a window or door.

This section provides examples of patented wall-form systems. All of these systems are designed to perform the same basic function, but each of them has different framework designs, attaching hardware, and accessories. Keep in mind that the examples shown in this section are a small sampling of the many types of patented wall forms available. There are many manufacturers of wall forms, and each manufacturer may have several product lines within the broad category of wall forms. The purpose of this material is to familiarize you with some of the different designs. It is your responsibility to obtain and read the manufacturer's instructions for assembly, stripping, and storage of the particular forms used at the job site.

The examples in *Figures 10* and *11* are lightweight panel systems designed to be hand-set; they do not require a crane. Other systems, such as the gang form shown in *Figure 12*, are designed to be assembled at the site assembly area, and then moved to their final position by a crane. Such forms are larger and much heavier than the assembled-in-place type. Some of these form panels can weigh up to 500 pounds, in comparison to the system shown in *Figure 10*, where the largest panel weighs about 50 pounds.

Figure 13 shows a heavy-duty wall-form system. Note that it is has larger frame members and is assembled with bolts and nuts rather than the quick-connect hardware used with the light-duty form shown in *Figure 11*.

Figure 14 shows three different levels of hardware from one manufacturer: light duty, medium duty, and heavy duty. The proper hardware and other components must be used for the type of forms being assembled and erected.

> **NOTE:** Gang forms are never pinned. They are either bolted or connected with proprietary locking devices.

2.1.0 Curved Wall Forms

Several manufacturers offer flexible wall forms that can be used to form tanks, retaining walls, and a variety of other curved surfaces (*Figure 15*). Several different styles of these forms are available. Some are made of flexible steel or aluminum panels shaped by curved ribs. Others use curved walers combined with metal or wood strongbacks.

Other curved wall forms use panels with flexible joints. These panels are placed into special templates, and their tension bolts are adjusted to form the panel to the curvature of the template (*Figure 16*).

Although patented forms for curved walls are readily available, curved forms can also be made from wood. In a curved wood frame, the top and bottom plates of the form panel, as well as the panel itself, are made of plywood (*Figure 17*). In cases where the weight of the concrete demands heavier-gauge plates and plywood, sections can be cut out and laminated to obtain the necessary thickness. Radius walers are cut the same way.

In order to bend plywood form panels into small-radius curves, it may be necessary to cut kerfs in the plywood with a circular saw (*Figure 18*). It is generally better to use two thinner panels than to kerf a thick sheet, as scoring the panel may reduce its strength. It makes a difference how the bend is made in relation to the face grain of the plywood. A bend made with the face grain can achieve a tighter radius than one made against the face grain.

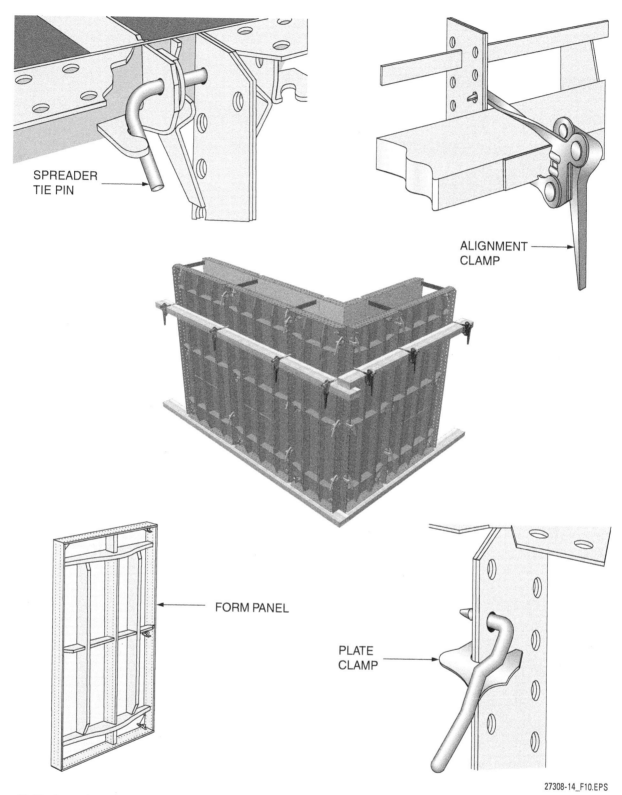

Figure 10 Steel panel system components.

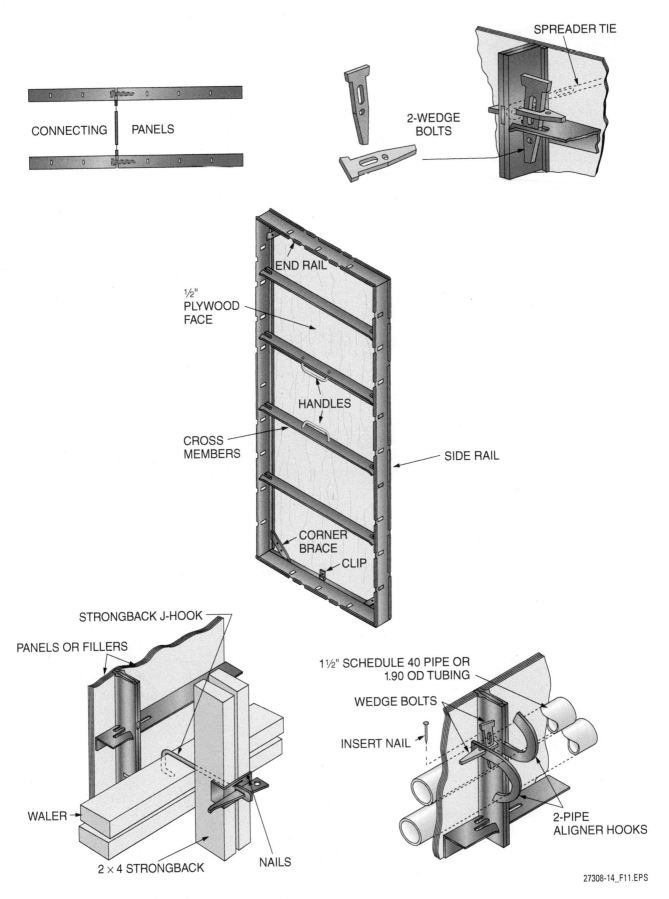

Figure 11 Steel-framed wood panel system components.

> **WARNING!**
>
> While erecting wall forms, craftworkers may be exposed to elevations that will require the use of a personal fall arrest system (PFAS) and/or fall protection. It is important that you understand OSHA, local, and company requirements in regard to working at elevations, such as on wall forms. A job hazard analysis (JHA) should be completed for the work performed that details the requirements to protect the worker from falls from elevations. Gang-form manufacturers may also have information about tie-off points that may be used during erection of gang forms; familiarize yourself with these tie-off points during training on the wall system. Safety is the first priority when forming vertical wall systems. Once the wall is erected, an elevated work platform or walkway with railing will be required for placement of concrete if the top of wall is above the maximum height allowed by code.

frame itself can be set into the form and used as a buck. In such cases, the frame must be plumbed, aligned, and properly braced to ensure that proper placement is maintained throughout concrete placement.

If small openings or recesses are needed for ducts, vents, and large pipes, small wood box frames, rigid foam plastic blocks, or frames made of sheet metal or fiber can be used. A vibrator hole is needed in the lower spreader of a window buck to allow the concrete below the buck to be properly vibrated. A plug or patch is secured in place over the hole after the concrete is vibrated. Openings may be placed in the formwork to provide access for vibrators or to remove debris from between the form walls.

Figure 12 Gang wall form.

2.2.0 Framing Wall Openings

Openings for windows and doors can be formed using bucks, which are wood or steel frames installed between the inner and outer wall forms. A typical buck consists of a plywood or lumber frame reinforced by spreaders and braces (*Figure 19*). In some cases, a metal window or door

Form Faces

The steel frame and plywood on some form systems will leave marks on the concrete; therefore, these forms cannot be used for architectural finishes. For example, when a plywood panel has defects in it, those defects will be transferred to the concrete finish. Some newer systems have a plastic resin-coated form face. If it gets damaged, the defect in the panel can be covered by applying a plastic filler and sanding it smooth.

Inspection Windows in Patented Forms

Most manufactured straight and curved form systems are designed so an individual form panel can be removed and replaced at any time from an installed formwork assembly without disturbing the other form panels. This provides a temporary window in the formwork that can be used to clean, inspect, or place concrete in the form.

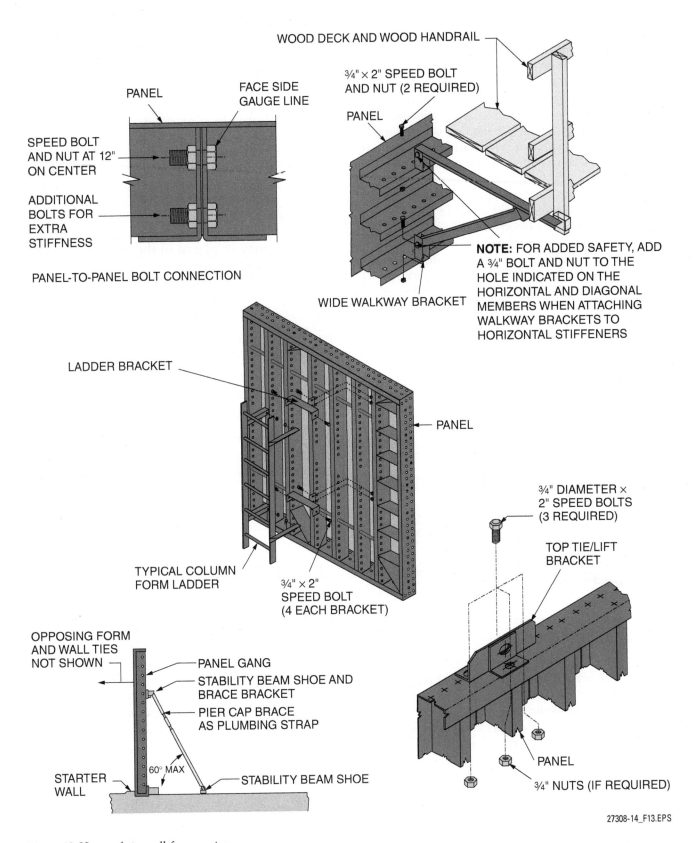

Figure 13 Heavy-duty wall-form system.

LIGHT DUTY

MEDIUM DUTY

HEAVY DUTY

Figure 14 Light-, medium-, and heavy-duty hardware.

Figure 15 Example of curved wall forms.

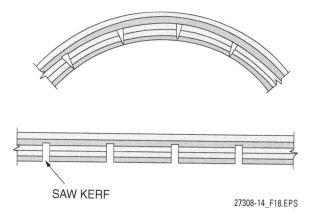

Figure 17 Cutting a curved form plate from a plywood panel.

Figure 16 Radius form template.

Figure 18 End view of plywood panel showing how to cut saw kerfs to bend panels.

27308-14 Vertical Formwork

Module Two 15

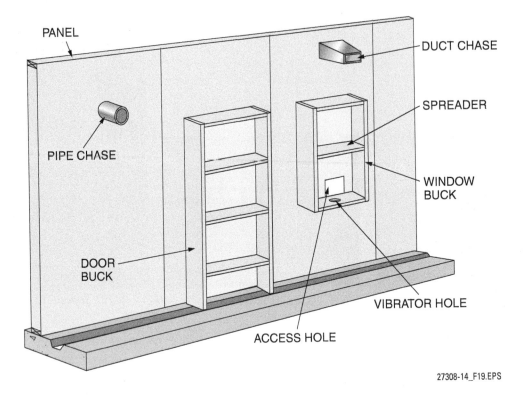

Figure 19 Wall-form openings.

Framing Wall Openings

When framing a wood buck for a door, window, or other opening in a wall form, verify the rough-opening size from the plans or manufacturer's specifications, and check the layout work. It is important that all such openings be made the correct size and be level and plumb, because it will be too late to make corrections after the concrete hardens. This photo shows a window buck being placed into an insulating concrete form (ICF) wall.

2.0.0 Section Review

1. When bending plywood panels to form small-radius curves, it is generally better to use two thinner panels rather than kerfing a thicker panel.
   a. True
   b. False

2. The lower spreader of a window buck requires a _____.
   a. wall tie
   b. brace
   c. vibrator hole
   d. strongback

SECTION THREE

3.0.0 ASSEMBLING AND SETTING FORMS

Objective

Explain how to properly assemble and set forms.
a. Explain how to assemble forms.
b. Explain how to set forms.

Performance Task

Erect, plumb, and brace an instructor-selected wall form.

Trade Terms

Bulkhead: Form component fastened vertically inside a form to stop the flow of concrete at a certain location.

Form liner: Plastic or wood liners used on the inside of wall forms to add a design or special feature to the concrete.

Lifting eye: Load-lifting device attached to a heavy form panel; used to rig the panel with a crane or other lifting equipment.

Spud wrench: Tool to align holes in adjoining form panels; it has a long, tapered steel handle with an open-end wrench on one end.

When erecting forms, it is critical to have the proper materials and equipment in the right place at the right time. Any delays can result in a heavy cost burden in wasted manpower, as well as the rental cost for equipment and materials. To start off, formwork needs to be properly staged in preparation for construction. This includes the following tasks:

- Formwork must be received from the storage yard or from the formwork leasing company.
- As the formwork components and attaching hardware are unloaded, they must be checked against the order to make sure everything is present.
- Tools needed to construct the forms must be gathered. This includes spud wrenches, impact wrenches, pry bars, and any proprietary tools or equipment.
- The wall must be laid out on the foundation and the sill prepared to receive the form.
- Someone must shoot the grades and place shims to level the form.

3.1.0 Assembling Forms

Gang forms are usually assembled on the ground or on work tables, then moved by crane or other lifting equipment to the erection location.

- The formwork is first laid out in accordance with the supplier's shop drawings. The shop drawings should be verified against the construction drawings.
- The form sections are then assembled per the manufacturer's instructions. During the assembly, the forms must be checked to ensure that they are true and straight.
- Lifting eyes must be placed in accordance with the drawings and must be installed correctly.

> **WARNING!**
> It is very important to determine the weight of the finished form and to make sure the lifting eyes will support that weight.

- Once the form section is assembled, make sure the braces are properly installed.
- The form section is flipped over and the form liners, reveals, and other materials are placed.
- Before lifting the form, ensure the lifting hardware and releases are lubricated.

3.2.0 Setting Forms

After the form panels are assembled, they are lifted and moved into place on the foundation. A crane is generally used for this purpose, but smaller lifting equipment such as a forklift may suffice, depending on the size of the form.

> **WARNING!**
> This is a critical time in the form erection process. Make sure tag lines are set and the wind is not going to be a problem once the form is lifted off the ground.

The following is a typical sequence:

Step 1 The form panel is flown in, set into place, and braced. Pins made of rebar are often set into the foundation so that forms can be pushed up against them to ensure proper placement, as shown in *Figure 20*. Note the rebar mat already in place. There are other methods of positioning the form, such as placing the form against a prepositioned wood sill.

Step 2 The form panel must be checked for level and plumb, and the braces adjusted as necessary.

Step 3 Once the form is properly braced, the form panel is unhooked from the crane. At that time, blockouts and embedments are installed.

Step 4 Release agent is applied to the form.

Step 5 Other trades such as rebar installers and electricians complete their work.

Step 6 *Bulkheads* are attached to the ends of the form to prevent concrete from escaping.

Step 7 The opposing form sections are then placed. Form ties and spreaders are installed.

Step 8 The concrete is placed into the forms. While concrete is being placed, the form must be checked for plumb and level again; the weight of the concrete can cause the form to move.

> **NOTE**
> Contract documents must be checked to determine when forms can be stripped. The structural engineer provides information to determine when the forms can be safely stripped.

Step 9 When it is time to strip the form, ties, braces, and spreaders are removed, but one tie is left in place.

Step 10 The crane is hooked up to the form.

> **NOTE**
> At this point, it is important to have an experienced crane operator and signal person to ensure that the right amount of tension is applied to the form.

Figure 20 Pins for positioning new gang-form sections.

Step 11 The final tie is removed and the form is pried loose from the concrete using wedges.

Step 12 The form is flown out, cleaned, lubricated, and reset for the next pour.

Step 13 Tie holes in the finished concrete are filled, and other necessary patching is done.

When the formwork is no longer needed, it is returned to the staging area. All the materials must be cleaned and lubricated at that time.

An inventory of all materials should be performed and verified against the received inventory. If the formwork is rented, the rental company will have a detailed inventory. If any material is not returned, they will charge your company for it.

Form Maintenance

Concrete and debris can be cleaned from form-panel face sheets manually by using a power grinder equipped with a 6" wire brush. Concrete buildup can be removed from the form-panel flanges with a putty knife or scraper. After cleaning, oil the form-panel face and flanges before reuse or storage. Commercial form oil or release agents can be applied with a common paint roller or can be sprayed on. Power-operated form cleaning machines are also available. Their use enables large quantities of form panels to be cleaned and oiled in a day, making them immediately available for reuse.

3.0.0 Section Review

1. Formwork is first laid out as indicated in the _____.
    a. floor plan
    b. supplier's shop drawings
    c. manufacturer's assembly instructions
    d. building elevations

2. After a form panel is flown in, it is checked for level and plumb as the _____
    a. form ties are installed
    b. bulkheads are attached
    c. release agent is applied
    d. braces are adjusted

SECTION FOUR

4.0.0 COLUMN FORMS

Objective

Identify the types of column forms.
a. List applications for fiber and steel column forms.
b. List applications for job-built column forms.

Performance Task

Erect, plumb, and brace an instructor-selected column form.

Trade Terms

Batten: Strip of lumber laid flat against the form panel to provide reinforcement.

Bracing collar: Metal strap used to brace a round column form; it is attached near the top of the form and extends to the ground.

Fiberglass-reinforced plastic (FRP): A durable material used for column forms and as an overlay on plywood panels.

Column forms are made from a variety of materials (*Figure 21*). They may be round, oval, square, or rectangular in cross section. Column forms require tight joints and strong tie supports.

Columns can also be formed using the same forms that may be used for wall forms. *Figure 22* is an example of a square column form constructed using the same panels and hardware used for built-in-place wall forms.

Round column forms are made of fiber, fiberglass, fiberglass-reinforced plastic (FRP), or steel. Although it is possible to build round column forms entirely from lumber, it is very labor intensive and is therefore not commonly done.

Figure 23 shows a one-piece form made of reinforced fiberglass. The bracing collar is attached near the top and then the unit is aligned with a plumb bob. Braces are installed to keep the form in position. When the concrete is dry, the bolts are taken off and the form is spread apart at the flange and removed. It can also be lifted off the column by a crane.

4.1.0 Fiber and Steel Column Forms

Columns are vertical structural members that are used to support imposed loads. Square or round columns may be required, depending on the construction drawings.

4.1.1 Fiber Forms

Fiber forms (*Figure 24*) are manufactured by wrapping layers of fiber in a spiral to produce tubes of the desired diameter. The fiber layers then receive a water-resistant coating to minimize moisture absorption while the concrete sets and hardens. Fiber forms are easily sawed to any length and can be erected quickly.

Fiber forms are made to be used once. If there is a delay in placing concrete after a fiber form has been set in position, the form must be protected against rain and snow by covering the top with waterproof sheathing and by keeping snow and rain from accumulating around the base.

4.1.2 Steel Column Forms

Steel forms consisting of sheet metal attached to prefabricated steel shapes are frequently used for round and square columns when the number of reuses justifies the initial cost or where special conditions require their use. For round columns, two semicircular form sections are fastened together with steel bolts. Sections may be stacked on top of each other and bolted together to provide forms of any desired height. *Figure 25* shows an all-steel square column form.

Round column forms are available in diameters from 12" to 48" and are furnished in 180-degree sections. Forms of larger diameters (up to 120") are furnished in either 90-degree or 60-degree sections. These standard forms are available in 1', 2', 4', and 12' lengths. The column form sections are connected at horizontal and vertical joints by bolts made for use with this system.

Form sections of the same diameter are interchangeable, and various lengths may be stacked to reach the desired height. Column form sections have lifting eyes at the top of each section for easy lifting with a crane. A combination of round and flat form sections can be used to build the form for an oval (bull nose) pier (*Figure 26A*). The same types of components can be used to build the girders that sit on top of piers (*Figure 26B*).

FIBER

STEEL

FIBERGLASS

PLYWOOD

Figure 21 Materials commonly used for column forms.

Custom-Made Concrete Forms

This is one example of the special shapes that can be formed using custom-made concrete forms. Special forms like these are available from several manufacturers.

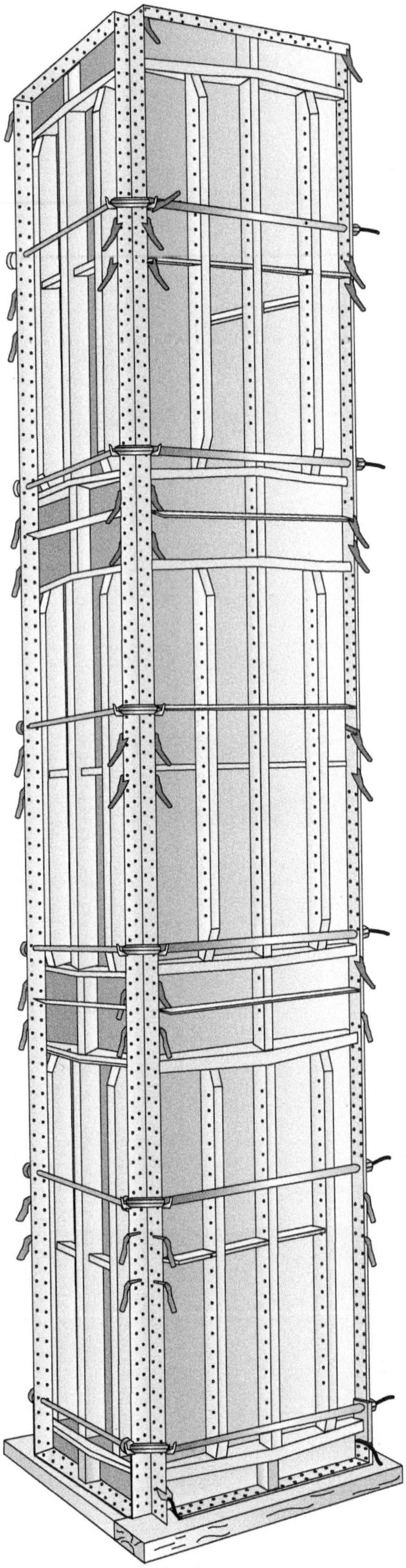

Stripping Fiber Column Forms

The typical procedure for stripping fiber forms is to first set a portable saw to a depth equal to the thickness of the form, then make two vertical cuts on opposing sides to divide the form in half, much like removing a cast. The two halves can then be removed from the concrete. The stripped form may be put back onto the column to protect it during construction.

Plastic Form Systems

Form manufacturers are constantly working to make forms lighter, easier to handle, and easier to assemble and disassemble. Any improvement that can be made in these activities increases productivity. One advance that has been made in column forms is the introduction of a plastic form section that can be assembled with a simple locking handle. This system is available for forming both round and rectangular columns. In addition to their simplicity, these forms are easy to clean because concrete will not stick to them. The smooth face of the form results in a smooth concrete finish.

Figure 22 Square column form constructed using wall-form panels.

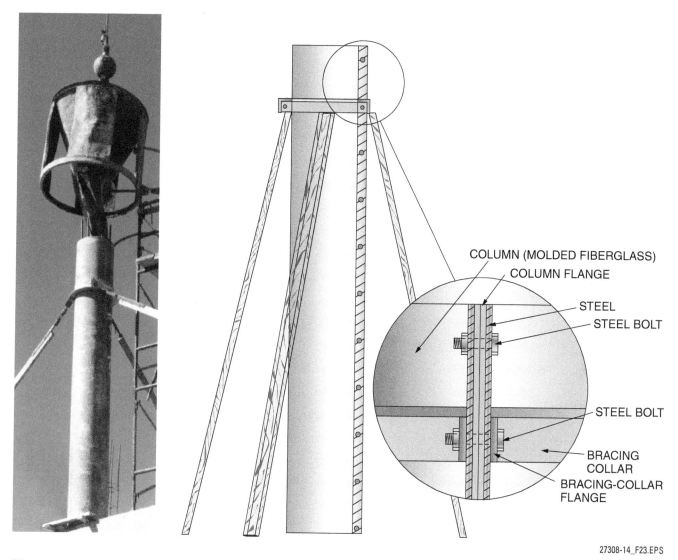

Figure 23 One-piece reinforced-fiberglass round column form.

Figure 24 Fiber column forms.

Figure 25 All-steel square column form.

27308-14 Vertical Formwork

Module Two 23

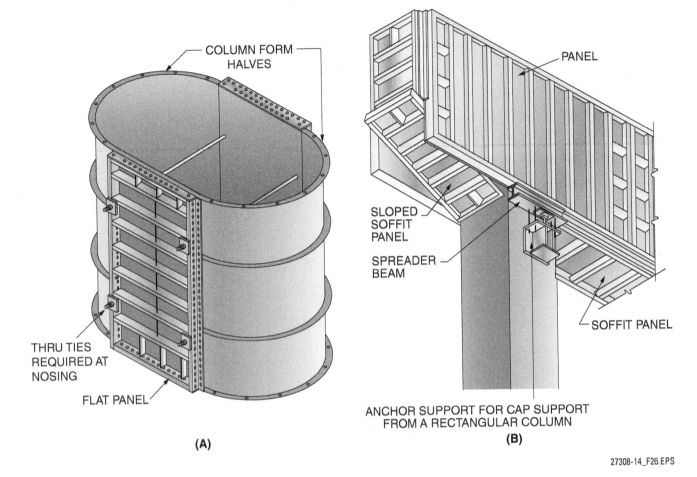

Figure 26 Heavy-duty form components.

4.2.0 Job-Built Column Forms

Although any type of column form, including round forms, can be made from lumber and plywood panels, usually only square and rectangular forms are assembled as job-built forms. Round and oval columns are more likely to be made from prefabricated (manufactured) forms.

Square and rectangular column forms are made of plywood and batterns, which are 2 × 4s placed flat against the plywood to act as studs (*Figure 27*). Braces are used to support and plumb the forms. The use of battens reduces the number of column clamps that would otherwise be needed to prevent the sheathing from bulging. Note that the clamps are spaced closer together at the bottom of the form because that area is subjected to the greatest pressure from plastic concrete. *Figure 27* shows manufactured clamps, but 2 × 4 walers are also used.

Figure 28 shows the typical clamp spacing for 20' and 10' round forms. If the form is more than 36" in diameter, or if there is a rapid pour rate, additional clamps may be needed. If manufactured clamps are used, the manufacturer will specify the spacing.

Figure 29 shows examples of the kinds of clamps used to hold column forms together. The clamps are used in place of walers.

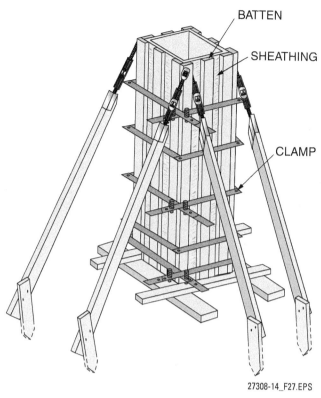

Figure 27 Job-built column form.

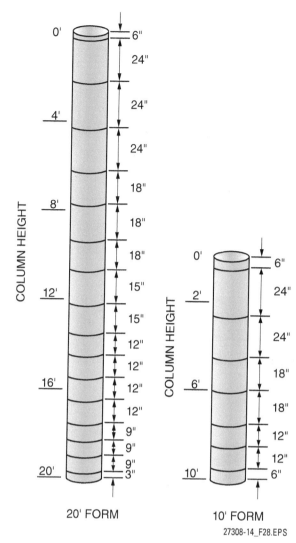

Figure 28 Typical spacing of round-column form clamps.

Common Causes of Form Failure

The most common points of form failure are joints and corners. Some common causes of form failure include the following:

- Exceeding the form design working pressure as a result of:
  - Excessive rate of concrete placement
  - Concrete mixture design not taken into account
  - Improper vibration of the concrete
  - Temperature not taken into account
- Form ties incorrectly placed or improperly fastened
- Improper form construction, especially when layout plans are not provided
- Form fillers, corners, and bulkheads not adequately designed and/or constructed
- Lack of or improper bracing of the form
- Connecting hardware not installed
- Not inspected by authorized qualified personnel to see if the form layout has been interpreted correctly

HINGED, DOUBLE-BAR CLAMP

SINGLE-BAR CLAMP

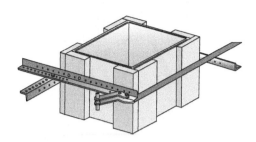
GATES LOK-FAST COLUMN CLAMP

SINGLE BAR CLAMP

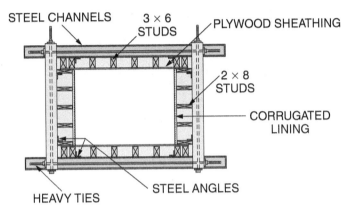
STEEL CHANNEL AND TIE BOLTS (ALL THREAD)

SCISSORS CLAMP

LIGHT-DUTY CLAMP

Figure 29 Column form clamps.

4.0.0 Section Review

1. A material that is typically *not* used to build round column forms is _____.

   a. lumber
   b. fiberglass-reinforced plastic
   c. steel
   d. fiber

2. Fiber forms may be quickly stripped from round columns with the use of a _____.

   a. linoleum knife
   b. pry bar
   c. hammer and chisel
   d. portable saw

Section Five

5.0.0 Vertical Slipforming

Objective

List applications of vertical slipforming and describe each.
  a. Identify slip-form components.
  b. Describe applications of climbing forms.

Trade Terms

Climbing form: A form used to construct vertical walls in successive pours. It is raised to a new level for each pour.

Yoke assembly: Clamping and support device used in slip forms.

Figure 30 Slip-form construction.

In slipforming, plastic concrete is placed or pumped into the forms and the forms are continuously moved to shape the concrete. The rate of movement of the forms is regulated so the forms leave the concrete after it is strong enough to retain its shape while supporting its own weight. The slip forms may be moved a couple of inches per hour or several inches per hour, depending on factors such as the air temperature and the properties of the concrete being placed.

Vertical slip-form construction is used primarily where the height of the structure is 40' or more and where there are few projections or obstacles to impede the movement of the form. Slipforming typically requires an experienced crew.

Vertical slipforming was once used primarily to form silos, bridge piers, storage bins, and similar facilities. Today, slipforming is used for a wide range of other applications, including the construction of tall structures (*Figure 30*). Another popular slip-form application is the forming of the shaft cores for stairs and elevators in buildings (*Figure 31*).

A slip-form system consists of the form itself, as well as upper and lower work decks as shown in *Figure 32*. The form is made with form panels and walers contained in a yoke assembly that is lifted by the jacking system. *Figure 32* provides a perspective on the entire slip-form assembly.

5.1.0 Slip-Form Components

Slip forms consist of common form components such as form panels and walers, as well as the yoke-and-jack assemblies that are used to lift the

Figure 31 Building-core construction.

slip form while concrete is being placed. Slip-form components include the following:

- *Form panels* – Form panels can be made of 1" board, ⅝" or ¾" plywood, or sheet steel (usually 10 gauge). If boards are used, they should be straight grained and center matched. Plywood panels should be installed with the face running vertically. When using plywood or boards, remember that they will be in constant contact with wet concrete, resulting in swelling. To minimize swelling that occurs during forming, either presoak the plywood or boards

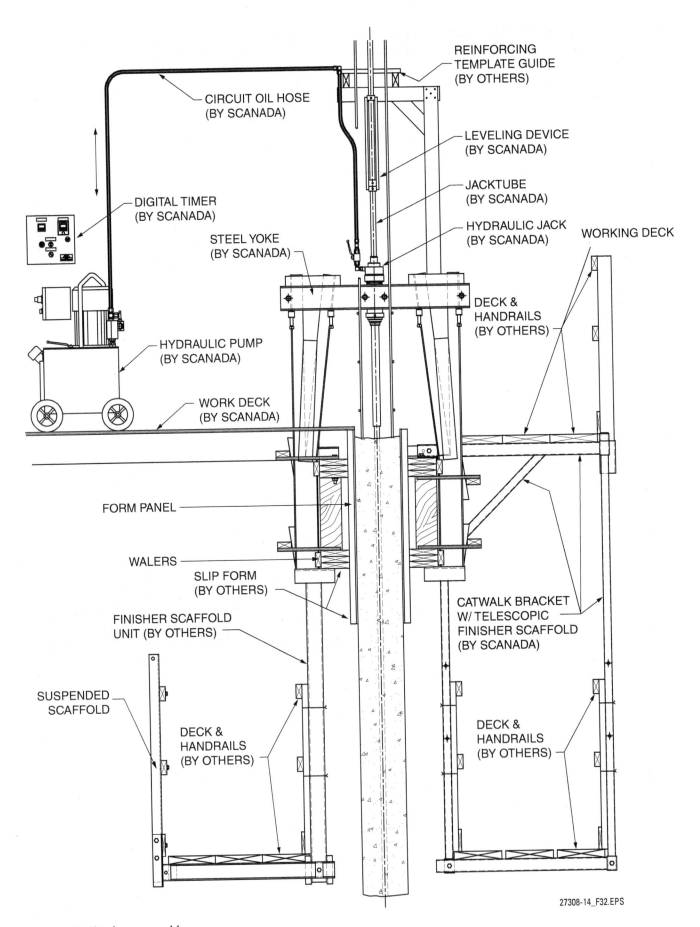

Figure 32 Slip-form assembly.

in water or apply a waterproofing solution. When constructing the form, leave a small gap to allow for any swelling once forming is underway. Also, ensure that the outside panel on the exterior wall form is a little higher than the inside panel, in order to provide a splashboard.

- *Walers or ribs* – Walers or ribs support and hold the form panels in place, support scaffolds, and transmit the lifting forces from the yokes to the form system. The steel members or timbers used as walers must be stiff enough horizontally to withstand the lateral pressure of the concrete. This pressure has a tendency to make the forms bulge out between the yokes. As shown in *Figure 32*, the form panels are held in alignment by two or more rows of walers on each side. The walers or ribs should be braced to improve the rigidity when the span between jacks reaches about 6'. Timber walers should be 4 × 6 or 4 × 8; however, 2 × 6s or 2 × 8s sawed to the proper curvatures can be used.
- *Yoke assembly* – Each yoke assembly consists of a horizontal cross member connected to a jack, plus a yoke for each set of form panels or walers/ribs. The top of each leg is attached to the cross member and the lower end is attached to the bottom waler. The yokes transmit the lifting forces from the jacks to the walers. Since form ties cannot be used to maintain the required spacing between the form panels, the yoke legs must also hold the panel in the required position.
- *Working deck* – The working deck usually consists of 1"-thick boards or ¾" plywood supported by joists. The joists may be supported at the ends only by the walers, or for long spans, they may require intermediate supports of wood or steel trusses, steel joists, or steel beams attached to the walers.

If the structure is to be finished with a concrete roof or cap, the working deck can be used as a form to support the concrete. For this purpose, pointed steel pins are driven through the form panels into the concrete wall under the walers, and then the yokes are removed.

- *Suspended scaffold* – The scaffold suspended under the form allows the finishers to have access to the concrete surfaces. The scaffold should be assembled in sections before concrete placement is started. When concrete placement has progressed sufficiently, the form raising is stopped temporarily and the scaffold is attached to the form assembly.

- *Form jacks* – Any of the three types of jacks (screw, hydraulic, and pneumatic) may be used to lift the forms.
  - The screw jack is a smooth-steel jack rod, usually 1" in diameter, with its lower end embedded in the concrete. It passes through a hollow square-threaded jack. The assembly is attached to the jack rod during the lifting operation. As the jack is rotated manually, a nut moves upward, lifting a yoke with it.
  - The hydraulic jack is a smooth-steel jack rod with its lower end embedded in the concrete. It passes upward through the hollow jack, which is attached to a yoke. When pressure is applied to the jack, one element of it grips the jack rod and another element moves upward, lifting the yoke with it. When the pressure is released temporarily, the jack resets itself automatically for another upward movement. Since all jacks are interconnected to a centrally located pump, the oil pressure on each jack will be the same. This should ensure uniform upward movement.
  - The pneumatic jack system is basically the same as the hydraulic system, but instead of oil pressure, air pressure is used.

Once a slip-form assembly is constructed completely on a concrete base, the forms are slowly filled with concrete. When the concrete in the bottom of the forms has gained sufficient rigidity, the upward movement of the forms is started. This motion is continued at a speed controlled by the rate at which the concrete sets. Lifting rates may vary from 2" to 3" per hour to more than 12" per hour, depending on the ambient temperature and the properties of the concrete. The reinforcing steel is placed as the form moves upward. For most slipforming operations, concrete is continually placed until the structure is completed.

5.2.0 Climbing Forms

Climbing forms, sometimes called jump forms, are different from slip forms. While slip forms are moved slowly up as concrete is placed, climbing forms are lifted directly from one level to the next using cranes or other lifting methods. Some systems use hydraulic lifting mechanisms to raise the form. A climbing form is usually supported by anchor bolts or rods embedded in the top of the previous lift.

27308-14 Vertical Formwork

Figure 33 shows one type of climbing form system. The entire assembly is supported by devices that are bolted to inserts placed in the previous lift. An 8'-wide work platform is used to strip and clean the forms. A trailing platform hangs below the work platform. The trailing platform is used to patch tie holes, remove inserts, and perform necessary finishing work after the climbing form has been lifted to the next level. The trailing platform is added to the form assembly before the third lift.

One type of climbing form is the shaft form used for elevator and stair shafts in building cores. *Figure 34* shows the interior wall section of a shaft form being moved to the next level. A major factor in a shaft form is that there must be a collapsible panel or corner to allow the form to be stripped. *Figure 35* shows an example of a collapsible corner.

Figure 34 Shaft form.

Figure 33 Climbing form.

Figure 35 Shaft-form corner section.

5.0.0 Section Review

1. A slip form consists of forms that are contained in a _____.
   a. jacking system
   b. lifting frame
   c. yoke assembly
   d. working deck

2. Jacks used to lift slip forms include all the following *except* the _____.
   a. screw jack
   b. lever jack
   c. pneumatic jack
   d. hydraulic jack

Section Six

6.0.0 Stair Forms

Objective
Describe how to construct stair forms.

Performance Task
Erect, plumb, and brace a stair form.

Trade Terms

Flight: Continuous series of steps from one floor to another or from a floor to a landing.

Landing: Horizontal slab or platform in a stairway to break the run of stairs.

Figure 36 shows a basic concrete stair form. Note that the size and position of the riser boards establish the height, depth, and spacing of the stairs. When the stairs are wide, a center strongback is needed for additional support.

The *International Building Code®* (*IBC®*) specifies minimum and/or maximum requirements for commercial stairways and handrails. The following are examples of the *IBC* requirements:

- *Stairway width* – 44 inches minimum (exception: 36 inches minimum for occupancies serving fewer than 50 people).
- *Stair tread depth* – 11 inches minimum. All treads on a given flight of stairs must be the same depth.
- *Stair riser height* – 7 inches maximum, 4 inches minimum. The riser height on a given flight of stairs must be the same. As the riser height increases, the tread depth must decrease proportionally.
- *Headroom* – 80 inches minimum.
- *Nosing* – 1¼ inches maximum.
- *Vertical rise* – 12 feet maximum between floor levels or landings.
- *Handrail height* – 34 inches minimum, 38 inches maximum from top of rail to nosing.

The *International Residential Code®* (*IRC®*) specifies minimum and/or maximum requirements for residential stairways and handrails. The following are examples of the IRC requirements:

- *Stairway width* – 36 inches minimum.
- *Stair tread depth* – 10 inches minimum. All treads on a given flight of stairs must be the same depth.

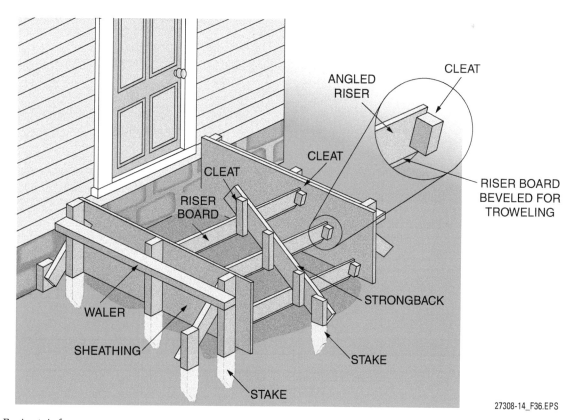

Figure 36 Basic stair form.

- *Stair riser height* – 7¾ inches maximum. The riser height on a given flight of stairs must be the same. As the riser height increases, the tread depth must decrease proportionally.
- *Headroom* – 80 inches minimum.
- *Nosing* – ¾ inch minimum, 1¼ inches maximum.
- *Vertical rise* – 12 feet maximum between floor levels or landings.
- *Handrail height* – 34 inches minimum, 38 inches maximum from top of rail to nosing.

Always refer to the local building code for stairway requirements for the particular type of work you are performing in your area. *Figure 37* shows a basic stair layout.

The general procedure for constructing formwork for a solid concrete stairway is as follows:

Step 1 Establish landing elevation(s).

Step 2 Set outside forms. Plumb and brace the forms using the same procedure as used for plumbing and bracing wall forms.

Step 3 Determine riser height to tread depth.

Step 4 Lay out risers on the forms (*Figure 38*). The figure also shows tie holes in the forms. Form ties can be placed through these holes to tie the two form panels together.

Step 5 Fasten cleats for riser boards on the inside face of the forms.

Step 6 Cut risers to the correct height. Bevel the bottom edge of the riser boards so concrete finishers can properly finish the treads.

Step 7 Install reinforcement, as needed, including the nosing bar.

Step 8 Fasten risers to cleats so they can be easily removed.

Step 9 Brace risers by installing strongbacks connecting the risers together (minimum of one strongback every 3 feet).

Step 10 Place concrete.

Step 11 When concrete starts to set, strip risers for final finishing.

Strip remaining forms according to the engineering specifications.

> **NOTE**
> Check local and national building codes before constructing stairways.

Figures 39 through *44* show different types of stairways and their associated forms.

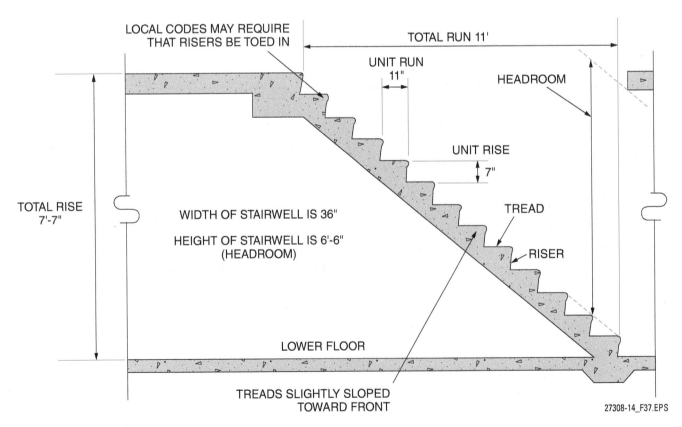

Figure 37 Basic stair layout.

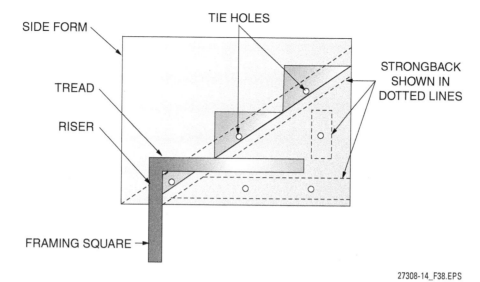

Figure 38 Laying out treads and risers on form sides.

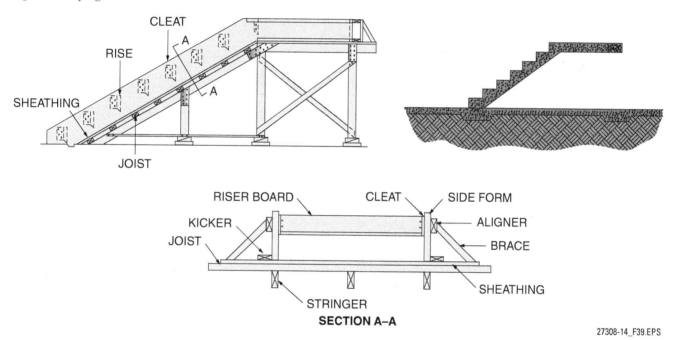

Figure 39 Open-sided suspended stairs and form.

Stair Forms

When building stair forms, make sure to check the drawings and specifications to see if handrails are required. Also, make sure that riser boards are tapered inward at the bottom to allow for finishing.

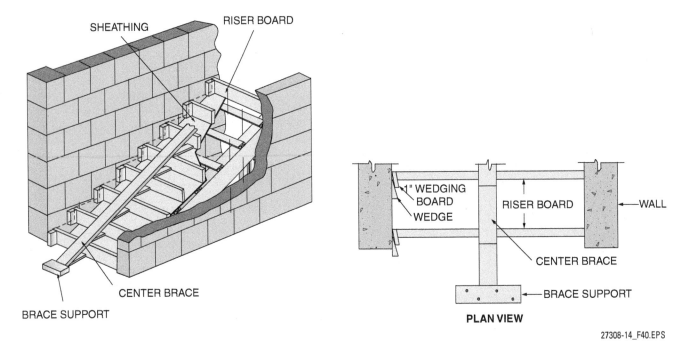

Figure 40 Form for suspended stairs between walls.

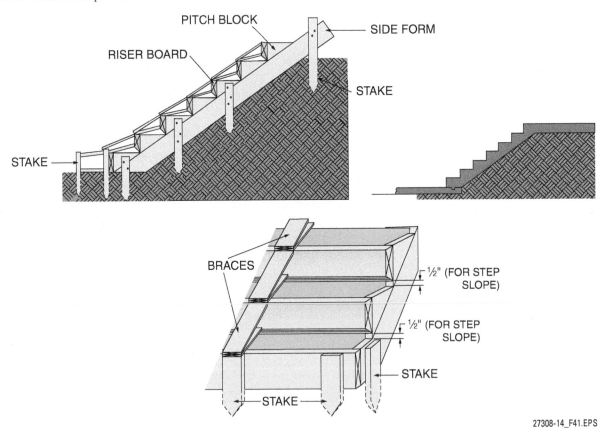

Figure 41 Earth-supported stairs and form.

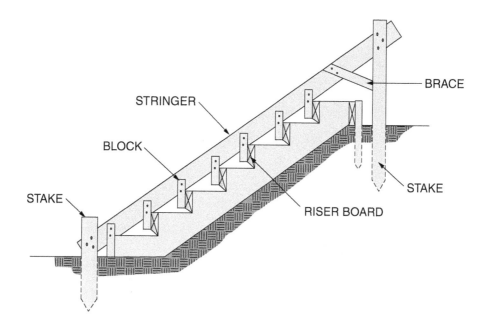

PLAN VIEW

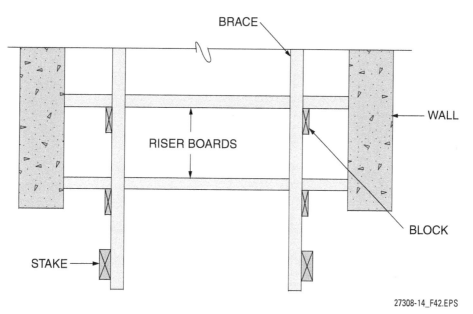

Figure 42 Form for earth-supported stairs between walls.

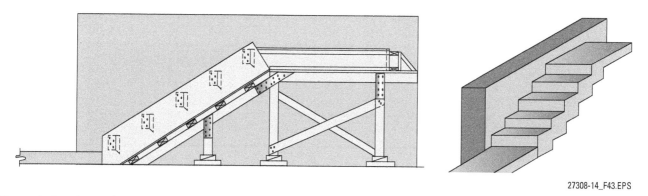

Figure 43 Cantilevered stairs and form.

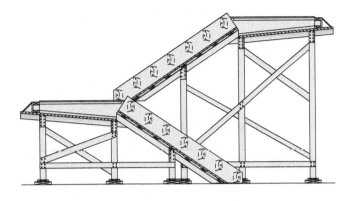

Figure 44 Freestanding stairs and form.

6.0.0 Section Review

1. On a given flight of stairs, the stair riser height must be _____.
    a. the same for each step
    b. six inches minimum
    c. equal to the tread depth
    d. in proportion to the handrail height

SECTION SEVEN

7.0.0 VERTICAL ARCHITECTURAL AND SPECIALTY FORMS

Objective

List various vertical architectural and specialty forms, and describe applications for each.
  a. Describe how smooth finishes are created.
  b. Describe how textured surfaces are created.
  c. Explain the use of insulating concrete forms (ICFs).

Trade Terms

Rustication line: Indentation made in a concrete panel by attaching a thin strip of tapered wood or plastic to the form.

Shiplap: Boards that have been rabbeted on both edges so that they overlap when they are placed together.

In some cases, structural concrete is covered with some type of veneer, such as brick. Architectural concrete, on the other hand, not only serves as structural material, but it also becomes an architectural medium by providing the desired surface quality and appearance. Therefore, special care is needed in the construction of the formwork involved. Mistakes or imperfections in the formwork will show as light conditions vary, as will form-panel joint marks, crooked corners, bulges, and other irregularities. Problems other than those caused by formwork can affect the quality of the exposed concrete surface. Other factors that can cause imperfections are as follows:

- Different types of cement used in the concrete mixture
- Mixture of coarse and fine aggregates
- Lack of uniformity in mixing and placement
- Curing methods
- Curing time and temperature
- Form-release agents

All of these factors can affect the color and/or texture of the exposed surfaces no matter how much care is taken in preparing the formwork.

When reading the construction drawings for any job, you will find a section on design. This section will be accompanied by design drawings, which refer to the location and item number, as well as to such items as openings, control joints, rustication lines, and expansion joints. These drawings provide detailed descriptions of how special areas are to be built. Recommendations are included for accessories, closure techniques, concealment of joints, and the sealing of the forms to make them watertight for smooth or lined surfaces.

A part of the need for watertightness is due to a condition caused by variations in the water content of the concrete and also by the moisture movement within the concrete during setting. Water-related discoloration penetrates concrete to a considerable depth and cannot ordinarily be concealed by abrasive blasting or tooling the surface. It is important to complete this part of the construction as carefully as possible to avoid water damage.

Contemporary design calls for a wide range of surface textures and treatments. A surface design may vary from a glass-like, smooth finish to a finish needing special sculptured ornamentation. These surfaces require different types of form panels and form liners. Chemical retarders applied to the form surface make it possible to remove surface mortar and expose the aggregate after forms have been removed. Variations in finish can also be achieved by grinding, chiseling, hammering, and sandblasting after the concrete is set. Precast panels used as forms or liners may also provide the desired surface finish. A nonstaining form oil or coating should be used for the architectural forms to prevent uneven coloring of the concrete.

7.1.0 Creating Smooth Finishes

Smooth finishes are commonly cast against materials such as steel, plywood, reinforced or unreinforced plastic, and hardboard. Contact surfaces of the formwork should be carefully installed to produce neat and symmetrical joint patterns unless otherwise specified. In building formwork for smooth finishes, large gang forms are ordinarily used in order to minimize the marks left by form panels. Gang forms can also be built horizontally, which minimizes voids called bug holes or blowholes. These voids are caused by bubbles of air or water trapped at the form surface while the concrete is hardening.

If lumber is used to build the smooth-face form, dressed material is a necessity; tongue-and-groove or shiplap is desirable. Dressed lumber is more easily and accurately placed.

When plywood or hardboard serves as the form panel, it is best and most economical if full panels are used. Joints should fit tightly and be filled or taped for the best appearance.

Metal forms and liners produce a smooth surface, but there is difficulty in concealing the joints. This is particularly true at breaks between smooth and textured surfaces.

27308-14 Vertical Formwork

7.2.0 Creating Textured Surfaces

Textured surfaces are used for several reasons, including their appearance and their ability to cover conspicuous defects (*Figure 45*). Textured surfaces can be achieved by the use of form sheathing or by a liner attached to the forms. These textured liners or textured forms can be made of fiberglass-reinforced plastics, plywood, polystyrene, polyvinyl chloride, rough-sawn lumber, rubber, aluminum, and other materials.

7.3.0 Insulating Concrete Forms

Insulating concrete forms (ICFs) are now commonly used for forming concrete structures (*Figure 46*). ICFs, which remain in place after the concrete has hardened, consist of a layer of concrete that is sandwiched between two expanded polystyrene (EPS) foam panels. These forms have grown in popularity as an alternative to more labor-intensive concrete block foundations and walls in residential and light commercial applications. More recently, ICFs are being used for heavy commercial construction work (*Figure 47*) and high-rises. Although these forms are very light, their design allows them to withstand the pressure of cast-in-place concrete. In addition, they provide substantial insulation, so separate insulation may be unnecessary. Finished ICF walls can be finished on the interior and exterior using the same type of finish as used for block walls.

Figure 46 Insulating concrete forms (ICFs).

Figure 47 ICFs used for commercial structures.

Figure 45 Textured surface.

Architectural Concrete

When reusing patterned architectural-concrete form liners to produce an architectural finish on a concrete wall like the one shown here, the key to acquiring a high-quality finish is the careful cleaning of the form liner between uses.

7.0.0 Section Review

1. To avoid water-related discoloration, architectural-concrete forms must be _____.
   a. plastic-lined
   b. watertight
   c. provided with drain holes
   d. coated with polyurethane sealer

2. Between uses, patterned architectural-concrete form liners must be _____.
   a. sandblasted
   b. rolled for storage
   c. carefully cleaned
   d. stretched out flat

3. A forming material that remains in place after concrete hardens is the _____.
   a. form liner
   b. biodegradable form
   c. fiberglass form
   d. insulating concrete form

Summary

Carpenters are required to construct or assemble a variety of concrete forms for walls, stairs, and columns. Job-built forms can be constructed almost entirely from lumber and plywood or lumber sheathing. Most of the attaching components, such as form ties, are made of metal.

When constructing or assembling formwork, safety is paramount. If forms are improperly built or are not adequate for the concrete pressure they must handle, people can be injured or killed as a result of structural failure. Therefore, it is very important to follow the plans and specifications established by the building designers, especially when supporting and bracing the forms.

Patented concrete forms and form hardware are widely used in commercial construction. They are very durable and can be reused many times. They are also easy to assemble and disassemble. Patented forms are available in a variety of sizes and shapes to suit any construction need. They range from small, lightweight forms for light construction to large, steel-reinforced forms that must be assembled and moved using a crane.

Many companies manufacture concrete forms and related accessories. In order to safely assemble and use these forms, a carpenter must be thoroughly familiar with the particular forms being used, especially the limitations of those forms. The instructions supplied with the forms must be followed. In some cases, it may be necessary to take a training course in the proper assembly and use of the forms.

Review Questions

1. The amount of formwork needed for a job can be influenced by _____.
   a. the size of the forming crew
   b. the project budget
   c. materials used for the forms
   d. the curing time needed for the concrete

2. Formwork must support the weight of the concrete and the _____.
   a. pressure from winds blowing across the job site
   b. weight of people working on the forms
   c. weight of the forms themselves
   d. weight of the steel reinforcement

3. Forms built on the job site from lumber and plywood are fastened with _____.
   a. 16d cement-coated nails
   b. lag screws
   c. duplex nails
   d. bolts and nuts

4. Accessories used to align the forms and hold them in place are the _____.
   a. stakes and braces
   b. spreaders
   c. strongbacks
   d. form ties

5. Panel form systems can be grouped into _____.
   a. three types
   b. four types
   c. five types
   d. six types

6. The usual panel size for all-steel forms is _____.
   a. 2' × 2'
   b. 2' × 4'
   c. 3' × 6'
   d. 4' × 8'

7. If properly cleaned and handled, some manufactured concrete forms may be reused _____.
   a. twice
   b. 20–30 times
   c. 50–75 times
   d. hundreds of times

8. An assembled gang form may have a width of as much as _____.
   a. 50 feet
   b. 45 feet
   c. 30 feet
   d. 25 feet

9. A crane usually is needed to set gang forms, which can weigh in excess of _____.
   a. 350 pounds
   b. 500 pounds
   c. 875 pounds
   d. 1,200 pounds

10. For a curved form, a sheet of plywood can be bent after using a saw to cut shallow parallel slots called _____.
    a. channels
    b. rabbets
    c. relievers
    d. kerfs

11. When a window buck is inserted in a form, the bottom spreader must include a _____.
    a. lifting handle
    b. stripping hook
    c. vibrator hole
    d. weep channel

12. When formwork components are unloaded at the job site, they must be _____.
    a. thoroughly cleaned
    b. checked against the order to be sure none are missing
    c. stacked in the order they will be used
    d. carefully measured to verify the ordered sizes

13. When setting a gang form, rebar pins placed in the foundation are used to _____.
    a. anchor the form
    b. aid in stripping after the concrete has cured
    c. properly position the form
    d. align the reinforcement

14. To keep concrete from escaping, the open ends of a form are closed with _____.
    a. bulkheads
    b. form seals
    c. end panels
    d. barriers

15. After forms are stripped, concrete finishers fill in the _____.
    a. stress cracks
    b. anchor voids
    c. cavities
    d. tie holes

16. Fiber forms are used for _____.
    a. walls
    b. round columns
    c. footings
    d. elevator pits

17. Round and flat steel form sections can be combined when building a type of structure known as a _____.
    a. half-round pier
    b. bullhead pilaster
    c. bull-nose pier
    d. combination column

18. Usually, only forms for square and rectangular columns are _____.
    a. reused
    b. job-built
    c. rented or leased
    d. made from lumber and plywood

19. Plastic concrete exerts the greatest pressure at the _____.
    a. bottom of the forms
    b. ends of the forms
    c. top of the forms
    d. center of the forms

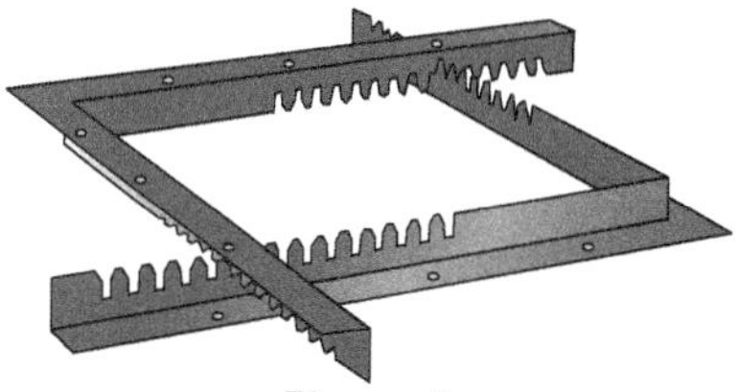

Figure 1

20. The column clamp shown in Review Question *Figure 1* is a _____.
    a. single-bar clamp
    b. scissors clamp
    c. hinged double-bar clamp
    d. light-duty clamp

21. Slip forms are raised at a rate measured in _____.
    a. inches per minute
    b. inches per hour
    c. feet per hour
    d. feet per day

22. Structures in tall buildings used for elevator shafts and stairwells are called _____.
    a. backbones
    b. raceways
    c. shaft cores
    d. utilities shafts

23. As concrete is placed in the slip form, the entire form unit is steadily raised by _____.
    a. hoisting gear
    b. a yoke-and-jack assembly
    c. hydraulic rams
    d. a rack-and-pinion system

24. Plywood used to form panels in the slip form should be presoaked in water to prevent _____.
    a. concrete sticking to the form
    b. splitting
    c. surface blemishes in the concrete
    d. swelling

25. The forms used in slip-form construction are aligned and braced on each side by two or more rows of _____.
    a. brace beams
    b. walers
    c. strongbacks
    d. girders

26. The trailing platform used by concrete finishers is attached to the bottom of a climbing form after the _____.
    a. first lift
    b. second lift
    c. third lift
    d. fourth lift

27. Building codes specify a maximum rise between floors or stair landings of _____.
    a. 10'-6"
    b. 12'
    c. 13'-6"
    d. 15'

28. The first step in constructing formwork for a concrete stairway is to _____.
    a. establish the landing elevation(s)
    b. erect the outside forms
    c. determine the rise and run
    d. determine riser height to tread depth

29. To allow access for finishing, riser forms should be stripped _____.
    a. in four hours or less
    b. as soon as the concrete starts to set
    c. after 24 hours
    d. when the concrete is fully cured

30. When using lumber to build a smooth-face form, it is necessary to use _____.
    a. edge-matched boards
    b. seasoned lumber
    c. dressed lumber
    d. wide boards

Trade Terms Quiz

Fill in the blank with the correct term that you learned from your study of this module.

1. A form can be reinforced against the weight of concrete by a diagonal supporting member called a(n) _____.

2. A(n) _____ is raised to a new level for each pour when constructing vertical walls.

3. A continuous series of steps from one floor to the next, or from a floor to a landing, is a(n) _____.

4. A design or other feature can be added to the surface of the concrete by attaching a plastic or wood _____ to the inside of the form.

5. A(n) _____ is a device attached to a heavy form to allow hoisting by a crane or other lifting equipment.

6. _____ is the material, such as plywood, that makes up the surface of a concrete form.

7. A(n) _____ is a type of form that is moved continuously as the concrete is being placed.

8. The long, tapered handle of the _____ is used to align holes on adjoining panels when constructing forms.

9. To stiffen a wall form, a wood or metal _____ is attached to the outside of the form.

10. A form panel can be reinforced by applying a(n) _____, which is a strip of lumber laid flat against the form.

11. To stop the flow of concrete at a desired location within a form, a(n) _____ is fastened vertically between the facing form panels.

12. _____ is a durable material used for column forms or applied as an overlay on plywood panels.

13. To smoothly overlap when placed together, _____ boards are rabbeted in both edges.

14. A(n) _____ is a vertical reinforcing member applied to the back of forms, especially around door and window openings.

15. Used with slip forms, the _____ is a clamping and support device.

16. _____ is exposed on the exterior of a structure, serving as a finish material.

17. A window or door opening is made in a concrete form by inserting a frame called a(n) _____.

18. A run of stairs is interrupted by a(n) _____ that is formed by a slab or platform.

19. The opposing sides of a concrete form are kept at the desired distance apart by inserting a wood or metal _____.

20. A bracing device known as a(n) _____ is attached near the top of a round column form.

21. A(n) _____ consists of two expanded polystyrene foam panels that are left in place after the concrete has cured.

22. A(n) _____ is an indentation, used as an architectural feature, that is created by fastening a thin, tapered strip of wood or plastic to the inside of a form.

Trade Terms

Architectural concrete	Buck	Flight	Lifting eye	Spreader
Batten	Bulkhead	Form liner	Rustication line	Spud wrench
Brace	Climbing form	Insulating concrete form (ICF)	Sheathing	Strongback
Bracing collar	Fiberglass-reinforced plastic (FRP)	Landing	Shiplap	Waler
			Slip form	Yoke assembly

27308-14 Vertical Formwork

Module Two 43

Trade Terms Introduced in This Module

Architectural concrete: Concrete that serves as the architectural finish material.

Batten: Strip of lumber laid flat against the form panel to provide reinforcement.

Brace: A diagonal supporting member used to reinforce a form against the weight of the concrete.

Bracing collar: Metal strap used to brace a round column form; it is attached near the top of the form and extends to the ground.

Buck: A frame placed inside a concrete form to provide an opening for a window or door.

Bulkhead: Form component fastened vertically inside a form to stop the flow of concrete at a certain location.

Climbing form: A form used to construct vertical walls in successive pours. It is raised to a new level for each pour.

Fiberglass-reinforced plastic (FRP): A durable material used for column forms and as an overlay on plywood panels.

Flight: Continuous series of steps from one floor to another or from a floor to a landing.

Form liner: Plastic or wood liners used on the inside of wall forms to add a design or special feature to the concrete.

Insulating concrete form (ICF): Concrete form system in which concrete is cast between two expanded polystyrene (EPS) foam panels.

Landing: Horizontal slab or platform in a stairway to break the run of stairs.

Lifting eye: Load-lifting device attached to a heavy form panel; used to rig the panel with a crane or other lifting equipment.

Rustication line: An indentation made in a concrete panel by attaching a thin strip of tapered wood or plastic to the form.

Sheathing: Plywood, planks, or sheet metal that make up the surface of a form.

Shiplap: Boards that have been rabbeted on both edges so that they overlap when they are placed together.

Slip form: A form that is moved continuously while the concrete is being placed.

Spreader: A wood or metal device used to hold the sides of a form apart.

Spud wrench: Tool to align holes in adjoining form panels; it has a long, tapered steel handle with an open-end wrench on one end.

Strongback: An upright supporting member attached to the back of a form to stiffen or reinforce it, especially around door and window openings.

Waler: Horizontal wood or metal members installed on the outside of form walls to strengthen the sheathing and stiffen the walls.

Yoke assembly: Clamping and support device used in slip forms.

Additional Resources

This module presents thorough resources for task training. The following resource material is suggested for further study.

OSHA 29 *CFR* 1926, *Safety and Health Regulations for Construction*, latest edition. Washington, D.C.
Principles and Practices of Commercial Construction. Upper Saddle River, NJ: Prentice Hall.
Scaffold, Shoring, and Forming Institute. **www.ssfi.org**

Figure Credits

Haskell, CO01, SA02
Courtesy of Symons by Dayton Superior, Figure 1, Figure 7b, Figures 12–13, Figure 21b, Figure 33
Courtesy of Western Forms, Figure 2, Figure 7a
Award Metals, Figure 5f, E01
Courtesy of EFCO Corp., Figure 7c, Figure 8, Figure 10c, Figure 22
Courtesy of Gates & Sons, Figure 9, Figure 14
PERI GmbH, Figure 16, Figure 30, Figure 34
Portland Cement Association, SA03, Figure 21d
Brasfield and Gorrie LLC, Figure 20, Figure 35

SONOCO, Figure 21a
MFG Construction Products, Figure 21c
Symons, Figure 26
Ellis Manufacturing Co. Inc., **www.Ellisok.com**, Figure 29d, Figure 29f
Courtesy of Scanada International, Inc., Figures 31–32
Western Forms, Figure 45
Greenstreak, Inc., SA05
Courtesy of Quad-Lock Building Systems Ltd. **www.quadlock.com**, Figures 46–47

Section Review Answer Key

Answer	Section Reference	Objective Reference
Section One		
1. True	1.0.0	1
2. c	1.1.0	1a
3. b	1.2.1	1c
4. d	1.4.0	1d
Section Two		
1. a	2.1.0	2a
2. c	2.2.0	2b
Section Three		
1. b	3.1.0	3a
2. d	3.2.0	3b
Section Four		
1. a	4.0.0	4b
2. d	4.1.1	4a
Section Five		
1. c	5.0.0	5a
2. b	5.1.0	5a
Section Six		
1. a	6.0.0	6
Section Seven		
1. b	7.0.0	7
2. c	7.2.0	7b
3. d	7.3.0	7c

NCCER CURRICULA — USER UPDATE

NCCER makes every effort to keep its textbooks up-to-date and free of technical errors. We appreciate your help in this process. If you find an error, a typographical mistake, or an inaccuracy in NCCER's curricula, please fill out this form (or a photocopy), or complete the online form at **www.nccer.org/olf**. Be sure to include the exact module ID number, page number, a detailed description, and your recommended correction. Your input will be brought to the attention of the Authoring Team. Thank you for your assistance.

Instructors – If you have an idea for improving this textbook, or have found that additional materials were necessary to teach this module effectively, please let us know so that we may present your suggestions to the Authoring Team.

NCCER Product Development and Revision
13614 Progress Blvd., Alachua, FL 32615

Email: curriculum@nccer.org
Online: www.nccer.org/olf

❏ Trainee Guide   ❏ Lesson Plans   ❏ Exam   ❏ PowerPoints   Other _____

Craft / Level: _____   Copyright Date: _____

Module ID Number / Title: _____

Section Number(s): _____

Description: _____

_____

_____

_____

Recommended Correction: _____

_____

_____

_____

Your Name: _____

Address: _____

Email: _____   Phone: _____

27309-14

Horizontal Formwork

Overview

Horizontal concrete formwork is used in the construction of floors, bridge decks, culverts, beams, girders, and other horizontal concrete structures. When a deck is being formed and the concrete is placed, it must be supported from underneath by shoring systems until the concrete hardens. Anyone doing heavy commercial or industrial carpentry work can expect to work with these special forming and shoring systems.

Module Three

Trainees with successful module completions may be eligible for credentialing through the NCCER Registry. To learn more, go to **www.nccer.org** or contact us at **1.888.622.3720**. Our website has information on the latest product releases and training, as well as online versions of our *Cornerstone* magazine and Pearson's product catalog.

Your feedback is welcome. You may email your comments to **curriculum@nccer.org**, send general comments and inquiries to **info@nccer.org**, or fill in the User Update form at the back of this module.

This information is general in nature and intended for training purposes only. Actual performance of activities described in this manual requires compliance with all applicable operating, service, maintenance, and safety procedures under the direction of qualified personnel. References in this manual to patented or proprietary devices do not constitute a recommendation of their use.

Copyright © 2015 by NCCER, Alachua, FL 32615, and published by Pearson Education, Inc., New York, NY 10013. All rights reserved. Printed in the United States of America. This publication is protected by Copyright, and permission should be obtained from NCCER prior to any prohibited reproduction, storage in a retrieval system, or transmission in any form or by any means, electronic, mechanical, photocopying, recording, or likewise. To obtain permission(s) to use material from this work, please submit a written request to NCCER Product Development, 13614 Progress Blvd., Alachua, FL 32615.

From *Construction Craft Laborer, Level Two, Trainee Guide*, Third Edition. NCCER.
Copyright © 2015 by NCCER. Published by Pearson Education. All rights reserved.

27309-14
Horizontal Formwork

Objectives

When you have completed this module, you will be able to do the following:

1. Identify safety hazards associated with elevated deck formwork.
2. Identify the various types of structural-concrete floor and roof slabs.
   a. Describe how one-way solid slabs are constructed.
   b. Describe how two-way flat slabs are constructed.
   c. Explain the difference between two-way flat plate slabs and two-way flat slabs.
   d. Describe how one-way joist slabs are constructed.
   e. Describe how two-way joist slabs are constructed.
   f. Describe how composite slabs are constructed.
   g. Describe how posttensioned concrete slabs are constructed.
3. Describe the different types of form systems.
   a. Describe applications for pan forms.
   b. Describe applications for I-joist pan forms.
   c. Describe applications for one- and two-way beam and slab forms.
   d. Describe applications for flat-slab or flat-plate forms.
   e. Describe applications for composite-slab deck forms.
4. Identify types of elevated decks.
   a. List the materials used for deck surfaces.
   b. Explain the use of hand-set multicomponent decks.
   c. Describe applications for hand-set panelized decks.
   d. Explain the use of outriggers.
   e. Describe applications for flying decks.
   f. Describe elevated-slab edge forms, blockouts, embedments, and jointing.
5. Identify the different types of shores and describe applications for each.
   a. Explain how adjustable wood shores are installed.
   b. Explain how manufactured shores are installed.
6. Identify specialty form systems.
   a. Explain how bridge decks are formed.
   b. Explain how tunnels and culverts are formed.

Performance Tasks

Under the supervision of your instructor, you should be able to do the following:

1. Erect, plumb, brace, and level a hand-set deck form.
2. Install edge forms, including instructor-selected blockouts, embedments, and bulkheads.

Trade Terms

Baseplate	Drop-head	Mudsill	Reshoring
Capital	Haunch	Phosphatized	Shoring
Corrugated	Live load	Plyform®	Stirrup
Dead load	Monolithically	Profile	Stringer

Industry-Recognized Credentials

If you're training through an NCCER-accredited sponsor you may be eligible for credentials from NCCER's Registry. The ID number for this module is 27309-14. Note that this module may have been used in other NCCER curricula and may apply to other level completions. Contact NCCER's Registry at 888.622.3720 or go to **www.nccer.org** for more information.

Code Note

Codes vary among jurisdictions. Because of the variations in code, consult the applicable code whenever regulations are in question. Referring to an incorrect set of codes can cause as much trouble as failing to reference codes altogether. Obtain, review, and familiarize yourself with your local adopted code.

Contents

Topics to be presented in this module include:

1.0.0 Elevated Deck Formwork Safety 1
2.0.0 Structural Concrete Floor and Roof Decks 4
   2.1.0 One-Way Solid Slabs 4
   2.2.0 Two-Way Flat Slabs 4
   2.3.0 Two-Way Flat Plate Slabs 5
   2.4.0 One-Way Joist Slabs 5
   2.5.0 Two-Way Joist Slabs 6
   2.6.0 Composite Slabs 6
   2.7.0 Posttensioned Concrete Slabs 7
3.0.0 Elevated Deck Formwork 9
   3.1.0 Pan Forms 10
   3.2.0 I-Joist Pan Forms 11
   3.3.0 One- and Two-Way Beam and Slab Forms 13
      3.3.1 Site-Built Beam and Slab Forms 13
      3.3.2 Manufactured Beam and Slab Forms 17
   3.4.0 Flat-Slab or Flat-Plate Forms 17
   3.5.0 Composite-Slab Deck Forms 17
4.0.0 Elevated Decks 22
   4.1.0 Deck Surfaces 22
   4.2.0 Hand-Set Multicomponent Decks 23
   4.3.0 Hand-Set Panelized Decks 24
   4.4.0 Outriggers 25
   4.5.0 Flying Decks 26
      4.5.1 Flying-Deck Safety 26
      4.5.2 Truss and Beam Tables 27
      4.5.3 Column-Mounted Tables 31
5.0.0 Shoring 34
   5.1.0 Adjustable Wood Shores 35
      5.1.1 Grading Elevated Slab Decks 36
   5.2.0 Manufactured Shores 36
      5.2.1 Frame Shores 36
      5.2.2 Steel and Aluminum Post Shores 36
6.0.0 Specialty Form Systems 41
   6.1.0 Bridge Decks 41
   6.2.0 Culverts and Tunnels 43

Figures

Figure 1 Shores supporting an elevated work surface 2
Figure 2 Types of structural-concrete floor and roof slabs 5
Figure 3 Monolithic beam-and-slab roof with posttensioning cables 5
Figure 4 A portion of a one-way joist slab 6
Figure 5 Underside of a two-way joist slab 6
Figure 6 Corrugated steel deck 7
Figure 7 Tensioning posttensioned tendons 7
Figure 8 Scaffold-supported shoring in place for a
          floor under construction 9
Figure 9 Beam/girder system 9
Figure 10 Drop-panel support system 10
Figure 11 Dome pans 10
Figure 12 Standard-length narrow and wide pan forms 11
Figure 13 Long-length pan forms 12
Figure 14 Wood-lattice joist supports 12
Figure 15 Wood pan forms 12
Figure 16 Partially stripped slab 12
Figure 17 Beam-and-slab floor 13
Figure 18 Types of beams 14
Figure 19 Steel girder or beam form without intermediate shores 15
Figure 20 Form for a small beam 15
Figure 21 Beam-and-slab forms supported by intermediate shores 15
Figure 22 Form for a large beam 15
Figure 23 Spandrel-beam form 16
Figure 24 Intersection of beam and column forms 16
Figure 25 Typical long-span minimum-column parking structure 18
Figure 26 A manufactured steel and aluminum beam-and-slab
          form system 18
Figure 27 Site-built beam and slab formwork using
          engineered-wood form components 18
Figure 28 Typical flat-slab form with drop panels 19
Figure 29 Underside of flat-slab form with drop panel 19
Figure 30 Typical composite floor-deck profiles 20
Figure 31 Anchor studs through a panel 20
Figure 32 Cross sections of typical aluminum stringers and joists 23
Figure 33 Engineered-wood I-joist framing members 23
Figure 34 Proprietary multicomponent aluminum deck framing 24
Figure 35 Drop-panel framing 24
Figure 36 Completed deck framing with plywood-decking
          and dome-pan installation in progress 24
Figure 37 Panelized decking under construction 24
Figure 38 Typical erection of panelized decking 25
Figure 39 Small panels with nailer strips used
          for fill-in around a column 26
Figure 40 Material-staging platform with outriggers 26
Figure 41 Outrigger beam bracing 26

Figure 42 Truss and beam table systems 27
Figure 43 Typical truss-table components 28
Figure 44 Truss table being lifted by a crane equipped with
           an auxiliary electric hoist 29
Figure 45 Truss tables partially rolled out of floor bays 29
Figure 46 Preparing a truss table for removal 29
Figure 47 Tilting a truss table 29
Figure 48 Truss table being guided into place on the next floor level 30
Figure 49 Truss table on jacks 31
Figure 50 Filler panels and shoring between tables 31
Figure 51 Typical adjustable-kicker installation 31
Figure 52 A column-mounted table 32
Figure 53 Column-mounted table being removed from floor bay 33
Figure 54 Column-mounted table being guided into position 33
Figure 55 Reshoring in a multistory building 34
Figure 56 Shoring 34
Figure 57 Pad-type mudsills 35
Figure 58 Adjustable post-type wood shores 35
Figure 59 Checking deck grading with a rotating-beam laser level 36
Figure 60 Typical frame shories 37
Figure 61 Stringer installation 37
Figure 62 Steel post shore 37
Figure 63 Lowering formwork 38
Figure 64 Integral post-mounted hammer-driven drop-pin 38
Figure 65 Shore-mounted drop-head accessory 38
Figure 66 Aluminum post shore 39
Figure 67 Final adjustment 39
Figure 68 Typical wood bridge-deck formwork 41
Figure 69 Modular deck panels 42
Figure 70 Example of a culvert form system 43
Figure 71 Culvert form setup 43
Figure 72 Precast-concrete tunnel segments 44

Section One

1.0.0 Elevated Deck Formwork Safety

Objective

Identify safety hazards associated with elevated deck formwork.

Trade Terms

Baseplate: Bottom component on a shore, typically a 6"-square steel plate with nail holes for fastening to a mudsill.

Mudsill: Hardwood planks or square pads used to support baseplates and prevent shores from sinking into the soil and causing the floor to collapse.

Reshoring: Process performed for multistory construction when shoring equipment is removed from a partially cured slab. As shores for forms are removed, the elevated deck is reshored by placing shores against the bottom of the slab.

Shoring: Construction of a temporary support system to carry the total load of horizontal decks and slabs, including the live loads that occur during construction added to the dead load (fresh cast-in-place concrete, reinforcing steel, and forms).

Stringer: Timber placed at the tops of shores to support joists and deck form panels.

This module provides information on various types of horizontal form systems used widely in heavy commercial or industrial construction. Horizontal form systems are used primarily in the construction of elevated structural floor and deck slabs in buildings and garages, in bridge-deck construction, and in tunnel and culvert construction. Also described in this module are the shoring methods used in conjunction with horizontal form systems and the safety considerations related to working with shoring and horizontal form systems.

Similar to vertical formwork, horizontal formwork has many safety hazards that must be identified. These hazards include, but are not limited to, the following items:

- Rigging
- Floor openings
- Materials handling
- Embedments
- Shoring
- Struck by/caught between
- Electrical and utility hazards

These hazards should be identified through the use of a job hazard analysis (JHA) and appropriate action should be taken to minimize the hazards.

> **WARNING!**
> Serious injury can result if safe practices are not followed when erecting, dismantling, or using forming and/or shoring equipment. Only persons familiar with current safety practices and the regulations concerning shores should engage in the erection, dismantling, or use of shoring equipment. OSHA regulations require the use of fall protection equipment, such as guardrail systems, and/or personal fall arrest devices at all working levels, open sides, and all other openings on platforms and work areas above certain heights. In all cases, where a worker is exposed to a fall hazard, guardrail systems or other fall protection equipment devices must be used.

Occupational Safety and Health Administration (OSHA) 29 *Code of Federal Regulations* (*CFR*) 1926.703 requires the following with regard to shoring and reshoring:

- All shoring equipment, including equipment used in reshoring operations, must be inspected prior to erection to determine that the equipment meets the requirements specified in the formwork drawings.
- Shoring equipment found to be damaged such that its strength is reduced to less than that required by 1926.703(a)(1) shall not be used for shoring.
- Erected shoring equipment shall be inspected immediately prior to, during, and immediately after concrete placement.
- Shoring equipment that is found to be damaged or weakened after erection, such that its strength is reduced to less than that required by 1926.703(a)(1), shall be immediately reinforced.
- The mudsills for shores shall be sound, rigid, and capable of carrying the maximum intended load.
- All baseplates, shore heads, extension devices, and adjustment screws shall be in firm contact, and secured when necessary, with the foundation and the form (*Figure 1*).

- Forms and shores, except those used for slabs-on-grade and slip forms, shall not be removed until the employer determines that the concrete has gained sufficient strength to support its weight and superimposed loads. Such determination shall be based on compliance with one of the following: the plans and specifications stipulate conditions for removal of forms and shores, and such conditions exist; or the concrete has been properly tested with an appropriate American Society for Testing and Materials (ASTM) standard test method designed to indicate the concrete compressive strength, and the test results indicate that the concrete has gained sufficient strength to support its weight and superimposed loads.
- Reshoring shall not be removed until the concrete being supported has attained adequate strength to support its weight and all loads in place upon it.

Other safety considerations specific to forming, shoring, and reshoring are:

- Use lumber rated as specified on the layout drawing taking special note of the stress, species, grade, and size. Use only lumber that is in good condition. Do not splice timber members between their supports.
- Provide a proper foundation below the baseplates for the distribution of leg loads to concrete slabs or ground. If installed on ground, mudsills must be used. The foundation must be level and thoroughly compacted prior to erection of shoring to prevent settlement. The area must be cleared of all obstructions and debris. Consideration must be given to potential adverse weather conditions such as washouts, and freezing and thawing of ground. Consult a qualified soil engineer to determine the proper-size foundation required for existing ground conditions.
- Do not make unauthorized changes or substitutions of equipment.
- Provide guardrail systems on all open sides and openings in formwork and slabs.
- Access must be provided to all deck levels being formed. If it is not available from the structure, access ladders or stair towers must be provided. Access ladders must extend at least 3 feet above formwork.
- If motorized concrete placement equipment is to be used, be sure that lateral loads, vibration, and other forces have been considered and adequate precautions taken to ensure stability.

Figure 1 Shores supporting an elevated work surface.

- Plan concrete placement methods and sequences to make sure that unbalanced loading of the shoring equipment does not occur.
- Fasten all braces securely.
- Check to see that all clamps, screws, pins, and all other components are in a closed or engaged position.
- Make certain that all baseplates and shore heads are in firm contact with the foundation and forming material.
- Avoid eccentric loads on U-heads and top plates by centering stringers on these members.
- Avoid shock or impact loads for which the shoring was not designed.
- Do not place additional temporary loads on erected formwork or cast-in-place slabs without checking the capacity of the shoring and/or structure to safely support such additional loads.
- Make sure that the completed shoring setup has the specified bracing to provide lateral stability.
- Make sure the erection of shoring is under the supervision of an experienced and competent person.
- When constructing frame shoring, follow the shoring layout drawing and do not omit required components. Do not exceed the shore frame spacings or tower heights as shown on the shoring layout. The shoring load must be carried on all legs, and all shoring frames must be made plumb and level as the erection proceeds. Also make sure to recheck the plumb and level of shoring towers just prior to concrete placement.

Additional Resources

OSHA 29 *CFR* 1926, *Safety and Health Regulations for Construction,* Latest Edition. Washington, D.C.: Occupational Safety and Health Administration.

1.0.0 Section Review

1. On platforms and work areas above a certain height, the minimum fall protection requirements are mandated by _____.

   a. local codes
   b. union contracts
   c. OSHA
   d. code inspectors

SECTION TWO

2.0.0 STRUCTURAL CONCRETE FLOOR AND ROOF DECKS

Objective

Identify the various types of structural-concrete floor and roof slabs.
a. Describe how one-way solid slabs are constructed.
b. Describe how two-way flat slabs are constructed.
c. Explain the difference between two-way flat plate slabs and two-way flat slabs.
d. Describe how one-way joist slabs are constructed.
e. Describe how two-way joist slabs are constructed.
f. Describe how composite slabs are constructed.
g. Describe how posttensioned concrete slabs are constructed.

Trade Terms

Capital: A flared section at the top of a concrete column.

Corrugated: Material formed with parallel ridges or grooves.

Dead load: The actual weight of the deck itself.

Live load: The carrying capacity of the deck including the dead load of the deck.

Monolithically: Used to describe concrete that is placed in forms continuously and without construction joints.

Stirrup: Rebar bent into a loop; used to provide shear reinforcement for concrete.

A number of elevated, structurally reinforced concrete floor and roof slabs are used for buildings (*Figure 2*). The type of floor or roof slab used depends on the loads that will be imposed on the slab, the span of the slab between supports, the type of frame used for the building, and the number of stories. Structural slabs differ from slabs-on-grade in that they must support their own weight (dead load) as well as any weight that will be placed on the slab (live load). Slabs-on-grade and the loads placed on them are supported by compacted material below the slab.

2.1.0 One-Way Solid Slabs

One-way solid slabs are also called one-way beam-and-slab floors or roofs. One-way solid slabs span parallel lines of support formed by beams or walls. In most cases, the concrete for beams and slabs is placed at the same time (*Figure 3*). Sometimes, the concrete for supporting columns is placed at the same time as the beams and slabs. The slab reinforcement is placed across the span to beams or walls. One-way solid slabs can support relatively high loads over short spans and are generally used for spans of less than 20' because their mass becomes too expensive for wider applications. Slab thickness ranges from 4" to 10". Structural engineers must design the shoring system supporting the formwork for the slabs as well as the beams because of the massive weight of the concrete being placed all at once.

Instead of standard beams with a depth of two or three times the width, beams that are wider than they are deep (slab bands) are sometimes used. The slab band is thinner than the outside beams. It also reduces the width of the slab, which, in turn, reduces the slab thickness and the amount of slab reinforcement that is required.

2.2.0 Two-Way Flat Slabs

Two-way flat slabs are reinforced in both directions at 90 degrees to each other (two ways). Except at the perimeter, beams and walls are not normally required to support two-way flat slabs because building columns with thickened drop panels (heads) centered above each column support the slab. Two-way flat slabs are commonly used for storage buildings and parking garages. For very heavily loaded industrial floors, a two-way slab is sometimes placed over a grid of beams and girders supported by columns to create two-way beam-and-slab floors. Two-way flat slabs are cast on temporary plywood decks supported by shores and temporary wood or metal beams and stringers. In many cases, removable (flying) decks covered with plywood are used to eliminate having to tear down and re-erect the decks for each floor. Shores and decking, including flying decks, are covered in the *Elevated Decks* and *Shoring* sections.

Two-way flat slabs range in thickness from 4" to 12", depending on the span width, which is usually less than 34'. The minimum size of the drop panel is ⅓ of the span and its thickness is normally 1¼ times the slab thickness. At one time, decorative tapered capitals at the top of the columns were used in addition to the drop panels;

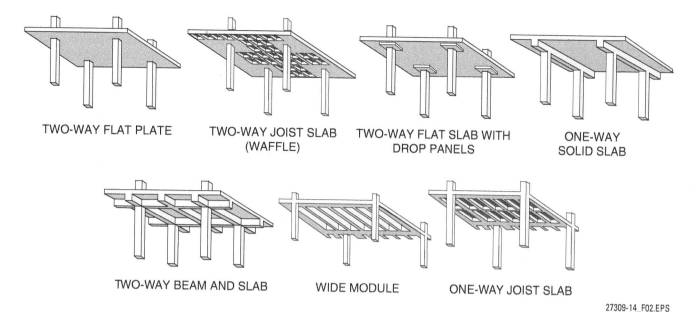

Figure 2 Types of structural-concrete floor and roof slabs.

however, they are not used much anymore because they are expensive to construct.

2.3.0 Two-Way Flat Plate Slabs

Two-way flat plate slabs are lighter-duty versions of two-way solid flat slabs without drop panels. They vary in thickness from 5" to 10" with spans up to 32'. Because two-way flat plate slabs are easily formed, they are generally used in high-rise apartment and office buildings with light floor loads. Both two-way solid flat slabs and flat plate slabs are generally cantilevered beyond the last row of columns for a distance of about 30 percent of the interior span, to take advantage of the slab's structural continuity. If the cantilever is not used, additional reinforcement is added to the slab edge to carry the higher stresses that will result. Like two-way flat slabs, flat plate slabs are cast using decks constructed with individual components or flying decks.

2.4.0 One-Way Joist Slabs

One-way joist slabs, also called ribbed slabs, are composed of reinforced concrete ribs or joists that are monolithically cast with a thin slab on top (*Figure 4*). The joists (ribs) are formed into the supporting beams or joist bands. The joist ends that are formed into supporting beams or joist bands are sometimes broadened. Because stirrups are usually not used in the joists due to space limitations, the broadened concrete at the ends provides sufficient resistance to diagonal tension forces. Joist bands, which are the same depth as the slab, are very wide beams placed over supporting columns and take the place of structural beams that would be deeper and would complicate the formwork. Joist bands are cast using the same formwork supporting the slab. Sometimes, a distribution rib may be used at midspan to spread live floor loads across the joists.

There is very little nonworking concrete in a one-way joist slab, reducing the dead load and allowing wider spans. Because the reinforcing steel is concentrated in the ribs, the slab can be much thinner and requires only shrinkage and temperature reinforcement. Slab thickness varies between 2½" and 4½". Joist spacing varies between 20" and 30" with depths of 6" to 20". The spans can be up to 45' wide. Overlapping manufactured steel or fiberglass-reinforced plastic (FRP) pan forms fas-

Figure 3 Monolithic beam-and-slab roof with posttensioning cables.

tened to plywood decking can be used for casting these types of slabs. If required, a tapered pan is used to form the broadened joist ends. Site-built wood pan forms using engineered-wood joists and plywood can also be used.

2.5.0 Two-Way Joist Slabs

Two-way joist slabs, also called waffle slabs, are economically similar to one-way joist slabs. They are composed of reinforced concrete ribs or joists that run at 90 degrees to each other (*Figure 5*). The waffle slabs can be formed using manufactured metal or FRP dome pans fastened to plywood decking. From an aesthetic viewpoint, the undersides of waffle slabs are more desirable than one-way joist slabs. In addition, they allow wider spans. In most cases, dome pans are not installed around columns. This allows solid reinforced sections, called heads, which are the same thickness as the joists to be cast along with the slab. These heads accomplish the same function as drop panels in two-way flat slabs.

Although other sizes are available, standard-size domes (30" × 30" or 19" × 19") result in 6"-wide joists on 3' centers or 5"-wide joists on 2' centers with depths from 8" to 20". Slab thickness usually ranges from 3" to 5" and spans can be 25' to 50'. If the slab is not cantilevered at an edge, a reinforced perimeter beam is required at that edge.

2.6.0 Composite Slabs

Composite slabs combine the compressive strength of concrete and the tensile strength of steel to form a bonded combination that is stronger than either material by itself. Corrugated steel deck that is fastened to structural steel beams and

Figure 4 A portion of a one-way joist slab.

Figure 5 Underside of a two-way joist slab.

Wide-Module Joist Slabs

A popular version of the one-way slab is the wide-module joist slab, also called a skip-joist slab. These slabs are formed with deep, wide metal or site-built pans that result in deeper joists with a wide spacing of 4' to 6'. Wide-module joist slabs are used when a slab thickness of 4½" or more is required for fireproofing. The thicker slab allows the use of wider joist spacing. The formwork for the slab in the figure is being assembled on a flat temporary plywood deck. Note the wide joist band in the foreground of the figure. The reinforcement for the joists and joist band has not yet been placed. Loads are higher in wide-module joists than in normal one-way joists. This usually requires that stirrups be used near the end of each joist. Because joist space is constricted, U-stirrups must be installed at an angle or single-leg stirrups must be used. The figure shows site-built skip-joist pans.

joists of a building is used as the slab form (*Figure 6*). The concrete and the deck, which is left in place, form the composite slab. The deck bonds to the slab and acts as the slab reinforcement after the concrete cures.

Two methods of composite-slab construction are used. One requires that the structural steel of the building be strong enough to support the deck and concrete without additional support while the concrete cures and bonds to the deck. The other method requires temporary support of lighter structural steel until the concrete cures and bonds to the deck.

In either case, the deck must be securely fastened to the structural steel by plug welding or spot welding. Welded stud anchors are also used to provide additional slab anchorage between the concrete and the structural steel. A wide variety of decks is available for use in floor or roof slabs in various gauges and rib heights with specific loadbearing capacities. The strength of the deck typically specified by the structural engineers is such that no deck support is required between structural steel spans when the deck concrete is placed and while the concrete cures. Lightweight concrete is often used for composite slabs.

2.7.0 Posttensioned Concrete Slabs

Posttensioning tendons made of very high-strength stranded-steel cable that is sheathed in plastic can be used in any of the previously described site-cast slabs and any supporting beams, girders, or joists. The tendons are tightened after the concrete has set properly (*Figure 7*). Posttensioning is used to reduce the size of beams, girders, or joists; eliminate the need for temperature reinforcement; reduce the amount of standard reinforcement required; reduce slab cracking; and extend the span of a slab. Two-way flat plate slabs are commonly posttensioned, and placement of posttensioning tendons in these slabs is different from the placement of conventional reinforcement.

In a typical installation, tendons are placed close together between columns in one direction across the slab. These closely spaced tendons are called banded tendons. Other tendons, called distributed tendons, are placed evenly in the other direction across the slab. However, the same number of tendons is used in both directions. Rebar can also be used at the column to provide additional shear resistance for the slab. The banded tendons are set high at the columns and low between the columns and at the edges of the slab unless a column is at the slab edge. The distributed tendons are set high over the banded tendons and low between the banded tendons and at the edges of the slab. After curing of the concrete and tensioning of the tendons, this draping causes upward pressure on the slab at the low points of the tendons and downward pressure at the high points. For slabs supported by structural steel beams, girders, and columns (such as noncomposite slabs on a metal deck or removable form slabs), the tendons are evenly distributed in both directions and are set high across the beams and girders and low between the beams and girders.

Structural engineers calculate the dead and live loads as well as short- and long-term tensioning losses to determine the amount of tensioning required. The losses include elastic shortening of the concrete, friction of the tendons, initial movement (set) of the anchorages, steel relaxation, concrete shrinkage, and creep. Certified personnel must perform placement and tensioning of posttensioning tendons.

Figure 6 Corrugated steel deck.

Figure 7 Tensioning posttensioned tendons.

Welded Stud Anchors

Besides composite slabs, welded studs are also used with other types of slabs that are placed in structural-steel buildings. In these cases, the studs provide the only anchorage for the slabs.

Posttensioned Slabs-on-Grade

Posttensioning of slabs-on-grade is common in some areas of the country, especially in the South and Southwest, where soil required for support of the slab may be poor. In these applications, the posttensioning tendons are equally spaced across the slab in both directions. This virtually eliminates slab cracking and the need for welded-wire reinforcement.

2.0.0 Section Review

1. Both structural slabs and slabs-on-grade must support both their own weight plus a live load.
   a. True
   b. False

2. One-way solid slabs are usually used for spans of less than _____.
   a. 10 feet
   b. 16 feet
   c. 20 feet
   d. 24 feet

3. Beams that are wider than they are deep are called _____.
   a. slab bands
   b. ribbed bands
   c. wide bands
   d. flat bands

4. A structural concrete slab frequently used for parking decks is the _____.
   a. one-way joist slab
   b. two-way flat plate slab
   c. posttensioned concrete slab
   d. two-way flat slab

5. The amount of tensioning required for posttensioned concrete slabs is determined by _____.
   a. OSHA
   b. the AHJ
   c. structural engineers
   d. certified personnel

6. One-way joist slabs can have a span with a width of up to _____.
   a. 45 feet
   b. 50 feet
   c. 65 feet
   d. 70 feet

7. Two-way joist slabs are also known as _____.
   a. corrugated decks
   b. waffle slabs
   c. dome pans
   d. reinforced decking

SECTION THREE

3.0.0 ELEVATED DECK FORMWORK

Objective

Describe the different types of form systems.
a. Describe applications for pan forms.
b. Describe applications for I-joist pan forms.
c. Describe applications for one- and two-way beam and slab forms.
d. Describe applications for flat-slab or flat-plate forms.
e. Describe applications for composite-slab deck forms.

Trade Terms

Phosphatized: Used to describe metal that has been treated with a phosphoric acid to prepare it for a finish coat. In composite floor decking, this unpainted surface will contact the concrete.

Profile: The appearance of a floor deck viewed cross-sectionally.

A variety of form systems are used to cast structural concrete slabs in heavy construction. These include systems for:

- Concrete slabs with concrete beams and girders
- Concrete slabs with no beams (flat slabs)
- Concrete slabs and joists using fiberglass-reinforced plastic (FRP) or metal pan forms
- Composite concrete slabs using corrugated-steel decking forms

Elevated slab forms require shoring for support (*Figure 8*) during the casting and curing of the concrete. Shoring is a temporary structure designed to carry the weight of plastic concrete, reinforcing steel, and formwork, as well as any live loads (people, equipment, etc.) that may be imposed on the formwork during construction. The shoring is not removed until the concrete has reached a specified strength, typically 75 or 80 percent.

A concrete floor must be properly supported. Several methods are used for this purpose. For floors that will be required to bear heavy loads, a beam/girder system is often used (*Figure 9*). In this type of system, beams and girders are tied to and rest on columns. In some cases, the concrete for the columns is cast first, then the beams and girders are formed. In other applications, the col-

Figure 8 Scaffold-supported shoring in place for a floor under construction.

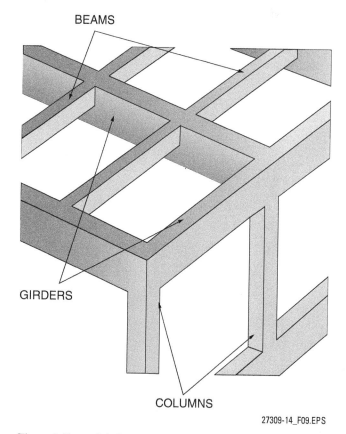

Figure 9 Beam/girder system.

umns, beams, and girders are formed monolithically, which means that the concrete is placed for all three components at the same time to form one solid piece. The beams and girders are reinforced with rebar that ties into the rebar of the columns above and below the floor.

Where lighter loads are involved, the slab receives its primary support from the columns. In some cases, a drop panel (*Figure 10*) is placed between the column and the slab. In the arrangement shown, a capital is added to the top of the column. The capital is an architectural, rather than a structural, feature.

3.1.0 Pan Forms

Commercial pan forms are molded steel or FRP forms that are used to construct concrete slab-and-beam joist systems. Two types are used: dome or square pans and long pans. Dome or square pans are used for two-way joist slab, or waffle slab, construction. Edge forms are retainers used to limit the horizontal spread of plastic concrete for slabs.

Standard manufactured dome or square pans are available in 19" × 19" and 30" × 30" sizes (*Figure 11*). Larger sizes with deeper depths are available to support spans up to 50'. The slab shown in the figure was formed using large dome pans. In this application, beams were formed over the columns by spacing the pans farther apart at the beam locations.

In a typical installation, the edge flanges on the pans are butted up against each other to form the joists. Butting the pans produces a clean, smooth surface on the bottom of the joist for appearance purposes. The pans are normally secured to the deck with nails along each edge, especially if the deck and pans will be stripped and used repeatedly for other floors on the same job. Where a beam or column head is to be formed, the edges of the pans facing them are nailed or screwed down more securely to resist the pressure of concrete that will be placed for the beam or head.

Once the slab has gained the specified strength, the temporary deck is lowered with all the pans securely attached. Sometimes, the pans are only minimally secured between each other, and when the deck is lowered, the pans remain in the slab. Some dome pans are equipped with a valve (blowhole) on the underside to aid in stripping the pans.

Manufactured steel pans are also available to form one-way joist slabs. They are available in a number of lengths and widths. Standard-length narrow and wide pan sections are shown in *Figure 12*. These pan sections, available in 2', 3', and 4' lengths, are overlapped end-to-end to form the slab and joists. Because the numerous overlaps are re-

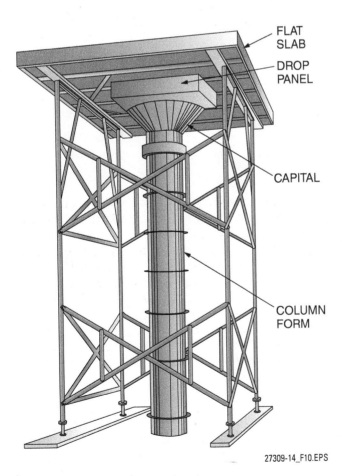

Figure 10 Drop-panel support system.

Figure 11 Dome pans.

NARROW WIDE

Figure 12 Standard-length narrow and wide pan forms.

flected in the undersurface of the slab, these forms are not suitable where appearance is important. Note the overlap indentations on the underside of the slab and the flange indentations on the bottom edges of the slab joists. When secured to a plywood deck, the flanges of these pans are not butted up to adjoining pans because the flanges are not wide enough to allow proper joist width. Overlapping end caps are used on the pans where they terminate for a beam, distribution rib, or column head.

Long versions (8', 12', and 16') of the narrow and wide pan forms are available to eliminate most of the lap joints and improve the appearance of the slab underside (*Figure 13*). As shown in the figure, separate support panels are installed on the deck before long wide pans are placed and secured. One manufacturer fabricates pans to specified lengths and flange widths, and joints are made by butt-welding the pans together. This eliminates any lap-joint and flange indentations on the underside of the slab and joists.

Like dome pans, pan forms for one-way joist slabs must be fastened to the deck with nails or screws. End caps must be secured with additional nails or screws because of the higher pressure caused by concrete in these areas. Like other forms, steel pans and exposed deck sections should be coated with a release agent (form oil) before the concrete is placed; however, reinforcement must not be coated.

3.2.0 I-Joist Pan Forms

Long pans or wide module pans can be site-built using wood I-joists covered with plywood. *Figure 14* shows the partial construction of wide module

Figure 13 Long-length pan forms.

pans with wood I-joist supports secured to strips of plywood. The plywood strips, resting on shoring framework, are the bottom forms for the slab joists that will be cast along with the slab. *Figure 15* shows plywood sheeting secured to the top and sides of the joists and the rebar being placed. Note the wide section at the left for a joist band that is the same depth as the slab joists. The band will be formed around the vertical column ties at the rear of the slab when the concrete for the slab and slab joists is placed. Also note the tapered sides of the pans along the slab joists to allow removal of the pans after the slab has gained sufficient strength for stripping. *Figure 16* shows a partially stripped slab with the slab-joist bottom form strips removed and with the pans ready to be stripped.

Figure 15 Wood pan forms.

Figure 14 Wood-lattice joist supports.

Figure 16 Partially stripped slab.

3.3.0 One- and Two-Way Beam and Slab Forms

Due to their expense, beam-and-slab floors (*Figure 17*) are typically used only for special applications involving heavy live loads. This is because they are very expensive to construct. The beam, girder, and slab forms can be site-built, or manufactured forms can be used.

3.3.1 Site-Built Beam and Slab Forms

Beams and girders are heavily reinforced with rebar that tie into the rebar of the columns on which they rest. *Figure 18* shows four types of beams:

- Interior beam
- Cantilever beam
- Inverted (or T-) beam
- Spandrel beam

Beam or girder forms may be set onto a column or across a series of columns (*Figure 19*). If the beam or girder is not directly supported by a column, it must be supported by shores (*Figure 20*). If the beam is set across columns and the columns are widely separated, shores may be needed between the columns (*Figure 21*).

Large beams require a different form configuration (*Figure 22*). With large beam forms, stiffeners (studs), walers, braces, and form ties may be used as they are with wall forms. For example, walers extend across the studs of a beam form to support the form sides. In addition, the deck may be extended to provide a work platform on either side of the form.

Beam and girder forms are often integral with the slab form, as in the spandrel beam shown in *Figure 23*. In this case, the concrete for the slab and beam are placed together.

Beam and column forms may intersect, with the beam form resting on and supported by the column form, as shown in *Figure 24*.

Beam form bottoms are usually constructed of ¾" (minimum) form plywood with 2 × 4 supporting members. Beam sides are constructed of either ¾" (minimum) plywood or lumber. If forms are constructed of lumber, 2 × 4 vertical cleats will be used. The cleats should be on 2' to 2'-6" centers. If form sides are constructed of plywood, the vertical cleats will be eliminated and 2 × 4 vertical stiffeners will be used. Vertical stiffeners will be placed according to plans and specifications.

The ledger is nailed to the beam form side at a certain point below the top to allow for the depth of joists that may have to be supported on the beam sides. Another design method places the ledger low enough on the beam form side to allow for the slab sheathing thickness.

The kicker is nailed flush with the bottom of the beam form side and later nailed to shore heads. In some cases, the kicker is not nailed to the beam form side until after the beam form side is in place. This allows for ease of installation. However, installing the kicker at the time of construction adds stiffness to the beam form side.

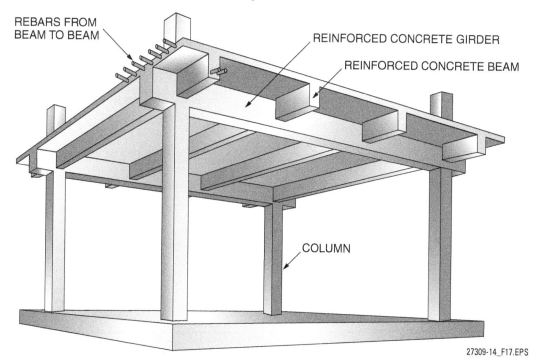

Figure 17 Beam-and-slab floor.

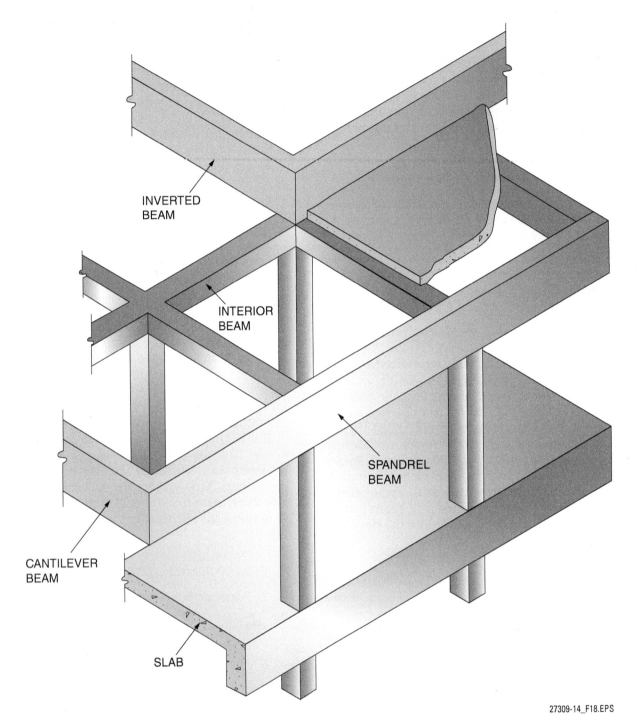

Figure 18 Types of beams.

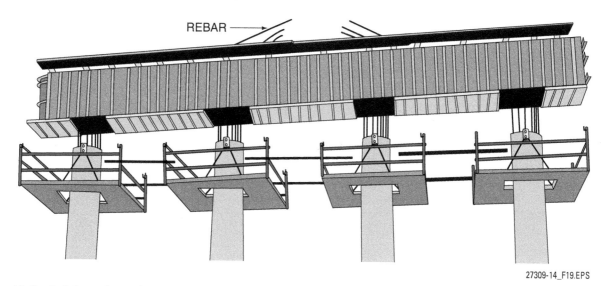

Figure 19 Steel girder or beam form without intermediate shores.

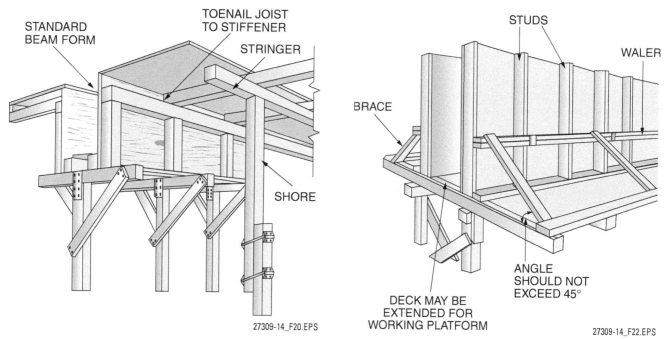

Figure 20 Form for a small beam.

Figure 22 Form for a large beam.

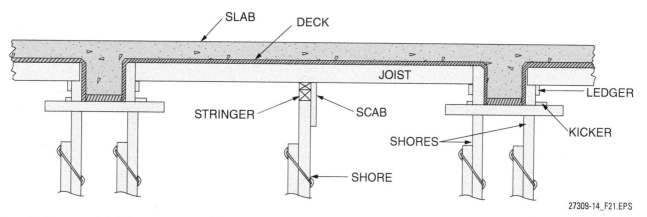

Figure 21 Beam-and-slab forms supported by intermediate shores.

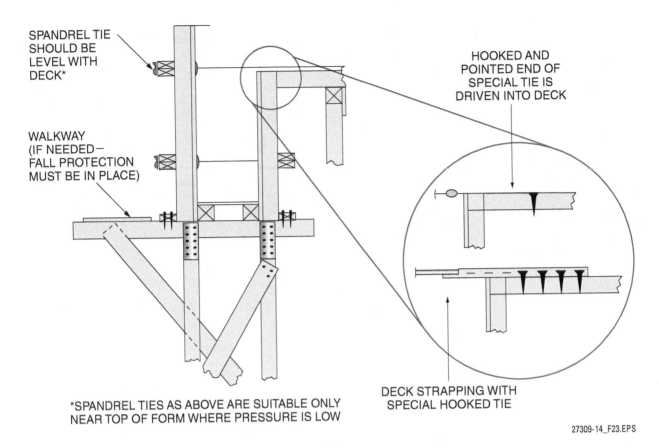

Figure 23 Spandrel-beam form.

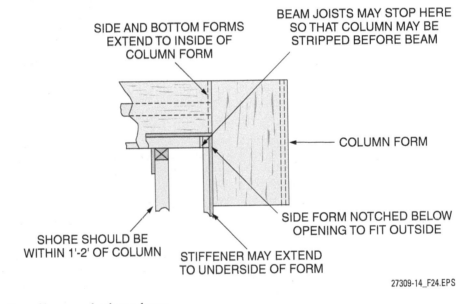

Figure 24 Intersection of beam and column forms.

Both ledgers and kickers are normally cut shorter than the beam form side to allow for framing of intersecting members.

Vertical stiffeners are often nailed in place when the beam form side is fabricated. When deep beam form sides are constructed, it may be necessary to predrill for form ties. If vertical members act as transfer points to transfer slab loads to the shores, many contractors prefer to cut the required number of vertical pieces and attach them loosely to the beam form side. They are positioned when the beam form side is placed. This permits the blocking to be set directly over shore heads in the event there is a discrepancy in placement of shoring. When off-site form assembly is being done, care should be taken to code all pieces for ease in identification and installation.

The lengths of the beam form sides will depend on the framing methods used, the method of transportation, and the concrete-placing technique employed at the job site. Forms may be constructed the same length as the clear span, minus the thickness of sheathing on the members the beam intersects, as well as a specific allowance for each end (1½", for example). The remaining gap is filled with a beveled 2 × 4 or other material suitable for a larger opening.

When framing beam pockets in girder sides, two methods are commonly used:

- The beam pocket may be constructed to the exact size of the final dimension of the beam, leaving a small allowance for give in the forms. The beam form is butted up against the opening in the girder form.
- In the second method, the beam form is made the size of the beam plus the thickness of the beam sides and bottom. The opening in the girder is made to accept the beam form, making sure that the beam form does not extend beyond the inside edge of the girder form.

3.3.2 Manufactured Beam and Slab Forms

Manufactured beam and slab forms reduce the amount of time and number of carpenters required to construct beam and slab forms. One such application is the long-span, posttensioned, one-way beam-and-slab floor that is popular for minimum-column parking structures (*Figure 25*). One version of a manufactured beam-and-slab form system is shown in *Figure 26*. In this system, the only shoring used is directly under the beams. In this example, frame shoring was used, but flying decks can also be used. In either case, mechanical lifting is required to set and strip the beam and deck forms.

Steel-beam form lengths of up to 60' with up to 31" of depth are available in a fixed 14" width or as 16"-wide split-bottom forms to accommodate wider beam requirements. To reduce project costs, beam depths are usually held constant and beam widths are varied to accommodate different loading within the project. For projects that cannot accommodate standard beam and slab widths and depths, site-built forms can also be used (*Figure 27*).

3.4.0 Flat-Slab or Flat-Plate Forms

Flat-slab or flat-plate forms can be either hand-set forms that are usually surfaced with plywood and supported by post shores with stringer and joist framing or some other type of deck. Manufactured panelized forms with post shoring are also used. Flat-slab formwork is more costly to build because of the drop panels required at each column. *Figure 28* shows the top of the formwork for a typical flat slab with drop panels. *Figure 29* shows the bottom of the drop-panel formwork. Note the mixture of aluminum joists and stringers for the deck framing along with the wood stringers and formwork used for the drop panel. Also note the mixture of metal-frame and wood-post shores used to support the formwork.

3.5.0 Composite-Slab Deck Forms

Steel deck-form material, called composite floor deck or structural deck, is made from corrugated sheets of galvanized steel, phosphatized/painted steel, or stainless steel. Supported by steel joists, steel beams, or precast concrete joists, the deck form material is installed as a floor deck over which concrete is placed to form a concrete floor (or roof) slab. The composite floor deck serves several purposes: it acts as a working platform, stabilizes the frame, serves as a concrete form for the slab, and reinforces the concrete slab to carry the design loads specified for the building floor.

Commonly used composite steel deck is made from 16-, 18-, 20-, and 22-gauge steel in 24"- or 36"-wide panels. Deck panels are made in various lengths ranging up to about 24' long. Deep-ribbed

Figure 25 Typical long-span minimum-column parking structure.

Figure 27 Site-built beam and slab formwork using engineered-wood form components.

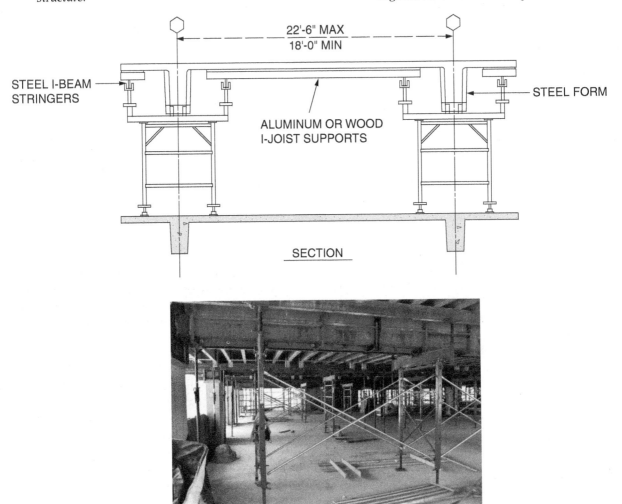

Figure 26 A manufactured steel and aluminum beam-and-slab form system.

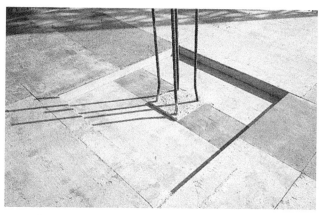

Figure 28 Typical flat-slab form with drop panels.

dimples, raised patterns, or rods crossing the corrugations are manufactured into the deck panels along with interlocking panel side laps to create a physical bond with the concrete, thus forming a strong floor slab. *Figure 30* shows typical profiles for composite floor-deck and cellular composite floor-deck panels. The raised portion of the deck-panel profile is made in different heights, with 1½", 2", and 3" heights being typical. Generally, the higher the profile height used, the deeper and stronger the resulting concrete slab. Cellular panels have a flat bottom welded to the corrugated panel that provides utility conduits and presents a smooth ceiling for the rooms below.

When using composite steel deck with a phosphatized/painted finish, the bare (phosphatized) top surface is the one that makes contact with the concrete. The bottom side of the deck panel is coated with a primer. In comparison, galvanized steel deck has a zinc coating on both sides. Floor deck is installed in accordance with the manufacturer's guidelines and the job erection drawings. During installation, the deck panels should be overlapped and installed in the opposite direction of the concrete placement to prevent loss of concrete through the joints. It is important to maintain rib alignment across the structure in order to achieve continuous concrete ribs across abutting panel ends.

Floor deck panels should be fastened in place as soon as possible after their placement. Spot welding or electric-arc plug welding is the best and most efficient method for fastening composite deck panels to structural supports. In many cases, studs (often called Nelson studs) are arc-welded through the panels to the structural steel to anchor the concrete slab (*Figure 31*).

Figure 29 Underside of flat-slab form with drop panel.

After the deck is installed, edge forms are placed to contain the concrete for the slab. Where penetrations are needed, one forming method is to block out concrete from the floor locations. After the concrete is sufficiently cured, a cutting torch is used to cut the penetration-area deck away. Concrete should be placed in the deck form from a low level in a uniform manner over the supporting structure and spread toward the center of the deck span. If necessary to store deck on site prior to its installation, it should be stored off the ground with one end elevated and protected with a tarpaulin or other weatherproof covering. It also should be ventilated to prevent condensation.

> **CAUTION**
> Concrete containing chloride admixtures or admixtures containing chloride salts should not be used with composite steel deck; such admixtures can deteriorate the steel.

Some specific safety considerations when working with composite decking include the following:

- Unfastened composite deck should not be used as a work or storage platform.
- Caution must be used with construction loading so that deck capacity is not exceeded and the deck is not damaged.
- When placing concrete in the deck form, care must be taken so that the deck is not subjected to impact that exceeds the design capacity of the deck.

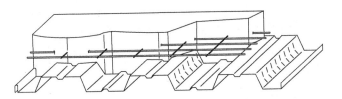 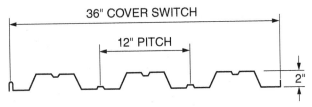

COMPOSITE FLOOR DECK

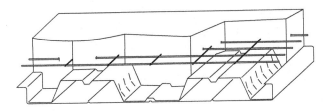 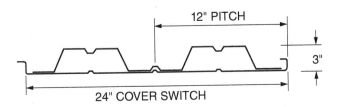

CELLULAR COMPOSITE FLOOR DECK

Figure 30 Typical composite floor-deck profiles.

Figure 31 Anchor studs through a panel.

Metal Deck Panels

Corrugated metal deck panels without deep-ribbed profiles or embossments are sometimes used to construct noncomposite floor slabs. When such metal deck is used solely as a form, slab reinforcement and concrete must be designed to carry the total slab load.

3.0.0 Section Review

1. Shoring is not removed from beneath a slab until the concrete has reached _____.
   a. its full design strength
   b. a specified strength, usually 75 or 80 percent
   c. 50 percent of its design strength
   d. sufficient strength to avoid sagging

2. A column capital is an architectural rather than a structural feature.
   a. True
   b. False

3. To aid in form stripping, some dome pans are equipped on the underside with a _____.
   a. handle
   b. hook
   c. valve
   d. release lever

4. When preparing slab or beam joist forms for concrete placement, form oil should be applied to all components *except* _____.
   a. dome pans
   b. reinforcement
   c. edge forms
   d. decks

5. Long versions (up to 16') of pan forms are used to _____.
   a. simplify forming work
   b. allow the use of less shoring material
   c. lower materials costs
   d. improve the finished appearance by minimizing lap joints

SECTION FOUR

4.0.0 ELEVATED DECKS

Objective

Identify types of elevated decks.
a. List the materials used for decking surfaces.
b. Explain the use of hand-set multicomponent decks.
c. Describe applications for hand-set panelized decks.
d. Explain the use of outriggers.
e. Describe applications for flying decks.
f. Describe elevated-slab edge forms, blockouts, embedments, and jointing.

Performance Tasks

Erect, plumb, brace, and level a hand-set deck form.

Install edge forms, including instructor-selected blockouts, embedments, and bulkheads.

Trade Terms

Drop-head: Shoring hardware that allows formwork to be released and dropped a few inches prior to stripping the forms.

Plyform®: An APA performance-rated panel used for concrete forms.

Many decks are erected in a labor-intensive multicomponent fashion using post or frame shores with separate stringer and joist framing members and separate plywood panels for the deck form. Other decks are constructed using hand-set panelized systems that are composed of shoring systems and panelized deck forms with integral framing and plywood deck surfaces or, in some cases, plastic surfaces. These types of decks are less labor intensive to erect. In many high-rise projects, flying decks or column tables are used as decks. The tables are assembled once at a job site and then flown from floor-to-floor by a crane as construction progresses. Often called flying forms, the tables are the least labor-intensive type of deck. They only require hand-set filler panels with appropriate shoring around columns and across adjacent decks. All deck framing and deck tables must be carefully placed so that they are properly positioned for the edges of the slab and aligned with the supporting columns or walls. Once set, hand-set deck framing or panelized decks as well as tables must be graded (leveled).

4.1.0 Deck Surfaces

The form materials most commonly used for all types of decks are exterior (waterproof) grades of plywood. Any exterior-grade plywood can be used, but Plyform® is recommended for formwork. Plyform® is limited to certain wood species and veneer (outer face surfaces) grades for high performance. Plyform® is available in three basic grades. All three grades are available with or without medium-density overlay (MDO) or high-density overlay (HDO) thermoset resins.

- *Structural I* – This grade has high-strength plies throughout and is the strongest of the grades. It is recommended that the face grain run parallel with the support framing.
- *Class I* – This grade has high-strength veneer faces for high strength and stiffness.
- *Class II* – This grade has lower-strength veneer faces, but still provides adequate strength for formwork.

MDO and HDO surfaces are bonded to plywood faces under intense heat and pressure. These overlays add stability, reduce the amount of concrete adhering to the surface, and provide a smooth and durable forming surface. Regular MDO is intended for painted surfaces and should not be used for concrete forming; special MDO plywood must be used for concrete forming. The MDO is usually applied to one surface only.

HDO is a tougher, more abrasion-resistant surface that is usually applied to both faces of the plywood. This type of overlay can produce a nearly polished concrete surface on the underside of the slab. However, scratches and dents on the back side of the panel, caused by fastening the panels to framing members, may make the use of the back side impractical for a polished concrete appearance. If reasonable care is used, an HDO surface will normally produce 20 to 50 reuses or more. In some cases, 200 or more reuses have been achieved.

Another type of Plyform® product that is not overlaid is B-B Plyform®. It is made with different veneer faces that are sanded on both sides and treated with a release agent at the mill (mill oiled). B-B Plyform® is available as Structural I, Class I, and Class II and can be reused 5 to 10 times. With Plyform® products, an edge sealer must be applied before first usage (if not applied at the mill). A release agent must be applied to the slab side surface before concrete placement to allow easy stripping. The release agent must also be applied after stripping and cleaning to preserve the surface.

For most slab formwork, ¾" (or thicker) × 4 × 8 plywood panels are used. Normally, the grain of the face veneer is parallel with the long sides of the sheet. For maximum strength, the face grain (long side of the sheet) should be placed so that it crosses over the supporting framing members directly under the sheet. An exception is the Structural I plywood where the face grain can be parallel with the framing members. To prevent skewing of plywood panels and resultant fit-up problems across an entire deck, a row of panels should be laid first along the longest edge of the slab and checked to ensure that they are square with the support framing.

4.2.0 Hand-Set Multicomponent Decks

Hand-set multicomponent decks can be cost effective for small- or medium-size projects. Stringers are normally placed atop the shoring and the joists are then laid across the stringers. Cross sections of typical aluminum stringers (also called beams) and joists are shown in *Figure 32*. They have a wood or plastic insert at the top and some have holes in the base flange for nailing purposes. Others have attachment clamps available to secure stringers to post or frame shores. In other cases, engineered-wood I-joists are used as stringers and joists as shown in *Figure 33*. The I-joists shown in the figure are specially designed for formwork framing and are available in various lengths. In many cases, standard engineered-wood I-joists used in construction are substituted for the specially designed I-joists. In the figure, note the use of tripods to support some of the post shores while the framing is being installed.

Deck framing that uses a proprietary multi-component aluminum system is shown in *Figure 34*. This type of framing is more expensive than individual aluminum framing stringers and joists. The particular manufacturer of this system refers to the stringers as main beams and the joists as secondary beams. The main and secondary beams are supplied in only two standard lengths, 5'-7" and 3'-9", making the framing very modular and uniform when installed. The secondary beams interlock with a lip at the bottom of

Figure 33 Engineered-wood I-joist framing members.

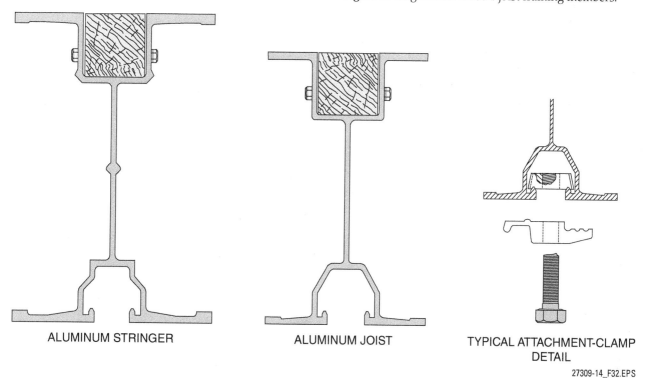

Figure 32 Cross sections of typical aluminum stringers and joists.

Figure 34 Proprietary multicomponent aluminum deck framing.

Figure 36 Completed deck framing with plywood-decking and dome-pan installation in progress.

the main beams at any point, and the main beams can interlock with each other and with proprietary drop-heads used on post shores, as shown in *Figure 35*. In the figure, note the drop-panel framing using short (filler) main beams and supported by hangers from the lip of short main beams used under the slab decking. *Figure 36* shows completed deck framing in place, with plywood decking and dome pans being installed.

4.3.0 Hand-Set Panelized Decks

Many of these types of deck systems use only two components: shoring systems and proprietary deck panels (*Figure 37*). The deck panels are manufactured in various sizes and shapes, with the supporting framing and decking material supplied as a complete unit. The replaceable deck material is usually plywood. One manufacturer supplies deck panels made entirely of plastic.

These deck panels can be interlocked with other panels for ganged placement.

Manufacturers of panelized deck systems require proprietary drop-heads for shoring and stripping purposes. These types of decks are initially expensive, but they are the fastest to hand-set and strip of any available deck system. Contractors for large projects use panelized deck systems to shorten construction time and reduce labor. *Figure 38* shows typical erection of a panelized deck. The 6 × 6 panels in this installation weigh about 100 pounds, so two workers can lift and hook them onto proprietary drop-heads. Then the panel is raised and propped into position with an erec-

Figure 35 Drop-panel framing

Figure 37 Panelized decking under construction.

tion pole. Finally the shores, set to a predetermined length, are placed at the unsupported corners of the panel. This process is repeated for each panel.

When each panel is being stripped and the slab reshored, the process is reversed. The drop-heads are released, the panel is propped, and the shores are removed from one end. Then the panel is swung down and removed. After the panel is removed, the drop-heads are removed, and the same shores can be reset against the slab. For fill-in around columns, curves, and other objects, specially shaped panels, adjustable panels, and small panels are available that are used with nailer strips as shown in *Figure 39*.

4.4.0 Outriggers

When stripping and reshoring slabs constructed with hand-set decks, many contractors use two or more outriggers, sometimes called poke-outs (*Figure 40*), to support material-staging platforms. The outriggers support the material-staging platform and allow the stripped formwork to be placed outside the slab for transfer to the next slab being constructed or to the ground. In the figure, note the side guardrails and the netting at the end of the outrigger. These prevent material and workers from falling to the ground below. The side guardrails also extend into the building. Outriggers like the one shown have two horizontal beams that extend into the building.

When a slab is going to be stripped, some of the shoring and formwork is removed at the edge of

Figure 38 Typical erection of panelized decking.

Site-Built Panelized Decks

Some contractors fabricate their own deck panels using standard engineered-wood I-joists and HDO plywood as shown in the figure. Like proprietary panelized decks, they are faster to erect, but require mechanical lifting to set or strip because of their weight. For posttensioned 6" slabs, the contractor claims that longer spans were possible with the I-joists than with aluminum joists. The panels shown averaged 24 reuses before the plywood surface had to be replaced.

Figure 39 Small panels with nailer strips used for fill-in around a column.

Figure 41 Outrigger beam bracing.

Flying decks are generally used for forming flat slabs and the deck is usually plywood. However, pan forms and cast-in-place beam forms can be mounted on truss or beam tables, depending on the height between slabs. Truss tables can also be ganged to save time in stripping and resetting.

4.5.1 Flying-Deck Safety

In addition to general safety guidelines, flying-deck forms also have inherent safety hazards. To ensure your own safety as well as that of others, strictly adhere to the manufacturer's instructions, applicable codes, and the following guidelines:

- Follow all state, local, and federal codes, ordinances, and regulations pertaining to shoring systems.
- Inspect all equipment before using it. Never use damaged equipment.
- Ensure that a shoring system layout is on the job site at all times.
- Inspect erected shoring systems and formwork:
  - Immediately prior to the concrete being placed
  - During concrete placement
  - After concrete placement until the concrete is set
- Consult your shoring equipment supplier when in doubt. Shoring is their business. Never take chances.
- Do not exceed the manufacturer's recommended safe working load for the equipment being used.
- All flying-deck forms must be assembled, moved, and maintained in accordance with the supplier's recommended procedures.
- If motorized concrete equipment is used, be sure that the shoring-system layout has been

Figure 40 Material-staging platform with outriggers.

the slab, and the outrigger is placed by crane so that the beams are inside the building, as shown in *Figure 41*. The beams are braced against the slab below and the slab being stripped by post shores set into cups on the beam. Once braced, the crane lifting slings are removed. The beams function as cantilever supports for the outrigger. Note the bottom portion of a guardrail fastened to the post shoring inside the building.

4.5.0 Flying Decks

Flying decks are complete slab-forming units with their own framing members, deck, and shores. They are tables that are moved (flown) by crane from one floor slab to the next slab location to be constructed at a project. There are three basic types of flying decks in general use: truss tables, beam tables, and column-mounted tables.

designed for use with this equipment and that this is noted on the layout.

- A method of adjustment should be provided on all flying-deck-form supporting members for form leveling, vertical positioning, ease of stripping, and adjusting to uneven grade conditions where applicable.
- Make certain that all supporting members are in firm contact with the flying form stringer/ledger and that supports are located in position as shown on the shoring-system layout.
- Use special precautions when placing shores from or to sloped surfaces.
- The reshoring procedure, if applicable, must be approved by the engineer of record.
- Use deck-form materials with properties as stated on the shoring-system layout. Do not splice joists or ledgers between supports unless details are given on the shoring-system layout.
- Do not release forms until the proper authority has given approval to do so.
- All field operations must be under the direct authority of a supervisor who is qualified and familiar with the procedures for assembly, erection, flying, and horizontal movement of the flying-deck form system being used.
- Make certain that a positively controlled method of tieback or braking is used when moving the deck form. The system must never be allowed to have free, uncontrolled horizontal movement.
- Ledgers/stringers and joists must be stabilized and laterally braced to ensure that the deck form system is stable against any foreseeable lateral loads.
- The crane used to fly the deck form must not be used to pull the deck form out of the building bay. A controlled and independent device or force must provide for horizontal movement of the deck form.
- Slings and rigging used in flying the deck form system must comply with all safe practices, applicable codes, and regulations governing their use.
- Do not make unauthorized changes or substitutions of equipment; always consult your supplier prior to making changes required by job-site conditions.
- All riggers must be qualified.
- Safety measures must be taken for all personnel involved in the rigging of flying deck forms.
- During concrete placement and deck-form rigging, the free-end cantilever of a deck form may not exceed the amount recommended by the supplier. Follow the recommended flying procedure as provided by the supplier.
- Any and all loose components of the deck form system (bulkheads, beam sides, filler strips, etc.), if flown with the form, must be securely fastened to the deck form prior to moving.
- Consult your supplier if a weatherproof covering or similar material is to be attached to the flying system.
- All personnel in the area must be advised and protected during all flying operations.
- All attached perimeter guardrails, midrails, and toeboards must conform to applicable codes and regulations.
- The weight of the flying-deck form must not exceed the capacity of the crane over the full range of the working radius.

4.5.2 Truss and Beam Tables

Truss and beam tables use an aluminum or steel truss or beam system to support aluminum deck framing and the plywood deck (*Figure 42*). Truss tables up to 30' wide and exceeding 35' in length can

ONE OF FOUR KNOCKOUT DECK PANELS REMOVED FOR SLING CONNECTION TO TRUSSES
TRUSS TABLE

BEAM TABLE

Figure 42 Truss and beam table systems.

be assembled and flown as a single unit. *Figure 43* shows the components of a typical truss table.

Various methods for extracting and flying truss tables can be utilized. One method involves tilting the table outside the building to allow pickup by a crane. Another method involves the use of an auxiliary electric hoist attached to a crane hook (*Figure 44*) so that the table does not have to be tilted. The auxiliary-hoist method for column-mounted tables is described in the next section.

For a typical truss table, all filler-panel shores are removed, and the jacks supporting the trusses are lowered until the table is placed on anchored tilt rollers at the outer edge of the slab, with rollers or positioning dollies under other portions of the trusses. All filler panels around columns and across adjacent tables are removed as required. The table is then tethered to an anchor point or points (columns) with 1"-diameter rope(s) that are only long enough to allow all the knockout panel openings on the table to be outside the building when the table is pushed out to the end of the tether. With the table pushed by two to four workers a short distance out of the building (*Figure 45*), relatively long equal-length slings from a crane hook located outside the building are connected through the knockout panel openings to the trusses (*Figure 46*).

The table is then pushed farther out of the building until the center of gravity (CG) is beyond the tilt rollers at the edge of the slab. This causes the table to tilt up gently until the rear of the table rests against the upper slab (*Figure 47*). After the table tilts, the crane hook, positioned over the outer knockout panels, lifts the table slightly and moves slowly away from the building while lowering the hook to maintain the tilt, until table movement is limited by the tether rope(s). During this operation, the rear slings remain slack due to the tilt of the ta-

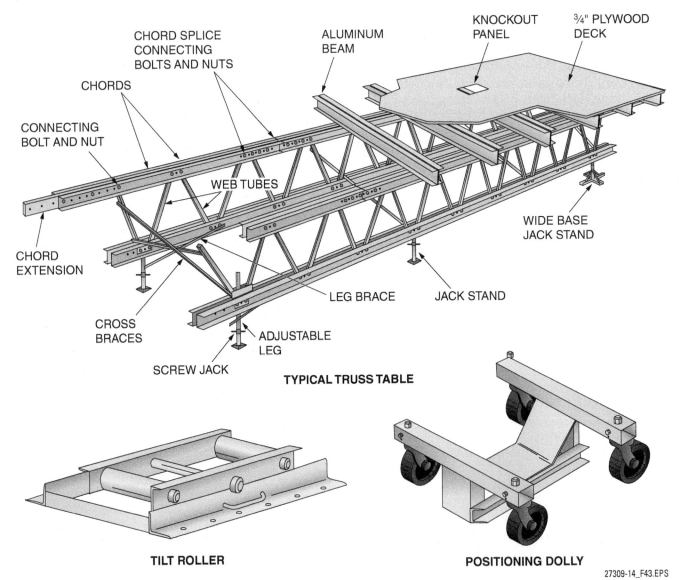

Figure 43 Typical truss-table components.

Figure 44 Truss table being lifted by a crane equipped with an auxiliary electric hoist.

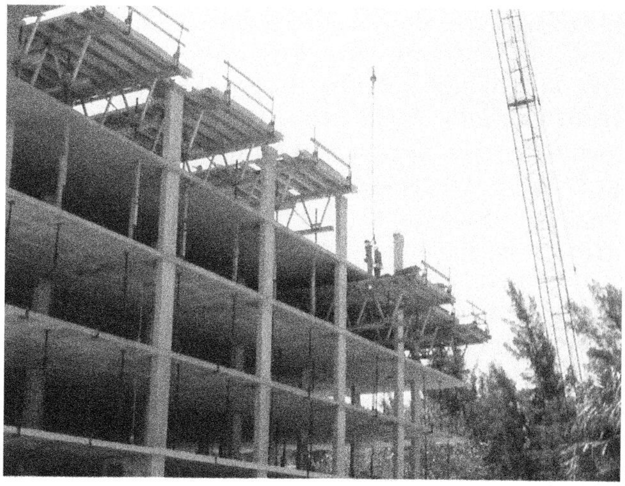

Figure 45 Truss tables partially rolled out of floor bays.

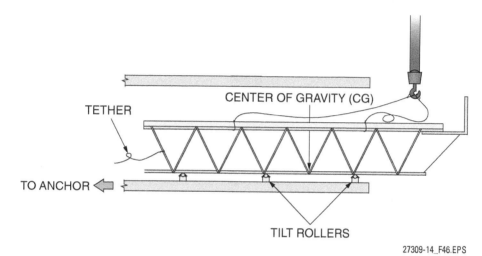

Figure 46 Preparing a truss table for removal.

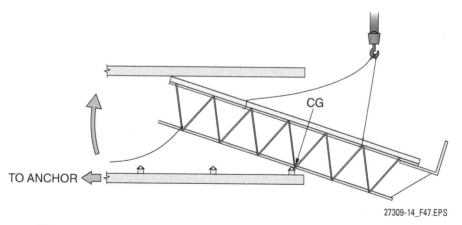

Figure 47 Tilting a truss table.

ble. The crane then lifts the front of the table until it is rotated to a level position and the rear slings are also taut. The table is then lifted slightly so most of its weight is off the rollers.

The crane moves the table into the building slightly, and the tether rope(s) are disconnected. A long guide rope is then tied to the rear of the table. The guide rope runs from the area where the table will be positioned after it is removed from the building. The crane lifts the table slightly to take the table weight off the rollers. Then, with help from workers in the building, the crane moves away from the building to remove the table.

When the table is clear of the building, workers use the guide rope to eliminate horizontal rotation. The crane lifts the table to the next floor location to be constructed (or to the ground) while workers guide it into position (*Figure 48*). Once the table is in position on jacks (*Figure 49*), it is raised to the correct level for the slab and graded (leveled). Then the filler panels, previously removed, are placed across the adjacent tables and around any columns. They are supported by additional shoring (*Figure 50*).

Before concrete can be placed on a graded slab deck, edge forms, blockouts, embedments, and any construction joints must be installed along with the required slab reinforcement.

- *Elevated-slab edge forms* – Conventional site-built wood edge forms, with bottom-edge kickers and wood braces fastened to the deck, are used extensively to contain elevated-slab concrete at the perimeter of a slab and within the slab to form blockouts for large openings such as stairwells. To save on wood, many contractors are using adjustable metal braces, sometimes called adjustable kickers, to replace the wood braces and kickers normally used. These devices, which can be used many times over, are shown in *Figure 51*. They can be fastened to a wood deck or concrete surface, or staked in the ground for slab-on-grade applications. The vertical stiffener is hinged so that it can be adjusted to any angle. For repetitive use on flying decks, the kicker can be fastened to the deck using a bracket that allows the entire device, including the form, to be moved back a distance for form stripping, as shown in the figure. When the deck is repositioned for a new slab, the forms can be returned to the same position and secured for placement of the next slab. Brackets are also available for the vertical stiffener to allow the attachment of metal edge forms instead of wood edge forms.
- *Blockouts* – If possible, blockouts for any required elevated-slab penetrations, including piping, conduit, or ducts, should be placed prior to placing concrete. Preplacement prevents having to drill through the slab and possibly damage reinforcement after the slab has cured.

> **NOTE**
> All blockouts and penetrations must be approved by the structural engineer.

Figure 48 Truss table being guided into place on the next floor level.

- *Embedments* – Before concrete placement, the elevation of the bottom of the slab deck, the elevation of the reinforcement used, and any embedments such as piping, electrical conduits, and outlet boxes must be checked. Deck forms that are too high can result in slab reinforcement and embedments being above the desired elevation of the slab. Screed rails or guides should be set to compensate for any initial downward movement of cambered deck forms during concrete placement. They also may be set to compensate for downward deflection of the structure after concrete placement and stripping of the forms.
- *Jointing* – Construction joints, if required for elevated slabs, should be specified by the designers of the building. Because joints can introduce weak vertical planes in a monolithic concrete member, they are located where shear stresses are low. Under most conditions, shear stresses are low in the middle of a span. As a guideline for slabs on removable forms, construction joints in floors are normally located within the middle third of spans of slabs, beams, and primary beams. Joints in girders should be offset a minimum distance of two times the width of any intersecting beams.

For composite slabs, construction joint location can affect the deflection of the floor framing near the joint. Construction joints that parallel secondary structural steel beams should normally be placed near the midpoint of the slab between the beams. Construction joints that parallel primary structural steel beams and cross secondary steel beams should be placed near the primary beam. The primary beam should not be included in the initial

Figure 49 Truss table on jacks.

Figure 50 Filler panels and shoring between tables.

Figure 51 Typical adjustable-kicker installation.

concrete placement. Placing the joint a distance of 4' from the primary beam allows nearly the full dead load from concrete placement to deflect the secondary beams included in the initial placement. This action also partially loads and deflects the primary beam. The second placement at the construction joint should include the primary beam. This will fully load the primary beam and allow complete deflection of the primary beam before the concrete at the primary beam hardens. Construction joints that cross a primary steel beam should be placed near a support at one end of the primary beam. This allows the beam to deflect completely before the concrete hardens. The placement of construction joints in noncomposite slabs on a metal deck follow the same guidelines as for removable forms.

4.5.3 Column-Mounted Tables

Column-mounted tables, sometimes called roller decks, are supported by jacking attachments that are anchored by bolts to the columns or walls for the floor under construction (*Figure 52*). Commonly used jacks are rated to support a working load of 50,000 lbs. The tables are constructed of heavy-duty aluminum or steel side stringers, aluminum joists or wood I-joists, and a plywood deck that can support clear spans up to 26' wide with an 8" slab. The advantage of these types of tables is that the area underneath the table allows free access to other trades, finishing operations, and space for material storage. Once the slab cast on the table as well as any columns/walls for the next higher or an adjacent floor level have sufficient strength and their vertical formwork has been stripped, the table can be stripped and moved by crane to the floor location above or to another location.

Table removal and transfer to another location or to the ground is similar to the method used for truss tables. If the transfer is to the next higher or an adjacent floor level, jack attachments are attached at the proper positions on the columns/walls for that location. Any shores for filler panels around the table to be stripped are removed. Each jack attachment for the table is lowered, one at a time, to allow a roller assembly to be inserted between the jack head and the stringer (refer to *Figure 52*). Then all the jacks are incrementally lowered to strip the table from the slab. Any filler panels are removed as required. A crane with an auxiliary electric hoist attached to the hook, along with a pair of slings from the hook and a pair of slings from the auxiliary-hoist hook, are positioned outside the building. The slings are lowered so that there is enough slack to attach them to the stringers through knockout panels in the table decking. The slings from the crane hook are connected through the knockouts at the end of the

Figure 52 A column-mounted table.

table nearest to the slab edge. The slings from the auxiliary hoist are connected through the knockouts at the other end of the table. Tether rope(s) are tied to the table and to anchor(s) within the building. A long guide rope from the intended location is also tied to the back of the table. Then workers with push poles or a forklift push the table part way out of the building to allow the crane to apply a light tension to the front slings at the outer edge of the table. The auxiliary-hoist slings remain slack. As the table is moved farther out of the building, the crane maintains the table in a horizontal position with the front slings (*Figure 53*). Once the rear slings are clear of the upper slab edge, the auxiliary hoist is operated to tighten them and center the crane hook over the middle of the table. As the rear slings are tightened, the crane hook is moved downward to maintain the table in a horizontal position. After the hook is centered, the table is raised slightly to take some of the weight off the bottom rollers. The tether rope(s) are removed and the crane hook is moved away from the building as workers push the table the rest of the way out of the building. Workers use the guide rope to keep the table from rotating horizontally when it clears the building. The crane lifts the table to the next floor location to be constructed (or to the ground) while workers guide it into position over the retracted jack assemblies that were previously placed (*Figure 54*). Once the table is in position on the jacks, it is raised by the jacks to the correct slab level and graded. Then filler panels with their shoring, removed from the previous table position, are placed across the adjacent tables and around any columns.

Figure 53 Column-mounted table being removed from floor bay.

Figure 54 Column-mounted table being guided into position.

4.0.0 Section Review

1. With reasonable care, plywood forms with an HDO surface often can be can be reused more than _____.
    a. 50 times
    b. 75 times
    c. 150 times
    d. 200 times

2. The fastest type of decking to hand-set and strip is the _____.
    a. flying deck
    b. multicomponent deck
    c. panelized deck system
    d. column-mounted table

3. Platforms cantilevered outside the building for use as material-staging areas are supported with _____.
    a. lookouts
    b. outriggers
    c. perches
    d. bumpouts

4. All blockouts and deck penetrations must be approved by the _____.
    a. architect
    b. building inspector
    c. structural engineer
    d. general contractor

5. Jacks most commonly used with column-mounted tables are rated to support a working load of _____.
    a. 25,000 pounds
    b. 50,000 pounds
    c. 75,000 pounds
    d. 100,000 pounds

Section Five

5.0.0 Shoring

Objective

Identify the different types of shores and describe applications for each.
a. Explain how adjustable wood shores are installed.
b. Explain how manufactured shores are installed.

Figure 55 Reshoring in a multistory building.

Figure 56 Shoring.

Shoring and reshoring are the most important aspects of constructing elevated slabs. Shoring refers to the construction of a temporary support system to carry the total load of horizontal decks and slabs, including the live loads that occur during construction added to the dead load (fresh cast-in-place concrete, reinforcing steel, and forms). Reshoring is the process performed for multistory construction when shoring equipment is removed from a partially cured slab. As the shores for the forms are removed, the elevated deck is reshored by placing shores against the bottom of the slab. As the construction of multistory concrete buildings moves up, shores for the lowest shored slab are stripped and placed on top of the highest shored slab in order to cast the next floor slab (*Figure 55*). The stripped lower slabs, normally at least two floor slabs below, must remain reshored to support the partially cured concrete and construction loads.

Unless specified otherwise, reshoring is typically accomplished on a leg-for-leg basis as a safety precaution. This means that the vertical supports for each reshored floor are placed directly under each of the supports for the floor above. When reshoring, shores for these slabs must be released and reset after placement of concrete on the floor being cast. Sometimes, preshoring (also called backshoring) is used when formwork must be stripped from a floor that has been cast, but has not yet achieved sufficient strength to support itself. Vertical supports are installed through the formwork while the formwork is carefully removed. Some manufactured formwork systems allow the formwork to be removed without disturbing the existing shores (*Figure 56*). Backshoring must not be confused with reshoring. Reshoring is used to transfer loads from shoring or backshoring to the floors below.

If a slab-on-grade has not yet been placed and cured, mudsills (*Figure 57*) must be placed on the soil under the shores for the first floor. They are used to prevent shores from sinking into the soil and causing the floor to collapse when the concrete for the first and any succeeding floors is placed. Mudsills are usually made of hardwood and can be planks or square pads, depending on the soil.

Both safety and cost must be taken into consideration in order to determine the appropriate method of shoring or reshoring. Factors that must be considered include: the time duration between the concrete placement of consecutive concrete slabs, the strength of the concrete, the configuration of the building, and the dimensions of the building and shores. Inadequate or improperly installed deck formwork, deck framing, shoring, and reshoring can result in catastrophic collapse of one or more floors of a building when concrete is being placed on the floor under construction.

It is essential that all shores and reshores be installed by qualified individuals in accordance with the building plans and the shoring manufac-

Figure 57 Pad-type mudsills.

Figure 58 Adjustable post-type wood shores.

turer's instructions. Except in certain applications involving edge support of cantilevered slabs and work decks, all shoring and reshoring members must be vertically plumbed when installed to prevent lateral loads on the shoring member and any decking member above the shoring. Lateral loading can significantly reduce the vertical loadbearing capacity of the shoring member as well as any deck framing member above it.

When concrete is being placed for an elevated slab, qualified individuals must constantly check the underside of the deck, deck framing, shoring, and the reshoring below for any signs of cracking, sagging, or bending that could signal impending failure. If any such signs are observed, immediate communications must occur so that concrete placement stops and workers leave the area. Structural engineers or other qualified individuals must then evaluate the deck, deck framing, shoring, and reshoring to determine the proper course of action. For this reason, both shoring and reshoring for a building must be designed and planned in advance by a qualified structural engineer, approved by the project architect/engineer, and installed and monitored by qualified individuals.

Vertical shores can be constructed from wood using patented manufactured metal components. Other types of patented manufactured steel and aluminum shores are also used.

5.1.0 Adjustable Wood Shores

One adjustable post-type wood shore widely used is made from two 4 × 4 (nominal) wood posts and two patented 4 × 4 shore clamps (*Figure 58*). This type of shore can also be made using 4 × 6 or 6 × 6 wood posts (nominal) with the appropriate-size shore clamps. The typical length of the lower shore post is 5'-6" or 6', with the upper shore post long enough to reach the desired height plus provide for a 24" overlap of the two wood posts.

The two shore clamps are attached to the lower shore post 12" apart on center, with the top clamp placed 2" below the post top. The shore clamps are then nailed through the holes in the clamp castings. Following this, the upper shore post is placed alongside the lower shore post and slid through the clamps until it is at the desired height. Final height adjustment is then made using a removable jack wrench especially designed for use with the shore clamps. The jack grips the wood of the lower shore post, and the upper shore post can be raised about 1" per stroke by a cam on the wrench lever. After the final height is obtained, the clamps on the upper shore post are tapped down to seat them, and a nail is driven into the upper shore post above each of the two clamp castings to prevent the clamps from vibrating loose. Loosening and removal of the shores is done by removing the safety nails from the upper shore post only, then loosening each clamp casting by tapping them up with a hammer one at a time.

The advantages of adjustable post-type wood shores are that they are inexpensive and can be fabricated on the job site as necessary. However, in comparison to steel or aluminum post shores, they are heavy to handle and more of them are required because of their lower loadbearing capacity.

When used for reshoring, pole-type wood shores can be equipped with reshoring springs that are nailed to the top of the upper post. The spring holds the shore snugly in place against a slab during the initial setting operation and any subsequent release/resetting operations. When the shore is in place, the reshoring spring is compressed flat between the upper shore pole and slab when 200 lbs or more of force is applied. When removed from compression, it returns to its original shape and can be used over and over again.

5.1.1 Grading Elevated Slab Decks

Prior to placing concrete on a hand-set or flying deck, the deck must be graded (leveled). This is usually accomplished by checking the bottom of the deck at periodic points along stringers and long-span joists across the entire deck and adjusting the shores as required. Grading can be performed by one person and a rotating-beam laser level set up at various spots near the center of the deck, as shown in *Figure 59*. Two workers and a builder's level can also perform the activity. Immediately after concrete placement and before the concrete sets, the deck should be rechecked for grade to determine if any appreciable movement has occurred that may require height adjustment of some shores or some spot addition of shores.

5.2.0 Manufactured Shores

Vertical shoring systems for most large commercial jobs are commonly constructed using patented manufactured metal components and systems. There are many configurations, depending on the manufacturer and application. These typically involve steel-frame or single-post shores used with wood, laminated, and/or aluminum joists and stringers.

5.2.1 Frame Shores

Typical frame shores are shown in *Figure 60*. Note the cross bracing used between the frames to support and maintain the frame positions. Also note the screw jacks at the bottom of the frames that are used to level the formwork for the floor slab above and for stripping the formwork.

Typically, the frame sections are 2' or 4' wide and 3', 4', 5', or 6' feet high, and are made with varying load capacities. Load capacities typically range from 20,000 to 50,000 pounds per frame. Top and bottom screw jacks provide for shore height and plumb adjustment. Braces connect the frames for tower rigidity. Connecting pins are used when stacking the frames. Stringers (beams) are often used in conjunction with frame shores by placing them in U-head screw jacks installed at the top of the uppermost frame members (*Figure 61*).

5.2.2 Steel and Aluminum Post Shores

Steel post shores (*Figure 62*) are made in a variety of designs, sizes, and load capacities. Most steel and aluminum post shores consist of two tube sections, upper and lower, and a threaded collar with a handle that provides for height adjustment. Rough height adjustments of the shore are made

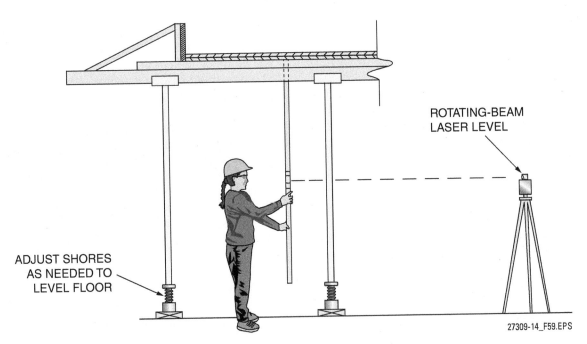

Figure 59 Checking deck grading with a rotating-beam laser level.

Figure 60 Typical frame shores.

by sliding the upper tube upward and inserting a pin in the appropriate slot spaced at regular intervals along the tube length. Fine adjustment is provided by turning the adjustment collar using the attached handle. Loosening the collar with a hammer (*Figure 63*) and rotating it with the handle lowers the formwork.

Many types of steel and aluminum post shores have some form of quick-release mechanism for stripping purposes. Some use a hammer-driven drop-pin that is part of the lower post tube (*Figure 64*). When driven out of position, the drop-pin causes the upper post tube to drop about ⅛" so formwork and posts can be removed.

Standard-duty versions of steel and aluminum post shores are capable of supporting 4,000 to 7,000 pounds, depending on height. Heavy-duty versions can support from 3,000 to 11,000 pounds, depending on height.

A variety of accessories may be used with steel and aluminum post shores. When installing shoring systems, tripods may be used to support groups of individual posts until sufficient stringers and joists are placed and secured to allow a deck under construction to be stable.

Figure 62 Steel post shore.

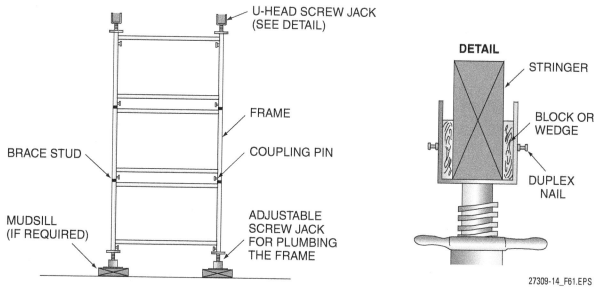

Figure 61 Stringer installation.

A U-head is commonly installed at the top of the shore to support aluminum or wood stringers. The stringers extend across several sections of deck forms to provide alignment and support. Other shores use a separate drop-head accessory. The drop-head is released by driving a slotted bar or pin on the drop-head out of position. *Figure 65* shows one type of drop-head that has been released, causing the formwork framing members to lower several inches. This allows removal of the framing members without disturbing the shores. With this particular version, the plywood deck is retained against the slab by the flat plate on top of the drop-head. Other types allow the framing members, as well as the deck panels, to be removed without disturbing the posts.

Aluminum post shores of various heights are also available with similar load capacities. These types of shores weigh much less than their steel counterparts. *Figure 66* shows one version of an aluminum post shore with an integrated measuring tape. For initial placement, the adjustment collar is pre-positioned to the desired height for the post using the measuring tape (*Figure 66*). Once the posts and formwork are set, a bar is used to rotate the collar when leveling the formwork (*Figure 67*). A drop-head, normally used with these types of shores, can be mounted on either end of the pole. For high poles, this allows the adjustment collar to be at the bottom for easier adjustment when leveling the formwork.

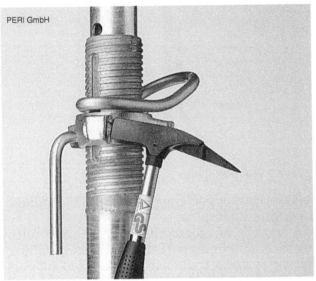

Figure 63 Lowering formwork.

DROP-HEAD RELEASED

Figure 64 Integral post-mounted hammer-driven drop-pin.

FRAMING MEMBERS REMOVED

Figure 65 Shore-mounted drop-head accessory.

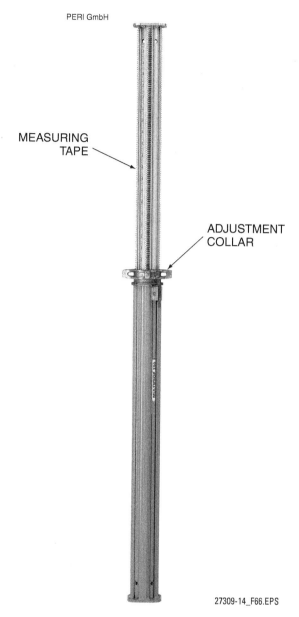

Figure 66 Aluminum post shore.

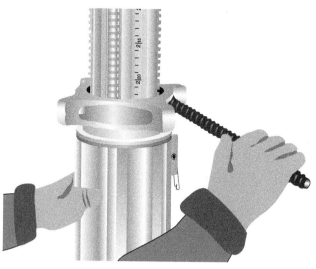

Figure 67 Final adjustment.

Adjustable Horizontal Shores

Adjustable horizontal shores usually do not require intermediate support and are used to support deck framing without the use of post shoring. These shores are available in a number of lengths to cover spans from 4'-6" to 20'. Constructed of double-lattice steel sections that telescope, they can be set up at floor level to the prescribed length, locked, and then installed on stringers or on structural-steel building beams using a hanger at each end. Most have a predetermined camber to accommodate loading sag. Wedge-type locks on the shore maintain the shore length and any camber.

5.0.0 Section Review

1. To prevent lateral loads from being imposed upon shores, they must be _____.
    a. installed in pairs
    b. vertically plumbed
    c. placed less than 10' from the deck-form edges
    d. diagonally braced

2. Indicators that may signal impending deck failure include all of the following *except* _____.
    a. sagging
    b. bending
    c. discoloration
    d. cracking

Section Six

6.0.0 Specialty Form Systems

Objective

Identify specialty form systems.
a. Explain how bridge decks are formed.
b. Explain how tunnels and culverts are formed.

Trade Term

Haunch: A structure provided on all bridges with steel girders or prestressed concrete I-beams. They provide a means of final adjustment of the deck-slab elevation to match the design roadway profile and cross slope. The haunch allows this adjustment without having the top flange of the girder project into the structural deck.

Horizontal structures such as bridge decks, culverts, and tunnels need to be formed if they are constructed of concrete and are not precast. While the forms for these structures were traditionally constructed using lumber and plywood panels, many forms now used for these structures are proprietary systems.

6.1.0 Bridge Decks

A bridge deck is the riding surface of a bridge. Forms for bridge decks support the concrete between adjacent structural members until it hardens sufficiently to stand alone. Deck forms must be concrete tight and sufficiently rigid to support the concrete without distorting under load. Deck forms are either removable or permanent. Most removable forms are made of wood. Permanent forms are usually made of metal or prestressed concrete. Removable wood forms used with conventional reinforced concrete decking are typically made of ¾" exterior-grade plywood panels for the deck (*Figure 68*). It is common for the deck to be supported by wood or metal joists and stringers hung from adjustable metal joist hangers.

Preformed metal bridge-deck panels made of corrugated steel are commonly used as permanent forms over which concrete is placed. They are similar to the metal forms described earlier for composite floor decks. Like composite floor decks, they serve several purposes:

- Used as a working platform
- Stabilize the frame
- Serve as concrete forms for the bridge deck slab
- Reinforce the concrete slab to carry the design loads specified for the bridge deck

Preformed metal bridge-deck panels are typically supported by metal angles that are welded or strapped to the top flanges of the supporting steel beams or girders. However, it is important that no support angles are welded onto flanges of steel beams or girders that are in tension. Metal deck panels are overlapped and installed in the opposite direction of the concrete placement to prevent loss of concrete through the joints. The panels should be welded in place as soon as they are placed and aligned to prevent them from being misaligned or blown off by wind.

Today, a decking method widely used for bridges involves the use of modular bridge-deck panels. Each of the modular deck panels consists

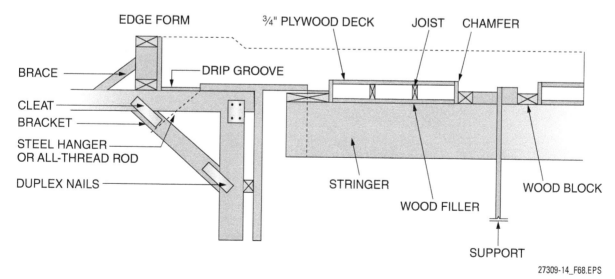

Figure 68 Typical wood bridge-deck formwork.

of a reinforced concrete slab placed on top of an unfilled steel grid (*Figure 69*). Using this construction method, the strength of the steel grid takes the place of the bottom half of a conventional reinforced concrete slab. The result is that the deck panel normally weighs about 35 to 50 percent less than a conventional reinforced concrete deck of the same size and span, without sacrificing strength or stiffness. By reducing the deck weight, the deck dead load is reduced, thus a higher live-load rating for the bridge can be achieved. The concrete for the bridge deck can be precast by a licensed manufacturer before the panels are placed on the bridge, or the concrete can be cast in place at the job site. Because of their modular construction, use of both precast and cast-in-place panels reduces construction time compared to the construction of conventional reinforced concrete decks.

Before placing either precast or cast-in-place deck panels on the bridge, haunches are usually formed using self-adhesive foam clamps, galvanized sheet steel, or structural angles connected with straps or welded to the supporting beam or timber. Haunches provide a means for final adjustment of the deck slab elevation. Precast deck-panel construction provides for blockouts in the panels so there is no concrete over the grid in the areas that are located directly over the top flanges of stringers, girders, or floor beams in the bridge superstructure.

Installing precast panels involves positioning the panels in place on the bridge superstructure and setting them to the correct elevation. The panels are attached to the bridge superstructure by welding headed shear studs onto the stringers, girders, or floor beams below that are accessible through the blockout areas in the panel. Following this, the blockout areas are filled to full depth with rapid-setting concrete to make the deck an integral part of the superstructure.

When the concrete is cast in place, precut, preformed, steel grid panels are placed on the bridge support structure to serve as a permanent form. The panels are set to the required elevation using built-in leveling clamps and are fastened to the steel superstructure of the bridge by welding headed shear studs through the grid to stringers, floor beams, and main girders below. Rebar is then placed on the grid. Following this, the headed studs in the haunch areas are embedded in concrete at the same time as the concrete is placed for the deck. This procedure makes the deck an integral part of the superstructure.

Bridge-deck panels made of fiber-reinforced plastic (FRP) material are also available. Use of FRP deck panels increases the ability of bridge decks to resist the corrosive effects of weather and road de-icing salts over that of decks made with concrete and steel materials. This increases deck longev-

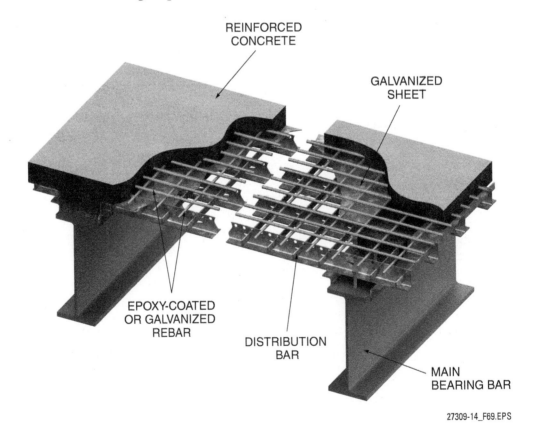

Figure 69 Modular deck panels.

ity and reduces deck maintenance. FRP panels are normally prefabricated into large, lightweight modules that can be installed easily and with light-duty equipment. FRP decks are increasingly being used as replacement solid-surface decks for existing deteriorated concrete decks of older bridges and drawbridges. This is because their use reduces the deck dead load, thus increasing the bridge's live-load capacity. For these reasons, their use can sometimes enable the bridge deck to be widened. FRP deck-panel construction is somewhat similar to that of precast concrete deck panels. This allows the deck-to-beam connections to be made using shear studs encased in concrete in the same manner as done in conventional precast-concrete deck construction.

6.2.0 Culverts and Tunnels

Culverts and tunnels provide unique forming challenges as the interior and exterior surfaces are typically formed at the same time. Culverts are commonly used for storm drainage, sanitary sewers, utilities, and irrigation canals. Tunnels are larger in diameter than culverts and are used to transport goods and people.

Culvert forms are patented forms that can be easily moved from one concrete placement to the next because the entire inside of the traveler frame is set on wheels and can be moved without disassembly (*Figure 70*). The headers and legs of the traveler frame are adjustable so that more than one box size can be formed with a single traveler frame.

In a typical setup, a ratchet jack is attached to several headers (*Figure 71*). The ratchet jack will retract the headers approximately 2" to pull the form away from the wall. A pipe brace between the leg assemblies on the bottom pulls the form away from the wall at the bottom of the setup. Secondary vertical adjustment is made with hydraulic hand jacks on both ends of the setup. Base jacks are used under each frame section to maintain position when the setup is raised off the wheels.

Figure 70 Example of a culvert form system.

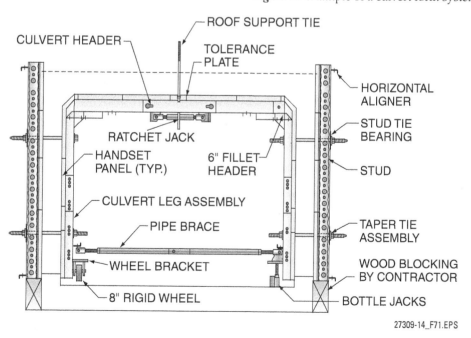

Figure 71 Culvert form setup.

Tunnels are created using several methods. They may be created using precast tunnel segments or the tunnel walls can be cast in place. For precast tunnels, the segments are transported inside the tunnel opening (*Figure 72*) and installed using heavy equipment. In one cast-in-place method, the tunnel opening is bored using a tunnel-boring machine, which is immediately followed by slipforming equipment that is used to create the tunnel walls. Other cast-in-place tunnels are formed using separate boring and forming operations.

Figure 72 Precast-concrete tunnel segments.

6.0.0 Section Review

1. Permanent forms used for bridge decks are made of metal or _____.
   a. prestressed concrete
   b. HDO plywood
   c. composite materials
   d. posttensioned concrete

2. Culvert forms are easily moved because they are _____.
   a. light enough to hand carry
   b. mounted on wheels
   c. quickly disassembled and reassembled
   d. motorized

SUMMARY

This module has provided an overview of horizontal concrete formwork used for heavy commercial or industrial construction. It also focused on the shoring and reshoring methods and equipment used to support the various types of horizontal forms. Floor forming typically involves the use of pan forms, composite-deck forms, or flying-deck forms. Forms are constructed for interior beams, cantilevered beams, inverted or T-beams, and spandrel beams. Bridge decks can be constructed using conventional reinforced concrete formwork. Newer methods for forming bridge decks involve the use of modular precast deck panels or metal cast-in-place deck panels. Modular bridge-deck panels made of fiber-reinforced plastic (FRP) material are also coming into increased use, especially for replacement decks on older bridges. Patented movable culvert forms are used when constructing storm drainage systems, sanitary sewers, and similar structures. Horizontal forms can be constructed either of wood or by assembling appropriate manufactured metal forms and equipment. Manufactured forms are more widely used today because they are easy to assemble and disassemble and can be reused many times. In order to safely assemble and use patented forms, the carpenter must be thoroughly familiar with the particular forms being used, especially the limitations of those forms. The instructions supplied with the forms must be followed.

Review Questions

1. All open sides and openings in formwork and slabs must be provided with _____.
   a. hazard warning signs
   b. safety nets
   c. guardrails
   d. fall protection equipment

2. Shoring must be erected under the supervision of a(n) _____.
   a. engineer
   b. experienced and competent person
   c. architect
   d. master carpenter

3. The weight of the concrete and reinforcement in a structural slab is the _____.
   a. combined load
   b. live load
   c. structural load
   d. dead load

4. Beams that are a part of a one-way solid slab are typically deeper than they are wide by _____.
   a. 15 percent
   b. two to three times
   c. 50 percent
   d. four to six times

5. Two-way flat slabs are often cantilevered beyond the last row of columns by a distance equal to _____.
   a. 15 percent of the interior span
   b. 25 percent of the interior span
   c. 30 percent of the interior span
   d. 40 percent of the interior span

6. Thickened ends on the joists of one-way joist slabs provide resistance to _____.
   a. compression forces
   b. diagonal tension forces
   c. internal stress forces
   d. perpendicular tension forces

7. Two-way joist slabs allow the use of spans as wide as _____.
   a. 50 feet
   b. 65 feet
   c. 70 feet
   d. 98 feet

8. Composite slabs are often built with _____.
   a. prestressed concrete
   b. shotcrete
   c. lightweight concrete
   d. corrugated aluminum decks

9. Elevated concrete slabs that do not include beams are referred to as _____.
   a. flat slabs
   b. solid slabs
   c. lightweight slabs
   d. monolithic slabs

10. Floors that will bear heavy loads are often supported by a _____.
    a. joist bands
    b. web of tensioned cables
    c. network of joists
    d. beam/girder system

11. Commercial pan forms used in concrete beam-and-slab joist systems are available in _____.
    a. six types
    b. five types
    c. three types
    d. two types

12. Form pans are usually secured to the deck with _____.
    a. thin battens
    b. nails
    c. drywall screws
    d. panel adhesive

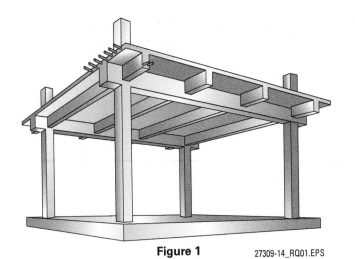

Figure 1 27309-14_RQ01.EPS

13. The floor shown in the illustration in Review Question *Figure 1*, used to support heavy live loads, is a _____.
    a. wide module
    b. two-way joist slab
    c. two-way flat with drop panels
    d. beam and slab

14. If a beam or girder form is not directly supported by a column, it must be supported by _____.
    a. shores
    b. hangers
    c. joists
    d. pilasters

15. Long-span posttensioned one-way beam-and-slab floors are a favored method for constructing _____.
    a. high-rise office buildings
    b. minimum-column parking structures
    c. heavy industrial buildings
    d. multilevel highway interchanges

16. Because drop panels are used for each column, flat-slab formwork _____.
    a. can be constructed quickly
    b. is less costly to build
    c. requires fewer form panels
    d. is more costly to build

17. Dimples or raised patterns on steel composite-deck panels are used to _____.
    a. provide a nonslip surface for worker safety
    b. create an architectural feature
    c. create a physical bond with the concrete
    d. prevent the formation of air pockets in the concrete

18. The most efficient method for fastening composite deck panels to structural supports is _____.
    a. welding
    b. sheet-metal screws
    c. riveting
    d. adhesive bonding

19. An important reason why hand-set panelized deck form systems are popular is that _____.
    a. the components are light and easy to handle
    b. they are less labor intensive to erect
    c. the panels can be reused many times
    d. they can be cleaned easily

20. The most commonly used facing material for concrete deck forms is _____.
    a. sealed interior-grade plywood
    b. medium-density fiberboard
    c. FRP panels
    d. exterior-grade plywood

21. For large decks, a single row of panels is laid on the longest side of the slab and checked for squareness with the support framework to prevent _____.
    a. swelling
    b. splitting
    c. skewing
    d. rippling

22. Hand-set multicomponent deck systems use engineered-wood _____.
    a. panel faces
    b. shores
    c. mudsills
    d. I-joists

23. A beam table is a form of _____.
    a. shoring system
    b. flying deck
    c. manufactured deck panel
    d. support joist

24. Horizontal movement of a deck form out of a building bay must be done by _____.
    a. an independent device or force
    b. the lifting crane
    c. hand labor
    d. forklift trucks

25. To save time in stripping and resetting, truss forms can be _____.
    a. gathered
    b. galleyed
    c. ganged
    d. grouped

Figure 2   27309-14_RQ02.EPS

26. The metal bracing devices shown in Review Question *Figure 2* are adjustable _____.
    a. kickers
    b. walers
    c. edge forms
    d. hold-downs

27. Openings framed into the slab form to allow penetrations such as conduits or piping are called _____.
    a. knockouts
    b. through-vents
    c. blockouts
    d. embedments

28. Column-mounted tables can support an 8-inch slab with clear spans as wide as _____.
    a. 26 feet
    b. 32 feet
    c. 47 feet
    d. 55 feet

29. The term *grading* is used to describe the process of _____.
    a. shoring forms
    b. leveling forms
    c. releasing forms
    d. relocating forms

30. Modular bridge-deck panels consist of a reinforced concrete slab _____.
    a. bonded to a corrugated steel panel
    b. fastened to the bridge's structural framework
    c. equipped with posttensioning tendons
    d. placed atop an unfilled steel grid

Trade Terms Quiz

Fill in the blank with the correct term that you learned from your study of this module.

1. A _____ is usually a steel plate, 6" square, used as the bottom component of a shore.

2. The weight of a structure, such as a bridge deck, as opposed to a load placed upon it, is the _____.

3. Made using hardwood planks or pads, a _____ supports the baseplate for a shore to prevent the shore from sinking into the soil.

4. The _____ is a combination of a bridge deck's weight and the weight that it can carry.

5. _____ is the trade name for an APA performance-rated plywood panel used in constructing concrete forms.

6. A temporary support system erected for horizontal decks and slabs, _____ is designed to carry both the dead load of the concrete, forms, and reinforcement and the live load of construction workers and equipment.

7. To support joists and deck form panels, a timber _____ is placed at the top of shores.

8. Sheet material formed with parallel ridges and grooves is described as _____.

9. A _____ is a type of shoring hardware that allows formwork to be released and dropped a few inches before being stripped.

10. Concrete that is placed in a continuous operation without construction joints is said to be formed _____.

11. A cross-sectional view of a floor deck is its _____.

12. In multistory construction, _____ is done by removing shoring equipment from beneath a partially-cured slab and replacing it with shores to bear the load.

13. At the top of a concrete column, a flared section forms its _____.

14. To provide shear reinforcement for concrete, rebar is bent into a loop referred to as a _____.

15. _____ reinforcement is metal that has been treated with phosphoric acid to prepare it for a finish coat or for contact with concrete.

16. A _____ is an adjustment structure, used on bridges with prestressed concrete I-beams or steel girders, that permits final alignment of the deck slab to match the design roadway profile and cross slope.

Trade Terms

Baseplate	Drop-head	Mudsill	Reshoring
Capital	Haunch	Phosphatized	Shoring
Corrugated	Live load	Plyform®	Stirrup
Dead load	Monolithically	Profile	Stringer

Trade Terms Introduced in This Module

Baseplate: Bottom component on a shore, typically a 6"-square steel plate with nail holes for fastening to a mudsill.

Capital: A flared section at the top of a concrete column.

Corrugated: Material formed with parallel ridges or grooves.

Dead load: The actual weight of the deck itself.

Drop-head: Shoring hardware that allows formwork to be released and dropped a few inches prior to stripping the forms.

Haunch: A structure provided on all bridges with steel girders or prestressed concrete I-beams. They provide a means of final adjustment of the deck-slab elevation to match the design roadway profile and cross slope. The haunch allows this adjustment without having the top flange of the girder project into the structural deck.

Live load: The carrying capacity of the deck including the dead load of the deck.

Monolithically: Used to describe concrete that is placed in forms continuously and without construction joints.

Mudsill: Hardwood planks or square pads used to support baseplates and prevent shores from sinking into the soil and causing the floor to collapse.

Phosphatized: Used to describe metal that has been treated with a phosphoric acid to prepare it for a finish coat. In composite floor decking, this unpainted surface will contact the concrete.

Plyform®: An APA performance-rated panel used for concrete forms.

Profile: The appearance of a floor deck viewed cross-sectionally.

Reshoring: Process performed for multistory construction when shoring equipment is removed from a partially cured slab. As shores for forms are removed, the elevated deck is re-shored by placing shores against the bottom of the slab.

Shoring: Construction of a temporary support system to carry the dead load of fresh cast-in-place concrete for horizontal decks and slabs, including the dead load of the reinforcing steel and forms as well as the live loads that occur during construction.

Stirrup: Rebar bent into a loop; used to provide shear reinforcement for concrete.

Stringer: Timber placed at the tops of shores to support joists and deck form panels.

Additional Resources

This module presents thorough resources for task training. The following resource material is suggested for further study.

American Concrete Institute (ACI). **www.concrete.org**

Cement Association of Canada. **www.cement.ca**

OSHA 29 *CFR* 1926, *Safety and Health Regulations for Construction*, latest edition. Washington, D.C.: Occupational Safety and Health Administration.

Portland Cement Association. **www.cement.org**

Figure Credits

Courtesy of Portland Cement Association, CO01, Figures 3–4, Figure 6, SA02, Figure 7, Figure 11a, Figure 26b, Figure 31, Figure 60, Figure 72

Ceco Concrete Construction, L.L.C., Figure 2, Figure 5, Figure 11b, Figures 12–13, Figure 25, Figure 26a, Figures 27–28

Brasfield and Gorrie LLC, SA01, Figures 14–16, Figure 29, Figure 41, Figure 50, Figure 57

Symons, Figure 8

Courtesy of Metal Dek Group®, a unit of CSI®, Figure 30

Symons, Figure 32, Figures 37–39, Figure 42a, Figure 43, Figure 45, Figures 48–49, Figure 52, Figure 54, Figure 64, SA04

PERI GmbH, Figure 33, Figure 42b, Figure 56, Figures 62–63, Figure 66

Titan Formwork Systems, LLC, Figures 34–36, Figure 65

APA – The Engineered Wood Association, SA03

Courtesy of EFCO Corp., Figure 44, Figure 53, Figures 70–71

Dayton Superior, Figure 51, RQ02

Titan Formwork Systems, LLC, Figure 55

Ellis Manufacturing Co. Inc. **www.Ellisok.com**, Figure 58

Courtesy of the D.S. Brown Company, Figure 69

Section Review Answer Key

Answer	Section Reference	Objective Reference
Section One		
1. c	1.0.0	1
Section Two		
1. False	2.0.0	2
2. c	2.1.0	2a
3. a	2.1.0	2a
4. d	2.2.0	2b
5. c	2.7.0	2g
6. a	2.4.0	2d
7. b	2.5.0	2e
Section Three		
1. b	3.0.0	3
2. True	3.0.0	3
3. c	3.1.0	3a
4. b	3.1.0	3a
5. d	3.1.0	3a
Section Four		
1. a	4.1.0	4a
2. c	4.3.0	4b
3. b	4.4.0	4d
4. c	4.5.2	4e
5. b	4.5.3	4e
Section Five		
1. b	5.0.0	5
2. c	5.0.0	5
Section Six		
1. a	6.1.0	6a
2. b	6.2.0	6b

NCCER CURRICULA — USER UPDATE

NCCER makes every effort to keep its textbooks up-to-date and free of technical errors. We appreciate your help in this process. If you find an error, a typographical mistake, or an inaccuracy in NCCER's curricula, please fill out this form (or a photocopy), or complete the online form at **www.nccer.org/olf**. Be sure to include the exact module ID number, page number, a detailed description, and your recommended correction. Your input will be brought to the attention of the Authoring Team. Thank you for your assistance.

Instructors – If you have an idea for improving this textbook, or have found that additional materials were necessary to teach this module effectively, please let us know so that we may present your suggestions to the Authoring Team.

NCCER Product Development and Revision
13614 Progress Blvd., Alachua, FL 32615

Email: curriculum@nccer.org
Online: www.nccer.org/olf

❏ Trainee Guide ❏ Lesson Plans ❏ Exam ❏ PowerPoints Other _____

Craft / Level: _____ Copyright Date: _____

Module ID Number / Title: _____

Section Number(s): _____

Description: _____

_____

_____

_____

_____

Recommended Correction: _____

_____

_____

_____

_____

Your Name: _____

Address: _____

_____

Email: _____ Phone: _____

75123-13

Heavy Equipment, Forklift, and Crane Safety

Overview

Motorized equipment operators must develop safe working habits and be able to recognize hazardous conditions in order to protect themselves and others from death and injury. This module covers the safety hazards and precautions necessary when working near heavy equipment. It also covers the general safety requirements for the use of forklifts and cranes.

Module Four

Trainees with successful module completions may be eligible for credentialing through the NCCER Registry. To learn more, go to **www.nccer.org** or contact us at **1.888.622.3720**. Our website has information on the latest product releases and training, as well as online versions of our *Cornerstone* magazine and Pearson's product catalog.

Your feedback is welcome. You may email your comments to **curriculum@nccer.org**, send general comments and inquiries to **info@nccer.org**, or fill in the User Update form at the back of this module.

This information is general in nature and intended for training purposes only. Actual performance of activities described in this manual requires compliance with all applicable operating, service, maintenance, and safety procedures under the direction of qualified personnel. References in this manual to patented or proprietary devices do not constitute a recommendation of their use.

Copyright © 2015 by NCCER, Alachua, FL 32615, and published by Pearson Education, Inc., New York, NY 10013. All rights reserved. Printed in the United States of America. This publication is protected by Copyright, and permission should be obtained from NCCER prior to any prohibited reproduction, storage in a retrieval system, or transmission in any form or by any means, electronic, mechanical, photocopying, recording, or likewise. To obtain permission(s) to use material from this work, please submit a written request to NCCER Product Development, 13614 Progress Blvd., Alachua, FL 32615.

 From *Construction Craft Laborer, Level Two, Trainee Guide*, Third Edition. NCCER. Copyright © 2015 by NCCER. Published by Pearson Education. All rights reserved.

75123-13
HEAVY EQUIPMENT, FORKLIFT, AND CRANE SAFETY

Objectives

When you have completed this module, you will be able to do the following:

1. Explain the general guidelines for working safely around heavy equipment.
    a. State the general guidelines for job-site safety.
    b. State the guidelines for the safe operation of heavy equipment.
2. Explain the general guidelines for forklift safety.
    a. Describe the safe operation of a forklift.
    b. State the general guidelines for safe load handling.
3. Explain the general guidelines for crane safety.
    a. State the guidelines for working safely around power lines.
    b. Describe various site hazards and restrictions.

Performance Tasks

This is a knowledge-based module; there are no performance tasks.

Trade Terms

Anti-two-blocking devices
Center of gravity
Combined center of gravity
Equipment maintenance
Equipment operator

Hydraulic
Momentum
Operator's compartment
Outrigger

Industry-Recognized Credentials

If you are training through an NCCER-accredited sponsor, you may be eligible for credentials from NCCER's Registry. The ID number for this module is 75123-13. Note that this module may have been used in other NCCER curricula and may apply to other level completions. Contact NCCER's Registry at 888.622.3720 or go to www.nccer.org for more information.

Contents

Topics to be presented in this module include:

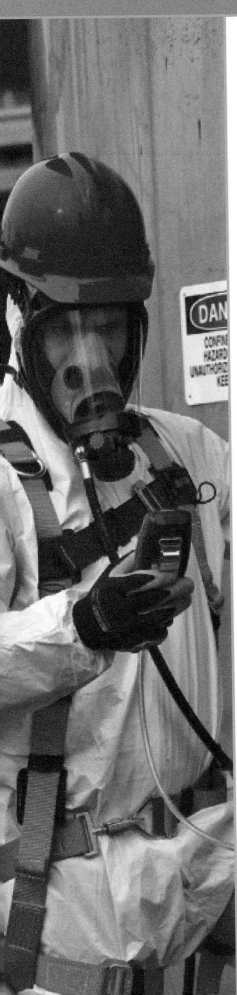

1.0.0 Heavy Equipment Safety 1
   1.1.0 General Guidelines for Job-Site Safety 2
      1.1.1 Personal Safety 4
      1.1.2 Soil and Demolition Dust 5
      1.1.3 Soil Contamination 6
      1.1.4 Fire and Explosion Hazards 6
   1.2.0 Guidelines for the Safe Operation of Heavy Equipment 6
      1.2.1 Moving Heavy Equipment 7
      1.2.2 Communication 7
      1.2.3 Maintenance 9
      1.2.4 Fueling 12
2.0.0 Forklift Safety 15
   2.1.0 Safe Operation of a Forklift 15
   2.2.0 General Guidelines for Safe Load Handling 16
      2.2.1 Traveling with Loads 17
      2.2.2 Tipping 17
      2.2.3 Obstructing the View 18
3.0.0 Crane Safety 20
   3.1.0 Working Around Power Lines 20
   3.2.0 Site Hazards and Restrictions 22

Figures

Figure 1	Pinch/crush points	2
Figure 2	Types of heavy equipment	3
Figure 3	Traffic-control devices	4
Figure 4	Personal protective equipment	5
Figure 5	Misting systems reduce dust	6
Figure 6	Example of an equipment operator's daily checklist	8
Figure 7	Standard hand signals	10–11
Figure 8	Forklifts	16
Figure 9	Forklift stability triangle	18
Figure 10	Mobile crane with outriggers	21
Figure 11	Prohibited zone and avoidance zone	21
Figure 12	Wind and lightning hazards	22

Section One

1.0.0 Heavy Equipment Safety

Objectives

Explain the general guidelines for working safely around heavy equipment.
  a. State the general guidelines for job-site safety.
  b. State the guidelines for the safe operation of heavy equipment.

Trade Terms

Anti-two-blocking devices: Devices that provide warnings and prevent two-blocking from occurring. Two-blocking occurs when the lower load block or hook comes into contact with the upper load block, boom point, or boom point machinery. The likely result is failure of the rope or release of the load or hook block.

Equipment maintenance: The care, cleaning, inspection, and proper use of machinery and equipment.

Equipment operator: A person skilled in operating certain equipment.

Hydraulic: Tools and equipment that are powered or moved by liquid under pressure.

Momentum: A physical force that causes an object in motion to stay in motion.

Motorized equipment is used in many different jobs including construction, mining, plant maintenance and operations, road maintenance, equipment transportation, and snow removal. Working with motorized equipment can be extremely dangerous. Workers can be crushed by falling loads, fall from equipment, be electrocuted by power lines, or be struck or trapped by vehicles. The swing radius of equipment can also be a hazard if the job is not carefully planned and properly barricaded. *Figure 1* shows some of the pinch point hazards that are accessible without barricades.

Dangers exist for both equipment operators and other workers on the site. In one example, a contractor was operating a backhoe when another employee attempted to walk between the swinging back end of the backhoe and a concrete wall. As the employee approached the backhoe from the operator's blind side, the back end hit the victim, crushing him against the wall.

Because working with or near motorized equipment is dangerous, you must understand that your first responsibility on a job is safety. This includes your own safety, the safety of others on the site, and the safe use of equipment on the site. You must know the hazards and safety procedures of every job you are on, regardless of the work you are doing.

Those who work around motorized equipment need a basic knowledge of all types of equipment, as well as the hazards and safeguards of this equipment. The types of jobs performed by equipment operators vary. The following are just a few of the jobs that use motorized equipment:

- Civil, residential, and industrial construction work
- Mining
- Snow and ice control operations
- Maintaining road surfaces, such as patching potholes and cracks (surface failures), raising depressed concrete, resurfacing asphalt, painting stripes, sweeping highways, inspecting, and repairing
- Loading, lashing, and unloading equipment and materials
- Minor equipment maintenance and repair
- Transporting equipment over public highways

Case History

Consider all of the Dangers

Two employees were attempting to adjust the brakes on a backhoe. The victim told the backhoe operator to raise the wheels off the ground with the front bucket and the outriggers so that he could get to the brakes. The victim then crawled under the machine and began to adjust the brakes. He did this without considering that there was only a 36" space from the ground to the drive shaft. While adjusting the brakes, the hood of his rain jacket wrapped around the drive shaft and broke his neck. He died instantly.

   The Bottom Line: Loose clothing can be caught in moving parts of machinery. You must consider all possible dangers when working on equipment.

Source: The Occupational Safety and Health Administration (OSHA)

75123-13 Heavy Equipment, Forklift, and Crane Safety

Figure 1 Pinch/crush points.

Construction workers may be expected to operate many different types of motorized equipment with proper training (*Figure 2*). This equipment can vary greatly in size and weight. Some of the heavy equipment used in construction includes the following:

- Trucks
- Cranes
- Compacting equipment
- Backhoes
- Backhoe loaders
- Scrapers
- Bulldozers
- Excavators
- Motor graders
- Skid steer loaders
- Forklifts
- Concrete paving equipment
- Concrete plant equipment
- Asphalt plant equipment
- Pug mill mixers
- Fine grade trimmers
- Cold milling machines
- Stabilizers
- Trenchers and rock saws
- Breaking equipment

It's important to know how to use this equipment safely and to know the hazards and safeguards on your job site.

1.1.0 General Guidelines for Job-Site Safety

You may not be an equipment operator, but you will probably work on a site with heavy equipment. It is important to remember that it may be

---

Case History

Always Wear Your Seat Belt

An employee was driving a front-end loader up a dirt ramp onto a lowboy trailer. The tractor tread began to slide off the trailer. As the tractor began to tip, the operator, who was not wearing a seat belt, jumped from the cab. As he hit the ground, the tractor's rollover cage fell on top of him, crushing him.

The Bottom Line: Always wear your seat belt when operating heavy equipment.

Source: OSHA

 (A) BULLDOZER

 (B) ARTICULATED OFF-ROAD DUMP TRUCK

 (C) EXCAVATOR

 (D) BACKHOE LOADER

 (E) SKID STEER

 (F) MOTOR GRADER

Figure 2 Types of heavy equipment.

difficult for the operator to see you. The operator may be concentrating on the load and may not be paying attention to pedestrians. Therefore, always assume that the operator doesn't see you. The most common type of heavy equipment accident is hitting a pedestrian.

Safety on the job is everyone's responsibility. The equipment operator has control over equipment application, operation, inspection, lubrication, and maintenance. Therefore, it is primarily the operator's responsibility to use good safety practices in these areas. It's important to understand, however, that even though the operator has much of this responsibility, everyone working on the site is responsible for working safely. They are also responsible for being aware of their co-workers and their actions. Follow these general job-safety rules to make sure everyone on the site is safe:

- Be alert. Watch out for moving equipment. Assume the operator cannot see you.
- Use spotters to guide equipment or loads whenever possible.
- Construction equipment is heavy and usually carries large loads. This results in a large amount of momentum, which means that it will probably take a distance equal to the length of the equipment to stop it. Never risk your safety on the equipment's ability to stop in time.
- Make sure that you're never positioned between any moving equipment and an immovable object.
- Some heavy equipment is designed to carry heavy loads high overhead. Never stand under a raised load. Objects may fall or, if there is a sudden loss of hydraulic pressure, the load may drop suddenly, crushing any people or objects beneath them.
- Use retaining guards and safety devices on all equipment.
- Report all defective tools, machines, or other equipment to your supervisor and report all accidents and near-accidents.
- Disconnect power and lock and tagout machines before performing maintenance.
- Use properly fitting wrenches on nuts and bolts.
- Keep your work area clear of scraps and litter.
- Dispose of combustible materials properly.
- Clean up any spilled liquids immediately.
- Store oily rags in self-closing metal containers.
- Never use compressed air to clean yourself or your clothing.
- Make sure all others on the site are at a safe distance from the equipment.

- Use traffic-control devices where required (*Figure 3*).
- Follow manufacturer's safety rules and limitations for all equipment.

1.1.1 Personal Safety

Personal safety is as important as site safety. Your actions affect everyone on the site. Follow these guidelines to prevent accidents and injury:

- Wear close-fitting clothing that is appropriate for the activity being performed.
- Tie back and secure long hair underneath your hard hat before operating equipment.
- Wear safety glasses, a suitable hard hat, goggles, hearing protection, and respiratory equipment where required (*Figure 4*).
- Fasten loose sleeves when working around machine tools or rotating equipment.
- Remove rings and other jewelry when working.

Figure 3 Traffic-control devices.

- Be alert.
- Do not jump on or off of moving equipment.
- Equipment must be at a full stop and the parking brake engaged before entering or exiting the operator's seat. Maintain three points of contact when entering or exiting equipment.
- Know the location of first-aid equipment, fire extinguishers, and emergency telephone numbers.

1.1.2 Soil and Demolition Dust

Excavation work generates a great deal of dust. It is a best practice to use a dust control method, such as a misting system (*Figure 5*). Before demolition work begins, a company safety inspector should evaluate the site for any harmful materials. Respiratory protection may be required.

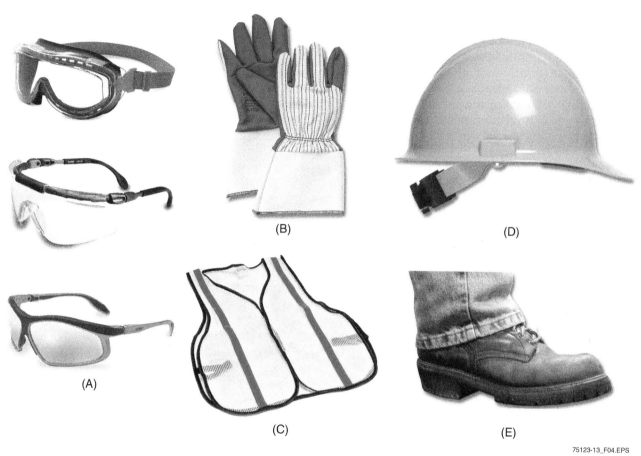

Figure 4 Personal protective equipment.

Case History
Know the Dangers of Your Work Site

Two employees were placing concrete as it was being delivered by a concrete pumper truck boom. The truck was parked across the street from the work site. Overhead power lines ran above the boom on the pumper truck.

One employee was moving the hose to pour the concrete when the boom of the pumper truck came in contact with the overhead power lines. These lines carry 7,200 volts. The employee was electrocuted and died immediately. He then fell on the employee who was assisting him. The second employee received a massive electrical shock and burns. No one on the site had received the proper safety training. Otherwise, they would have known how dangerous it is to work under power lines.

The Bottom Line: Always check the job site for hazards before you begin work. Overhead and underground power lines can be deadly.

Source: OSHA

CAUTION: Many demolition sites contain asbestos materials. These materials may only be removed by properly licensed companies.

1.1.3 Soil Contamination

When working at a new construction site or performing demolition on a building, there is always the possibility of uncovering live sewers, septic tanks, or a contaminated site. Contaminants can be in the soil, buried underground, or stored in some container such as a barrel. Identifying contaminants is a complicated and time-consuming procedure that requires advanced education and training. Heavy equipment operators are not expected to identify contaminants, but are required to report any potential contaminations to a supervisor. Some warning signs of potential contamination are as follows:

- Puddles of fluid on the ground, especially fluid with an unusual color or odor
- Any fluid seeping out of the ground
- Unusual or foul odors
- Unmarked barrels or tanks—often buried or otherwise camouflaged

1.1.4 Fire and Explosion Hazards

All types of motorized equipment use fuels such as gasoline, liquid propane (LP) gas, and diesel fuel. All of these fuels are capable of causing a fire or explosion if not handled properly. In addition, LP gas is stored in cylinders under pressure, creating an explosion hazard if the cylinder is exposed to extreme heat or fire. It is extremely important to keep these fuels away from any source of fire, and to keep the areas in which the equipment is used free of any flammable materials. There are specific precautions that must be taken to avoid the possibility of a fire or explosion.

1.2.0 Guidelines for the Safe Operation of Heavy Equipment

The identification of safety hazards begins with a daily machine-safety inspection. Before daily operations begin, the operator should inspect around and under the equipment for visible evidence of potential safety hazards or operational problems. Any safety hazards must either be corrected immediately or reported, and equipment operations should not proceed until the conditions are safe.

The following is a list of basic items that should be checked daily:

- Engine performance and gauges
- Oil pressure and levels
- Housekeeping tasks
- Fire extinguishers
- Audible warning devices
- Anti-two-blocking devices
- Hydraulic lines and fluid
- Windshield wipers, mirrors, and lights
- Batteries and charging system

Figure 5 Misting systems reduce dust.

- Visible hand signal and load charts
- Guards and clutches
- Tracks, wheels, and tires
- Brakes, locks, and safety devices
- Drive chains, steering, and all rollers
- Sheaves, drums, and all cables
- Blocks and hooks
- Boom, gantry, jib, and extension
- Carrier assembly
- All controls, including hoisting, swing, travel, and boom

The more thorough you are when inspecting the equipment, the safer and more productive you will be during your shift. *Figure 6* shows a sample checklist covering the basic items that need to be checked during a pre-shift inspection of a forklift.

If any problems are noted during the preshift inspection, notify the supervisor or maintenance manager immediately. The equipment should be locked out and tagged. It cannot be used until all problems are corrected.

1.2.1 Moving Heavy Equipment

All motorized equipment is moved during the course of a job. It is moved to the site, on the site, and away from the site. Many injuries and deaths happen during the movement of equipment. It is important to be especially safety conscious whenever equipment is moving. When operating motorized equipment, be responsible. A responsible operator must be a qualified and authorized operator. To be qualified, you must understand the instruction manual supplied by the manufacturer of the equipment, have training in the actual operation of the equipment, and know the safety rules and regulations for the work site. Never operate any equipment if under the influence of alcohol or drugs. If taking prescription or over-the-counter drugs, seek medical advice about whether you can safely operate machinery. Never knowingly allow anyone to operate equipment when he or she is impaired.

Travel on public highways requires a commercial driver's license (CDL) and possibly an escort. Always observe the following guidelines when driving equipment on public roads:

- Drive slowly and never speed.
- Know the distance it takes to stop safely.
- Allow extra time to enter traffic.
- Know and follow the traffic control pattern.
- Travel with lights on.
- Use proper warning signs and flags.
- Secure all attachments and loose gear.
- Turn cautiously; allow for extensions or attachments and for structural clearances. Some equipment is top-heavy and will tip over if a turn is made too fast.
- Know the turning radius of the equipment.
- Know the swing radius of the equipment and make sure the swing path is clear of people and objects.
- Know and obey all state and local laws.

Always follow these guidelines for driving equipment on the job site:

- Never drive a machine in a congested area, or around people, without a spotter or flagger. The spotter or flagger is responsible for determining and directing the driver's speed.
- Be sure everyone is in the clear while backing up, hooking up, or moving attachments.
- Never move buckets or shovels over the heads of other workers.
- If you cannot see the surrounding area clearly from the operator's seat, get a spotter or do not operate the equipment.
- Wait for an all-clear signal before moving.
- Signal a forward move with two blasts of the horn.
- Signal a reverse move with three blasts of the horn.
- Yield the right-of-way to loaded equipment on construction sites.
- Maintain a safe distance from all other vehicles.
- When moving, keep the equipment in gear at all times; never coast.
- Maintain a ground speed consistent with ground conditions and posted speed limits.
- Know where overhead electrical power lines are located.
- Ensure that buried pipes and power and gas lines have been located and marked.

Each piece of equipment has its own unique set of hazards. Be aware of its pinch (crush) points. Pinch points occur when there is motion between equipment parts. Serious injury or death can result from getting part of your body or your clothing caught in a pinch point.

1.2.2 Communication

Operation of motorized equipment requires concentration by the operator. Any distractions may have disastrous results. At times, the operator will not have a clear view of the site or other equipment. This means that a signal person will have to act as the operator's eyes. To accomplish this, there must be a standard form of communication between the operator and the signal person.

75123-13 **Heavy Equipment, Forklift, and Crane Safety**

Operator's Daily Checklist – Internal Combustion Engine Industrial Truck – Gas/LPG/Diesel Truck

Record of Fuel Added

Date		Operator		Fuel	
Truck #		Model #		Engine Oil	
Department		Serial #		Radiator Coolant	
Shift		Hour Meter		Hydraulic Oil	

SAFETY AND OPERATIONAL CHECKS (PRIOR TO EACH SHIFT)
Have a **qualified** mechanic correct all problems.

Engine Off Checks	OK	Maintenance
Leaks – Fuel, Hydraulic Oil, Engine Oil or Radiator Coolant		
Tires – Condition and Pressure		
Forks, Top Clip Retaining Pin and Heel – Check Condition		
Load Backrest – Securely Attached		
Hydraulic Hoses, Mast Chains, Cables and Stops – Check Visually		
Overhead Guard – Attached		
Finger Guards – Attached		
Propane Tank (LP Gas Truck) – Rust Corrosion, Damage		
Safety Warnings – Attached (Refer to Parts Manual for Location)		
Battery – Check Water/Electrolyte Level and Charge		
All Engine Belts – Check Visually		
Hydraulic Fluid Level – Check Level		
Engine Oil Level – Dipstick		
Transmission Fluid Level – Dipstick		
Engine Air Cleaner – Squeeze Rubber Dirt Trap or Check the Restriction Alarm (if equipped)		
Fuel Sedimentor (Diesel)		
Radiator Coolant – Check Level		
Operator's Manual – In Container		
Nameplate – Attached and Information Matches Model, Serial Number and Attachments		
Seat Belt – Functioning Smoothly		
Hood Latch – Adjusted and Securely Fastened		
Brake Fluid – Check Level		
Engine On Checks – Unusual Noises Must Be Investigated Immediately	**OK**	**Maintenance**
Accelerator or Direction Control Pedal – Functioning Smoothly		
Service Brake – Functioning Smoothly		
Parking Brake – Functioning Smoothly		
Steering Operation – Functioning Smoothly		
Drive Control – Forward/Reverse – Functioning Smoothly		
Tilt Control – Forward and Back – Functioning Smoothly		
Hoist and Lowering Control – Functioning Smoothly		
Attachment Control – Operation		
Horn and Lights – Functioning		
Cab (if equipped) – Heater, Defroster, Wipers – Functioning; Glass		
Gauges: Ammeter, Engine Oil Pressure, Hour Meter, Fuel Level, Temperature, Instrument, Monitors – Functioning		

Figure 6 Example of an equipment operator's daily checklist.

There are many ways for the operator and the signal person to communicate. One is a radio. This involves the signal person using a radio to tell the operator how to control the load. Disadvantages of this type of communication are interference from other radio or electronic equipment and the inability to hear the radio over background machinery noise.

Hardwired communication circuits are another way to set up communication. This involves a dedicated circuit. With a dedicated circuit, the interference from other radios is removed. The disadvantage of this type of communication is not being able to hear because of loud noise on the site.

To get rid of the problem of noise in voice and radio communications, a standard set of hand signals can be used (*Figure 7*). They are not affected by noise or interference. The only disadvantage is that when using hand signals, the signal person and the operator must be in sight of each other. This can be difficult if there is a lot of equipment and workers on the site. Still, it is generally the safest and most effective way to communicate.

1.2.3 Maintenance

Heavy equipment breakdowns can often be traced to improper maintenance or lack of regular maintenance. These malfunctions can cause unexpected failure of the equipment, resulting in scheduling delays and accidents. An operator's minimum responsibility is to ensure that the equipment is in safe working order before using it. However, operators sometimes perform periodic maintenance.

> **NOTE**
> Only persons qualified to do so and authorized by the employer may perform maintenance on a vehicle or equipment.

Manufacturers supply detailed operating and maintenance manuals for equipment. Workers qualified and authorized to perform maintenance on the equipment should study the manufacturer's manuals, including maintenance schedules, lubrication charts, and parts lists, before starting the equipment. Maintenance can be dangerous unless performed properly. Workers should have the necessary skills, information, tools, and equipment to do the job correctly. Use parts, lubricants, and service techniques recommended by the manufacturer. If you do not know what you are doing—stop!

While performing maintenance on equipment, keep the following in mind:

- Disconnect power, then lock and tagout machines before performing maintenance.
- Block any hydraulic arms/lifts.
- Use properly fitting wrenches on nuts and bolts.
- Keep the work area clear of scraps and litter.
- Dispose of combustible materials properly.
- Clean up any spilled liquids immediately.
- Store oily rags in self-closing metal containers.
- Never use compressed air to clean yourself or your clothing.

Before servicing equipment, prepare the service area. Begin by emptying your pockets of any items that could fall into machinery. Then collect any PPE that may be needed, such as aprons and gloves; safety glasses, goggles, or shields; safety shoes; welding protection equipment; and filter mask or respirator. Wear all the protective clothing that the job requires.

If the work site does not have a designated service area, select a clean, level area out of the flow of traffic. Make sure there is enough room to work, as well as adequate light and ventilation. Check overhead clearances if necessary. Before

---

Ergonomics

Ergonomics is the science that deals with identifying and reducing stress on equipment users. In the past, jarring movements, uncomfortable seats, tight compartments, deafening noise, and hard-to-use controls were common. In the long term, operators were faced with a variety of physical problems, including bad backs and hearing loss, because of the harshness of their working environment.

In recent years, equipment designers have focused on eliminating these problems. Their efforts have resulted in such improvements as the following:

- Adjustable suspension seats
- Roomier cabs with better visibility
- More leg and foot room
- Easy-to-operate joystick controls
- Quieter operation
- Easier-to-reach service components, such as oil dipsticks and fuel tank nozzles

FORKLIFT HAND SIGNALS

Figure 7 Standard hand signals. (1 of 2)

CRANE HAND SIGNALS

Figure 7 Standard hand signals. (2 of 2)

starting, clean the working surface, removing oil, grease, and water to eliminate slippery areas. Put sand or other absorbent material on these areas.

After moving the equipment to the service area, follow the necessary shutdown procedures. Remove the ignition key and use a lockout device if possible. After the equipment is shut down, always attach a Do Not Operate tag or similar warning tag to the starter switch or steering lever before performing maintenance. Many machines have ignition lockout switches in or near the engine compartment for this purpose.

Remember that hydraulic systems can hold pressure even after shutdown. Cycle the controls to relieve the remaining pressure in the lines. If the engine must be started during maintenance, ensure that there is proper ventilation in the service area. Exhaust fumes can be deadly.

When performing maintenance, always use the correct tools. If guards or covers need to be removed during servicing, remove only those necessary to gain access to other parts and replace them as soon as possible. Never leave guards off while the equipment is unattended. Make sure that any devices that are used during maintenance, like lifting and supporting devices, are strong enough to perform their jobs.

1.2.4 Fueling

Stop the engine and shut off electrical equipment while filling the fuel tank. Let a hot engine cool before fueling it. Always ground the fuel nozzle or container spout against the filler neck to avoid sparks. Never overfill the fuel tank. If any fuel is spilled, clean it up immediately. Keep sparks and open flames away from fuel. Never smoke while handling fuel or working on the fuel system. The fumes in an empty container are explosive. Never cut or weld on fuel lines, tanks, or containers. Do not walk away when fueling equipment. If the fill spout is elevated, climb up and then have someone hand up the fuel hose.

> **WARNING!**
> Do not grind, flame-cut, braze, or weld any fuel tanks or lines. Doing so can cause residual vapors to explode, causing serious injuries or even death.

Case History

Study Your Service Manual

An employee was repairing and testing a hydraulic pump. He had his hand over the open valve to the pump on the outflow side when a high-pressure stream of hydraulic fluid struck it. The stream punctured his hand and forced hydraulic fluid under the skin, resulting in injuries that required hospitalization. The cause of this accident was an open hydraulic line capable of developing a pressure of 2,000 psi.

The Bottom Line: Pressurized fluids can cause injuries. Study the service manual before servicing equipment, and follow all instruction and warnings.

Source: OSHA

Case History

Double Check Your Equipment

A 49-year old bulldozer operator was crushed to death while compacting earthen fill during the construction of an oil exploration island. The victim was operating the bulldozer on level ground, alternately in forward and reverse at half throttle. With the blade completely down, the victim shifted the transmission control lever toward neutral. The transmission was partially disengaged, but not fully in neutral.

He assumed, without checking, that the bulldozer transmission was in a stable neutral position. He exited the cab on the right side. When he did this, the transmission slipped back into the first gear of reverse, causing the bulldozer to suddenly move. As a result, the victim was pulled between the underside of the fender and the top of the track cleats. As the bulldozer continued in reverse, the victim was fatally crushed beneath the track cleats.

The Bottom Line: Always use parking brakes, and double-check settings before exiting the equipment. Never leave equipment when it is still running.

Source: The National Institute for Occupational Safety and Health (NIOSH)

---

12   NCCER – *Construction Craft Laborer Level Two*   75123-13

Case History
Communicate With Co-Workers

A radiological control technician was providing support for maintenance employees performing work in a facility. While other workers prepared the work site, the manlift operator began traveling toward the site. The control technician was walking along next to the manlift. The route used to approach the job site involved navigating through a narrow contamination control zone (CCZ) corridor. This required the operator's full attention. As they approached the job site, the technician moved ahead of the manlift and out of the direct view of the operator. The technician knelt down to pick up some surveying equipment that was in the path. He lost eye contact with the operator. Therefore, the operator did not realize that the technician was in the pathway of the manlift. As a result, the front wheel of the manlift ran over the technician's left foot. When the operator heard the technician screaming, he immediately stopped and backed up. Luckily, the technician did not have any broken bones. He did, however, miss 14 days of work. Using hand signals and maintaining eye contact could have spared this technician his injured foot.

The Bottom Line: Equipment operators have limited visibility. Pedestrians must be aware of and stay away from moving equipment.

Source: US Department of Energy

Don't Get Burned

The following are some facts about gasoline:

- One gallon of gasoline contains the same explosive force as 14 sticks of dynamite.
- Gasoline vapors are heavier than air, can travel a number of feet to an ignition source, and can ignite at temperatures of 45°F. To be safe, keep open gasoline containers well removed from all potential ignition sources.
- Gasoline has a low electrical conductivity. As a result, a charge of static electricity builds up on gasoline as it flows through a pipe or hose. Getting into and out of a vehicle during refueling can build up a static charge, especially during dry weather. That charge can cause a spark that can ignite gasoline vapors if it occurs near the fuel nozzle.

Source: US Department of Energy

Case History
Think It Through

An employee was using a MIG welder to repair a 100-gallon tank containing diesel fuel. The tank ruptured, and the fuel ignited and spilled onto the employee. The employee was engulfed in flames, suffering third- and fourth-degree burns from which he died.

The Bottom Line: Do not start a job until you have thought it through. Check material safety data sheets for information.

Source: OSHA

Additional Resources

Many equipment manufacturers have videos and literature on heavy equipment safety, including virtual walkarounds for a variety of heavy equipment.

Construction Safety, Second Edition, 2006. Jimmie Hinze. Englewood Cliffs, NJ: Prentice Hall.

Handbook of OSHA Construction Safety and Health, Second Edition, 2006. Charles D. Reese and James V. Eidson. Boca Raton, FL: CRC Press.

1.0.0 Section Review

1. Which of the following is true regarding work near heavy equipment?
   a. Always leave the power on when performing maintenance in order to check for proper operation.
   b. It is okay to stand under a load as long as it is rigged properly.
   c. Always assume the operator cannot see you.
   d. Store oily rags out in the open so any fumes can dissipate.

2. Moving heavy equipment on public highways requires that the operator have a(n) _____.
   a. CDL
   b. police escort
   c. permit
   d. minimum of two pilot vehicles

3. Which of the following is true regarding equipment maintenance?
   a. Wait until the job is completed before cleaning up spills and other work area clutter.
   b. Store oily rags in open areas to allow the fumes to dissipate.
   c. To avoid cross contamination, use the same lubrication on all pieces of equipment.
   d. Lock and tagout machines before performing maintenance.

Section Two

2.0.0 Forklift Safety

Objectives

Explain the general guidelines for forklift safety.
  a. Describe the safe operation of a forklift.
  b. State the general guidelines for safe load handling.

Trade Terms

Center of gravity: The point around which all of an object's weight is evenly distributed.

Combined center of gravity: When the weight of two items is combined, the center of gravity shifts to one point for both items.

Operator's compartment: The portion of a forklift where the operator is positioned to control the forklift.

Forklifts are common on many work sites. They are useful for lifting and moving heavy or awkward loads of materials, supplies, and equipment (*Figure 8*). While extremely useful and relatively easy to operate, these machines can be very dangerous. They present several risks including hitting other workers, dropping loads, tipping over, and causing fires and explosions.

While mechanical and hydraulic problems can cause accidents, the most common cause of forklift accidents is human error. That means most forklift accidents can be avoided if the operator and other workers in the area stay alert and use caution and common sense. In fact, research by Liberty Mutual Insurance Company shows that drivers with more than a year of experience operating a forklift are more likely to have an accident than someone with little experience. This is because operators tend to become too comfortable and less attentive after they gain experience on the equipment. The same study showed that the most common type of forklift accident is one in which a pedestrian is hit by the truck.

2.1.0 Safe Operation of a Forklift

Before you can begin operating a forklift, you must be trained and certified on that particular piece of equipment. Once you are trained and certified, you must thoroughly inspect your forklift before you begin each shift.

It is a common misconception that if you can drive a car, truck, or piece of heavy equipment, you can just hop on a forklift and start operating it. However, the Occupational Safety and Health Administration (OSHA) requires forklift operators to be trained and certified on each piece of equipment before they operate it on the job site. The operator's card only applies to the specific piece of equipment on which the operator is trained. Operators of powered forklifts must be qualified as to sight, hearing, and physical and mental ability to operate the equipment safely. Personnel who have not been trained in the operation of forklifts may only operate them for the purpose of training. The training must be conducted under the direct supervision of a qualified trainer.

Safe operation is the operator's responsibility. Operators must develop safe working habits and be able to recognize hazardous conditions in order to protect themselves and others from death and injury. They must always be aware of unsafe conditions to protect the load and the forklift from damage. They must also understand the operation and function of all controls and instruments before operating any forklift. Operators must read and fully understand the operator's manual for each piece of equipment being used.

The following safety rules are specific to forklift operation:

- Always check the capacity or load chart mounted on the machine before operating any forklift.
- Never put any part of the body into the mast structure or between the mast and the forklift.
- Never put any part of the body within the reach mechanism.
- Understand the limitations of the forklift.
- Do not permit passengers to ride in the forklift unless a safe place to ride has been provided by the manufacturer.
- Never leave the forklift running unattended.
- Never carry passengers on the forks.

Forklift operators must pay special attention to the safety of any pedestrians on the job site. Safeguard pedestrians at all times by observing the following rules:

- Always look in the direction of travel.
- Do not drive the forklift up to anyone standing in front of an object or load.
- Make sure that personnel stand clear of the rear swing area before turning.
- Exercise particular care at cross aisles, doorways, and other locations where pedestrians may step into the travel path.

75123-13 Heavy Equipment, Forklift, and Crane Safety

(A) FIXED-MAST ROUGH-TERRAIN FORKLIFT

(B) ROUGH-TERRAIN TELEHANDLER FORKLIFTS

Figure 8 Forklifts.

- Always use a spotter when moving an elevated load with a telescoping-boom forklift.

When working in a storage room or warehouse with racking, never work in the aisle on the other side of a racking unit where a forklift is working. Occasionally, the operator may push a load into the rack, causing it to fall off the other side. If you are unlucky enough to be on the other side, there is a good chance that you will be struck or crushed by a falling object.

Never hitch a ride on a forklift. Forklifts are designed for one operator and a load. Riding on the forks creates a high risk of falling and being run over by the forklift. Riding on the tractor of the forklift is also not permitted. It is easy to fall and/or be crushed between the forklift and other objects.

2.2.0 General Guidelines for Safe Load Handling

A forklift's main use is to transport large, heavy, or awkward loads. If not handled correctly, loads can fall from the forks, obstruct the operator's view, or cause the forklift to tip. The most important factor to consider when using any forklift is its capacity. Each forklift is designed with an intended capacity, and this capacity must never be exceeded. Exceeding the capacity jeopardizes not only the machine and the load, but also the safety of everyone on or near the forklift. Every manufacturer supplies a capacity chart for each forklift. The operator must be aware of the capacity of the machine before being allowed to operate the forklift.

2.2.1 Traveling with Loads

Keep the following precautions in mind when traveling with a load on a forklift:

- Always travel at a safe rate of speed with a load. Never travel with a raised load. Keep the load as low as possible (no higher than 6" above the travel surface unless the road is very rough) and tilt the mast rearward to cradle the load.
- Stay alert and pay attention. Watch the load and the conditions ahead, and alert others to your presence. Avoid sudden stops and abrupt changes in direction. Be careful when downshifting because sudden deceleration can cause the load to shift or topple. Watch the machine's rear clearance when turning.
- If you are traveling with a telescoping-boom forklift, be sure the boom is fully retracted. If you have to drive on a slope, keep the load as low as possible. Do not drive across steep slopes. If you have to turn on an incline, make the turn wide and slow.
- Travel the smoothest route possible because rough surfaces can cause bouncing and loss of the load.
- Traveling with long loads presents special hazards, particularly if the load is flexible and subject to damage. Traveling multiplies the effect of bumps over the length of the load. A stiffener may be added to the load to give it extra rigidity.
- To prevent slippage, secure long loads to the forks. This may be done in one of several ways. A field-fabricated cradle may be used to support the load. While this is an effective method, it requires that the load be jacked up.
- The forklift may be used to carry pieces of rigging equipment. This method requires the use of slings and a spreader bar.
- In some cases, long loads may be snaked through openings that are narrower than the load itself. This is done by approaching the opening at an angle and carefully maneuvering one end of the load through the opening first. Avoid making quick turns because abrupt maneuvers will cause the load to shift.
- Position the forklift at the landing point so that the load can be placed where you want it. Be sure everyone is clear of the load. The area under the load must be clear of obstructions and must be able to support the weight of the load. If you cannot see the placement, use a signaler to guide you.
- With the forklift in the unloading position, lower the load and tilt the forks to the horizontal position. When the load has been placed and the forks are clear from the underside of the load, back away carefully to disengage the forks.
- Special care needs to be taken when placing elevated loads. Some forklifts are equipped with a leveling device that allows the operator to rotate the fork carriage to keep the load level during travel. When placing elevated loads, it is extremely important to level the machine before lifting the load.
- One of the biggest potential safety hazards during elevated load placement is poor visibility. There may be workers in the immediate area who cannot be seen. The landing point itself may not be visible. Your depth perception decreases as the height of the lift increases. To be safe, use a signal person to help you position the load.
- Use tag lines to tie off long loads. Tie off loads to the mast of the forklift. Drive the forklift as closely as possible to the landing point with the load kept low. Set the parking brake. Raise the load slowly and carefully while maintaining a slight rearward tilt to keep the load cradled. Under no circumstances should the load be tilted forward until the load is over the landing point and ready to be set down.
- If the forks start to move, sway, or lean, stop immediately but not abruptly. Lower the load slowly; reposition it, or break it down into smaller components if necessary. If ground conditions are poor at the unloading site, it may be necessary to reinforce the ground with planks to provide greater stability. As the load approaches the landing point, slow the lift speed to a minimum. Continue lifting until the load is slightly higher than the landing point.
- All attachments used on a forklift must be approved by the forklift manufacturer.
- Never attempt to rig an unstable load with a forklift. Be especially mindful of the load's center of gravity when rigging loads with a forklift.
- When carrying cylindrical objects, such as oil drums, keep the mast tilted rearward to cradle the load. If necessary, secure the load to keep it from rolling off the forks.

2.2.2 Tipping

There are three main causes for a forklift tipping:

- The load is too heavy.
- The load is placed too far forward on the forks.
- The operator is not driving safely.

To avoid tipping, you need to understand what the center of gravity is and how it applies to fork-

75123-13 Heavy Equipment, Forklift, and Crane Safety   Module Four   17

lifts. The center of gravity is the point around which all of an object's weight is evenly distributed. Your forklift has a center of gravity and the load you're moving will have its own center of gravity. When the forklift picks up the load, the center of gravity shifts to the combined center of gravity. Understanding a concept known as the stability triangle helps operators to better evaluate a lift situation. The stability triangle is represented by three points on the chassis of the forklift (*Figure 9*). The center of the two front wheels and the center of the rear axle form the three points. The center of the triangle is the center of gravity (CG) of the forklift with no load, the forks down, and the boom retracted. Adding a fourth point—straight up from the CG to the boom—creates a pyramid. Imagine that there is a weight hanging from the boom at this point. As long as the weight is hanging inside the lines of the triangle below, the forklift remains stable. As the boom is raised, the CG rises within the pyramid. As a result, the triangle begins to shrink in size. The smaller the triangle, the more difficult it becomes to keep the imaginary weight hanging within it. If the forklift is not level, especially if it is tilting left or right, the imbalance can easily cause the machine to tip over as the load rises.

The forklift will tip if the center of gravity moves too far forward, backward, right, or left. Putting too heavy a load on the forklift, or placing the load too far forward on the forks, will cause the combined center of gravity to be too far forward, causing the forklift to tip forward. Turning too sharply or quickly can cause the forklift to sway or swing to the left or right, causing the combined center of gravity to veer far enough off center to tip the forklift over. These types of accidents can easily result in the operator or a bystander being crushed by the forklift. The forklift operator must read, understand, and follow the load charts for the equipment in use.

2.2.3 Obstructing the View

It is common to try to move as much as possible in the fewest number of trips. Sometimes this causes forklift operators to stack loads too high. This may exceed the safe weight limit of the forklift, and it may also block the operator's vision. To operate a forklift safely, the operator must be able to clearly see what is in front of and behind the forklift. This should be done without the operator leaning outside of the operator's compartment. Leaning outside the operator's compartment can result in serious injuries.

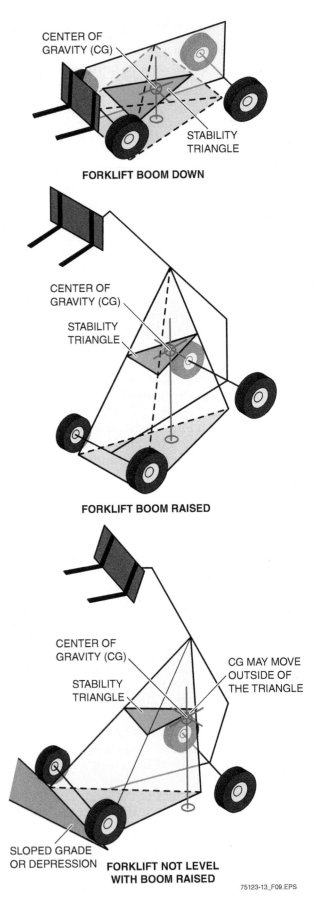

Figure 9 Forklift stability triangle.

Case History

Deadly Overload

A forklift operator was carrying a load that was stacked too high. The load obstructed his view. To make up for it, he stuck his head out the side of the operator's compartment to see around the load. Unfortunately, as he was preparing to drive forward, another forklift was backing up and sideswiped his machine, decapitating him.

The Bottom Line: Never carry a load that obstructs your view. Always keep all of your body parts inside the operator's compartment.

Source: OSHA

Case History

Chock the Wheels

As a forklift passes from a dock to an over-the-road truck and back, the force can slowly move the truck forward if the wheels are not properly chocked. In one case, a forklift operator incorrectly assumed that the truck driver had chocked the wheels of the truck. Every time the forklift passed from the dock onto the truck, the truck moved away from the dock. Because the operator was concentrating on his job, he did not notice that the gap between the dock and the truck was growing. After several trips in and out of the truck, the gap was wide enough for the forklift's front wheels to fall into the gap. This caused the forklift to tip forward into the truck and tip the truck trailer backward, crushing the operator between the truck and the forklift.

The Bottom Line: Never assume that the wheels are chocked properly. Always chock the wheels yourself or check to make sure it was done correctly.

Source: OSHA

Additional Resources

OSHA provides safety, inspection, and maintenance information on powered industrial trucks (forklifts) at **www.osha.gov**.

2.0.0 Section Review

1. A forklift operator card provides certification on all types of forklifts, regardless of manufacturer.
    a. True
    b. False

2. When travelling with a forklift, the forks should be no higher than _____.
    a. 3" from the travel surface
    b. 6" from the travel surface
    c. 9" from the travel surface
    d. 12" from the travel surface

Section Three

3.0.0 Crane Safety

Objectives

Explain the general guidelines for crane safety.
a. State the guidelines for working safely around power lines.
b. Describe various site hazards and restrictions.

Trade Terms

Outrigger: An extension that projects from the main body of a crane to add stability and support.

The most important crane safety precaution is determining the weight of all loads before attempting to lift them. When the assessment of the weight load is difficult, safe-load indicators or weighing devices should be attached to the rigging equipment. It is equally important to rig the load so that it is stable and the center of gravity is below the hook.

The personal safety of riggers and hoisting operators depends on common sense. The following safety practices should be observed:

- Always read the manufacturer's literature for all equipment with which you work. This literature provides information on required startup checks and periodic inspections, as well as inspection guidelines. Also, this literature provides information on configurations and capacities in addition to many safety precautions and restrictions of use.
- Determine the weight of all loads before rigging. Site management should provide this information if not known.
- Know the safe working load of the equipment and rigging and never exceed the limit.
- Examine all equipment and rigging before use. Destroy all defective components.
- Immediately report defective equipment or hazardous conditions to the supervisor. Someone in authority must issue orders to proceed after safe conditions have been ensured.
- Stop hoisting or rigging operations when weather conditions present hazards to property, workers, or bystanders, such as when winds exceed 25 to 30 miles per hour (mph); when the visibility of the rigger or hoist crew is impaired by darkness, dust, fog, rain, or snow; or when the temperature is cold enough that hoist or crane steel structures could fracture upon shock or impact.
- Safe working loads of all hoisting and rigging equipment are based on ideal conditions. These conditions are seldom achieved under working conditions and it is important to recognize factors that can reduce equipment capacity.
- Avoid rapid swinging of suspended loads. This subjects the equipment to additional side loading which can cause a collapse. Keep the load directly below the boom. Remove any equipment from the work line if damaged. Safe working loads apply only to undamaged structural members and unkinked lines.
- Safe working loads of hoisting equipment are applicable only to freely suspended loads and plumb hoist lines. Side loads and hoist lines that are not plumb can stress equipment beyond design limits and structural failure can occur without warning.
- Ensure that the safe working load of equipment is not exceeded if it is exposed to wind. Avoid sudden snatching, swinging, and stopping of suspended loads. Rapid acceleration and deceleration greatly increases the stress on equipment and rigging.
- Equipment load ratings only extend to the hook. Blocks, hooks, slings, equalizer beams, lifting components, and other equipment from the hook down must be taken into consideration when computing equipment load-rating factors.
- Most modern cranes are equipped with outriggers. To be properly stabilized, the outriggers must be used to take the crane's weight off the tires. Follow all manufacturer's guidelines and consult applicable standards to properly stabilize mobile cranes (*Figure 10*).

3.1.0 Working Around Power Lines

A crane working within a boom's length of any energized power line must have a competent signal person stationed at all times to warn the operator when any part of the machine or load

Figure 10 Mobile crane with outriggers.

is approaching the minimum safe distance from the power line. The signal person must be in full view at all times. *Figure 11* shows the minimum safe distance from power lines. Power line clearances are determined by values supplied by the local utility and the requirements of all applicable safety standards. Minimum safe power line clearances for cranes in operation range from 10 feet for lines carrying voltages up to 50,000 volts (50kV) to 45 feet for lines carrying voltages between 750 and 1,000kV. Consult your local utility for additional information.

WARNING! The most frequent cause of death of riggers and material handlers is electrocution caused by contact of the crane's boom, load lines, or load with electric power lines. To prevent personal injury or death, stay clear of electric power lines. Even though the boom guards, insulating links, or proximity warning devices may be used or required, these devices do not alter the precautions given in this section.

The preferred working condition is to have the owner of the power lines de-energize and provide grounding of the lines that are visible to the crane operator.

When working near energized power lines, observe the following:

- Erect non-conductive barricades to restrict access to the work area.
- Load control, when required, shall use tag lines of a non-conductive type.
- A qualified signal person(s), whose sole responsibility is to verify that the proper clearances are established and maintained, shall be in constant contact with the crane operator.
- The person(s) responsible for the operation shall alert and warn the crane operator and all persons working around or near the crane about the hazards of electrocution or serious injury, and instruct them on how to avoid these hazards.

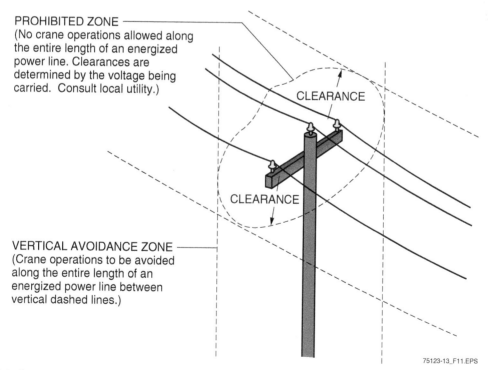

Figure 11 Prohibited zone and avoidance zone.

Heavy Equipment, Forklift, and Crane Safety

- All non-essential personnel shall be removed from the crane work area.
- No one shall be permitted to touch the crane or load unless the signal person indicates it is safe to do so.

If a crane or load comes in contact with or becomes entangled in power lines, the operator must assume that the power lines are energized unless the lines are visibly grounded. Any other assumption could be fatal. The following guidelines should be followed if the crane comes in contact with an electrical power source:

- Stay in the cab of the crane.
- Do not allow anyone to touch the crane or the load.
- If possible, reverse the movement of the crane to break contact with the energized power line.
- If staying in the cab is not possible due to fire or arcing, the operator should jump clear of the crane, landing with both feet together on the ground. Once out of the crane, take very short steps or hops to keep your feet together until you are well clear of the crane.
- Call the local power authority or owner of the power line.
- Have the lines verified secure and properly grounded within the operator's view before allowing anyone to approach the crane or the load.

3.2.0 Site Hazards and Restrictions

It takes a combined effort by everyone involved in crane operations to make sure no one is injured or killed. There are many site hazards and restrictions related to crane operations. These hazards include the following:

- Underground utilities such as gas, oil, electrical, and telephone lines; sewage and drainage piping; and underground tanks
- Electrical lines or high-frequency transmitters
- Structures such as buildings, excavations, bridges, and abutments

> **WARNING!**
> Power lines and environmental issues such as weather are common causes of injury and death during crane operations.

The operator should inspect the work area and identify hazards or restrictions that may affect the safe operation of the crane. This includes the following actions:

- Ensuring the ground can support the crane and the load
- Checking that there is a safe path to move the crane around on site
- Making sure that the crane can rotate in the required quadrants for the planned lift

The operator is required to follow the manufacturer's recommendations and any locally established restrictions placed on crane operations, such as traffic considerations or time restrictions for noise abatement.

High winds and lightning may cause severe problems on the job site (*Figure 12*). The crane operator needs to be prepared for these types of situations to avoid accidents. It is rare that high winds or lightning arrive without some warning. Lightning is a major weather hazard for crane operation. A crane's boom is a great target for lightning because it is made of metal and extends far into the sky. The distance between an observer and lightning can be estimated by the sound of thunder. If you hear thunder, the associated lightning is within hearing range; that is, within six to eight miles of the observation point. The distance from lightning strike A to lightning strike B can also be six to eight miles. In high-risk areas, it is a best practice to use lightning proximity sensors that will provide a warning when lightning occurs within 20 miles. When a warning is given, crane operations should be halted as soon as possible and the crane secured against high winds per the crane manufacturer's instructions. Wait a minimum of 30 minutes from the last observed lightning or thunder before resuming activities.

> **WARNING!**
> Even with the boom in the lowest position, it may be taller than surrounding structures and could still be a target for lightning strikes.

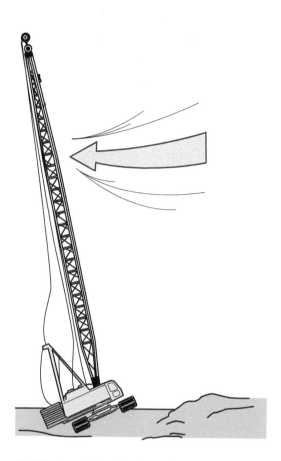 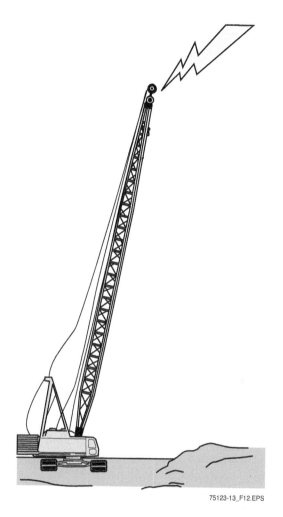

Figure 12 Wind and lightning hazards.

Additional Resources

OSHA provides safety, inspection, and maintenance information on powered industrial trucks (forklifts) at **www.osha.gov**.

3.0.0 Section Review

1. The most frequent cause of crane-related deaths is _____.

    a. crane upset
    b. rigging failure
    c. two-blocking accidents
    d. electrocution

2. When operating a crane in a high-risk area, a warning should be given when lightning is spotted within _____.

    a. 5 miles
    b. 10 miles
    c. 15 miles
    d. 20 miles

Summary

Motorized equipment is used in a variety of different construction jobs. The size and use of the equipment will vary with each job, but the dangers won't. Working around or on motorized equipment can be extremely dangerous. Serious injury or death can occur if the equipment is used improperly or unsafely, or is poorly maintained. It is your responsibility to be aware of the hazards on the job site and the safety procedures required to keep you and your co-workers safe.

Forklifts and cranes are the workhorses of a construction site. They can move, lift, and lower large, heavy, and awkward loads. They can get equipment off rail cars and out of trucks. They are heavy and powerful machines that take special training to operate. The best way to work safely around this equipment is to follow all safety procedures and stay alert to any possible hazards.

Review Questions

1. The most common type of moving heavy equipment accident is _____.
   a. a rollover
   b. striking a pedestrian
   c. backing into an excavation
   d. striking a building

2. Which of the following is true regarding asbestos?
   a. It is commonly used in the manufacture of new insulation materials.
   b. You are not likely to encounter it on a demolition site.
   c. Its removal may only be performed by properly licensed companies.
   d. It poses no real health hazards.

3. Any fluid seeping out of the ground should be reported as a potential contaminant.
   a. True
   b. False

4. A heavy equipment operator signals a reverse movement using _____.
   a. one blast of the horn
   b. two blasts of the horn
   c. three blasts of the horn
   d. four blasts of the horn

5. The crane hand signal shown by extending one arm with the thumb pointed up means _____.
   a. raise the boom
   b. nice work
   c. raise the load
   d. use the main hoist

6. In order to operate a forklift, you must _____.
   a. have a Class E driver's license
   b. be specifically trained on how to operate it
   c. be trained in the use of other motorized equipment
   d. supply your employer with a copy of your driver's license

7. A forklift is likely to tip over if the _____.
   a. load is lighter than anticipated
   b. load is too far back on the forks
   c. forklift is not level
   d. forks are lowered

8. The CG of a forklift is determined with _____.
   a. a full load, forks down, and boom extended
   b. a full load, forks up, and boom retracted
   c. no load, forks down, and boom retracted
   d. no load, forks up, and boom extended

9. Leaning your head out of the operator's compartment _____.
   a. is often necessary to see around a tall load
   b. requires the proper personal protective equipment
   c. makes you vulnerable to serious injuries
   d. is only acceptable when working outdoors

10. Rapid swinging of suspended loads subjects the equipment to collapse caused by _____.
    a. power failure
    b. decreased swing radius
    c. unsafe rigging
    d. additional side loading

Trade Terms Introduced in This Module

Anti-two-blocking devices: Devices that provide warnings and prevent two-blocking from occurring. Two-blocking occurs when the lower load block or hook comes into contact with the upper load block, boom point, or boom point machinery. The likely result is failure of the rope or release of the load or hook block.

Center of gravity: The point around which all of an object's weight is evenly distributed.

Combined center of gravity: When the weight of two items is combined, the center of gravity shifts to one point for both items.

Equipment maintenance: The care, cleaning, inspection, and proper use of machinery and equipment.

Equipment operator: A person skilled in operating certain equipment.

Hydraulic: Tools and equipment that are powered or moved by liquid under pressure.

Momentum: A physical force that causes an object in motion to stay in motion.

Operator's compartment: The portion of a forklift where the operator is positioned to control the forklift.

Outrigger: An extension that projects from the main body of a crane to add stability and support.

Figure Credits

Komatsu America Corp., Module opener, Figures 2B, 2D

Courtesy of Atlas Copco, Figure 1

Courtesy of Deere & Company, Figures 2A, 2C, 2E, 2F

Carolina Bridge Co., Figure 3

Bacou-Dalloz, Figure 4A

Honeywell Safety Products, Figure 4B

MSA The Safety Company, Figure 4C

Photo courtesy of Bullard, Figure 4D

Topaz Publications, Inc., Figure 4E

Photo courtesy of Dust Control Technology, Figure 5

U.S. Department of Labor, Figure 6

Courtesy of Sellick Equipment Limited, Figure 8A

Courtesy of JLG Industries, Inc., Figure 8B

Link-Belt Construction Equipment Co., Figure 10

Section Review Answer Key

Answer	Section Reference	Objective
Section One		
1. c	1.1.0	1a
2. a	1.2.1	1b
3. d	1.2.3	1b
Section Two		
1. b	2.1.0	2a
2. b	2.2.1	2b
Section Three		
1. d	3.1.0	3a
2. d	3.2.0	3b

NCCER CURRICULA — USER UPDATE

NCCER makes every effort to keep its textbooks up-to-date and free of technical errors. We appreciate your help in this process. If you find an error, a typographical mistake, or an inaccuracy in NCCER's curricula, please fill out this form (or a photocopy), or complete the online form at **www.nccer.org/olf**. Be sure to include the exact module ID number, page number, a detailed description, and your recommended correction. Your input will be brought to the attention of the Authoring Team. Thank you for your assistance.

Instructors – If you have an idea for improving this textbook, or have found that additional materials were necessary to teach this module effectively, please let us know so that we may present your suggestions to the Authoring Team.

NCCER Product Development and Revision
13614 Progress Blvd., Alachua, FL 32615

Email: curriculum@nccer.org
Online: www.nccer.org/olf

❏ Trainee Guide ❏ Lesson Plans ❏ Exam ❏ PowerPoints Other _____

Craft / Level: _____ Copyright Date: _____

Module ID Number / Title: _____

Section Number(s): _____

Description: _____

_____

_____

_____

_____

Recommended Correction: _____

_____

_____

_____

_____

Your Name: _____

Address: _____

Email: _____ Phone: _____

75110-13

Steel Erection

OVERVIEW

A steel-erection job site presents a variety of safety hazards. This module covers common safety precautions related to steel-erection work, including controlled decking zones, hazardous materials and equipment precautions, tool safety, and appropriate personal protective equipment.

Module Five

Trainees with successful module completions may be eligible for credentialing through the NCCER Registry. To learn more, go to **www.nccer.org** or contact us at **1.888.622.3720**. Our website has information on the latest product releases and training, as well as online versions of our *Cornerstone* magazine and Pearson's product catalog.

Your feedback is welcome. You may email your comments to **curriculum@nccer.org**, send general comments and inquiries to **info@nccer.org**, or fill in the User Update form at the back of this module.

This information is general in nature and intended for training purposes only. Actual performance of activities described in this manual requires compliance with all applicable operating, service, maintenance, and safety procedures under the direction of qualified personnel. References in this manual to patented or proprietary devices do not constitute a recommendation of their use.

Copyright © 2015 by NCCER, Alachua, FL 32615, and published by Pearson Education, Inc., New York, NY 07458. All rights reserved. Printed in the United States of America. This publication is protected by Copyright, and permission should be obtained from NCCER prior to any prohibited reproduction, storage in a retrieval system, or transmission in any form or by any means, electronic, mechanical, photocopying, recording, or likewise. To obtain permission(s) to use material from this work, please submit a written request to NCCER Product Development, 13614 Progress Blvd., Alachua, FL 32615.

From *Construction Craft Laborer, Level Two, Trainee Guide*, Third Edition. NCCER. Copyright © 2015 by NCCER. Published by Pearson Education. All rights reserved.

75110-13
STEEL ERECTION

Objectives

When trainees have completed this module, you will be able to do the following:

1. Identify the safety concerns related to steel erection.
   a. Identify common safety hazards associated with steel-erection jobs.
   b. Explain the safeguards that are required to prevent injury and equipment/property damage.

Performance Tasks

This is a knowledge-based module; there are no performance tasks.

Trade Terms

Capacity
Connector
Controlled Decking Zone (CDZ)

Industry-Recognized Credentials

If you are training through an NCCER-accredited sponsor, you may be eligible for credentials from NCCER's Registry. The ID number for this module is 75110-13. Note that this module may have been used in other NCCER curricula and may apply to other level completions. Contact NCCER's Registry at 888.622.3720 or go to **www.nccer.org** for more information.

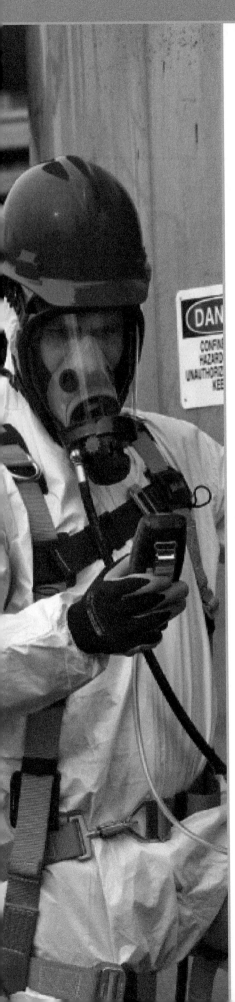

Contents

Topics to be presented in this module include:

1.0.0 Safety Concerns Related to Steel Erection 1
   1.1.0 Common Safety Hazards Associated with Steel-Erection Jobs........ 1
   1.2.0 Safeguards to Prevent Injury and Equipment/Property Damage....... 2
      1.2.1 Controlled Decking Zones 3
      1.2.2 Warning Signs 4
      1.2.3 Contaminated and Dangerous Areas 4
      1.2.4 Hazardous Materials 4
      1.2.5 Falling and Flying Objects 5
      1.2.6 Rigging Safety 5
      1.2.7 Tool Safety 5
      1.2.8 Welding Safety 6
      1.2.9 Crane Safety Precautions 6
      1.2.10 Material Handling and Storage 7

Figures

Figure 1   PPE 2
Figure 2   Horizontal safety cable 3
Figure 3   Safety harness in use on test dummies 4
Figure 4   Pinch points 7
Figure 5   Cab swing 7
Figure 6   Power line hazards 8

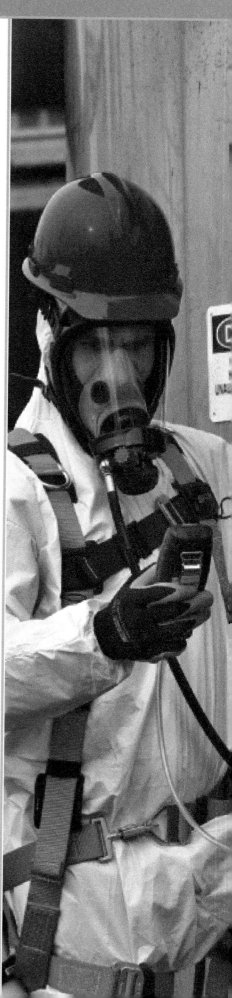

SECTION ONE

1.0.0 SAFETY CONCERNS RELATED TO STEEL ERECTION

Objectives

Identify the safety concerns related to steel erection.

a. Identify common safety hazards associated with steel-erection jobs.
b. Explain the safeguards that are required to prevent injury and equipment/property damage.

Trade Terms

Capacity: The total amount of weight capable of being lifted. This includes personnel, tools and materials, and/or equipment.

Connector: An employee who, working with hoisting equipment, is placing and connecting structural members and/or components.

Controlled Decking Zone (CDZ): An area in which certain work (for example, initial installation and placement of metal decking) may take place without the use of guardrail systems, personal fall arrest systems, fall restraint systems, or safety net systems and where access to the zone is controlled.

Steel erection is used during the construction of many types of structures including high-rise buildings, metal decking, bridges, office buildings, schools, medical facilities, and retail stores. Steel-erection jobs can be extremely dangerous. Common types of steel-erection accidents include hand injuries, back injuries, and falling from elevations. Workers can also be struck or crushed by falling loads, electrocuted by power lines, or run over or trapped by cranes.

Hazards at the job site vary with the size and complexity of the job. However, there are certain fundamental principles for job site safety that should be observed by every worker. Most accidents in the field result from inadequate maintenance of equipment, insufficient instruction in safe practices, or failure to follow safety regulations. Pay special attention to the hazards involved when working above the ground. Work operations and the moving of equipment and materials pose an increased risk.

1.1.0 Common Safety Hazards Associated with Steel-Erection Jobs

A wide range of conditions can fall under the classification of a job site hazard. Examples of hazardous conditions on the job-site include the following:

- Poor housekeeping
- Defective or unsafe tools and equipment
- Improper ventilation
- Inadequate lighting
- High noise levels
- Unsafe storage of flammable fluids
- Slippery floors
- Poorly constructed and rigged scaffolding
- Improperly stacked construction materials
- Unmarked low ceilings

As a trainee, it is important to have an awareness of safety. Always understand the hazards that exist and the reasons for the safety rules. Practice following safety rules until they become second nature. The safe way is always the right way to perform any activity. The following is a partial list of hazards that may be encountered on steel erection job sites:

- Falling from heights
- Being caught between a stationary object and a moving truck or railway car when unloading steel
- Being struck by falling materials, tools, and equipment or swinging hooks, chains, and cables
- Hearing damage from continuous exposure to high noise levels
- Burns and eye damage from welding, cutting, and chipping
- Sprains and strains from improper lifting and carrying methods
- Electrical shock from poorly grounded or defective tools and equipment
- Injuries due to defective tools and equipment
- Injuries due to improper tool or equipment use

Stay focused on the job being performed. If tired or distracted, immediately stop what you are doing. Just a moment's lapse in attentiveness when working above the ground can cause a serious accident. The same guidelines should be observed when approaching co-workers. Wait until they have finished a particular operation before attracting their attention.

The proper use of fall protection equipment will prevent most fatalities and injuries from falls. The most common circumstance of falls is from a roof; this is probably based on the fact that many roofs are sloped and do not provide good anchorages for fall protection. Ladders and scaffoldings or staging locations are other common circumstances of falls; the relative instability of both supports makes them places where great caution should be employed. Some risk can be avoided by tying off ladders and properly installing fall protection systems.

In the case of material and equipment collapses, avoid the areas under crane booms and under unsecured or hoisted material. When the worker has to be in a risky position, constant vigilance necessary to prevent accidents. Pieces being moved by any machine can escape control, but if the worker is constantly alert, it is often possible to dodge threats.

The same rules apply to all forms of threats. Keep a watch on all directions, especially overhead. If a crane brings an object close, keep an eye on it. If a machine is operating nearby, be aware of the hazards that could occur. A spinning wheel or drum that catches a worker's sleeve or pants leg can pull the worker into it. An object being moved can strike a column or other obstruction and shift suddenly. Always be ready to move out of the way of danger and have an escape route in mind. Follow all safety procedures, and remain alert at all times.

1.2.0 Safeguards to Prevent Injury and Equipment/Property Damage

Steel erection involves grinding, cutting, and welding. Wear clothing that will protect against flying debris. Proper clothing includes long-sleeved shirts, full-length pants with no cuffs, and high-topped leather work boots (*Figure 1*).

Most job sites and plant operations require additional safety clothing to be worn. Among these items are hard hats or helmets, aprons, goggles or safety glasses, and regulation ear plugs or muffs for excessive noise levels. The proper use of safety equipment is determined by the specific hazards of the job. OSHA has specific requirements for the use of safety equipment. For example, proper eye protection should be worn when cutting, chipping, or welding

The following items should be considered before starting any steel erection activities:

- *Loose clothing and accessories* – Do not wear loose or torn clothing or rings or other jewelry on the job. Clothing and accessories that do not fit tightly to the body can become caught in moving machinery.
- *Clothing materials* – Because of the flying sparks and intense heat created by grinding, cutting,

Figure 1 PPE.

and welding activities, work clothing cannot be made of polyester or other synthetic fibers. Sparks or intense heat will melt these materials, possibly causing severe burns. Because wool and cotton are more resistant to sparks, choose work clothing made from these materials.

- *Cuffs and flaps* – A shirt with long sleeves protects the upper body, and a pair of full-length pants (with no cuffs) shields the lower half of the body. The cuffless pants should hang straight through the leg and over the top of the work boots. Pants with cuffs must be avoided. Cuffs can catch sparks, which could start a fire. Pockets on shirts should be covered with flaps that can be buttoned. Always keep the pocket flaps closed and buttoned and the shirt buttoned all the way to the top collar button. This will keep flying debris or sparks from getting down inside a shirt or shirt pockets. Long sleeves and a closed shirt collar, along with long pants, also provide protection from exposure to the flash and ultraviolet rays resulting from cutting and welding.

- *Worn clothing* – Clothing must not be worn-out, frayed, or have material hanging down from it. Clothing cannot protect the body if it is worn to the point of being ragged. Clothing that is frayed or ragged could become snagged on machinery or could catch a stray spark and ignite.
- *Footwear* – Footwear should be sturdy leather work boots, with safety (reinforced) toes recommended. The shanks of the work boots should be tall enough to protect against sparks and any arc flash generated by welding. Low-top shoes should never be worn when conducting steel-erection activities.
- *Headgear* – The type of headgear worn depends on the work area. Hard hats are required in areas where falling or moving objects could cause head injuries. Hard hats also protect workers from injuries resulting from impact with stationary objects. When welding, some workers wear a solid material (non-mesh) cap with the bill pointing toward the back. Wear this cap under a hard hat and welding helmet. This protects the back of the neck and will help prevent sparks from going down a collar. Safety glasses are worn in most situations.
- *Gloves* – Ironworkers are almost always handling metal objects. Because of this, they must wear good leather work gloves to protect themselves from cuts and burns. Workers performing or assisting in welding activities must wear welding gloves instead of basic work gloves.

Weather conditions also affect the type of clothing needed. In the summer heat, it is tempting to open shirt collars and roll up sleeves. For proper protection, keep shirt collars closed and sleeves rolled down and buttoned. Winter weather usually results in extra clothing, such as coats and coveralls. These are worn over normal work clothes. For safety reasons, additional clothing items, such as outerwear, should also be kept buttoned in the same manner as normal work clothing.

While OSHA's general fall protection requirements are six feet or more, *OSHA Section 1926.760* describes the fall protection requirement during steel-erection activities. This regulation is intended to provide each worker with 100 percent fall protection above 15 feet when performing decking or detail operations and 30 feet during initial construction. This protection includes establishing perimeter safety cable systems (*Figure 2*), guardrail systems, safety nets, and personal fall arrest systems (PFAS) (*Figure 3*) for every worker who is working or walking on a surface that is 15 feet or more above the next lower level. Remember that a safety harness is useless unless it is used correctly. It should be buckled and worn tightly enough so that there is no danger of slipping out if a fall occurs. Fall protection equipment is covered in detail in the *Working from Elevations* module.

> **WARNING!**
> Body belts can result in severe injuries during a fall and may not be used.

1.2.1 Controlled Decking Zones

Controlled decking zones (CDZs) may be established for areas where metal decking is being installed more than 15 feet but less than 30 feet above the next lower level. Access to the CDZ is restricted to those employees, such as connectors, who are performing leading-edge work and who have been trained for CDZ work.

Figure 2 Horizontal safety cable.

Figure 3 Safety harness in use on test dummies.

A CDZ is a substitute for fall protection in areas where decking is initially being installed. Only workers trained for CDZ work are permitted in the areas. The CDZ must be clearly marked. The zone can be no wider than 90 feet wide and 90 feet deep from any leading edge.

All openings in floors or roofs must be covered with materials capable of supporting twice the weight of workers and equipment that may be placed on the cover. These covers must be painted with high-visibility paint or marked with the word HOLE or COVER to provide warning of the hazard.

OSHA requires training to be provided on the proper use of each of the fall protection devices used at the work site. The employer must also provide special training for all employees engaged in installing and maintaining fall protection equipment, establishing CDZs, and connecting structural members.

1.2.2 Warning Signs

Observe all warning signs. Use tools and equipment as prescribed, and maintain good housekeeping. Safety conscious workers can greatly reduce risks by anticipating hazards and observing general safety precautions such as the following:

- Comply with OSHA requirements for head and foot protection in hard hat zones.
- Locate and observe all warning and accident-prevention signs, and confine movements only to the work area.
- Only enter restricted areas with permission from the superintendent or foreman.
- Anticipate the movement paths of all motorized equipment.
- Keep out from under raised loads, unless duties require it.
- Never ride on a load or crane hook.
- Walk around scaffolding areas. If it is necessary to work under raised platforms, be sure that the work site is covered and protected.

1.2.3 Contaminated and Dangerous Areas

Check the immediate work area for major health hazards. If work must be done in contaminated and dangerous areas, always use protective clothing and equipment. Keep track of exposure time and spend as little time as possible in these areas. OSHA has identified the following as major hazards:

- Air contaminants
- Asbestos
- Coal tar pitch volatiles
- Cancer-causing agents
- Occupational noise
- Ionizing radiation
- Nonionizing radiation

1.2.4 Hazardous Materials

Check the job site for the location of highly flammable, volatile, corrosive, and explosive materials. The hazardous materials workers may come into contact with include the following:

Case History

Guard Floor Openings

A worker had just completed bolting up a vertical beam at a manufacturing facility. He was attempting to disconnect the hoisting line when he fell 60' through an unguarded floor opening in the metal decking on which he was standing.

The Bottom Line: This worker would not have fallen if guardrails had been in place. Always ensure that the area is safe before you begin working.

Source: The Occupational Safety and Health Administration (OSHA)

- Compressed gases (acetylene, oxygen, propane)
- Hydrogen
- Nitrous oxide
- Gasoline
- Diesel

Always check the Safety Data Sheet (SDS) and Hazcom for the material being used.

1.2.5 Falling and Flying Objects

Falling and flying objects are a common hazard during steel erection. Workers are at risk from falling objects when they are beneath cranes, scaffolds, forms, or where overhead work is being performed. There is a danger from flying objects when power tools or activities such as pushing, pulling, or prying cause objects to become airborne. Injuries can range from minor abrasions to concussions, blindness, or death. Follow these safeguards to avoid getting struck by falling or flying objects:

- Never walk under a load.
- Always wear a hard hat.
- Use toeboards, screens, and guardrails to prevent falling objects from striking those below.
- Use debris nets, catch platforms, or canopies to catch or deflect falling objects.
- Use safety glasses, goggles, or face shields around machines or tools that can cause flying particles.
- Inspect tools to ensure that protective guards are in good condition.
- Do not use powder-actuated tools unless qualified and properly trained.
- Avoid working underneath loads being moved.
- Barricade hazardous areas and post warning signs.
- Inspect cranes and hoists to see that all components, such as wire rope, lifting hooks, and chains are in good condition.

- Do not exceed the lifting capacity of cranes and hoists.
- Do not use buckets or pails to hoist materials. Use bolt bags designed for this type of work.
- Secure tools using tool holders/lanyards.
- Secure loose materials to prevent them from falling on anyone below.

1.2.6 Rigging Safety

Safe rigging and hoisting of steel members and materials are essential parts of the steel-erection process. Rigging operations can be extremely complicated and dangerous, especially when moving heavy material like steel. Some rigging operations require cranes, while others use a loader to move materials around the job site. Regardless of whether rigging operations involve a simple vertical lift, a powered hoist, or a crane, only qualified workers who have been properly trained can do it without supervision.

1.2.7 Tool Safety

Specific safety rules for using hand and power tools are discussed in the modules covering tools, but there are some general rules that apply to all tools and processes. These rules should become part of a positive attitude toward safety. Always use the right tool for the job. This requires knowing the capabilities of each tool and understanding how each tool is used. Also, always use tool holders and lanyards when working at elevation. Dropped tools present a serious hazard to anyone walking or working below.

Selecting the wrong tool for the job may result in injury. A wrench that is the wrong size can slip or break; a file used as a pry bar may break; and a poorly selected cable for lifting a heavy load could snap.

Case History

Pay Attention

A rigger was in the process of rigging up three beams in a triangular shape. He had put the sling around the first beam, and tightened it by pulling on the leg of the sling. This pushed up the top end of the sling, so the eye of the sling rode up onto or over the point of the hook, where it caught on the latch. The rigger was unaware that the beam was not properly secured. He then signaled the crane operator to raise the load. When the beam was overhead, the rigger began rigging the second beam. Almost instantly, the eye of the sling slipped off the latch, and the rigged beam dropped onto the victim. Although the rigger survived the accident, he spent several weeks in the hospital.

The Bottom Line: Ensure that your work area is safe before beginning any job, and stay alert for potential safety hazards.

Source: The Occupational Safety and Health Administration (OSHA)

75110-13 Steel Erection

Portable power tools present special hazards to their users. Because they are handheld and can be easily moved, the cutting, shaping, or penetrating part of the tool may contact the operator's body. This can cause burns, cuts, scrapes, or even more serious injuries. Another source of danger is the possibility of dropping a handheld power tool. Because the source of the tool's power is close to the operator, this can create electrical shock, explosion, and fire hazards. Be aware that the size and mobility of these tools make them difficult to guard properly.

Proper care of tools will prolong their use and may prolong a worker's life. Caring for tools properly requires that a number of checks are carried out. Always be sure of the following:

- Tools are free of grease and oil before using them.
- Tools are properly dressed, free of defects, and sharpened where appropriate.
- Electrical hand tools are equipped with an automatic shut-off control.
- Portable electric saws and drills are properly guarded and grounded.
- Damaged tools are repaired before use.
- Never perform any welding on tools.

All damaged equipment must be removed from service and immediately reported to the job supervisor. Take steps to see that the equipment is not used until it has been checked out. Always disconnect the power source, and post a sign to prevent others from using the equipment until it can be repaired. Rules for the care and safe use of power tools include the following:

- Do not attempt to operate any power tool without being properly trained on that particular tool.
- Always wear appropriate safety equipment and protective clothing for the job being done. For example, wear safety glasses and close-fitting clothing that cannot become caught in the moving parts of a tool. Roll up long sleeves, tuck in your shirt tail, and tie back long hair.
- Never leave a power tool energized and unattended.
- Assume a safe and comfortable position before starting a power tool.
- Do not distract others or let anyone distract you while operating a power tool.
- Be sure that any electric power tool is properly grounded before using it.
- Be sure that power is not connected before performing maintenance or changing accessories.
- Do not use dull or broken accessories.
- Use power tools only for their intended purpose.
- Do not use a power tool with the guards or safety devices removed.
- Use a proper extension cord of sufficient size to service the particular electric tool being used.
- Become familiar with the correct operation and adjustments of a power tool before attempting to use it.
- Be sure there is proper ventilation before operating gasoline-powered equipment indoors.
- Keep a fire extinguisher nearby when filling and operating gasoline-powered equipment.
- Store tools properly when not in use.

1.2.8 Welding Safety

Welding and cutting operations are common in steel erection work. Some basic safety precautions must be observed at all times when welding or cutting. Special eye protection (safety glasses) should be worn at all times, even under a welding helmet. Precautions must be taken for spark containment, and any workers below the welding area must be protected.

Additional appropriate personal protective equipment that must be worn while welding or cutting includes welding leathers, such as aprons, sleeves or jackets; welding gloves; and welding helmets or shields. The gases, dust, and fumes produced during welding can be hazardous if the appropriate precautions are not observed. Welding should only be done in well-ventilated areas. Avoid inhaling welding fumes and smoke. Respirators may be needed if welding is to be performed in a space with less than 10,000 cubic feet of air per worker. Additional precautions are necessary when welding metals that are galvanized or are painted with epoxy paint. Hot work permits must be obtained before any welding or cutting is performed, if this is required on the job site.

1.2.9 Crane Safety Precautions

Cranes are a potential danger that can be minimized by taking preventive action. The first step in prevention is knowing how and where serious accidents occur. The area close to the crane is especially dangerous. As is the case with all accidents, most crane accidents can be prevented. The trainee working in the crane area can prevent a serious accident by observing the following safety procedures:

- Do not walk under raised loads unless specifically instructed to do so.

- Always be aware of the position of the load when hooking on, arranging loads, or moving across the construction site. Treat all lifting operations as if equipment failures could occur.
- Keep hands out of pinch points when assisting with hooking or unhooking a load. Use tag lines whenever possible. *Figure 4* shows typical pinch points to avoid.
- Do not use a crane to pull a sling out from under a load.
- Always stand clear of the rear of the rotating super structure (*Figure 5*).
- Keep clear of power lines. Contact with power lines is the largest single cause of fatalities associated with cranes. *Figure 6* shows the hazardous locations created when a crane contacts a power line.
- Learn the crane signals, especially the emergency stop signal. Use it only when it may prevent an accident from happening. The crane signals should always be posted on the job site and/or crane.

> **NOTE**
> Crane operations should stop when the crane operator cannot see or hear the signal person. Anyone can use the emergency stop signal, but only a qualified person can signal the crane operator for lifts.

1.2.10 Material Handling and Storage

Many occupational injuries occur when manually handling or moving construction materials. Most nonmechanized moving, stacking, loading, and unloading tasks are performed by less experienced craftworkers. In many cases, these are trainees. Strains, sprains, fractures, and crushing injuries can be minimized by having the proper knowledge of safe lifting and handling procedures and by using good body mechanics. General points to keep in mind when handling or moving construction materials include the following:

- Inspect materials for slivers and for rough or sharp edges.
- Determine the weight of the load before attempting to move it.
- Know your own capacity for lifting objects.
- Be sure that the pathway is free from obstacles.
- Keep hands free of oil and grease.
- Take a firm grip on the object before it is moved, being careful to keep fingers from being pinched.
- Use good body mechanics.
- Use gloves when handling sharp-edged materials.
- Get help when the load is too large or heavy.

Figure 4 Pinch points.

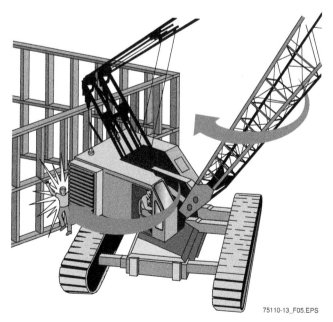

Figure 5 Cab swing.

Figure 6 Power line hazards.

Additional Resources

The complete *OSHA Safety and Health Regulations for Construction* can be found at **www.osha.gov**.

1.0.0 Section Review

1. It is safe to walk under a moving load as long as you are wearing the appropriate PPE.
   a. True
   b. False

2. The hand protection worn by ironworkers should be made of _____.
   a. plain cotton
   b. rubberized cotton to provide a good grip
   c. good-quality leather
   d. special rubber designed to protect against electrical contact

Summary

Safety is, and should be the primary concern at every level of responsibility in construction work. Employers are responsible for providing the means, support, and training for safe work. Employees are responsible for maintaining safety awareness at all times. All forms of steelwork require close attention to the appropriate safety procedures for that aspect of the work. Working around cranes requires awareness of the location and movement of loads and cranes.

Working with power tools demands that the worker know the hazards associated with those tools. Working at heights demands constant attention to the environment and to the activities going on around the worker. Even the clothing a worker wears is a safety factor. Take all possible practical precautions.

Falls can cause serious injuries or death when no fall protection equipment or the wrong kind of equipment is used. The three types of fall protection commonly used when working at elevated levels are guardrails, personal fall arrest systems, and safety nets.

Review Questions

1. Which of the following is appropriate to wear on a steel-erection job site in the heat of summer?
    a. Cutoff shorts
    b. Knee length shorts
    c. Light, loose pants
    d. Straight pants without cuffs

2. The same headgear will be required on all job sites.
    a. True
    b. False

3. Which of the following is *true* regarding safety on a steel erection site?
    a. The OSHA general fall protection requirements apply to steel erection activities.
    b. Steel workers are never required to wear fall protection.
    c. Fall protection is required above 15 feet when performing decking or detail operations.
    d. Fall protection is required above 6 feet during initial steel construction.

4. A CDZ may be established where metal decking is being installed _____.
    a. up to 10 feet above the next lower level
    b. from 5 to 15 feet above the next lower level
    c. from 10 to 20 feet above the next lower level
    d. from 15 to 30 feet above the next lower level

5. All of the following are considered major health hazards by OSHA, *except* _____.
    a. asbestos
    b. ionizing radiation
    c. nonionizing radiation
    d. fuel oil

Trade Terms Introduced in This Module

Capacity: The total amount of weight capable of being lifted. This includes personnel, tools and materials, and/or equipment.

Connector: An employee who, working with hoisting equipment, is placing and connecting structural members and/or components.

Controlled Decking Zone (CDZ): An area in which certain work (for example, initial installation and placement of metal decking) may take place without the use of guardrail systems, personal fall arrest systems, fall restraint systems, or safety net systems and where access to the zone is controlled.

Additional Resources

This module presents thorough resources for task training. The following resource material is suggested for further study.

The complete *OSHA Safety and Health Regulations for Construction* can be found at **www.osha.gov**.

Figure Credits

LPR Construction, Module opener, Figures 1–3

Section Review Answer Key

Answer Section One	Section Reference	Objective
1. b	1.1.0	1a
2. c	1.2.0	1b

NCCER CURRICULA — USER UPDATE

NCCER makes every effort to keep its textbooks up-to-date and free of technical errors. We appreciate your help in this process. If you find an error, a typographical mistake, or an inaccuracy in NCCER's curricula, please fill out this form (or a photocopy), or complete the online form at **www.nccer.org/olf**. Be sure to include the exact module ID number, page number, a detailed description, and your recommended correction. Your input will be brought to the attention of the Authoring Team. Thank you for your assistance.

Instructors – If you have an idea for improving this textbook, or have found that additional materials were necessary to teach this module effectively, please let us know so that we may present your suggestions to the Authoring Team.

NCCER Product Development and Revision
13614 Progress Blvd., Alachua, FL 32615

Email: curriculum@nccer.org
Online: www.nccer.org/olf

❏ Trainee Guide ❏ Lesson Plans ❏ Exam ❏ PowerPoints Other _____

Craft / Level: _____ Copyright Date: _____

Module ID Number / Title: _____

Section Number(s): _____

Description: _____

_____

_____

_____

_____

Recommended Correction: _____

_____

_____

_____

_____

Your Name: _____

Address: _____

_____

Email: _____ Phone: _____

75121-13

Electrical Safety

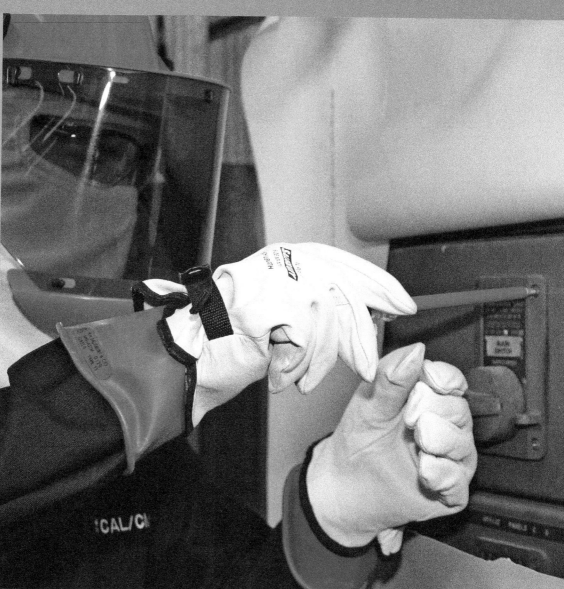

OVERVIEW

Everyone on a job site is exposed to potential electrical safety hazards. This module covers common electrical hazards and the safeguards used to protect against them. It also provides an overview of a basic lockout/tagout procedure.

Module Six

Trainees with successful module completions may be eligible for credentialing through the NCCER Registry. To learn more, go to **www.nccer.org** or contact us at **1.888.622.3720**. Our website has information on the latest product releases and training, as well as online versions of our *Cornerstone* newsletter and Pearson's product catalog.

Your feedback is welcome. You may email your comments to **curriculum@nccer.org**, send general comments and inquiries to **info@nccer.org**, or fill in the User Update form at the back of this module.

This information is general in nature and intended for training purposes only. Actual performance of activities described in this manual requires compliance with all applicable operating, service, maintenance, and safety procedures under the direction of qualified personnel. References in this manual to patented or proprietary devices do not constitute a recommendation of their use.

Copyright © 2015 by NCCER, Alachua, FL 32615, and published by Pearson Education, Inc., New York, NY 10013. All rights reserved. Printed in the United States of America. This publication is protected by Copyright, and permission should be obtained from NCCER prior to any prohibited reproduction, storage in a retrieval system, or transmission in any form or by any means, electronic, mechanical, photocopying, recording, or likewise. To obtain permission(s) to use material from this work, please submit a written request to NCCER Product Development, 13614 Progress Blvd., Alachua, FL 32615.

From *Construction Craft Laborer, Level Two, Trainee Guide*, Third Edition. NCCER.
Copyright © 2015 by NCCER. Published by Pearson Education. All rights reserved.

75121-13
ELECTRICAL SAFETY

Objectives

When you have completed this module, you will be able to do the following:

1. Identify the risks associated with working around electricity.
   a. Describe the effects of electrical shock, arc, and blast.
   b. Describe common power cord hazards.
   c. Describe the hazards of working near overhead lines.
   d. Explain how to minimize the risks associated with work around electricity.
2. Describe the lockout/tagout procedure for all energy sources associated with a device or process.
   a. Identify the steps in a typical lockout/tagout procedure.
   b. Identify situations under which emergency removal of a lockout may be required.

Performance Task

Under the supervision of the instructor, you should be able to do the following:

1. Demonstrate how to properly use a lockout/tagout device.

Trade Terms

Arc blast
Arc fault
Arc fault circuit interrupter (AFCI)
Arc flash boundary
Assured equipment grounding conductor program
Bolted fault
Breakdown voltage
Energy source
Energy-isolating device
Equipment grounding conductor

Fibrillation
Ground fault circuit interrupter (GFCI)
Grounding
Insulation
Lockout
Lockout device
Shock hazard
Tagout
Tagout device

Industry Recognized Credentials

If you are training through an NCCER-accredited sponsor, you may be eligible for credentials from NCCER's Registry. The ID number for this module is 75121-13. Note that this module may have been used in other NCCER curricula and may apply to other level completions. Contact NCCER's Registry at 888.622.3720 or go to **www.nccer.org** for more information.

Contents

Topics to be presented in this module include:

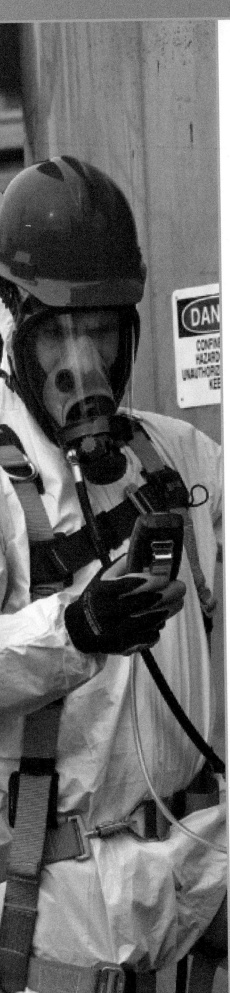

1.0.0 Risks Associated with Working Around Electricity 1
  1.1.0 The Effects of Electrical Shock, Arc, and Blast 1
    1.1.1 How Shock Occurs 3
    1.1.2 Burns 4
    1.1.3 Arc Flash and Blast Hazards 5
    1.1.4 Emergency Response 5
  1.2.0 Common Power Cord Hazards 5
  1.3.0 Hazards of Working Near Overhead Power Lines 6
  1.4.0 Minimizing the Risks Associated with Work Around Electricity 7
    1.4.1 Ground Fault Circuit Interrupters 8
    1.4.2 Arc Fault Circuit Interrupters 9
    1.4.3 Assured Equipment Grounding Conductor Program 10
    1.4.4 Selecting a Meter with the Appropriate Category Rating 10
    1.4.5 Recognizing Hazard Boundaries 10
2.0.0 Performing Lockout/Tagout Procedures for All Energy
       Sources Associated with A Device or Process 16
  2.1.0 Steps in a Typical Lockout/Tagout Procedures 19
    2.1.1 Preparation for Lockout/Tagout 19
    2.1.2 Sequence for Lockout/Tagout 20
    2.1.3 Restoration of Energy 20
  2.2.0 Situations Under Which Emergency Removal of a Lockout
       May Be Required 22

Figures and Tables

Figure 1   Power distribution system 2
Figure 2   Typical body resistance and currents 2
Figure 3   Defibrillation equipment 3
Figure 4   Arc flash diagram 6
Figure 5   Electrical grounding conductor 6
Figure 6   Receptacle configurations 7
Figure 7   Double-insulated electric drill 9
Figure 8   Extension with a GFCI 9
Figure 9   Test equipment 11
Figure 10 Category rating shown on test equipment 11
Figure 11 Approach limits 12
Figure 12 Arc flash in process 13
Figure 13 Worker using appropriate PPE 14
Figure 14 Lockout/tagout device 17
Figure 15 Typical safety tags 17
Figure 16 Lockout devices 18
Figure 17 Multiple lockout/tagout device 19
Figure 18 Placing a lockout/tagout device 19

Table 1   Effects of Current on the Human Body 3
Table 2   Overvoltage Installation Categories 11

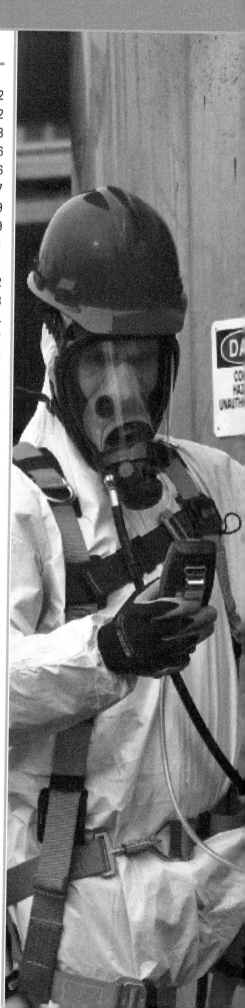

SECTION ONE

1.0.0 RISKS ASSOCIATED WITH WORKING AROUND ELECTRICITY

Objectives

Identify the risks associated with working around electricity.
  a. Describe the effects of electrical shock, arc, and blast.
  b. Describe common power cord hazards.
  c. Describe the hazards of working near overhead power lines.
  d. Explain how to minimize the risks associated with work around electricity.

Trade Terms

Arc blast: An explosion similar to the detonation of dynamite that occurs during an arc flash incident.

Arc fault: A high-energy discharge between two or more conductors.

Arc fault circuit interrupter (AFCI): A device intended to provide protection from the effects of arc faults by recognizing characteristics unique to arcing and by functioning to de-energize the circuit when an arc fault is detected.

Arc flash boundary: An approach limit at a distance from exposed energized electrical conductors or circuit parts within which a person could receive a second-degree burn if an electrical arc flash were to occur.

Assured equipment grounding conductor program: A detailed plan specifying an employer's required equipment inspections and tests and a schedule for conducting those inspections and tests.

Bolted fault: A short circuit or electrical contact between two conductors at different potentials, in which the impedance or resistance between the conductors is essentially zero.

Breakdown voltage: The voltage at which an insulator has a breakdown and ceases to act as a resistor.

Equipment grounding conductor: A wire that connects metal enclosures and containers to ground.

Fibrillation: Very rapid, irregular contractions of the muscle fibers of the heart that result in the heartbeat and pulse going out of rhythm with each other.

Ground fault circuit interrupter (GFCI): A fast-acting circuit breaker that senses small imbalances in the circuit caused by current leakage to ground and, in a fraction of a second, shuts off the electricity.

Grounding: The process of directly connecting an electrical circuit to a known ground to provide a zero-voltage reference level for the equipment or system.

Insulation: The practice of placing nonconductive material such as plastic around the conductor to prevent current from passing through it.

Shock hazard: A dangerous condition associated with the possible release of energy caused by contact or approach to energized electrical conductors or circuit parts.

Electrical accidents are the third leading cause of death in the workplace. In fact, each year in the United States there are approximately 700 electricity related deaths in the workplace.

> **WARNING!**
> This module is intended to provide an overview of the basic concepts of electrical safety. Only qualified individuals may work on electrical circuits and equipment.

Electricity is generated at a power station and transmitted through power distribution lines to residential, commercial, and industrial users (*Figure 1*). At each point on the power distribution route, maintenance personnel must install, connect or disconnect, adjust, and maintain the equipment used to generate, transmit, and consume electricity. Each point along this route presents hazards that can injure or kill personnel, whether or not they are working directly on the equipment. Anyone working on or near electrical equipment may encounter one of the following electrical hazards:

- Shock hazard
- Arc flash
- Arc blast

1.1.0 The Effects of Electrical Shock, Arc, and Blast

An electrical circuit consists of an electrical power source tied through a designated path to a point of utilization. As long as the current travels the intended path, the circuit is safe. When a person

75121-13 Electrical Safety

Module Six 1

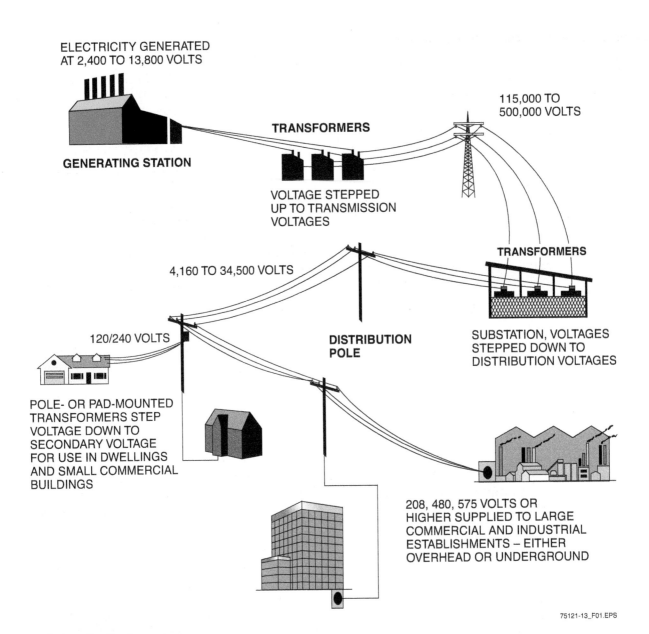

Figure 1 Power distribution system.

in contact with ground (or grounded objects) contacts an energized part, that person receives an electrical shock. Electrical shock can also occur when an internal fault causes the cases or enclosures of tools, appliances, or electrical equipment to become energized. The danger from electrical shock depends on the amount of current flowing through the person, which is a function of voltage and body resistance and may be calculated using Ohm's law. *Figure 2* shows the resistance of the human body to electrical voltage and current. *Table 1* shows the effects of current on the human body. Current is measured in amps (A) or milliamps (mA). A milliamp is $\frac{1}{1,000}$ of an amp.

As shown in *Table 1*, a minor shock of 5mA results in an involuntary movement away from the source. This can result in injuries as the shocked worker jumps back from the electrical shock source, only to rip open or break an arm or hand on the way out of a cabinet or work area. When the current is between 6mA and 30mA, the shock causes loss of muscular control. This may result in the worker falling from an elevated position or cause the worker to fall into a more dangerous

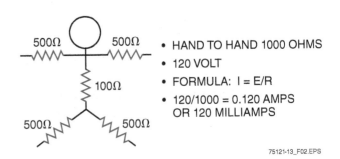

Figure 2 Typical body resistance and currents.

Table 1 Effects of Current on the Human Body

Current Value	Common Item/Tool	Typical Effects
1mA	Watch battery	Perception level. Slight tingling sensation.
5mA	9-volt battery	Slight shock. Involuntary reactions can result in other injuries.
6 to 30mA	Christmas tree light bulb	Painful shock, loss of muscular control.
50 to 150mA	Small electric radio	Extreme pain, respiratory arrest, severe muscular contractions. Death possible.
1000mA to 4300mA	Jigsaw (4 amps); Sawsall® or Port-a-Band® saw (6 amps); portable drill (3–8 amps); ShopVac® (15-gallon); circular saw	Ventricular fibrillation, severe muscular contractions, nerve damage. Typically results in death.

electrical source. As the current levels increase beyond about 20mA, muscular contractions can prevent the victim from pulling away. At 50mA, respiratory paralysis may result in suffocation. Current levels above 150mA may cause the heart to go into a state of fibrillation (abnormal rhythm). This condition is fatal unless the heart rhythm is corrected using special defibrillation equipment (*Figure 3*). Current levels of 4A or more may stop the heart, resulting in death unless immediate medical attention is provided.

Other effects of electrical shock may include immediate entry and exit wounds from high-voltage contact, heart stoppage and thermal burns. Thermal burns are often not apparent at first; however, the tissue in the current path may be destroyed and necrotize (rot away) from the inside over time. In addition, the body's reaction to the shock can cause a fall or other accident.

1.1.1 How Shock Occurs

The current enters the body at one point and leaves at another. Shock normally occurs in one of three ways: the person must come in contact with both wires of the electric circuit, one wire of the electric circuit and the ground, or a metallic part that is in contact with an energized wire while the person is in contact with the ground. To fully understand the harm done by electrical shock, you need to understand something about the human body and its parts: the skin, the heart, and muscles.

Skin covers the body and is made up of three layers. The most important layer, as far as electric shock is concerned, is the outer layer of dead cells referred to as the horny layer. This layer is composed mostly of a protein called keratin, and it is the keratin that provides the largest percentage of the body's electrical resistance. When it is dry, the outer layer of skin may have a resistance of several thousand ohms, but when it is moist, there is

> **Did you know?**
> Approximately 30,000 nonfatal and 1,000 fatal electrical shock accidents occur each year. However, the majority of hospital admissions due to electrical accidents are from arc flash burns, not from shocks.
>
> Source: *NFPA 70E®, Electrical Safety in the Workplace, Informative Annex K*

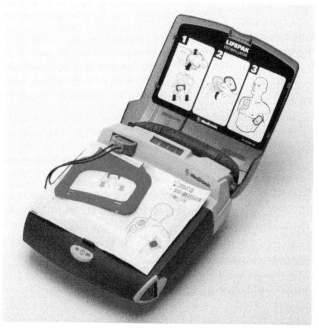

Figure 3 Defibrillation equipment.

a radical drop in resistance. This is also the case if there is a cut or abrasion that pierces the horny layer. The amount of resistance provided by the skin varies widely from individual to individual. A worker with a thick horny layer will have a much higher resistance than someone with a thin horny layer. The resistance also varies at different parts of the body. For instance, a worker with high-resistance hands may have low-resistance skin on the back of his calf.

The skin, like any insulator, has a breakdown voltage at which it ceases to act as a resistor and is simply punctured, leaving only the lower-resistance body tissue to impede the flow of current. The breakdown voltage varies with the individual, but is in the area of 600 volts (V). Since most industrial power-distribution systems operate at 480V or higher, technicians working at these levels need to be aware of the shock potential.

The heart is the pump that generates blood flow throughout the body. The blood flow is caused by the contractions of the heart muscle, which is controlled by electrical impulses. The electrical impulses are delivered by an intricate system of nerve tissue with built-in timing mechanisms, which make the chambers of the heart contract at exactly the right time. An outside electric current measuring as little as 75mA can upset the rhythmic, coordinated beating of the heart by disturbing the nerve impulses. When this happens, the heart is said to be in fibrillation, and the pumping action stops. Death will occur quickly if the normal beat is not restored. Remarkable as it may seem, what is needed to defibrillate the heart is a shock of an even higher intensity.

The other muscles of the body are also controlled by electrical impulses delivered by nerves. Electric shock can cause loss of muscular control, resulting in the inability to let go of an electrical conductor. Electric shock can also cause injuries of an indirect nature in which involuntary muscle reaction from the electric shock can cause bruises, fractures, and even deaths resulting from collisions or falls.

The severity of shock received when a person becomes a part of an electric circuit is affected by three primary factors: the amount of current flowing through the body (measured in amperes), the path of the current through the body, and the length of time the body is in the circuit. Other factors that may affect the severity of the shock are the frequency of the current, the phase of the heart cycle when shock occurs, and the general health of the person prior to the shock. Effects can range from a barely perceptible tingle to immediate cardiac arrest. A difference of only 100mA exists between a current that is barely perceptible and one that can be fatal.

A severe shock can cause considerably more damage to the body than is visible. For example, a person may suffer internal hemorrhages and destruction of tissues, nerves, and muscle. In addition, shock is often only the first injury in a chain of events. The final injury may well be from a fall, cuts, burns, or broken bones.

1.1.2 Burns

The most common electrical injury is a burn. Burns suffered in electrical accidents may be of three types: electrical burns, arc burns, and thermal contact burns.

Electrical burns are the result of electric current flowing through the tissues or bones. Tissue damage is caused by the heat generated by the current flow through the body. An electrical burn is one of the most serious injuries you can receive, and should be given immediate attention. Since the most severe burning is likely to be internal, what may appear to be a small surface wound could, in fact, be an indication of severe internal burns.

Arc burns make up a large portion of the injuries from electrical malfunctions. The electric arc between metals can be up to 35,000°F, which is about four times hotter than the surface of the sun. Workers several feet from the source of the arc can receive severe or fatal burns. Since most electrical safety guidelines recommend safe working distances based on shock considerations, workers following these guidelines can still be at risk for arc burns. Electric arcs can occur due to poor electrical contact or failed insulation. Electrical arcing is caused by the passage of substantial amounts of current through vaporized terminal material, usually metal or carbon.

Case History
The Added Risk of Wearing Jewelry

People who work around electricity must not wear any jewelry, including wedding rings, bracelets, and necklaces. One unfortunate worker lost his ring finger when his gold wedding band came in contact with an electrical capacitor.

The Bottom Line: Jewelry is usually made of materials that are excellent conductors, such as gold, silver, or platinum. Wearing these items greatly increases the risk of severe injury or death from electrical shock.

A thermal contact burn is caused by contact with objects thrown during the blast associated with an electric arc. This blast comes from the pressure developed by the near-instantaneous heating of the air surrounding the arc, and from the expansion of the metal as it is vaporized. (Copper expands by a factor in excess of 65,000 times when vaporized.) These pressures can be great enough to propel people, switchgear, and cabinets considerable distances. Another hazard associated with the blast is the explosion of molten metal droplets, which can also cause thermal contact burns and associated damage. A possible beneficial side effect of the blast is that it could throw a nearby person away from the arc, thereby reducing the effect of arc burns.

1.1.3 Arc Flash and Blast Hazards

Electrical faults may occur when safe work practices are not followed, because of equipment failure, or both. Two common types of faults are bolted faults and arc faults. A bolted fault occurs when a low-resistance connection (short) is made between two conductors at different potentials. For example, if the conductors of two electrical power phases are crossed during initial installation or repair, the result is a bolted fault. The current flowing between bolted components is called bolted fault current. Electrical equipment that is installed per applicable codes and is in good condition should withstand a bolted fault within its rating without damage. Properly sized and applied circuit protective devices are expected to interrupt a bolted fault before it causes damage to equipment or conductors. The high current of a bolted fault enables fast response by these devices.

An arc fault occurs when an electrical current jumps or arcs the air gap between two energized sources with different potentials, or between an energized electrical circuit and ground. The current is called arc fault current.

Electric arcs occur in the normal operation of nearly all electrical equipment in situations ranging from plugging a cord into an electrical outlet to operating a switch. This also occurs when meter leads contact energized parts. Any time contact is made or broken between energized components and non-energized components, it creates a small arc/spark. Whenever that small arc is not interrupted and grows into a big arc, it creates an event called an arc flash. This is why it is important to use appropriately rated test equipment.

The duration of the arc and the amount of arc current flow determine the intensity of the arc. If the current level is low and the arc time is short, minimal damage will occur. Any increase in time or current results in greater energy release and destruction and may cause a fire or explosion. An arc flash generates a very bright light and can reach temperatures up to 35,000°F. These temperatures melt most metals and all plastics. Blast pressure from an arc may be several hundred pounds per square foot and can blow enclosures apart and people away from the source.

When an arc flash occurs in an enclosed area, the resulting chain of events can cause a fire, explosion, toxic fumes, intense light, and shock waves that damage or destroy anything near the center of the arc flash (*Figure 4*).

The hazardous effects of arc flashes must be evaluated to determine the type of personal protective equipment (PPE) required when exposed to an arc hazard. All equipment in which arc flashes can occur must be labeled with warnings indicating the arc hazard and the anticipated energy exposure and/or required category of PPE. The labeling must be large enough to be readable from outside the hazard area. Anyone working on or near the equipment must be trained to recognize and avoid hazards.

1.1.4 Emergency Response

If someone near you is receiving an electrical shock, do not touch that person. Instead, immediately turn off the power source. If that is not possible, try using nonconductive material such as a blanket, rope, coat, or piece of dry wood to separate the person from the electrical source. Never use anything wet or damp. If you touch the person with your body, use a metal object to move the person, or use wet or damp material on or near the person, you could also become a victim.

Once the person has been separated from the shock source, immediately call for medical help. If the victim is not breathing, a trained person should immediately begin artificial respiration.

1.2.0 Common Power Cord Hazards

Flexible power cords (extension cords) used to supply power to tools and equipment during construction are very common on a job site. Inside a power cord are conductors that carry potentially lethal voltages. The insulation on the outside of the power cord protects against shock. However, power cord insulation is subject to damage because the cords are often given rough use and are exposed to foot and vehicle traffic, sharp edges, and strain from being pulled. Over time, the pro-

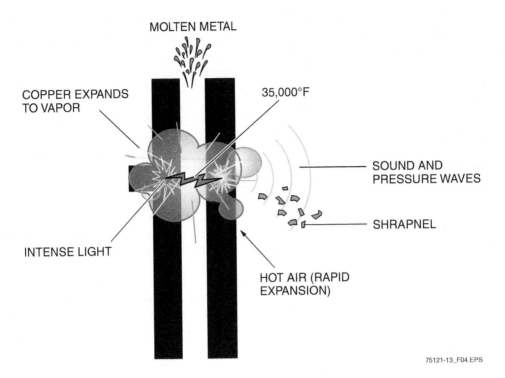

Figure 4 Arc flash diagram.

tective insulation can wear away or be cut. Once the insulation becomes damaged, you can receive a shock by touching the exposed conductor. You can also be shocked if you and the exposed conductor are touching the same conductive surface, such as metal or water. For example, if you are standing on a metal roof, and an exposed conductor touches the roof, you can receive a shock.

The following are important tips that can protect you from electrical shock:

- Never use tape to repair a damaged power cord. It is no substitute for the original insulation, and may not prevent a shock.
- Never use a cord from which the grounding conductor (*Figure 5*) has been removed. Grounded plugs may only be used with three-prong receptacles (*Figure 6*).
- Always connect power tools to a circuit protected by a ground fault circuit interrupter or GFCI (described later in this module).
- Be extremely careful when using a power cord in a wet or damp area. Keep the cord out of any water. Electricity from an exposed conductor can travel along the outside of the cord, or along a wet surface, and give you a shock.

1.3.0 Hazards of Working Near Overhead Power Lines

High-voltage electricity, defined as 600V or more, is 10 times more likely to kill you than normal household voltages, which are around 120V or 240V.

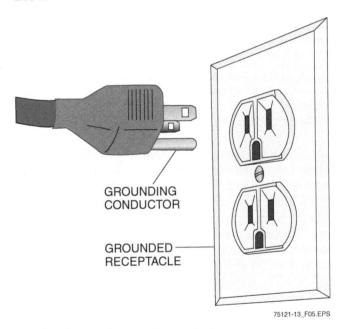

Figure 5 Electrical grounding conductor.

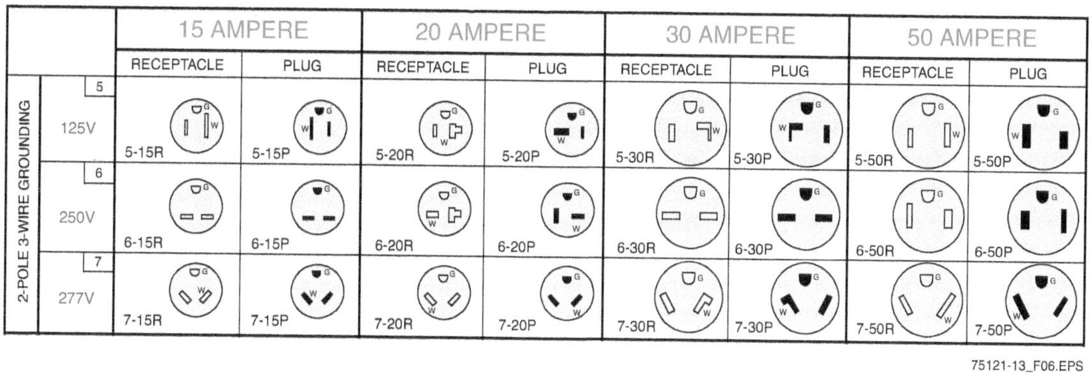

		15 AMPERE		20 AMPERE		30 AMPERE		50 AMPERE	
		RECEPTACLE	PLUG	RECEPTACLE	PLUG	RECEPTACLE	PLUG	RECEPTACLE	PLUG
2-POLE 3-WIRE GROUNDING	5 125V	5-15R	5-15P	5-20R	5-20P	5-30R	5-30P	5-50R	5-50P
	6 250V	6-15R	6-15P	6-20R	6-20P	6-30R	6-30P	6-50R	6-50P
	7 277V	7-15R	7-15P	7-20R	7-20P	7-30R	7-30P	7-50R	7-50P

Figure 6 Receptacle configurations.

High-voltage hazards come from several sources; however, the most common are contacting overhead power lines with ladders, scaffolds, or equipment with booms, such as cranes; or contacting underground power lines with backhoes and other digging equipment. Every community has laws that require contractors to contact the local power company or independent agency before digging. The agency will mark the locations of power lines and other utilities so they can be avoided.

Main power lines carry thousands of volts. Touching them will result in severe burns or death. Even getting too close is dangerous, because high voltages can arc to a grounded conducting source, such as a person.

> **NOTE**
> Anyone who cuts power, gas, or communications lines is subject to the cost of repairing the lines and may also be charged for the equipment downtime. Imagine what the loss-of-service cost would be if you cut a cable carrying thousands of phone lines.

Before beginning any job, survey the site for overhead power lines. Remain at a minimum distance of 25 feet, then follow the guidelines for the specific voltage. Maintain an approach distance of 25 feet or greater until the line voltage has been verified. Other regulations or requirements may apply. Contact the utility company to shut down or insulate the lines if you must work near them. Treat all power lines as energized until the utility company has verified that it is safe to proceed. Equipment should be used at slower-than-normal speeds to maintain safe distances. Use warning signs. A spotter may be needed to prevent accidental contact with power lines.

1.4.0 Minimizing the Risks Associated with Work Around Electricity

The amount of current that passes through a body determines the outcome of an electrical shock. The higher the voltage, the greater the chance exists for a fatal shock.

High-voltage currents are primarily found in power lines. Electrical equipment, building wiring, and portable generators generally use low voltages

---

Did you know?
Birds on a Wire

Have you ever wondered why a bird can perch safely on an electrical wire, but other animals, like squirrels, commonly get electrocuted on power lines or at substations? It is because birds never complete the circuit by touching the power lines and ground at the same time. Squirrels may touch the charged conductors and ground (the power pole) simultaneously and be electrocuted.

Case History
Know What's Over Your Head

Five employees were constructing a chain-link fence in front of a house and directly below a 7,200V energized power line. They were installing 21-foot sections of metal top rail on the fence. One employee picked up a 21-foot section of top rail and held it up vertically. The top rail contacted the 7,200V line, and the employee was electrocuted.

The Bottom Line: Following its inspection, OSHA determined that the employee who was killed had never received any safety training from his employer nor any specific instructions for avoiding the hazards posed by overhead power lines.

Source: OSHA

(under 600V). Due to the frequency of contact, most electrocution deaths actually occur at low voltages. A lax attitude about low voltages can kill you.

Understanding the risks involved in working with electrical equipment is the first step in learning to work safely.

Insulation and grounding are two recognized means of preventing injury from electrical equipment. Conductor insulation may be provided by placing nonconductive material such as plastic around the conductor. Grounding may be achieved through the use of a direct connection to a known ground, such as a metal cold water pipe.

Consider, for example, the metal housing or enclosure around a motor or on a portable tool or generator. Such enclosures protect the equipment from dirt and moisture and prevent accidental contact with exposed wiring. However, there is a hazard associated with housings and enclosures. A malfunction within the equipment, such as damaged insulation, may increase the risk of electrical shock. Many metal enclosures are connected to a ground to eliminate this hazard. If a hot wire contacts a grounded enclosure, it results in a ground fault, which will normally trip a circuit breaker or blow a fuse. Metal enclosures and housings are usually grounded by a wire connecting them to ground. This is called the equipment grounding conductor. Most portable electric tools and appliances are grounded this way. There is one disadvantage to grounding: a break in the grounding system may occur without the user's knowledge.

Insulation on a power tool may be damaged by dropping the tool, hard use on the job, or simply by aging. If this damage causes a break in the grounding system, a shock hazard exists and the tool must be taken out of service. Double insulation may be used as additional protection on the live parts of a tool, but double insulation does not protect against defective cords and plugs, or against heavy-moisture conditions. *Figure 7* shows an example of a double-insulated ungrounded tool.

The use of a ground fault circuit interrupter (GFCI) is one method used to overcome grounding and insulation problems.

1.4.1 Ground Fault Circuit Interrupters

A ground fault circuit interrupter (GFCI) is a fast-acting circuit breaker that senses small imbalances in the circuit caused by current leakage to ground. A GFCI continually matches the amount of current going to an electrical device against the amount of current returning from the device. Whenever the two values differ by more than 5 milliamps, the GFCI interrupts the electric power within of a second. *Figure 8* shows an extension cord with a GFCI.

A GFCI will not protect you from line-to-line contact hazards such as holding either two hot wires or a hot and a neutral wire in each hand. It does provide protection against the most common form of electrical shock, which is a ground fault. It also provides protection from fires, overheating, and wiring insulation deterioration.

GFCIs can be used successfully to reduce electrical hazards on construction sites. Tripping of GFCIs—interruption of circuit flow—is sometimes caused by wet connectors and tools. Limit the amount of water that tools and connectors come into contact with by using watertight or sealable connectors. Tripping may also be caused by cumulative leakage from several tools or from extremely long circuits.

Did you know?
Current Kills

Current, measured in amps, increases if the resistance, measured in ohms, decreases and the voltage, measured in volts, remains the same. For example, a shock from a 110V plug across dry skin will be uncomfortable; however, the same 110V shock across sweaty skin can be fatal because sweaty skin offers less resistance and increases the amperage of the shock.

Case History
Grounding and Insulation Saves Lives

One worker was climbing a metal ladder to hand an electric drill to the journeyman installer on a scaffold about 5 feet above him. When the worker climbing the ladder reached the third rung from the bottom of the ladder, he received a fatal shock. He died because the cord of the drill he was carrying had an exposed wire that made contact with the conductive ladder.

The Bottom Line: Inspect all equipment before using it. Never use electrical equipment with a damaged power cord. Use fiberglass ladders when working with electrical equipment.

Source: OSHA

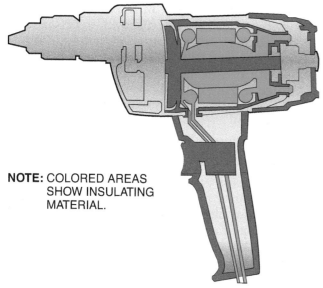

NOTE: COLORED AREAS SHOW INSULATING MATERIAL.

Figure 7 Double-insulated electric drill.

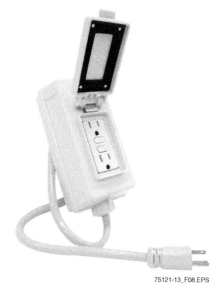

Figure 8 Extension with a GFCI.

 Do not plug a GFCI-protected device into a GFCI-protected circuit.

OSHA has determined that to minimize the risk of electrical shock at a construction site, it is the employer's responsibility to provide either of the following:

- Ground fault circuit interrupters are needed for all 120V, single-phase, 15A and 20A receptacle outlets being used if they are not part of the permanent wiring of the building or structure. That might include receptacles at the end of extension cords, even if the extension cord is plugged into the building's permanent wiring.
- A scheduled and recorded assured equipment grounding conductor program is needed to cover cord sets, receptacles that are not part of the permanent wiring of the building, and equipment connected by cords and plugs that employees may use.

1.4.2 Arc Fault Circuit Interrupters

Arc faults can be caused by many common activities, such as driving nails in walls that pierce

Working Safely Around High Voltage

If you are working near high-voltage equipment, Danger signs like the one shown here should be posted on the job site. Overhead power lines don't have any signs to remind you of the voltage hazard. Whenever you are working with ladders, scaffolds, or equipment with booms, make sure there are no power lines nearby.

There may not be any warning about buried power lines either. To prevent accidental contact with underground lines, you are required by law to have the designated authority locate and mark any buried utilities before digging in the area.

hidden wiring, loose electrical connections, or electrical cords that have been damaged by contact with furniture. An arc fault circuit interrupter (AFCI) is designed to de-energize the circuit when an arc fault is detected. AFCIs are now required by the *National Electrical Code®* in new construction in many areas of the home, including family rooms, dining rooms, living rooms, parlors, libraries, dens, bedrooms, sunrooms, recreation rooms, closets, hallways, or similar rooms or areas.

1.4.3 Assured Equipment Grounding Conductor Program

The assured equipment grounding conductor program covers all cord sets, receptacles that are not part of the permanent wiring, and equipment connected by a cord or plug that is available for employee use. OSHA requires that a written description of the employer's assured equipment grounding conductor program, including the specific procedures used, be kept at the job site. This program should outline the employer's specific procedures for the required equipment inspections, tests, and test schedule.

The required tests must be recorded, and the record maintained at the job site. The written program description and the test results must be made available to any employee upon request. The employer must designate one or more competent people to implement the program.

Electrical equipment must be visually inspected for damage or defects before each day's use. Any damaged or defective equipment must not be used until it is repaired. If you come across a questionable cord, outlet, or piece of equipment, do not use it. Show it to your supervisor immediately.

Two tests are required for the assured equipment grounding conductor program. One is a continuity test to ensure that the equipment grounding conductor is electrically continuous. It must be performed on all cord sets, nonpermanent receptacles, and cord- and plug-connected equipment that is required to be grounded. Many simple continuity testers are available, including the following:

- Lamp and battery
- Bell and battery
- Multimeter
- Receptacle tester

Figure 9 shows a continuity tester and a multimeter. This equipment can also be used to conduct the second test. In the second test, plugs and receptacles are checked to make sure that the equipment grounding conductor is connected to its proper terminal. Any equipment that fails these tests cannot be used until it is repaired and passes these tests. If you discover an unsafe tool, machine, or piece of equipment, tell your supervisor. Nicked or broken extension cords, power tools with broken housings, or equipment with modified wiring can all pose electrical hazards. They must be tagged and replaced or repaired.

1.4.4 Selecting a Meter with the Appropriate Category Rating

Safety systems are built into test equipment to protect the user from transient power spikes. The International Electrotechnical Commission (IEC) has developed a standard that defines four overvoltage installation categories: CAT I, CAT II, CAT III, and CAT IV (*Table 2*). These categories identify the hazards posed by transients; the higher the category, the greater the risk. A CAT IV meter provides the greatest degree of protection. Workers must be aware of and select meters for use in the appropriate category for the work environment. Category ratings are identified on the meter face, as shown in *Figure 10*.

1.4.5 Recognizing Hazard Boundaries

Ideally, work on or near electrical equipment would always be performed with no electrical power applied (also known as an electrically safe work condition), but that is not always possible. *National Fire Protection Association (NFPA) 70E®, Electrical Safety in the Workplace* requires special safety procedures when working on or near circuits having voltage levels of more than 50V line-to-line.

Distance is the best protection against electrical hazards. *NFPA 70E®* establishes specific limits of approach to exposed energized parts. These limits are for shock protection and are called approach boundaries. The approach boundaries and

---

Case History
GFCIs Can Save Your Life

A self-employed builder was using a metal cutting tool on a metal carport roof and was not using GFCI protection. The male and female plugs of his extension cord partially separated, and the active pin touched the metal roofing. When the builder grounded himself on the gutter of an adjacent roof, he received a fatal shock.

The Bottom Line: Always use GFCI protection and be aware of potential hazards.

Source: OSHA

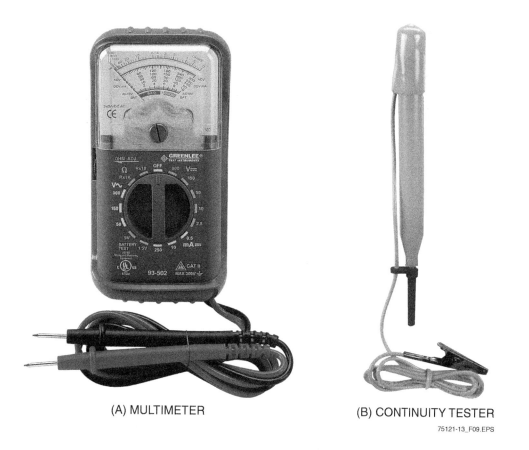

(A) MULTIMETER (B) CONTINUITY TESTER

Figure 9 Test equipment.

required PPE and other tools or equipment are determined by performing a shock hazard analysis.

When working with electrical equipment, assume that the equipment is energized until personally verifying otherwise. As noted earlier, electrical equipment presents a potential shock hazard, flash hazard, and blast hazard. *Figure 11* shows a diagram indicating the approach limits defined by *NFPA 70E®*.

Table 2 Overvoltage Installation Categories

Overvoltage Category	Installation Examples
CAT I	Electronic equipment and circuitry
CAT II	Single-phase loads such as small appliances and tools, outlets at more than 30 feet from a CAT III source or 60 feet from a CAT IV source
CAT III	Three-phase motors, single-phase commercial or industrial lighting, switchgear, busduct and feeders in industrial plants
CAT IV	Three-phase power at meter, service-entrance, or utility connection, any outdoor conductors

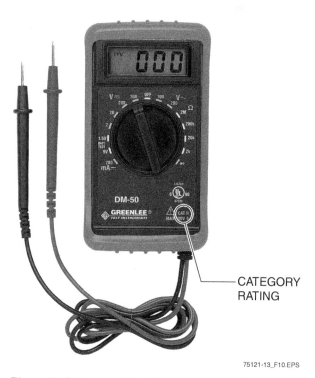

Figure 10 Category rating shown on test equipment.

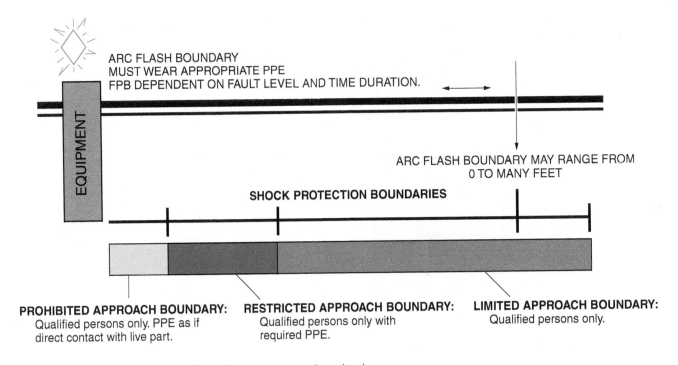

Figure 11 Approach limits.

Note that the arc flash boundary is determined independently and may be at a greater or lesser distance than shock protection boundaries. Only qualified persons are allowed within these boundaries. Unqualified personnel must be trained to stay away from potentially dangerous electrical equipment and processes.

The exposed energized component can be a wire or a mechanical component inside the electrical equipment. All boundary distances are measured from that point. When establishing boundaries, exposed movable conductors are treated differently than exposed fixed circuit parts or conductors.

Every possible electrical hazard within a work area must be analyzed and documented. Specific PPE for a given situation are based on the information gathered from the analysis of a given hazard. That documented data includes all the electrical hazards (arc flash, blast, and shock).

Temporary Grounds

Temporary grounding jumpers, such as those shown here, are used to safely conduct the maximum fault current expected at a location for the amount of time necessary for protective devices to operate to clear the fault. The application and removal of temporary grounds is considered work on energized parts and is hazardous. Many electrical shock and arc flash events have occurred during the application and removal of temporary grounds. Only trained and authorized individuals may perform this task.

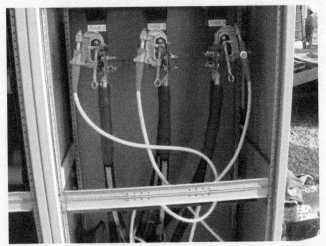

After all hazards have been documented, all personnel (qualified and unqualified) working in the area must be trained to recognize and avoid the identified hazards. Only qualified persons using all required PPE are allowed to enter and work inside the electrical hazard boundaries.

Arc Flash Boundary – When an arc flash hazard is present, an arc flash boundary must be established. This boundary is determined by how far away a person would need to be located to avoid receiving serious burns in the event of an arc flash. Anyone within the arc flash boundary is exposed to the possibility of second-degree (blistering) burns or worse and must therefore use arc-rated protective apparel as determined through arc hazard analysis and other PPE identified by *NFPA 70E*®.

Depending on the incident energy, an arc flash boundary might be within a shock protection boundary or outside of (exceed) a shock protection boundary. Many electrical safety programs establish both the arc flash boundary and the outer shock protection boundary at whatever distance is greater, as determined by the hazard analysis.

When an electrical fault causes an arc flash, the explosion produces both a fireball and a shock wave extending away from the arc flash location (the conductor or circuit). Anyone within the arc flash boundary will be exposed to searing heat as well as an extremely bright light that may cause temporary loss of vision and pain. The heat from arc flashes is often hot enough to melt metal fixtures inside the enclosure. *Figure 12* shows an arc flash.

The arc flash boundary is determined for thermal energy, but blast accompanies the arc flash. The blast creates a shock wave that can blow equipment apart and people away from the blast. Shrapnel, toxic gases, and copper vapor explode in all directions. The blast also creates sound waves that can damage hearing. The amount of current flowing through the fault affects the size of the arc flash.

Shock Protection Boundaries – There are three electrical shock protection boundaries or limits of approach. *NFPA 70E*® identifies these electrical shock boundaries as follows:

- *Limited approach boundary* – A shock protection boundary at a specified distance from an exposed energized part that can be crossed only by qualified persons. Unqualified persons may cross a limited approach boundary for purposes of on-the-job training and when escorted by a qualified person. All unqualified personnel in the area must be made aware of the hazards and warned not to cross the boundary.

Figure 12 Arc flash in process.

- *Restricted approach boundary* – A shock protection boundary which, due to its proximity to exposed energized parts, requires the use of shock protection techniques and equipment when crossed. The restricted approach boundary may be crossed only by qualified persons using the required PPE (*Figure 13*) and authorized by an energized electrical work permit. Work within the restricted approach boundary requires that rubber-insulating equipment be used within that boundary. It also requires the use of insulated tools for voltages of 1,000V and below, or live-line tools above 1,000V. Restricted approach boundaries include an added safety margin to compensate for inadvertent movement of the worker. The interaction of exposed energized parts and test equipment or hand tools is the initiator of many shock and arc flash incidents when the tool or test lead becomes part of the circuit path. Some estimates are that 75 percent or more of arc incidents begin in this manner.
- *Prohibited approach boundary* – A shock protection boundary at a specified distance from an exposed energized part, within which work is considered the same as making contact with

the energized part. Any part of the body crossing the prohibited approach boundary must be suitably insulated and protected. There is no margin for inadvertent movement in the prohibited approach boundary.

Shock protection approach boundaries to exposed energized parts are listed in *NFPA 70E*®.

WARNING! Any worker exposed to electrical hazards must be qualified to manage the hazard and using appropriate PPE. If they are not exposed to an electrical hazard but have the potential to be, they must be informed of the hazard and instructed in how to avoid it.

Figure 13 Worker using appropriate PPE.

Additional Resources

IEEE C2-2012, National Electrical Safety Code. Institute of Electrical and Electronics Engineers: New York, NY.

NFPA 70E®-2012, Standard for Electrical Safety in the Workplace®. National Fire Protection Association: Quincy, MA.

1.0.0 Section Review

1. During an electrical shock, loss of muscular control typically occurs when the current is _____.

    a. less than 5mA
    b. between 6mA and 30mA
    c. between 31mA and 50mA
    d. above 50mA

2. Approved electrical tape is a safe and acceptable method of reinsulating damaged power cords.

    a. True
    b. False

3. Accidental contact with high-voltage power lines is commonly caused by excavating equipment.

    a. True
    b. False

4. All electrical equipment must be inspected _____.

    a. when it is purchased
    b. before each day's use
    c. weekly
    d. monthly

Section Two

2.0.0 Performing Lockout/Tagout Procedures for All Energy Sources Associated with a Device or Process

Objectives

Describe the lockout/tagout procedure for all energy sources associated with a device or process.
a. Identify the steps in a typical lockout/tagout procedure.
b. Identify situations under which emergency removal of a lockout may be required.

Performance Task 1

Demonstrate how to properly use a lockout/tagout device.

Trade Terms

Energy-isolating device: Any mechanical device that physically prevents the transmission or release of energy. These include, but are not limited to, manually operated electrical circuit breakers, disconnect switches, line valves, and blocks.

Energy source: Any source of electrical, mechanical, hydraulic, pneumatic, chemical, thermal, or other energy.

Lockout: The placement of a lockout device on an energy-isolating device, in accordance with an established procedure, ensuring that the energy-isolating device and the equipment being controlled cannot be operated until the lockout device is removed.

Lockout device: Any device that uses positive means such as a lock to hold an energy-isolating device in a safe position, thereby preventing the energizing of machinery or equipment.

Tagout: The placement of a tagout device on an energy-isolating device, in accordance with an established procedure, to indicate that the energy-isolating device and the equipment being controlled may not be operated until the tagout device is removed.

Tagout device: Any prominent warning device, such as a tag and a means of attachment, that can be securely fastened to an energy-isolating device in accordance with an established procedure. The tag indicates that the machine or equipment to which it is attached is not to be operated until the tagout device is removed in accordance with the energy-control procedure.

Many accidents on a job site involve machinery and equipment. These accidents often happen because there is an uncontrolled release of energy. Uncontrolled releases of energy can cause machinery to start up while being serviced or maintained. Failure to lockout and tagout machinery before working on it is a major cause of injury and death on job sites. Some of those injuries include electrocution, amputation of body parts, and severe crushing injuries. In one instance, a maintenance worker was killed when he was knocked off balance by equipment that moved unexpectedly, knocking him into the spinning blade of a 36-inch saw.

The proper implementation of an effective lockout/tagout system can eliminate this type of hazard. It safeguards workers from unexpected releases from various energy sources. An energy source is any source of electrical, mechanical, hydraulic, pneumatic, chemical, thermal, or other energy. Energy can be stored in a system or machine even after the power is shut off. Stored energy in a system must be released or grounded to prevent unexpected operation of the machine while it is being serviced. For example, although the power has been shut off, residual air pressure remains in a compressor, which can activate pneumatic tools.

There is an additional danger if the machine contains acids, chemicals, flammable liquids, or high-temperature liquids or gases. Stored energy may cause the explosive release of these hazardous materials, resulting in serious injuries.

When anyone is working on or around any of these hazards, mechanical and other systems must be shut down, drained, or de-energized. Energy-isolating devices, such as switches, circuit breakers, valves, or other components, are used with tags and locks (*Figure 14*) to make sure that motors aren't started, valves aren't opened or closed, and any other changes that would endanger workers cannot be made.

(A) ELECTRICAL LOCKOUT

(B) PNEUMATIC LOCKOUT

Figure 14 Lockout/tagout device.

WARNING! Electrical equipment is not the only source of lockout/tagout accidents.

NOTE Some disconnect devices are equipped with keyed interlocks for protection during operation. These locks are called kirklocks and are relied upon to ensure proper sequence of operation only. They are not to be used for the purpose of locking out a circuit or system. Where disconnects are installed for use in isolation, they should never be opened under load. When opening a disconnect manually, it should be done quickly using a positive force. Again, lockouts should be used when the disconnects are open.

Each lock has its own key, and the person who puts the lock on keeps the key. That person is the only one who can remove the lock. Tags typically have the word Danger on them (*Figure 15*).

Lockout/tagout procedures protect workers from all possible uncontrolled releases of energy. In a lockout, an energy-isolating device such as a disconnect switch or circuit breaker is placed in the OFF position and locked. *Figure 16* shows some examples of the types of lockout devices that can

Figure 15 Typical safety tags.

be used with key or combination locks. Multiple lockout devices (*Figure 17*) are used when more than one person has access to the equipment. In some instances, as with work involving valves, a chain and lock can be used to hold the valve in place and keep it from being turned.

In a tagout, components that power up equipment and machinery, such as switches, are set in a safe position and a written warning or tagout device is attached to them (*Figure 18*).

It is important to know how to protect yourself and your co-workers on the job. Everyone must be aware of what activities are being done on the site and understand how to perform them safely. These are the most common safeguards that should be used to keep the job site safe.

- Never operate any device, valve, switch, or piece of equipment that has a lock or a tag attached to it.
- Use only tags that have been approved for your job site.
- If a device, valve, switch, or piece of equipment is locked out, make sure the proper tag is attached.
- Lock out and tag all electrical systems.
- Lock out and tag pipelines containing acids, explosive fluids, and high-pressure steam.
- Lock and tag motorized vehicles and equipment when they require repair or are being replaced. Also, disconnect or disable any starting devices.

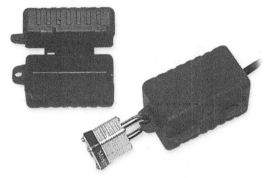

ELECTRICAL PLUG LOCKOUT

CIRCUIT BREAKER LOCK

BALL VALVE LOCKOUT

ELECTRICAL SWITCH LOCKOUT

Figure 16 Lockout devices.

Case History
Near Miss

While maintenance workers were de-coupling an actuator from a 100-psi steam control valve, the packing gland was removed from the valve. The valve was not under lockout/tagout control. As a result, the wrong fasteners were removed. This caused the packing gland follower fasteners and the packing bore components to be expelled from the valve due to residual steam pressure. Fortunately, no one was injured, but the valve was destroyed and had to be replaced.

The Bottom Line: A lockout/tagout program can prevent accidents from the unexpected release of steam, spring, or hydraulic energy. Lockout/tagout is not limited to electrical safety.

Source: US Department of Energy

Figure 17 Multiple lockout/tagout device.

2.1.0 Steps in a Typical Lockout/Tagout Procedures

The exact procedures for lockout/tagout may vary at different companies and job sites. Ask your supervisor to explain the lockout/tagout procedure on your job site. You must know this procedure and follow it. This is for your safety and the safety of your co-workers. If you ever have questions about lockout/tagout procedures, ask your supervisor. An effective lockout/tagout program should include the following tasks:

- An inspection of equipment by a trained individual who is thoroughly familiar with the equipment operation and associated hazards.
- Identification and labeling of lockout devices.
- The purchase of locks, tags, and blocks.
- A standard written operating procedure that is followed by all employees.

The following is an example of a typical lockout/tagout procedure. It is made up of the following four components:

Figure 18 Placing a lockout/tagout device.

- Preparation for lockout/tagout
- Sequence for lockout/tagout
- Restoration of energy
- Emergency removal authorization

2.1.1 Preparation for Lockout/Tagout

When preparing for a lockout/tagout, it is important to follow these steps.

Step 1 Check the procedures to ensure that no changes have been made since you last used a lockout/tagout.

75121-13 Electrical Safety

Module Six 19

Case History

Flood Warning

The threaded male end of a ½-inch pressure-relief valve (PRV) on a sprinkler system riser broke off during an attempt to re-pipe the discharge side of the valve. This caused approximately 550 gallons of water to be discharged into a utility closet and surrounding office areas. There were no injuries as a result of this event, but there was significant property damage.

The Bottom Line: A lockout/tagout should have been performed on the sprinkler system.

Source: US Department of Energy

Step 2 Identify all authorized and affected employees involved with the pending lockout/tagout.

> **NOTE**
> There is a difference between turning off a machine and actually disengaging a piece of equipment. When turning off a control switch, you are opening a circuit; however, there is still electrical energy at the switch. A short in the switch or someone turning on the machine may start it running again.

2.1.2 Sequence for Lockout/Tagout

To perform a lockout/tagout, follow these steps:

Step 1 Notify all authorized and affected personnel that a lockout/tagout is to be used and explain why it is necessary.

Step 2 Shut down the equipment or system using the normal Off or Stop procedures.

Step 3 Lock out all energy sources and test disconnects to be sure they cannot be moved to the On position. Next, open the control cutout switch. If there is no cutout switch, block the magnet in the switch open position before working on electrically operated equipment or apparatus such as motors or relays. Remove the control wire.

Step 4 Lock and tag the required switches or valves in the open position. Each authorized employee must affix a separate lock and tag, if necessary.

Step 5 Dissipate any stored energy by attaching the equipment or system to ground.

Step 6 Verify that the test equipment is functional using a known power source.

Step 7 Confirm that all switches are in the open position and use test equipment to verify that all components are de-energized.

Step 8 If it is necessary to temporarily leave the area, upon returning, retest to ensure that the equipment or system is still de-energized.

2.1.3 Restoration of Energy

After work is done on the machinery or equipment, use these steps to restore energy.

Step 1 Completely reassemble and secure the equipment or system.

Step 2 Confirm that all equipment and tools, including shorting probes, are accounted for and removed from the equipment or system.

Step 3 Replace and reactivate all of the safety controls.

Case History

Harmful Energy

An employee was attempting to correct an electrical problem involving two non-operational lamps. He did not shut off the power at the circuit breaker panel nor did he test the wires to see if they were live. He was killed instantly when he fell from the ladder after grabbing the two live wires with his left hand.

The Bottom Line: Do not work on electrical circuits unless they are positively de-energized and tagged out.

Source: OSHA

Lockout Method

Think About It

Which lockout method would you rather use if your life was at stake?

(A) TAPE

(B) CIRCUIT BREAKER LOCK

Step 4 Remove the locks and tags from the isolation switches. Each employee must remove his or her own lock and tag.

Step 5 Notify all affected personnel that the lockout/tagout has ended and the equipment or system will be re-energized.

Step 6 Operate or close the isolation switches to restore energy.

Case History

Testing Can Save Your Life

On a site in Georgia, electricians found energized switches after the lockout of a circuit panel in an older system that had been upgraded several times. The existing wiring did not match the current site drawings. A subsequent investigation found many such situations in older facilities.

The Bottom Line: Never rely solely on drawings. The circuit must be tested after lockout to verify that it is de-energized.

Source: OSHA

P.R.O.P.E.R. Lockout/Tagout Procedures

Using P.R.O.P.E.R lockout/tagout procedures is the best way to prevent accidents, injury, and death.

P – Process shut down.
R – Recognize energy sources.
O – Off (shut energy sources off).
P – Place locks and tags.
E – Release stored energy.
R – Return controls to neutral.

2.2.0 Situations Under Which Emergency Removal of a Lockout May Be Required

There are times when emergency removal of a lockout/tagout device is required. The device can only be removed by an authorized supervisor. Follow these guidelines if a lockout/tagout device is left secured, and the authorized employee is absent, or the key is lost. Follow company-specific guidelines for lockout removal.

- The entire line must be examined to ensure that all hazards have been mitigated and it is safe to remove the lock. All efforts must be exhausted to locate the employee and/or key.
- The authorized employee must be informed that the lockout/tagout device has been removed.
- Written verification of the action taken, including informing the authorized employee of the removal, must be recorded in the job journal.

Additional Resources

IEEE C2-2012, National Electrical Safety Code. Institute of Electrical and Electronics Engineers: New York, NY.
   OSHA 29 CFR, Part 1910, Occupational Safety and Health Standards for General Industry. Occupational Safety and Health Administration: Washington, DC.

2.0.0 Section Review

1. The first step in a lockout/tagout procedure is _____.
    a. removing the power to the work area
    b. powering down the equipment
    c. ensuring that the procedure has not changed
    d. applying tags to the equipment to be affected

2. If an employee involved a lockout/tagout leaves for the day without removing their lock, _____.
    a. all work must stop until the next scheduled shift
    b. the lock cannot be removed until the employee returns to the job site
    c. an authorized supervisor may remove the lock
    d. they will be fired immediately

Summary

Safety must be your primary concern at all times so that you do not become either the victim of an accident or the cause of one. Safety requirements and safe work practices are provided by OSHA and your employer. You must adhere to all safety requirements and follow your employer's safe work practices and procedures. Also, you must be able to identify the potential safety hazards at your job site. The consequences of unsafe job-site conduct can be expensive, painful, and even deadly.

Here is a summary of the important things to remember about electrical safety:

- Use three-wire extension cords and protect them from damage. Never use damaged cords.
- Make sure that panels, switches, outlets, and plugs are grounded.
- Never use metal ladders near any source of electricity.
- Always inspect electrical power tools before using them.
- Never operate any piece of electrical equipment that has a lockout device attached to it.
- Use a GFCI.

Lockouts and tagouts protect workers from all possible sources of energy. In a lockout, an energy-isolating device such as a disconnect switch or circuit breaker is placed in the Off position and locked.

In a tagout, components that power up equipment and machinery, such as switches, are set in a safe position and a written warning or tagout device is attached to them

Failure to follow your company's lockout/tagout procedures can result in serious injury or death.

Review Questions

1. The heart's normal pumping rhythm can be interrupted by an electric current of as small as _____.
   a. 75mA
   b. 5A
   c. 7.5A
   d. 50A

2. Shock can occur by coming into contact within any of the following, *except* _____.
   a. water and a ground wire
   b. both wires of the electrical circuit
   c. one wire of a circuit and the ground
   d. both the ground and a metallic object energized through contact with the circuit

3. A type of burn in which the injuries may not be immediately apparent is a(n) _____.
   a. arc burn
   b. electrical burn
   c. powder burn
   d. thermal contact burn

4. In an electrical accident, the higher the voltage passing through the body, the _____.
   a. greater the resistance of the body's horny layer
   b. greater the chance for a fatal shock
   c. greater the chance the worker will be thrown out of harm's way
   d. lower the amperage required to kill you

5. Contact with low-voltage conductors causes the majority of electrocution deaths.
   a. True
   b. False

6. A ground fault circuit interrupter is a fast-acting circuit breaker that senses small imbalances in the circuit and quickly shuts off the electricity.
   a. True
   b. False

7. A GFCI will provide protection from all of the following, *except* _____.
   a. ground faults
   b. line-to-line contact hazards
   c. overheating
   d. wiring insulation deterioration

8. Lockout/tagout systems only protect workers from electrical hazards.
   a. True
   b. False

9. Circuit breakers are examples of _____.
   a. energy-isolating devices
   b. energy-removal devices
   c. lockout/tagout devices
   d. multiple lockout devices

10. The final step in the energy resoration procedure is to _____.
    a. restore the energy
    b. account for all tools
    c. secure the equipment
    d. remove the locks and tags

Trade Terms Introduced in This Module

Arc blast: An explosion similar to the detonation of dynamite that occurs during an arc flash incident.

Arc fault: A high-energy discharge between two or more conductors.

Arc fault circuit interrupter (AFCI): A device intended to provide protection from the effects of arc faults by recognizing characteristics unique to arcing and by functioning to de-energize the circuit when an arc fault is detected.

Arc flash boundary: An approach limit at a distance from exposed energized electrical conductors or circuit parts within which a person could receive a second-degree burn if an electrical arc flash were to occur.

Assured equipment grounding conductor program: A detailed plan specifying an employer's required equipment inspections and tests and a schedule for conducting those inspections and tests.

Bolted fault: A short circuit or electrical contact between two conductors at different potentials, in which the impedance or resistance between the conductors is essentially zero.

Breakdown voltage: The voltage at which an insulator has a breakdown and ceases to act as a resistor.

Energy-isolating device: Any mechanical device that physically prevents the transmission or release of energy. These include, but are not limited to, manually operated electrical circuit breakers, disconnect switches, line valves, and blocks.

Energy source: Any source of electrical, mechanical, hydraulic, pneumatic, chemical, thermal, or other energy.

Equipment grounding conductor: A wire that connects metal enclosures and containers to ground.

Fibrillation: Very rapid, irregular contractions of the muscle fibers of the heart that result in the heartbeat and pulse going out of rhythm with each other.

Ground fault circuit interrupter (GFCI): A fast-acting circuit breaker that senses small imbalances in the circuit caused by current leakage to ground and, in a fraction of a second, shuts off the electricity.

Grounding: The process of directly connecting an electrical circuit to a known ground to provide a zero-voltage reference level for the equipment or system.

Insulation: The practice of placing nonconductive material such as plastic around the conductor to prevent current from passing through it.

Lockout: The placement of a lockout device on an energy-isolating device, in accordance with an established procedure, ensuring that the energy-isolating device and the equipment being controlled cannot be operated until the lockout device is removed.

Lockout device: Any device that uses positive means such as a lock to hold an energy-isolating device in a safe position, thereby preventing the energizing of machinery or equipment.

Shock hazard: A dangerous condition associated with the possible release of energy caused by contact or approach to energized electrical conductors or circuit parts.

Tagout: The placement of a tagout device on an energy-isolating device, in accordance with an established procedure, to indicate that the energy-isolating device and the equipment being controlled may not be operated until the tagout device is removed.

Tagout device: Any prominent warning device, such as a tag and a means of attachment, that can be securely fastened to an energy-isolating device in accordance with an established procedure. The tag indicates that the machine or equipment to which it is attached is not to be operated until the tagout device is removed in accordance with the energy-control procedure.

Additional Resources

This module presents thorough resources for task training. The following resource material is suggested for further study.

IEEE C2-2012, National Electrical Safety Code. Institute of Electrical and Electronics Engineers: New York, NY.

NFPA 70E®-2012, Standard for Electrical Safety in the Workplace®. National Fire Protection Association: Quincy, MA.

OSHA 29 CFR, Part 1910, Occupational Safety and Health Standards for General Industry. Occupational Safety and Health Administration: Washington, DC.

Figure Credits

Salisbury by Honeywell, Module opener, Figures 12, 13
Physio-Control, Inc., Figure 3
Flexon Ind. - Division of U.S. Wire & Cable Corporation, Figure 8
Greenlee / A Textron Company, Figures 9A, 10
Amprobe, Figure 9B

Jim Mitchem, SA02
Topaz Publications, Inc., Figure 14A
Panduit Corp., Figure 14B
Accuform Signs, Figure 15
Honeywell Safety Products, Figures 16, 18, SA03B
Mike Powers, SA03A

Section Review Answer Key

Answer	Section Reference	Objective
Section One		
1. b	1.1.0	1a
2. b	1.2.0	1b
3. a	1.3.0	1c
4. b	1.4.3	1d
Section Two		
1. c	2.1.1	2a
2. c	2.2.0	2b

NCCER CURRICULA — USER UPDATE

NCCER makes every effort to keep its textbooks up-to-date and free of technical errors. We appreciate your help in this process. If you find an error, a typographical mistake, or an inaccuracy in NCCER's curricula, please fill out this form (or a photocopy), or complete the online form at **www.nccer.org/olf**. Be sure to include the exact module ID number, page number, a detailed description, and your recommended correction. Your input will be brought to the attention of the Authoring Team. Thank you for your assistance.

Instructors – If you have an idea for improving this textbook, or have found that additional materials were necessary to teach this module effectively, please let us know so that we may present your suggestions to the Authoring Team.

NCCER Product Development and Revision
13614 Progress Blvd., Alachua, FL 32615

Email: curriculum@nccer.org
Online: www.nccer.org/olf

❏ Trainee Guide ❏ Lesson Plans ❏ Exam ❏ PowerPoints Other _____

Craft / Level: _____ Copyright Date: _____

Module ID Number / Title: _____

Section Number(s): _____

Description: _____

_____

_____

_____

_____

Recommended Correction: _____

_____

_____

_____

_____

Your Name: _____

Address: _____

_____

Email: _____ Phone: _____

27406-14

Introduction to Construction Equipment

Overview

Backhoes, forklifts, aerial lifts, generators, and compressors are commonly required on a job site. While some of this equipment requires specially trained operators, other equipment is often used by carpenters and other personnel working on the site. In all cases, knowing how to safely and properly operate a piece of equipment in accordance with the manufacturer's instructions is extremely important.

Module Seven

Trainees with successful module completions may be eligible for credentialing through the NCCER Registry. To learn more, go to **www.nccer.org** or contact us at **1.888.622.3720**. Our website has information on the latest product releases and training, as well as online versions of our *Cornerstone* magazine and Pearson's product catalog.

Your feedback is welcome. You may email your comments to **curriculum@nccer.org**, send general comments and inquiries to **info@nccer.org**, or fill in the User Update form at the back of this module.

This information is general in nature and intended for training purposes only. Actual performance of activities described in this manual requires compliance with all applicable operating, service, maintenance, and safety procedures under the direction of qualified personnel. References in this manual to patented or proprietary devices do not constitute a recommendation of their use.

Copyright © 2015 by NCCER, Alachua, FL 32615, and published by Pearson Education, Inc., New York, NY 10013. All rights reserved. Printed in the United States of America. This publication is protected by Copyright, and permission should be obtained from NCCER prior to any prohibited reproduction, storage in a retrieval system, or transmission in any form or by any means, electronic, mechanical, photocopying, recording, or likewise. To obtain permission(s) to use material from this work, please submit a written request to NCCER Product Development, 13614 Progress Blvd., Alachua, FL 32615.

From *Construction Craft Laborer, Level Two, Trainee Guide*, Third Edition. NCCER.
Copyright © 2015 by NCCER. Published by Pearson Education. All rights reserved.

27406-14
INTRODUCTION TO CONSTRUCTION EQUIPMENT

Objectives

When you have completed this module, you will be able to do the following:

1. State the safety precautions associated with construction equipment.
    a. Identify safety precautions when transporting construction equipment.
    b. Identify safety precautions related to interlocking and hydraulic systems.
    c. Identify safety precautions to observe when fueling construction equipment.
    d. Identify safety precautions related to batteries of construction equipment.
2. Identify and explain the safe operation and use of various pieces of construction equipment.
    a. Explain the safe operation of aerial lifts.
    b. Explain the safe operation of skid-steer loaders.
    c. Explain the safe operation of generators.
    d. Explain the safe operation of compressors.
    e. Explain the safe operation of compactors.
    f. Explain the safe operation of forklifts.
    g. Explain the safe operation of backhoes.

Performance Tasks

This is a knowledge-based module; there are no performance tasks.

Trade Terms

Aerial lift
Boom lift
Circuit breaker
Compactor
Compressor
Drive/steer controller
Forklift
Generator
Powered industrial trucks
Proportional control
Safety interlock
Scissor lift
Skid-steer loader
Twist-lock outlet
Whip check

Industry-Recognized Credentials

If you're training through an NCCER-accredited sponsor, you may be eligible for credentials from NCCER's Registry. The ID number for this module is 27406-14. Note that this module may have been used in other NCCER curricula and may apply to other level completions. Contact NCCER's Registry at 888.622.3720 or go to www.nccer.org for more information.

WARNING! The Core Curriculum *Basic Safety* module and the Safety Review Questions at the back of this module must be successfully completed before operating any equipment. Additionally, appropriate personal protective equipment must be worn while operating or working near any equipment. Every operation must be performed under the direct supervision of your instructor or other qualified personnel.

Code Note

Codes vary among jurisdictions. Because of the variations in code, consult the applicable code whenever regulations are in question. Referring to an incorrect set of codes can cause as much trouble as failing to reference codes altogether. Obtain, review, and familiarize yourself with your local adopted code.

Contents

Topics to be presented in this module include:

1.0.0 Construction Equipment Safety Precautions 1
  1.1.0 Transporting Equipment 1
  1.2.0 Safety Interlocks and Hydraulic Systems 1
  1.3.0 Fueling Safety 2
  1.4.0 Battery Safety 2
2.0.0 Safe Operation and Use of Construction Equipment 4
  2.1.0 Aerial Lifts 4
    2.1.1 Aerial Lift Assemblies 5
    2.1.2 Aerial Lift Operator Qualifications 6
    2.1.3 Typical Aerial Lift Controls 6
    2.1.4 Aerial Lift Safety Precautions 7
    2.1.5 Aerial Lift Operating Procedure 7
    2.1.6 Aerial Lift Operator's Maintenance Responsibility 8
  2.2.0 Skid-Steer Loader 10
    2.2.1 Skid-Steer Loader Assemblies 10
    2.2.2 Skid-Steer Loader Operator Qualifications 11
    2.2.3 Skid-Steer Loader Controls 12
    2.2.4 Skid-Steer Loader Safety Precautions 12
    2.2.5 Skid-Steer Loader Operating Procedure 13
    2.2.6 Skid-Steer Loader Operator's Maintenance Responsibility 14
  2.3.0 Generators 14
    2.3.1 Generator Assemblies 14
    2.3.2 Typical Generator Controls 14
    2.3.3 Generator Safety Precautions 16
    2.3.4 Generator Operating Procedure 17
    2.3.5 Generator Operator's Maintenance Responsibility 18
  2.4.0 Air Compressors 20
    2.4.1 Compressor Assemblies 20
    2.4.2 Typical Compressor Controls 20
    2.4.3 Compressor Safety Precautions 21
    2.4.4 Compressor Operating Procedure 22
    2.4.5 Air Compressor Operator's Maintenance Responsibility 23
  2.5.0 Compaction Equipment 23
    2.5.1 Compaction Equipment Assemblies 24
    2.5.2 Compaction Equipment Operator Qualifications 24
    2.5.3 Typical Compaction Equipment Controls 24
    2.5.4 Compaction Equipment Safety Precautions 25
    2.5.5 Compaction Equipment Operating Procedure 25
    2.5.6 Compactor Operator's Maintenance Responsibility 25

Contents (continued)

2.6.0 Forklifts 26
    2.6.1 Forklift Assemblies 26
    2.6.2 Forklift Operator Qualifications 26
    2.6.3 Typical Forklift Controls 26
    2.6.4 Forklift Safety Precautions 28
    2.6.5 Forklift Operation 28
    2.6.6 Forklift Operator's Maintenance Responsibility 29
2.7.0 Backhoe 30
    2.7.1 Backhoe Operator Qualifications 30
    2.7.2 Backhoe Controls 30
    2.7.3 Backhoe Safety Precautions 33
    2.7.4 Backhoe Operating Procedure 34
    2.7.5 Backhoe Operator's Maintenance Responsibility 35
Appendix Safety Review Questions 39

Figures

Figure 1   Aerial lifts 5
Figure 2   Aerial lift components 5
Figure 3   Example of an aerial lift operator's checklist 8
Figure 4   Example of an aerial lift maintenance and inspection schedule 9
Figure 5   Skid-steer loader 10
Figure 6   Major assemblies of a typical skid-steer loader 11
Figure 7   Skid-steer loader instrument panel 12
Figure 8   Skid-steer loader operator compartment controls 13
Figure 9   Example of a skid steer loader service schedule 15
Figure 10 Generators 16
Figure 11 Tow-behind generator/light tower 16
Figure 12 Generator engine control panel 17
Figure 13 Generator control panel 17
Figure 14 Example of a generator preventive maintenance schedule 19
Figure 15 Types of compressors 20
Figure 16 Tow-behind compressor components 20
Figure 17 Compressor control panel 21
Figure 18 Properly securing air hoses 22
Figure 19 Types of compactors 24
Figure 20 Flat-plate compactor components 24
Figure 21 Fixed-mast rough terrain forklifts 26
Figure 22 Telescoping-boom forklift 27
Figure 23 Example of a forklift operator's daily checklist 31
Figure 24 Backhoe 32

Section One

1.0.0 Construction Equipment Safety Precautions

Objective

State the safety precautions associated with construction equipment.

a. Identify safety precautions when transporting construction equipment.
b. Identify safety precautions related to interlocking and hydraulic systems.
c. Identify safety precautions to observe when fueling construction equipment.
d. Identify safety precautions related to batteries of construction equipment.

Trade Term

Safety interlock: A safety mechanism used to prevent incorrect control operation.

Safety is the primary concern regardless of the type of equipment being operated. Most of the types of equipment discussed in this module share common safety precautions. These safety precautions are discussed in this section. As each piece of construction equipment is discussed, reefer back to this overview and apply these precautions to the equipment being covered. Safety precautions that are unique to a particular type of equipment will also be discussed.

The manufacturer for each piece of construction equipment provides safety precautions with the equipment. These must be must be observed whenever you are operating the equipment. All of the equipment presented in this module can cause serious injury or death if applicable safety precautions are not followed.

> **NOTE**
> Site-specific training for construction equipment may be required by the authority having jurisdiction (AHJ). Always consult with the AHJ prior to operating any construction equipment.

1.1.0 Transporting Equipment

When transporting equipment between jobs, ensure that these safety guidelines are followed:

- Park, unload, and load the equipment from a trailer on level ground.
- To prevent tipping, connect the trailer to the tow vehicle before loading or unloading.
- Ensure the trailer hitch is properly engaged and locked down. Make sure safety chains are properly attached to the tow vehicle.
- Follow the manufacturer's requirements for towing and transporting a piece of equipment.
- Explosive separation of a tire and/or rim parts can cause injury or death. Ensure that a properly trained technician removes and services tires on construction equipment.
- Always maintain the correct tire pressure for the trailer and tow vehicle.
- Inspect the tires and wheels before towing. Do not tow with high or low tire pressures, cuts, excessive wear, bubbles, damaged rims, or missing lug bolts or nuts.
- Do not inflate tires with flammable gases or from systems using an alcohol injector.

> **NOTE**
> Be careful and plan ahead. Always think safety. Create a job hazard analysis before undertaking any task. Accidents do not just happen; they are generally caused by carelessness and unsafe practices.

1.2.0 Safety Interlocks and Hydraulic Systems

Some equipment comes with safety interlocks to prevent unsafe operation. Observe the following guidelines when using this type of equipment:

- Interlock systems must not be modified. Interlock systems help prevent incorrect control operation.
- If an interlock system fails, contact the manufacturer's repair representative immediately.
- Runaway equipment is always possible; learn how to use all the controls and emergency procedures.

27406-14 Introduction to Construction Equipment

WARNING!

Anyone operating power equipment must be properly trained in its use and the associated safety precautions. Some localities may require the operator to be trained and qualified in the use of certain equipment. Always adhere to the following general safety practices:

- Read and fully understand the operator's manual and follow the instructions and safety precautions for the equipment.
- Complete a job hazard analysis (JHA) prior to working on an assigned task.
- Gasoline, diesel, and liquefied petroleum (LP) engines produce deadly carbon monoxide. If they are used in a closed environment, the exhaust must be vented to the outdoors.
- Know the capacity and operating characteristics of the equipment.
- Inspect the equipment before each use to make sure everything is in proper working order. Have any defects repaired before using it. Never modify or remove any part of the equipment unless authorized to do so by the manufacturer.
- Check for hazards above, below, and all around the job site. Always maintain a safe distance from power lines and other electrical hazards.
- Learn as much about the work area as you can before beginning work.
- Fasten seat belt or operator restraints before starting.
- Set up warning barriers and keep others away from the equipment and job site.
- Know where to get assistance, if needed.
- Learn how to use a first-aid kit and fire extinguisher and know where they are located.

When operating equipment that uses a hydraulic system, observe the following precautions:

- Before disconnecting any hydraulic lines, relieve system pressure by cycling controls.
- Before pressurizing the system, be sure all connections are tight and the lines are undamaged.
- Do not perform any work on the equipment unless you are authorized and qualified to do so.
- Check for leaks using a piece of cardboard or wood; never use bare hands.

1.3.0 Fueling Safety

Fueling safety precautions are applicable to all construction equipment using petroleum-based fuels, such as gasoline, diesel fuel, or liquefied petroleum (LP). Make sure you follow the applicable manufacturer's instructions before fueling the equipment. The following are general safety precautions associated with fueling construction equipment:

- Never fill the fuel tank with the motor running, while smoking, or near an open flame.
- Be careful not to overfill the tank or spill fuel. If fuel is spilled, clean it up immediately.
- Use clean fuel only.
- Do not operate the equipment if fuel has been spilled inside or near the unit.
- Ground the fuel funnel or nozzle against the filler neck by keeping them in contact to prevent sparks.
- Replace the fuel tank cap after refueling.

1.4.0 Battery Safety

Battery safety precautions are applicable to all construction equipment equipped with a battery for starting, or for operating electronic components. General safety precautions associated with batteries are as follows:

- Chargers can ignite flammable materials and vapors. They should not be used near fuels, grain dust, solvents, or other flammables.
- To reduce the possibility of electric shock, a charger should only be connected to a properly grounded single-phase outlet. Do not use an extension cord longer than 25'.

Hydraulic Fluid Leaks Can Be Dangerous

Hydraulic fluid escaping through pinholes in hoses and hose fittings can be almost invisible. Escaping hydraulic fluid under pressure can penetrate the skin or eyes, causing serious injury. When inspecting light construction equipment, hydraulic hoses, and hose couplings for fluid leaks, always wear protective clothing and eye protection. Do not trace for leaks along hydraulic hoses and hydraulic system components with bare hands; instead, use a piece of cardboard or wood. In the event that hydraulic fluid penetrates the skin, seek immediate medical attention from a doctor familiar with this type of injury.

- If the battery is frozen, do not charge or attempt to jump-start the equipment, as the battery may explode.
- Battery and fuel fumes could ignite and cause explosions and burns. Keep batteries away from flames or sparks and do not smoke.

> **NOTE**
> Be careful and plan ahead. Always think safety. Create a job hazard analysis before undertaking any task. Accidents do not just happen; they are generally caused by carelessness and unsafe practices.

1.0.0 Section Review

1. Tracing along hydraulic hoses for pinhole leaks should be done using ____.
   a. nitrile gloves
   b. a piece of wood or cardboard
   c. bare hands
   d. a UV light source

2. A mechanism designed to prevent incorrect control operation is called a safety ____.
   a. override
   b. backup
   c. interlock
   d. monitor

3. A deadly exhaust product produced by diesel, gasoline, or LP-gas engines is ____.
   a. carbon dioxide
   b. nitrous oxide
   c. hydrogen peroxide
   d. carbon monoxide

4. If a battery is frozen, it is best to jumpstart the equipment.
   a. True
   b. False

Section Two

2.0.0 Safe Operation and Use of Construction Equipment

Objective

Identify and explain the safe operation and use of various pieces of construction equipment.
a. Explain the safe operation of aerial lifts.
b. Explain the safe operation of skid-steer loaders.
c. Explain the safe operation of generators.
d. Explain the safe operation of compressors.
e. Explain the safe operation of compactors.
f. Explain the safe operation of forklifts.
g. Explain the safe operation of backhoes.

Trade Terms

Aerial lift: A mobile work platform designed to transport and raise personnel, tools, and materials to overhead work areas.

Boom lift: Aerial lift with a single arm that extends a work platform/enclosure capable of holding one or two workers; some models may have an articulated joint.

Circuit breaker: A device designed to protect circuits from overloads and to enable them to be reset after tripping.

Compactor: A machine used to compact soil to prevent the settling of the soil after construction.

Compressor: A motor-driven machine used to supply compressed air for pneumatic tools.

Drive/steer controller: One-handed joystick lever control used to control the speed and steering of an aerial lift.

Forklift: A machine designed to facilitate the movement of bulk items around the job site.

Generator: A machine designed to generate electricity.

Powered industrial trucks: See *Forklift*.

Proportional control: A control that increases speed in proportion to the movement of the control.

Scissor lift: Aerial lift used to raise a work enclosure vertically by means of crisscrossed supports.

Skid-steer loader: A small, highly maneuverable machine equipped with a bucket for moving stone, dirt, and bulk items around a job site.

Twist-lock outlet: A type of electrical connector.

Whip check: A cable attached to the ends of an air supply hose and a tool, or to the ends of two hoses that are connected to each other, to prevent the hose from whipping uncontrollably when charged with compressed air.

Construction equipment, such as aerial lifts, skid-steer loaders, generators, air compressors, compactors, forklifts, and backhoes, are frequently used on job sites for a wide variety of applications. While it is important to have a general understanding of the safe operation of this equipment, always review the operator's manual prior to operation of any specific piece of equipment.

2.1.0 Aerial Lifts

Aerial lifts are used to raise and lower workers to and from elevated work areas. There are two main types of lifts: scissor lifts and boom lifts. Various models of both types of lifts are available. Some are transported on a vehicle to a job site where they are unloaded. Others are trailer-mounted and towed to the job site by a vehicle, or they may be permanently mounted on a vehicle. Depending on the design, they can be used for indoor work, outdoor work, or both. *Figure 1* shows two common types of aerial lifts.

Boom lifts have a single arm that extends a work platform/enclosure capable of holding one or two workers. Some models have a jointed (articulated) arm that allows the work platform to be positioned both horizontally and vertically. Scissor lifts raise a work enclosure vertically by means of crisscrossed supports.

Most models of aerial lifts are self-propelled, allowing workers to move the platform as work is performed. The power to move these lifts is provided by electric motors or gasoline, diesel, and LP engines.

Scissor-Lift Tires

To prevent surfaces from being marked or scuffed, nonmarking tires are available for scissor lifts used indoors. This feature is particularly desirable when using the lift on wood or tile floor surfaces.

BOOM LIFT   SCISSOR LIFT

Figure 1 Aerial lifts.

2.1.1 Aerial Lift Assemblies

Aerial lifts typically consist of three major assemblies: the platform, a lifting mechanism, and the base. *Figure 2* shows these components for a scissor lift.

The platform of an aerial lift is constructed of a tubular steel frame with a skid-resistant deck surface, railings, toeboard, and midrails. Entry to the platform is normally from the rear. The entry opening is closed either with a chain or a spring-returned gate with a latch. The work platform may also be equipped with a retractable extension platform.

The lifting mechanism is raised and lowered either by electric motors and gears, or by one or more single-acting hydraulic lift cylinder(s). A pump, driven by either an AC (alternating current) or DC (direct current) motor, provides hydraulic power to the cylinder(s).

The base provides a housing for the electrical and hydraulic components of the lift. These components are normally mounted in swing-out access trays. This allows easy access when performing maintenance or repairs to the unit. The

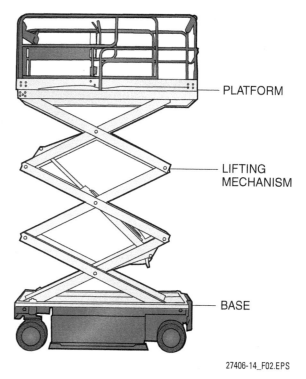

Figure 2 Aerial lift components.

base also contains the axles and wheels for moving the assembly. In the case of a self-propelled platform, electrical or hydraulic motors drive two or more of the wheels to allow movement of the lift from one location to another. Brakes are incorporated into one or more of the wheels to prevent inadvertent movement of the lift.

2.1.2 Aerial Lift Operator Qualifications

Only trained and authorized workers may use an aerial lift. Safe operation requires the operator to understand all limitations and warnings, operating procedures, and operator requirements for maintenance of the aerial lift. The operator must meet the following requirements:

- Understand and be familiar with the operator's manual for the lift being used.
- Understand all procedures and warnings within the operator's manual, and those posted on decals on the aerial lift.
- Be familiar with the employer's work rules and all related Occupational Safety and Health Administration (OSHA) and other government safety regulations.
- Demonstrate this understanding and operate the specific model of aerial lift during training in the presence of a qualified trainer.

2.1.3 Typical Aerial Lift Controls

Aerial lifts typically have two sets of controls: platform controls and ground controls. The ground controls include a switch to select whether the operation will be controlled from the platform or the ground. The ground controls override the platform controls.

The worker on the platform is designated as the operator who will use the platform controls. In some aerial lifts, operation from the platform is limited to raising and lowering the platform. On others, the aerial platform is designed to be driven by the operator using controls provided on the platform. In either case, the platform controls are located within the platform assembly.

The platform controls of a scissor lift typically consist of the following switches and controls:

- *Drive/steer controller* – This is a one-handed joystick lever control for controlling the speed and steering of the aerial lift. This is usually a deadman switch that returns to neutral and locks when released. The handle is moved forward to drive the aerial platform forward. The platform speed is determined by how far forward the handle is moved. The handle is moved backward to drive the aerial platform backward; the speed is adjusted by the extent of the backward movement. Releasing the stick stops all motion. Steering is performed by depressing a rocker switch on the top of the stick in the desired direction of travel, either right or left.
- *Thumb rocker switch* – Allows the lift to turn in the direction the switch is depressed. Without the switch depressed, the lift moves straight ahead.
- *Diagnostic panel* – Provides information on hours the lift has been in service and provides diagnostic capabilities for ease in troubleshooting.
- *Platform-up/platform-down buttons* – When depressed, these buttons move the platform vertically up or down, respectively. When released, the switch stops the movement of the platform.
- *Emergency Stop button* – When the button is pushed in, all functions of the lift cease. To restore power, pull the button out.
- *Low- and high-speed lift-enable buttons* – When pushed and held down, these buttons enable the lift circuit. The appropriate button must be held down to raise or lower the platform. Releasing the button stops the motion of the platform.
- *Drive-speed selector* – Allows the lift to move forward or backward at a slower pace.
- *Horn button* – Allows the operator to sound the horn.

The ground controls for the scissor lift contain the following switches and controls:

- *Emergency Stop button* – When the button is pushed in, all functions of the lift cease. To restore power, pull the button out.
- *Breaker* – For the electrical circuits in the lift.
- *Key switch* – Turn the key to the Platform position, and the platform controls will operate. Turn the key to the Ground position, and the ground controls will operate.
- *Diagnostic panel* – Displays the hours the lift has been in service, and provides diagnostic capabilities for ease in troubleshooting.
- *Menu buttons* – Allow the operator to diagnose problems and view information about hours in service, etc.

The basic controls for an aerial lift with a hydraulic system generally include the following:

- *Emergency lowering valve* – Allows for platform lowering in the event of an electrical/hydraulic system failure.
- *Free-wheeling valve* – Opening this valve allows hydraulic fluid to flow through the wheel motors. This allows the aerial lift to be pushed by hand, and prevents damage to the motors when

the aerial lift is moved between job-site locations. There are usually strict limits as to how fast the aerial lift may be moved without causing damage to hydraulic system components.

Other types of controls that may be available on aerial lifts include:

- *Parking-brake manual release* – Allows manual release of the parking brake. This control should only be used when the aerial lift is located on a level surface.
- *Safety bar* – Used to support the platform lifting hardware in a raised position during maintenance or repair.

Outriggers (stabilizing hardware) are required on some aerial lifts. At a minimum, this hardware includes a stabilizer leg, stabilizer lock pin and cotter key, and a stabilizer jack at each corner of the aerial lift. Outriggers help to support the lift and minimize the potential for the lift to tip, which could cause injury or death.

2.1.4 Aerial Lift Safety Precautions

Safety precautions unique to aerial lifts are listed here. Remember that the safety precautions discussed earlier also apply. Each manufacturer provides specific safety precautions in the operator's manual provided with the equipment. OSHA standard 1926.453 defines and governs the safe use of aerial lifts. Important safety guidelines include the following:

- Avoid using the lift outdoors in stormy weather or in strong winds.
- Erect barricades to prevent people from walking beneath the lift.
- Use personal fall arrest equipment (body harness and lanyard) as required for the type of lift being used. Use approved anchorage points. Do not attach belts to adjacent poles, structures, or equipment.
- Do not use an aerial lift on uneven ground.
- Lower the lift and lock it into place before moving the equipment. Also, lower the lift, shut off the engine, set the parking brake, and remove the key before leaving it unattended.
- Stand firmly on the floor of the basket or platform. Do not lean over the guardrails of the platform, and never stand on the guardrails. Do not sit or climb on the edge of the basket or use planks, ladders, or other devices to attain additional height.
- Lift controls must be tested each day prior to use to verify that the controls are in safe working order.

2.1.5 Aerial Lift Operating Procedure

This section describes operations for a scissor lift. Operating procedures vary from model to model. The operator must be familiar with all operating procedures, controls, safety features, and associated safety precautions before operating any type of aerial lift.

Aerial Lift Operator Controls

Operator controls for aerial lifts differ depending on the manufacturer and model of the lift. Shown here are the ground control station and platform control box for an electric-drive scissor lift produced by one major lift manufacturer.

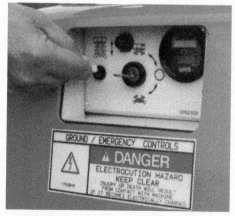

GROUND CONTROL STATION

PLATFORM CONTROL BOX

A drive/steer controller is a type of *proportional control* that causes an aerial lift to move. The farther the control is moved, the more power is applied to the motor and the faster the aerial lift will move.

The following tasks must be performed before operating the aerial lift:

- Carefully read and fully understand the operating procedures in the operator's manual and all warnings and instruction decals on the work platform.
- Do not exceed the load capacity of the aerial lift.
- Check for obstacles around the work platform and in the path of travel, such as holes, drop-offs, debris, ditches, and soft fill. Operate the aerial lift only on firm surfaces.
- Check overhead clearances. Make sure to stay at least 10' away from overhead power lines.
- Use aerial lifts only for their intended purpose. Do not use aerial lifts to hoist joists, beams, or any other materials.
- Make sure batteries are fully charged (if applicable).
- Make sure all guardrails are in place and locked in position.
- Perform an inspection and complete an operator's checklist.
- Never make unauthorized modifications to the components of an aerial lift.

Figure 3 shows an example of an operator's checklist.

To drive the aerial lift forward, the operator selects the drive position with the drive/steer controller and moves it forward. The speed can be adjusted by continuing to move the controller forward until the desired speed is reached. By releasing the controller, the forward motion of the lift is stopped. To drive in reverse, the lever is moved in the opposite direction (backward).

2.1.6 Aerial Lift Operator's Maintenance Responsibility

The operator must ensure that the work platform has been properly maintained before using it.

WARNING! Death or injury to workers, and damage to equipment, can result if the platform is not maintained in good working condition. Inspection and maintenance should be performed by personnel who are authorized to perform such procedures.

OPERATOR'S CHECKLIST

INSPECT AND/OR TEST THE FOLLOWING DAILY OR AT BEGINNING OF EACH SHIFT

___ 1. OPERATING AND EMERGENCY CONTROLS
___ 2. SAFETY DEVICES
___ 3. PERSONNEL PROTECTIVE DEVICES
___ 4. TIRES AND WHEELS
___ 5. OUTRIGGERS (IF EQUIPPED) AND OTHER STRUCTURES
___ 6. AIR, HYDRAULIC, AND FUEL SYSTEM(S) FOR LEAKS
___ 7. LOOSE OR MISSING PARTS
___ 8. CABLES AND WIRING HARNESSES
___ 9. DECALS, WARNINGS, CONTROL MARKINGS, AND OPERATING MANUALS
___ 10. GUARDRAIL SYSTEM
___ 11. ENGINE OIL LEVEL (IF SO EQUIPPED)
___ 12. BATTERY FLUID LEVEL
___ 13. HYDRAULIC RESERVOIR LEVEL
___ 14. COOLANT LEVEL (IF SO EQUIPPED)

Figure 3 Example of an aerial lift operator's checklist.

If the operator is not responsible for the maintenance of the aerial lift, the operator must at least perform the daily checks. *Figure 4* shows an example of a maintenance and inspection schedule for a typical aerial lift. The daily checks should be performed at the beginning of each shift or at the beginning of the day if only one shift is worked. The aerial lift must never be used until these checks are completed satisfactorily. Any deficiencies found during the daily inspection must be corrected before using the lift.

Figure 4 is divided into four major categories. Each category is further divided into the components that should be checked daily, weekly, monthly, every three months, every six months, and yearly. Footnotes (the numbers in parentheses) following each component note the type of inspection required. For example, under the Electrical category, battery fluid level should be checked daily or at the beginning of each shift. The (1) refers the operator to the Notes portion of the schedule. Note 1 tells the operator to perform a visual inspection of the battery fluid level.

	Daily	Weekly	Monthly	3 Months	6 Months	*Annually
Mechanical						
Structural damage/welds (1)	✓					✓
Parking brakes (2)	✓					✓
Tires and wheels (1)(2)(3)	✓					✓
Guides/rollers/slides (1)	✓					✓
Railings/entry chain/gate (2)(3)	✓					✓
Bolts and fasteners (3)	✓					✓
Rust (1)			✓			✓
Wheel bearings (2) King pins (1)(8)	✓					✓
Steer cylinder ends (8)				✓		✓
Electrical						
Battery fluid level (1)	✓					✓
Control switches (1)(2)	✓					✓
Cords and wiring (1)	✓					✓
Battery terminals (1)(3)	✓					✓
Terminals and plugs (3)	✓					✓
Generator and receptacle (2)	✓					✓
Limit switches (2)	✓					✓
Hydraulic						
Hydraulic oil level (1)	✓					✓
Hydraulic leaks (1)	✓					✓
Lift/lowering time (10)				✓		✓
Hydraulic cylinders (1)(2)		✓				✓
Emergency lowering (2)	✓					
Lift capacity (7)				✓		✓
Hydraulic oil/filter (9)					✓	✓
Miscellaneous						
Labels (1)(11) Manual (12)	✓					✓

Notes:
(1) Visually inspect. (3) Check tightness. (5)(6) N/A. (8) Lubricate.
(2) Check operation. (4) Check oil level. (7) Check relief-valve setting. Refer to serial-number nameplate. (9) Replace.

(10) General specifications.
(11) Replace if missing or illegible.
(12) Proper operating manual *must* be in the manual tube.
*Record inspection date

Figure 4 Example of an aerial lift maintenance and inspection schedule.

Different models of aerial lifts have different maintenance requirements. Always refer to the operator's manual to determine exactly what checks are required before operation.

General maintenance rules applicable to any aerial lift include the following:

- Disconnect the battery ground negative (–) lead before performing any maintenance.
- Ensure the safety bar is in the proper position before performing maintenance with the platform in the raised position.

Preventive maintenance is easier and less expensive than corrective maintenance.

2.2.0 Skid-Steer Loader

Skid-steer loaders (*Figure 5*), also called skid loaders or skidsters, are small, highly maneuverable machines designed for moving dirt, crushed stone, and gravel around the job site. The standard configuration skid-steer loader is equipped with a bucket for handling these materials. Additional functions can be performed with the skid-steer loader using optional attachments available from the manufacturer.

Skid-steer loaders are equipped with gasoline, diesel, or LP engines. The lift and bucket attachment is operated using a self-contained hydraulic system. A single operator performs the operations.

2.2.1 Skid-Steer Loader Assemblies

Figure 6 shows the major assemblies of a typical skid-steer loader.

Protection of Underground Utilities

In order to avoid damage or interruptions to underground utilities caused by digging, it is mandatory to contact the various utilities for a utility stake-out prior to any digging. Most states have a One-Call Notification System center that makes the digging notification process easy by calling a single phone number or by contacting them on the internet. Typically, the call must be made at least two or three working days before the digging is to begin.

Common skid-steer loader components include the following:

- Seat bar
- Grab handles
- Joystick or steering levers
- Safety tread
- Lift and tilt cylinders
- Operator seat
- Operator cab/roll-over protective structure/falling-object protective structure (ROPS/FOPS)
- Lift-arm support device
- Lift arms
- Tires

(A) TRACKED

(B) WHEELED

Figure 5 Skid-steer loader.

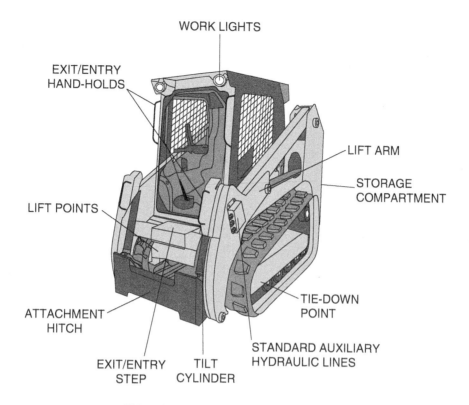
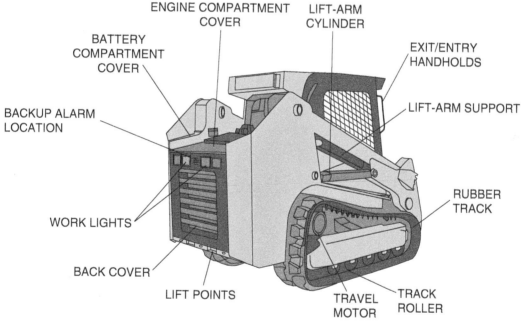

Figure 6 Major assemblies of a typical skid-steer loader.

2.2.2 Skid-Steer Loader Operator Qualifications

The operator is responsible for reading and understanding the safety manual and operator's manual provided with the equipment. You must be qualified and authorized before operating a skid-steer loader. Before operation, the following requirements must be met:

- The operator must have a complete understanding of the written instructions provided by the manufacturer.
- The operator must successfully complete a training program that includes actual operation of the skid-steer loader equipment.
- The operator must know the safety rules and regulations for the job site.

2.2.3 Skid-Steer Loader Controls

Figure 7 shows the typical instrument panel controls located on a skid-steer loader. The panel typically displays the following items:

- *Ambient-light sensor* – Senses the amount of light and adjusts the screen contrast accordingly.
- *Engine oil-pressure warning indicator* – Lights when the engine oil pressure is too low, indicating that the loader should be immediately shut off.
- *Coolant-temperature warning indicator* – Lights when the coolant temperature is too high.
- *Battery-voltage warning indicator* – Lights when the alternator is not charging the battery.
- *Hydraulic oil-temperature warning indicator* – Lights when the hydraulic temperature is too high.
- *High-speed travel range indicator* – Lights when the high-speed travel range is activated.
- *Preheat indicator* – Lights when the ignition switch is in the Run position and engine preheat is required; goes out when the engine preheat is complete.
- *Coolant temperature indicator* – Displays the coolant temperature.
- *Display select button* – Used for screen selection and display/operation configuration.
- *Lift-arm float indicator* – Lights when the lift-arm float is activated.
- *Hydraulic oil-filter warning indicator* – Lights when the hydraulic oil filter requires service.
- *Engine air-filter restriction indicator* – Lights when the engine air filter requires service.
- *Parking-brake indicator* – Lights when the parking brake is applied.
- *Seat-belt reminder indicator* – Lights when the engine is started as a reminder to fasten the seat belt.
- *Fuel gauge* – Displays the amount of fuel in the fuel tank.
- *Display screen* – Displays status/configuration information.

Figure 8 shows the typical operator compartment controls. The typical operator compartment controls include the following:

- *Control joysticks* – Control direction of travel and turning.
- *Safety bars* – When in the raised position, applies the parking brake, locks out work hydraulics, and prevents the starting of the engine.
- *Throttle knob* – Controls the engine speed.
- *Throttle pedal* – Provides supplemental speed control.
- *Work-lights switch* – Controls the work lights.
- *Ignition switch* – Controls the ignition to start and run the loader.
- *Electrical accessory socket* – Provides 12-volt accessory outlet.

2.2.4 Skid-Steer Loader Safety Precautions

The following is a list of safety rules specific to the operation of a skid-steer loader. Remember that the safety precautions discussed earlier also apply.

- Use the seat belt, and if provided, the driver-restraint bar.
- Lower the lift arms and put the bucket or attachment flat on the ground before exiting the skid-steer loader.
- Never permit riders on the equipment.
- Never overload the bucket with material that could injure someone if spilled.
- Never attempt to work the controls unless properly seated.
- Keep the load bucket level as the lift arms are moved and as the loader moves up or down on slopes or ramps.
- Carry the bucket low for stability and visibility when traveling.
- Operate at a slow enough speed to make sure you have control of the load at all times.
- Travel slowly over rough or slippery surfaces and on hillsides.
- Never ram the bucket into a material pile; most loaders have more force at slow speeds.

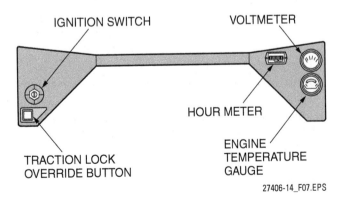

Figure 7 Skid-steer loader instrument panel.

Attachments Used With a Skid-Steer Loader

A skid-steer loader can be used to perform the functions of many other machines when equipped with optional attachments available from the manufacturer.

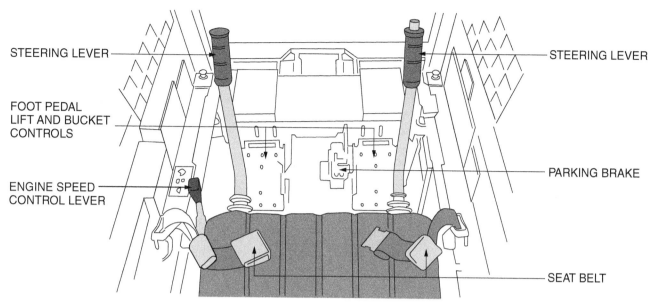

Figure 8 Skid-steer loader operator compartment controls.

- Always have the lift arms down when traveling or turning.
- Raise the load slowly at an even rate, and be ready to lower the load quickly if the loader becomes unstable.
- When traveling up or down a slope, keep the load as low as possible. Keep the heavy end of the loader pointed uphill. For a loaded bucket, keep the bucket and load facing up the slope. For an empty bucket, keep the bucket facing down the slope.
- Never undercut a steep bank.
- Use caution when handling materials such as rocks or site debris, which could fall into the operator's compartment if the lift arms are raised too high and the bucket is rolled back too far.
- Be familiar with your work area and use the proper equipment for the job.

Skid-Steer Loaders

A skid-steer loader does not have a steering wheel, and the wheels do not actually turn to change the direction of vehicle movement. Instead, a joystick controls the wheels on each side of the machine. If one joystick is pushed harder than the other, the skid loader will turn in the direction with the most force applied. It is possible to do a complete turn in place by pushing one joystick forward and pulling the other one back.

Skid-Loader Buckets

The skid loader can have either a smooth bucket or a toothed bucket. The smooth bucket can be used to pick up loose sand, gravel, and similar material, as well as for scraping. The toothed bucket is good for digging and for spreading material.

2.2.5 Skid-Steer Loader Operating Procedure

This section covers the basic operation of a skid-steer loader. Each manufacturer provides detailed operating instructions with its equipment. Refer to the specific operator's manual for the equipment. The general outline given here is intended to explain the operator controls.

The operator should perform a complete inspection of the skid-steer loader and correct any deficiencies before beginning operation. Begin by checking the machine for broken, missing, or damaged parts. Any problems found must be corrected before beginning operation. Inspect the track for damage and wear. If equipped with tires, check them for cuts, bulges, and correct pressure, if required.

The following items must also be checked to make sure they are working properly:

- Check the parking brake for proper operation.
- Check the hydraulic system for leaks.
- Check all fluid levels.
- Check all safety features.

Ensure that the operator's area, control joysticks, pedal, steps, and grab handles are working correctly, and are free of debris.

The following is a discussion of the controls used to start a skid-steer loader. The exact starting procedure for a machine is contained in the operator's manual. This generic discussion will help you understand the functions of the controls.

For safety reasons, the operator must be aware of the following when starting the skid-steer loader:

- When getting on and off the equipment, the operator must maintain three-point contact with the steps and grab handles provided. The operator should face the machine when getting on or off. Never jump on or off the machine, or attempt to get on or off a moving machine. Do not use control joysticks as handholds when getting on or off the machine.
- Before starting, walk completely around the machine to make sure no one is under, on, or close to the machine. Notify others in the area that you are starting the machine and do not start the engine until everyone is clear. Sit in the operator's seat and adjust the seat so that all controls can be reached for proper operation. Fasten the seat belt and any other operator-restraining device the machine is equipped with. Engage the parking brake and put all controls in the Neutral or Park position.
- Start the engine following the procedure in the operator's manual. Observe the gauges, instruments, and warning lights for proper operation and readings. Operate each control pedal or joystick to verify that the lift and bucket functions are satisfactory.
- Make sure that both speed and direction are controllable before moving the machine. Operate the travel control lever(s) to verify both forward and reverse motion. Test the steering (both right and left) while moving slowly in a clear safe area.

The following controls are used to shut down a skid-steer loader. The exact shutdown procedure for a specific machine is contained in the operator's manual. This is a generic description to explain the functions of the skid-steer loader controls. For safety reasons, the operator must be aware of and perform the following when shutting down the skid-steer loader:

- To stop the machine, lower the bucket and other attachments so they are flat on the ground. Position the operator controls in Neutral. Engage the parking brake.
- Let the engine idle for a short time to allow the engine to cool down. Turn the engine off. Cycle the hydraulic controls to release system pressure.
- Unbuckle the seat belt or other operator restraint. Remove the ignition key and lock covers and closures.
- Exit the loader safely, using the same guidelines as presented for safe entry.

2.2.6 Skid-Steer Loader Operator's Maintenance Responsibility

Maintenance work must be done at regular intervals. Failure to do so will result in excessive wear and early failure. The manufacturer supplies a service schedule with the skid-steer loader. The schedule contains required service checks and the time intervals when these checks should be performed. *Figure 9* is an example of a typical skid-steer loader service schedule. Following the manufacturer's service schedule will extend the life of the equipment and improve its performance.

2.3.0 Generators

Generators are used to provide electrical power and/or lighting at the job site. Generators are available in many different sizes and configurations (*Figures 10 and 11*). They range from small portable generators to large installed backup power systems for emergency power generation. Construction-site requirements for power vary depending on the scale of the work being performed. This section describes a typical tow-behind generator used to provide electrical power to a job site.

2.3.1 Generator Assemblies

A tow-behind generator is comprised of the following main components:

- Towing assembly and frame
- Protective covers and doors
- Engine and generator assemblies
- Operator control panels

2.3.2 Typical Generator Controls

Tow-behind generators have operator controls for the engine and an electrical control panel to control and monitor the power produced by the generator. *Figure 12* shows a typical engine control panel. This panel provides the operator with the controls required to start and stop the motor and monitor critical motor functions. Controls and indicators include the following:

SERVICE SCHEDULE		HOURS					
ITEM	SERVICE REQUIRED	8-10	50	100	250	500	1000
Engine Air Cleaner	Empty the dust cup & replace the filter element as needed.	✓					
Engine Oil	Check the oil level & add oil as needed.	✓					
Engine Cooling System	Clean debris from oil cooler, radiator, & grill. Check coolant level in recovery tank and add as needed.	✓					
Track	Check for damage and wear.	✓					
Indicators, Gauges & Lights	Check for correct operation of all indicators, gauges, & lights.	✓					
Seat Belt and Seat Bar	Check the condition of seat belt. Check the seat belt & pedal interlocks for correct operation. Clean dirt and debris from moving parts.	✓					
Safety Signs (Decals) and Safety Tread	Check for damaged signs and safety treads. Replace any signs or safety treads that are damaged or worn.	✓					
Lift Arm and Pivot Pin and Wedges	Lubricate with multipurpose lithium-based grease.	✓					
Engine Fuel Filter	Remove the water from the filter.	✓					
Hyd. Fluid, Hose & Tube Lines	Check fluid level & add as needed. Check for damage & leaks. Repair & replace as needed.	✓					
Battery	Check cables & electrolyte level.		✓				
Control Pedals & Steering	Check for correct operation repair or adjust as needed.		✓				
Wheel Nuts	Check for loose wheel nuts & tighten to 40–45 ft.-lbs. (54-61 Nm) torque.		✓				
Parking Brake	Check operation.		✓				
Alternator Belt	Check tension & adjust as needed.		✓				
Engine Fuel Filter	Remove the water from the filter.		✓				
Engine Oil &	Replace oil & filter. Use CD or better grade oil and Melroe filter.				✓		
Hydra...	Replace filter element				✓		
	Clean the sp...				✓		

Figure 9 Example of a skid-steer loader service schedule.

- *Engine overspeed indicator* – Monitors engine speed. If speed exceeds the manufacturer's set limit, the engine will shut down automatically.
- *High engine-temperature indicator* – Indicates when the engine coolant temperature has exceeded the manufacturer's set limit. The engine will shut down automatically.
- *Low engine-oil pressure indicator* – Indicates when the engine oil pressure has fallen below the manufacturer's preset limit. The engine will shut down automatically.
- *Alternator not charging* – Indicates the engine alternator is not outputting enough energy to charge the unit's battery.
- *Ignition switch* – Place this switch in the On position to run; place it to the Off position to stop.
- *Start switch* – Pressing this switch activates the engine starting motor.
- *Safety-circuit bypass* – Pressing this switch bypasses automatic shutdowns when starting the engine.

- *Emergency stop* – Pressing this switch causes the engine to shut down with no other operator action required.
- *Engine tachometer gauge* – Indicates engine speed in revolutions per minute (rpm).
- *Engine oil pressure gauge* – Displays engine oil pressure.
- *Engine coolant temperature* – Indicates engine coolant temperature.
- *Ammeter gauge* – Indicates the charge rate of the engine alternator.
- *Fuel-level gauge* – Indicates the level of fuel in the fuel tank(s).
- *Hour meter* – Records total engine operating hours. It is used to determine the maintenance schedule on the unit.

The second control panel in a tow-behind generator is the generator control panel. This panel contains meters, monitor switches, a voltage regulator, circuit breakers, and receptacles to control

27406-14 Introduction to Construction Equipment

(A) TOW-BEHIND GENERATOR

(B) PORTABLE GENERATOR

27406-14_F10.EPS

Figure 10 Generators.

27406-14_F11.EPS

Figure 11 Tow-behind generator/light tower.

- *20A circuit breakers* – Circuit breakers for the 120V ground fault circuit interrupter (GFCI) outlets.
- *120V GFCI duplex outlets* – Used for connecting additional loads or equipment to the generator.
- *50A circuit breakers* – For the 120V/240V twist-lock outlets.
- *120V/240V twist-lock outlets* – Used for connecting additional loads or equipment to the generator.

2.3.3 Generator Safety Precautions

The safe operation of a generator is the operator's responsibility. Remember that the general safety precautions presented earlier also apply.

and monitor the output of the generator. *Figure 13* shows a typical generator control panel with the following controls and indicators:

- *Generator ground-connection lug* – For connecting a good earth ground per any local, state, or *National Electrical Code*® (*NEC*®) guidelines before starting the generator.
- *Generator output-connection lugs* – Allow appropriate loads to be wired directly to the generator.
- *Battery charger connection* – Allows for 120VAC (voltage alternating current) input to power an onboard battery charger.
- *Main circuit breaker* – Disconnects power to the connection lugs; will not disconnect power to the convenience outlets when the engine is running.
- *Auxiliary light switches* – These switches operate control panel and interior lights.
- *Air-filter meter* – Shows the condition of the air filter when the engine is running.

> **WARNING!** High voltages are present when the generator is operating. Exercise caution to avoid electric shock.

The following are general safety precautions that must be observed when operating a generator:

- Do not operate electrical equipment when standing in water or on wet ground, or if your hands or shoes are wet.
- Grounding should be performed in compliance with local electrical codes and in accordance with the manufacturer's manual.
- Use a generator as an alternate power supply only after the main feed at the service entrance panel has been opened and locked.

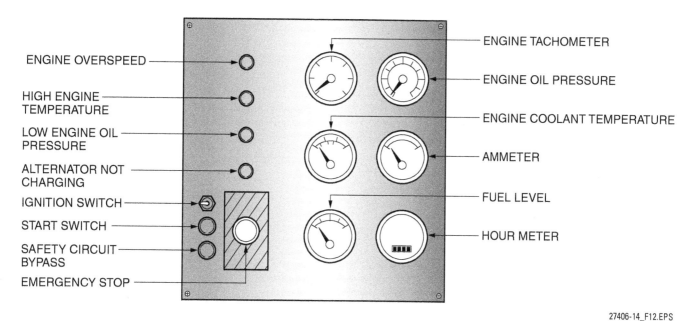

Figure 12 Generator engine control panel.

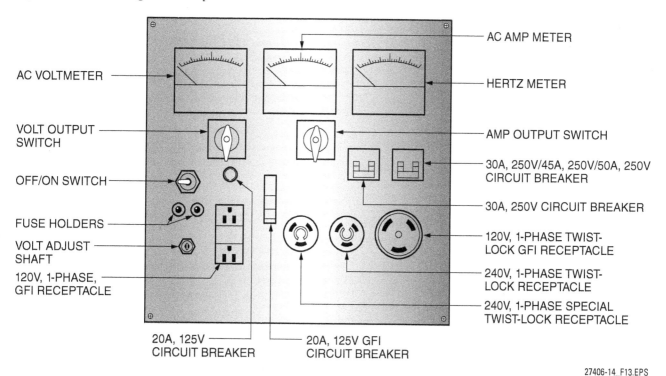

Figure 13 Generator control panel.

- Do not change voltage selection while the engine is running. Voltage selection, adjustment, and electrical connections may only be performed by qualified personnel.
- Do not exceed the generator power (watt) rating during operation.
- Never make electrical connections with the unit running.
- If welding is required on the unit, follow the manufacturer's instructions to prevent damage to the circuitry.

2.3.4 Generator Operating Procedure

Before operating the generator, local grounding requirements must be investigated and followed. The generator set can produce high voltages, which can cause severe injury or death to personnel and damage to equipment. The generator should have proper internal and external grounds as required by the *National Electrical Code*® (*NEC*®). A qualified, licensed electrical contractor, knowledgeable in local codes, should be consulted. Follow all manufacturer's requirements to ensure that the generator is connected properly before operation.

The following steps describe the basic operation of a generator. Each manufacturer provides detailed operating instructions with its equipment. Refer to the operator's manual provided with the generator for specific instructions. The general guidelines given here are intended to help you understand the operator controls. The operator should perform a complete inspection of the generator and correct any deficiencies before beginning operation.

Generator Setup and Pre-Operational Inspection

A typical procedure for setting up a generator for use at the job site is as follows:

Step 1 Place the unit as level as possible. Follow the manufacturer's directions concerning equipment placement and any special considerations for equipment location.

Step 2 Disconnect the generator from the towing vehicle.

Step 3 Chock the wheels of the generator.

Step 4 Unlock the jack and lower it to the service position.

Step 5 Lock the jack in the service position.

Step 6 Disconnect the safety chains and crank the jack to raise the coupling off the hitch.

Before putting the generator into operation, check it for evidence of arcing on or around the control panel. If any arcing is noted, the problem must be located and repaired before beginning operation. Check for loose wiring or loose routing clamps within the housing.

Perform the following additional checks before operating:

- Check for excessive engine fluid leaks on or under the engine.
- Check the engine oil and coolant levels.
- Check for frayed or loose fan belts, hoses, or wiring insulation.
- Check for obstructions in air vents.
- Check the fuel-tank level. Fill the tank at the end of each day to prevent condensation.
- Check the engine air filter.

Maintain the engine in accordance with the manufacturer's recommendations.

Generator Setup, Startup, and Shutdown

A typical procedure for setting up and starting a generator is as follows:

Step 1 Open the voltage selector door and position the voltage selector switch to the correct voltage. Close and latch the voltage selector door. Verify that the voltage regulator switch is on. Verify that all external electrical power requirements are turned off and all receptacle breakers and main breakers are on. Ensure that the direct hookup terminal door is closed.

Step 2 Position the ignition switch to On and start the engine. Allow the engine to warm up for three to five minutes. Check the control panel readouts for proper voltages. Power is now present and available for use. Close the access doors for maximum cooling while the engine is running.

Step 3 When shutting down the generator, verify that all external electrical loads are turned off. Place the ignition switch to Off. The engine will stop running.

2.3.5 Generator Operator's Maintenance Responsibility

Figure 14 shows a typical preventive maintenance schedule for a generator. The schedule is laid out with the performance period across the top and the items to be checked listed down the side of the page. For example, under the Daily column, the following items need to be checked:

- Evidence of arcing around electrical terminals
- Loose wire-routing clamps
- Engine oil and coolant levels
- Grounding circuit
- Instruments
- Fan belts, hoses, and wiring insulation
- Air vents
- Fuel/water separator
- Service air indicator

Proper maintenance will extend the life and performance of the equipment.

Grounding Portable and Vehicle-Mounted Generators

The grounding requirements for portable and vehicle-mounted generators are listed in the *National Electrical Code®*, *NEC®* Section 250.34. *NEC®* Section 250.20 governs the grounding of portable generators for applications that supply fixed wiring systems.

PREVENTIVE MAINTENANCE SCHEDULE						
	DAILY	WEEKLY	MONTHLY/ 150 HOURS	3 MONTHS/ 250 HOURS	6 MONTHS/ 500 HOURS	YEARLY 1000 HOURS
Evidence of Arcing around Electrical Terminals	✓					
Loose Wire-Routing Clamps	✓					
Engine Oil and Coolant Levels	✓					
Proper Grounding Circuit	✓					
Instruments	✓					
Frayed/Loose Fan Belts, Hoses, and Wiring Insulation	✓					
Obstructions in Air Vents	✓					
Fuel/Water Separator (drain)	✓					
Service Air Indicator	✓					
Precleaner Dumps		✓				
Tires		✓				
Battery Connections		✓				
Engine Radiator (exterior)			✓			
Air Intake Hoses and Flexible Hoses			✓			
Fasteners (tighten)			✓			
Emergency Stop Switch Operation			✓			
Engine Protection Shutdown System			✓			
Diagnostic Lamps			✓			
Voltage Selector and Direct Hookup Interlock Switches				✓		
Air-Cleaner Housing				✓		
Control Compartment (interior)					✓	
Fuel Tank (fill at end of each day)					Drain	
Fuel/Water Separator Element					Replace	
Wheel Bearings and Grease Seals					Repack	
Engine Shutdown System Switches (settings)						✓
Exterior Finish				(as needed)		
Engine			Refer to Engine Operator			
Decals			Replace decals if removed, damaged, or missing.			
✓ = Check or Clean (and Adjust or Replace, if necessary)						

Figure 14 Example of a generator preventive maintenance schedule.

Extension Cords

It is important that the extension cords used to supply power from a portable generator to electric power tools or other devices have an adequate current-carrying capacity for the job. If undersized extension cords are used, excessive voltage drops will result, causing excessive heating of the extension cords and the portable tools, as well as additional generator loading.

 27406-14 Introduction to Construction Equipment   Module Seven 19

2.4.0 Air Compressors

Compressors are used to provide compressed air for tools at the job site. Compressors are available in many different sizes and configurations (*Figure 15*). They range from portable units to large installed units for industrial applications. Construction-site requirements for compressed air vary depending on the scale of the work being performed. This section describes a typical tow-behind compressor.

2.4.1 Compressor Assemblies

Figure 16 shows a tow-behind compressor. Its primary components include:

- Towing frame assembly
- Protective cover and doors
- Engine and compressor assembly
- Operator control panel located behind the access door

(A) TOW-BEHIND COMPRESSOR

(B) SMALL, PORTABLE COMPRESSOR

Figure 15 Types of compressors.

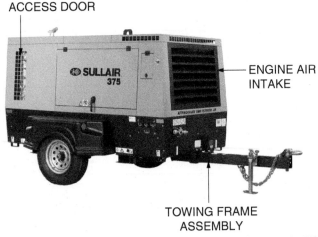

Figure 16 Tow-behind compressor components.

2.4.2 Typical Compressor Controls

Compressors normally have operator controls for the engine and compressor (*Figure 17*). This section discusses controls associated with a tow-behind compressor. This panel provides the operator with the controls required to start and stop the motor and monitor critical motor and compressor functions.

Controls and indicators typically include the following:

- *Engine diagnostic display panel* – Displays information related to the compressor engine. Information such as engine overspeed, high temperature, low engine oil, alternator not charging, and similar information is displayed.
- *Compressor diagnostic display panel* – Displays information related to the compressor assembly.
- *Engine power switch* – Placed in the On position to run; placed in the Off position to stop.
- *Warm-up controls* – Activates the engine starting motor.
- *Diagnostic service port* – Allows a cable to be inserted to provide additional diagnostic capabilities.
- *Air-pressure gauge* – Indicates the output air-pressure level.
- *Pressure selector switch* – Allows high or low pressure to be selected for use with different equipment.

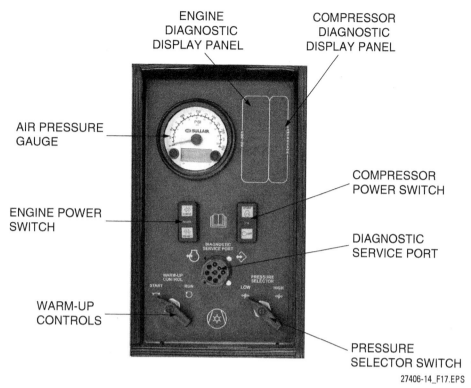

Figure 17 Compressor control panel.

2.4.3 Compressor Safety Precautions

The safe operation of the air compressor is the operator's responsibility. The following are general safety precautions that are applicable to air compressors. Remember the safety precautions discussed earlier also apply. Refer to the operator's manual for the particular compressor being used.

- Wear eye protection.
- Compressed-air venting to the atmosphere can cause hearing damage. Make sure proper hearing protection is worn by everyone in the area.
- Ether is highly flammable. Do not inject ether into a hot engine or an engine equipped with a glow-plug type of preheater.
- Do not inject ether into the compressor air filter or a common air filter for the engine and compressor.
- Do not attempt to move the compressor or lift the drawbar without adequate personnel or equipment to handle the weight.
- Do not exceed the machine's air-pressure rating.
- Do not use compressed air for breathing.
- Never direct compressed air at anyone.
- Ensure no pressure is in the system before removing the compressor filler cap.
- Do not use tools that are rated for a pressure lower than that provided by the compressor.
- Be sure all connections are securely made and hoses under pressure are secured to prevent whipping. Use appropriate safety devices to secure hoses.
- Do not weld or perform any modifications on the air-compressor receiver tank.
- Use a safe, nonflammable solvent when cleaning parts.

The air hose must be connected properly and securely. An unsecured air hose can come loose and whip around violently, causing serious injury. To prevent this, some fittings require the use of whip checks to keep them from coming loose (see *Figure 18A*). A whip check is a strong steel cable with loops on either end. The loops are attached to the ends of the air supply hose and the tool, or to the ends of two hoses that are connected to each other. OSHA regulations require the use of whip checks or other means of preventing air supply hoses from whipping if compressed air is flowing through them when they are disconnected.

Tools that require very high air pressure, such as jackhammers, use hoses with a larger diameter than those used for other types of air tools. These hoses are fitted with cam-operated twist locks secured by a removable pin (see *Figure 18B*). Be sure to follow the manufacturer's instructions when using these locks to connect high-pressure hoses

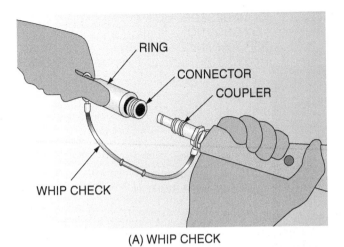

(A) WHIP CHECK

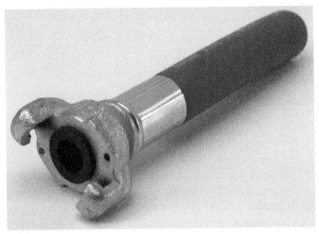

(B) CAM-OPERATED TWIST LOCK

Figure 18 Properly securing air hoses.

with tools. Always check to ensure that pressurized air has been bled completely from these hoses before disconnecting them from the tool.

> **WARNING!**
> Whip checks are rated by the maximum working pressure of the compressed-air system. Always use a properly rated whip check. Using a whip check rated for less pressure than the system can produce can result in failure of the whip check in the event of an accident, which can result in serious injury or even death.

2.4.4 Compressor Operating Procedure

This section covers the general operation of a compressor. Each manufacturer provides detailed operating instructions with its equipment. Refer to the operator's manual provided with the compressor for specific instructions. The general outline provided here is intended to explain the operator controls. The operator should perform a complete inspection of the compressor and correct any deficiencies before beginning operation.

The following is a general discussion of the type of inspection the operator performs before beginning work. Each manufacturer provides specific inspection procedure(s) in the operator's manual.

Compressor Setup and Pre-Operational Inspection

Before beginning operation, the operator must perform a setup and pre-operational inspection. Each manufacturer provides a checklist with the equipment. This section covers the basic setup and pre-operational checks for compressors.

Place the unit in as level a position as possible. Follow the manufacturer's directions concerning equipment placement and any special considerations for equipment location. Disconnect the compressor from the towing vehicle. Chock the wheels of the compressor. Unlock the jack and lower it to the service position. Lock the jack in the service position. Disconnect the safety chains and crank the jack to raise the coupling off the hitch.

Vent all internal pressure before opening any air line, fitting, hose, valve, drain plug, connector, oil filler, filler caps, and before refilling anti-ice systems. Be sure all hoses, shutoff valves, flow-limiting valves, and other attachments are connected according to the manufacturer's instructions. If the air compressor is to be used along with other sources of air, be sure there is a check valve installed at the service valve.

The following checks should also be performed on the compressor:

- Check for loose wiring and routing clamps.
- Check for excessive engine fluid leaks.
- Check the engine oil and coolant levels.
- Check for frayed or loose fan belts, hoses, or wiring insulation.
- Check for obstructions in the air vents.
- Check the fuel tank level. The tank should be filled at the end of each day to prevent condensation.
- Check the engine and compressor air cleaner(s).

Maintain the engine in accordance with the manufacturer's recommendations.

Compressor Startup and Shutdown

The operator must be familiar with the correct starting procedure for the equipment. Start the compressor in accordance with the manufacturer's detailed instructions. Follow these general steps:

Sizing Compressors

Do not incorrectly size compressors based on the horsepower rating. Selection of a larger-horsepower compressor motor/engine allows the compressor to run at faster speeds or larger displacement to produce greater airflow at a rated pressure. The size of a compressor should be based on the total air requirements of the various tools or equipment to be used with the compressor. The total airflow is expressed in cubic feet per minute (cfm) and the required air pressure in pounds per square inch (psi). Pneumatic-tool manufacturers provide this information for each of their tools. For example, a typical framing nailer gun operates at a pressure of between 70 and 100 psi and consumes an average airflow of 9.6 cfm.

Step 1 Be familiar with warning devices, gauges, and operating controls before starting the air compressor. Before starting the engine, fully understand the emergency shutdown procedure.

Step 2 After starting the engine, observe gauges, instruments, and warning lights to be sure they are functioning and their readings are within normal operating range.

Step 3 While operating the compressor, listen for any unusual noises and check for unusual vibrations that may indicate a problem. No equipment should be operated if any part is not in proper working condition. Ensure that any unsafe condition has been corrected before operating the equipment. If an unsafe condition exists, tag the control panel with a Do Not Operate tag and disconnect the battery until the problem is corrected.

The operator must know the correct shutdown procedure for the equipment in accordance with the manufacturer's detailed instructions. Follow these general steps to shut down a compressor:

Step 1 Shut off all air service valves at the compressor.

Step 2 Allow the equipment to cool down at idle speed.

Step 3 Turn the ignition switch to Off to stop the air compressor engine.

Step 4 Using the appropriate tool, relieve the residual air pressure.

Step 5 Install and lock all locking devices and remove the ignition key, if applicable.

2.4.5 Air Compressor Operator's Maintenance Responsibility

The manufacturer's manual provides a schedule and procedures to be followed when performing maintenance on the air compressor. Before performing any maintenance, the operator must be trained, authorized, and have the proper tools to perform these procedures. The following are examples of the types of maintenance that may need to be performed on an air compressor:

- Changing the engine oil and filter
- Changing the engine and compressor air filters
- Changing the engine coolant
- Performing battery maintenance
- Lubricating parts
- Inspecting all guards and safety devices

2.5.0 Compaction Equipment

Compactors are used to eliminate soil settlement. If the soil under a structure is not compacted, it will eventually settle and cause damage to the structure. This settling will cause the structure to deform. If the settling occurs to a greater degree at one side or corner, cracks or structural failures can occur.

Compactors are available in a variety of styles and sizes, including upright and flat-plate compactors and ride-on rollers, such as sheepsfoot rollers. Each type of compactor is recommended for different soil conditions and activities. Compactors can also be installed on other types of equipment as an attachment. *Figure 19* shows three common types of compactors.

The compactors commonly used for various soil conditions are as follows:

- *Cohesive soils (silts and clay)* – Use upright tampers or sheepsfoot rollers.
- *Granular soils (sand or gravel)* – Use flat-plate compactors or upright tampers, or smooth-drum vibratory rollers.
- *Mixed soil* – Use an upright or flat-plate compactor, or a sheepsfoot roller or smooth-drum vibratory roller.

This section covers one type of flat-plate compactor commonly used by carpenters. All compactors come with an operator's manual. Refer to the operator's manual before operating any compactor.

27406-14 Introduction to Construction Equipment

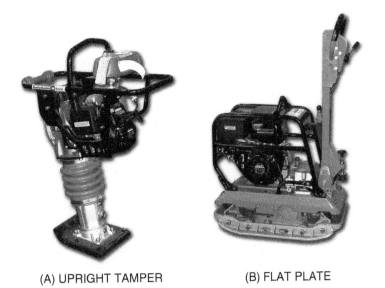

(A) UPRIGHT TAMPER (B) FLAT PLATE (C) SHEEPSFOOT ROLLER

Figure 19 Types of compactors.

2.5.1 Compaction Equipment Assemblies

The flat-plate compactor shown in *Figure 20* includes the following components:

- Vibratory plate
- Engine bed
- Pulley cover
- Operating handle
- Clutch control lever
- Throttle lever
- Engine guard

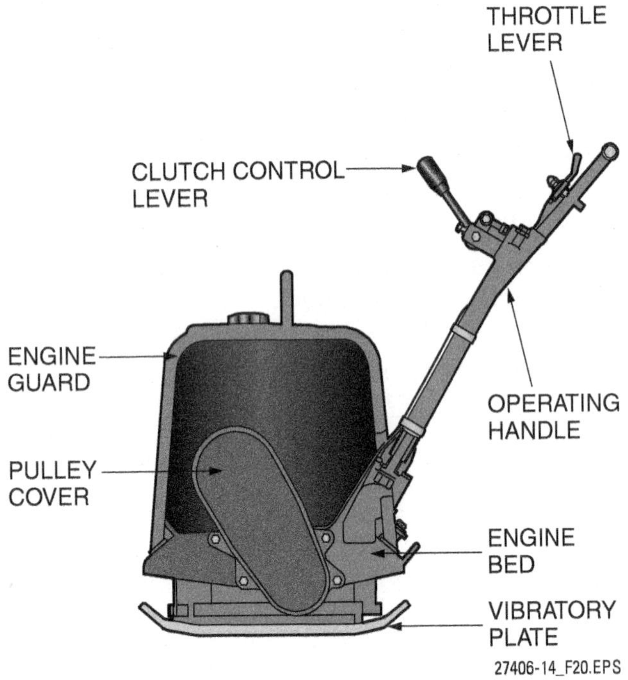

2.5.2 Compaction Equipment Operator Qualifications

Only trained, authorized persons are permitted to operate a compactor. Personnel who have not been trained in the operation of compactors may only operate them for the purpose of training. The training must be conducted under the direct supervision of a qualified trainer.

2.5.3 Typical Compaction Equipment Controls

Compactors have operator controls for the engine and vibratory plate (*Figure 21*). This section discusses typical controls associated with a flat-plate compactor:

Controls associated with the compactor include the following:

- *Operating handle* – Provides the operator with a handle for controlling the compactor and mounting other controls.
- *Clutch control lever* – Used to engage the clutch for the compacting plate.
- *Throttle lever* – Controls the engine speed.

Figure 20 Flat-plate compactor components.

Soil Compaction Specifications

Earthwork specifications will determine the depth of fill that can be placed and compacted in each layer. This helps prevent excessive settling. The job specifications will also dictate the frequency of soil compaction testing.

- *Engine guard* – Protects the engine during use and transportation.
- *Pulley cover* – Protects the operator from the rotating pulleys used to drive the vibratory plate.
- *Vibratory plate* – Provides the compaction capability.
- *Engine bed* – Used as an engine mount and provides weight for the compactor.

2.5.4 Compaction Equipment Safety Precautions

The safe operation of the compactor is the operator's responsibility. The following are general safety precautions that are applicable to compactors. Remember that the safety precautions discussed earlier also apply. Refer to the manufacturer's manual for specific safety precautions.

- After performing maintenance, make certain all guards are correctly installed and all safety devices are functioning.
- Be familiar with the safety devices and proper operation of the compactor.
- Do not perform any maintenance work on the compactor unless authorized and qualified to do so.

2.5.5 Compaction Equipment Operating Procedure

The following discussion covers generic operation of a compactor. Each manufacturer provides detailed operating instructions with its equipment. Refer to the operator's manual for the specific instructions. This general outline is intended to help you understand the operator controls. The operator should perform a complete inspection of the compactor and correct any deficiencies before beginning operation.

Compaction Equipment Pre-Operational Inspection

Each manufacturer provides specific inspection procedure(s) in the operator's manual. Before beginning operation, the operator must perform a pre-operational inspection. Each manufacturer provides a checklist with the equipment. The inspection normally includes the following items:

- Inspect the machine to verify that all guards and safety items are properly installed.
- Check the fuel level and add fuel, if required.
- Check the engine oil level and add oil, if required.
- Check the engine air filter and clean or replace it following the manufacturer's procedures.

Compaction Equipment Starting, Operating, and Stopping

Before beginning operation, the operator must perform a startup. Each manufacturer provides detailed startup instructions with the equipment. Follow these general steps for startup:

Step 1 Verify that the clutch control lever is in the disengaged position.

Step 2 Engage the choke if the engine is cold.

Step 3 Pull on the recoil starter handle until the engine starts. Allow the engine to run until warm, then release the choke.

Step 4 Use the throttle lever to control the engine speed during operation.

Each manufacturer provides detailed instructions for operation of the equipment. General steps are as follows:

Step 1 Adjust the throttle until the desired engine speed is reached.

Step 2 Slowly engage the clutch using the clutch control lever.

Step 3 Use the operating handle to control the direction and movement of the compactor.

Each manufacturer provides detailed instructions for shutting down the equipment. Follow these general steps for shutdown:

Step 1 Use the throttle lever to decrease the engine's speed.

Step 2 Use the clutch control lever to disengage the clutch.

Step 3 Press the engine shutdown control to turn off the engine.

Emergency shutdown of the compactor is accomplished by pressing the engine shutdown control to stop the engine.

2.5.6 Compactor Operator's Maintenance Responsibility

The manufacturer's manual provides a schedule and procedures to be followed when performing maintenance on the compactor. Before performing any maintenance, the operator must be trained, authorized, and have the proper tools to

27406-14 Introduction to Construction Equipment

perform these procedures. The following are examples of maintenance that may need to be performed on a flat-plate compactor:

- Changing the engine oil
- Changing the engine air filters
- Lubricating parts
- Inspecting all guards and safety devices

2.6.0 Forklifts

Forklifts are used to move, unload, and place material (such as pallets of shingles or stacks of plywood) at required locations on the job site. Forklifts are also sometimes referred to as **powered industrial trucks**. In this module, the term *forklift* is used since this term is most commonly used on a job site.

This section introduces you to safe forklift operation. A firm foundation in forklift safety will ensure greater productivity and a safer working environment.

2.6.1 Forklift Assemblies

Many types of forklifts are available to meet different needs. Most forklifts used in construction are gasoline- or diesel-powered and fall into one of two broad categories: fixed mast and telescoping boom. Within these categories, forklifts are further differentiated by their drive train, steering, and capacity.

There are two types of fixed-mast forklifts: rough terrain and warehouse. Rough terrain forklifts are by far the most common type of fixed-mast forklift used in construction (*Figure 21*). These forklifts are made for outside use. They have higher ground clearances than warehouse forklifts, larger tires, and leveling devices. Warehouse forklifts are designed for inside use. They have hard rubber tires that are usually all the same size.

When the term *fixed-mast forklift* is used in this module, it is assumed to refer to the rough terrain forklift.

2.6.2 Forklift Operator Qualifications

Only trained, authorized persons are permitted to operate a forklift. Operators of forklifts must be qualified as to visual, auditory, physical, and mental ability to operate the equipment safely. Training must be conducted under the direct supervision of a qualified trainer.

Figure 21 Fixed mast rough terrain forklifts.

2.6.3 Typical Forklift Controls

Forklifts have either two-wheel drive or four-wheel drive. The distinction is the same as that used with other vehicles. A two-wheel-drive forklift has a drive train that transmits power to the front wheels. Many times, the front wheels are larger than the rear wheels and do not steer the forklift. These forklifts are good for general use, but may lack power in rough terrain. A four-wheel-drive forklift has a drive train that transmits power to all four wheels. These forklifts are well suited to rough terrain and other conditions that might require additional traction.

The rear wheels of a forklift with two-wheel drive move when the steering wheel is turned. This feature is standard on most forklifts and provides increased maneuverability. The rear wheels are generally smaller than the front wheels.

When four-wheel drive is engaged on a forklift with this feature, all the wheels move. However, some forklifts offer the following three options:

- The rear wheels may be locked to allow the front wheels to move.
- All wheels may move in the same direction. This is called crab steering or oblique steering.
- The front and rear wheels may move in opposite directions. This is called articulated steering or four-wheel steering.

Fixed-Mast Forklifts

The upright member along which the forks travel is called the mast. A typical fixed-mast forklift has a mast that may tilt as much as 19 degrees forward and 10 degrees rearward. These types of forklifts are suitable for placing loads vertically and traveling with loads, but their horizontal reach is limited to how close the machine can be driven to the pick-up or landing point. For example, a fixed-mast forklift cannot place a load of purlins beyond the edge of the roof because of its limited reach.

The mast can be either two-stage or three-stage. This designation refers to the number of telescoping channels built in the mast. A two-stage mast has one telescoping channel and a three-stage mast has two telescoping channels. The purpose of the telescoping channels is to provide greater lift height.

Telescoping-Boom Forklifts

Telescoping-boom forklifts (*Figure 22*) provide more versatility in horizontal and vertical placement than a fixed-mast forklift. A telescoping-boom forklift is really a combination of a telescoping-boom crane and a forklift.

Some models of telescoping-boom forklifts have a level-reach fork carriage, sometimes called a squirt boom. A squirt boom allows the fork carriage to be moved horizontally while the boom remains in a stationary position.

Soil Compaction

Some form of soil compaction is needed before placing concrete for foundations, slabs, or floors. Improperly prepared soil can settle, causing foundations or slabs to crack, and even causing the structure to lean. The most famous example of settling is the Leaning Tower of Pisa. It was built plumb, but now leans at an angle of about 12 degrees.

Figure 22 Telescoping-boom forklift.

2.6.4 Forklift Safety Precautions

Safe operation is the responsibility of the operator. Operators must develop safe working habits and recognize hazardous conditions to protect themselves and others from death or injury. Always be aware of unsafe conditions to protect the load and the forklift from damage. Be familiar with the operation and function of all controls and instruments before operating a forklift. Read and fully understand the operator's manual for the equipment.

The following safety rules are specific to forklift operation. Remember the safety precautions discussed earlier also apply.

- Never put any part of your body into the mast structure or between the mast and the forklift.
- Never put any part of your body within the reach mechanism.
- Understand the limitations of the forklift.
- Do not permit passengers to ride unless a safe place to ride has been provided by the manufacturer.
- Never leave the forklift running unattended.

Safeguard pedestrians at all times by observing the following rules:

- Always look in the direction of travel.
- Do not drive the forklift up to anyone standing in front of an object or load.
- Make sure that personnel stand clear of the rear swing area before turning.
- Exercise particular care at blind spots, cross aisles, doorways, and other locations where pedestrians may step into the travel path.
- Use a spotter when landing an elevated load with a telescoping-boom forklift.

2.6.5 Forklift Operation

This section contains general guidelines for operation of a forklift. The most important factor to consider when using any forklift is its capacity. Each forklift is designed with an intended capacity, and this capacity must never be exceeded. Exceeding the capacity jeopardizes not only the equipment but also the safety of everyone on or near the equipment. Each manufacturer supplies a capacity chart for each forklift model. Be sure to read and follow the capacity chart.

Picking Up a Load

Some forklifts are equipped with a side-shift device that allows the operator to horizontally shift the load several inches in either direction with respect to the mast. A side-shift device enables more precise placing of loads, but it changes the center of gravity and must be used with caution. If the forklift being used is equipped with a side-shift device, be sure to return the fork carriage to the center position before attempting to pick up a load Follow these general steps to safely pick up a load:

Step 1 Check the position of the forks with respect to each other. They should be centered on the carriage. If the forks have to be moved, check the operator's manual for the proper procedure. Usually, there is a pin at the top of each fork that, when lifted, allows each fork to be slid along the upper backing plate until the fork centers over the desired notch.

Step 2 Travel to the area at a safe speed. Always keep the forks lowered when traveling with a forklift.

Step 3 Before picking up a load with a forklift, make sure the load is stable. If it looks like the load might shift when picked up, secure the load. Knowing the center of gravity is crucial, especially when picking up tapered sections. Make a trial lift, if necessary, to determine and adjust the center of gravity.

Step 4 Approach the load so that the forks straddle the load evenly. It is important that the weight of all loads be distributed evenly on the forks. Overloading one fork can damage both forks. In some cases, it may be advisable to measure the load and mark its center of gravity.

Step 5 Drive up to the load with the forks straight and level. If the load is on a pallet, be sure the forks are low enough to clear the pallet boards.

Step 6 Move forward until the leading edge of the load rests squarely against the back of both forks. If you cannot see the forks engage the load, ask someone to signal for you. This prevents expensive damage and injury.

Step 7 Raise the carriage, then tilt the mast rearward until the forks contact the load. Raise the carriage until the load safely clears the ground. Then tilt the mast fully rearward to cradle the load. This minimizes the chance that the load may slip during travel.

Traveling with a Load

Always travel with a load at a safe speed. Never travel with a raised load. Keep the load as low as possible and be sure the mast is tilted rearward to cradle the load.

As you travel, keep your eyes open and stay alert. Watch the load and the conditions ahead of you, and alert others of your presence. Avoid sudden stops and abrupt changes in direction. Be careful when downshifting because sudden deceleration can cause the load to shift or topple. Be aware of front and rear swing when turning.

If you are traveling with a telescoping-boom forklift, be sure the boom is fully retracted.

If you have to drive on a slope, keep the load as low as possible. Do not drive across steep slopes. If you have to turn on an incline, make the turn wide and slow.

Placing a Load

Position the forklift at the landing point so that the load can be placed where you want it. Be sure everyone is clear of the load.

The area under the load must be clear of obstructions and must be able to support the weight of the load. If you cannot see the placement, use a signaler to guide you.

With the forklift in the unloading position, lower the load and tilt the forks to the horizontal position. When the load has been placed and the forks are clear from the underside of the load, back away carefully to disengage the forks or retract the boom on variable-reach units.

Placing Elevated Loads

Special care needs to be taken when placing elevated loads. Some forklifts are equipped with a leveling device that allows the operator to rotate the fork carriage to keep the load level during travel. When placing elevated loads, it is extremely important to level the machine before lifting the load.

One of the biggest potential safety hazards during elevated load placement is poor visibility. There may be workers in the immediate area that cannot be seen. The landing point itself may not be visible. Your depth perception decreases as the height of the lift increases. To be safe, use a signaler to help you spot the load.

Use tag lines to tie off long loads.

Drive the forklift as closely as possible to the landing point with the load kept low. Set the parking brake. Raise the load slowly and carefully while maintaining a slight rearward tilt to keep the load cradled. Under no circumstances should the load be tilted forward until the load is over the landing point and ready to be set down.

If the forklift's rear wheels start to lift off the ground, stop immediately, but not abruptly. Lower the load slowly and reposition it, or break it down into smaller components, if necessary. If surface conditions are bad at the unloading site, it may be necessary to reinforce the surface conditions to provide more stability.

Traveling with Long Loads

Traveling with long loads presents special problems, particularly if the load is flexible and subject to damage. Traveling multiplies the effect of bumps over the length of the load. A stiffener may be added to the load to give it extra rigidity.

To prevent slippage, secure long loads to the forks. This may be done in one of several ways. A field-fabricated cradle may be used to support the load. While this is an effective method, it requires that the load be jacked up.

The forklift may be used to carry pieces of rigging equipment. This method requires the use of slings and a spreader bar.

In some cases, long loads may be snaked through openings that are narrower than the load itself. This is done by approaching the opening at an angle and carefully maneuvering one end of the load through the opening first. Avoid making quick turns because abrupt maneuvers cause the load or its center of gravity to shift.

Using the Forklift to Rig Loads

The forklift can be a very useful piece of rigging equipment if it is properly and safely used. Loads can be suspended from the forks with slings, moved around the job site, and placed.

All the rules of careful and safe rigging apply when using a forklift to rig loads. Be sure not to drag the load or let it swing freely. Use tag lines to control the load.

Never attempt to rig an unstable load with a forklift. Be especially mindful of the load's center of gravity when rigging loads with a forklift.

When carrying cylindrical objects, such as oil drums, keep the mast tilted rearward to cradle the load. If necessary, secure the load to keep it from rolling off the forks.

2.6.6 Forklift Operator's Maintenance Responsibility

This section of the module is intended to present general preventive maintenance procedures required of forklift operators, including the following:

27406-14 Introduction to Construction Equipment   Module Seven 29

Parking the Forklift

At the end of the day, park the forklift on level ground. Lower the forks to the ground and set the parking brake. If you must park on a slope, position the forklift at right angles to the slope and block the wheels.

- Check the operator's manual for lubrication points and suggested lubrication periods. Lubrication periods are expressed in terms of the number of hours of service.
- Check fluid levels frequently and use the recommended fluids when refilling.
- Because forklifts often operate in dusty and dirty environments, it is important to keep the moving parts of the machine well-greased. This reduces wear, prolongs the life of the machine, and ensures safe operation. Follow the manufacturer's recommendations for lubricants.

Forklift manufacturers provide a daily checklist. *Figure 23* shows an example of a daily checklist and the items that are usually checked. These items may differ from unit to unit.

2.7.0 Backhoe

A backhoe (*Figure 24*) is a dual-purpose, highly maneuverable machine used to dig trenches, foundations, and similar excavations. It is also used to move dirt, crushed stone, gravel, and other materials around the job site.

Backhoes are equipped with either a gasoline or diesel engine. The backhoe bucket and loader bucket attachments are hydraulically operated under the control of a single operator who performs all backhoe operations.

2.7.1 Backhoe Operator Qualifications

Before operating a backhoe, the following requirements must be met:

- The operator must successfully complete a training program that includes actual operation of the backhoe.
- The operator is responsible for reading and understanding the safety manual and operator's manual provided with the equipment.
- The operator must know the safety rules and regulations for the job site.

2.7.2 Backhoe Controls

Depending on the manufacturer and model of backhoe, there can be several different arrangements in the operating controls and their locations. Always refer to the operator's manual for the specific equipment you are operating to familiarize yourself with the controls and their locations. The basic controls used to operate any backhoe fall into three categories:

- Vehicle engine and movement-related controls
- Backhoe bucket and backhoe stabilizer controls
- Loader bucket controls

Vehicle Engine and Movement-Related Controls

Most of the backhoe engine and movement-related controls and their functions are similar to those found in other construction equipment, including the engine start switch, transmission lock control and direction control lever, steering wheel, steering-column tilt control, accelerator pedal, parking brake, and service brake. Gauges used to monitor engine performance can include battery, engine coolant, tachometer, oil pressure, transmission oil temperature, and service-hour meter. Some engine and movement-related controls and their functions unique to some backhoes are:

- *Accelerator lever* – Used to control the engine speed between low and high idle during backhoe operation.
- *Differential lock pedal* – Used to engage the differential lock to prevent wheel slippage when moving on soft or wet ground.

Some backhoes are equipped with an all-wheel drive and/or an all-wheel steering capability. The all-wheel drive capability enables the operator to switch from two-wheel drive to all-wheel drive anytime additional traction is needed. All-wheel steering capability provides three modes of steering to suit various job-site conditions: two-wheel steering, all-wheel steering, and circle steering. Two-wheel steering is normally used to operate the machine on the road or other surfaces when additional maneuvering ability is not needed. In this mode, only the front axle is used to steer the backhoe. All-wheel steering allows the operator to choose independent rear-axle operation so that that the rear wheels will move the back of the backhoe either to the left or right when the unit is moving forward. This enables the operator to position the front and back wheels in opposite directions or in the same direction. Turning the front and back wheels in opposite directions

OPERATOR'S DAILY CHECKLIST

Check Each Item before Start of Each Shift   Date:_____

Check One: ☐ Gas/LGP/Diesel Truck   ☐ Electric Sit-down   ☐ Electric Stand-up   ☐ Electric Pallet

Truck Serial Number:_____   Operator:_____   Supervisor's OK: _____

Hour Meter Reading: _____

Check each of the following items before the start of each shift. Let your supervisor and/or maintenance department know of any problem. DO NOT OPERATE A FAULTY TRUCK. Your safety is at risk.

After checking, mark each item accordingly. Explain below as necessary.

Check boxes as follows: ☐ OK   ☐ NG, needs attention, or repair. Circle problem and explain below.

OK	NG	Visual Checks	OK	NG	Visual Checks
		Tires/Wheels: wear, damage, nuts tight			Steering: loose/binding, leaks, operation
		Head/Tail/Working Lights: damage, mounting, operation			Service Brake: linkage loose/binding, stops OK, grab
		Gauges/Instruments: damage, operation			Parking Brake: loose/binding, operational, adjustment
		Operator Restraint: damage, mounting, operation oily, dirty			Seat Brake (if equipped): loose/binding, operational, adjustment
		Warning Decals/Operators' Manual: missing, not readable			Horn: operation
		Data Plate: not readable, missing adjustment			Backup Alarm (if equipped): mounting, operation
		Overhead Guard: bent, cracked, loose, missing			Warning Lights (if equipped): mounting, operation
		Load Back Rest: bent, cracked, loose, missing			Lift/Lower: loose/binding, excessive drift, leaks
		Forks: bent, worn, stops OK			Tilt: loose/binding, excessive drift, "chatters," leaks
		Engine Oil: level, dirty, leaks			Attachments: mounting, damaged, operation, leaks
		Hydraulic Oil: level, dirty, leaks			Battery Test (electric trucks only): indicator in green
		Radiator: level, dirty, leaks			
		Fuel: level, leaks			
		Battery: connections loose, charge, electrolyte low			Battery: connections loose, charge, electrolyte low while holding full forward tilt
		Covers/Sheet Metal: damaged, missing			Control Levers: loose/binding, freely return to neutral
		Brakes: linkage, reservoir fluid level, leaks			Directional Control: loose/binding, find neutral OK
		Engine: runs rough, noisy, leaks			

Explanation of problems marked above: _____

_____

_____

_____

_____

Figure 23 Example of a forklift operator's daily checklist.

27406-14 Introduction to Construction Equipment   Module Seven   31

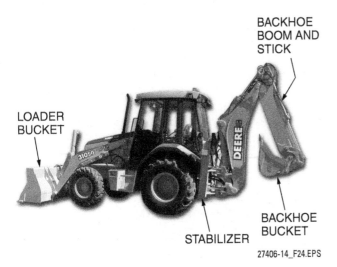

Figure 24 Backhoe.

is typically used when it is required to move the backhoe around tight corners. Turning the wheels in the same direction is done when crab steering is desired. The operator can determine the position of the rear axle by looking at a rear-axle position gauge provided for this purpose. Selection of the circle mode reduces the turning radius of the backhoe, allowing for better steering in confined spaces.

> **CAUTION**
> Before changing from one all-wheel steering mode to another on a backhoe, always center both the front and rear wheels.

Backhoe Bucket and Stabilizer Controls

The stabilizer controls are used to lower and raise the left and right stabilizer arms. Before operating the backhoe, the stabilizer arms must be lowered firmly to the ground in order to provide a firm and level base for backhoe operation. The left and right stabilizers are independently controlled by left and right stabilizer levers, respectively. Each lever has three positions: Down, Hold, and Up. Moving the lever in the Down direction lowers the associated stabilizer. When the stabilizer firmly contacts the ground, lowering it further causes the rear of the backhoe to be raised. Moving the lever in the Up position raises the stabilizer, thus lowering the rear of the backhoe. Releasing the stabilizer lever either from the Down or Up position causes the lever to return to the Hold (center) position, stopping stabilizer movement.

Backhoe-bucket operation is controlled by the backhoe boom and backhoe stick-and-bucket controls. Depending on the machine and model, the arrangement of these controls varies. Some machines use a two-lever control arrangement, others use a three-lever scheme with a foot swing control, and still others use a four-lever control scheme. The two-lever control scheme is briefly described here.

A backhoe with a two-lever control scheme has two levers, one to control the backhoe boom and the other to control the stick and bucket. The backhoe boom lever is a five-position lever having two positions to lower and raise the boom and two positions to swing the boom to the left or right. Movement of the boom is stopped when the lever is in the Hold (center) position. The lever returns to the Hold position when released from any of the other four positions.

The backhoe stick-and-bucket lever is a five-position lever. Two positions are used to extend and retract the stick and two positions to load and dump the bucket. Movement of the stick and/or the bucket is stopped when the lever is in the Hold (center) position. The lever returns to the Hold position when released from any of the other four positions.

A boom-lock lever is used to enable or prevent boom operation. When in the locked position, a boom lock is engaged that prevents the backhoe from moving and swinging into objects or into traffic. When in the released position, movement of the backhoe is enabled for digging operations.

Loader Bucket Controls

The loader bucket for the equipment being described here is controlled by an eight-position lever. These positions are:

- *Float* – Allows the loader bucket to move freely following the contour of the ground.
- *Lower* – Lowers the loader bucket.
- *Hold* – Stops the movement of the loader bucket. The lever returns to the Hold position when it is released from any other position, except for the float position. The lever stays in the float position until it is moved by the operator to the Hold position.
- *Raise* – Raises the loader bucket.
- *Tilt back* – Tilts back the loader bucket.
- *Return-to-dig* – Returns the loader bucket to the dig position. The lever stays in this position until the bucket is level; then it automatically returns to the Hold position.
- *Dump* – Used to empty the loader bucket.
- *Quick dump* – Shortens the time that is required to tilt the loader bucket.

2.7.3 Backhoe Safety Precautions

The following is a list of safety rules specific to the operation of a backhoe. Remember that the safety precautions discussed earlier also apply.

- Mount the backhoe only at locations that are equipped with steps and/or handgrips.
- Face the backhoe while entering and leaving the operator's compartment.
- Do not mount a backhoe while carrying tools or supplies.
- Do not mount a moving backhoe.
- Do not use controls as handgrips when entering the operator's compartment.
- Do not obstruct your vision when traveling or working.
- Never lift, move, or swing a load over a truck cab or over workers.
- When traveling, operate at speeds slow enough so that you have complete control of the backhoe at all times, especially when traveling over rough or slippery ground and when on hillsides. Never place the transmission in neutral in order to allow the backhoe to coast.
- Never approach overhead power lines with any part of the backhoe unless you are in strict compliance with all local, state, and federal required safety precautions.
- Make sure that you know the underground location of all gas and water pipelines and of all electrical or fiber-optic cables.

When operating the backhoe the following safety precautions apply:

- Never enter or allow anyone to enter the backhoe swing area.
- Operate the backhoe from the correct backhoe operating position. Never operate the backhoe controls from the ground. Never allow riders in or on the backhoe.
- To prevent cave-ins and the possibility of the backhoe falling into the excavation, do not dig under the backhoe or its stabilizers.
- When operating the backhoe on a slope, swing to the uphill side to dump the load, if possible. If necessary to dump downhill, swing only as far as required to dump the bucket.
- Always dump the soil far enough away from the trench to prevent cave-ins.

When operating the loader bucket, the following safety precautions apply:

- Carry the bucket low for maximum stability and visibility.
- Keep the backhoe in gear when traveling downhill, and use the same gear range as you would for traveling up a grade.
- When on a steep slope, drive up or down the slope; do not drive across the slope. If the bucket is loaded, drive with the bucket facing uphill. If empty, drive with the bucket pointed downhill.
- When operating the backhoe at a fixed position, make sure that the backhoe is in the transport-lock position to prevent backhoe movement.
- When working at the base of a bank or overhang, never undercut a high bank and/or operate the backhoe close to the edge of an overhang or ditch.

Safety Manuals

A series of easy-to-read and easy-to-use equipment safety manuals, such as those shown here, are available from the Association of Equipment Manufacturers (AEM), formerly the Equipment Manufacturers Institute (EMI). These manuals were produced by the EMI to provide equipment owners, operators, service personnel, and mechanics with the basic safety procedures and precautions that are pertinent in the day-to-day operation and maintenance of the specific equipment. These safety manuals are generic in nature and are to be used in conjunction with the operator/user manuals provided by manufacturers for a specific piece of equipment.

27406-14_SA03.EPS

2.7.4 Backhoe Operating Procedure

This section covers basic operations performed with a backhoe. Each manufacturer provides detailed operating instructions with its equipment. Refer to the operator's manual for specific operating instructions.

Backhoe Inspection

Before operating the backhoe, the operator should inspect the equipment and attachments and perform the recommended operator's daily maintenance procedures in accordance with the manufacturer's instructions. Examples of the daily maintenance tasks are given later in this section. A backhoe with defective or missing parts must be repaired before being operated.

General Backhoe Operating Guidelines

Startup and warm-up of the backhoe engine should be done in accordance with the manufacturer's instructions. The duration of the engine warm-up period is determined by the prevailing temperature. The colder the temperature, the longer the required warm-up period. During the warm-up period, check the gauges and controls for proper indications/operation.

Follow these general guidelines pertaining to backhoe operation and travel:

- Raise all attachments during travel high enough to clear any unexpected obstacles.
- Make changes in the direction of travel by putting the transmission direction control lever to the desired speed and direction. When making changes in direction, it is recommended to reduce the backhoe speed.
- Select the best gear speed for conditions before starting down a hill, and do not change gears while going down the hill. When going down grade, the rule of thumb is to use the same gear speed that would be used if going up-grade. Do not allow the engine to overspeed when going downhill. If equipped, activate the all-wheel drive function when operating on a hill or when additional traction and/or braking are needed.
- Avoid turning on a slope.
- When the load will be pushing the backhoe, put the transmission lever in the first speed position before you start downhill. Engage the all-wheel drive function (if equipped).

WARNING! Never use the Float position to lower a loaded bucket. Damage to the machine can result from the bucket falling too fast.

When the work is completed and the backhoe is to be shut down, lower all the attachments to the ground. Shift the controls to neutral/park and lock (if equipped) and set the parking brake. Before turning off the engine, allow it to idle for a short time in order to cool down. After turning off the engine, cycle the hydraulic controls. If the backhoe is being parked on an incline, block the wheels.

Backhoe Bucket Operating Guidelines

The backhoe bucket should be prepared for operation and operated in accordance with the manufacturer's instructions. When digging with the backhoe bucket, follow these general guidelines:

- When digging, close the bucket slowly and move the stick inward at the same time for maximum performance.
- Keep the bucket teeth at an angle that gives the best ground penetration. This helps prevent the bucket from just scraping the ground.
- Apply downward pressure with the boom to increase bucket penetration in hard-packed ground conditions.
- Move the boom downward to close the bucket completely. Move the stick and the bucket outward from the backhoe slightly while lifting the bucket from the excavation. Then, swing the bucket to the side and dump the load as you approach the pile.

CAUTION When digging with the backhoe, do not swing the bucket against the sides of the excavation and/or spoils. Repeated stopping of the bucket by the sides of the excavation and/or spoils can cause structural damage to the boom and/or result in premature wear of the boom pin and bushings.

- After the bucket is emptied, the stick and bucket should be operated to the closed dig position and returned to the excavation.
- When backfilling, lift the bucket over the spoils; then pull the bucket inward and lift the boom evenly.

Loader Bucket Operating Guidelines

The loader bucket should be prepared for operation and operated in accordance with the manufacturer's instructions. When using the loader bucket, follow these general guidelines:

- Carry an empty or loaded bucket low to the ground to achieve equipment stability and better vision.
- When carrying a load downhill, travel in reverse. When carrying a load uphill, travel forward.
- To load the bucket, skim the ground as you move the backhoe forward. Move the control lever in order to lift the bucket and tilt back the bucket. Fully tilt back the bucket to avoid spillage.
- When dumping the load into a truck, position the backhoe at a 45-degree angle to the truck. This helps reduce the amount of backhoe turning and travel.
- Move the loader bucket as close as possible to the truck before dumping the load. Dump the load in the center of the truck bin. On successive dumps, load the truck starting from the front of the truck bin and working toward the rear.
- When possible, keep the wind to your back when picking up and dumping loads. This helps keep dust and debris from getting on you or the equipment.

2.7.5 Backhoe Operator's Maintenance Responsibility

The manufacturer's manual provides a schedule and procedures to be followed when performing maintenance on a backhoe. Before performing any maintenance, the operator must be trained, authorized, and have the proper tools to perform these procedures. The following are examples of the types of maintenance a backhoe operator may perform on a daily basis:

- Clean the radiator core.
- Clean the windows.
- Check the engine fuel level.
- Check the cooling-system level.
- Check the engine oil level.
- Check the hydraulic-system oil level.
- Check the transmission oil level.
- Check the brake-reservoir oil level.
- Check the engine air-filter service indicator.
- Check the tire inflation.
- Check the wheel-nut torque.
- Check for the proper installation of access covers and the guards.
- Drain the fuel-system water separator.
- Lubricate the backhoe boom, stick, bucket, and cylinder bearings.
- Lubricate the front kingpin bearings.
- Lubricate the loader-bucket, cylinder, and linkage bearings.
- Lubricate the stabilizer and cylinder bearings.
- Lubricate the swing-frame and cylinder bearings.

2.0.0 Section Review

1. An aerial lift has both platform controls and _____.
   a. cab controls
   b. ground controls
   c. base controls
   d. remote controls

2. Utilities should be asked to stake out underground lines 24 hours before digging begins.
   a. True
   b. False

3. Each of the following is a main component of a tow-behind generator, *except* the _____.
   a. engine and generator assemblies
   b. hydraulic reservoir
   c. towing assembly and frame
   d. operator control panels

4. Whip checks are rated by the _____.
   a. rpm of the equipment
   b. physical size of the equipment
   c. maximum working pressure of the compressed-air system
   d. diameter of the orifice

5. Types of compactors include upright, ride-on, and _____.
   a. flat plate
   b. clutch control
   c. pulley-driven
   d. vibratory plate

6. The most common type of fixed-mast forklift used in construction is the _____.
   a. all-weather forklift
   b. high-capacity forklift
   c. rough-terrain forklift
   d. warehouse forklift

7. When placing elevated loads with a forklift, use a signaler to spot the load.
   a. True
   b. False

Summary

As a carpenter, it is important for you to be familiar with the different types of construction equipment that might be encountered on the job. Recognizing the equipment and fully understanding the operation of each type will allow you to perform your work more efficiently. Safety has been emphasized throughout this module. The safe operation of any construction equipment is paramount to your health and well-being and that of your co-workers.

Review Questions

1. If a safety interlock fails when using any type of construction equipment, you should _____.
   a. bypass it and continue operating
   b. complain to your supervisor
   c. post an "out of order" sign on it
   d. contact the manufacturer's repair representative immediately

Figure 1

2. The piece of equipment shown in Review Question *Figure 1* is a _____.
   a. scissor lift
   b. forklift
   c. boom lift
   d. backhoe

3. The lifting mechanism of a scissor lift is raised and lowered by a hydraulic cylinder or by _____.
   a. electric motors and gears
   b. pneumatic cylinders
   c. diesel engine and chain drive
   d. a screw mechanism

4. To restore power to an aerial lift after the Emergency Stop button is pressed, you must _____.
   a. turn the ignition key off, then on again
   b. release (pull out) the button
   c. call your supervisor to reset the mechanism
   d. reset the circuit breaker

5. The travel direction and turning of a skid-steer loader is controlled with two _____.
   a. joysticks
   b. thumbwheels
   c. trackballs
   d. rocker switches

6. When traveling up or down a slope with a loaded skid-steer bucket, _____.
   a. move across the slope diagonally
   b. keep the load facing upslope
   c. raise the bucket at least 3 feet to clear obstacles
   d. keep the load facing downslope

7. A device that can automatically shut down a generator is the _____.
   a. high engine-temperature indicator
   b. ammeter gauge
   c. safety-circuit bypass
   d. engine tachometer

8. A generator should be used as an alternate power supply only _____.
   a. when authorized by the contractor
   b. with OSHA certification
   c. when approved by the local electrical inspector
   d. after the main feed at the service entrance panel has been opened and locked

9. If you find evidence of arcing on the control panel of a tow-behind generator, _____.
   a. continue with operation of the generator
   b. replace the control panel
   c. locate and repair the problem before operating
   d. call the manufacturer for additional information

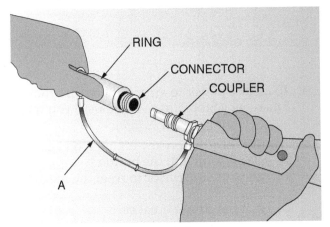

Figure 2

10. The device labeled A in Review Question *Figure 2* is a _____.
    a. whip check
    b. connector lock
    c. safety cable
    d. security link

11. Air hoses used with tools such as jackhammers are connected with _____.
    a. clamp-equipped couplings
    b. cam-operated twist locks
    c. threaded couplers
    d. bayonet connectors

Figure 3

12. The device shown in Review Question *Figure 3* is a(n) _____.
    a. flat-plate compactor
    b. jackhammer
    c. upright tamper
    d. vibratory consolidator

13. For compacting cohesive soils, use a sheepsfoot roller or a(n) _____.
    a. smooth-drum vibratory roller
    b. upright tamper
    c. flat-plate compactor
    d. segmented-drum roller

14. The engine bed of a compactor _____.
    a. protects the operator from moving parts
    b. acts as the vibratory plate
    c. requires periodic adjustment
    d. provides weight for the compactor

15. Emergency shutdown of a flat-plate compactor is done by _____.
    a. closing the throttle lever
    b. disengaging the clutch
    c. pressing the engine shutdown control
    d. closing the fuel-line valve

16. Forklifts are sometimes referred to as _____.
    a. loaders
    b. powered industrial trucks
    c. pallet jacks
    d. hoist trucks

17. Fixed-mast lift trucks are divided into _____.
    a. two types
    b. three types
    c. four types
    d. five types

18. When a forklift has crab (oblique) steering, _____.
    a. only the front wheels can be used to steer
    b. front and rear wheels move in opposite directions
    c. all wheels move in the same direction
    d. only the rear wheels can be used to steer

19. To provide a firm and level base for operation, the backhoe is equipped with _____.
    a. leveling struts
    b. baseplates
    c. outriggers
    d. stabilizer arms

20. When dumping a load into a truck, position the backhoe to the truck at an angle of _____.
    a. 30 degrees
    b. 45 degrees
    c. 90 degrees
    d. 180 degrees

Appendix

Safety Review Questions

1. When loading a piece of equipment onto a trailer, _____ to prevent the trailer from tipping.
   a. make sure the trailer jack is down and locked in position
   b. make sure the wheels of the trailer are blocked
   c. weigh down the front of the trailer
   d. have the trailer hitched to the towing vehicle before loading

2. The operator should check for leaks in a pressurized hydraulic system by _____.
   a. checking the hydraulic oil level in the supply tank
   b. feeling all the hoses and connectors
   c. using a piece of wood or cardboard
   d. using a flashlight and an inspection mirror

3. Before disconnecting any hydraulic system hoses, the operator should shut down the engine and _____.
   a. check the hydraulic fluid level
   b. cycle the hydraulic controls
   c. set the parking brake
   d. chock the wheels

4. When refueling a piece of equipment, keep the nozzle or fuel funnel in contact with the filler neck to _____.
   a. prevent the possibility of sparks
   b. lower the possibility of a fuel spill
   c. protect you from breathing the fuel fumes
   d. make it easier to see when the fuel tank is full

5. When a battery is frozen, _____.
   a. disconnect the negative lead before attaching the charger
   b. do not charge it or attempt to jump-start it
   c. jump-start the equipment following the manufacturer's instructions
   d. replace the battery and charge the old battery

6. During the pre-operational check of a piece of equipment, the operator notices a problem with a safety shield on the equipment. The operator should _____.
   a. note the problem and begin working with the equipment
   b. remove the shield and begin working with the equipment
   c. have the problem corrected before operating the equipment
   d. tell the manufacturer of the problem and request a modified operating procedure

7. Before operating a skid-steer loader, the operator must _____.
   a. be qualified and authorized
   b. be licensed and certified
   c. pass a mental and physical examination
   d. meet height and weight requirements

8. When carrying a full load on a skid-steer loader, _____.
   a. carry the load high to improve vision
   b. travel as fast as possible over rough or slippery surfaces and on hillsides
   c. always have the lift arms up when traveling or turning
   d. carry the bucket low for stability and visibility when traveling

9. When setting up a compressor, place the unit in as level a position as possible, disconnect the compressor from the towing vehicle, and _____.
   a. start up the compressor
   b. chock the wheels
   c. ground the compressor to a grounding rod
   d. elevate the tongue of the compressor trailer for greater stability

10. When setting an elevated load down on a roof using a telescoping-boom forklift, _____.
    a. keep the equipment away from oil, grease, wet paint, mud, or any slippery material
    b. look for oil spills, wet spots, and slippery surfaces
    c. approach the drop point with the load fully raised
    d. use a spotter

27406-14   Introduction to Construction Equipment

Trade Terms Quiz

Fill in the blank with the correct term that you learned from your study of this module.

1. A(n) _____ is a machine used to prevent the settling of soil after construction.

2. A(n) _____ uses a crisscrossed support to raise a work enclosure.

3. A(n) _____ is a safety device designed to protect circuits from electrical overloads.

4. Another term for forklift is _____.

5. A(n) _____ is a general term referring to a mobile elevated work platform.

6. Bulk materials such as joists and beams are moved around a job site using a(n) _____.

7. The more a(n) _____ is moved, the faster the equipment moves.

8. A(n) _____ is attached to the ends of an air supply hose and a tool, or to the ends of two hoses that are connected to each other, to prevent the hose from moving uncontrollably when charged with compressed air.

9. Pneumatic tools are powered by air delivered by a(n) _____.

10. A(n) _____ produces electricity.

11. The lever used to control the speed and steering of an aerial lift is the _____.

12. A(n) _____ is equipped with a bucket for moving stone, dirt, and bulk items around a job site.

13. A mechanism used to prevent incorrect control operation is the _____.

14. A type of electrical connector is the _____.

15. A(n) _____ has a single arm that extends a work platform/enclosure.

Trade Terms

Aerial lift
Boom lift
Circuit breaker
Compactor
Compressor

Drive/steer controller
Forklift
Generator
Powered industrial trucks
Proportional control

Safety interlock
Scissor lift
Skid-steer loader
Twist-lock outlet
Whip check

Trade Terms Introduced in This Module

Aerial lift: A mobile work platform designed to transport and raise personnel, tools, and materials to overhead work areas.

Boom lift: Aerial lift with a single arm that extends a work platform/enclosure capable of holding one or two workers; some models may have an articulated joint.

Circuit breaker: A device designed to protect circuits from overloads and to enable them to be reset after tripping.

Compactor: A machine used to compact soil to prevent the settling of the soil after construction.

Compressor: A motor-driven machine used to supply compressed air for pneumatic tools.

Drive/steer controller: One-handed joystick lever control used to control the speed and steering of an aerial lift.

Forklift: A machine designed to facilitate the movement of bulk items around the job site.

Generator: A machine designed to generate electricity.

Powered industrial trucks: See *Forklift*.

Proportional control: A control that increases speed in proportion to the movement of the control.

Safety interlock: A safety mechanism used to prevent incorrect control operation.

Scissor lift: Aerial lift used to raise a work enclosure vertically by means of crisscrossed supports.

Skid-steer loader: A small, highly maneuverable machine equipped with a bucket for moving stone, dirt, and bulk items around a job site.

Twist-lock outlet: A type of electrical connector.

Whip check: A cable attached to the ends of an air supply hose and a tool, or to the ends of two hoses that are connected to each other, to prevent the hose from whipping uncontrollably when charged with compressed air.

Additional Resources

This module presents thorough resources for task training. The following resource material is suggested for further study.

Construction Equipment Guide. New York: John Wiley & Sons.
Heavy Equipment Operations Levels 1, 2, and 3. Latest Edition. NCCER. Upper Saddle River, NJ: Pearson.

Figure Credits

Courtesy of **GeneralContractor.com**, Module Opener
JLG Industries, Inc., Figure 1, SA01, RQ01
Courtesy of S4Carlisle Publishing Services, Figure 5A, SA02
Deere & Company, Figure 5B
Ingersoll Rand Construction Technologies, Figure 10A
Courtesy of Multiquip Inc., Figure 10B
Topaz Publications, Inc., Figure 11, Figure 24
Courtesy of Sullair, LLC, Figure 15A, Figures 16-17
Campbell Hausfeld, Figure 15B
AIR POWER Jackhammer Hose Photos courtesy of HBD/Thermoid, Inc., Figure 18B
Caterpillar, Inc., Figure 19C
Sellick Equipment Ltd., Figure 21
Manitou North America, Inc., Figure 22
Courtesy of Association of Equipment Manufacturers, SA03

Section Review Answer Key

Answer	Section Reference	Objective
Section One		
1. b	1.2.0	1b
2. c	1.2.0	1b
3. d	1.2.0	1b
4. b	1.4.0	1d
Section Two		
1. b	2.1.3	2a
2. b	2.2.0	2b
3. b	2.3.1	2c
4. c	2.4.3	2d
5. a	2.5.0	2e
6. c	2.6.0	2f
7. a	2.6.5	2f

NCCER CURRICULA — USER UPDATE

NCCER makes every effort to keep its textbooks up-to-date and free of technical errors. We appreciate your help in this process. If you find an error, a typographical mistake, or an inaccuracy in NCCER's curricula, please fill out this form (or a photocopy), or complete the online form at **www.nccer.org/olf**. Be sure to include the exact module ID number, page number, a detailed description, and your recommended correction. Your input will be brought to the attention of the Authoring Team. Thank you for your assistance.

Instructors – If you have an idea for improving this textbook, or have found that additional materials were necessary to teach this module effectively, please let us know so that we may present your suggestions to the Authoring Team.

NCCER Product Development and Revision
13614 Progress Blvd., Alachua, FL 32615

Email: curriculum@nccer.org
Online: www.nccer.org/olf

❏ Trainee Guide   ❏ Lesson Plans   ❏ Exam   ❏ PowerPoints   Other _____

Craft / Level: _____   Copyright Date: _____

Module ID Number / Title: _____

Section Number(s): _____

Description: _____

_____

_____

_____

_____

Recommended Correction: _____

_____

_____

_____

_____

Your Name: _____

Address: _____

_____

Email: _____   Phone: _____

22206-13

Rough-Terrain Forklifts

OVERVIEW

Forklifts, also called lift trucks, are motorized vehicles equipped primarily with a fork assembly that allows the machine to lift and carry materials. Forklifts are used inside and outside of buildings. Those used inside are built low to the ground and are usually equipped with slick tires. The tires used on inside forklifts may be solid rubber, or inflated. Those forklifts used outside are usually built higher off the ground and are equipped with treaded tires that have more traction in dirt or other natural surfaces. Outdoor forklifts normally have inflated rubber tires. The forklifts found on most construction sites are usually referred to as rough-terrain forklifts.

Module Eight

Trainees with successful module completions may be eligible for credentialing through the NCCER Registry. To learn more, go to **www.nccer.org** or contact us at **1.888.622.3720**. Our website has information on the latest product releases and training, as well as online versions of our *Cornerstone* magazine and Pearson's product catalog.

Your feedback is welcome. You may email your comments to **curriculum@nccer.org**, send general comments and inquiries to **info@nccer.org**, or fill in the User Update form at the back of this module.

This information is general in nature and intended for training purposes only. Actual performance of activities described in this manual requires compliance with all applicable operating, service, maintenance, and safety procedures under the direction of qualified personnel. References in this manual to patented or proprietary devices do not constitute a recommendation of their use.

Copyright © 2015 by NCCER, Alachua, FL 32615, and published by Pearson Education, Inc., New York, NY 10013. All rights reserved. Printed in the United States of America. This publication is protected by Copyright, and permission should be obtained from NCCER prior to any prohibited reproduction, storage in a retrieval system, or transmission in any form or by any means, electronic, mechanical, photocopying, recording, or likewise. To obtain permission(s) to use material from this work, please submit a written request to NCCER Product Development, 13614 Progress Blvd., Alachua, FL 32615.

From *Construction Craft Laborer, Level Two, Trainee Guide,* Third Edition. NCCER.
Copyright © 2015 by NCCER. Published by Pearson Education. All rights reserved.

22206-13
ROUGH-TERRAIN FORKLIFTS

Objectives

When you have completed this module, you will be able to do the following:

1. Identify and describe the components of a rough-terrain forklift.
   a. Identify and describe chassis components.
   b. Identify and describe the controls.
   c. Identify and describe the instrumentation.
   d. Identify and describe the attachments.
2. Describe the prestart inspection requirements for a rough-terrain forklift.
   a. Describe prestart inspection procedures.
   b. Describe preventive maintenance requirements.
3. Describe the startup and operating procedures for a rough-terrain forklift.
   a. State rough-terrain forklift-related safety guidelines.
   b. Describe startup, warm-up, and shutdown procedures.
   c. Describe basic maneuvers and operations.
   d. Describe related work activities.

Performance Tasks

Under the supervision of your instructor, you should be able to do the following:

1. Complete a proper prestart inspection and maintenance on a rough-terrain forklift.
2. Perform proper startup, warm-up, and shutdown procedures.
3. Execute basic maneuvers with a rough-terrain forklift.
4. Interpret a forklift load chart.
5. Perform basic lifting operations with a rough-terrain forklift.
6. Demonstrate proper parking of a rough-terrain forklift.

Trade Terms

Crab steering
Four-wheel steering
Fulcrum
Oblique steering
Powered industrial trucks
Power hop
Telehandler
Tines

Industry Recognized Credentials

If you are training through an NCCER-accredited sponsor, you may be eligible for credentials from NCCER's Registry. The ID number for this module is 22206-13. Note that this module may have been used in other NCCER curricula and may apply to other level completions. Contact NCCER's Registry at 888.622.3720 or go to www.nccer.org for more information.

Contents

Topics to be presented in this module include:

1.0.0 Rough-Terrain Forklift Components 1
  1.1.0 Chassis Components 1
    1.1.1 Axles and Steering 1
    1.1.2 Fixed-Mast Forklifts 3
    1.1.3 Telescoping-Boom Forklifts 6
    1.1.4 Engine Area 7
    1.1.5 Transmissions 7
    1.1.6 Loader Hydraulic System Components 7
    1.1.7 Operator Cab 8
    1.1.8 Outriggers 8
  1.2.0 Controls 8
    1.2.1 Disconnect Switches 9
    1.2.2 Operator Seat and Steering Wheel Adjustments 10
    1.2.3 Engine Start Switch 10
    1.2.4 Vehicle Movement Controls 11
    1.2.5 Boom and Lift Attachment Controls 11
    1.2.6 Control Switches 13
    1.2.7 Special Features 13
    1.2.8 Operator Comfort and Other Controls 14
  1.3.0 Instruments 14
    1.3.1 Fuel Level Gauge or Indicator 14
    1.3.2 Engine Coolant Temperature Gauge or Indicator 15
    1.3.3 Transmission Oil Temperature Gauge or Indicator 15
    1.3.4 Hydraulic Oil Temperature Gauge or Indicator 15
    1.3.5 Service Hour Meter 16
    1.3.6 Speedometer/Tachometer 16
    1.3.7 Other Indicators 16
    1.3.8 Frame Level Indicator and Longitudinal Stability Indicator 16
  1.4.0 Attachments 17
    1.4.1 Forks 17
    1.4.2 Booms and Hooks 18
    1.4.3 Personnel Platform 21
    1.4.4 Other Attachments 21
2.0.0 Inspection and Maintenance 24
  2.1.0 Prestart Inspections 24
    2.1.1 Tires 27
    2.1.2 Fuel Service 26
  2.2.0 Preventive Maintenance 26
    2.2.1 Weekly (50-Hour) Maintenance 26
    2.2.2 Long Interval Maintenance 29
    2.2.3 Preventive Maintenance Records 29

3.0.0	Startup and Operating Procedures	30
3.1.0	Safety Guidelines	30
3.1.1	Operator Training and Certification	30
3.1.2	Operator Safety	31
3.1.3	Safety of Co-Workers and the Public	31
3.1.4	Equipment Safety	32
3.1.5	Spill Containment and Cleanup	32
3.1.6	Hand Signals Used with Forklifts	33
3.2.0	Startup, Warm-Up, and Shutdown Procedures	33
3.2.1	Preparing to Work	35
3.2.2	Startup	35
3.2.3	Warm Up	36
3.2.4	Shutdown	36
3.3.0	Basic Maneuvering	36
3.3.1	Moving Forward	36
3.3.2	Moving Backward	37
3.3.3	Steering and Turning	37
3.3.4	Leveling the Forklift	37
3.4.0	Work Activities	38
3.4.1	Load Charts	39
3.4.2	Picking Up a Load	41
3.4.3	Picking Up an Elevated Load	42
3.4.4	Traveling with a Load	42
3.4.5	Placing a Load	43
3.4.6	Placing an Elevated Load	44
3.4.7	Unloading a Flatbed Truck	44
3.4.8	Using Special Attachments	44
3.4.9	Using a Bucket	45
3.4.10	Clamshell Bucket	46
3.4.11	Working in Unstable Soils	48
Appendix A	Typical Periodic Maintenance Requirements	54
Appendix B	OSHA Inspection Checklist	57

Figures

Figure 1	Examples of rough-terrain forklifts	2
Figure 2	Steering modes for forklifts	3
Figure 3	Fixed-mast rough-terrain forklift	4
Figure 4	Fixed-mast rough-terrain forklift specifications	5
Figure 5	Telescoping-boom forklift	6
Figure 6	Example of squirt boom in action	6
Figure 7	Locations of engine and operator cab on a telehandler	7
Figure 8	Components of a telehandler boom	8
Figure 9	Operator cab of a telehandler	9
Figure 10	Telehandler with outriggers spread and set	9
Figure 11	Telehandler operator seat	10

Figures (continued)

Figure 12 F-N-R transmission control on steering column 12
Figure 13 Forklift boom controls 12
Figure 14 Control switches 13
Figure 15 Instrument panel 14
Figure 16 Instrument panel with LCD display 15
Figure 17 Frame level and load stability indicators 17
Figure 18 Forklift with bale handling attachment 18
Figure 19 Cube forks 18
Figure 20 Extension boom with load chart 19
Figure 21 Adjustable angle extension boom with load chart 20
Figure 22 Lifting hook 20
Figure 23 Fork-mounted lifting hook 21
Figure 24 Personnel platform 22
Figure 25 General-purpose bucket 22
Figure 26 Clamshell, or grappler, attachment 23
Figure 27 Concrete hopper for forklift use 23
Figure 28 Fuel/water separator 26
Figure 29 Electrical master switch 27
Figure 30 Coolant fill and overflow tank 27
Figure 31 Wear pads 28
Figure 32 Refuel safely 32
Figure 33 Check all fluids daily 33
Figure 34 Basic hand signals for forklifts 34
Figure 35 Parking brake 35
Figure 36 Forklift level indicator 38
Figure 37 Outriggers in use; front wheels raised 39
Figure 38 Load charts 40
Figure 39 The stability triangle changes due to the terrain and boom position 41
Figure 40 Approach the load squarely 42
Figure 41 Raise the boom to align the forks with the load 42
Figure 42 Tilt the forks back to cradle the load 42
Figure 43 Keep the load low 42
Figure 44 Traveling uphill with a load 43
Figure 45 Place elevated loads carefully 43
Figure 46 Quick coupler 44
Figure 47 Forklift with boom extension and lifting hook 45
Figure 48 Shorter rigging reduces swinging 45
Figure 49 I-pattern for loading 45
Figure 50 Y-pattern for loading 47

Section One

1.0.0 Rough-Terrain Forklift Components

Objective 1

Identify and describe the components of a rough-terrain forklift.
  a. Identify and describe chassis components.
  b. Identify and describe the controls.
  c. Identify and describe the instrumentation.
  d. Identify and describe the attachments.

Trade Terms

Crab steering: A steering mode where all wheels may move in the same direction, allowing the machine to move sideways on a diagonal, also known as oblique steering.

Four-wheel steering: A steering mode where the front and rear wheels may move in opposite directions, allowing for very tight turns, also known as independent steering or circle steering.

Oblique steering: A steering mode where all wheels may move in the same direction, allowing the machine to move sideways on a diagonal. Also known as crab steering.

Powered industrial trucks: An OSHA term for several types of light equipment that include forklifts.

Telehandler: A type of powered industrial truck characterized by a boom with several extendable sections known as a telescoping boom. Another name for a shooting boom forklift.

Tines: A prong of an implement such as a fork. For forklifts, tines are often called forks.

The Occupational Safety and Health Administration (OSHA) uses the term powered industrial trucks for forklifts. OSHA uses the American Society of Mechanical Engineers (ASME) definition that describes a powered industrial truck as being "a mobile, power-propelled truck used to carry, push, pull, lift, stack, or tier material." Powered industrial trucks are classified by their manufacturers and the Industrial Truck Association (ITA) according to the individual characteristics of the trucks. Those classifications are as follows:

- *Class 1* – Electric motor, sit-down rider, counterbalanced trucks (on solid or pneumatic tires)
- *Class 2* – Electric motor, narrow-aisle, trucks (on solid tires)
- *Class 3* – Electric motor, hand trucks or hand/rider trucks (on solid tires)
- *Class 4* – Internal combustion engine trucks (on solid tires)
- *Class 5* – Internal combustion engine trucks (on pneumatic tires)
- *Class 6* – Electric and internal combustion engine trucks (on solid and pneumatic tires)
- *Class 7* – Rough-terrain forklift trucks (on pneumatic tires)

Manual lift hand trucks moved by humans are classified as Class 8 fork trucks. This module focuses only on the Class 7 rough-terrain forklift trucks. They are referred to as either rough-terrain forklifts, or simply forklifts. The module also focuses only on forklifts powered by combustion engines. OSHA and the ASME further separate forklifts into the following groups, based on where they are being used:

- General Industry
- Shipyards
- Marine Terminals
- Longshoring
- Construction

The rough-terrain forklifts grouped under the construction banner are frequently used in logging, agriculture, and general construction. In those environments, the surfaces over which they are driven are often uneven with foreign objects possibly present.

Generally, rough-terrain forklifts are further divided into two broad categories: fixed-mast and telescoping-boom forklifts. A telescoping boom forklift is normally called a telehandler. *Figure 1* shows examples of a rough-terrain fixed-mast forklift and of rough-terrain telehandlers that are often seen on construction sites.

1.1.0 Chassis Components

The frame used in any vehicle is often called its chassis. The frame or chassis is the backbone of the vehicle. The engine, transmission, axles, operator cab, and the main body parts are attached directly to the chassis.

1.1.1 Axles and Steering

Within the fixed-mass forklift and telehandler categories, the forklifts are further differentiated by their drive train, steering, and capacity. Both types have either two-wheel or four-wheel drive. A two-wheel-drive forklift has a drive train that transmits power to the axle and wheels closest to the tines, or forks. That axle and those wheels

FIXED-MAST ROUGH-TERRAIN FORKLIFT

ROUGH-TERRAIN TELEHANDLER FORKLIFTS

Figure 1 Examples of rough-terrain forklifts.

In The Beginning...

The Pennsylvania Railroad put battery-powered platform trucks to work for the first time, using them to move luggage in an Altoona train station in 1906. Between the years of 1917 and 1920, several companies entered the market. Clark built this unit in 1917 and placed it in service at a Michigan plant. It was manually loaded and unloaded, but others quickly saw the value in it. About 75 units were built for sale in 1919. Both Towmotor and Yale & Towne entered the market with their own products around the same time. According to Clark, about 350,000 of their units are presently in use worldwide.

are considered to be the front wheels. The front wheels are often larger than the rear wheels. Two-wheel-drive forklifts typically have rear-wheel steering. The rear wheels move when the steering wheel is turned. Rear-wheel steering provides greater maneuverability. This is especially useful when backing up with a load. Some rear-wheel-drive forklifts may have only one rear wheel. These three-wheeled forklifts are good for general use, but they lack power and stability on rough-terrain.

A four-wheel-drive forklift has a drive train that transmits power to both the front and rear axles. These forklifts are well suited for rough-terrain and other conditions that require additional traction. When four-wheel drive is engaged, power is transmitted to all four wheels.

Rough-terrain forklifts may also have different steering modes, as illustrated in *Figure 2*. Different steering modes enable the forklift to maneuver in smaller spaces. Some models can be operated in the following steering modes:

- The rear wheels can be locked so only the front wheels steer. This is called two-wheel steering.
- All wheels may move in the same direction. This is called crab steering or oblique steering.
- The front and rear wheels may move in opposite directions. This is called four-wheel steering, circle steering, or independent steering. Some manufacturers may also refer to it as round steering.

1.1.2 Fixed-Mast Forklifts

The mast is the upright structure mounted to the front of the forklift chassis. It is a frame on which the fork assembly travels. A typical fixed-mast forklift has a mast that tilts 20 degrees forward and 10 degrees rearward. These types of forklifts are suitable for placing loads vertically and traveling with the load. However, their reach is limited to how close the machine can be driven to the pickup or landing point. For example, a fixed-mast forklift cannot place a load beyond the leading edge of a roof because of its limited reach. For this reason, telescoping-boom forklifts or telehandlers are frequently used in construction. *Figure 3* shows a typical fixed-mast rough-terrain forklift. *Figure 4* shows the specifications for the same forklift.

On a fixed-mast forklift, a fork carriage rides up and down the mast. Most carriages are driven by chain drives. Some forklifts are equipped with one or more hydraulic cylinders that adjust the overall length of the mast to allow the fork car-

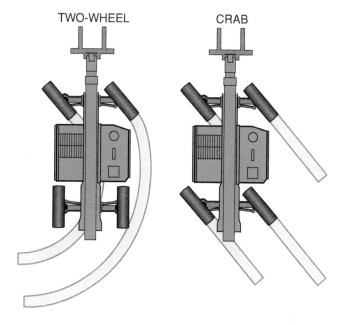

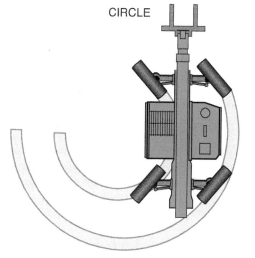

Figure 2 Steering modes for forklifts.

riage to be lifted higher than a forklift without the hydraulic cylinder(s).

> **CAUTION**
> Fixed-mast forklift operators must be careful and watch the top of the mast when passing under any low passageways. If the top of the mast comes in contact with a solid overhead object, the mast and the overhead object may be damaged.

The forklift's forks are anchored on a heavy horizontally mounted rod built into the fork carriage. On most fixed-mast forklifts, the forks must be manually adjusted along the anchor rod to whatever width of separation is desired. On some higher-end fixed-mast forklifts, a hydraulically

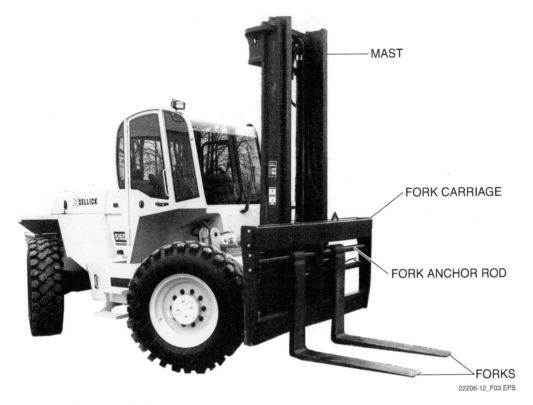

Figure 3 Fixed-mast rough-terrain forklift.

Rotating Telehandlers

This telehandler from Genie has the capability of rotating the entire cab and boom assembly 360 degrees. This allows the machine to set up in a fixed position and rotate around a vertical axis to quickly handle repetitive lifts and similar work activities. The unit is also equipped with four independently controlled outriggers.

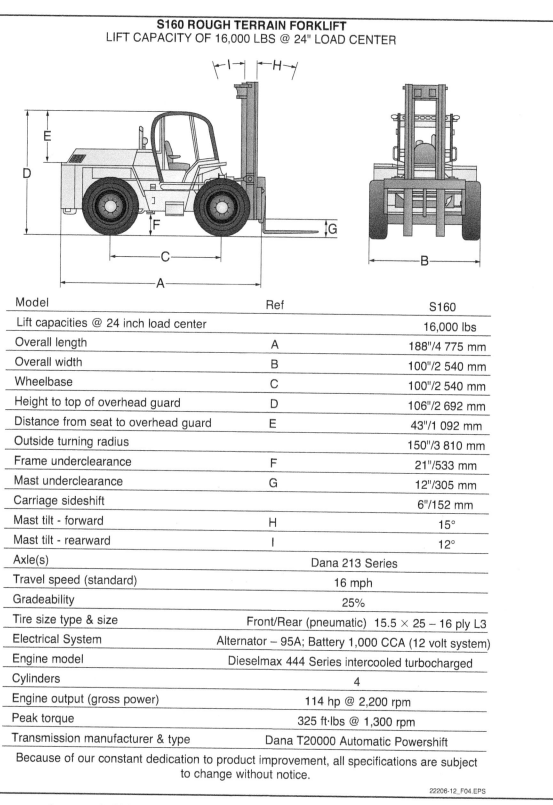

Figure 4 Fixed-mast rough-terrain forklift specifications.

driven device on the carriage allows the operator to spread or narrow the forks without getting off the operator seat. This module focuses on the telescoping-boom forklifts more often found on a construction site.

1.1.3 Telescoping-Boom Forklifts

Telescoping-boom forklifts have more versatility in placing loads. They are also called shooting-boom forklifts or telehandlers. They are a combination of a crane and a forklift. They can easily place loads on upper floors, well beyond the leading edge (*Figure 5*).

Some models of telehandlers have a level-reach fork carriage, sometimes called a squirt boom. A squirt boom allows the fork carriage to be moved forward or backward while keeping the load level and at the same elevation (*Figure 6*). Squirt booms can also extend as the load is being raised, again keeping the load level and rising in a straight vertical plane.

This module describes the operations of telehandlers similar to the SkyTrak and JLG rough-terrain telehandlers often used on construction sites. These machines are typically used for light and medium work, including staging of materials and equipment.

Figure 6 Example of squirt boom in action.

> NOTE: Always read the operator manual before operating equipment. Follow all safety and startup procedures.

Most telescoping forklifts, or telehandlers, have similar parts. To help balance the machine, the engine is mounted along one side of the chassis, and the operator cab is mounted on the opposite side. *Figure 7* shows the engine and operator cab locations on a SkyTrak telehandler's right and left sides.

The boom assembly is centered on and mounted above the main chassis. The boom is attached at a pivot point toward the rear of the chassis. Large hydraulic cylinders lift and lower the boom as needed. Some telehandlers have a single section boom, but most have booms made up of three or more telescoping sections. Hydraulic cylinders and chain drives extend or retract the boom sections. At the end of the last boom section is an attachment tilt cylinder that allows the operator to level or tilt (up or down) an attachment such as a

Figure 5 Telescoping-boom forklift.

Figure 7 Locations of engine and operator cab on a telehandler.

fork assembly. *Figure 8* shows the basic parts of a three-section telehandler boom.

1.1.4 Engine Area

The engine on most telehandlers is mounted on the side of the chassis opposite the operator cab. The engine on later telehandler models has a cover that can be raised to allow a person standing on the ground to inspect and service the engine. Obviously, larger telehandlers may require the operator or service person to use a short ladder or climb up onto the machine to reach the engine area. However, most can be reached while standing on the ground. To locate specific engine components, such as filters, dipsticks, and fill ports, refer to the operator manual for the machine being used. Most telehandlers are powered by diesel engines.

1.1.5 Transmissions

A telehandler's engine provides power for all the machine's operations. The power from the engine is transmitted through a transmission to the forklift's drive axle(s). The engine and transmission systems are similar to those found in most tractors. Most modern forklifts use an electronically controlled hydrostatic transmission. It provides infinitely variable speed within the speed range of the machine. Many have a load-sensing feature that automatically adjusts the speed and power to changing load conditions.

Some forklifts are driven only by a single axle (two-wheel drive) while others are driven by both axles (four-wheel drive). As noted earlier, most forklifts are steered by their rear axle and wheels. Because the forklifts are steered primarily by their rear wheels, it is critical that they not be put into situations where the load is either too heavy or extended too far out in front. This relieves too much weight from the rear wheels, and the forklift may become difficult or impossible to steer.

1.1.6 Loader Hydraulic System Components

The engine of a forklift also powers the hydraulic system that provides power to the machine's steering and lifting devices. The engine supplies power to one or more pumps that build pressure in the hydraulic system. The output pressure of the hydraulic pumps varies with the engine speed. Therefore, hydraulic components operate faster when the engine speed is increased. In some cases, a hydraulic component may not move at all if the machine is only operating at a low idle. The operational controls open and close valves to the hydraulic lines connected to the machine's steering and lifting hydraulic cylinders and pistons. When the hydraulic fluid is applied to the lifting cylinders, the pistons extend and raise the boom. When the flow of hydraulic fluid is reversed, the cylinders retract their pistons and the boom lowers. The hydraulic cylinder(s) located nearest the attachment end of the last boom section tilts the attachment (fork assembly or other) up or down. Ensuring that the hydraulic cylinders and hoses are not damaged

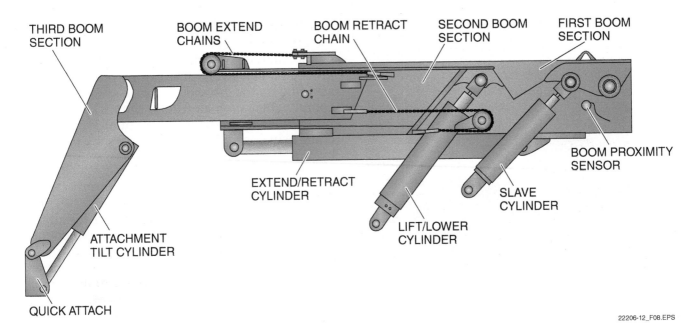

Figure 8 Components of a telehandler boom.

is critically important to the safe operation of the forklift. The same goes for all the hydraulic lines connected to the hydraulic devices.

1.1.7 Operator Cab

The operator cab is normally on the left side of most telehandlers. The cabs are normally installed low along the side of the chassis so that the operator can better see the movements of the boom. Every forklift is somewhat different, but most share similar locations for things like the steering wheel, brake pedal, accelerator pedal, and the joysticks that control the boom and the installed attachment. *Figure 9* shows the general layout of a SkyTrack telehandler's operator cab.

The operator's cab of a forklift serves several purposes. The cab provides some operator protection in case the forklift was to overturn through the rollover protective structure (ROPS). The ROPS identifies a structural integrity and cab safety model that heavy equipment such as rough-terrain forklifts must incorporate into the design to protect the operator. It may provide a controlled environment in which the operator can work comfortably on some models. The cab also serves as the central hub for forklift operations. An operator must understand the controls and instruments before operating a forklift.

Big or small, the operator cabs on most rough-terrain forklifts are laid out in a similar manner. The steering wheel is in the center of the cab. The operator's seat is at the back of the cab and centered on the steering wheel. A dash panel with indicators and some controls is below the steering wheel. The brake and accelerator pedals are near the floor and under the steering wheel. A controls console, usually on the right of the seat, houses most of the controls used on the forklift. Most modern operator cabs also have good heating and cooling systems, as well as adjustable and comfortable seating for the operator.

1.1.8 Outriggers

Telehandlers are equipped with outriggers for stability when it is picking up or dropping off a load (*Figure 10*). To use the outriggers, the forklift must be stopped and put into neutral with its brakes set. When the outriggers are extended, they must be set on firm terrain. After they are properly set, a boom and outrigger interlock allows the operator to extend the boom to its full extension. Without the outriggers in position, boom extension is intentionally limited for safety. The outriggers can be individually controlled to level the telehandler. The specifications indicate the range of correction. For example, the Skytrak Model 6036 can adjust the angle of the forklift up to 10 degrees. The load charts for the telehandler indicate how much load it can handle with and without the outriggers in use.

1.2.0 Controls

All the controls for a forklift or telehandler are located within reach of the operator inside the operator cab. These include the controls for acceleration, braking, steering, and the movement of the boom and any installed attachments. If the machine is

Figure 9 Operator cab of a telehandler.

Figure 10 Telehandler with outriggers spread and set.

equipped with outriggers, they are also controlled from the operator's position in the cab.

1.2.1 Disconnect Switches

Some forklifts are equipped with disconnect switches located outside of the cab. These switches disconnect critical functions so that the machine cannot be operated. Refer to the operator manual to determine where these disconnect switches are located, and what they control.

The disconnect switches must be activated before the machine can be started. Turn them on before mounting the machine. They offer an additional level of safety and security. Unauthorized users are usually unaware of, or cannot locate, these switches and will not be able to operate the equipment. Some of the newer machines may even require a security code to be entered into the machine's computer system before the machine will start.

Some models have a fuel shutoff valve or switch that must be turned to the On position before the machine can be operated. The fuel shutoff valve or switch physically prevents fuel from flowing from the fuel tank into the supply lines. This prevents unwanted fuel flow during idle periods or when transporting the machine, significantly reducing the potential for fuel leaks.

Some machines may also have a battery disconnect switch. When the battery disconnect switch is turned off, the entire electrical system is disabled. This switch should be turned off when the machine is idle overnight or longer to prevent a short circuit or active components from draining the battery. Before mounting the machine, check that the switch is in the On position.

1.2.2 Operator Seat and Steering Wheel Adjustments

Every forklift is slightly different when it comes to adjusting the position of the operator's seat and the steering wheel. Not all forklifts or telehandlers have adjustable steering wheels and seats. The seat in *Figure 11* can be moved up or down, and forward or backward. The operator needs to review the operator manual to see what adjustments can be made on the machine being used. When the seat is in the proper position, the operator should be able to reach all the foot pedals comfortably. When any pedal is fully depressed, there should still be a slight bend at the knee.

After the operator gets the seat and steering wheel positioned to the most comfortable position, he or she needs to make sure that the seat belts are in good condition and usable. After that, the operator can move on to the other controls.

1.2.3 Engine Start Switch

The ignition switch functions can vary widely between makes and models of forklifts. Some only activate the starter and ignition system. Others may activate fuel pumps, fuel valves, and starting aids. In some cases, the starter and starting aids are engaged by other manual controls. The engine start switch on most forklifts is located somewhere near the dash and steering wheel. Most engine start switches have the following four positions:

- *Off* – Turning the key to this position stops the engine. It also disconnects power to electrical circuits in the cab. However, several lights remain active when the key is in the Off position, including the hazard warning light, the interior light, and the parking lights.
- *Auxiliary* – If available, turning the key to this position only applies power to accessory power outlets located on top of the front console, or to devices such as radios that may be connected to accessory power. Power is not supplied to the instrument panel or other electrical controls, other than a battery gauge or indicator.

> **CAUTION**
> Leaving the key in the Auxiliary position without the engine running will drain the battery.

- *On* – Turning the key to the On position activates all of the electrical circuits except the starter motor circuit. When the key is first turned to the On position, it may initiate a momentary instrument panel and indicator bulb check.
- *Start* – The key is turned to the Start position to activate the starter, which starts the engine. This position is spring-loaded to return to the On position when the key is released. If the engine fails to start, the key must be returned to the Off position before the starter can be activated again. To reduce battery load during starting, the ignition switch of some forklifts may be configured to shut off power to accessories and lights when the key is in the start position.

Figure 11 Telehandler operator seat.

> **CAUTION**
> Activate the starter for a maximum of 30 seconds. If the machine does not start, turn the key to the off position and wait 2 minutes before activating the starter again. Overuse of the starter will cause it to overheat.

1.2.4 Vehicle Movement Controls

Forklift movement is controlled in a manner similar to that of cars and trucks. The throttle and brakes are operated by foot pedals. A steering wheel is used to turn the vehicle. Rotating the steering wheel counterclockwise guides the machine to the left. Rotating the steering wheel clockwise guides the machine to the right. The steering on modern forklifts is aided by computer systems built into the machine. The computer system(s) detects the ground speed of the machine and then adjusts the reaction speed of the steering to ensure that the steering is smooth and controlled. This helps prevent oversteering when the forklift is moving too fast to safely execute the maneuver.

Some steering wheels are equipped with a knob. This gives the operator greater leverage and faster control of the wheel when steering with one hand. This is important when the operator is using one hand to steer and the other to operate the boom and forks with the joystick.

One of the things that a forklift operator must always watch when steering is the swing of the forks and the rear of the machine. Since the forklift can be moved with the boom elevated, the operator must also watch where the top of the boom is in relationship to anything overhead.

As noted earlier, the power of a loader's engine is sent through the machine's transmission to the drive axle(s). A hydrostatic transmission is controlled by a forward/neutral/reverse (F-N-R) control that is normally mounted on the steering wheel shaft. *Figure 12* shows a typical F-N-R transmission control mounted to the left of the steering wheel. The selected transmission direction may be indicated on one of the instrument panels in the cab, depending on the model. Models without a hydrostatic transmission likely have a manual gear shift with multiple gears from which to choose.

Forklifts have transmissions with multiple speeds. On most forklifts, the first two or three speeds can be used for either forward or reverse movements. The highest gear is reserved for forward motion only. Do not skip gears when downshifting. For which gears to use, refer to the operator manual for the forklift being used.

> **CAUTION**
> Always come to a complete stop before changing from Forward to Reverse, or vice versa. Changing travel direction while the machine is moving can damage some machines. Although it is possible to change gears while in motion on machines with a hydrostatic transmission, it is not recommended. Changing directions suddenly can also dislodge the load.

1.2.5 Boom and Lift Attachment Controls

The boom and forks on a forklift can be controlled with levers or with a joystick. The forks can be moved in several directions. The forks can be tilted up and down and the boom can be raised and lowered. Some forks can be moved side to side and others can be tilted or angled from side to side. Telescopic booms can be extended and retracted.

Older forklifts are controlled with a series of levers. Each lever controls one aspect of the motion of the forks. For example, some forklifts have three levers. One lever moves the forks up and down, while another tilts them forward and backward. The third lever moves the forks from side to side. Because different makes and models have different controls, review the operator manual to become familiar with the controls before operating any forklift. This module describes joystick controls for fork and boom movement.

The joystick is used in conjunction with switches, triggers, or buttons to control the boom and fork movement. The six ways to move the fork and boom are as follows:

- Boom raise
- Boom lower
- Boom extend
- Boom retract
- Tilt forward
- Tilt back

The speed of the movement is usually determined by both how far the joystick is moved and the engine speed. Increase the engine speed and then move the joystick slowly until the forks are moving at the desired speed. The maximum speed of movement occurs when the joystick is moved to its limit in some direction, and the engine speed is high. When the joystick is released it returns to the central or neutral position and movement stops. Avoid sudden movement with the forks, as that can dislodge the load or cause accidents.

A typical control arrangement is shown in *Figure 13*. It is important to note that joystick control layouts can vary dramatically. Even though joy-

22206-13 Rough-Terrain Forklifts

Figure 12 F-N-R transmission control on steering column.

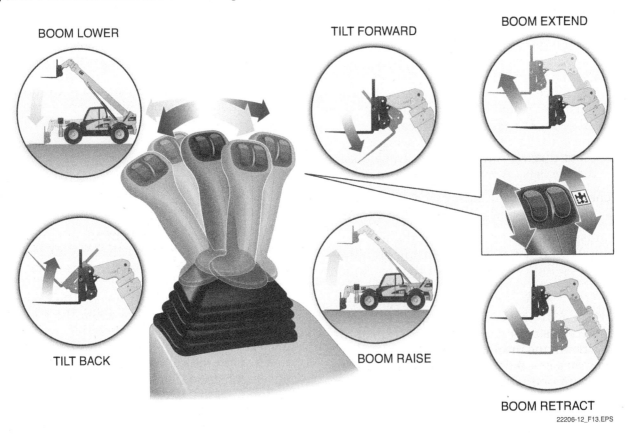

Figure 13 Forklift boom controls.

sticks may appear to be the same, they may have different button and switch assignments from one model to another. Make sure to review and clearly understand the controls shown in the operator manual of the machine being used. Never make assumptions about joystick controller actions.

For the joystick shown, right and left movement of the joystick tilts the forks forward and back. Backward and forward movement of the joystick raises and lowers the boom. On this joystick, there are two sliding switches shown on top. The sliding switch on the left retracts and extends the telescopic boom. The switch on the right raises and lowers the stabilizers. For these two switch-actuated movements, the speed at which the boom extends and retracts depends only on the engine speed.

With a joystick, multiple movements can be made simultaneously. Moving the joystick diagonally tilts the forks while raising or lowering the boom. The switch can also be activated to retract or extend the boom while the joystick is off-center. With practice, an operator will be able to move the boom and forks smoothly in several directions at once.

The carriage on some forklifts can be shifted from side to side or tilted up to 10 degrees to either side. The tilt feature allows the operator to position the forks to pick up a load that is not level. This is particularly useful on rough terrain. The side-shift feature allows the forks to move side to side horizontally while the boom remains stationary. The operator can precisely position the forks under a load without repositioning the forklift.

1.2.6 Control Switches

Each model of forklift has its own set of control switches (*Figure 14*). In most cases, the switches are located on the dash in front of the operator, off to the operator's right or left side. They activate the features and functions of the forklift. Some switches are used in conjunction with other controls, including the steering mode select and the quick coupler. The following are some examples of control switches:

- *Service brake or parking brake* – A control that locks down the machine's brakes.
- *Quick coupler* – Allows the operator to disconnect from one attachment and hook onto another attachment.
- *Steering mode select* – Allows the operator to change the speed of the machine's steering. This option is especially important when working in close quarters.

Figure 14 Control switches.

- *Fog lights* – Allows the operator to have some additional lighting when needed.
- *Left stabilizer* – Allows the operator to operate only the left stabilizer.
- *Right stabilizer* – Allows the operator to operate only the right stabilizer.
- *Work lights* – Allows the operator to control the lights on the front or rear of the operator cab, and any lights that may be mounted on the boom.
- *Rotating beacon* – Some jobs require forklifts to have a rotating beacon on and operating when the machine is in motion.

1.2.7 Special Features

There are several features that override the normal functions of a forklift. These include the transmission neutralizer and the differential lock. These features are an important part of operating the forklift. They offer added control of the machine in specialized situations.

The transmission neutralizer switch is a two-position switch. With the switch activated, the transmission is disengaged (neutralized) when the service brake is applied. With the switch turned off, the transmission remains engaged when the service brakes are applied. It can be considered a clutch that is interlocked with the brake pedal.

The differential lock control overrides the normal operation of the front axle differential. The differential normally allows the drive wheels to rotate at different speeds when turning the forklift. Locking the differential disables this feature.

Torque is then transmitted equally to both wheels, even though one wheel may not have good traction. This helps maintain traction when ground conditions are soft or slippery.

If one or both wheels start to spin, release pressure on the accelerator until the wheels stop spinning. Engage the differential lock and increase pressure on the accelerator. Once clear of the area, release pressure on the accelerator, and release the switch to disengage the differential lock. When using the differential lock, the forklift should be driven straight ahead or in reverse; only very mild and shallow turns should be made.

> **CAUTION**
> Limit steering maneuvers while the differential lock is engaged. Steering maneuvers with the differential lock engaged can damage the machine.

1.2.8 Operator Comfort and Other Controls

There are several controls designed to adjust the seat and controls for maximum operator comfort. Adjust the seat, armrests, mirrors, and cab climate controls while the machine is warming up or parked. Do not move or operate the machine before adjusting the seat and mirrors for comfort and visibility. The seat belt should be worn any time the engine is running.

> **WARNING!**
> The seat belt must be worn at all times. Failure to use a seat belt may result in serious personal injury or death.

Seat and mirror adjustments affect safety as well as operator comfort. Operators who are adjusting the seat or mirrors after beginning operations are not giving their full attention to machine operations. If an operator cannot reach all of the controls, the machine may be hard to control and cause injury or property damage. Properly setting the climate controls reduces operator fatigue and increases alertness. Failure to properly set the climate controls can also reduce visibility if the windshield is fogged.

If the machine will be used at night, the operator must become familiar with all of the light switches. Once it is dark, it is too late to look for the light switches. Make sure that all lights are functioning properly. Adjust the lights so that the work area is properly illuminated.

1.3.0 Instruments

Many of the instruments on a forklift are very similar to those found on a car or truck driven on the street. Among other things, the instruments tell the operator the status of the vehicle's engine and cooling system. They also let the operator know when something is wrong with the vehicle. Forklifts also have instruments that tell the operator how level the machine is, or where the boom is positioned.

An operator must pay attention to the instrument panel of any machine being operated. There are several warning lights and indicators that must also be monitored. An operator can seriously damage the equipment by ignoring the instrument panel.

The instrument panel can vary dramatically on different makes and models of forklifts. Generally, they include indicators for the engine coolant temperature, transmission oil temperature, fuel level, and service hour meter (*Figure 15*). Other types of forklifts may have different gauges, such as the one shown in *Figure 16*. Refer to the operator manual for the specific gauges on the machine you are operating.

1.3.1 Fuel Level Gauge or Indicator

A fuel gauge indicates the amount of fuel in the forklift's fuel tank. On diesel engine forklifts, the gauge may contain a low-fuel warning zone. On forklifts with analog gauges (those with needle

Figure 15 Instrument panel.

Figure 16 Instrument panel with LCD display.

movements), the warning zone is usually yellow. On models with a digital readout, a low fuel warning light may turn on when the fuel is too low. Avoid running out of fuel on diesel engine forklifts because the fuel lines and injectors must be bled of air before the engine can be restarted.

1.3.2 Engine Coolant Temperature Gauge or Indicator

The engine coolant temperature gauge indicates the temperature of the coolant flowing through the engine cooling system. Refer to the operator manual to determine the correct operating range for normal forklift operations. Temperature gauges normally read left to right, with cold on the left and hot on the right. If the gauge is in the white zone, the coolant temperature is in the normal range. Most gauges have a section that is red. If the needle is in the red zone, the coolant temperature is excessive. Some machines may also activate warning lights if the engine overheats.

> **CAUTION**
> Operating equipment when temperature gauges are in the red zone may severely damage it. Stop operations, determine the cause of the problem, and resolve it before continuing to operate the equipment.

If the engine coolant temperature gets too high, stop immediately. Get out of the cab and investigate the problem. There are three primary causes of the engine overheating: low coolant level, a nonfunctional cooling fan, or blocked air flow through the radiator. An operator needs to check them all before turning the machine in for service. These three issues can often be resolved easily on site with a fan belt, water, and/or an antifreeze solution.

Never try to open the cap on a radiator while the coolant is still hot. While waiting for the coolant to cool, check the radiator fins to ensure that they are not fouled. If they are fouled or clogged, clean them as necessary.

> **WARNING!**
> Engine coolant may be extremely hot and under pressure. Check the operator manual and follow the procedure to safely check and fill engine coolant. Accessing the coolant system while pressures and temperatures are high can result in severe burns or scalding.

Next, check to see if the fan belt is loose or broken. Replace it if necessary. If the radiator fins are clear and the fan blade and belt are good, check the radiator and its hoses for leaks. If no leaks can be seen, and the radiator has cooled, open the radiator cap and check the level of coolant inside the radiator. Add coolant and/or an antifreeze solution as needed. Restart the engine and monitor the temperature gauge. If the temperature goes up again, stop operations and take the machine out of service. If the issue is not related to these three common maintenance issues, then an internal problem, such as a thermostat or water pump, is likely.

1.3.3 Transmission Oil Temperature Gauge or Indicator

The transmission oil temperature gauge indicates the temperature of the oil flowing through the transmission. This gauge also reads left to right in increasing temperature. It has a red zone that indicates excessive temperatures. When the weather is colder, allow the transmission oil to warm up sufficiently before operating the machine. Some models may not provide a gauge showing the present transmission oil temperature. Instead, only a warning light may be provided, or a digital display may need to be changed to a different display mode to view the temperature. Most rough-terrain models have a transmission fluid cooler, constructed similar to the radiator and located in the same area.

1.3.4 Hydraulic Oil Temperature Gauge or Indicator

As noted earlier, forklifts use hydraulic cylinders to lift and lower the boom, and to control the tilt of the attachment at the end of the boom. Hydraulic system pressure is also used to help steer the forklift. Any time the fluid in a hydraulic system is being used to perform work, its temperature increases. The hydraulic fluid returning to the hydraulic system's reservoir is filtered and cooled before it is reused. Sensors in the hydraulic system show the temperature of the hydraulic fluid. Some forklifts may not have a gauge or indicator for the hydraulic oil temperature, but many will. Others may have only a warning light when the fluid temperature rises too high. When the weather is colder, the forklift operator may need to operate the boom and fork assembly several times to build up some heat in the hydraulic fluid. Like the transmission, the hydraulic system also has a radiator-like cooling coil, usually in the same area as the radiator.

1.3.5 Service Hour Meter

The service hour meter on any machine is important because it indicates the total hours of operation. It indicates the period of time the machine has been running over the life of the machine. It cannot be reset, since the total number of hours needs to be tracked for a variety of reasons.

The number of hours a machine has been in service determines when the machine needs to have its next periodic maintenance performed. The operator must keep track of the hours since the last service using an hour meter and a service log. Periodic maintenance is covered in more detail later.

1.3.6 Speedometer/Tachometer

Some machines also have a tachometer and speedometer. A tachometer indicates the engine speed in revolutions per minute (rpm). A speedometer shows the machine's ground speed. Typically they can be set for either miles per hour (mph) or kilometers per hour (kph).

1.3.7 Other Indicators

Some forklifts have a series of lights above the steering wheel or on the dash behind the steering wheel. These lights show the operator when various features are activated under normal operating conditions, or they may alert the operator to a problem. Indicator lights show the status of a component or system, while warning lights indicate a dangerous or undesirable condition. Two warning lights, for example, are the alternator light and oil pressure light. If the alternator light is on, the battery is not receiving a charge. If the oil pressure light is on, the oil pressure is too low. Stop the machine and resolve the problem before continuing operations. Typical indicator lights show the activation of the following features:

- Left and right turn signals
- High-beam headlights
- Circular steering
- Crab steering

> **CAUTION**
> A flashing warning light requires immediate attention by the operator. If a warning light illuminates, stop working and investigate the problem. Ignoring such indicators may cause serious damage to the equipment.

1.3.8 Frame Level Indicator and Longitudinal Stability Indicator

Just above the front windshield or front opening of most forklifts is an indicator that shows the level of the forklift's frame (*Figure 17*). This is normally called the frame level indicator. When its pointer moves too far to the left or right of zero (0), the forklift has the potential of rollover.

Just below and to the right of the frame level indicator is a load stability indicator that is actually mounted on the side of the boom. Load stability is an important consideration in forklift operations. The longitudinal stability indicator is located on the right side of the front windshield or front opening in the operator cab. This way, the operator can easily monitor the indicator and the load at the same time. This indicator determines the forward stability of the load. It produces visual (and possibly audible signals) to indicate the limits of forward stability of the machine. The indicator alerts the operator to make adjustments to the machine or how it is being operated. Carefully monitor this indicator as the load is being raised to maintain load stability. If the machine senses that a dangerous load stability level is being approached, an audible alert may sound. Such issues often include audible warnings, since the operator's eyes are typically focused on the load or the surroundings.

For example, if the forklift is equipped with a digital readout system, the digital system may show the machine's forward stability went from 20 to 80 percent of its range while the boom was being extended. When the stability reaches 85 percent of its range, an indicator light comes on and an alarm may sound intermittently. At this point, the operator must not increase the outreach of the boom any further. In addition, the boom may need to be retracted some before the load is lowered. If the stability reaches 100 percent of the maximum allowed, red lights illuminate and the

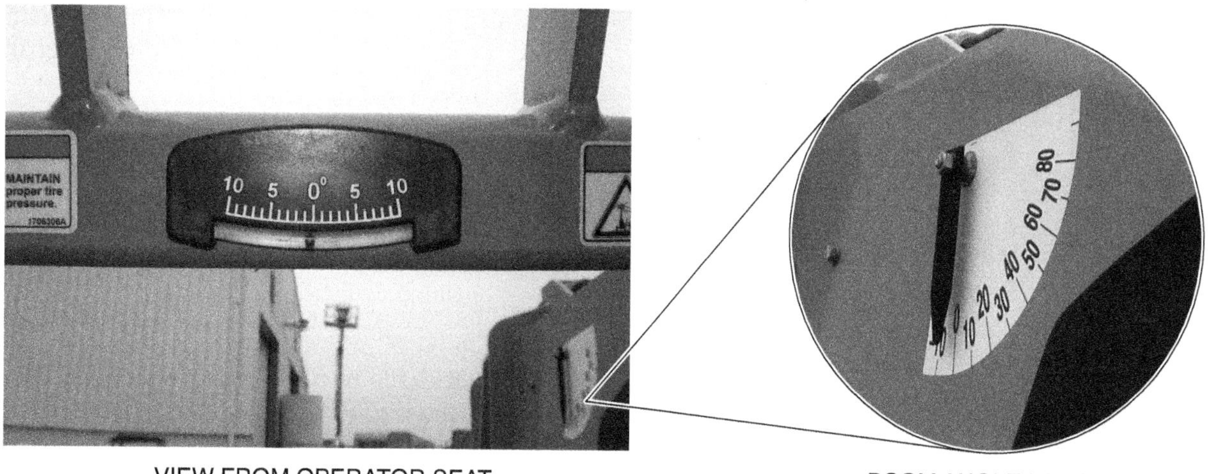

VIEW FROM OPERATOR SEAT

BOOM ANGLE INDICATOR

Figure 17 Frame level and load stability indicators.

alarm sounds continuously. The operator must immediately lower the load or retract the boom to avoid an accident.

> **WARNING!**
> Extending the boom or lowering a raised boom increases the outreach of the load and reduces forward stability. If you are approaching the machine's stability limit, retract the boom before lowering it. Otherwise the machine could tip over and cause personal injury or death.

If the forklift being used has a digital system, it must be tested daily. The load stability indicator can be adjusted for operations on tires or with the stabilizers lowered. The visual and audible alarms may operate momentarily if the machine is carrying a load close to its maximum capacity, especially when traveling over rough terrain.

1.4.0 Attachments

The main attachment on a forklift is the forks. However, on many types of forklifts, the forks can be detached and replaced with different forks or other attachments. The length and configuration of the forks can be changed to maximize productivity when handling certain materials. Some forks are designed to handle specific materials like pallets, cubes of brick and block, or sheet metal. Forks are also available in different lengths.

> **CAUTION**
> The lifting capacity of the forklift changes any time a different set of forks or another attachment is used. Load charts must be rechecked for the different weights. Also, ensure that any new set of forks or attachment is securely fastened before use.

In addition to lifting and moving, forklifts can be used for other operations. Different attachments, such as sweepers, buckets, or hoppers, can be fitted onto some forklifts. A forklift can be used as a raised workstation when an access platform is attached. These attachments greatly expand the capabilities of the forklift and increase its usefulness on the construction site. Use the proper attachment to lift loads safely. Although a sling can be attached to the forks, it is safer to use the boom and hook attachment to lift loads from the top. Make sure that the attachment is certified by the manufacturer for the intended lift.

> **WARNING!**
> Only use attachments that are designed for the machine. Using attachments that were not designed for the specific model forklift can cause a serious accident and/or equipment failure.

Some forklift attachments are designed for use in specific industries such as farming or waste handling. A bale handler is attached to the forklift shown in *Figure 18*. The forklift can easily move bales of scrap even though they are not palletized. The attachment is hooked up to the forklift's hydraulic system. The bale grabber is operated using auxiliary switches on the forklift controls.

1.4.1 Forks

Forks are available in various sizes to handle different types of materials, including standard forks, cube forks, and lumber forks. Standard forks are 48 inches long, 2 inches thick at the head, and 4 inches wide. Cube forks (*Figure 19*) have narrower, 2-inch wide tines. They are used in sets of four or more, rather than two. Lumber

Figure 18 Forklift with bale-handling attachment.

Figure 19 Cube forks.

forks are normally longer and wider than the others, at 60 inches long and 7 inches wide.

Some carriages have fixed forks. On other models, the space between the forks can be adjusted. To manually adjust the forks, lift the tip of the fork and tilt it upward. Slide it along the carriage until it reaches the desired position. Lower the tip of the fork to lock it into place.

> **WARNING!**
> Manually adjusting the forks poses a severe pinch hazard. Read the operator manual and follow all safety precautions. Do not drop forks into place. Lower them carefully to avoid crushing your fingers or hands.

The forklift's hydraulic system can be used to power an attachment installed over the forks. For example, some forks are designed to swing from side to side. The carriage is connected to the forklift's hydraulic system. The auxiliary hydraulic switches are used to rotate the carriage up to 45 degrees to either side. This provides better maneuverability and allows loads to be lifted or placed when the forklift cannot be directly perpendicular to the load. On rough terrain, this is a common problem.

1.4.2 Booms and Hooks

Some loads are not palletized or easily lifted from the bottom. These loads must be lifted from the top using slings and a lifting hook. The lifting hook and boom extension allow the forklift to operate like a crane. This arrangement is often used in construction to place trusses or to set pipe. The operator should always remain at the controls while a load is suspended from either the boom extension or the lifting hook.

Boom extensions can be either fixed angle (no angle adjustments) or adjustable angle. *Figure 20* shows a Caldwell fixed angle telescoping (extension) boom with its associated load chart. Note that the boom of the attachment must be extended manually and a pin is placed through aligned holes at the desired point. It is not operated hydraulically. This is typical for this attachment. If the boom extension is used on a telehandler, the telehandler boom can still extend or retract.

The forks of the forklift are inserted into the boom slots of the boom extension's main body. After the forks are fully engaged with the extension, one or more chains or straps are used to secure the extension to the mass of the forklift.

The boom of such extension booms can be extended by one-foot segments. With this Caldwell boom fully retracted (Model FB-30), the boom is capable of lifting 3,000 pounds (assuming that the forklift has that capability). Note that the load chart shown in *Figure 20* is for the extension boom only, and not the forklift itself. For this discus-

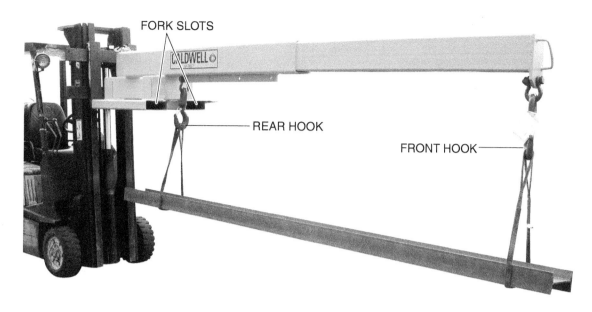

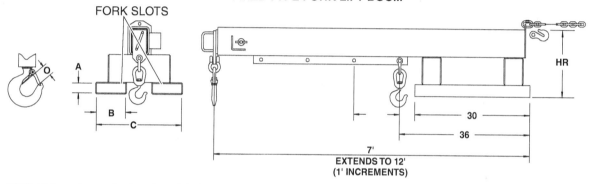

NOTE: All dimensions on drawings shown in inches unless stated.

SPECIFICATIONS

Model Number	Dimensions (in.)					Maximum Capacity at Hook Position (lb.)							Weight (lb.)
	A	B	C	HR	0	3'–6'	7'	8'	9'	10'	11'	12'	
FB–30	2½	7½	22	16	1.00	3,000	3,000	2,600	2,200	1,900	1,600	1,500	340
FB–40				16	1.09	4,000	3,200	2,600	2,200	1,900	1,600	1,500	340
FB–60	2½	7½	22	17	1.36	6,000	5,000	4,200	3,500	3,000	2,700	2,500	390
FB–80				18	1.61	8,000	7,000	5,700	4,800	4,100	3,600	3,100	520

Capacities are for boom only and are dependent on fork lift used. Check with forklift manufacturer for load capacities before use.

Figure 20 Extension boom with load chart.

sion, assume that the forklift can lift the weights shown. When the boom is extended out to 7 feet, the boom is still capable of lifting 3,000 pounds, but extend it another foot and the lifting capacity is reduced to 2,600 pounds. Extending the boom out to 10 feet results in a maximum capacity of 1,900 pounds. When the boom is fully extended to 12 feet, it can only lift 1,500 pounds.

> **WARNING!**
> When an extension boom is used on the forks of a forklift, the forklift's overall center of gravity is changed proportionally. Also, the additional weight of the extension boom and any rigging materials (clevises, hooks, and straps) must be calculated into the overall amount of weight that the forklift can lift. Refer to the operator manual, or to the necessary tables and charts, to see how an extension boom changes the forklift's center of gravity and capacity.

As mentioned earlier, some extension booms can be adjusted to different elevation angles. Such booms have anchor pins at their base that allow the angle of the boom to be moved. The pin is removed so that the boom can be manually moved up or down to a new position. After the boom is reset, the pin is replaced and the boom is secured in its new location. *Figure 21* shows an adjustable extension boom from the Vestil Manufacturing Corporation. Note that the capacity of the boom is always higher when the load is closest to the forklift. This occurs when the boom is fully retracted and angled up to its maximum angle. = max cap.

Vestil extension booms come in different lengths and weights. A boom capable of lifting 8,000 pounds weighs more than the one capable of lifting 4,000 pounds. The weight of the extension boom must be taken into consideration when determining how much total weight the forklift can actually lift.

When the Vestil 8K boom is lowered to 0 degrees and fully retracted (at the 36-inch point), the boom can lift up to 8,000 pounds at its outer hook, assuming that the forklift is capable of lifting that much weight. The further the boom is extended, the less weight the boom can handle at its outer hook. If the boom is fully extended and then moved to a higher elevation angle, the amount it can lift changes proportionally. When the boom is elevated, anything being lifted by its outer hook comes closer to the boom's base. The closer the load is to the boom's base, the more load the boom can lift. Refer to the load charts in the owner manual of the boom being used.

> **WARNING!**
> Do not exceed the rated lift capacity of the forklift, the extension boom, or any rigging component used to lift the load.

Any time that a forklift is equipped with an attachment that allows it to be used as a crane, such as the extension booms, OSHA guidelines may require that only certified riggers rig the load being lifted. Check with the local safety representative to clarify who is allowed to rig loads lifted by forklifts equipped with extension booms and lifting hooks.

Another lifting device is shown in *Figure 22*. The hook is mounted on a truss boom that is attached to the end of the forklift boom. Only use the boom extension and lifting hook with loads that can be rigged using chains or slings. This particular model also has a hydraulic winch to raise and lower the load without moving the boom. It uses the hydraulic pressure source normally used to tilt or rotate the forks to power the hydraulic winch motor. A different style of lifting hook is shown in *Figure 23*. This attachment simply slips over the forks and provides a proper rigging point.

> **WARNING!**
> Never rig directly to a set of forks. The rigging could slip off or the fork edges could cut the rigging. Always use the appropriate rigging methods to secure a load to the hook. Improper rigging can fail unexpectedly, causing the load to fall and cause injury or property damage. Use tag lines to prevent the load from swinging.

1.4.3 Personnel Platform

Some forklifts can be equipped with an articulating personnel platform (*Figure 24*). This provides an elevated workstation for two workers and their tools and materials. The platform uses the forklift's hydraulic system and is operated with controls located on the platform itself. Once the machine is positioned and the stabilizers lowered, the occupant controls the movement of the platform. However, unlike boom lifts designed specifically for this purpose, the forklift cannot be driven from the platform.

As an additional safeguard, other machine controls are locked out with a key switch. The lockout switch disables the stabilizers, frame level, transmission control, and quick coupler. This prevents the machine from being moved while the platform is raised. Moving the machine while workers are on the platform is dangerous and must be prevented.

> **WARNING!**
> Always use the proper fall prevention and arrest equipment when working on an elevated workstation. Follow all safety procedures for operating the access platform. Improper operation of the platform could result in injury or death.

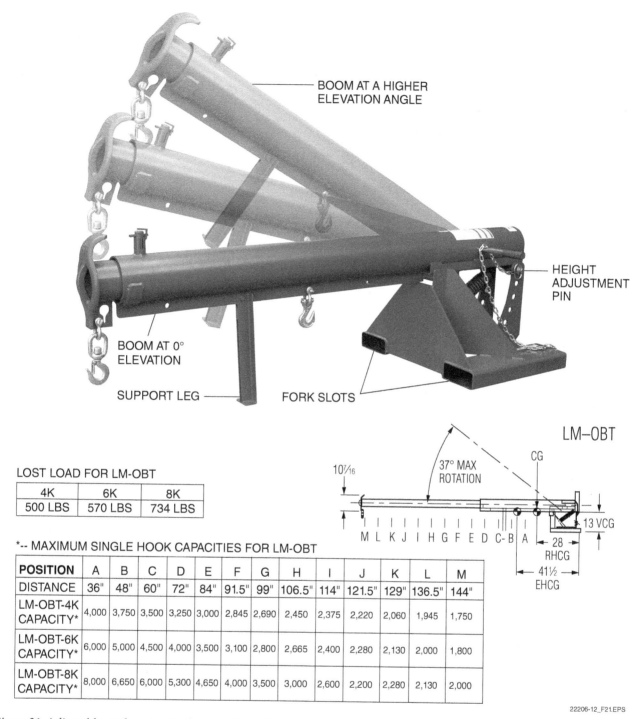

Figure 21 Adjustable angle extension boom with load chart.

Review the operator manual or other literature before operating the access platform. Follow all safety precautions to minimize hazards from slips and falls. Keep the platform clear of power lines, and do not allow the platform to contact any structure. An operator should not move the forklift unless the access platform is lowered to 18 inches or less off the ground, and is unoccupied.

Note that not every personnel platform is articulating. A simple personnel basket can be mounted and secured to the forks for easy aerial tasks that require very limited movement of the basket itself. Remember to always use proper fall arrest and/or fall prevention PPE while in a personnel basket.

Figure 22 Lifting hook.

Figure 24 Personnel platform.

Figure 23 Fork-mounted lifting hook.

1.4.4 Other Attachments

Other types of attachments for a forklift include concrete buckets, hoppers, utility buckets, multipurpose buckets, augers, and sweepers. These attachments enable the forklift to be used for a wide variety of tasks on the construction site.

Several types of buckets can be used with the forklift. General-purpose buckets (*Figure 25*) are used for a variety of activities from digging to loading loose material, such as dirt or gravel. There are several bucket sizes for different types of materials. Light material buckets are generally larger than the standard bucket. Light material buckets are used for low-density materials such as wood chips.

A multipurpose bucket allows the bucket to function several different ways. It is also known as a clamshell or a grappler (*Figure 26*). It can be used

Figure 25 General-purpose bucket.

as a standard bucket, a clamp, or a controlled-discharge bucket. It is also connected to the machine's hydraulic system and is operated with the auxiliary switches on the joystick controls.

Concrete hoppers and buckets (*Figure 27*) allow the forklift to carry loose or wet materials. Hoppers can be manually dumped. They can lift ½ or 1 cubic yard of concrete over columns or other forms. The concrete bucket has a clamshell gate on the bottom that is opened to place the concrete.

Figure 26 Clamshell, or grappler, attachment.

WARNING! When moving a concrete hopper, the contents can easily shift as the machine moves. This could result in an imbalance that may cause the machine to tip over, especially when the hopper is high in the air.

The function and operation of an attachment is described in the manufacturer's literature and the operator manual. Always review the instructions before using any equipment or attachments. Improper operation can result in personal injury, machine damage, or property damage.

Figure 27 Concrete hopper for forklift use.

WARNING! Only use attachments that are compatible with the machine. Using nonstandard attachments could damage the equipment or cause serious personal injury.

1.0.0 Section Review

1. The specific steering mode that allows the front and back wheels to move in different directions is known as _____.
   a. crab steering
   b. locked steering
   c. oblique steering
   d. four-wheel steering

2. The maximum amount of time that the starter should be engaged is _____.
   a. 15 seconds
   b. 20 seconds
   c. 30 seconds
   d. 45 seconds

3. The service hour meter is reset after each maintenance action is performed.
   a. True
   b. False

4. Cube forks are different from standard forks in that the tines are _____.
   a. wider
   b. narrower
   c. longer
   d. shorter

Section Two

2.0.0 Inspection and Maintenance

Objective 2

Describe the prestart inspection requirements for a rough-terrain forklift.
a. Describe prestart inspection procedures.
b. Describe preventive maintenance requirements.

Performance Task 1

Complete proper prestart inspection and maintenance for a rough-terrain forklift.

Trade Term

Power hop: Action in heavy equipment that uses pneumatic tires to create a bouncing motion between the fore and aft axles. Once started, the oscillation back and forth usually continues until the operator either stops or slows down significantly to change the dynamics.

Preventive maintenance is an organized effort to regularly lubricate and service the machine in order to avoid poor performance and breakdowns at critical times. Performing preventive maintenance on a forklift keeps it operating more efficiently and safely, and may help avoid costly and untimely failures.

The preventive maintenance requirements for forklifts are essential, but most tasks are quite simple, if you have the right tools and equipment. Preventive maintenance should become a habit, and be performed on a regular basis. Inspect and lubricate the machine on a daily basis. Be aware of hours of service and have the machine serviced at the appropriate intervals.

Many maintenance activities are performed after certain intervals of service. Others, however, must be done daily. Daily maintenance is typically done before the machine is started at the beginning of the workday. This group of activities done at the beginning of the day is usually referred to as a prestart inspection. The operator manual includes lists of inspections and servicing activities required for each time interval. *Appendix A* is an example of typical periodic maintenance requirements.

Maintenance time intervals for most machines are established by the Society of Automotive Engineers (SAE) and adopted by most equipment manufacturers. Instructions for preventive maintenance are usually in the operator manual for each piece of equipment. Typical time intervals are: 10 hours (daily); 50 hours (weekly); 100 hours, 250 hours, 500 hours (quarterly); and 1,000 hours (semi-annually). Note that the comparison of operating hours to days, weeks, and months is based on a typical 40 to 50 hour workweek. Forklifts may operate around the clock at some sites, although rough-terrain models are less likely to operate this way. When they do, an inspection that is typically done semi-annually can quickly become one that is required monthly, based on the operating hours accumulated.

2.1.0 Prestart Inspections

The first thing to do each day before beginning work is to conduct the prestart, or daily, inspection. Some companies provide an inspection checklist that must be completed daily. A model inspection checklist for forklifts that was developed by OSHA is included in *Appendix B*. The prestart inspection obviously should be done before starting the engine. It will identify any potential problems that could cause a breakdown and indicate whether the machine can be operated safely.

Note that the prestart inspection should not be the only time that the condition of the equipment is considered. Operators should always be attentive to the condition of the forklift while it is in use, and after it is shut down as well.

The prestart inspection is sometimes called a walk around. The operator should walk completely around the machine checking various items. Many manufacturers provide a diagram of the suggested walk-around path, noting items to check as the machine is circled. The following items are typically checked and serviced on a prestart inspection:

> **WARNING!**
> Do not check for hydraulic leaks with your bare hands. Use cardboard or a similar material. Pressurized fluids can cause severe injuries to unprotected skin. Long-term exposure to these fluids can cause cancer or other chronic diseases.

- Look around and under the machine for leaks, damaged components, or missing bolts or pins. Fluids that can leak include battery electrolyte, fuel, hydraulic fluid, engine oil, and engine coolant.
- Inspect the cooling system for leaks or faulty hoses. Remove any debris from the radiator. If

a leak is observed, locate the source of the leak and fix it. If leaks are suspected, check fluid levels more frequently.

- Inspect all attachments and implements for wear and damage. Make sure there is no damage that would create unsafe operating conditions or cause an equipment breakdown.
- Inspect all hydraulic cylinders and actuators. Wipe surfaces clean of any debris that could damage the seals.
- Inspect the chassis and main structure for cracks in welds or parent metals.
- Carefully inspect any and all sensors, limit switches, and other electrical devices. These devices provide the information that initiates alarms or status indicator lights. If they have been damaged or dislodged, they may not function which creates an unsafe condition.
- Check to ensure safety decals are in place, as well as placards that provide operating information, load capacity charts, and travel limits for moving components.
- Inspect and clean steps, walkways, and handholds.
- Inspect the engine compartment and remove any debris. Clean access doors.
- Inspect the ROPS for obvious damage.
- Inspect the hydraulic system for leaks, faulty hoses, or loose clamps.
- Inspect the lights and replace any broken bulbs or lenses.
- Inspect the axles, differentials, wheel brakes, and transmission for leaks.
- Inspect tires for damage and replace any missing valve caps. Check that the tires are inflated to the correct pressure and with the correct substance.
- Check the condition and adjustment of drive belts on the engine.
- Inspect the operator's compartment and remove any trash.
- Inspect the windows for visibility and clean them if needed.
- Adjust the mirrors.

Most manufacturers require that daily maintenance be performed on specific parts or systems. These items are usually identified as 10-hour maintenance items. Checking the engine oil level, using the dipstick, is a common daily requirement. Although dipstick markings vary, the indication of a Safe level and a Full level are marked in some manner. The oil level should be between those markings, but not above the Full level.

Another common 10-hour inspection item for forklifts is the hydraulic fluid level. Most machines have a vertical sight glass to view the fluid level. The sight glass usually has markings to indicate the safe operating level. If the fluid level is low, locate the fill cap and fill the reservoir to the proper level.

> **CAUTION**: Operating the forklift with a low hydraulic fluid level could cause a sudden failure of hydraulically operated components, resulting in equipment or load damage.

2.1.1 Tires

Forklift tires can be inflated with one of several substances including air, nitrogen, or a liquid mixture. Tires inflated with liquid ballast provide added machine stability. The liquid may be a mixture of water, corrosion inhibitors, and calcium chloride; the latter ingredient provides antifreeze protection.

> **CAUTION**: Using a liquid ballast product in tires that lack corrosion-inhibiting ingredients (such as plain water) can cause severe and rapid corrosion of the wheel, from the inside out. Ensure that only products designed for this purpose are used.

Heavy equipment sometimes suffers from a condition known as power hop. Rough-terrain forklifts, as well as tractors, loaders, and other heavy equipment using inflatable tires have the potential for power hop to occur. Heavy equipment with a short wheelbase tend to experience its ill effects more so than longer machines. The machine may bounce or hop during operation, causing a poor ride, load instability, a loss of traction, and excessive wear and tear on the equipment. Operators usually have to slow down and operate in a lower gear to reduce the effects. The conditions that are likely to cause power hop are: dry soil, especially when it is a loose mixture over a hard sub-layer; pulling soil implements that require a great deal of torque; and improperly adjusted or connected implements.

Liquid ballast has been used in heavy equipment tires for many years. Ballasting helps prevent slipping by increasing traction, and also may improve fuel economy. A liquid ballast remains near the bottom of the tire as it turns, concentrating weight in the traction area. Adjusting the percentage of inner tire area filled with liquid helps to resolve power hop.

Power hop can often be controlled by stiffening front tires using higher inflation pressures and/or adding more liquid ballast. Rear tires are softened by reducing the inflation pressure and/or removing liquid ballast in some cases. The combination of changes to front and rear tires can significantly reduce power hop.

The operator needs to know if liquid ballast is being used, and what the appropriate tire pressure should be. Regardless of what is used inside the tires, check the pressure daily to make sure it meets the recommended pressure in the operator manual. While checking the inflation pressure, also inspect the tires for significant damage and excessive wear. Tires that have suffered cuts or tears that expose the underlying cords should be considered unsafe and be taken out of service. Also check to ensure that all the lug nuts are present and tight on the wheels.

2.1.2 Fuel Service

Most rough-terrain forklifts operate on diesel fuel, but some require gasoline. Although no engine benefits when the fuel supply is allowed to run out, diesel engines are especially sensitive to damage or operating problems. It is essential that diesel engines not be allowed to run out of fuel while in operation.

It is best to refuel the machine at the end of the day. Leaving the tank only partially filled overnight can lead to water condensing in the fuel tank as temperatures change overnight. To prevent water from leaving the tank and entering the fuel lines, a fuel/water separator (*Figure 28*) is typically provided. Since water is heavier than diesel fuel, it collects in the bottom of the separator. To eliminate the water, a drain cock on the bottom is opened, and the water is allowed to drain into a proper container until only fuel is clearly draining. Draining the fuel/water separator may be identified as a 10-hour or a 50-hour maintenance requirement.

2.2.0 Preventive Maintenance

When servicing a forklift, follow the manufacturer's recommendations and service chart. Any special servicing for a particular forklift model is highlighted in the manual. Normally, the service chart recommends specific intervals, based on hours of operation, for such things as changing oil, filters, and coolant.

Since inspection items identified as 10-hour requirements are considered daily or prestart inspection tasks, preventive maintenance tasks are typically those that are conducted at longer intervals.

DRAIN COCK

Figure 28 Fuel/water separator.

The first step in the maintenance process is to prepare the forklift. The following steps should be taken:

- Park the forklift on firm, level ground and apply the parking brake.
- Shift the transmission into Neutral.
- Lower the forks or other attachment to the ground; the ground should support any attachment, rather than relying on the hydraulic system to hold it up.
- Operate the engine at idle speed for three to five minutes.
- Shut the engine down and take the key out of the ignition.
- Turn off the electrical master switch (*Figure 29*) and the fuel system switch (if equipped).
- Firmly chock the wheels to prevent any movement. when doing maintence

2.2.1 Weekly (50-Hour) Maintenance

There are a number of items that are typically part of a 50-hour maintenance program. However, it is essential that the manufacturer's recommenda-

Figure 29 Electrical master switch.

> **WARNING!** Eye protection should be worn at all times when working near batteries. The electrolyte solution is a strong acid that can cause severe burns or blindness.

- *Engine coolant* – The engine coolant may be checked one of two ways, depending on the engine and forklift manufacturer. Some engines may have a transparent overflow reservoir (*Figure 30*). The tank allows the level to be checked easily, without removing any caps. Water and/or antifreeze products can also be added here. If the engine is not equipped this way, then the radiator cap must be removed.

> **WARNING!** The radiator may be under pressure, and the coolant may be at or near the boiling point. When the cap is removed and the radiator returns to atmospheric pressure, steam and/or extremely hot water may be expelled. Use extreme caution and the proper PPE (safety glasses, heavy rubber gloves) when removing the cap from a radiator.

tions for the specific model of forklift in use be followed.

It is important to note that new forklifts may start out with a special maintenance schedule. For example, replacement of the axle oil, wheel-end oil, and the engine oil is often required after the first 50 hours of operation. These components, including the new engine, typically produce significantly more wear to metals during the first 50 hours of operation than they will in the future. Replacing the lubricant in these components after this short interval removes these extra metal bits and flakes, preventing them from causing damage or plating other surfaces. Fluid filters are also changed the first time after a short interval. If a new forklift is being driven, be sure to follow the required maintenance schedule to ensure the warranty remains valid.

The following are common examples of 50-hour maintenance tasks.

- *Batteries* – The battery should be inspected every 50 hours of operation. During this inspection, the battery case should be inspected for any signs of damage from an external source, casing cracks, or buckling. Sealed batteries require only a visual inspection.

Radiator caps have three positions. When it is in the closed position, it forms a tight seal that allows pressure to build in the radiator. The pressure is controlled by a pressure relief valve in the cap. If the pressure exceeds the relief setting, pressure will be relieved. By allow-

Figure 30 Coolant fill and overflow tank.

ing pressure to build in the cooling system, the boiling point of the coolant will be increased. Cooling systems under pressure can operate somewhat above 212°F (100°C) without boiling over.

The second position of the radiator cap allows pressure to escape, but the cap is still locked to the radiator fill neck. This prevents it from being blown off by the pressure behind it as a safety measure. This position is usually reached with a quarter- to half-turn counter-clockwise. With slight downward pressure, the cap can again be turned counter-clockwise for complete removal. The coolant level should then be clearly visible. If coolant needs to be added, be sure to maintain the appropriate ratio of antifreeze and water.

- *Boom wear pads* – The sectional booms are equipped with wear pads that provide a smooth contact surface for the boom sections as they slide in and out. If they are not maintained, the mounting hardware may eventually make contact with and damage the boom section(s). The wear pads may also require periodic lubrication, even though some are made of slick materials. Some pads may be made with silicone or Teflon® to provide an inherently slick surface. As the name implies, the wear pads experience wear, protecting the metal boom surfaces from being damaged. Wear pads can be in locations that are difficult to access for lubrication. It is not unusual for a telehandler to have 30 to 50 individual wear pads. The manufacturer may provide a dimension or other criteria to determine if the wear pads need to be replaced. The pad shown in *Figure 31* is considered functional until the top is worn down beyond the beveled edge shown on either end. Worn out, missing, or poorly lubricated wear pads are usually evidenced by chattering as the boom moves in and out.
- *Lubrication* – The forklift manual provides a lubrication schedule, which generally points out the exact locations that require attention. It is not unusual for some locations to require attention during the 50-hour inspection, while others may require lubrication at longer intervals. Note that several grease products, each with different characteristics, may be required. It is very important to ensure that correct grease is being applied at each point. Using the wrong grease may lead to component failure, especially if it is a component that operates at high speeds or is exposed to significant stress.

> **NOTE**
> In cold weather, it may be preferable to lubricate pivot points at the end of a work shift when the mechanism is warm. Warm the grease gun before using it for better grease flow and penetration.

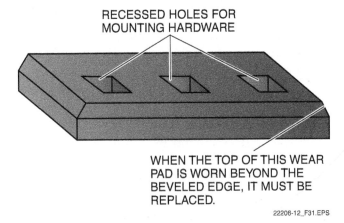

Figure 31 Wear pad inspection.

Hydraulic System Assassins

There are a number of things that can destroy hydraulic systems and components. Aeration creates tiny air bubbles that collapse violently and can erode metal surfaces over time. Hydraulic fluid in this condition appears milky and makes crackling noises as the pressure rises. Cavitation has much the same effect, with air bubbles expanding and exploding as they approach the low-pressure zone of the pump inlet. However, contamination is credited with causing roughly 70 percent of all system failures. Particles of dirt and debris also erode metal surfaces, increasing the level of contamination and wear. These problems can be avoided through periodic fluid sampling and testing.

2.2.2 Long Interval Maintenance

A number of tasks are required at longer intervals. Generally, the tasks become more complex and are done in the maintenance shop. Tasks that are typically done at longer intervals include:

- 250 hours:
  - Replacement of engine oil and oil filter
  - Sampling and testing of engine oil and hydraulic fluid
  - Checking axle and wheel end lubricant levels
  - Replacement of air filter elements
  - Inspecting and lubricating the boom drive chain
- 500 hours:
  - Replacement of the fuel filter
  - Verification of wheel lug nut torque values
- 1,000 hours:
  - Changing the transmission fluid and filter
  - Changing the hydraulic fluid and filter
  - Changing the differential oil
  - Checking and adjusting the boom chain tension
  - Inspecting and/or testing brake systems
  - Changing axle and wheel end lubricants
- 1,500 to 2,000 hours:
  - Changing the engine coolant

Remember that the above list is only a single example of a maintenance schedule. Make sure that the schedule for the specific forklift model in use is followed.

Hydraulic fluid maintenance is addressed in all maintenance schedules. However, the hydraulic fluid should be changed whenever it becomes dirty or breaks down due to overheating. Continuous operation of the hydraulic system in hot environments can heat the hydraulic fluid to the boiling point and cause it to break down. In dusty areas, the fluid cooler may become clogged or blocked, causing the fluid to remain hot. As a result, the hydraulic fluid condition should be observed consistently, and replaced if it shows signs of overheating and breakdown. Filters should also be replaced whenever the fluid is changed.

2.2.3 Preventive Maintenance Records

Accurate, up-to-date maintenance records are essential for knowing the history of your equipment. Each machine should have a record that describes any inspection or service that is to be performed and the corresponding time intervals. Typically, an operator manual and some sort of inspection sheet are kept with the equipment at all times. Actions taken, along with the date, are recorded on the log. Various operators can then share information about fluids that have been added, for example, so that patterns in fluid loss are noted.

The operator manual usually has detailed instructions for performing periodic maintenance. If you find any problems with your machine that you are not authorized to fix, inform the foreman or field mechanic before operating the machine.

2.0.0 Section Review

1. The hydraulic fluid level is typically checked by _____.
   a. removing a dipstick
   b. observing a vertical sight glass
   c. observing a floating indicator arm
   d. removing the fill cap and observing the level

2. Which of the following fluids is most likely to be replaced on a new forklift after 50 hours of service?
   a. engine oil
   b. brake fluid
   c. hydraulic fluid
   d. transmission fluid

SECTION THREE

3.0.0 STARTUP AND OPERATING PROCEDURES

Objective 3

Describe the startup and operating procedures for a rough-terrain forklift.
  a. State rough-terrain forklift-related safety guidelines.
  b. Describe startup, warm-up, and shutdown procedures.
  c. Describe basic maneuvers and operations.
  d. Describe related work activities.

Performance Tasks 2 through 6

Perform proper startup, warm-up, and shutdown procedures.

Interpret a forklift load chart.

Execute basic maneuvers with a rough-terrain forklift.

Perform basic lifting operations with a rough-terrain forklift.

Demonstrate proper parking of a rough-terrain forklift.

Trade Term

Fulcrum: A point or structure on which a lever sits and pivots.

Now that all the basic rough-terrain forklift components have been covered along with the operator-performed inspections and preventive maintenance activities, the next step is to actually start and operate the machine. Before that can happen, operators must fully understand the safety issues associated with this versatile piece of equipment. The operator manual for a given forklift contains safety information about that specific machine. Forklifts and telehandlers in general are similar in their operation. However, they all have slight differences. Some may be a bit clumsier to operate than others are. The type of steering modes available and the terrain also change how a machine maneuvers.

3.1.0 Safety Guidelines

Safe forklift operation is the responsibility of the operator. Operators must develop safe working habits and recognize hazardous conditions to protect themselves and others from injury or death. Always be aware of unsafe conditions to protect the load and the forklift from damage. Become familiar with the operation and function of all controls and instruments before operating the equipment. Read and fully understand the operator manual. Operators must be properly trained and certified.

3.1.1 Operator Training and Certification

Not everyone is allowed to operate a forklift. Due to liability and insurance guidelines, only trained and qualified personnel are allowed to operate forklifts on the job. Operators receive a license certifying them to operate specific forklifts. They also have to be re-evaluated at different times over the years. OSHA recommends that operators be recertified every three years. Since operators often move from one job site to another or from one company to another, they may be required to requalify at each new location. The following explanation of operator training requirements was extracted from OSHA's *Occupational Safety & Health Administration (OSHA) Directive CPL 02-01-028*, which deals with Compliance Assistance for the Powered Industrial Truck Operator Training Standards:

> The training requirement found in *29 CFR 1910.178(l)* for operators of powered industrial trucks and the same requirement for operators of powered industrial trucks in the construction *[1926.602(d)]* and maritime *[1915.120, 1910.16(a)(2)(x), 1910.16(b)(2)(xiv), 1917.1(a)(2)(xiv), 1918.l(b)(10)]* industries specify that the employer must develop a complete training program. OSHA requires that operators of powered industrial trucks be trained in the operation of such vehicles before they are allowed to operate them independently. The training must consist of instruction (both classroom-type and practical training) in proper vehicle operation, the hazards of operating the vehicle in the workplace, and the requirements of the OSHA standard for powered industrial trucks. Operators who have completed training must then be evaluated while they operate the vehicle in the workplace. Operators must also be periodically evaluated (at least once every three years) to ensure that their skills remain at a high level and must receive refresher training whenever there is a demonstrated need. To maximize the effectiveness of the training, OSHA will not require training that is duplicative of other training the employee has previously received if the operator has been evaluated and found competent to operate the truck safely. Finally, the training provisions require that the employer certify that the training and evaluations have been conducted.

Forklift-Related Fatality

A 39-year-old supply motorman was helping a forklift operator place crib blocks on a mine flat car. The motorman was using a shovel to remove ice and snow from the bottom of a crib block while it was raised on the forklift. The load slipped and crushed the motorman. He later died from his injuries.

The Bottom Line: Always maintain a safe distance from the load, and do not approach a load that is unstable for any reason. Ice and snow represent significant hazards in the work place, changing many situations from acceptably safe to unpredictable.

Individual companies may have their own guidelines that specify how they comply with these OSHA guidelines. It is in the individual's best interest that he or she maintains copies of any operator certifications received. In some cases, it may keep the operator from having to recertify at a new job site.

3.1.2 Operator Safety

There are a number of things workers can do to protect themselves and those around them from getting hurt on the job.

Know and follow your employer's safety rules. Your employer or supervisor will provide you with the requirements for proper dress and safety equipment. The following are recommended safety procedures for all occasions:

- Operate the machine from the operator's cab only.
- Wear the seat belt and tighten it firmly. The effectiveness of the ROPS depends heavily on the operator remaining in the seat in the event of an accident.
- Mount and dismount the equipment carefully.
- Wear a hard hat, safety glasses, safety boots, and gloves when operating the equipment.
- Do not wear loose clothing or jewelry that could catch on controls or moving parts.
- Keep the windshield, windows, and mirrors clean at all times.
- Never operate equipment under the influence of alcohol or drugs.
- Never smoke while refueling.
- Do not use a cell phone and avoid other sources of static electricity while refueling. Cell phones should also never be used while operating the machine.
- Never remove protective guards or panels.
- Never attempt to search for leaks with your bare hands. Hydraulic and cooling systems operate at high pressure. Fluids under high pressure can cause serious injury.
- Always lower the forks or other attachments to the ground before performing any service or when leaving the forklift unattended.

3.1.3 Safety of Co-Workers and the Public

You are not only responsible for your personal safety, but also for the safety of other people who may be working around you. Sometimes, you may be working in areas that are very close to pedestrians or motor vehicles. In these areas, take time to be aware of what is going on around you. Create a safe work zone using cones, tape, or other barriers. Remember, it is often difficult to hear when operating a forklift. Use a spotter and a radio in crowded conditions.

The main safety points when working around other people include the following:

- Walk around the equipment to make sure that everyone is clear of the equipment before starting and moving it.
- Always look in the direction of travel.
- Do not drive the forklift up to anyone standing in front of an object or load.
- Make sure that personnel are clear of the rear area before turning (*Figure 32*).
- Know and understand the traffic rules for the area you will be operating in.
- Use a spotter when landing an elevated load or when you do not have a clear view of the landing area.
- Exercise particular care at blind spots, crossings, and other locations where there is traffic or where pedestrians may step into the travel

---

Did You Know?

Each year in the United States nearly 100 workers are killed and another 20,000 are seriously injured in forklift-related incidents. The most frequent type of accident is a forklift striking a pedestrian. This accounts for 25 percent of all forklift accidents. However, these figures include all forklifts. Rough-terrain models have a lower incidence of accidents, especially in striking pedestrians, since they are generally used outdoors where visibility is better. The tight confines and blind corners of a warehouse significantly increase the accident rate.

 22206-13 Rough-Terrain Forklifts

Figure 32 Checking the rear before turning.

path. Sound the horn to communicate your presence at such locations.
- Do not swing loads over the heads of workers. Make sure you have a clear area to maneuver.
- Travel with the forks no higher than 12 to 18 inches above the ground. Always drop the forks or attachment to the ground when parking the machine.
- Do not allow workers to ride in the cab or on the forks.

When working around pedestrians or in other public areas, create a safe work zone for forklift operations. Use barrels, cones, tape, or barricades to keep others out of your work area. This protects both you and bystanders.

3.1.4 Equipment Safety

The forklift has been designed with certain safety features to protect you as well as the equipment. For example, it has guards, canopies, shields, rollover protection, and seat belts. Know your equipment's safety devices and be sure they are in working order.

Forklift overturns are the leading cause of deadly forklift accidents. They represent 25 percent of all forklift-related deaths. In too many cases, the seat belt was not in use. Know the weight of the load and the limits of your machine. Don't take risks and overload the machine. Be especially careful moving suspended loads in windy conditions.

Use the following guidelines to keep your equipment in good working order:
- Perform prestart inspection and lubrication daily (*Figure 33*).
- Look and listen to make sure the equipment is functioning normally. Stop the equipment if it is malfunctioning. Correct or report trouble immediately.
- Always travel with the forks or load low to the ground.
- Never exceed the manufacturer's limits for speed, lifting, or operating on inclines.
- Know the weight of all loads before attempting to lift them. Review the appropriate load chart. Do not exceed the rated capacity of the forklift. It is important to remember that a forklift can rarely lift its maximum load capacity. Especially when a mast is raised and tilted forward, or a telescoping boom is extended, the actual lifting capacity is usually a small percentage of the maximum load capacity.
- Always lower the forks, engage the parking brake, turn off the engine, and secure the controls before leaving the equipment.
- Never park on an incline.
- Maintain a safe distance between your forklift and other equipment that may be on the job site.

Know your equipment. Learn the purpose and use of all gauges and controls as well as your equipment's limitations. Never operate your machine if it is not in good working order.

3.1.5 Spill Containment and Cleanup

As noted in earlier sections of this text, forklifts can be powered by batteries or combustible fuels. Fuels must be contained during refueling activities. Most companies have designated refueling areas specifically built to contain any fuel spills that may occur.

Forklifts also use one or more hydraulic systems to steer the forklift and to extend, retract, tilt, lift, and lower the forklift mast and forklift assembly. With so many of the forklift operational controls relying on the hydraulic system, it is critical that hydraulic fluid be contained without spills or leaks. Anyone who has ever worked around any hydraulic system knows that it is almost impossible to stop all leaks. Leaks need to be cleaned and repaired whenever reasonably possible. Hydraulic fluid spills, on the other hand, are much more of a problem. If a hydraulic hose ruptures under pressure, gallons of hydraulic fluid are spilled. To protect the environment, companies must maintain an oil-spill kit that is easily acces-

Figure 33 Check all fluids daily.

sible to the forklift operator. A typical oil spill kit may include the following:

- A large (20-gallon or more) salvage drum/container to store spill kit materials until needed and for disposal after materials are used
- An emergency response guide for spills of petroleum-based products
- Personal protective equipment (goggles and unlined rubber gloves)
- One or more 3" by 10' absorbent socks for containment
- Half dozen 3" by 4' absorbent socks for containment
- About 20 absorbent pads for cleanup
- About 50 wiper towels or rags for cleanup
- 3 to 5 pounds of powdery absorbent material
- Several disposable bags with ties

NOTE: Although a flat-tipped shovel may not fit into a spill kit, having one handy would be a good idea for spill cleanups.

3.1.6 Hand Signals Used with Forklifts

Forklift operators often work alone. They operate their forklift controls based on what they can see out of the forklift operator's area. When they are placing or picking up a load and cannot see exactly everything they need to see, they must use a spotter or signal person. A spotter uses electronic communication devices, voice commands, or hand signals to tell the forklift operator what moves are needed. Most companies have published examples of the hand signals they want their people to use when performing forklift operations. *Figure 34* shows a few basic hand signals used with forklifts.

3.2.0 Startup, Warm-Up, and Shutdown Procedures

Before starting forklift operations, make sure that you are familiar with the load and the area of operations. Check the area for both vertical and horizontal clearances. Make sure that the path is clear of electrical power lines and other obstacles.

The following suggestions can help improve operating efficiency:

- Observe all safety rules and regulations.
- Determine the weight of the load, review the load charts, and plan operations before starting.
- Use a spotter if you cannot see the area where the load will be placed.

 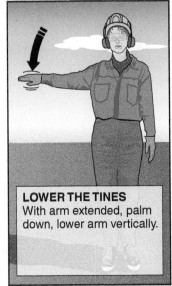 
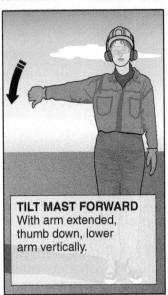  

Figure 34 Basic hand signals for forklifts.

3.2.1 Preparing to Work

Preparing to work involves getting organized in the cab, fastening the seat belt, and starting the machine. Mount your equipment using the grab rails and foot rests. Getting in and out of equipment can be dangerous. Always face the machine and maintain three points of contact when mounting and dismounting a machine. That means you should have three out of four of your hands and feet on the equipment. That can be two hands and one foot or one hand and two feet.

> **WARNING!** OSHA requires that approved seat belts and a ROPS be installed on virtually all heavy equipment. Old equipment must be retrofitted and the seat belt must be used at all times. Do not use heavy equipment that is not equipped with these safety devices.

Adjust the seat to a comfortable operating position. The seat should be adjusted to allow full pedal travel with your back firmly against the seat back. This will permit the application of maximum force on the brake pedals. The knees should still be slightly bent when each pedal is pushed to its maximum position. Make sure you can see clearly and reach all the controls.

The startup and shutdown of an engine is very important. Proper startup lengthens the life of the engine and other components. A slow warm up is essential for proper operation of the machine under load. Similarly, the machine must be shut down properly to cool the hot fluids circulating through the system. These fluids must cool so that they can cool the metal parts of the engine before it is switched off.

3.2.2 Startup

There may be specific startup procedures for the piece of equipment you are operating. But in general, the startup procedure should follow this sequence:

Step 1 Be sure the transmission control is in neutral.

Step 2 Engage the parking brake (*Figure 35*). This is done with either a lever or a knob, depending on the forklift make and model.

> **NOTE** When the parking brake is engaged, an indicator light on the dash should light up or flash. If it does not, stop and correct the problem before operating the equipment.

Step 3 Depress the throttle control slightly.

Step 4 Turn the ignition switch to the start position. The engine should turn over. Never operate the starter for more than 30 seconds at a time. If the engine fails to start, wait two to five minutes before cranking again. The lesser amount of time typically applies in cold weather.

Step 5 As soon as the engine starts, release the key; it should return to the On position.

Adjust the engine speed to one-third to one-half throttle for the warm-up period. Keep the engine speed low until the oil pressure shows on the gauge. The oil pressure light should initially light and then go out. If the oil pressure light does not turn off within 10 seconds, stop the engine, investigate, and correct the problem.

If the machine you are using has a diesel engine, there are special procedures for starting the engine in cold temperatures. Many diesel engines have glow plugs that heat up the engine for ignition. Some units are also equipped with ether starting aids. Review the operator manual so that you fully understand the procedures for using these aids. Follow the manufacturer's instructions for starting the machine in cold weather.

Figure 35 Parking brake.

WARNING!
Ether is a highly flammable liquid. It mixes with the air in the carburetor. As the pistons compress the gases, ignition takes place. The flame from the ether produces enough heat to overcome the compression heat lost through the cold metal of the engine. Ether should only be used when temperatures are below 0°F (-17.7°C) and when the engine is cold.

3.2.3 Warm Up

Warm up a cold engine for at least five minutes. Warm up the machine for a longer period in colder temperatures. While allowing time for the engine to warm up, complete the following tasks:

Step 1 Check all the gauges and instruments to make sure they are working properly. Watch the temperature gauge as the coolant temperature climbs.

Step 2 Shift the transmission into forward and rotate the gear control (if any) to low range.

Step 3 Release the parking brake and depress the service brakes to ensure they work as the forklift begins to roll.

Step 4 Check the steering for proper operation.

Step 5 After a few minutes, manipulate the controls to be sure all components are operating properly.

CAUTION
Do not pick up a load if the hydraulics are sluggish. Allow the machine to warm up until the hydraulic components function normally. The hydraulics can fail if not warmed up completely.

Step 6 Shift the gears to neutral and lock.

Step 7 Reset the parking brake while the machine continues to warm up.

Step 8 Make a final visual check for leaks, unusual noises, or vibrations.

If there are any problems that have no obvious cause, shut down the machine and investigate or get a mechanic to look at the problem.

3.2.4 Shutdown

Shutdown should also follow a specific procedure. Proper shutdown reduces engine wear and possible damage to the machine.

Step 1 Find a dry, level spot to park the forklift. Stop the forklift by decreasing the engine speed, depressing the clutch, and bring the machine to a full stop.

Step 2 Place the transmission in neutral and engage the parking brakes.

Step 3 Release the service brake and make sure that the parking brake is holding the machine.

Step 4 Make sure that the boom is fully retracted. Lower the forks or other attachment so that they are resting on the ground.

Step 5 Place the speed control in low idle and let the engine run for approximately five minutes.

CAUTION
Failure to allow the machine to cool down can cause excessive temperatures in the engine or systems. The engine oil, hydraulic fluid, or transmission fluid may overheat or even boil.

Step 6 Turn the engine start switch to the Off position.

Step 7 Release hydraulic pressures by moving the control levers until all movement stops.

Step 8 Turn the engine switch to Off and remove the key. If you must park on an incline, chock the wheels.

Some machines have security panels or vandalism caps for added security. The panels cover the controls and can be locked when the machine is not in use. Lock the cab door, and secure and lock the engine enclosure. Always engage any security systems when leaving the forklift unattended.

3.3.0 Basic Maneuvering

To maneuver the forklift, you must be able to move it forward, backward, and turn.

3.3.1 Moving Forward

The first basic maneuver is learning to drive forward. To move forward, follow these steps:

Step 1 Before starting to move, use the joystick to raise the forks to roughly 15 inches above the ground. This is the travel position.

Step 2 Put the shift lever in low forward. Release the parking brake, and press the accelerator pedal to start moving the forklift.

Step 3 Steer the machine using the steering wheel.

Step 4 Once underway, shift to a higher gear to drive on a smooth road. To shift from a lower to a higher gear, rotate the collar of the shift lever.

> **WARNING!** Always travel with the forks low to the ground (12 to 18 inches). Be aware of the tips of the forks and avoid hitting things with them. On very rough terrain, an even higher travel position may be required.

3.3.2 Moving Backward

To back up or reverse direction, always come to a complete stop first. Then move the shift lever to reverse. Once in reverse gear, you can apply some acceleration and begin to move backwards.

Depending on the steering modes available or in use, backing up can be challenging. The operator should always be looking in the direction of travel, and using the proper steering inputs while looking backwards requires some practice.

3.3.3 Steering and Turning

How a forklift is steered depends on the make and model. However, most rough-terrain models have a steering wheel. Some forklift steering wheels can be operated with one hand. This allows the operator to use the other hand to control the forks and boom. The steering wheels on a forklift operate in the same manner as steering wheels on cars and trucks. Moving the wheel to the right turns the forklift to the right. Turning the wheel to the left moves the wheels to the left.

Some forklifts have different steering modes, including two-wheel, four-wheel, and crab steering. When the machine is in a two-wheel steering mode, only the front wheels may move; if the unit is front-wheel drive, then the rear wheels will likely move. Use two-wheel steering when you are traveling, especially on a smooth road. Use four-wheel steering for work activities and close-quarter operations. When the machine is in four-wheel steering mode, the front and back wheels turn in opposite directions. This allows the machine to make tight turns in a circle. When working in a confined area, crab steering is probably the best choice. In crab steering, the front and back wheels turn in the same direction. The machine will travel forward and to one side, or backward and to one side. The work area and tasks to be done determine which steering mode is the most advantageous.

> **CAUTION** Always straighten the wheels before switching modes. You can damage the steering if the wheels are not centered before operating the machine in two-wheel steer.

> **NOTE** It is possible for the steering to go out of synchronization if the correct procedures are not followed when changing steering modes. Review the operator manual before changing steering modes. Follow the procedure in the operator manual to synchronize the wheels if they become unsynchronized.

If the forklift is equipped with a differential lock, it should be disengaged before making any turns. The differential lock ensures that an equal amount of torque is sent to the drive wheels. As a result, they will attempt to rotate at the same speed. The differential lock is best used when in slippery or muddy conditions, and one drive wheel may be spinning while the other is nearly or completely motionless. Locking the differential rebalances the torque to each drive wheel and helps get the forklift moving. When turning the forklift, the drive wheels on the inside of the turn must rotate more slowly than those on the outside of the turn. Turning with the differential lock engaged does not allow this to occur, causing a clumsy turn and unnecessary stress on the drive train.

3.3.4 Leveling the Forklift

Keeping the forklift level is a vital part of safe operations. The forklift must be level and balanced before significant loads are lifted. If the forklift is not level when a load is lifted, it could tip over. Remember that an overturned forklift is the leading cause of death in forklift operations.

Most rough-terrain forklifts have some type of level indicator. On many telehandlers, the level indicator is located on the frame above the front windshield as shown in *Figure 36*. An air bubble or small bead is contained within a small-arced sight glass. When the bubble or bead is centered in the middle of the sight glass, the forklift frame is level. A scale on the side of the glass is marked zero in the center and increases to either side. The numbers represent the number of degrees off level.

22206-13 Rough-Terrain Forklifts

Module Eight 37

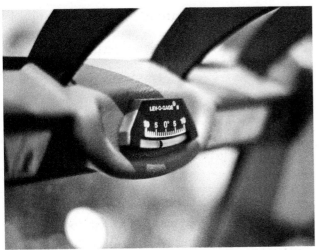

Figure 36 Forklift level indicator.

Note that a forklift must always be leveled before a load is raised. Although leveling can be done with a load on the forks, the forks should be very near the ground before any leveling is attempted.

> **WARNING!**
> Never attempt to level a forklift with a load raised in the air. Using any method of leveling with the load raised could result in a serious imbalance leading to serious personal injury and/or property damage.

There are two ways to level the forklift. One is to lower and adjust the outriggers until the machine is level. When the forklift must be leveled while the machine is on its wheels, use the frame-leveling control to level the machine (if the machine is so equipped). This feature rotates the cab and boom assembly on the chassis to level the machine. The frame level controls cannot be used after the outriggers are lowered.

The frame-leveling controls are usually located on the console near the joystick. The default position for the three-position switch is center or hold. Depressing the switch to either side rotates the frame. Pressing the right side of the switch lowers the right side of the machine. Pressing the left side of the switch lowers the left side.

Use the frame-leveling controls when a load must be lifted from an uneven surface. The controls tilt the frame 10 degrees to the left or right. Lower the boom before using the level controls. The boom attachment should be close to, but not touching, the ground. Depress the appropriate switch until the machine is level. When you release the switch, it returns to the hold position and the machine stops rotating.

> **CAUTION**
> Do not operate the frame-leveling controls when the boom is raised. Doing so could damage the machine.

The outriggers can also be used to level the machine. The outrigger controls are used to lower and raise the left and right outriggers. The outriggers are controlled with switches located on the instrument panel or the joystick. Lowering the outriggers firmly to the ground provides a firm and level base for lifting. When the stabilizers are lowered, the forklift can handle heavier loads and the boom can be fully extended.

> **WARNING!**
> Check that the ground will support the stabilizers before they are lowered. If the soil collapses from the weight of the machine, the forklift could overturn and cause injury or death. Make sure that all personnel stand clear when operating the stabilizers. Moving stabilizers poses a significant crushing hazard.

Before lowering the stabilizers, check that the area is free of obstructions and that the ground will support the stabilizers. Lower and retract the boom. The boom should be close to, but not on, the ground. Use the frame-leveling controls to level the machine. Run the engine at sufficient speed to supply enough hydraulic pressure to the stabilizers. Check that all personnel are clear of the area. Depress or slide the switch for the right outrigger to lower it. Release it when the outrigger has reached the desired position. Repeat the procedure for the left outrigger. When both outriggers are lowered, the front wheels of the forklift should be off the ground (*Figure 37*). Adjust the positions of the outriggers until the frame is level as indicated in the sight glass.

Before raising and retracting the outriggers, fully retract and lower the boom. Check that all personnel are clear of the area before retracting the outriggers. Make sure that both outriggers are fully raised and retracted before moving the forklift.

3.4.0 Work Activities

Operation of the forklift is fairly straightforward, but it requires attention to detail. With proper planning, you will have no trouble operating the forklift. The basic work activities performed with a forklift are described in this section. One thing that is different with these machines is the terrain. The terrain is very predictable in a warehouse en-

Figure 37 Outriggers in use; front wheels raised.

vironment. The environment where rough-terrain models are used presents an entirely different set of hazards and surprises.

> **NOTE:** The controls on specific forklifts may be different from those described in the procedures. Check your operator manual for information about the controls and limitations of your equipment.

3.4.1 Load Charts

The most important factor to consider when operating any forklift is its lifting capacity. Each forklift is designed with an intended capacity that must never be exceeded. Exceeding the capacity jeopardizes the equipment, the load, and the safety of everyone near the equipment. Capacity information is provided in the load charts, which are included in the operator manual or posted in the forklift. Be sure to read and follow the load chart.

The ability of a forklift to lift a load without tipping is called the rated capacity. The capacity of a forklift varies depending on the angle and height of the boom. Load charts, like the ones shown in *Figure 38*, provide data on the maximum capacity of the forklift. A forklift operator must be able to read load charts and make sure that the load does not exceed the capacity.

Different load charts are used for different machine configurations. For example, the capacity changes if the outriggers are lowered or raised. Using different attachments also requires that you use a different load chart. Make sure that you are using the correct load chart for the lift you are performing.

Lifting a load, regardless of how it is done, is a matter of leverage. Almost everyone has played on a playground seesaw. When nobody is on the seesaw, the board can be balanced on the pivot point, which is the fulcrum for the seesaw. When equal weights are placed on each end of the board, the board can still be balanced. With a forklift, the rear end provides a given amount of weight that acts as a counterweight to any load the forklift tries to lift. The front axle, aided by the wheels, is the fulcrum, or pivot point. Anything heavy placed on the forks is the load. If the forklift's rear weight is enough of a counterweight, the forklift can lift the load. As with any lifting device, the further the load gets away from the fulcrum horizontally, the more likely it is to pivot due to the added leverage. If it is too heavy and placed far enough from the fulcrum, the rear wheels of the forklift rise off the ground.

In summary, the closer the load is to the front axle horizontally, the more stable the lift. When the boom's elevation is kept at zero degrees, and the load is extended out further away from the axle, the forklift's balance is more unstable.

Understanding a concept known as the stability triangle helps operators to better evaluate a lift situation. The stability triangle is represented by three points on the chassis of the forklift (*Figure 39*). The center of the two front wheels and the center of the rear axle form the three points. The center of the triangle is the center of gravity (CG) of the forklift with no load, the forks down, and the boom retracted. Adding a fourth point—straight up from the CG to the boom—creates a pyramid. Imagine that there is a weight hanging from the boom at this point. As long as the weight is hanging inside the lines of the triangle below, the forklift remains stable. As the boom is raised, the CG rises within the pyramid. As a result, the triangle begins to shrink in size. The smaller the triangle, the more difficult it becomes to keep the imaginary weight hanging within it. If the forklift is not level, especially if it is tilting left or right, the imbalance can easily cause the machine to tip over as the load rises.

Forklifts are often selected based on their maximum load capacity. However, most can only achieve the maximum capacity under ideal conditions, on ideal terrain, and without any horizontal extension of the boom or forward-tilting of the forks. A 6,000-pound forklift can quickly become a 1,500-pound forklift as the position of the load in relation to the fulcrum changes.

Figure 38 shows load charts for a telehandler with and without outriggers being used. With the outriggers retracted (not in use or non-existent), the machine on the right can lift up to 8,000

OPERATING SPECS

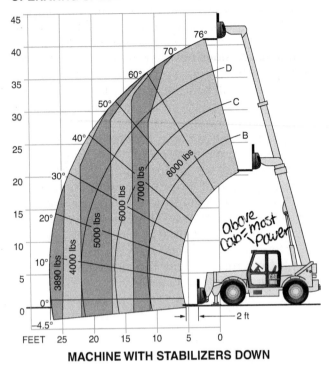
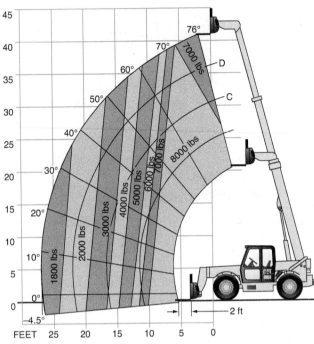

Figure 38 Load charts.

pounds when its boom is fully retracted (closest in to the axle) and lowered to 0 degrees. It can also lift that same load to a height of nearly 40 feet as long as the fork tips remain within 6 feet forward of the front wheels, which occurs at a boom angle of 70 degrees. If the boom is kept at zero degrees, with the fork tips extended out to 15 feet, you will see that the machine can only lift 4,000 pounds. If the fork tips are extended all the way out to 25 feet, while being kept at zero degrees, the machine can then lift only 1,800 pounds.

Now, still operating without the outriggers, see what the forklift can do when the boom is elevated to some angle above zero degrees. If the boom is fully retracted and then raised, the machine can lift an 8,000-pound load up to 70 degrees, which would put the load up to nearly 40 feet above the ground. If the boom is raised beyond 70 degrees, the amount of load the machine can lift is reduced to 7,000 pounds because the height of the load becomes unstable when lifted almost directly over the machine's front axle.

If the same machine, not using its stabilizers, has its boom extended and raised at the same time, the load chart shows that a different set of limitations must be applied. If the boom is extended to 15 feet and elevated to 20 degrees, the machine's lifting capacity is reduced to 4,000 pounds. If the boom is kept at 20 degrees, but extended out to 25

feet, you can see that the machine's lifting capacity is further reduced to only 1,800 pounds.

There are times when the machine operator has plenty of room to use the machine's stabilizers or outriggers. The chart on the left of *Figure 38* shows that the machine can lift heavier loads further away from its axle when the outriggers are used. This particular machine can lift up to 3,890 pounds when its boom is at zero degrees elevation and fully extended out to 25 feet. When the machine has its stabilizers in use and its boom retracted, it can safely elevate its boom with the full 8,000 pounds all the way up to the 76-degree point. The stabilizers allow for the increased lifting capacity.

The point of this discussion about load charts is that the machine operator must clearly understand how much of a load is to be lifted and what boom extensions and elevation angles are needed to move and position the load. Before beginning any forklift work, make sure that you clearly understand the rated capacities of that given machine. However, it is equally important to remember that, regardless of the operator's expertise in interpreting load charts, it is of no value at all if the actual weight of the load is not first determined. Ignoring or trying to guess the weight of a significant load can start a tragic series of events.

Load capacities for forklifts are calculated based on the assumption that the load's CG is 24 inches up from the forks and 24 inches forward

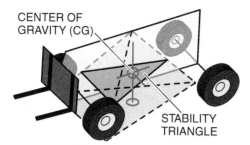
FORKLIFT BOOM DOWN

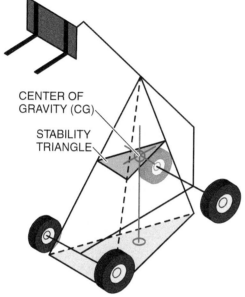
FORKLIFT BOOM RAISED

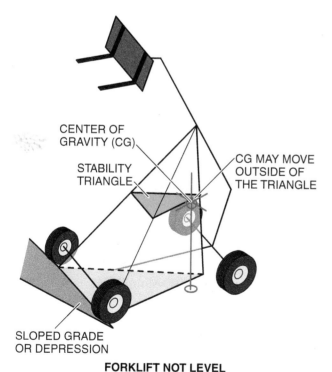
FORKLIFT NOT LEVEL WITH BOOM RAISED

Figure 39 The stability triangle changes due to the terrain and boom position.

of the back of the forks. Attachments can make a dramatic difference in determining the capacity. Operators must consider the added weight of the attachment at a minimum. However, they must also consider how the attachment changes the CG of the load. A good example of this is seen in the use of lifting hooks/boom extensions and fork-mounted lifting hooks. Fork-mounted lifting hooks, such as the one shown in *Figure 23*, do not usually change the center of gravity at all. Only the added weight of the attachment must be considered. However, with a lifting hook such as the one shown in *Figure 22* is used, the CG is moved far forward of the forks. Operators must then consider the location of the load CG on the load chart and disregard the location of the forks. Of course, the load capacity of the attachment itself must always be considered as well. The load capacity of the attachment and the forklift at a given set of conditions must be compared, and the lowest value of the two becomes the maximum capacity for the lift conditions.

3.4.2 Picking Up a Load

To pick up a load, first check the position of the forks. They should be centered on the carriage. If they must be adjusted, check the operator manual for the proper procedure. Usually, there is a pin at the top of each fork. When the fork is lifted, you can slide it along the upper backing plate until it reaches the desired position. Lock them into place.

Travel to the area at a safe rate of speed. Use the foot throttle and the steering wheel to maneuver. Always keep the forks 12 to 18 inches above the ground and tilted slightly back.

Before picking up the load, make sure it is stable. If it looks like the load might shift when picked up, secure the load with strapping or other tie-downs. The center of gravity is critical to load stability. If necessary, make a trial lift and adjust the load.

Approach the load so that the forks straddle the load evenly. It is important that the weight of the load is distributed evenly across the forks. Overloading one fork can damage it or cause the forklift or the load to overturn. In some cases, it may be advisable to measure and mark the load's center of gravity. When lifting critical, sensitive, or unusual equipment, check the equipment manufacturers documentation for weight and balance information.

Drive up to the load with the forks straight and level (*Figure 40*). Approach the load so that the forklift is square and level to the intended load. If necessary, level the forklift using the frame level controls or the stabilizers. If the load is on a pallet, make sure that the forks will clear the pallet board. If the load is on blocks, you may need to get out and check if the forks will clear.

Move forward slowly until the leading edge of

> **WARNING!**
> Set the parking brake and turn off the engine before exiting the cab. Do not leave the forklift running unattended.

the load rests against the back of both forks. If you cannot see the forks engage the load, ask someone to signal for you. This can avoid expensive damage and injury. Raise the carriage slowly until the forks contact the load. Continue raising the carriage until the load safely clears the ground. Tilt the mast fully rearward to cradle the load. This minimizes the possibility of the load slipping in transport.

> **CAUTION**
> Oil, ice, mud, grease, rain, or snow can cause the forks to become very slippery. Clean the forks whenever they become wet or fouled. Wet metal on wet metal is extremely slippery. Be extremely careful when picking up metal loads with wet forks.

3.4.3 Picking Up an Elevated Load

Picking up a load from an elevated position is more difficult than picking a load off the ground. First, it is more difficult to see an elevated load. Use a spotter if necessary. Second, the load can fall farther and cause other materials to fall. Use extreme caution when picking up elevated loads. Remember that the stability is diminished as load height increases. Use the load chart to make sure that you can safely lift and place the load.

Approach the elevated load square and level as previously described. Raise and extend the boom so that the forks are in line with the load (*Figure 41*). Extend and lower the boom to move the forks under the load. The forks will remain level. Be careful not to hit the stack.

Carefully raise the boom to lift the load, and then tilt the forks slightly backward to cradle the load (*Figure 42*) before continuing to raise the load. Once the load is cradled, retract the boom slowly until the load is clear. Lower the boom to the travel position before moving the forklift. Be careful not to hit the stack as you retract the load.

3.4.4 Traveling with a Load

Always travel at a safe rate of speed. Never travel with the load higher than necessary. Keep the load as low as possible without risking the forks or attachment hitting the ground (*Figure 43*). The terrain has a great deal to do with the choice of traveling height. Be sure that the carriage is level and tilted slightly rearward so the load does not slip off.

As you travel, keep your eyes open and stay alert. Watch the load and conditions ahead of you. Alert others to your presence. Avoid sudden stops and abrupt changes in direction. Be careful when downshifting because sudden deceleration can cause the load to shift. Be aware of front and rear swing when turning.

If you have to drive on a steep slope, keep the load as low as possible. Do not drive across steep slopes, because the forklift could easily overturn. Drive or back up and down a steep hill, but make sure the forklift is pointed uphill with a load (*Figure 44*) and downhill without one. If you have to turn on an incline, make the turn wide and slow. This minimizes the risk of tipping over.

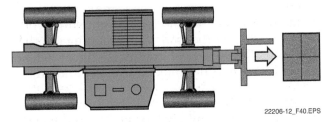

Figure 40 Approach the load squarely.

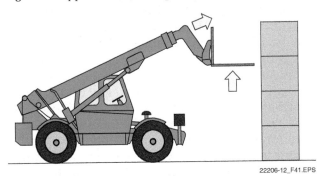

Figure 41 Raise the boom to align the forks with the load.

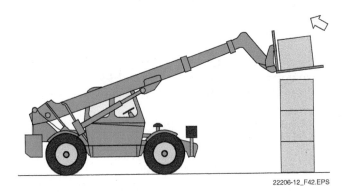

Figure 42 Tilt the forks back to cradle the load.

Figure 43 Keep the load low.

Figure 44 Traveling uphill with a load.

Driving over bumps and holes with a load can cause you to lose a load. If the machine bumps, the load can bounce off. Avoid driving over bumps and holes. If impossible to avoid, drive over them slowly and keep the load low.

3.4.5 Placing a Load

Position the forklift at the landing point so the load can be placed where needed. Remember that it is easier to reposition the forklift than to reposition the load after it is placed. If the load is not palletized, place blocking material where the load is to be placed. The blocking will create space under the load so that you can remove the forks. Make sure that the blocking will adequately support the load and be sure that everyone is clear of the area.

The area under the load must be clear of obstruction and must be able to support the weight of the load. If you cannot see the placement, use a spotter to guide you.

When the forklift is in position, tilt the forks forward to the horizontal position. Lower the load slowly. You can usually feel when the load is resting on the ground or blocking. When the load has been placed, lower the forks a little more to clear the underside of the load. Back carefully away from the load to disengage the forks.

> **NOTE**
> Do not lower the forks too much after placing the load. You can lower the forks too far and dig up the ground under the load when backing up. Lower the forks just enough to clear the bottom of the load but still remain off the ground.

3.4.6 Placing an Elevated Load

You must take extra precautions when placing elevated loads. It is extremely important to level the machine before lifting the load. Failure to do so can cause the machine, the load, or the existing stack to tip over.

One of the biggest safety hazards for elevated load placement is poor visibility (*Figure 45*). There may be workers in the area who cannot be seen. The landing point itself may not be visible. Your depth perception decreases as the height of the load increases. To be safe, use a signal person to help you spot the load. Use tag lines for long loads.

Drive the forklift as close as possible to the landing point with the load kept low. Set the parking brake. Lower the outriggers if necessary. The outriggers offer a significant advantage in stability with a raised load. Raise the load slowly and carefully while maintaining a slight rearward tilt to keep the load cradled against the back of the forks. Do not tilt the load forward until the load is over the landing point and ready to be set down.

If the forklift's rear wheels start to lift off the ground, stop immediately but not abruptly. Lower the load. Slowly reposition it, break it down into smaller loads, or use the outriggers. If the surface conditions are poor at the unloading site, it may be necessary to reinforce the surface to provide more stability.

3.4.7 Unloading a Flatbed Truck

The following steps are used to unload material from a flatbed truck:

Step 1 Position the forklift at either side of the truck bed.

Step 2 Manipulate the control levers in order to obtain the appropriate fork height and angle.

Step 3 Drive forks into the opening of the pallet or under the loose material. Use care not to damage any material.

Step 4 Adjust the controls as required to lift the material slightly off the bed.

Step 5 Tilt the forks back slightly to keep the pallet or other material from sliding off the front of the forks.

Step 6 Retract the boom or back the forklift away from the truck.

Step 7 Lower the forks to the travel position and move to the stockpile area.

Step 8 Position the forklift so that the material can be placed in the desired area.

Step 9 Lower or raise the boom until the material is set on the required surface.

Step 10 Adjust the forks with the boom lever in order to relieve the pressure under the pallet.

Step 11 Back the forklift away from the pallet.

Step 12 Repeat the cycle until the truck is unloaded.

3.4.8 Using Special Attachments

In addition to the various types of forks, there are several special attachments that expand the forklift's operational capability. The three main attachments are the hook, the boom extension, and the bucket. Always read the operator manual to make sure you follow the proper procedure for securing attachments to the forklift.

Some forklifts have a coupler system that allows the operator to easily change attachments. *Figure 46* shows the main features of the coupler. The coupler is activated with switches located on the instrument panel or on the joystick. On some models the coupler is controlled with the joystick after another switch is activated to change the joystick mode.

Before detaching the forks from the forklift, you need to disengage the hydraulics. Typically, there is a diverter valve on the hydraulic hoses that must be closed. Check the operator manual for the correct procedures. If the attachments are not secured properly, they could fail, causing property damage and significant injury.

Figure 45 Place elevated loads carefully.

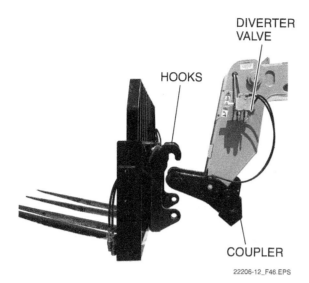

Figure 46 Quick coupler.

> **WARNING!**
> The hydraulic system is under pressure. Follow the safety procedures listed in the operator manual for relieving pressure before disconnecting hydraulic hoses. The release of fluids under pressure can cause significant injury.

To attach the coupler to the forks or other attachments, position the coupler in line with the attachment. Tilt the coupler forward so that it is below the levels of the hooks. Move the forklift forward or extend the boom until the coupler contacts the carriage. Tilt the coupler back until the lower part of the carriage contacts the coupler, and then secure the attachment to the coupler.

The boom extension and lifting hook allow the forklift to lift objects from above. These attachments feature a sturdy hook mounted on the carriage or the end of a boom extension (*Figure 47*). Loads must be securely rigged using approved lifting equipment. Each attachment has different lifting capacities. Operators must be aware that lifting attachments may have lower capacities than the forklift itself. Be sure to use the correct load chart when planning the lift, and consider the limitations of the attachment first. Once the capacity of the attachment is known, the operator can then determine if the forklift or the attachment represents the weakest link. Do not exceed the lifting capacity of the equipment or the attachment.

> **WARNING!**
> Only use approved chains, slings, hooks, and other rigging. Nonstandard rigging can fail and cause property damage and personal injury.

Figure 47 Forklift with boom extension and lifting hook.

Position the hook or lifting point directly above the load before lifting. If it is not directly above, the load could swing when it is lifted. Secure the load to the hook. Using shorter slings also reduces swinging (*Figure 48*). Use tag lines to control load swing and placement. Once the load is secured, use the boom controls to lift and position the load.

> **CAUTION**
> In extremely cold temperatures, the load can freeze to the ground. Free the load before attempting to lift it. Lifting a frozen load can cause a jolt that could affect the stability of the machine or dislodge the load.

3.4.9 Using a Bucket

Loading trucks, bins, and other containers can be done using a forklift with a bucket attachment. Usually, material is loaded from a forklift by tak-

Figure 48 Shorter rigging reduces swinging.

ing the material from a stockpile. The procedure for carrying out a loading operation from a stockpile is as follows:

Step 1 Travel to the work area with the bucket in the travel position.

Step 2 Position the bucket parallel to and just skimming the ground.

Step 3 Drive the bucket straight into the stockpile.

Step 4 Tilt the bucket backwards to fill it.

Step 5 Work the tilt control lever back and forth to move material to the back of the bucket. This is called bumping. When the bucket is full, move the tilt control lever to the tilt-back position.

Step 6 Shift the gears to reverse and back the forklift away from the stockpile.

Step 7 Place the bucket in the travel position and move the forklift to the truck.

Step 8 Center the forklift with the truck bed and raise the bucket high enough to clear the side of the truck.

Step 9 Move the bucket over the truck bed and shift the bucket control lever forward to dump the bucket.

Step 10 Pull the bucket control lever to retract the empty bucket, and back the forklift away from the truck as soon as the bucket is empty.

Step 11 Lower the bucket to the travel position and return to the stockpile.

Step 12 Repeat the cycle until the truck is loaded.

As the truck fills, the material needs to be pushed across the truck bed to even the load. As the leading edge of the bucket passes the sideboard of the truck, roll the bucket down quickly. Dump the material in the middle of the bed. The load is then pushed across the truck as the bucket is raised. By raising the bucket and backing up slowly, the material is distributed evenly across the bed.

> **CAUTION**
> Avoid hitting the side of the truck with the boom or bucket when you are unloading.

While there are many ways to maneuver a forklift, the two most common patterns for a truck loading operation are the I-pattern and the Y-pattern. For the I-pattern, both the forklift and the truck move in only a straight line, backward and forward (*Figure 49*). This is a good method for small, cramped areas. The forklift fills the bucket and backs approximately 20 feet away from the pile. The truck backs up between the machine and the pile. The forklift dumps the bucket into the truck. The truck moves out of the way and the cycle repeats.

To perform this I-pattern loading maneuver, position the forklift so that it is on the driver's side of the truck. That way, eye contact can be made with the driver. Fill the bucket, as shown in *Figure 49A*. Back far enough away from the pile to allow room for the truck to back in. Signal the truck driver with the horn or agreed hand signal. The truck backs up to a predetermined position, as shown in *Figure 49B*. Move the forklift forward and center it on the truck bed. Raise the bucket to clear the side of the truck and place it over the truck bed. Move the boom and bucket control lever to dump the bucket. At the same time, raise the forklift boom to make sure the bucket clears the truck bed. When the bucket is empty, move the boom and bucket control from side to side to shake out the last of the material. Back the forklift away from the truck and signal the truck driver to move. When the truck is out of the way, lower the bucket and position the forklift to return for another bucket of material.

The other loading pattern is the Y-pattern (*Figure 50*). This method is used when larger open areas are available. The dump truck remains stationary and as close as possible to the pile. The forklift does all the moving in a Y-shaped pattern.

To perform the Y-pattern, position the truck so that eye contact can be made with the driver. Fill the bucket with material. While backing up, turn the forklift to the right or left, depending on the position of the truck. Shift to a forward gear and turn the forklift while approaching the truck slowly. Stop when the forklift is lined up with the truck bed. Dump the bucket in the same way done for the I-pattern. When the bucket is empty, back away from the truck while turning toward the pile. Drive forward into the pile to repeat the pattern. Repeat the cycle until the truck is full.

3.4.10 Clamshell Bucket

The clamshell bucket has a hydraulically operated clamshell design. The bucket can perform four basic functions. It can be used as a clamshell bucket, scraper, dozer, and a regular forklift bucket.

The clamshell can be used for removing stumps and large rocks, as well as picking up de-

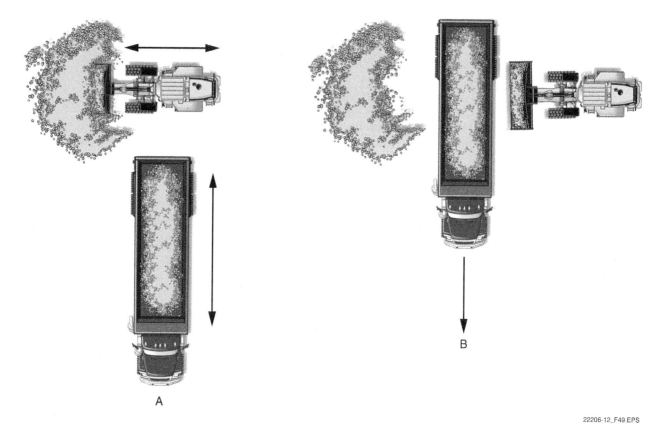

Figure 49 I-pattern for loading.

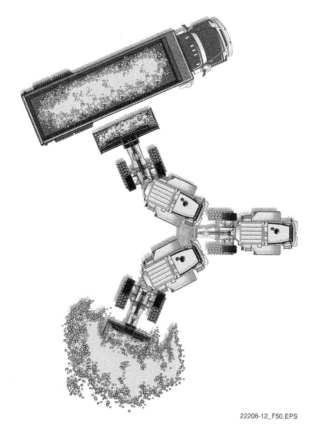

Figure 50 Y-pattern for loading.

bris and brush. To do this, the operator must open the bowl and position the bucket over the material to be loaded, then lower the bucket, and close the bowl to fill. Material can then be transported to the truck or stockpile for dumping. Use of the clamshell configuration gives the forklift added height for dumping and better handling of sticky material such as wet clay soils.

Using the bucket as a scraper requires the operator to open the bowl and use the backside of the bucket to cut material. When the material is filling the back of the bucket, close the bowl over the material, and raise the bucket for transporting to the dumpsite.

For use as a dozer or pusher, the operator must open the bowl fully and use the back of the bucket as a blade. Level cutting is maintained by the bucket lift control. The material can then be pushed into an area to create a stockpile. The multipurpose bucket is useful for roughing-in access roads.

With the added height of the multipurpose bucket, loading trucks becomes easier because the boom can remain higher and stay away from the side of the truck. The clamshell configuration also makes it easier to dump sticky material because the bucket does not compact the material.

3.4.11 Working in Unstable Soils

Working in mud or unstable soils that will not support the forklift can be aggravating and dangerous. This is a problem even for experienced operators.

When entering a soft or wet area, go very slowly. If the front of the machine feels like it is starting to settle, stop and back out immediately. That settling is the first indication that the ground is too soft to support the equipment. The engine will lug slightly and the front end of the forklift will start to settle.

After backing out, examine how deep the wheels or tracks sank into the ground. If they sink deep enough that the material hits the bottom of the machine, the ground is too soft to work in a normal way.

To work in soft or unstable material, consider this approach:

- Start from the edge and work forward slowly.
- Push the mud ahead of the bucket to test the consistency of the soil and be sure the ground below is firm.
- Don't try to move too much material on any one pass.
- Try to keep the wheels or tracks from slipping and digging in.

Partially stable material can also be a hazard because an operator may drive in and out over relatively firm ground many times, while it slowly gets softer because the weight of the wheels or tracks pumps more water to the surface. If this happens, the wheels or tracks will sink a little more each time until the forklift finally gets mired in the hole. Then the machine must be pulled out.

To keep this from happening, do not run in the same track each time entering or leaving an area. Move over slightly in one direction or the other so the same tracks are not pushed deeper into the unstable material each time.

3.0.0 Section Review

1. Which of the following accidents causes 25 percent of forklift fatalities?
    a. Overturning a forklift
    b. Striking a pedestrian with a forklift
    c. Unstable loads falling from the forklift
    d. Exceeding the safe lifting capacity of a forklift

2. If you are asked to drive a forklift and discover the seat belt strap is broken, _____.
    a. improvise by using a rope or other material and drive it
    b. repair it by securely fastening the ends together and drive it
    c. do not operate it and report the problem
    d. drive it as is and report the problem

3. Before making a turn with a forklift, the operator needs to ensure that the _____.
    a. differential lock is engaged
    b. differential lock is disengaged
    c. forks are 3 to 5 inches above the ground
    d. boom is at no more than a 10-degree angle

4. The two common driving patterns for loading a truck using a bucket attachment are the _____.
    a. I- and the Y-patterns
    b. A- and the B-patterns
    c. V- and the Y-patterns
    d. I- and the Z-patterns

Summary

Rough-terrain forklifts are used primarily for lifting and loading material on job sites. The forks are used to lift palletized loads and other bundled material. The forklift can be fitted with several attachments that enable it to lift and place loose or bulky material. These attachments include a lifting hook, a boom extension, a hopper, and a bucket.

Vehicle movement is controlled with the steering wheel, accelerator, and brake pedals. The forks and boom are controlled with either levers or a joystick and switches. Study the operator manual to become familiar with the machine you will be operating.

Safety considerations when operating a forklift include keeping the forklift in good working condition, obeying all safety rules, being aware of other people and equipment in the same area where you are operating, and not taking chances. Perform inspections and maintenance daily to keep the forklift in good working order. One of the primary safety considerations for forklift operations is the rated capacity of the forklift. Know the weight of the load and the capacity of the machine. You must be able to read and interpret a load chart. Exceeding the rated capacity of the machine can cause injury and death.

Always position the forklift so that it is square and level to the load. Pick a load up slowly and tilt the forks backward to cradle the load. Lower the load before traveling to the unloading area. Use extra caution when picking up elevated loads.

Review Questions

1. Two-wheel drive forklifts usually use the _____.
   a. rear wheels to both drive and steer
   b. front wheels to both drive and steer
   c. rear wheels to drive and the front wheels to steer
   d. front wheels to drive and the rear wheels to steer

2. The steering mode that allows all four wheels to turn and point in the same direction at the same time is called _____.
   a. articulated steering
   b. dynamic steering
   c. oblique steering
   d. circle steering

3. A hydrostatic transmission provides _____.
   a. only manual shifting, using a clutch
   b. highway speeds for a rough-terrain forklift
   c. more reverse gears than a common transmission
   d. infinitely variable speeds

4. A forklift fitted with an F-N-R control typically has a _____.
   a. squirt boom
   b. pivoting boom
   c. set of outriggers
   d. hydrostatic transmission

5. With a joystick in use, the operator can make multiple movements occur at the same time, such as raising the boom and tipping the forks.
   a. True
   b. False

6. The typical means of adjusting the length of a boom extension for lifting is _____.
   a. manually
   b. electrically
   c. hydraulically
   d. engine power take-off

7. Per the chart shown in *Figure 1*, the lifting capacity of the FB-40 boom attachment at an 11-foot extension is _____.
   a. 1,500 pounds
   b. 1,600 pounds
   c. 3,000 pounds
   d. 4,800 pounds

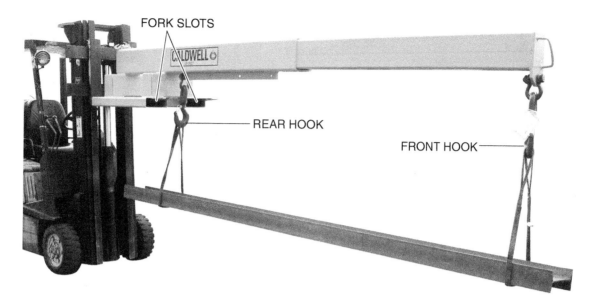

**MODEL FB
FIXED TYPE FORK LIFT BOOM**

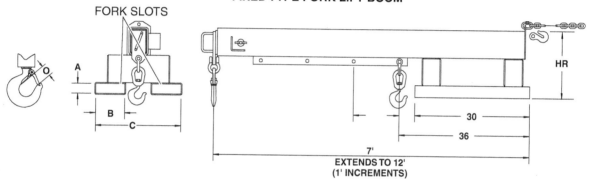

NOTE: All dimensions on drawings shown in inches unless stated.

SPECIFICATIONS

Model Number	Dimensions (in.)					Maximum Capacity at Hook Position (lbs.)							Weight (lbs.)
	A	B	C	HR	O	3'–6'	7'	8'	9'	10'	11'	12'	
FB-30	2½	7½	22	16	1.00	3,000	3,000	2,600	2,200	1,900	1,600	1,500	340
FB-40				16	1.09	4,000	3,200	2,600	2,200	1,900	1,600	1,500	340
FB-60	2½	7½	22	17	1.36	6,000	5,000	4,200	3,500	3,000	2,700	2,500	390
FB-80				18	1.61	8,000	7,000	5,700	4,800	4,100	3,600	3,100	520

Capacities are for boom only and are dependent on fork lift used. Check with forklift manufacturer for load capacities before use.

Figure 1

8. Maintenance time intervals for most machines that are adopted by manufacturers are established by _____.
   a. SAE
   b. OSHA
   c. NAMA
   d. ASHRAE

Figure 2

9. The movement being communicated to the operator in *Figure 2* is _____.
   a. lower the forks
   b. retract the boom
   c. tilt the mast forward
   d. raise one or more outriggers

10. When the machine is equipped with a ROPS and was built after 2008, the seat belt is not necessary.
    a. True
    b. False

11. What is the appropriate length of time for an operator to wait after startup for the oil pressure light to extinguish before shutting down the engine?
    a. 5 seconds
    b. 10 seconds
    c. 20 seconds
    d. 30 seconds

12. At shutdown, the forklift should be allowed to continue operating to cool the fluids for approximately _____.
    a. 1 minute
    b. 2 minutes
    c. 5 minutes
    d. 15 minutes

13. When an operator is in a slippery area and one wheel is spinning while the other is motionless, the component or feature that should be engaged to help is _____.
    a. the outriggers
    b. the differential lock
    c. two-wheel steering
    d. the transmission neutralizer

14. The purpose of a frame-leveling control is to _____.
    a. extend the outriggers
    b. raise and lower the outriggers
    c. level the forklift by rotating the chassis on its wheels and axles
    d. automatically sense an out-of-level condition and adjust the outriggers

15. If the forklift's rear wheels start to lift off the ground with a load, _____.
    a. stop raising the load and slowly lower it
    b. raise the outriggers higher on both sides
    c. drop the load to the ground immediately
    d. back a little farther away from the load and try again

Trade Terms Introduced in This Module

Crab steering: A steering mode where all wheels may move in the same direction, allowing the machine to move sideways on a diagonal; also known as oblique steering.

Four-wheel steering: A steering mode where the front and rear wheels may move in opposite directions, allowing for very tight turns; also known as independent steering or circle steering.

Fulcrum: A point or structure on which a lever sits and pivots.

Oblique steering: A steering mode where all wheels may move in the same direction, allowing the machine to move sideways on a diagonal; also known as crab steering.

Powered industrial trucks: An OSHA term for several types of light equipment that include forklifts.

Power hop: Action in heavy equipment that uses pneumatic tires to create a bouncing motion between the fore and aft axles. Once started, the oscillation back and forth usually continues until the operator either stops or slows down significantly to change the dynamics.

Telehandler: A type of powered industrial truck characterized by a boom with several extendable sections known as a telescoping boom; another name for a shooting boom forklift.

Tines: A prong of an implement such as a fork. For forklifts, tines are often called forks.

Appendix A

TYPICAL PERIODIC MAINTENANCE REQUIREMENTS

Maintenance Interval Schedule

Note: All safety information, warnings, and instructions must be read and understood before you perform any operation or maintenance procedure.

Before each consecutive interval is performed, all of the maintenance requirements from the previous interval must also be performed.

The normal oil change interval is every 500 service hours. If you operate the engine under severe conditions or if the oil is not Caterpillar oil, the oil must be changed at shorter intervals. Refer to the Operation and Maintenance Manual, "Engine Oil and Filter – Change" for further information. Severe conditions include the following factors: high temperatures, continuous high loads, and extremely dusty conditions.

Refer to the Operation and Maintenance Manual, "S.O.S. Oil Analysis" in order to determine if the oil change interval should be decreased. Refer to your Caterpillar dealer for detailed information regarding the optimum oil change interval.

The normal interval for inspecting and adjusting the clearance between the wear pads and the boom is 500 service hours. If the machine is working with excessively abrasive material then the clearance may need to be adjusted at shorter intervals. Refer to the Operation and Maintenance Manual, "Boom Wear Pad Clearance – Inspect/Adjust" for further information.

When Required

Battery – Recycle
Battery or Battery Cable – Inspect/Replace
Boom Telescoping Cylinder Air – Purge
Boom and Frame – Inspect
Engine Air Filter Primary Element – Clean/Replace
Engine Air Filter Secondary Element – Replace
Engine Air Filter Service Indicator – Inspect
Engine Air Precleaner – Clean
Fuel System – Prime
Fuel System Primary Filter – Replace
Fuel System Secondary Filter – Replace
Fuel Tank Cap and Strainer – Clean
Fuses and Relays – Replace
Oil Filter – Inspect
Radiator Core – Clean
Radiator Screen – Clean
Transmission Neutralizer Pressure Switch – Adjust
Window Washer Reservoir – Fill
Window Wiper – Inspect/Replace

Every 10 Service Hours or Daily

Backup Alarm – Test
Boom Retracting and Boom Lowering with Electric Power – Check
Braking System – Test
Cooling System Coolant Level – Check
Cooling System Pressure Cap – Clean/Replace
Engine Oil Level – Check
Fuel System Water Separator – Drain
Fuel Tank Water and Sediment – Drain
Indicators and Gauges – Test
Seat Belt – Inspect
Tire Inflation – Check
Transmission Oil Level – Check
Wheel Nut Torque – Check
Windows – Clean

Every 50 Service Hours or 2 Weeks

Axle Support – Lubricate
Bearing (Pivot) for Axle Drive Shaft – Lubricate
Boom Cylinder Pin – Lubricate
Boom Pivot Shaft – Lubricate
Brake Control Linkage – Lubricate
Carriage Cylinder Bearing – Lubricate
Carriage Pivot Pin - Lubricate
Compensating Cylinder Bearing – Lubricate
Cylinder Pin (Grapple Bucket) – Lubricate
Cylinder Pin and Pivot Pin (Bale Handler) – Lubricate
Cylinder Pin and Pivot Pin (Multipurpose Bucket) – Lubricate
Cylinder Pin and Pivot Pin (Utility Fork) – Lubricate
Fork Leveling Cylinder Pin – Lubricate
Frame Leveling Cylinder Pin – Lubricate
Pulley for Boom Extension Chain – Lubricate
Pulley for Boom Retraction Chain – Lubricate
Quick Coupler – Lubricate
Stabilizer and Cylinder Bearings – Lubricate

Initial 250 Service Hours (or after rebuild)

Boom Wear Pad Clearance – Inspect/Adjust
Service Brake – Adjust

Initial 250 Service Hours (or at first oil change)

Hydraulic System Oil Filter – Replace

Every 250 Service Hours or 3 Months

Axle Breathers – Clean/Replace
Belts – Inspect/Adjust/Replace
Boom Chain Tension – Check/Adjust
Differential Oil Level – Check
Drive Shaft Spline – Lubricate
Drive Shaft Universal Joint – Lubricate
Engine Oil Sample – Obtain
Final Drive Oil Level – Check
Hydraulic System Oil Level – Check
Longitudinal Stability Indicator – Test
Transfer Gear Oil Level – Check
Transmission Breather – Clean

Every 500 Service Hours

Differential and Final Drive Oil Sample – Obtain

Every 500 Service Hours or 6 Months

Boom Wear Pad Clearance – Inspect/Adjust
Engine Oil and Filter – Change
Fuel System Primary Filter – Replace
Fuel System Secondary Filter – Replace
Fuel Tank Cap and Strainer – Clean
Hydraulic System Oil Filter – Replace
Hydraulic System Oil Sample – Obtain
Hydraulic Tank Breather – Clean
Service Brake – Adjust
Transmission Oil Sample – Obtain

Every 1,000 Service Hours or 1 Year

Differential Oil – Change
Engine Valve Lash – Check
Final Drive Oil – Change
Rollover Protective Structure (ROPS) and Falling
  Object Protective Structure (FOPS) – Inspect
Transfer Gear Oil – Change
Transmission Oil – Change
Transmission Oil Filter – Replace

Every 2,000 Service Hours or 2 Years

Fuel Injection Timing – Check
Hydraulic System Oil – Change

Every 3 Years After Date of Installation or Every 5 Years After Date of Manufacture

Seal Belt – Replace

Every 3,000 Service Hours or 3 Years

Boom Chain – Inspect/Lubricate
Cooling System Coolant Extender (ELC) – Add
Cooling System Water Temperature Regulator –
  Replace
Engine Mounts – Inspect

Every 6,000 Service Hours or 6 Years

Cooling System Coolant (ELC) – Change

Appendix B

OSHA Inspection Checklist

Operator's Daily Checklist – Internal Combustion Engine Industrial Truck – Gas/LPG/Diesel Truck

Record of Fuel Added

Date		Operator		Fuel	
Truck #		Model #		Engine Oil	
Department		Serial #		Radiator Coolant	
Shift		Hour Meter		Hydraulic Oil	

SAFETY AND OPERATIONAL CHECKS (PRIOR TO EACH SHIFT)
Have a **qualified** mechanic correct all problems.

Engine Off Checks	OK	Maintenance
Leaks – Fuel, Hydraulic Oil, Engine Oil or Radiator Coolant		
Tires – Condition and Pressure		
Forks, Top Clip Retaining Pin and Heel – Check Condition		
Load Backrest – Securely Attached		
Hydraulic Hoses, Mast Chains, Cables and Stops – Check Visually		
Overhead Guard – Attached		
Finger Guards – Attached		
Propane Tank (LP Gas Truck) – Rust Corrosion, Damage		
Safety Warnings – Attached (Refer to Parts Manual for Location)		
Battery – Check Water/Electrolyte Level and Charge		
All Engine Belts – Check Visually		
Hydraulic Fluid Level – Check Level		
Engine Oil Level – Dipstick		
Transmission Fluid Level – Dipstick		
Engine Air Cleaner – Squeeze Rubber Dirt Trap or Check the Restriction Alarm (if equipped)		
Fuel Sedimentor (Diesel)		
Radiator Coolant – Check Level		
Operator's Manual – In Container		
Nameplate – Attached and Information Matches Model, Serial Number and Attachments		
Seat Belt – Functioning Smoothly		
Hood Latch – Adjusted and Securely Fastened		
Brake Fluid – Check Level		
Engine On Checks – Unusual Noises Must Be Investigated Immediately	OK	Maintenance
Accelerator or Direction Control Pedal – Functioning Smoothly		
Service Brake – Functioning Smoothly		
Parking Brake – Functioning Smoothly		
Steering Operation – Functioning Smoothly		
Drive Control – Forward/Reverse – Functioning Smoothly		
Tilt Control – Forward and Back – Functioning Smoothly		
Hoist and Lowering Control – Functioning Smoothly		
Attachment Control – Operation		
Horn and Lights – Functioning		
Cab (if equipped) – Heater, Defroster, Wipers – Functioning		
Gauges: Ammeter, Engine Oil Pressure, Hour Meter, Fuel Level, Temperature, Instrument, Monitors – Functioning		

22206-12_A03.EPS

Figure Credits

Courtesy of JLG Industries, Inc., Module opener, Figures 1B, 5–10, 12, 19, 22, 23, and 37

Courtesy of Sellick Equipment Limited, Figures 1A, 3, and 4

Courtesy of CLARK Material Handling Co., SA01

Courtesy of Terex Aerial Work Platforms, SA02

Reprinted courtesy of Caterpillar Inc., Figures 11, 13–16, 18, 24–26, 28–30, 32, 33, 35, 36, 38, 43–48, and Appendix A

Topaz Publications, Inc., Figure 17

Courtesy of The Caldwell Group, Inc., Figure 20

Vestil Manufacturing Corporation, Figure 21

Courtesy of Gar-Bro Manufacturing Co., Figure 27

US Department of Labor, Appendix B

Section Review Answers

Answer	Section Reference	Objectives
Section One		
1　d	1.1.1	1a
2　c	1.2.3	1b
3　b	1.3.5	1c
4　b	1.4.1	1d
Section Two		
1　b	2.1.0	2a
2　a	2.2.1	2b
Section Three		
1　a	3.1.4	3a
2　c	3.2.1	3b
3　b	3.3.3	3c
4　a	3.4.9	3d

NCCER CURRICULA — USER UPDATE

NCCER makes every effort to keep its textbooks up-to-date and free of technical errors. We appreciate your help in this process. If you find an error, a typographical mistake, or an inaccuracy in NCCER's curricula, please fill out this form (or a photocopy), or complete the online form at **www.nccer.org/olf**. Be sure to include the exact module ID number, page number, a detailed description, and your recommended correction. Your input will be brought to the attention of the Authoring Team. Thank you for your assistance.

Instructors – If you have an idea for improving this textbook, or have found that additional materials were necessary to teach this module effectively, please let us know so that we may present your suggestions to the Authoring Team.

NCCER Product Development and Revision
13614 Progress Blvd., Alachua, FL 32615

Email: curriculum@nccer.org
Online: www.nccer.org/olf

❏ Trainee Guide   ❏ Lesson Plans   ❏ Exam   ❏ PowerPoints   Other _____

Craft / Level: _____ Copyright Date: _____

Module ID Number / Title: _____

Section Number(s): _____

Description: _____

_____

_____

_____

Recommended Correction: _____

_____

_____

_____

Your Name: _____

Address: _____

Email: _____ Phone: _____

29102-15
Oxyfuel Cutting

OVERVIEW

Oxyfuel cutting is a method for cutting metal that uses an intense flame produced by burning a mixture of a fuel gas and pure oxygen. It is a versatile metal cutting method that has many uses on job sites. Because of the flammable gases and open flame involved, there is a danger of fire and explosion when oxyfuel equipment is used. However, these risks can be minimized when the operator is well-trained and knowledgeable about the function and operation of each part of an oxyfuel cutting outfit.

Module Nine

Trainees with successful module completions may be eligible for credentialing through the NCCER Registry. To learn more, go to **www.nccer.org** or contact us at **1.888.622.3720**. Our website has information on the latest product releases and training, as well as online versions of our *Cornerstone* magazine and Pearson's product catalog.

Your feedback is welcome. You may email your comments to **curriculum@nccer.org**, send general comments and inquiries to **info@nccer.org**, or fill in the User Update form at the back of this module.

This information is general in nature and intended for training purposes only. Actual performance of activities described in this manual requires compliance with all applicable operating, service, maintenance, and safety procedures under the direction of qualified personnel. References in this manual to patented or proprietary devices do not constitute a recommendation of their use.

Copyright © 2015 by NCCER, Alachua, FL 32615, and published by Pearson Education, Inc., New York, NY 10013. All rights reserved. Printed in the United States of America. This publication is protected by Copyright, and permission should be obtained from NCCER prior to any prohibited reproduction, storage in a retrieval system, or transmission in any form or by any means, electronic, mechanical, photocopying, recording, or likewise. To obtain permission(s) to use material from this work, please submit a written request to NCCER Product Development, 13614 Progress Blvd., Alachua, FL 32615.

From *Construction Craft Laborer, Level Two, Trainee Guide*, Third Edition. NCCER.
Copyright © 2015 by NCCER. Published by Pearson Education. All rights reserved.

29102-15
Oxyfuel Cutting

Objectives

When you have completed this module, you will be able to do the following:

1. Describe oxyfuel cutting and identify related safe work practices.
    a. Describe basic oxyfuel cutting.
    b. Identify safe work practices related to oxyfuel cutting.
2. Identify and describe oxyfuel cutting equipment and consumables.
    a. Identify and describe various gases and cylinders used for oxyfuel cutting.
    b. Identify and describe hoses and various types of regulators.
    c. Identify and describe cutting torches and tips.
    d. Identify and describe other miscellaneous oxyfuel cutting accessories.
    e. Identify and describe specialized cutting equipment.
3. Explain how to set up, light, and shut down oxyfuel equipment.
    a. Explain how to properly prepare a torch set for operation.
    b. Explain how to leak test oxyfuel equipment.
    c. Explain how to light the torch and adjust for the proper flame.
    d. Explain how to properly shut down oxyfuel cutting equipment.
4. Explain how to perform various oxyfuel cutting procedures.
    a. Identify the appearance of both good and inferior cuts and their causes.
    b. Explain how to cut both thick and thin steel.
    c. Explain how to bevel, wash, and gouge.
    d. Explain how to make straight and bevel cuts with portable oxyfuel cutting machines.

Performance Tasks

Under the supervision of your instructor, you should be able to do the following:

1. Set up oxyfuel cutting equipment.
2. Light and adjust an oxyfuel torch.
3. Shut down oxyfuel cutting equipment.
4. Disassemble oxyfuel cutting equipment.
5. Change empty gas cylinders.
6. Cut shapes from various thicknesses of steel, emphasizing:
    - Straight line cutting
    - Square shape cutting
    - Piercing
    - Beveling
    - Cutting slots
7. Perform washing.
8. Perform gouging.
9. Use a track burner to cut straight lines and bevels.

Trade Terms

- Backfire
- Carburizing flame
- Drag lines
- Dross
- Ferrous metals
- Flashback
- Gouging
- Kerf
- Neutral flame
- Oxidizing flame
- Pierce
- Soapstone
- Washing

Industry Recognized Credentials

If you are training through an NCCER-accredited sponsor, you may be eligible for credentials from NCCER's Registry. The ID number for this module is 29102-15. Note that this module may have been used in other NCCER curricula and may apply to other level completions. Contact NCCER's Registry at 888.622.3720 or go to **www.nccer.org** for more information.

Contents

Topics to be presented in this module include:

1.0.0 Oxyfuel Cutting Basics 1
  1.1.0 The Oxyfuel Cutting Process 1
  1.2.0 Oxyfuel Safety Summary 1
    1.2.1 Protective Clothing and Equipment 1
    1.2.2 Fire/Explosion Prevention 3
    1.2.3 Work Area Ventilation 4
    1.2.4 Cylinder Handling and Storage 4
2.0.0 Oxyfuel Cutting Consumables and Equipment 7
  2.1.0 Cutting Gases 7
    2.1.1 Oxygen 7
    2.1.2 Acetylene 9
    2.1.3 Liquefied Fuel Gases 13
  2.2.0 Regulators and Hoses 14
    2.2.1 Single-Stage Regulators 15
    2.2.2 Two-Stage Regulators 15
    2.2.3 Check Valves and Flashback Arrestors 15
    2.2.4 Gas Distribution Manifolds 17
    2.2.5 Hoses 18
  2.3.0 Torches and Tips 18
    2.3.1 One-Piece Hand Cutting Torch 18
    2.3.2 Combination Torch 19
    2.3.3 Cutting Torch Tips 19
  2.4.0 Other Accessories 24
    2.4.1 Tip Cleaners and Tip Drills 24
    2.4.2 Friction Lighters 24
    2.4.3 Cylinder Cart 24
    2.4.4 Soapstone Markers 25
  2.5.0 Specialized Cutting Equipment 25
    2.5.1 Mechanical Guides 25
    2.5.2 Motor-Driven Equipment 27
    2.5.3 Exothermic Oxygen Lances 29
3.0.0 Oxyfuel Equipment Setup and Shutdown 32
  3.1.0 Setting Up Oxyfuel Equipment 32
    3.1.1 Transporting and Securing Cylinders 32
    3.1.2 Cracking Cylinder Valves 32
    3.1.3 Attaching Regulators 33
    3.1.4 Installing Flashback Arrestors or Check Valves 34
    3.1.5 Connecting Hoses to Regulators 34
    3.1.6 Attaching Hoses to the Torch 35
    3.1.7 Connecting Cutting Attachments (Combination Torch Only) 35
    3.1.8 Installing Cutting Tips 35
    3.1.9 Closing Torch Valves and Loosening Regulator Adjusting Screws 35
    3.1.10 Opening Cylinder Valves 36
    3.1.11 Purging the Torch and Setting the Working Pressures 36

Contents (continued)

3.2.0 Testing for Leaks 37
    3.2.1 Initial and Periodic Leak Testing 38
    3.2.2 Leak-Down Testing of Regulators, Hoses, and Torch 39
    3.2.3 Full Leak Testing of a Torch 40
3.3.0 Controlling the Oxyfuel Torch Flame 40
    3.3.1 Oxyfuel Flames 40
    3.3.2 Backfires and Flashbacks 42
    3.3.3 Igniting the Torch and Adjusting the Flame 42
    3.3.4 Shutting Off the Torch 43
3.4.0 Shutting Down Oxyfuel Cutting Equipment 43
    3.4.1 Disassembling Oxyfuel Equipment 43
    3.4.2 Changing Cylinders 44
4.0.0 Performing Cutting Procedures 46
    4.1.0 Preparing for Oxyfuel Cutting with a Hand Cutting Torch 46
    4.1.1 Inspecting the Cut 46
    4.2.0 Cutting Steel 47
    4.2.1 Cutting Thin Steel 47
    4.2.2 Cutting Thick Steel 48
    4.2.3 Piercing a Plate 48
    4.3.0 Beveling, Washing, and Gouging 48
    4.3.1 Cutting Bevels 48
    4.3.2 Washing 50
    4.3.3 Gouging 50
    4.4.0 Operating Oxyfuel Track Burners 51
    4.4.1 Torch Adjustment 51
    4.4.2 Straight-Line Cutting 51
    4.4.3 Bevel Cutting 52
Appendix Performance Accreditation Tasks 57

Figures and Tables

Figure 1 Oxyfuel cutting 1
Figure 2 Personal protective equipment (PPE) for oxyfuel cutting 2
Figure 3 Welding blanket 3
Figure 4 Eliminating/minimizing oxygen in a container 4
Figure 5 Typical oxyfuel welding/cutting outfit 7
Figure 6 High-pressure oxygen cylinder markings and sizes 8
Figure 7 Oxygen cylinder valve 9
Figure 8 Oxygen cylinder with standard safety cap 9
Figure 9 Standard acetylene cylinder valve and fuse plugs 10
Figure 10 Acetylene cylinder markings and sizes 12
Figure 11 Acetylene cylinder with standard valve safety cap 13
Figure 12 Liquefied fuel gas cylinder 13
Figure 13 Oxygen and acetylene regulators 15
Figure 14 Two-stage regulator 16
Figure 15 Add-on check valve and flashback arrestor 16
Figure 16 Universal torch wrench 17
Figure 17 Gas distribution to stations through a manifold 17
Figure 18 Operator hookups for cutting gases 17
Figure 19 Heavy-duty three-tube one-piece positive-pressure hand cutting torch 20
Figure 20 Cutting torch attachment 21
Figure 21 Combination cutting torch 21
Figure 22 One- and two-piece cutting tips 21
Figure 23 Orifice-end views of one-piece acetylene torch cutting tips 22
Figure 24 Orifice-end view of one-piece cutting tip for liquefied fuel gases 23
Figure 25 Orifice-end view of two-piece cutting tip for liquefied fuel gases 23
Figure 26 Special-purpose torch cutting tips 23
Figure 27 Tip cleaner and drill kits 24
Figure 28 Friction lighters 25
Figure 29 Cylinder carts 26
Figure 30 Soapstone and graphite markers 26
Figure 31 Computer-controlled plate cutting machine 27
Figure 32 Track burner with oxyfuel machine torch 29
Figure 33 Track burner features 29
Figure 34 Track burner controls 29
Figure 35 Band-track pipe cutter/beveler 29
Figure 36 Ring gear pipe cutter/beveler 30
Figure 37 Oxygen lance holder 30
Figure 38 Cracking a cylinder valve 33
Figure 39 Checking connection fittings 34
Figure 40 Tightening regulator connection 34
Figure 41 Cleaning the regulator 34
Figure 42 Attaching a flashback arrestor 34
Figure 43 Connecting hose to regulator flashback arrestor 35

Figures and Tables (continued)

Figure 44 Connecting hoses to torch body 35
Figure 45 Installing a cutting tip 36
Figure 46 Torch valves and regulator adjusting screws............................. 36
Figure 47 Cylinder valve and gauges 36
Figure 48 Leak detection fluid use 38
Figure 49 Typical initial and periodic leak-test points 39
Figure 50 Blocking cutting tip for a leak test 39
Figure 51 Torch leak-test points 40
Figure 52 Acetylene and LP (propane) gas flames..................................... 41
Figure 53 Shutting down oxyfuel cutting equipment................................. 43
Figure 54 Typical empty cylinder marking 44
Figure 55 Examples of good and bad cuts.. 47
Figure 56 Cutting thin steel 48
Figure 57 Flame cutting with a hand torch... 49
Figure 58 Steps for piercing steel 49
Figure 59 Cutting a bevel 50
Figure 60 Washing 50
Figure 61 Gouging 51
Figure 62 Portable oxyfuel track burner 52

Table 1 Flame Temperatures of Oxygen With Various Fuel Gases 13
Table 2 Sample Acetylene Cutting Tip Chart... 22

SECTION ONE

1.0.0 OXYFUEL CUTTING BASICS

Objective

Describe oxyfuel cutting and identify related safe work practices.
 a. Describe basic oxyfuel cutting.
 b. Identify safe work practices related to oxyfuel cutting.

Trade Terms

Dross: The material (oxidized and molten metal) that is expelled from the kerf when cutting using a thermal process. It is sometimes called slag.

Ferrous metals: Metals containing iron.

Soapstone: Soft, white stone used to mark metal.

Figure 1 Oxyfuel cutting.

In order to perform oxyfuel cutting in a safe and effective manner, it is critical to understand the basic principles of oxyfuel cutting and be thoroughly familiar with safe work practices associated with the process.

1.1.0 The Oxyfuel Cutting Process

Oxyfuel cutting, also called flame cutting or burning, is a process that uses the flame and oxygen from a cutting torch to cut ferrous metals. The flame is produced by burning a fuel gas mixed with pure oxygen. The flame heats the metal to be cut to the kindling temperature (a cherry-red color); then a stream of high-pressure pure oxygen is directed from the torch at the metal's surface. This causes the metal to instantaneously oxidize or burn away. The cutting process results in oxides that mix with molten iron and produce dross, which is blown from the cut by the jet of cutting oxygen. This oxidation process, which takes place during the cutting operation, is similar to an accelerated rusting process. *Figure 1* shows an operator performing oxyfuel cutting.

The oxyfuel cutting process is usually performed only on ferrous metals such as straight carbon steels, which oxidize rapidly. This process can be used to quickly cut, trim, and shape ferrous metals, including the hardest steel.

Oxyfuel cutting can be used for certain metal alloys, such as stainless steel; however, the process requires higher preheat temperatures (white heat) and about 20 percent more oxygen for cutting. In addition, sacrificial steel plate or rod may have to be placed on top of the cut to help maintain the burning process. Other methods, such as carbon arc cutting, powder cutting, inert gas cutting, and plasma arc cutting, are much more practical for cutting steel alloys and nonferrous metals. Some of these methods are covered in other modules.

1.2.0 Oxyfuel Safety Summary

Cutting activities present unique hazards depending upon the material being cut and the fuel used to power the equipment. The proper safety equipment and precautions must be used when working with oxyfuel equipment because of the potential danger from the high-pressure flammable gases and high temperatures used. The following is a summary of safety procedures and practices that must be observed while cutting or welding. Keep in mind that this is just a summary. Complete safety coverage is provided in the *Welding Safety* module. Trainees who have not completed that module should do so before continuing.

1.2.1 Protective Clothing and Equipment

Oxyfuel cutting produces intense light and heat. It can also produce flying sparks and toxic fumes. To avoid injury, operators must wear appropriate personal protective equipment (PPE) when performing oxyfuel cutting operations. The following is a list of safety practices related to protective clothing and equipment that operators should follow:

- Always use safety glasses with a full face shield or a helmet (*Figure 2*). The glasses, face shield, or helmet lens must have the proper light-reducing tint for the type of cutting to be performed.
- Wear proper protective leather and/or flame-retardant clothing along with welding gloves that protect from flying sparks and molten metal as well as heat.
- Wear high-top safety shoes or boots. Make sure that the tongue and lace area of the footwear will be covered by a pant leg. If the tongue and lace area is exposed or the footwear must be protected from burn marks, wear leather spats under the pants or chaps and over the top of the footwear.
- Wear a 100-percent cotton cap with no mesh material included in its construction. The bill of the cap points to the rear. If a hard hat is required for the environment, use one that allows the attachment of rear deflector material and a face shield. A hard hat with a rear deflector is generally preferred when working overhead, and may be required by some employers and job sites.

WARNING! Do not wear a cap with a button in the middle. The conductive metal button beneath the fabric represents a safety hazard.

- Wear a face shield over safety glasses for cutting. Either the face shield or the lenses of the safety glasses must be an approved shade for the application. A welding hood equipped with a properly tinted lens is also acceptable. A shade 3 to 6 filter is recommended, depending on the thickness of the metal being cut, as required by *ANSI Z49.1*.
- Wear earplugs to protect ear canals from sparks. Wear hearing protection to protect against the consistent sound of the torch.

WARNING! Ear protection is essential to protect ears from the noise of the torch. Other personal protective equipment (PPE) must be worn to protect the operator from hot metal and slag.

Figure 2 Personal protective equipment (PPE) for oxyfuel cutting.

- Cutting operations involving materials or coatings containing cadmium, mercury, lead, zinc, chromium, and beryllium result in toxic fumes. For long-term cutting of such materials, always wear an approved full face, supplied-air respirator (SAR) that uses breathing air supplied externally of the work area. For occasional, very short-term exposure to zinc or copper fumes, a high-efficiency particulate arresting (HEPA)-rated or metal-fume filter may be used on a standard respirator.

1.2.2 Fire/Explosion Prevention

Most welding environment fires occur during oxyfuel gas welding or cutting. To minimize fire and explosion hazards, all cutting should be done in designated areas of the shop if possible. These areas should be made safe for cutting operations with concrete floors, arc filter screens, protective drapes, and fire extinguishers. A welding blanket (*Figure 3*) can be used to protect items in the area that would otherwise be damaged. No combustibles should be stored nearby. The work area should be kept neat and clean, and any metal scrap or dross must be cold before disposal.

Operators should be well-trained in the function and operation of each part of an oxyfuel gas welding or cutting station. In addition, it is often required that at least one fire watch be posted with an extinguisher to watch for possible fires.

The following list contains other steps that operators should follow to help prevent fires and explosions.

- Never carry matches or gas-filled lighters. Sparks can cause the matches to ignite or the lighter to explode, resulting in serious injury.
- Always comply with any site requirement for a hot-work permit and/or a fire watch.
- Never use oxygen to blow off clothing. The oxygen can remain trapped in the fabric for a time. If a spark hits the clothing during this time, the clothing can burn rapidly and violently out of control.
- Never release a large amount of oxygen or use oxygen in place of compressed air. Its presence around flammable materials or sparks can cause rapid and uncontrolled combustion. Keep pure oxygen away from oil, grease, and other petroleum products.
- Make sure that any flammable material in the work area is moved or shielded by a fire-resistant covering.

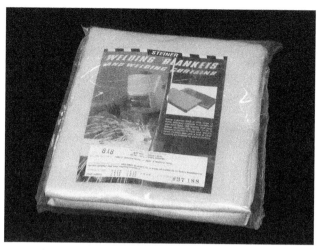

3,000°F (1,649°C) INTERMITTENT, 1,500°F (816°C) CONTINUOUS SILICON DIOXIDE CLOTH

Figure 3 Welding blanket.

- Approved fire extinguishers must be available before attempting any heating, welding, or cutting operations. Make sure the extinguisher is charged, the inspection tag is valid, and any individual that may be required to operate it knows how to do so.
- Never release a large amount of fuel gas, especially acetylene. Propylene and propane tend to concentrate in and along low areas and can ignite at a considerable distance from the release point. Acetylene is lighter than air but is even more dangerous. When mixed with air or oxygen, it will explode at much lower concentrations than any other fuel gas.
- To prevent fires, maintain a neat and clean work area, and make sure that any metal scrap or slag is cold before disposal.

Before cutting containers such as tanks or barrels, check to see if they have contained any explosive, hazardous, or flammable materials, including petroleum products, citrus products, or chemicals that decompose into toxic fumes when heated. Proper procedures for cutting or welding hazardous containers are described in the *American Welding Society (AWS) F4.1, Safe Practices for the Preparation of Containers and Piping for Welding and Cutting,* and *ANSI Z49.1*. As a standard practice, always clean and then fill any tanks or barrels with water, or purge them with a flow of inert gas such as nitrogen to displace any oxygen.

WARNING!
Welding or cutting must never be performed on drums, barrels, tanks, vessels, or other containers until they have been emptied and cleaned thoroughly, eliminating all flammable materials and all substances (such as detergents, solvents, greases, tars, or acids) that might produce flammable, toxic, or explosive vapors when heated. Do not assume that a container that has held combustibles is clean and safe until proven so by proper tests. Do not weld in places where dust or other combustible particles are suspended in air or where explosive vapors are present.

Containers must be cleaned by steam cleaning, flushing with water, or washing with detergent until all traces of the material have been removed.

WARNING!
Clean containers only in well-ventilated areas. Vapors can accumulate during cleaning, causing explosions or injury.

After cleaning the container, fill it with water (*Figure 4*) or a purge gas, such as carbon dioxide, argon, or nitrogen to displace the explosive fumes. Air, which contains oxygen, is displaced from inside the container by the water or inert gas. Without oxygen, combustion cannot take place.

A water-filled vessel is the best alternative. When using water, position the container to minimize the air space. When using an inert gas, provide a vent hole so the inert gas can push the air and other vapors out to the atmosphere. Keep in mind that these precautions do not guarantee the absence of flammable materials inside. For that reason, these types of activities should not be done without adequate supervision and the use of proper testing methods.

1.2.3 Work Area Ventilation

Cutting operations should always be performed in a well-ventilated area. This greatly reduces the risks associated with toxic fumes. This rule is especially true for confined spaces. Confined-space procedures must be followed before cutting begins. Proper ventilation must be provided before any cutting procedures take place, but never use oxygen for ventilation purposes.

Work area safety rules apply to all workers in the area. In a typical work area, there might be numerous workers performing various tasks. Always remain aware of other personnel in the area and take the necessary precautions to ensure their safety as well as your own.

1.2.4 Cylinder Handling and Storage

Operators must be aware of the hazards involved in the use of fuel gas, oxygen, and shielding gas cylinders and know how to store these cylinders are stored safely. One basic rule is that only compressed gas cylinders containing the correct gas for the process should be used. Regulators must be correct for the gas and pressure and must function properly. All hoses, fittings, and other parts must be suitable and maintained in good condition.

Any cylinder that leaks, bad valves, or damaged threads must be identified and reported to the supplier. Use a piece of soapstone or a tag to write the problem on the cylinder. If closing the cylinder valve cannot stop the leak, move the cylinder outdoors to a safe location, away from any source of ignition, and notify the supplier. Post a warning sign and then slowly release the pressure.

In its gaseous form, acetylene is extremely unstable and explodes easily. For this reason it must remain at pressures below 15 pounds per square inch (psi) or 103 kilopascals (kPa). If an acetylene cylinder is tipped over, stand it upright and wait at least an hour before using it. If the cylinder has spent a significant period of time laying on its side,

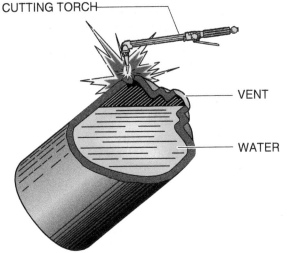

NOTE: ANSI Z49.1 AND AWS STANDARDS SHOULD BE FOLLOWED.

Figure 4 Eliminating/minimizing oxygen in a container.

Oxygen Consumption

Two-thirds of the oxygen in a neutral flame comes from the air you breathe inside the work space. This could be an issue in a heavily occupied, tight work space.

it is best to allow at least two hours to ensure all of the liquid has drained down to the base of the cylinder. Acetylene cylinders contain liquid acetone. If the liquid is withdrawn from a cylinder, it will foul the safety check valves and regulators and decrease the stability of the acetylene stored in the cylinder. For this reason, acetylene must never be withdrawn at a per-hour rate that exceeds one-seventh of the volume of the cylinder(s) in use. Acetylene cylinders in use should be opened no more than one and one-half turns and, preferably, no more than three-fourths of a turn.

Other precautions associated with gas cylinders include the following:

- Keep cylinders in the upright position and securely chained to an undercarriage or fixed support so that they cannot be knocked over accidentally. Even though they are more stable, cylinders attached to a gas distribution manifold should be chained or otherwise confined, as should cylinders stored in a special room used only for cylinder storage.
- Keep cylinders that contain combustible gases in one area of the building for safety. Cylinders must be at a safe distance from arc welding or cutting operations and any other source of heat, sparks, or flame.
- Cylinder storage areas must be located away from halls, stairwells, and exits so that in case of an emergency they will not block an escape route. Storage areas should also be located away from heat, radiators, furnaces, and welding sparks. The location of storage areas should be where unauthorized people cannot tamper with the cylinders. A warning sign that reads "Danger—No Smoking, Matches, or Open Lights", or similar wording, should be posted in the storage area.
- Oxygen and fuel gas cylinders or other flammable materials must be stored separately. The storage areas must be separated by 20 feet (6.1 m) or by a wall 5 feet (1.5 m) high with at least a 30-minute burn rating. The purpose of the distance or wall is to keep the heat of a small fire from causing the oxygen cylinder safety valve to release. If the safety valve releases the oxygen, a small fire could quickly become an inferno.
- Inert gas cylinders may be stored separately or with either fuel cylinders or oxygen cylinders. Empty cylinders must be stored separately from full cylinders, although they may be stored in the same room or area. All cylinders must be stored vertically and have the protective caps screwed on firmly.
- Never allow a welding electrode, electrode holder, or any other electrically energized parts to come in contact with the cylinder.
- When opening a cylinder valve, operators should stand with the valve stem between themselves and the regulator.
- If a cylinder is frozen to the ground, use warm water (not boiling) to loosen it.

Cylinders equipped with a valve protection cap must have the cap in place unless the cylinder is in use. The protection cap prevents the valve from being broken off if the cylinder is knocked over. If the valve of a full high-pressure cylinder (such as argon or oxygen) is broken off, the cylinder can take flight like a missile if it has not been secured properly. Never lift a cylinder by the safety cap or valve. The valve can easily break off or be damaged. When moving cylinders, the valve protection cap must be replaced, especially if the cylinders are mounted on a truck or trailer. Cylinders must never be dropped or handled roughly.

> **WARNING!**
> Using a wrench inserted through the cap can open the valve. If the cap is stuck, use a strap wrench to remove, or call the gas supplier.

Additional Resources

ANSI Z49.1, Safety in Welding, Cutting, and Allied Processes. Miami, FL: American Welding Society.

Uniweld Products, Inc. Numerous videos are available at **www.uniweld.com/en/uniweld-videos**. Last accessed: November 30, 2014.

The Harris Products Group, a division of Lincoln Electric. Numerous videos are available at **www.harrisproductsgroup.com/en/Expert-Advice/videos.aspx**. Last accessed: November 30, 2014.

1.0.0 Section Review

1. The oxyfuel cutting process creates oxides that mix with molten iron and produce dross, which is blown from the cut by a jet of ____ .
    a. pure nitrogen
    b. pure oxygen
    c. carbon dioxide
    d. cooling water

2. Carbon dioxide, argon, and nitrogen are all gases that are suitable for ____ .
    a. mixing with a fuel gas to enable oxyfuel cutting
    b. oxidizing dross to create reusable slag
    c. filling and pressurizing confined spaces
    d. purging explosive fumes from containers

Section Two

2.0.0 Oxyfuel Cutting Consumables and Equipment

Objective

Identify and describe oxyfuel cutting equipment and consumables.
a. Identify and describe various gases and cylinders used for oxyfuel cutting.
b. Identify and describe hoses and various types of regulators.
c. Identify and describe cutting torches and tips.
d. Identify and describe other miscellaneous oxyfuel cutting accessories.
e. Identify and describe specialized cutting equipment.

Trade Terms

Backfire: A loud snap or pop as a torch flame is extinguished.

Flashback: The flame burning back into the tip, torch, hose, or regulator, causing a high-pitched whistling or hissing sound.

Gouging: The process of cutting a groove into a surface.

Kerf: The gap produced by a cutting process.

Washing: A term used to describe the process of cutting out bolts, rivets, previously welded pieces, or other projections from the metal surface.

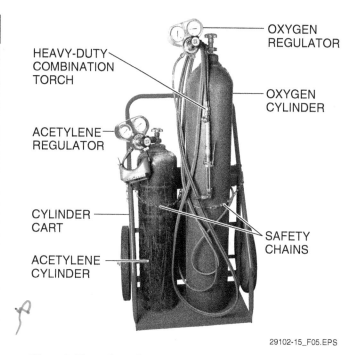

Figure 5 Typical oxyfuel welding/cutting outfit.

The equipment used to perform oxyfuel cutting includes oxygen and fuel gas cylinders, oxygen and fuel gas regulators, hoses, and a cutting torch. A typical movable oxyfuel (oxyacetylene) cutting outfit is shown in *Figure 5*.

2.1.0 Cutting Gases

Many oxyfuel cutting outfits use oxygen and acetylene. However, other fuel gases including natural gas and liquefied gases are also used with oxygen for cutting. This section examines some of the more common cutting gases.

2.1.1 Oxygen

Oxygen (O_2) is a colorless, odorless, tasteless gas that supports combustion. It is not considered a fuel gas, but is necessary for combustion. Combined with burning material, pure oxygen causes a fire to flare and burn out of control. When mixed with fuel gases, oxygen produces the high-temperature flame required to flame-cut metals.

Oxygen is stored at more than 2,000 psi (13,790 kPa) in hollow steel cylinders. The cylinders come in a variety of sizes based on different international standards. Specific details about the cylinders are typically regulated by the government agency for the country in which the cylinders are transported, such as the US Department of Transportation (DOT), Transport Canada (TC), and the Department for Transport (DfT) in Europe. *Figure 6* shows high-pressure oxygen cylinder markings and capacities in cubic feet based on US DOT specifications. The smallest standard cylinder holds about 85 cubic feet (2.4 cubic meters) of oxygen, and the largest ultra-high-pressure cylinder holds about 485 cubic feet (13.7 cubic meters). The most common oxygen cylinder size used for cutting operations is the 227-cubic foot (6.4-cubic meter) cylinder. It is more than 4' (1.2 m) tall and 9" (10.2 cm) in diameter. Regardless of their size and capacity, oxygen cylinders must be tested every 10 years in the United States. Other locations may have a different testing standard.

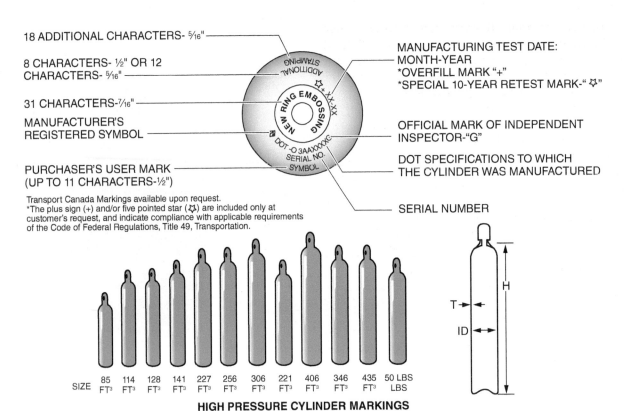

DOT SPECIFICATIONS	O₂ CAPACITY (FT³)		WATER CAPACITY (IN³)		NOMINAL DIMENSIONS (IN)			NOMINAL WEIGHT (LB)	PRESSURE (PSI)	
	AT RATED SERVICE PRESSURE	AT 10% OVERCHARGE	MINIMUM	MAXIMUM	AVG. INSIDE DIAMETER "ID"	HEIGHT "H"	MINIMUM WALL "T"		SERVICE	TEST
STANDARD HIGH PRESSURE CYLINDERS[1]										
3AA2015	85	93	960	1040	6.625	32.50	0.144	48	2015	3360
3AA2015	114	125	1320	1355	6.625	43.00	0.144	61	2015	3360
3AA2265	128	140	1320	1355	6.625	43.00	0.162	62	2265	3775
3AA2015	141	155	1630	1690	7.000	46.00	0.150	70	2015	3360
3AA2015	227	250	2640	2710	8.625	51.00	0.184	116	2015	3360
3AA2265	256	281	2640	2710	8.625	51.00	0.208	117	2265	3775
3AA2400	306	336	2995	3060	8.813	55.00	0.226	140	2400	4000
3AA2400	405	444	3960	4040	10.060	56.00	0.258	181	2400	4000
ULTRALIGHT® HIGH PRESSURE CYLINDERS[1]										
E-9370-3280	365	NA	2640	2710	8.625	51.00	0.211	122	3280	4920
E-9370-3330	442	NA	3181	3220	8.813	57.50	0.219	147	3330	4995
ULTRA HIGH PRESSURE CYLINDERS[2]										
3AA3600	347[3]	374	2640	2690	8.500	51.00	0.336	170	3600	6000
3AA6000	434[3]	458	2285	2360	8.147	51.00	0.568	267	6000	10000
E-10869-4500	435[3]	NA	2750	2890	8.813	51.00	0.260	148	4500	6750
E-10869-4500	485[3]	NA	3058	3210	8.813	56.00	0.260	158	4500	6750

1. Regulators normally permit filling these cylinders with 10% overcharge, provided certain other requirements are met.
2. Under no circumstances are these cylinders to be filled to a pressure exceeding the marked service pressure at 70°F.
3. Nitrogen capacity at 70°F.

All cylinders normally furnished with ¾" NGT internal threads, unless otherwise specified.
Nominal weights include neck ring but exclude valve and cap, add 2 lbs. (.91 kg) for cap and 1½ lb. (.8 kg) for valve.
Cap adds approximately 5 in. (127 mm) to height.
Cylinder capacities are approximately 5 in. (127 mm) to height.
Cylinder capacities are approximately at 70°F. (21°C).

Figure 6 High-pressure oxygen cylinder markings and sizes.

Oxygen cylinders have bronze cylinder valves on top (*Figure 7*). The cylinder valve controls the flow of oxygen out of the cylinder. A safety plug on the side of the cylinder valve allows oxygen in the cylinder to escape if the pressure in the cylinder rises too high. Although the escaping oxygen presents a hazard, the risk of explosion represents an even more significant hazard. Oxygen cylinders are usually equipped with Compressed Gas Association (CGA) Valve Type 540 valves for service up to 3,000 pounds per square inch gauge (psig), or 20,684 kPa. Some cylinders are equipped with CGA Valve Type 577 valves for up to 4,000 psig (27,579 kPa) oxygen service or CGA Valve Type 701 valves for up to 5,500 psig (34,473 kPa) service. Each CGA valve type is for a specific type of gas and pressure rating. Use care when handling oxygen cylinders because oxygen is stored at such high pressures. When it is not in use, always cover the cylinder valve with the protective steel safety cap and tighten it securely (*Figure 8*).

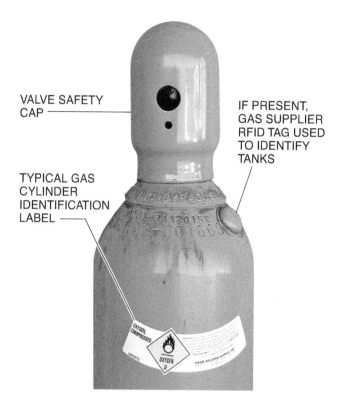

Figure 8 Oxygen cylinder with standard safety cap.

> **WARNING!**
> Do not remove the protective cap unless the cylinder is secured. If the cylinder falls over and the valve assembly breaks off, the cylinder will be propelled like a rocket, causing severe injury or death to anyone in its path.

2.1.2 Acetylene

Acetylene gas (C_2H_2), a compound of carbon and hydrogen, is lighter than air. It is formed by dissolving calcium carbide in water. It has a strong, distinctive, garlic-like odor, which is added to the gas intentionally so that it can be detected. In its gaseous form, acetylene is extremely unstable and explodes easily. Because of this instability, it cannot be compressed at pressures of more than 15 psi (103 kPa) when in its gaseous form. At higher pressures, acetylene gas breaks down chemically, producing heat and pressure that could result in a violent explosion. When combined with oxygen, acetylene creates a flame that burns hotter than 5,500°F (3,037°C), one of the hottest gas flames. Acetylene can be used for flame cutting, welding, heating, flame hardening, and stress relieving.

Because of the explosive nature of acetylene gas, it cannot be stored above 15 psi (103 kPa) in a hollow cylinder. To solve this problem, acetylene cylinders are specially constructed to store acetylene. The acetylene cylinder is filled with a porous material that creates a solid cylinder, instead of a hollow cylinder as used for all other common gases. The porous material is soaked with liquid acetone, which absorbs the acetylene gas, stabilizing it and allowing for storage at pressures above 15 psi (103 kPa).

Because of the liquid acetone inside the cylinder, acetylene cylinders must always be used in an upright position. If the cylinder is tipped over, stand the cylinder upright and wait at least one hour before using it. If liquid acetone is withdrawn from a cylinder, it will foul the safety check valves and regulators. It will also cause extremely unstable torch operation. Always take

Figure 7 Oxygen cylinder valve.

care to control the acetylene gas pressure leaving a cylinder at pressures less than 15 psig (103.4 kPa) and at hourly rates that do not exceed one-seventh of the cylinder capacity. This can easily happen if a torch with a large nozzle is connected to a cylinder that is too small for the task. High rates of discharge may cause liquid acetone to be caught up in the gas stream and be carried out with the gas.

Acetylene cylinders have safety fuse plugs in the top and bottom of the cylinder (*Figure 9*) that melt at 212°F (100°C). A fuse, or fusible, plug is a type of pressure relief device. It is not a valve however; once it is activated, it cannot be reclosed and must be replaced. Fuse plugs, also known as rupture disks, are often used in place of relief valves for low-pressure applications like this one. In the event of a fire, the fuse plugs will release the acetylene gas, preventing the cylinder from exploding.

As with oxygen cylinders, acetylene cylinders are available in a variety of sizes and are typically regulated by government agencies. *Figure 10* shows high-pressure acetylene cylinder markings and capacities in cubic feet based on US DOT specifications. The smallest standard cylinder holds about 10 cubic feet (0.28 cubic meters) of gas. The largest standard cylinder holds about 420 cubic feet (11.9 cubic meters) of gas. A cylinder that holds about 850 cubic feet (24.1 cubic meters) is also available. Like oxygen cylinders, acetylene cylinders used in the United States must be tested every 10 years.

Acetylene cylinders are usually equipped with a CGA 510 brass cylinder valve. The valve controls the flow of acetylene from the cylinder into a regulator. Some acetylene cylinders are equipped with an alternate CGA 300 valve. Some obsolete valves still in use require a special long-handled wrench with a square socket end to operate the valve.

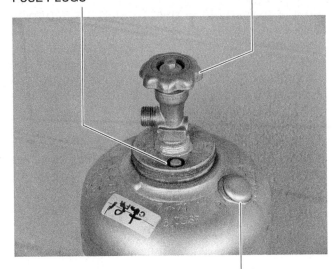
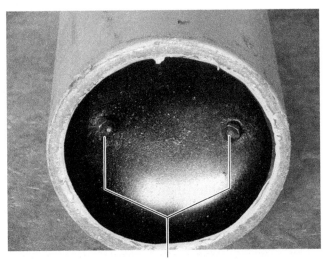

Figure 9 Standard acetylene cylinder valve and fuse plugs.

Lifetime Cylinder Management

Gas cylinders may be fitted with radio-frequency identification (RFID) tags so that they can be readily identified and tracked. The RFID tag is electronically scanned and a coded number is matched against the records for identification and tracking. This aids in quickly determining the identity of the purchaser or user of the cylinder, where it has been, and for determining the testing and maintenance records of the cylinder. The RFID tag may appear to be a button like the one shown here. However, it can also be concealed in the walls of the cylinder neck and covered for protection, or have a tight-fitting collar that snaps around the cylinder valve neck.

Alternate High-Pressure Cylinder Valve Cap

High-pressure cylinders can be equipped with a clamshell cap that can be closed to protect the cylinder valve with or without a regulator installed on the valve. This enables safe movement of the cylinder after the cylinder valve is closed. This type of cap is usually secured to the cylinder body cap threads when it is installed so that it cannot be removed. When the clamshell is closed, it can also be padlocked to prevent unauthorized operation of the cylinder valve.

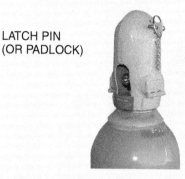

LATCH PIN (OR PADLOCK)

CLAMSHELL OPEN TO ALLOW CYLINDER VALVE OPERATION

CLAMSHELL CLOSED FOR MOVEMENT OR PADLOCKED TO PREVENT OPERATION OF CYLINDER VALVE

CLAMSHELL CLOSED FOR TRANSPORT

29102-15_SA02.EPS

Alternate Acetylene Cylinder Safety Cap

Acetylene cylinders can be equipped with a ring guard cap that protects the cylinder valve with or without a regulator installed on the valve. This enables safe movement of the cylinder after the cylinder valve is closed. This type of cap is usually secured to the cylinder body cap threads when it is installed so that it cannot be removed. Some other types of cylinders, such as propane cylinders, may also be fitted with this type of guard.

29102-15_SA03.EPS

29102-15 Oxyfuel Cutting   Module Nine   11

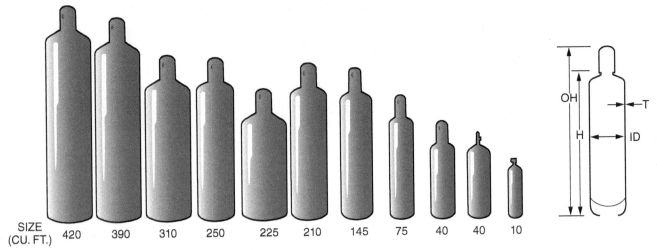

DOT SPECIFICATIONS	CAPACITY ACETYLENE (FT³)	MIN. WATER (IN³)	(LB.)	NOMINAL DIMENSIONS (IN) AVG. INSIDE DIAMETER "ID"	HEIGHT W/OUT VALVE OR CAP "H"	HEIGHT W/VALVE AND CAP "OH"	MINIMUM WALL "T"	ACETONE (LB - OZ)	APPROXIMATE TARE WEIGHT WITH VALVE WITHOUT CAP (LB)
8 AL[1]	10	125	4.5	3.83	13.1375	14.75	0.0650	1-6	8
8[1]	40	466	16.8	6.00	19.8000	23.31	0.0870	5-7	25
8[2]	40	466	16.8	6.00	19.8000	28.30	0.0870	5-7	28
8[3]	75	855	30.8	7.00	25.5000	31.25	0.0890	9-8	45
8	100	1055	38.0	7.00	30.7500	36.50	0.0890	12-2	55
8	145	1527	55.0	8.00	34.2500	40.00	0.1020	18-10	76
8	210	2194	79.0	10.00	32.2500	38.00	0.0940	25-13	105
8AL	225	2630	94.7	12.00	27.5000	32.75	0.1280	29-6	110
8	250	2606	93.8	10.00	38.0000	43.75	0.0940	30-12	115
8AL	310	3240	116.7	12.00	32.7500	38.50	0.1120	39-5	140
8AL	390	4151	150.0	12.00	41.0000	46.75	0.1120	49-14	170
8AL	420	4375	157.5	12.00	43.2500	49.00	0.1120	51-14	187
8	60	666	24.0	7.00	25.79 OH		0.0890	7-11	40
8	130	1480	53.3	8.00	36.00 OH		0.1020	17-2	75
8AL	390	4215	151.8	12.00	46.00 OH		0.1120	49-14	180

1. Tapped for 3/8" valve but are not equipped with valve protection caps.
2. Includes valve protection cap.
3. Can be tared to hold 60 ft³ (1.7 m³) of acetylene gas. Standard tapping (except cylinders tapped for 3/8") 3/4"-14 NGT.

Weight includes saturation gas, filler, paint, solvent, valve, fuse plugs. Does not include cap of 2 lb. (.91 kg).
Cylinder capacities are based upon commercially pure acetylene gas at 250 psi (17.5 kg/cm²), and 70°F (15°C).

Figure 10 Acetylene cylinder markings and sizes.

The smallest standard acetylene cylinder, which holds 10 cubic feet (0.28 cubic meters), is equipped with a CGA 200 small series valve, and 40 cubic foot (1.13 cubic meter) cylinders use a CGA 520 small series valve. As with oxygen cylinders, place a protective valve cap on the acetylene cylinders during transport (*Figure 11*).

WARNING! Do not remove the protective cap unless the cylinder is secured. If the cylinder falls over and the nozzle breaks off, the cylinder will release highly explosive gas.

2.1.3 Liquefied Fuel Gases

Many fuel gases other than acetylene are used for cutting. They include natural gas and liquefied fuel gases such as propylene and propane. Their flames are not as hot as acetylene, but they have higher British thermal unit (Btu) ratings and are cheaper and safer to use. Job site policies typically determine which fuel gas to use. *Table 1* compares the flame temperatures of oxygen mixed with various fuel gases.

Propylene mixtures are hydrocarbon-based gases that are stable and shock-resistant, making them relatively safe to use. They are purchased under trade names such as High Purity Gas (HPG™), Apachi™, and Prestolene™. These gases and others have distinctive odors to make leak detection easier. They burn at temperatures around 5,193°F (2,867°C), hotter than natural gas and propane. Propylene gases are used for flame cutting, scarfing, heating, stress relieving, brazing, and soldering.

Propane is also known as liquefied petroleum (LP) gas. It is stable and shock-resistant, and it has a distinctive odor for easy leak detection. It burns at 4,580°F (2,526°C), which is the lowest temperature of any common fuel gas. Propane has a slight tendency toward backfire and flashback and is used quite extensively for cutting procedures.

Natural gas is delivered by pipeline rather than by cylinders. Manifolds must be available on site for the connection of regulators and hoses. It burns at about 4,600°F (2,537°C). Natural gas is relatively stable and shock-resistant and has a slight tendency toward backfire and flashback. Because of its recognizable odor, leaks are easily detectable. Natural gas is used primarily for cutting on job sites with permanent cutting stations.

Liquefied fuel gases are shipped in hollow steel cylinders (*Figure 12*). When empty, they are much lighter than acetylene cylinders.

The hollow steel cylinders for liquefied fuel gases come in various sizes. They can hold from 30 to 225 pounds (13.6 to 102.1 kilograms) of fuel gas. As the cylinder valve is opened, the

Table 1 Flame Temperatures of Oxygen With Various Fuel Gases

Type of Gas	Flame Temperature
Acetylene	More than 5,500°F (3,038°C)
Propylene	5,130°F (2,832°C)
Natural gas	4,600°F (2,538°C)
Propane	4,580°F (2,527°C)

Figure 11 Acetylene cylinder with standard valve safety cap.

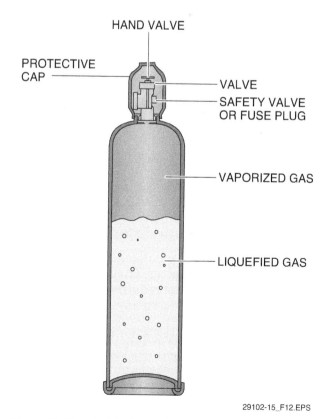

Figure 12 Liquefied fuel gas cylinder.

vaporized gas is withdrawn from the cylinder. The remaining liquefied gas absorbs heat and releases additional vaporized gas. The pressure of the vaporized gas varies with the outside temperature. The colder the outside temperature, the lower the vaporized gas pressure will be. If high volumes of gas are removed from a liquefied fuel gas cylinder, the pressure drops, and the temperature of the cylinder will also drop. A ring of frost can form around the base of the cylinder due to the cooling effect as the liquid vaporizes and absorbs heat. If high withdrawal rates continue, the regulator may also start to ice up. If high withdrawal rates are required, special regulators with electric heaters should be used.

> **WARNING!**
> Never apply heat directly to a cylinder or regulator. This can cause excessive pressure, resulting in an explosion.

The pressure inside a liquefied fuel gas cylinder is not an indicator of how full or empty the cylinder is. The weight of a cylinder determines how much liquefied gas is left. Liquefied fuel gas cylinders are equipped with CGA 510, 350, or 695 valves, depending on the fuel and storage pressures.

> **WARNING!**
> Do not remove the protective cap on liquefied fuel gas cylinders unless the cylinder is secured. If the cylinder falls over and the nozzle breaks off, the cylinder will release highly explosive gas. Cylinders containing a liquid such as propane must be kept in an upright position. If the valve is broken or is opened with the cylinder horizontal, the fuel can emerge as a liquid that will shoot a long distance before it vaporizes. If it is ignited, it produces an extremely dangerous, uncontrolled flame.

2.2.0 Regulators and Hoses

Regulators (*Figure 13*) are attached to the oxygen and fuel gas cylinder valves. They reduce the high cylinder pressures to the required lower working pressures and maintain a steady flow of gas from the cylinder.

A regulator's pressure-adjusting screw controls the gas pressure. Turned clockwise, it increases the pressure of gas. Turned counterclockwise, it reduces the pressure of gas. When turned counterclockwise until loose (released), it stops the flow of gas. When the adjusting screw feels very loose, it is an indication that the end of the screw is no longer in contact with the regulating spring. However, the regulator adjusting screw should never be considered a shut-off valve. When shut-off is desired, use the cylinder valve. The regulator can remain at its set position, allowing any gas remaining between the cylinder valve and the regulator to escape.

Most regulators contain two gauges. The high-pressure or cylinder-pressure gauge indicates the actual cylinder pressure (upstream); the low-pressure gauge indicates the pressure of the gas leaving the regulator (downstream).

Oxygen regulators differ from fuel gas regulators. Oxygen regulators may have green markings and always have right-hand threads on all connections. The oxygen regulator's high-pressure gauge generally reads up to 3,000 psi (20,684 kPa) and includes a second scale that shows the amount of oxygen in the cylinder in terms of cubic feet. The low-pressure or working-pressure gauge may read 100 psi (689 kPa) or higher.

Fuel gas regulators may have red markings and usually have left-hand threads on all the connections. As a reminder that the regulator has left-hand threads, a V-notch may be cut into the corners of the fitting nut. These notches are visible on the fitting nut of the fuel gas regulator shown in *Figure 13*. The fuel gas regulator's high-pressure gauge usually reads up to 400 psi (2,758 kPa). The low-pressure or working-pressure gauge may read up to 40 psi (276 kPa). Acetylene gauges, however, are always red-lined at 15 psi (103 kPa) as a reminder that acetylene pressure should not be increased beyond that point.

Single-stage and two-stage regulators will be discussed in the following sections.

Handling and Storing Liquefied Gas Cylinders

Liquefied fuel gas cylinders have a safety valve built into the valve at the top of the cylinder. The safety valve releases gas if the pressure begins to rise. Use care when handling fuel gas cylinders because the gas in cylinders is stored at significant pressures. Cylinders should never be dropped or hit with heavy objects, and they should always be stored in an upright position. When not in use, the cylinder valve must always be covered with the protective steel cap.

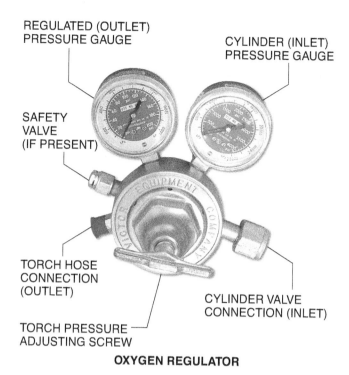
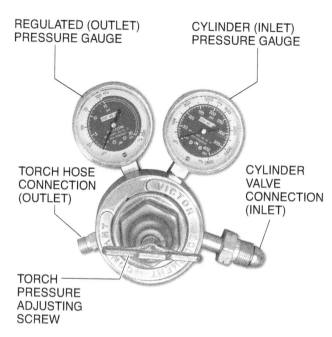

Figure 13 Oxygen and acetylene regulators.

> **WARNING!**
>
> To prevent injury and damage to regulators, always follow these guidelines:
> - Never subject regulators to jarring or shaking, as this can damage the equipment beyond repair.
> - Always check that the adjusting screw is fully released before the cylinder valve is turned on and when the welding has been completed.
> - Always open cylinder valves slowly and stand with the valve stem between you and the regulator.
> - Never use oil to lubricate a regulator. This can result in an explosion when the regulator is in use.
> - Never use fuel gas regulators on oxygen cylinders, or oxygen regulators on fuel gas cylinders.
> - Never work with a defective regulator. If it is not working properly, shut off the gas supply and have the regulator repaired by someone who is qualified to work on it.
> - Never use large wrenches, pipe wrenches, pliers, or slip-joint pliers to install or remove regulators.

2.2.1 Single-Stage Regulators

Single-stage, spring-compensated regulators reduce pressure in one step. As gas is drawn from the cylinder, the internal pressure of the cylinder decreases. A single-stage, spring-compensated regulator is unable to automatically adjust for this decrease in internal cylinder pressure. Therefore, it becomes necessary to adjust the spring pressure periodically to modify the output gas pressure as the gas in the cylinder is consumed. These regulators are the most commonly used because of their low cost and high flow rates.

2.2.2 Two-Stage Regulators

The two-stage, pressure-compensated regulator reduces pressure in two steps. It first reduces the input pressure from the cylinder to a predetermined intermediate pressure. This intermediate pressure is then adjusted by the pressure-adjusting screw. With this type of regulator, the delivery pressure to the torch remains constant, and no readjustment is necessary as the gas in the cylinder is consumed. Standard two-stage regulators (*Figure 14*) are more expensive than single-stage regulators and have lower flow rates. There are also heavy-duty types with higher flow rates that are usually preferred for thick material and/or continuous-duty cutting operations.

2.2.3 Check Valves and Flashback Arrestors

Check valves and flashback arrestors (*Figure 15*) are safety devices for regulators, hoses, and torches. Check valves allow gas to flow in one direction only. Flashback arrestors stop fire from being able to travel backwards through the hose.

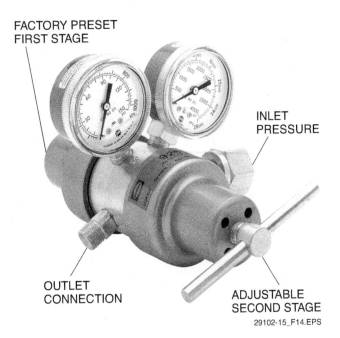

Figure 14 Two-stage regulator.

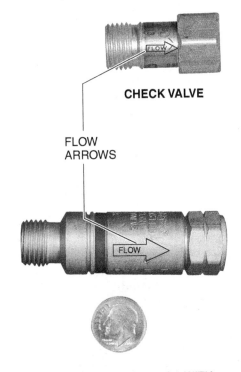

Figure 15 Add-on check valve and flashback arrestor.

oxygen hose or the entry and mixing of oxygen with acetylene in the acetylene hose.

Flashback arrestors prevent flashbacks from reaching the hoses and/or regulator. They have a flame-retarding filter that will allow heat, but not flames, to pass through. Most flashback arrestors also contain a check valve.

Add-on check valves and flashback arrestors are designed to be attached to the torch handle connections and to the regulator outlets. At a minimum, flashback arrestors with check valves should be attached to the torch handle connections. Both devices have arrows on them to indicate flow direction. When installing add-on check valves and flashback arrestors, be sure the arrow matches the desired gas flow direction.

The fittings for oxyfuel equipment are brass or bronze, and certain components are often fitted with soft, flexible, O-ring seals. The seal surfaces of the fittings or O-rings can be easily damaged by over-tightening with standard wrenches. For that reason, only a torch wrench (sometimes called a gang wrench) should be used to install regulators, hose connections, check valves, flashback arrestors, torches, and torch tips. Longer wrenches that provide more leverage should be avoided.

The universal torch wrench shown in *Figure 16* is equipped with various size wrench cutouts for use with a variety of equipment and standard CGA components. The length of a torch wrench is limited to reduce the chance of damage to fittings because of excessive torque. In some cases,

Check valves consist of a ball and spring that open inside a cylinder. The valve allows gas to move in one direction but closes if the gas attempts to flow in the opposite direction. When a torch is first pressurized or when it is being shut off, back-pressure check valves prevent the entry and mixing of acetylene with oxygen in the

Serious Cutting

The cutting power of oxyfuel equipment can be very surprising. This 6" (15.2 cm) carbon steel block is no match for a worker with the right torch.

NCCER – Construction Craft Laborer Level Two

Figure 16 Universal torch wrench.

manufacturers specify only hand-tightening for certain fitting connections of a torch set (tips or cutting/welding attachments, for example). In any event, follow the manufacturer's specific instructions when connecting the components of a torch set.

2.2.4 Gas Distribution Manifolds

In some applications, cutting gases are used on a large scale. Instead of using a cylinder for each operator, a large bank of cylinders connected to a common manifold is used. *Figure 17* is a representation of such an arrangement. High-pressure hoses called pigtails are used to connect the gas supply tanks to the manifold. Regulators are provided at both the source and the workstation hookups.

Figure 18 shows a manifold setup that might be used in a pipe fabrication shop. Operators would tie-in with their hoses at the drops on the manifolds. Each of these manifolds has four drops. Each drop would be provided with a pressure gauge like the one shown. One function of this gauge is to provide an indication if there is a leak

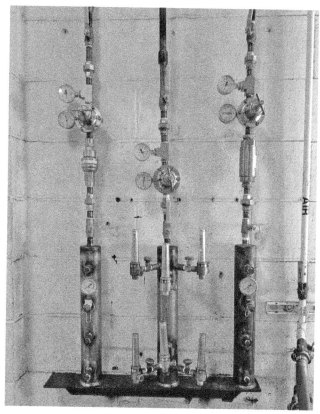

Figure 18 Operator hookups for cutting gases.

in a hose or in the torch. Hose length must be considered when using a manifold distribution system because the pressure drop increases with

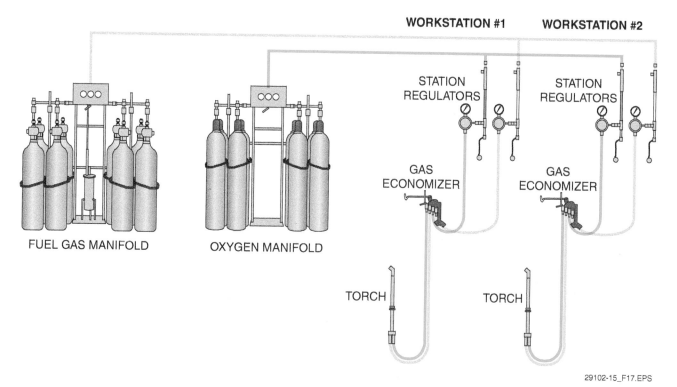

Figure 17 Gas distribution to stations through a manifold.

29102-15   Oxyfuel Cutting   Module Nine   17

the length of the hoses. Hoses with a larger inside diameter are needed for long runs.

OSHA provides the following safety precautions specifically for manifold systems:

- Fuel gas and oxygen manifolds must bear the name of the substance they contain.
- Fuel gas and oxygen manifolds must not be placed in confined spaces; they must be placed in safe, well ventilated, and accessible locations.
- Hose connections must be designed so that they cannot be interchanged between fuel gas and oxygen manifolds and supply header connections. Adapters may not be used to interchange hoses.
- Hose connections must be kept free of grease and oil.
- Manifold and header hose connections must be capped when not in use.
- Nothing may be placed on a manifold that will damage the manifold or interfere with the quick closing of the valves.

2.2.5 Hoses

Hoses transport gases from the regulators to the torch. Oxygen hoses are usually green with right-hand threaded connections. Hoses for fuel gas are usually red and have left-hand threaded connections. The fuel gas connection fittings are grooved as a reminder that they have left-hand threads.

Proper care and maintenance of the hose is important for maintaining a safe, efficient work area. Remember the following guidelines for hoses:

- Protect the hose from molten dross or sparks, which will burn the exterior. Although some hoses are flame retardant, they will burn.
- Do not place the hoses under the metal being cut. If the hot metal falls on the hose, the hose will be damaged. Keep hoses as far away from the cutting activity as possible.
- Frequently inspect and replace hoses that show signs of cuts, burns, worn areas, cracks, or damaged fittings. The hoses are tough and durable, but not indestructible.
- Never use pipe-fitting compounds or lubricants around hose connections. These compounds often contain oil or grease, which ignite and burn or explode in the presence of oxygen.

Propane and propylene require hoses designed for the mixture of hydrocarbons present in these fuels. Ensure that any hoses used are appropriate for the fuel gas. Hose and fuel gas providers can help provide the correct hoses.

2.3.0 Torches and Tips

Cutting torches mix oxygen and fuel gas for the torch flame and control the stream of oxygen necessary for the cutting jet. Depending on the job site, either a one-piece or a combination cutting torch may be used.

2.3.1 One-Piece Hand Cutting Torch

The one-piece hand cutting torch, sometimes called a demolition torch or a straight torch, contains the fuel gas and oxygen valves that allow the gases to enter the chambers and then flow

Fuel and Oxygen Cylinder Separation for Fixed Installations

For fixed installations involving one or more cylinders coupled to a manifold, fuel and oxygen cylinders must be separated by at least 20' (6.10 m) or be divided by a wall 5' (1.52 m) or higher with a 30-minute burn rating, per American National Standards Institute Z49.1. This also applies to cylinders in storage. Special wheeled cradles designed to distribute gas from multiple cylinders are available.

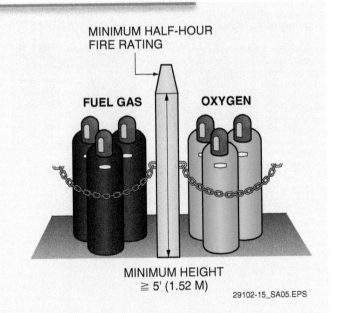

into the part of the torch where they are mixed. The main body of the torch is called the handle. The torch valves control the fuel gas and oxygen flow needed for preheating the metal to be cut. The cutting oxygen lever, which is spring-loaded, controls the jet of cutting oxygen. Hose connections are located at the end of the torch body behind the valves.

Figure 19 shows a three-tube, one-piece hand cutting torch in which the preheat fuel and oxygen are mixed in the tip. These torches are designed for heavy-duty cutting and little else. They have long supply tubes from the torch handle to the torch head to reduce radiated heat to the operator's hands. Cutting torches are generally available with sufficient capacity to cut steel up to 12" (≈30 cm) thick. Larger-capacity torches, with the ability to cut steel up to 36" (≈90 cm) thick, can also be obtained. Torches with this kind of capacity require a significant oxygen and fuel gas supply to perform.

Two different types of oxyfuel cutting torches are in general use. The positive-pressure torch (*Figure 19*) is designed for use with fuel supplied through a regulator from pressurized fuel storage cylinders. The injector torch is designed to use a vacuum created by the oxygen flow to draw the necessary amount of fuel from a very-low-pressure fuel source, such as a natural gas line or acetylene generator. The injector torch, when used, is most often found in continuous-duty, high-volume manufacturing applications. Both types may employ one of two different fuel-mixing methods:

- Torch-handle, or supply-tube, mixing
- Torch-head or tip mixing

The two methods can normally be distinguished by the number of supply tubes from the torch handle to the torch head. Torches that use three tubes from the handle to the head mix the preheat fuel and oxygen at the torch head or tip. This method tends to help eliminate any flashback damage to the torch head supply tubes and torch handle. One tube carries fuel gas to the head. The other two tubes carry oxygen; one carries the oxygen for the preheat flame while the other carries the oxygen for cutting.

The cutting torch with two tubes usually mixes the preheat fuel and oxygen in a mixing chamber in the torch body or in one of the supply tubes (*Figure 20*). Injector torches usually have the injector located in one of the supply tubes, and the mixing occurs in the tube between the injector and the torch head. Some older torches that have only two visible tubes are actually three-tube torches that mix the preheat fuel and oxygen in the torch head or tip. This is accomplished by using a separate preheat fuel tube inside a larger preheat oxygen tube.

2.3.2 Combination Torch

The combination torch consists of a cutting torch attachment that fits onto a welding torch handle. These torches are normally used in light-duty or medium-duty applications. Fuel gas and oxygen valves are on the torch handle. The cutting attachment has a cutting oxygen lever and another oxygen valve to control the preheat flame. When the cutting attachment is screwed onto the torch handle, the torch handle oxygen valve is opened all the way, and the preheat oxygen is controlled by an oxygen valve on the cutting attachment. When the cutting attachment is removed, brazing and heating tips can be screwed onto the torch handle. *Figure 21* shows a two-tube combination torch in which preheat mixing is accomplished in a supply tube. These torches are usually positive-pressure torches with mixing occurring in the attachment body, supply tube, head, or tip. These torches may be equipped with built-in flashback arrestors and check valves.

2.3.3 Cutting Torch Tips

Cutting torch tips, or nozzles, fit into the cutting torch and are either screwed in or secured with a tip nut. There are one- and two-piece cutting tips (*Figure 22*).

One-piece cutting tips are made from a solid piece of copper. Two-piece cutting tips have a separate external sleeve and internal section.

Torch manufacturers supply literature explaining the appropriate torch tips and gas pressures for various applications. *Table 2* shows a sample cutting tip chart that lists recommended tip sizes and gas pressures for use with acetylene fuel gas and a specific manufacturer's torch and tips.

> **CAUTION**
> Do not use the cutting tip chart from one manufacturer for the cutting tips of another manufacturer. The gas flow rate of the tips may be different, resulting in excessive flow rates. Different gas pressures may also be required. The cutting torch tip to be used depends on the base metal thickness and fuel gas being used. Special-purpose tips are also available for use in such operations as gouging and grooving.

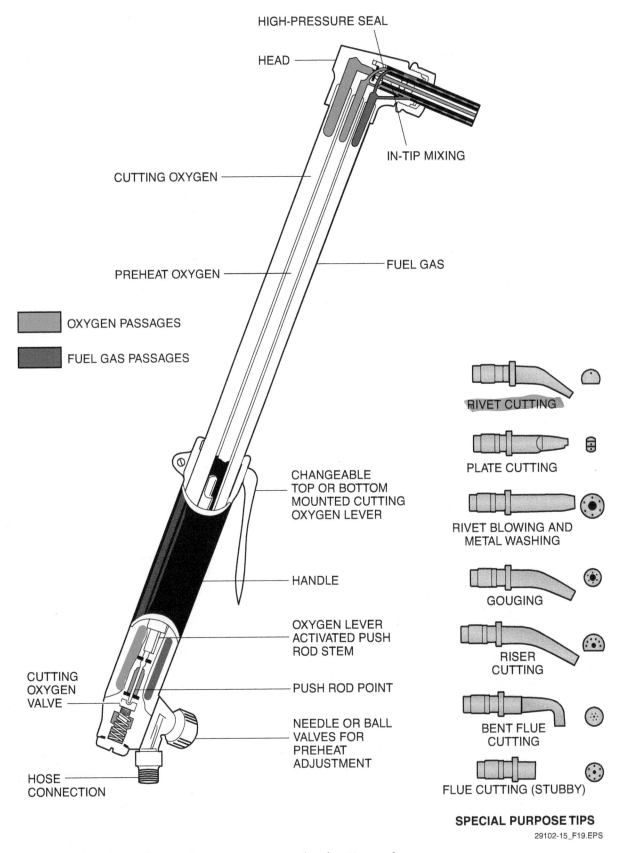

Figure 19 Heavy-duty three-tube one-piece positive-pressure hand cutting torch.

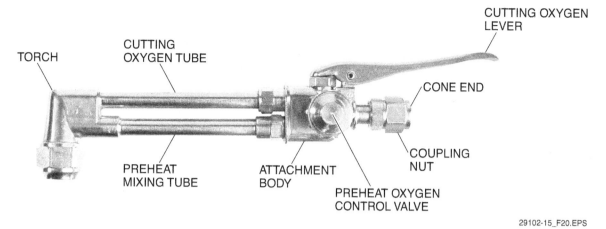

Figure 20 Cutting torch attachment.

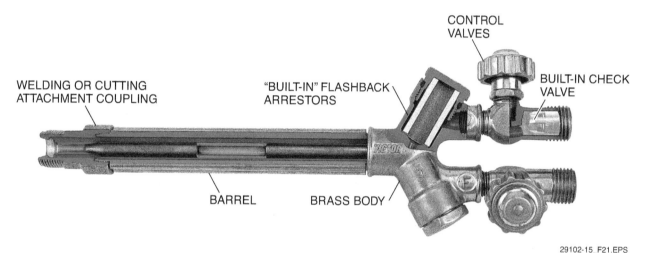

Figure 21 Combination cutting torch

The cutting torch tip to be used depends on the base metal thickness and fuel gas being used. Special-purpose tips are also available for such operations as gouging and grooving.

One-piece torch tips are generally used with acetylene cutting because of the high temperatures involved. They can have four, six, or eight preheat holes in addition to the single cutting hole. *Figure 23* shows the arrangement of typical acetylene torch cutting tips.

Tips used with liquefied fuel gases must have at least six preheat holes (*Figure 24*). Because fuel gases burn at lower temperatures than acetylene, more holes are necessary for preheating. Tips used with liquefied fuel gases can be one- or two-piece cutting tips. *Figure 25* shows a typical two-piece cutting tip used with liquefied fuel gases.

Special-purpose tips are available for special cutting jobs such as cutting sheet metal, rivets, risers, and flues, as well as washing and gouging. *Figure 26* shows special-purpose torch cutting tips, which are described as follows.

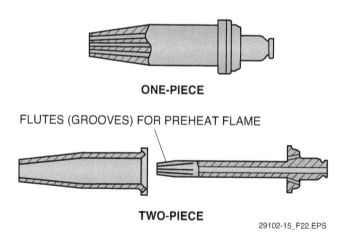

Figure 22 One- and two-piece cutting tips.

- The sheet metal cutting tip has only one preheat hole. This minimizes the heat and prevents distortion in the sheet metal. These tips are normally used with a motorized carriage, but can also be used for hand cutting.
- Rivet cutting tips are used to cut off rivet heads, bolt heads, and nuts.

Table 2 Sample Acetylene Cutting Tip Chart

Cutting Tip Series 1-101, 3-101, and 5-101												
Metal Thickness		Tip Size	Cutting Oxygen Pressure*		Preheat Oxygen*		Acetylene Pressure*		Speed		Kerf Width	
(in)	(mm)		(psig)	(kPa)	(psig)	(kPa)	(psig)	(kPa)	(in/min)	(cm/min)	(in)	(mm)
1/8	3.18	000	20-25	138-172	3-5	21-34	3-5	21-34	20-30	51-76	0.04	01.02
1/4	6.35	00	20-25	138-172	3-5	21-34	3-5	21-34	20-28	51-71	0.05	01.27
3/8	9.52	0	25-30	172-207	3-5	21-34	3-5	21-34	18-26	46-66	0.06	01.52
1/2	12.70	0	30-35	207-241	3-6	21-41	3-5	21-34	16-22	41-56	0.06	01.52
3/4	19.05	1	30-35	207-241	4-7	28-48	3-5	21-34	15-20	38-51	0.07	1.78
1	25.40	2	35-40	241-276	4-8	28-55	3-6	21-41	13-18	33-46	0.09	02.29
2	50.80	3	40-45	276-310	5-10	34-69	4-8	28-55	10-12	25-30	0.11	02.79
3	76.20	4	40-50	276-345	5-10	34-69	5-11	34-76	8-10	20-25	0.12	03.05
4	101.60	5	45-55	310-379	6-12	41-83	6-13	41-90	6-9	15-23	0.15	03.81
6	152.40	6	45-55	310-379	6-15	41-103	8-14	55-97	4-7	10-18	0.15	03.81
10	254.00	7	45-55	310-379	6-20	41-138	10-15	69-103	3-5	8-13	0.34	08.64
12	304.80	8	45-55	310-379	7-25	48-172	10-15	69-103	3-4	8-10	0.41	10.41

*The lower side of the pressure listings is for hand cutting and the higher side is for machine cutting.

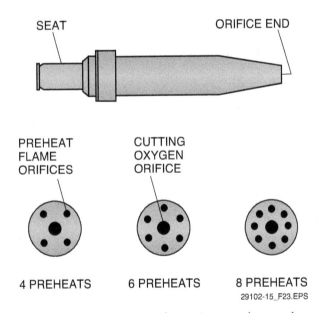

Figure 23 Orifice-end views of one-piece acetylene torch cutting tips.

- Riser cutting tips are similar to rivet cutting tips and can also be used to cut off rivet heads, bolt heads, and nuts. They have extra preheat holes to cut risers, flanges, or angle legs faster. They can be used for any operation that requires a cut close to and parallel to another surface, such as in removing a metal backing.
- Rivet blowing and metal washing tips are heavy-duty tips designed to withstand high heat. They are used for coarse cutting and for removing such items as clips, angles, and brackets.
- Gouging tips are used to groove metal in preparation for welding.
- Flue cutting tips are designed to cut flues inside boilers. They also can be used for any cutting operation in tight quarters where it is difficult to get a conventional tip into position.

Nearly all manufacturers use different tip-to-torch mounting designs, sealing surfaces, and diameters. In addition, tip sizes and flow rates are usually not the same between manufacturers even though the model number designations may be the same. This makes it impossible to safely interchange cutting tips between torches from different manufacturers. Even though some tips from different manufacturers may appear to be the same, do not interchange them. The sealing surfaces are very precise, and serious leaks may occur that

Acetylene Flow Rates

Manufacturers provide listings of the maximum fuel flow rate for each acetylene tip size in addition to recommended acetylene pressures. When selecting a tip, make sure that its maximum flow rate (in cubic feet or cubic meters per hour) does not exceed one-seventh of the total fuel capacity for the acetylene cylinder in use. Multiple cylinders must be manifolded together if the flow rate exceeds the cylinder(s) in use in order to prevent withdrawal of acetone along with acetylene.

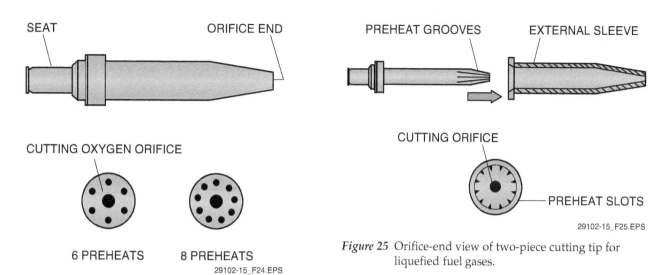

Figure 24 Orifice-end view of one-piece cutting tip for liquefied fuel gases.

Figure 25 Orifice-end view of two-piece cutting tip for liquefied fuel gases.

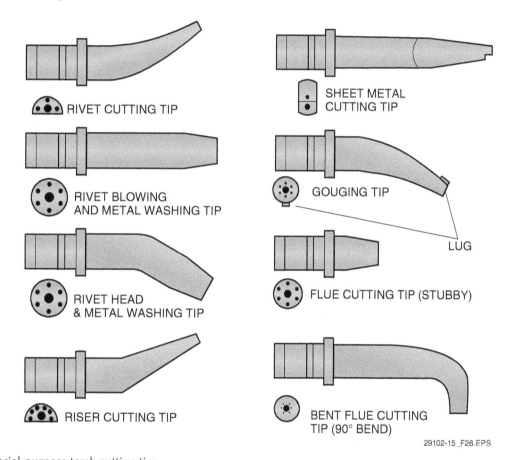

Figure 26 Special-purpose torch cutting tips.

could result in a dangerous fire or flashback. Each torch tip must be properly matched with the handle to which it is attached. Some manufacturers do make tips that are specifically designed for use with other manufacturer's torch handle. If these are used, ensure that the tip is listed as a precise match for the handle model in use.

> **CAUTION**
> Do not mix torch tips and handles from different manufacturers, unless they are specifically identified as being compatible in the manufacturer's documentation or catalog.

2.4.0 Other Accessories

In addition to the components that make up the actual oxyfuel cutting equipment, there are numerous accessories that are used along with the equipment. This section examines some accessories that are used for cleaning torch tips, lighting torches, transporting cylinders, and marking cylinders and metal workpieces.

2.4.1 Tip Cleaners and Tip Drills

With use, cutting tips become dirty. Carbon and other impurities build up inside the holes, and molten metal often sprays and sticks onto the surface of the tip. A dirty tip will result in a poor-quality cut with an uneven kerf and excessive dross buildup. To ensure good cuts with straight kerfs and minimal dross buildup, clean cutting tips with tip cleaners or tip drills (*Figure 27*). Tip cleaners are tiny round files. They usually come in a set with files to match the diameters of the various tip holes. In addition, each set usually includes a file that can be used to lightly recondition the face of the cutting tip. Tip cleaners are inserted into the tip hole and moved back and forth a few times to remove deposits from the hole. The small files are not made from hard metals. Torch tips are typically made of brass, which is also relatively soft. Using an aggressive tip file made from hard metals could easily damage the precise opening.

Tip drills are used for major cleaning and for holes that are plugged. Tip drills are tiny drill bits that are sized to match the diameters of tip holes. The drill fits into a drill handle for use. The handle is held, and the drill bit is turned carefully inside the hole to remove debris. They are more brittle than tip cleaners, making them more difficult to use. If a torch tip is properly cared for, tip drills are rarely needed.

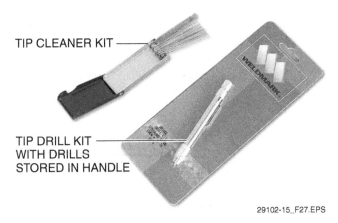

Figure 27 Tip cleaner and drill kits.

> **CAUTION**
> Tip cleaners and tip drills are brittle. Care must be taken to prevent these devices from breaking off inside a hole. Broken tip cleaners are difficult to remove. Improper use of tip cleaners or tip drills can enlarge the tip, causing improper burning of gases. If this occurs, the tip must be discarded. If the end of the tip has been partially melted or deeply gouged, do not attempt to cut it off or file it flat. The tip should be discarded and replaced with a new tip. This is because some tips have tapered preheat holes, and if a significant amount of metal is removed from the end of the tip, the preheat holes will become too large.

2.4.2 Friction Lighters

Always use a friction lighter (*Figure 28*), also known as a striker or spark-lighter, to ignite the cutting torch. The friction lighter works by rubbing a piece of flint on a steel surface to create sparks.

> **WARNING!**
> Do not use a match or a gas-filled lighter to light a torch. This could result in severe burns and/or could cause the lighter to explode.

2.4.3 Cylinder Cart

The cylinder cart, or bottle cart, is a modified hand truck that has been equipped with chains or straps to hold cylinders firmly in place. Bottle carts help ensure the safe transportation of gas cylinders. *Figure 29* shows two cylinder carts used for oxyfuel cylinders. Some carts are equipped with tool trays or boxes as well as rod holders.

TRIGGER OPERATED STRIKER   **COMMON CUP-TYPE STRIKER**

Figure 28 Friction lighters.

2.4.4 Soapstone Markers

Because of the heat involved in cutting operations, along with the tinted lenses that are required, ordinary pen or pencil marking for cutting lines or welding locations is not effective. The oldest and most common material used for marking is soapstone in the form of sticks or cylinders (*Figure 30*). Soapstone is soft and feels greasy and slippery. It is actually steatite, a dense, impure form of talc that is heat resistant. It also shows up well through a tinted lens under the illumination of an electric arc or gas cutting flame. Some welders prefer to use silver-graphite pencils (*Figure 30*) for marking dark materials and red-graphite pencils for aluminum or other bright metals. Graphite is also highly heat resistant. A few manufacturers also market heat-resistant paint/dye markers for cutting and welding.

2.5.0 Specialized Cutting Equipment

In addition to the common hand cutting torches, other types of equipment are used in oxyfuel cutting applications. This equipment includes mechanical guides used with a hand cutting torch, various types of motorized cutting machines, and oxygen lances. All of the motorized units use special straight body machine cutting or welding torches with a gear rack attached to the torch body to set the tip distance from the work.

2.5.1 Mechanical Guides

On long, circular, or irregular cuts, it is very difficult to control and maintain an even kerf with a hand cutting torch. Mechanical guides can help maintain an accurate and smooth kerf along the cutting line. For straight line or curved cuts, use a one- or two-wheeled accessory that clamps on the torch tip in a fixed position. The wheeled accessory maintains the proper tip distance while the tip is guided by hand along the cutting line.

The torch tip fits through and is secured to a rotating mount between two small metal wheels. The wheel heights are adjustable so that the tip distance from the work can be set. The radius of

---

Cup-Type Striker

When using a cup-type striker to ignite a welding torch, hold the cup of the striker slightly below and to the side of the tip, parallel with the fuel gas stream from the tip. This prevents the ignited gas from deflecting back from the cup and reduces the amount of carbon soot in the cup. Note that the flint in a striker can be replaced.

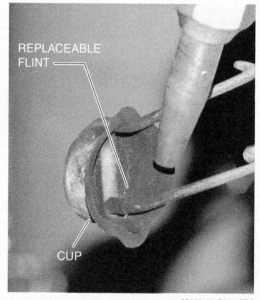

Figure 29 Cylinder carts.

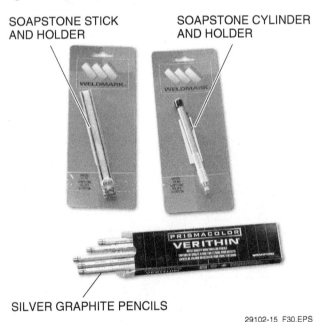

Figure 30 Soapstone and graphite markers.

the circle is set by moving the pivot point on a radius bar. After a starting hole is cut (if needed), the torch tip is placed through and secured to the circle cutter rotating mount. Then the pivot point is placed in a drilled hole or a magnetic holder at the center of the circle. When the cut is restarted, the torch can be moved in a circle around the cut, guided by the circle cutter. The magnetic guide is used for straight line cuts. The magnets hold it securely in place during cutting operations.

When large work with an irregular pattern must be cut, a template is often used. The torch is drawn around the edges of the template to trace the pattern as the cut is made. If multiple copies must be cut, a metal pattern held in place and designed to allow space for tip distance from the pattern is usually used. For a one- or two-time

Sharpening Soapstone Sticks

The most effective way to sharpen a soapstone stick marker is to shave it on one side with a file. By leaving one side flat, accurate lines can be drawn very close to a straightedge or a pattern.

copy, a heavily weighted Masonite or aluminum template that is spaced off the workpiece could be carefully used and discarded.

2.5.2 Motor-Driven Equipment

A variety of fixed and portable motorized cutting equipment is available for straight and curved cutting/welding. The computer-controlled plate cutting machine (*Figure 31*) is a fixed-location machine used in industrial manufacturing applications. The computer-controlled machine can be programmed to pierce and then cut any pattern from flat metal stock. There is also an optical pattern-tracing machine that follows lines on a drawing using a light beam and an optical detector. They both can be rigged to cut multiple items using multiple torches operated in parallel. Both units have a motor-driven gantry that travels the length of a table and a transverse motor-driven torch head that moves back and forth across the table. Both units are also equipped to use oxyfuel or plasma cutting torches.

Other types of pattern-tracing machines use metal templates that are clamped in the machine. A follower wheel traces the pattern from the template. The pattern size can be increased or decreased by electrical or mechanical linkage to a moveable arm holding one or more cutting torches that cut the pattern from flat metal stock.

Portable track cutting machines, or track burners, can be used in the field for straight or curved cutting and beveling. *Figures 32* and *33* show units driven by a variable-speed motor. The unit shown in *Figure 32* is available with track extensions for any length of straight cutting or beveling, along with a circle cutting attachment. Some models use a single, centered track.

Machine oxyfuel gas cutters or track burners are basic guidance systems driven by a variable speed electric motor to enable the operator to cut or bevel straight lines at any desired speed. The device (*Figure 33*) is usually mounted on a track or used with a circle-cutting attachment to enable the operator to cut various circle diameters up to 96" (≈244 cm). It consists of a heavy-duty tractor unit fitted with an adjustable torch mount and gas hose attachments. It is also equipped with an On/Off switch, a Low/High speed range switch, a Forward/Reverse directional switch, and a speed-adjusting dial calibrated in inches (or metric equivalent) per minute.

The device shown in *Figure 33* offers the following operational features:

- Makes straight-line cuts of any length
- Makes circle cuts up to 96" (≈244 cm) in diameter
- Makes bevel or chamfer cuts
- Has an infinitely variable cutting speed from 1" to 110" (2.5 cm to 279 cm) per minute
- Has dual speed and directional controls to enable operation of the machine from either end

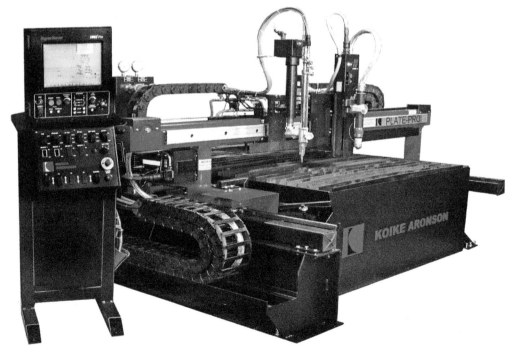

Figure 31 Computer-controlled plate cutting machine.

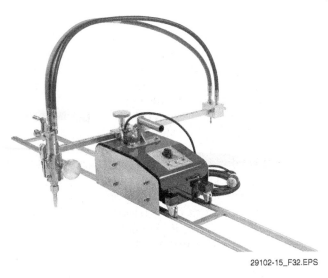

Figure 32 Track burner with oxyfuel machine torch.

Figure 34 shows the location of the following controls:

- *Power on/power off* – Turn the machine on and off by toggling the On/Off toggle switch.
- *Speed range control* – Set the machine's speed range by toggling the switch up or down. The use of two speed ranges allows for more precise speed control.
- *Directional control* – Set the machine's direction.
- *Speed control* – Turn the large knob to adjust the cutting speed based on the percentage of the selected speed range.

A portable, motor-driven band track or hand-cranked ring gear cutter/beveler can be set up in the field for cutting and beveling pipe with oxyfuel or plasma machine torches (*Figures 35*

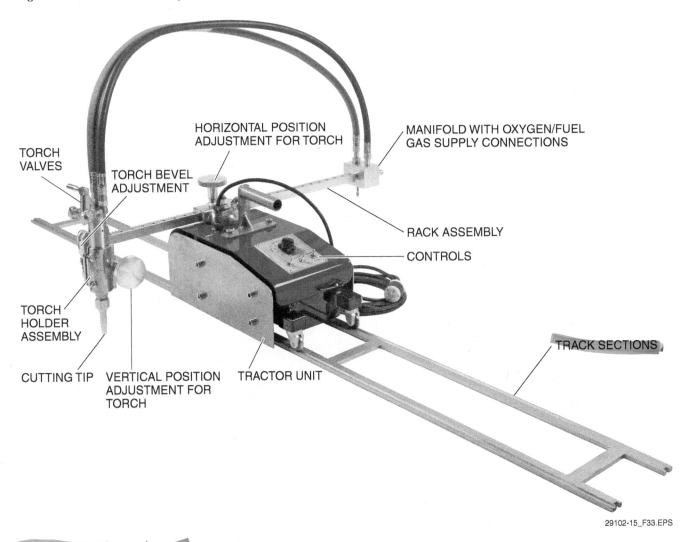

Figure 33 Track burner features.

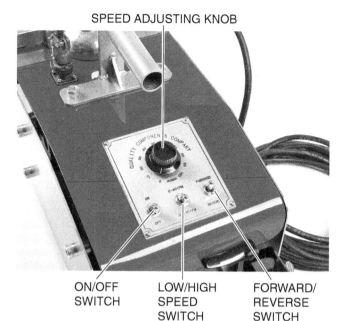

Figure 34 Track burner controls.

Figure 35 Band-track pipe cutter/beveler.

and *36*). The stainless steel band track cutter uses a chain and motor sprocket drive to rotate the machine cutting torch around the pipe a full 360 degrees. The all-aluminum ring gear type of cutter/beveler is positioned on the pipe, and then the saddle is clamped in place. In operation, the ring gear and the cutting torch rotate at different rates around the saddle for a full 360-degree cut.

2.5.3 Exothermic Oxygen Lances

Exothermic (combustible) oxygen lances are a special oxyfuel cutting tool usually used in heavy industrial applications and demolition work. The lance is a steel pipe that contains magnesium- and aluminum-cored powder or rods (fuel). In operation, the lance is clamped into a holder (*Figure 37*) that seals the lance to a fitting that supplies oxygen to it through a hose at pressures of 75 to 80 psi (517 to 552 kPa). With the oxygen turned on, the end of the lance is ignited with an acetylene torch or flare. As long as the oxygen is applied, the lance will burn and consume itself. The oxygen-fed flame of the burning magnesium, aluminum, and steel pipe creates temperatures approaching 10,000°F (5,538°C). At this temperature, the lance will rapidly cut or pierce any material, including steel, metal alloys, and cast iron, even under water. The lances for the holder shown in *Figure 37* are 10' (≈3 m) long and range in size from ⅜" (9.5 mm) to 1" (25.4 mm) in diameter. The larger sizes can be coupled to obtain a longer lance.

Shop-Made Straight-Line Cutting Guide

A simple solution for straight-line cutting is to clamp a piece of angle iron to the work and use a band clamp around the cutting torch tip to maintain the cutting tip distance from the work. When the cut is started, the band clamp rests on the top of the vertical leg of the angle iron, and the torch is drawn along the length of the angle iron at the correct cutting speed.

Figure 36 Ring gear pipe cutter/beveler.

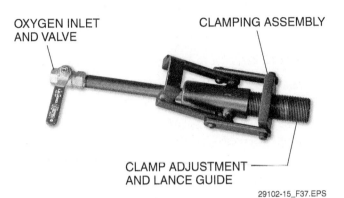

Figure 37 Oxygen lance holder.

A small pistol-grip heat-shielded unit that can be used with an electric welder is also available. This small unit uses lances from ¼" (6.35 mm) to ⅜" (9.52 mm) in diameter that are 22" to 36" (55.9 to 91.4 cm) long and that cut very rapidly at a maximum burning time of 60 to 70 seconds. The small unit is primarily used to burn out large frozen pins and frozen headless bolts or rivets. Like a large lance, it can be used to cut any material, including concrete-lined pipe. Both units are relatively inexpensive and can be set up in the field with only an oxygen cylinder, hose, and ignition device.

Additional Resources

ANSI Z49.1, Safety in Welding, Cutting, and Allied Processes. Miami, FL: American Welding Society.

Uniweld Products, Inc. Numerous videos are available at **www.uniweld.com/en/uniweld-videos**. Last accessed: November 30, 2014.

The Harris Products Group, a division of Lincoln Electric. Numerous videos are available at **www.harrisproductsgroup.com/en/Expert-Advice/videos.aspx**. Last accessed: November 30, 2014.

2.0.0 Section Review

1. A lighter-than-air compound of carbon and hydrogen that is formed by dissolving calcium carbide in water is ____.

    a. argon
    b. acetylene
    c. propane
    d. propylene

2. Safety devices used on regulators, hoses, and torches to prevent reverse gas flow and protect against fires and explosions are ____.

    a. check valves and flashback arrestors
    b. orifice plates and torch tips
    c. HEPA filters and flashback valves
    d. distribution manifolds and friction strikers

3. What basic type of oxyfuel cutting torch is designed for use with fuel that is supplied through a regulator from a pressurized fuel storage cylinder?

    a. Motor-controlled torch
    b. Vacuum-injector torch
    c. Neutral-flame torch
    d. Positive-pressure torch

4. The most common material used for marking cutting lines on metal is a form of heat-resistant talc called ____.

    a. soapstone
    b. Masonite
    c. graphite
    d. flint

5. A type of oxyfuel cutting tool that uses a steel pipe containing magnesium- and aluminum-cored powder or rods for heavy industrial cutting applications and demolition work is a(n) ____.

    a. circle cutting accessory
    b. plasma machine torch
    c. exothermic oxygen lance
    d. ring gear beveler

Section Three

3.0.0 OXYFUEL EQUIPMENT SETUP AND SHUTDOWN

Objective

Explain how to set up, light, and shut down oxyfuel equipment.
  a. Explain how to properly prepare a torch set for operation.
  b. Explain how to leak test oxyfuel equipment.
  c. Explain how to light the torch and adjust for the proper flame.
  d. Explain how to properly shut down oxyfuel cutting equipment.

Performance Tasks

1. Set up oxyfuel equipment.
2. Light and adjust an oxyfuel torch.
3. Shut down oxyfuel cutting equipment.
4. Disassemble oxyfuel cutting equipment.
5. Change empty gas cylinders.

Trade Terms

Carburizing flame: A flame burning with an excess amount of fuel; also called a reducing flame.

Neutral flame: A flame burning with correct proportions of fuel gas and oxygen.

Oxidizing flame: A flame burning with an excess amount of oxygen.

Operators should be trained and tested in the correct methods of safely preparing, starting, testing for leaks, and shutting down an oxyfuel cutting station. This part of the module examines procedures for setting up oxyfuel equipment, leak testing the equipment, lighting and adjusting the torch, and shutting down the equipment after use.

3.1.0 Setting Up Oxyfuel Equipment

When setting up oxyfuel equipment, follow procedures to ensure that the equipment operates properly and safely. The following sections explain the procedures for setting up oxyfuel equipment.

3.1.1 Transporting and Securing Cylinders

Cylinders should be transported to the workstation in an upright position on an appropriate hand truck or bottle cart. Once the cylinders are at the workstation, they must be secured with chain to a fixed support so that they cannot be knocked over accidentally. Leaving the cylinders in a proper cylinder cart is common; removal of the cylinders from the cart is not required. Then the protective cap from each cylinder can be removed and the outlet nozzles inspected to make sure that the seat and threads are not damaged. Place the protective caps where they will not be lost and where they will be readily available for reinstallation.

> **WARNING!** Always handle cylinders with care. They are under high pressure and should never be dropped, knocked over, rolled, or exposed to heat in excess of 140°F (60°C). When moving cylinders, always be certain that the valve caps are in place. Use a cylinder cage to lift cylinders. Never use a sling or electromagnet for cylinder lifting.

3.1.2 Cracking Cylinder Valves

To crack open a cylinder valve, start by ensuring that the cylinder is fully secured. Then crack open the cylinder valve momentarily to remove any dirt from the valve opening (*Figure 38*).

> **WARNING!** Operators should always stand with the valve stem between themselves and the regulator when opening valves to avoid injury from dirt that may be lodged in the valve. If a cloth is used during the cleaning process, it must not have any oil or grease on it. Oil or grease mixed with compressed oxygen can cause an explosion.

Hoisting Cylinders

Never attempt to lift a cylinder using the holes in a safety cap. Always use a lifting cage. Make sure that the cylinder is secured in the cage. Cages designed for storing and lifting cylinders are available in various sizes. The model shown here includes a partition to separate oxygen cylinders from fuel gas cylinders.

> **CAUTION**
> Use care not to over-tighten connections. The brass connections used will strip if over-tightened. Repair or replace equipment that does not seal properly. If the torch valves leak, try lightly tightening the packing nut at the valves. If a cutting attachment or torch tip leaks, disassemble and check the sealing surfaces and/or O-rings for damage. If the equipment or torch tip is new and the sealing surfaces do not appear damaged, over-tighten the connection slightly to seat the sealing surfaces; then loosen and retighten normally. If the leaks persist or the equipment is visibly damaged, replace the torch, cutting attachment (if used), or cutting tip as necessary.

OUTLET FACING AWAY

Figure 38 Cracking a cylinder valve.

3.1.3 Attaching Regulators

To attach the regulators, first check that the regulator is closed (adjustment screw is backed out and loose/turns with no resistance).

Check the regulator fittings to ensure that they are free of oil and grease (*Figure 39*).

Connect and tighten the oxygen regulator to the oxygen cylinder using a torch wrench (*Figure 40*).

> **WARNING!**
> Do not work with a regulator that shows signs of damage, such as cracked gauges, bent thumbscrews, or worn threads. Set it aside for repairs. Operators should never attempt to repair regulators. When tightening the connections, always use a torch wrench.

Figure 39 Checking connection fittings.

Figure 41 Cleaning the regulator.

Figure 40 Tightening regulator connection.

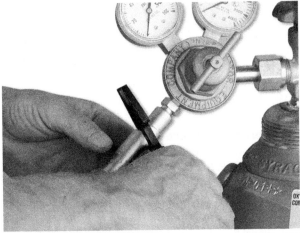

Figure 42 Attaching a flashback arrestor.

Connect and tighten the fuel gas regulator to the fuel gas cylinder. Remember that most fuel gas fittings have left-hand threads. Next, clean the outlet connection of the regulator. Crack the cylinder valve slightly and turn the pressure adjustment screw clockwise until you feel some resistance and the regulator begins to allow gas to pass through and expel any debris. Shut the cylinder valve and close the regulator (*Figure 41*).

3.1.4 Installing Flashback Arrestors or Check Valves

The installation of flashback arrestors or check valves is important and easy to accomplish if they are not already installed. Attach a flashback arrestor and/or check valve to the hose connection on the oxygen regulator or torch handle (*Figure 42*) and tighten with a torch wrench. Then attach and tighten a flashback arrestor and/or check valve to the fuel gas regulator or torch handle.

Once again, remember that fuel gas fittings have left-hand threads.

> **WARNING!** At least one flashback arrestor must be used with each hose. The absence of a flashback arrestor could result in flashback. Flashback arrestors can be attached either to the regulator, the torch, or both; however, flashback arrestors installed at the torch handle are preferred if only one is being used.

3.1.5 Connecting Hoses to Regulators

New hoses contain talc and possibly loose bits of rubber. These materials must be blown out of the hoses before the torch is connected. If they are not blown out, they will clog the tiny torch needle valves or tip openings.

To connect the hoses to the regulators, first inspect both the oxygen and fuel gas hoses for any damage, burns, cuts, or fraying. Replace any damaged hoses. Then connect the oxygen hose to the oxygen regulator flashback arrestor or check valve (*Figure 43*), and connect the fuel gas hose to the fuel gas regulator flashback arrestor and/or check valve. Complete the installation by opening the cylinder valves and regulators and purging the hoses until they are clear.

3.1.6 Attaching Hoses to the Torch

To attach the hoses to the torch, first attach flashback arrestors to the oxygen and fuel gas hose connections on the torch body—unless the torch has built-in flashback arrestors and check valves. Attach and tighten the oxygen hose to the oxygen fitting on the flashback arrestor or torch (*Figure 44*). Then attach and tighten the hose to the fuel gas fitting on the flashback arrestor or torch.

3.1.7 Connecting Cutting Attachments (Combination Torch Only)

If cutting attachments are being connected to a combination torch, be sure to check the torch manufacturer's instructions for the correct installation method. Then connect the attachment and tighten by hand as required.

3.1.8 Installing Cutting Tips

Before installing a cutting tip in a cutting torch, first identify the thickness of the material to be cut. Then identify the proper size cutting tip from the manufacturer's recommended tip size chart for the fuel being used.

> **WARNING!**
> If acetylene fuel is being used, make sure that the maximum fuel flow rate per hour of the tip does not exceed one-seventh of the fuel cylinder capacity. If a purplish flame is observed when the torch is operating, the fuel rate is too high and acetone is being withdrawn from the acetylene cylinder along with the acetylene gas.

Once the cutting tip has been selected, inspect the cutting tip sealing surfaces and orifices for damage or plugged holes. If the sealing surfaces are damaged, discard the tip. If the orifices are plugged, clean them with a tip cleaner or drill.

Check the torch manufacturer's instructions for the correct method of installing cutting tips. Then install the cutting tip and secure it with a torch wrench or by hand as required (*Figure 45*).

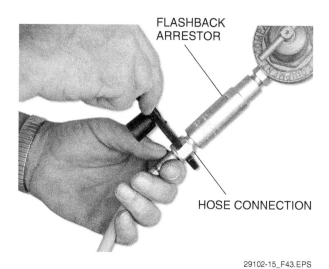

Figure 43 Connecting hose to regulator flashback arrestor.

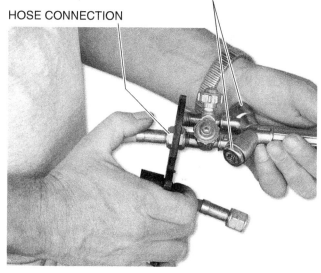

Figure 44 Connecting hoses to torch body.

3.1.9 Closing Torch Valves and Loosening Regulator Adjusting Screws

Closing the torch valves and loosening the regulator adjusting screws (*Figure 46*) are done before opening either cylinder valve. First, check the fuel and oxygen valves on the torch to be sure they are closed. Then check both the oxygen and fuel gas regulator adjusting screws to be sure they are loose (backed out).

> **CAUTION**
> Loosening regulator adjusting screws closes the regulators and prevents damage to the regulator diaphragms when the cylinder valves are opened.

Figure 45 Installing a cutting tip.

Figure 46 Torch valves and regulator adjusting screws.

3.1.10 Opening Cylinder Valves

To open cylinder valves (*Figure 47*), stand on the opposite side of the cylinder from the regulator and crack open the oxygen cylinder valve until the pressure on the regulator gauge rises and stops. The pressure in the cylinder and the pressure at the inlet of the regulator are now equal. Now open the oxygen cylinder valve all the way.

> **WARNING!**
> When opening a cylinder valve, keep the cylinder valve stem between you and the regulator. Never stand directly in front of or behind a regulator. The regulator adjusting screw can blow out, causing serious injury. Always open the cylinder valve gradually. Quick openings can damage a regulator or gauge or even cause a gauge to explode.

Figure 47 Cylinder valve and gauges.

Oxygen cylinder valves must be opened all the way until the valve seats at the top. Seating the valve at the fully open position prevents high-pressure leaks at the valve stem.

Once the oxygen cylinder valve is open, slowly open the fuel gas cylinder valve until the cylinder pressure gauge indicates the cylinder pressure, but no more than one and a half turns. This allows it to be quickly closed in case of a fire. This is especially important with acetylene.

3.1.11 Purging the Torch and Setting the Working Pressures

After the oxygen and fuel gas tank valves have been opened, the torch valves are opened to purge the torch and set the working pressures on the regulators.

Fully open the oxygen valve on the torch. Then depress and hold the cutting oxygen lever. Turn the oxygen regulator adjusting screw clockwise until the working pressure gauge shows the correct working pressure with the gas flowing. Allow the gas to flow for five to ten seconds to purge the torch and hoses of air or fuel gas. Then release the cutting lever and close the oxygen valve.

Note that, on single-stage regulators, the working pressure shown on the gauge will likely rise when the cutting lever is released. You will see the pressure gauge fall somewhat each time the lever is opened; this is normal operation. Set the pressure on the oxygen regulator while the gas is flowing to ensure it will be as desired when working.

Open the fuel valve on the torch about one eighth of a turn. Turn the fuel regulator adjusting screw clockwise until the working pressure gauge shows the correct fuel gas working pressure with the gas flowing. Allow the gas to flow for five to ten seconds to purge the hoses and torch of air. Then close the torch fuel valve. If acetylene is used, check that the acetylene static pressure does not rise above 15 psig (103 kPa). If it does, immediately open the torch fuel valve and reduce the regulator output pressure as needed. Because of its instability, acetylene cannot be used at pressures of more than 15 psig (103kPa) when in gaseous form. At higher pressures, acetylene gas breaks down chemically, producing heat and pressure that could result in a violent explosion inside the hose.

> **WARNING!**
> The working pressure gauge readings on single-stage regulators will rise after the torch valves are turned off. This is normal. However, if acetylene is being used as the fuel gas, make sure that the static pressure does not rise above 15 psig (103 kPa). Make sure that equipment is purged and leak tested in a well-ventilated area to avoid creating an explosive concentration of gases.

3.2.0 Testing for Leaks

Equipment must be tested for leaks immediately after it is set up and periodically thereafter. The torch should be checked for leaks before each use. Leaks could cause a fire or explosion. To test for leaks, apply a commercially prepared leak-testing formula (*Figure 48*) or a solution of detergent and water to each potential leak point. If bubbles form, a leak is present. Be aware though, that solutions made from common household soaps are not usually as effective as commercially prepared leak detection solutions.

> **WARNING!**
> If a detergent is used for leak testing, make sure the detergent contains no oil. In the presence of oxygen, oil can cause fires or explosions.

There are numerous leak points to test, including the following:

- Oxygen cylinder valve
- Fuel gas cylinder valve
- Oxygen regulator and regulator inlet and outlet connections
- Fuel gas regulator and regulator inlet and outlet connections

Portable Oxyacetylene Equipment

Most oxyfuel equipment used on large job sites and in shops is very heavy and is usually transported using a special hand truck. This type of equipment is typically used when extensive cutting must be accomplished. However, for small tasks in unusual locations, such as a commercial building rooftop, portable equipment that can be hand carried by one person is often used.

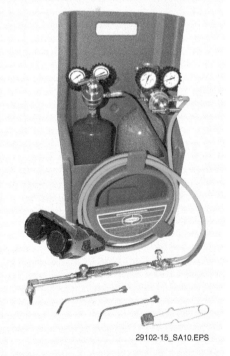

29102-15_SA10.EPS

APPLYING THE SOLUTION

REACTION TO A LEAK

Figure 48 Leak detection fluid use.

- Hose connections at the regulators, check valves/flashback arrestors, and torch
- Torch valves and cutting oxygen lever valve
- Cutting attachment connection (if used)
- Cutting tip

If there is a leak at the fuel gas cylinder valve stem, attempt to stop it by tightening the packing gland at the base of the stem. If this does not stop the leak, mark and remove the cylinder from service and notify the supplier. For other leaks, tighten the connections slightly with a wrench. If this does not stop the leak, turn off the gas pressure, open all connections, and inspect the fitting for damage.

> **WARNING!** Do not use Teflon® tape or pipe dope on these fittings since they do not provide a good seal. Make sure that equipment is purged and leak tested in a well-ventilated area to avoid creating an explosive concentration of gases.

3.2.1 Initial and Periodic Leak Testing

Initial and periodic leak testing is performed during initial equipment setup and periodically thereafter. First, set the equipment to the correct working pressures with the torch valves turned off. Then, using a leak-test solution, check for leaks at the cylinder valves, regulator relief ports, and regulator gauge connections (*Figure 49*). Also, check for leaks at hose connections, regulator connections, and check valve/flame arrestor connections up to the torch.

Fuel Cylinder Wrench

If the fuel cylinder is equipped with a valve requiring a T-wrench, always leave the wrench in place on the valve so that the fuel can be quickly turned off. This type of valve is obsolete but still in use.

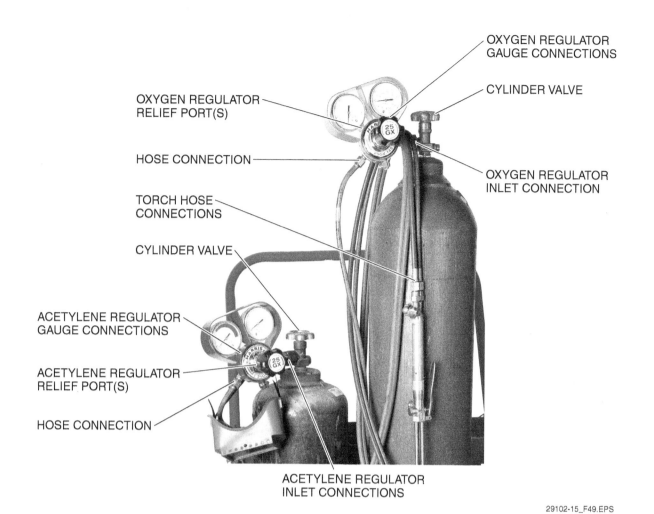

Figure 49 Typical initial and periodic leak-test points.

3.2.2 Leak-Down Testing of Regulators, Hoses, and Torch

Before the torch is ignited for use, the regulators, hoses, and torch should be quickly tested for leaks. First, set the equipment to the correct working pressures with the torch valves turned off. Then loosen both regulator adjusting screws. Check the working pressure gauges after a minute or two to see if the pressure drops. If the pressure drops, check the hose connection and regulators for leaks; otherwise, proceed with the test.

Place a thumb or finger over the cutting tip orifices and press tightly to block them (*Figure 50*). Turn on the torch oxygen valve and then depress and hold the cutting oxygen lever down. After the gauge pressure drops slightly, observe the oxygen working pressure gauge for a minute to see if the pressure continues to drop. If the pressure keeps dropping, perform the leak test described in the following section to determine the source of the leak. If the pressure does not change, close the torch oxygen valve and release the pressure at the cutting tip.

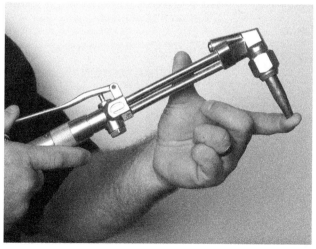

Figure 50 Blocking cutting tip for a leak test.

Next, block the tip again and turn on the torch fuel valve. After the gauge pressure drops slightly, carefully observe the fuel working pressure gauge for a minute. If the pressure continues to drop, perform the leak test described in the following section to determine the source of the

29102-15 Oxyfuel Cutting

leak. If the pressure does not change, close the torch fuel valve and release the pressure at the cutting tip.

If no leaks are apparent during the leak-down test, set the equipment to the correct working pressures.

3.2.3 Full Leak Testing of a Torch

Performing a full leak test involves testing for and isolating torch leaks in several places, as shown in *Figure 51*.

Start by setting the equipment to the correct working pressures with the torch valves turned off. Then, place a thumb or finger over the cutting tip orifices and press to block the orifices.

Turn on the torch oxygen valve and then depress the cutting oxygen lever. With the cutting tip blocked, check for leaks using a leak-test solution at the torch oxygen valve, cutting oxygen lever valve, cutting attachment connection to the handle (if used), preheat oxygen valve (if present), and cutting tip seal at the torch head. If no leaks are found, release the cutting oxygen lever, close the torch oxygen valve, and release the pressure at the cutting tip.

Next, with the cutting tip again blocked, open the torch fuel valve. Using a leak-test solution, check for leaks at the torch fuel valve, cutting attachment connection to the handle (if used), and cutting tip seal at the torch head. If no leaks are detected, close the torch fuel valve and release the pressure at the cutting tip.

3.3.0 Controlling the Oxyfuel Torch Flame

To be able to safely use a cutting torch, the operator must understand the flame and be able to adjust it and react to unsatisfactory conditions. The following sections will explain the oxyfuel flame and how to control it safely.

3.3.1 Oxyfuel Flames

There are three types of oxyfuel flames: neutral flame, carburizing flame, and oxidizing flame.

- *Neutral flame* – A neutral flame burns proper proportions of oxygen and fuel gas. The inner cones will be light blue in color, surrounded by a darker blue outer flame envelope that results when the oxygen in the air combines with the super-heated gases from the inner cone. A neutral flame is used for all but special cutting applications.
- *Carburizing flame* – A carburizing flame has a white feather created by excess fuel. The length of the feather depends on the amount of excess fuel present in the flame. The outer flame envelope is longer than that of the neutral flame, and it is much brighter in color. The excess fuel in the carburizing flame (especially acetylene) produces large amounts of carbon. The carbon will combine with red-hot or molten metal, making the metal hard and brittle. The carburizing flame is cooler than a neutral flame and is never used for cutting. It is used for some special heating applications.

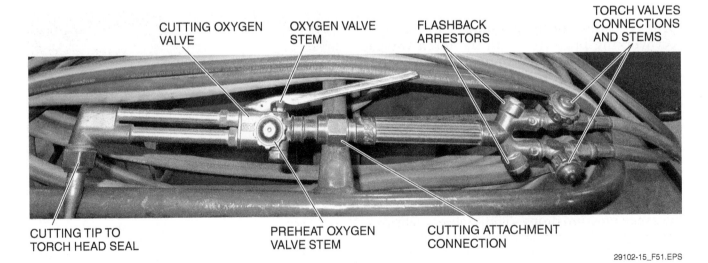

Figure 51 Torch leak-test points.

- *Oxidizing flame* – An oxidizing flame has an excess of oxygen. The inner cones are shorter, much bluer in color, and more pointed than a neutral flame. The outer flame envelope is very short and often fans out at the ends. An oxidizing flame is the hottest flame. A slightly oxidizing flame is recommended with some special fuel gases, but in most cases it is not used. The excess oxygen in the flame can combine with many metals, forming a hard, brittle, low-strength oxide. However, the preheat flames of a properly adjusted cutting torch will be slightly oxidizing when the cutting oxygen is shut off.

Figure 52 shows the various flames that occur at a cutting tip for both acetylene and LP gas.

Always Leak Test a Cutting Torch

Always take the time to perform a leak test on regulators, hoses, and the torch before using them the first time each day. A leak test should also be performed on a torch after tips have been changed or after converting the torch from one operation to another. Leaks, especially fuel gas leaks, can cause a fire or explosion after a cutting operation begins. A dangerous fire occurring at or near the torch may not be immediately noticed by the operator due to limited visibility through the tinted lenses.

Acetylene Burning in Atmosphere
Open fuel gas valve until smoke clears from flame.

Carburizing Flame
(Excess acetylene with oxygen)
Preheat flames require more oxygen.

Neutral Flame
(Acetylene with oxygen) Temperature 5589°F (3087°C).
Proper preheat adjustment when cutting.

Neutral Flame with Cutting Jet Open
Cutting jet must be straight and clean.
If it flares, the pressure is too high for the tip size.

Oxidizing Flame
(Acetylene with excess oxygen) Not recommended for average cutting. However, if the preheat flame is adjusted for neutral with the cutting oxygen on, then this flame is normal after the cutting oxygen is off.

OXYACETYLENE FLAME

LP Gas Burning in Atmosphere
Open fuel gas valve until flame begins to leave tip end.

Reducing Flame
(Excess LP-gas with oxygen) Not hot enough for cutting.

Neutral Flame
(LP-gas with oxygen) For preheating prior to cutting.

Oxidizing Flame with Cutting Jet Open
Cutting jet stream must be straight and clean.

Oxidizing Flame without Cutting Jet Open
(LP-gas with excess oxygen) The highest temperature flame for fast starts and high cutting speeds.

OXYPROPANE FLAME

Figure 52 Acetylene and LP (propane) gas flames.

3.3.2 Backfires and Flashbacks

When the torch flame goes out with a loud pop or snap, a backfire has occurred. Backfires are usually caused when the tip or nozzle touches the work surface or when a bit of hot dross briefly interrupts the flame. When a backfire occurs, relight the torch immediately. Sometimes the torch even relights itself. If a backfire recurs without the tip making contact with the base metal, shut off the torch and find the cause. Possible causes are the following:

- Improper operating pressures
- A loose torch tip
- Dirt in the torch tip seat or a bad seat

When the flame goes out and burns back inside the torch with a hissing or whistling sound, a flashback is occurring. Immediately shut off the oxygen valve on the torch; the flame is burning inside the torch. If the flame is not extinguished quickly, the end of the torch will melt off. The flashback will stop as soon as the oxygen valve is closed. Therefore, quick action is crucial. Flashbacks can cause fires and explosions within the cutting rig and, therefore, are very dangerous. Flashbacks can be caused by the following:

- Equipment failure
- Overheated torch tip
- Dross or spatter hitting and sticking to the torch tip
- Oversized tip (tip is too large for the gas flow rate being used)

After a flashback has occurred, wait until the torch has cooled. Then, blow oxygen (not fuel gas) through the torch for several seconds to remove soot that may have built up in the torch during the flashback before relighting it. If the torch makes a hissing or whistling sound after it is reignited or if the flame does not appear to be normal, shut off the torch immediately and have the torch serviced by a qualified technician.

3.3.3 Igniting the Torch and Adjusting the Flame

After the cutting equipment has been properly set up and purged, the torch can be ignited and the flame adjusted for cutting. The procedure for igniting the torch starts with choosing the appropriate cutting torch tip according to the base metal thickness being cut and the fuel gas being used. Always inspect the cutting tip sealing surfaces and orifices prior to installation. Attach the tip to the cutting torch by placing it on the end of the torch and tightening the nut. Some manufacturers recommend tightening the nut with a torch wrench, while others recommend tightening the nut by hand. Check the manufacturer's documentation for the equipment in use to ensure the tip is being installed correctly.

> **NOTE**
> Refer to the manufacturer's charts. Depending on the tip selected, the oxygen and fuel gas pressure may have to be adjusted.

Prior to opening the oxygen and fuel gas valves, be sure to put on the proper PPE. Also ensure that you are not depressing the oxygen cutting lever. If present, close the preheat oxygen valve and open the torch oxygen valve fully. Open the fuel gas valve on the torch handle about one-quarter turn. Then, holding the friction lighter near the side and to the front of the torch tip, ignite the torch.

> **WARNING!**
> Hold the friction lighter near the side of the tip, rather than directly in front of it, to prevent the ignited gas from being deflected backwards. Always use a friction lighter. Never use matches or cigarette lighters to light the torch because this could result in severe burns and/or could cause the lighter to explode. Always point the torch away from yourself, other people, equipment, and flammable material.

Once the torch is lit, adjust the torch fuel gas flame by adjusting the flow of fuel gas with the fuel gas valve. Increase the flow of fuel gas until the flame stops smoking or pulls slightly away from the tip. Decrease the flow until the flame returns to the tip. Open the preheat oxygen valve (if present) or the oxygen torch valve very slowly and adjust the torch flame to a neutral flame. Then press the cutting oxygen lever all the way down and observe the flame. It should have a long, thin, high-pressure oxygen cutting jet up to 8" (≈20 cm) long, extending from the cutting oxygen hole in the center of the tip. If it does not, do the following:

- Check that the working pressures are set as recommended on the manufacturer's chart.
- Clean the cutting tip. If this does not clear up the problem, change the cutting tip.

With the cutting oxygen on, observe the preheat flame. If it has changed slightly to a carburizing flame, increase the preheat oxygen until the flame is neutral. After this adjustment, the

NCCER – Construction Craft Laborer Level Two 29102-15

preheat flame will change slightly to an acceptable oxidizing flame when the cutting oxygen is shut off.

3.3.4 Shutting Off the Torch

Shutting off the torch itself is done by releasing the cutting oxygen lever and then closing the torch or preheat oxygen valves. After that, quickly close the torch fuel gas valve to extinguish the flame.

> **WARNING!**
> Always turn off the oxygen flow first to prevent a possible flashback into the torch.

3.4.0 Shutting Down Oxyfuel Cutting Equipment

When a cutting job is completed and the oxyfuel equipment is no longer needed, it must be shut down. *Figure 53* identifies the order in which various pieces of the oxyfuel cutting equipment are shut down.

Step 1 To begin the shutdown, first close the fuel gas and oxygen cylinder valves. Leave the regulators at their present setting.

Step 2 Open the fuel gas valve on the torch to allow all remaining gas to escape, and then close it. Next, open the oxygen valve on the torch to allow all remaining gas to escape, and then close it. These actions relieve the gas pressure in the hose and regulators, all the way back to the cylinder valve. Do not proceed to the next step until all pressure is released and all regulator gauges — both inlet and outlet — read zero.

Step 3 Turn the fuel gas and oxygen regulator adjusting screws counterclockwise to back them out, until they are loose.

Step 4 Coil and secure the hose and torch to prevent damage.

3.4.1 Disassembling Oxyfuel Equipment

In some situations, it may be necessary to disassemble the oxyfuel equipment after it has been used. Before any disassembly takes place, make sure that the equipment has been properly shut down. This includes checking that the cylinder valves are closed and all pressure gauges read zero.

Remove both hoses from the torch assembly and then detach the hoses from the regulators. Remove both regulators from the cylinders and reinstall the protective caps on the cylinders. The cylinders should now be returned to their proper storage place.

> **WARNING!**
> Always transport and store gas cylinders in the upright position. Be sure they are properly secured (chained) and capped. Regardless of whether the cylinders are empty or full, never store fuel gas cylinders and oxygen cylinders together without providing the required separation distance or fire-rated partition.

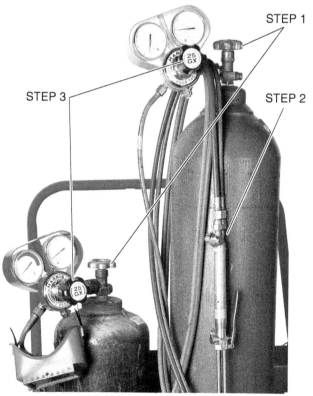

Figure 53 Shutting down oxyfuel cutting equipment.

Obtaining Maximum Fuel Flow

Increasing the fuel flow until the flame pulls away from the tip and then decreasing the flow until the flame returns to the tip sets the maximum fuel flow for the tip size in use.

3.4.2 Changing Cylinders

Empty is a relative term when discussing gas cylinders. These cylinders should never be completely emptied because reverse flow could occur. Oxygen tanks, for example, should never get below the required working pressure, or about 25 psi (172 kPa). Once the cylinder pressure drops near the working pressure value, the torch will stop performing properly and the cylinder will have to be replaced. As a result, some residual pressure is typically present. Follow these procedures to change a cylinder:

> **WARNING!** When moving cylinders, always be certain that they are in the upright position and the valve caps are secured in place. Never use a sling or electromagnet to lift cylinders. To lift cylinders, use a cylinder cage.

Step 1 Begin by making sure that the equipment has been properly shut down. This includes checking that the cylinder valves are closed and all pressure gauges read zero.

Step 2 Remove the regulator from the empty cylinder and replace the protective cap on the cylinder.

Step 3 Mark MT (empty) and the date (or the accepted site notation for indicating an empty cylinder) near the top of the cylinder using soapstone (*Figure 54*).

Figure 54 Typical empty cylinder marking.

Step 4 Transport the empty cylinder from the workstation to the storage area. Place the empty cylinder in the empty cylinder section of the storage area for the type of gas it contained.

Marking and Tagging Cylinders

Do not use permanent markers on cylinders; use soapstone or another temporary marker. If a cylinder is defective, place a warning tag on it.

Additional Resources

ANSI Z49.1, Safety in Welding, Cutting, and Allied Processes. Miami, FL: American Welding Society.

Uniweld Products, Inc. Numerous videos are available at **www.uniweld.com/en/uniweld-videos**. Last accessed: November 30, 2014.

The Harris Products Group, a division of Lincoln Electric. Numerous videos are available at **www.harrisproductsgroup.com/en/Expert-Advice/videos.aspx**. Last accessed: November 30, 2014.

3.0.0 Section Review

1. Gas cylinders can be lifted to height ____ .
   a. with a strong electromagnet
   b. using a cylinder cage
   c. with a sling routed through the openings in the cap
   d. with a cable routed through the openings in the cap

2. Immediately after setup and periodically thereafter, oxyfuel equipment must be tested for ____ .
   a. purging
   b. carburizing
   c. flashbacks
   d. leaks

3. An oxidizing torch flame is one that has a(n) ____ .
   a. excess of fuel gas
   b. proper oxygen/fuel gas mix
   c. excess of oxygen
   d. abnormally low temperature

4. When oxyfuel cutting equipment is being shut down, how should the fuel gas and oxygen regulator adjusting screws be positioned after bleeding the remaining gas pressure in the hoses?
   a. Both screws tight
   b. Fuel gas screw tight; oxygen screw loose
   c. Fuel gas screw loose; oxygen screw tight
   d. Both screws loose

Section Four

4.0.0 Performing Cutting Procedures

Objective

Explain how to perform various oxyfuel cutting procedures.

a. Identify the appearance of both good and inferior cuts and their causes.
b. Explain how to cut both thick and thin steel.
c. Explain how to bevel, wash, and gouge.
d. Explain how to make straight and bevel cuts with portable oxyfuel cutting machines.

Performance Tasks

6. Cut shapes from various thicknesses of steel, emphasizing:
   - Straight line cutting
   - Square shape cutting
   - Piercing
   - Beveling
   - Cutting slots
7. Perform washing.
8. Perform gouging.
9. Use a track burner to cut straight lines and bevels.

Trade Terms

Drag lines: The lines on the edge of the material that result from the travel of the cutting oxygen stream into, through, and out of the metal.

The following sections explain how to recognize good and bad cuts, how to prepare for cutting operations, and how to perform straight-line cutting, piercing, bevel cutting, washing, and gouging.

4.1.0 Preparing for Oxyfuel Cutting with a Hand Cutting Torch

Before metal can be cut, the equipment must be set up and the metal prepared. One important step is to properly lay out the cut by marking it with soapstone or punch marks. The few minutes this takes will result in a quality job, reflecting craftsmanship and pride. The following procedures describe how to prepare to make a cut.

Prepare the metal to be cut by cleaning any rust, scale, or other foreign matter from the surface. If possible, position the work so that it can be cut comfortably. Mark the lines to be cut with soapstone or a scriber. Then select the correct cutting torch tip according to the thickness of the metal to be cut, the type of cut to be made, the amount of preheat needed, and the type of fuel gas to be used. Ignite the torch and use the procedures outlined in the following sections for performing specific types of cutting operations.

4.1.1 Inspecting the Cut

Before attempting to make a cut, operators must be able to recognize good and bad cuts and know what causes bad cuts. This is explained in the following list and illustrated in *Figure 55*:

- A good cut features a square top edge that is sharp and straight, not ragged. The bottom edge can have some dross adhering to it but not an excessive amount. What dross there is should be easily removable with a chipping hammer. The drag lines should be near vertical and not very pronounced.
- When preheat is insufficient, bad gouging results at the bottom of the cut because of slow travel speed.
- Too much preheat will result in the top surface melting over the cut, an irregular cut edge, and an excessive amount of dross.
- When the cutting oxygen pressure is too low, the top edge will melt over because of the resulting slow cutting speed.
- Using cutting oxygen pressure that is too high will cause the operator to lose control of the cut, resulting in an uneven kerf.
- A travel speed that is too slow results in bad gouging at the bottom of the cut and irregular drag lines.
- When the travel speed is too fast, there will be gouging at the bottom of the cut, a pronounced break in the drag line, and an irregular kerf.
- A torch that is held or moved unsteadily across the metal being cut can result in a wavy and irregular kerf.
- When a cut is lost and then not restarted carefully, bad gouges will result at the point where the cut is restarted.

A square kerf face with minimal notching not exceeding 1/16" (1.6 mm) deep is expected and, in fact, required in the Performance Accreditation Tasks for this module.

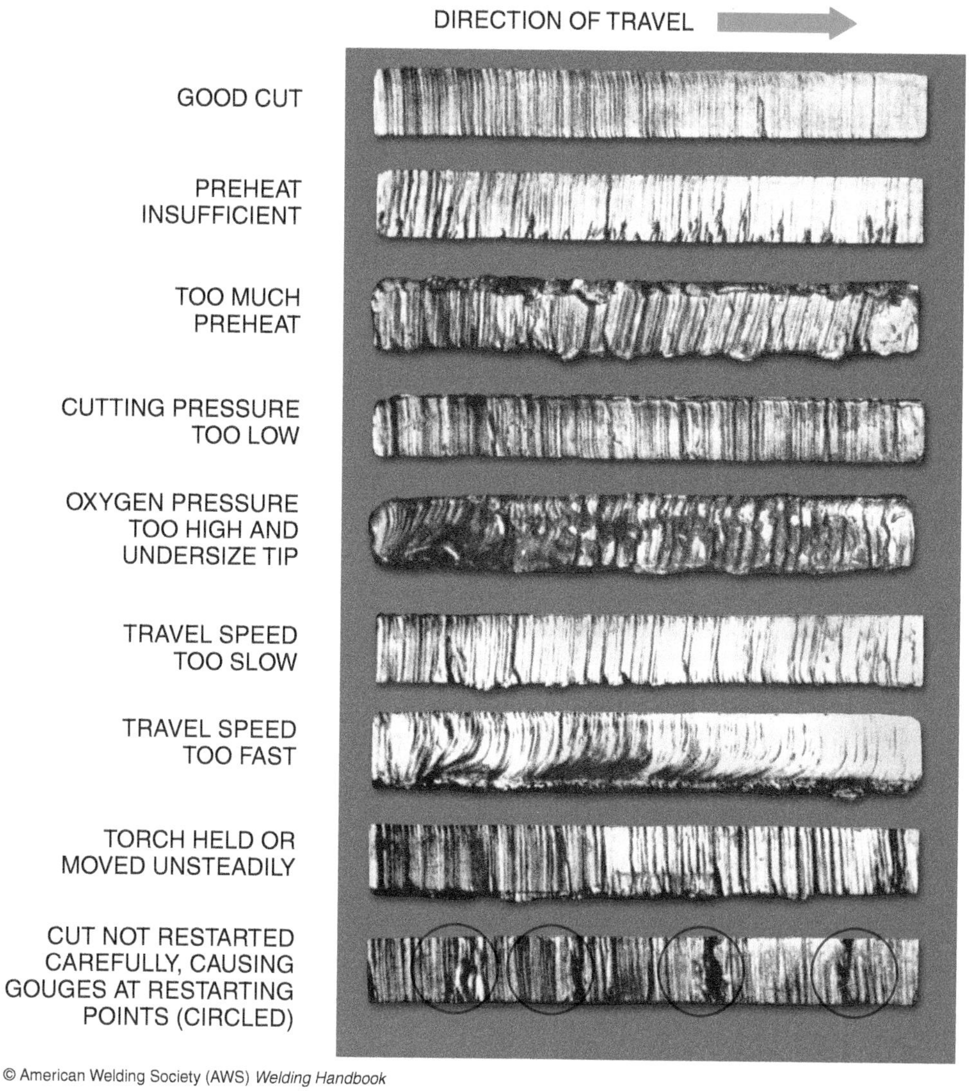

© American Welding Society (AWS) *Welding Handbook*

29102-15_F55.EPS

Figure 55 Examples of good and bad cuts.

The tasks in the sections that follow are designed to develop skills with a cutting torch. Each task should be practiced until there is thorough familiarity with the procedure. After each task is completed, it should be taken to the instructor for evaluation. Do not proceed to the next task until the instructor says to continue.

4.2.0 Cutting Steel

The effectiveness of cutting steel with an oxyfuel cutting outfit depends on factors such as the thickness of the steel, the cutting tip that is being used, and the skill of the operator.

4.2.1 Cutting Thin Steel

Thin steel is considered material 3⁄16" (≈5 mm) thick or less. A major concern when cutting thin steel is distortion caused by the heat of the torch and the cutting process. To minimize distortion, move as quickly as possible without losing the cut.

To begin the process for cutting thin steel, first prepare the metal surface. Then light the torch and hold it so that the tip is pointing in the direction the torch is traveling at a 15- to 20-degree angle. Make sure that a preheat orifice and the cutting orifice are centered on the line of travel next to the metal (*Figure 56*).

CAUTION

Holding the tip upright (perpendicular to the metal) when cutting thin steel will overheat the metal, causing distortion. Maintain the 15 to 20-degree push angle as shown.

Preheat the metal to a dull red. Use care not to overheat thin steel because this will cause distortion. The edge of the tip can be lightly rested on

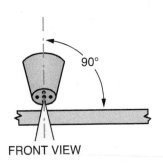

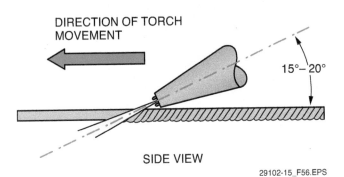

Figure 56 Cutting thin steel.

the surface of the metal being cut and then slid along the surface when making the cut in thin metal. Press the cutting oxygen lever to start the cut, and then move quickly along the line. To minimize distortion, move as quickly as possible without losing the cut.

4.2.2 Cutting Thick Steel

Most oxyfuel cutting is done on steel that is more than ³⁄₁₆" (≈5 mm) thick. Whenever heat is applied to metal, distortion is a problem, but as the steel gets thicker, it becomes less of a problem.

To cut thick steel with a cutting torch, start by preparing the metal surface. Then light the torch and adjust the torch flame. Follow the number sequence shown in *Figure 57* to perform the cut.

The torch can be moved from either right to left or left to right. Choose the direction that allows the best visibility of the cut. When cutting begins, the tips of the preheat flame should be held ¹⁄₁₆" to ⅛" (1.6 to 3.2 mm) above the workpiece. For steel up to ⅜" (10 mm) thick, the first and third procedures can usually be omitted.

4.2.3 Piercing a Plate

Before holes or slots can be cut in a plate, the plate must be pierced. Piercing puts a small hole through the metal where the cut can be started. Because more preheat is necessary on the surface of a plate than at the edge, choose the next-larger cutting tip than is recommended for the thickness to be pierced. When piercing steel that is more than 3" (≈8 cm) thick, it may help to first preheat the bottom side of the plate directly under the spot to be pierced. The following steps describe how to pierce a plate for cutting. *Figure 58* provides a visual reference.

Step 1 Start by preparing the metal surface and the torch for cutting.

Step 2 Ignite the torch and adjust the flame.

Step 3 Hold the torch tip ¼" to ⁵⁄₁₆" (6.4 mm to 7.9 mm) above the spot to be pierced until the surface is a bright cherry red.

Step 4 Raise the tip about ½" (12.7 mm) above the metal surface and tilt the torch slightly so that molten metal does not blow directly back into the tip as the oxygen lever is depressed. Depress the oxygen lever.

Step 5 Maintain the tipped position until a hole burns through the plate. Then rotate the torch back to the vertical position (perpendicular to the plate).

Step 6 Lower the torch back to the initial distance from the plate and continue to cut outward from the original hole to the line to be cut. Then follow the line.

4.3.0 Beveling, Washing, and Gouging

While oxyfuel cutting equipment is commonly associated with cutting through metal plate, it is also well suited for cutting angles in the edge of steel plate, removing bolts and rivets, and cutting groves in metal surfaces.

4.3.1 Cutting Bevels

Bevel cutting is often performed to prepare the edge of steel plate for welding. The procedure for bevel cutting is illustrated in *Figure 59*.

Step 1 Prepare the metal surface and the torch.

Step 2 Ignite the torch and adjust the flame.

Step 3 Hold the torch so that the tip faces the metal at the desired bevel angle. Using a piece of angle iron as a cutting guide as shown in *Figure 59* will result in a 45-degree bevel angle. Angle iron can be used as a guide for any angle, as long as the operator consciously maintains the torch at the proper bevel angle.

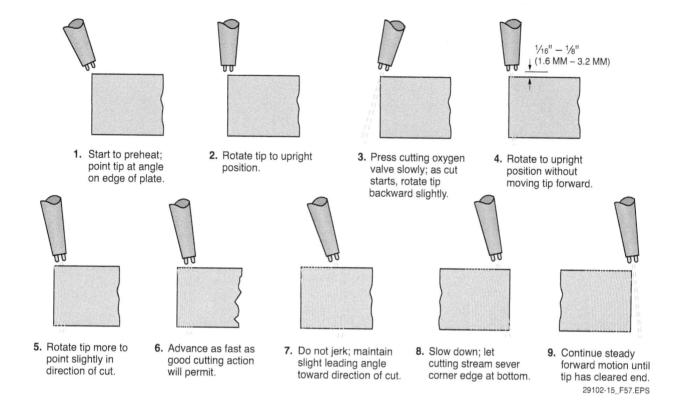

Figure 57 Flame cutting with a hand torch.

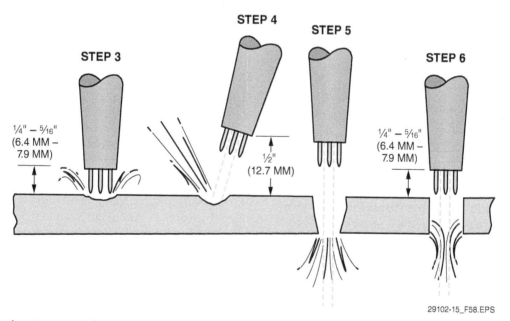

Figure 58 Steps for piercing steel.

Step 4 Preheat the edge to a bright cherry red.

Step 5 Press the cutting oxygen lever to start the cut.

Step 6 As cutting begins, move the torch tip at a steady rate along the line to be cut. Pay particular attention to the torch angle to ensure it creates a uniform bevel along the entire length of the cut.

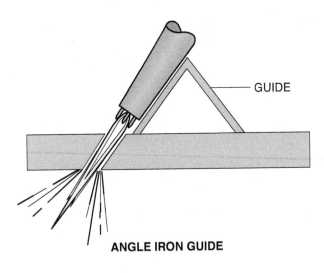
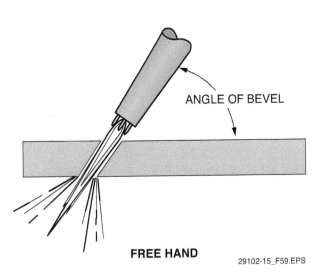

Figure 59 Cutting a bevel.

4.3.2 Washing

Washing is a term used to describe the process of cutting out bolts, rivets, previously welded pieces, or other projections from the surface. Washing operations use a special tip with a large cutting hole that produces a low-velocity stream of oxygen. The low-velocity oxygen stream helps prevent cutting into the surrounding base metal. *Figure 60* is a simplified illustration of a washing procedure.

Step 1 Prepare the metal surface and torch.

Step 2 Ignite the torch and adjust the flame.

Step 3 Preheat the metal to be cut until it is a bright cherry red.

Step 4 Rotate the cutting torch tip to roughly a 55-degree angle to the metal surface.

Step 5 At the top of the material, press the cutting oxygen lever to begin cutting the material to be removed. Continue moving back and forth across the material while

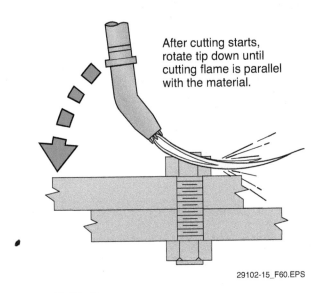

Figure 60 Washing.

rotating the tip to a position parallel with the material. Move the tip back and forth and down to the workpiece surface. Take care not to cut into it.

> **CAUTION**: As the surrounding metal heats up, there is a greater danger of cutting into it. Try to complete the washing operation as quickly as possible. If the surrounding metal gets too hot, stop and let it cool down.

4.3.3 Gouging

Gouging (*Figure 61*) is the process of cutting a groove into a surface. Gouging operations use a special curved tip that produces a low-velocity stream of oxygen that curves up, allowing the operator to control the depth and width of the groove. It is an effective means to gouge out cracks or weld defects for welding. Gouging tips can also be used to remove steel backing from welds or to wash off bolt or rivet heads. However, gouging tips are not as effective as washing tips for removing the shank of a bolt or rivet.

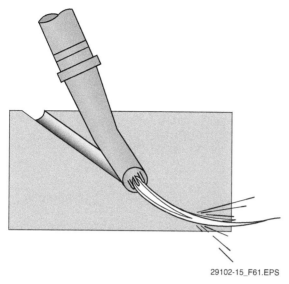

Figure 61 Gouging.

The travel speed and torch angle are very important when gouging. If the travel speed or torch angle is incorrect, the gouge will be irregular and there will be a buildup of dross inside the gouge. Practice until the gouge is clean and even, with a consistent depth.

Step 1 Prepare the metal surface and torch.

Step 2 Ignite the torch and adjust the flame.

Step 3 Holding the torch so that the preheat holes are pointed directly at the metal, preheat the surface until it becomes a bright cherry red.

Step 4 When the steel has been heated to a bright cherry red, slowly roll the torch away from the metal so that the holes are at an angle that will cut the gouge to the correct depth. While rolling the torch away, depress the cutting oxygen lever gradually.

Step 5 Continue to move the cutting torch along the line of the gouge while rocking it back and forth to create a gouge of the required depth and width.

4.4.0 Operating Oxyfuel Track Burners

Oxyfuel track burners, such as the one shown in *Figure 62*, provide a convenient way for operators to make straight cuts, curved cuts, and beveled cuts in the field. They can also enhance precision and uniformity in cuts. When a number of cuts are needed, track burners significantly increase productivity. Most models are portable, allowing them to be set up on the job site.

4.4.1 Torch Adjustment

The rack assembly on the track burner permits the torch holder assembly to move toward or away from the tractor unit. The torch holder allows vertical positioning of the torch. The torch bevel adjustment allows torch positioning at any commonly required angle. After adjusting the torch to the desired position, tighten all clamping screws to prevent the torch from making any unexpected movements.

4.4.2 Straight-Line Cutting

The following provides some basic information regarding the set-up of track burners. Although most track burners have a great deal in common, be sure to follow the manufacturer's operating procedures for the system in use.

To perform straight-line cutting with an oxyfuel track burner, first place the machine track on the workpiece and line it up before placing the machine on the track. Be sure the track is long enough for the cut to be made. If not, install additional track. Connect track sections carefully and ensure it is properly supported. When properly connected, the machine should travel smoothly from one track section to the next. If the cut is long, the track may have to be clamped at both ends beyond the cut to keep the track from moving during the cutting process.

WARNING! Many cutting machines are not designed to detect the end of their track or workpiece. Take care that an unattended machine does not fall from an elevated workpiece while in operation.

Once the track has been positioned, place the machine on the track. Be sure that the supply gas hoses and the power lines are long enough and free to move with the machine so that it can complete the cut properly. Move the machine to the approximate point where the cut will start. Then set the Low/High speed switch to the desired cutting speed or speed range. Set the On/Off switch to the Off position. Next, plug the power cord into an appropriate power supply outlet. Ensure that all clamping screws are properly tightened. Now ignite and properly adjust the torch, and preheat the start of the cut. Set the Forward/Reverse switch to the desired direction of travel. Simultaneously turn on the cutting oxygen and rotate the cutting speed control knob to the desired rate of travel. When the cut is completed, stop the machine and shut off the torch.

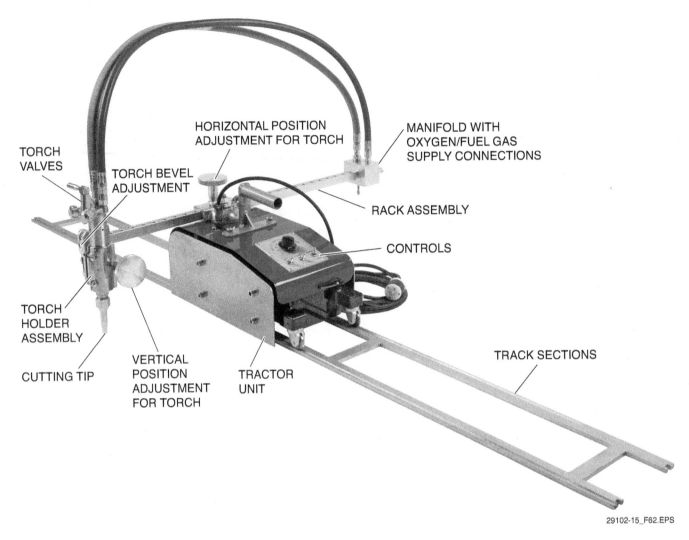

Figure 62 Portable oxyfuel track burner.

4.4.3 Bevel Cutting

Bevel cutting can also be performed with a portable oxyfuel track burner. As before, first place the machine track on the workpiece and line it up before placing the machine on the track. The track must be long enough for the cut to be made. If it is not, install additional track. Connect track sections carefully. Extend the track on both sides of the cut and support the track. When properly connected, the machine should travel smoothly from one track section to the next. If the cut is long, the track may have to be clamped at both ends beyond the cut to keep the track from moving during the cut.

With the track properly positioned, place the machine on the track. Make sure the supply gas hoses and the power lines are long enough and free to move with the machine so that it can complete the cut properly. Loosen the bevel adjusting knob, set the torch angle to the desired bevel angle, and then tighten the bevel adjusting knob.

Move the machine to the approximate point where the cut will start. Then set the Low/High speed switch to the desired cutting speed. Set the On/Off switch to the Off position. Next, plug the power cord into a 115 alternating current (AC), 60 Hertz (Hz) power outlet. Ensure that all clamping screws are properly tightened. Now ignite and properly adjust the torch, and preheat the start of the cut. Set the Forward/Reverse switch to the desired direction of travel. Simultaneously turn on the cutting oxygen and rotate the cutting speed control knob to the desired rate. When the cut is completed, stop the machine and shut off the torch.

Additional Resources

ANSI Z49.1, Safety in Welding, Cutting, and Allied Processes. Miami, FL: American Welding Society.

Uniweld Products, Inc. Numerous videos are available at www.uniweld.com/en/uniweld-videos. Last accessed: November 30, 2014.

The Harris Products Group, a division of Lincoln Electric. Numerous videos are available at www.harrisproductsgroup.com/en/Expert-Advice/videos.aspx. Last accessed: November 30, 2014.

4.0.0 Section Review

1. An oxyfuel cut that has gouging at the bottom, a pronounced break in the drag line, and an irregular kerf is most likely to occur when the ____.

    a. travel speed is too fast
    b. preheat is insufficient
    c. cutting oxygen pressure is too low
    d. cutting oxygen pressure is too high

2. The process of burning a small hole through the metal where an oxyfuel cut can be started is called ____.

    a. gouging
    b. beveling
    c. piercing
    d. washing

3. A washing operation uses a special torch tip with a large cutting hole that produces a ____.

    a. high-velocity stream of oxygen
    b. grooved cut in the base metal surface
    c. cleansing layer of dross in the cut
    d. low-velocity stream of oxygen

4. On a portable oxyfuel track burner, the cutting torch can be angled in a plane that is perpendicular to the track by using the torch ____.

    a. clamping screws
    b. bevel adjustment
    c. kerf positioner
    d. holder assembly

SUMMARY

Oxyfuel cutting has many uses on many different job sites. It can be used to cut metal plate and shapes to size, prepare joints for welding, clean metals or welds, and disassemble structures. Oxyfuel cutting equipment can range in size from small, portable sets to large, automated, fixed-position machines. In all cases, high pressures and flammable gases are involved. For that reason, there is always the danger of fire and explosion when using oxyfuel equipment. These risks can be minimized when the operator is well trained and knowledgeable. By understanding the safety precautions and equipment fundamentals presented in this module, operators will be better prepared to use oxyfuel cutting equipment and cutting techniques in their workplaces.

Review Questions

1. Oxyfuel cutting is best suited for use on ____.
   a. steel alloys
   b. ferrous metals
   c. nonferrous metals
   d. stainless steel

2. The stream of high-pressure cutting oxygen that is directed from an oxyfuel torch causes the metal to instantaneously ____.
   a. solidify
   b. de-kerf
   c. magnetize
   d. oxidize

3. The recommended range of tinting for either the face shield or the safety glass lenses used during an oxyfuel cutting operation is ____.
   a. 1 to 2
   b. 3 to 6
   c. 7 to 8
   d. 9 to 10

4. Most welding environment fires occur during ____.
   a. oxyfuel gas equipment transporting
   b. carbon arc welding activities
   c. oxyfuel gas welding or cutting
   d. acetylene tank installation

5. When preparing a tank or vessel that might have contained flammable materials to be cut with an oxyfuel cutting torch, the best approach is to ____.
   a. fill it with water
   b. purge it with oxygen
   c. fill it with air
   d. purge it with acetylene

6. When pure oxygen is combined with a fuel gas, it produces a ____.
   a. high-pressure jet of cutting oxygen
   b. non-explosive, non-flammable vapor
   c. colorless, odorless, and tasteless gas
   d. high-temperature flame for cutting

7. The most common size of oxygen cylinder used in oxyfuel cutting applications is ____.
   a. 85 cubic feet (2.4 cubic meters)
   b. 227 cubic feet (6.4 cubic meters)
   c. 350 cubic feet (9.9 cubic meters)
   d. 485 cubic feet (13.7 cubic meters)

8. If an acetylene cylinder is found lying on its side, what should be done before it is used?
   a. Contact the supplier for instructions.
   b. Release a small amount of the gas to atmosphere.
   c. Stand it upright and wait at least one hour.
   d. Add a liquid stabilizer to the gas.

9. The maximum hourly rate at which acetylene gas can be withdrawn from a cylinder is ____.
   a. one half of the cylinder's capacity
   b. one third of the cylinder's capacity
   c. one fifth of the cylinder's capacity
   d. one seventh of the cylinder's capacity

10. The fuel gas that burns with the lowest flame temperature is ____.
    a. propane
    b. propylene
    c. acetylene
    d. butane

11. Most gas pressure regulators contain two gauges—one that indicates the cylinder pressure and one that indicates the ____.
    a. ideal cylinder pressure for the conditions
    b. pressure of the gas at the regulator outlet
    c. pressure in the accompanying cylinder
    d. maximum pressure for the cylinder

12. Oxyfuel cutting equipment that has left-hand threads and a V-notch in the nut are most likely to be ____.
    a. oxygen fittings
    b. purging gas fittings
    c. aftermarket fittings
    d. fuel gas fittings

13. The type of cutting torch that uses a vacuum created by oxygen flow to draw in fuel from a very-low-pressure fuel source is a(n) ____.
    a. siphon torch
    b. inert torch
    c. injector torch
    d. suspension torch

14. A friction lighter produces sparks when its steel surface is rubbed with a piece of ____.
    a. flint
    b. soapstone
    c. graphite
    d. steatite

15. A computer-controlled plate cutting machine and an optical pattern-tracing machine are examples of ____.
    a. portable oxyfuel cutters
    b. exothermic machines
    c. manual guide cutters
    d. fixed-location machines

16. Cracking the cylinder valves during the setup of oxyfuel equipment is a way of ____.
    a. leak testing the valve regulators
    b. equalizing the cylinder pressures
    c. removing dirt from the valves
    d. venting residual gas from the cylinders

17. Before opening cylinder valves, verify that the adjusting screws on the oxygen and fuel gas regulators have been ____.
    a. tightened
    b. closed
    c. loosened
    d. purged

18. Oxyfuel cutting equipment is typically leak tested using a(n) ____.
    a. supervisor's sense of smell
    b. solution that produces bubbles
    c. lit match or a candle
    d. ultrasonic detector

19. An oxyfuel cutting flame that has an excess of fuel is called a(n) ____.
    a. hot lean flame
    b. neutral flame
    c. oxidizing flame
    d. carburizing flame

20. When disassembling oxyfuel equipment, verify that all pressure gauges are ____.
    a. reading zero
    b. open and showing atmospheric pressure
    c. chained securely to each other
    d. marked MT for storage

21. Gas cylinders should never be completely emptied because of the risk of ____.
    a. reverse flow
    b. valve cracking
    c. disproportional mixing
    d. cylinder implosion

22. When a cut has been made with oxyfuel cutting equipment, the drag lines of the cut should be close to ____.
    a. thirty degrees with minimal notching
    b. forty-five degrees with a wavy kerf
    c. horizontal and notched at the kerf
    d. vertical and not very pronounced

23. A major concern when cutting thin steel with an oxyfuel cutting torch is ____.
    a. flashback
    b. sparking
    c. distortion
    d. backfire

24. An oxyfuel cutting process that is often used for cutting off bolts, rivets, and other projections is called ____.
    a. beveling
    b. gouging
    c. piercing
    d. washing

25. What part of an oxyfuel track burner allows the vertical positioning of the torch to be controlled?
    a. The clamping screws
    b. The torch holder
    c. The kerf regulator
    d. The bevel adjustment

Trade Terms Quiz

Fill in the blank with the correct term that you learned from your study of this module.

1. The gap produced by a cutting process is called a(n) _____ .
2. A loud snap or pop that can be heard as a torch flame is extinguished is called a(n) _____ .
3. The soft, white material that is commonly used to mark metal is _____ .
4. Metals that contain iron are called _____ .
5. Creating a groove in the surface of a workpiece is a process called _____ .
6. A flame that is burning with too much fuel is called a(n) _____ .
7. A flame that is burning with too much oxygen is called a(n) _____ .
8. When correct portions of fuel gas and oxygen are fed to a flame, the flame is said to be a(n) _____ .
9. When a thermal process is used for cutting, the material expelled from the kerf is called _____ .
10. Cutting off projections such as bolts, rivets, and previous welded pieces is a process referred to as _____ .
11. The name used to describe what occurs when an oxyfuel cutting torch penetrates a metal plate is _____ .
12. When a flame burns back into the tip of a torch and causes a high-pitched whistling sound, the condition is called _____ .
13. The lines on the edge of a cut that result from the cutting oxygen streaming into, through, and out of the metal are called _____ .

Trade Terms

Backfire
Carburizing flame
Drag lines
Dross
Ferrous metals

Flashback
Gouging
Kerf
Neutral flame
Oxidizing flame

Pierce
Soapstone
Washing

Appendix

Performance Accreditation Tasks

The American Welding Society (AWS) School Excelling through National Skills Standards Education (SENSE) program is a comprehensive set of minimum Standards and Guidelines for Welding Education programs. The following performance accreditation is aligned with and designed around the SENSE program.

The Performance Accreditation Tasks (PATs) correspond to and support the learning objectives in *AWS EG2.0, Guide for the Training and Qualification of Welding Personnel: Entry-Level Welder*.

Note that in order to satisfy all learning objectives in *AWS EG2.0*, the instructor must also use the PATs contained in the second level of the NCCER Welding curriculum.

PATs 1 and 2 correspond to *AWS EG2.0, Module 8 – Thermal Cutting Processes, Unit 1 – Manual OFC Principles*, Key Indicators 5, 6, and 7.

PAT 3 corresponds to *AWS EG2.0, Module 8 – Thermal Cutting Processes, Unit 1 – Manual OFC Principles*, Key Indicators 3 and 4.

PATs provide specific acceptable criteria for performance and help to ensure a true competency-based welding program for students.

The following tasks are designed to test your competency with an oxyfuel cutting torch. Do not perform these cutting tasks until directed to do so by your instructor.

Performance Accreditation Tasks Module 29102-15

SETTING UP, IGNITING, ADJUSTING, AND SHUTTING DOWN OXYFUEL EQUIPMENT

Using oxyfuel equipment that has been completely disassembled, demonstrate how to:

- Set up oxyfuel equipment
- Ignite and adjust the flame
  - Carburizing
  - Neutral
  - Oxidizing
- Shut off the torch
- Shut down the oxyfuel equipment

Criteria for Acceptance:

- Set up the oxyfuel equipment in the correct sequence   _____
- Demonstrate that there are no leaks   _____
- Properly adjust all three flames   _____
- Shut off the torch in the correct sequence   _____
- Shut down the oxyfuel equipment   _____

Copyright © 2015 NCCER. Permission is granted to reproduce this page provided that copies are for local use only and that each copy contains this notice.

CUTTING A SHAPE

Using a carbon steel plate, lay out and cut the shape and holes shown in the figure. If available, use a machine track cutter to straight cut the longer dimension.

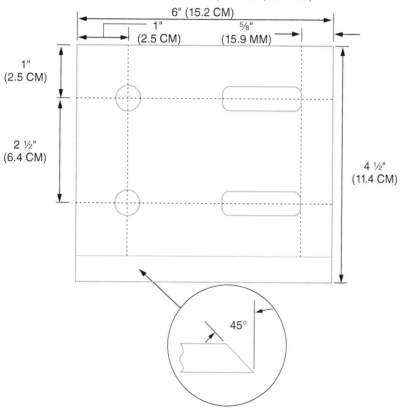

NOTE: MATERIAL – CARBON STEEL ¼" (>6 MM) THICK OR GREATER
HOLES ¾" (19.1 MM) DIAMETER
SLOTS ¾" (19.1 MM) × 1 ½" (38.1 MM)

Criteria for Acceptance:

- Perform this task in the flat position (1G)
- Outside dimensions ±⅛" (3.2 mm)
- Inside dimensions (holes and slots) ±⅛" (3.2 mm)
- Square ±5 degrees
- Minimal amount of dross sticking to plate which can be easily removed
- Square kerf face with minimal notching not exceeding 1/16" (1.6 mm) deep

Performance Accreditation Tasks

Module 29102-15

CUTTING A SHAPE

Using a carbon steel plate, lay out and cut the shape and holes shown in the figure. If available, use a machine track cutter to bevel and straight cut the longer dimension.

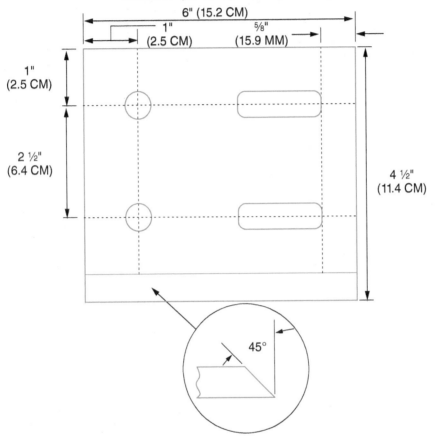

Criteria for Acceptance:

- Perform this task in the horizontal position (2G)
- Outside dimensions ±⅛" (3.2 mm)
- Inside dimensions (holes and slots) ±⅛" (3.2 mm)
- Square ±5 degrees
- Bevel ±2 degrees
- Minimal amount of dross sticking to plate which can be easily removed
- Square kerf face with minimal notching not exceeding ¹⁄₁₆" (1.6 mm) deep

Copyright © 2015 NCCER. Permission is granted to reproduce this page provided that copies are for local use only and that each copy contains this notice.

NCCER – *Construction Craft Laborer Level Two* 29102-15

Trade Terms Introduced in This Module

Backfire: A loud snap or pop as a torch flame is extinguished.

Carburizing flame: A flame burning with an excess amount of fuel; also called a reducing flame.

Drag lines: The lines on the edge of the material that result from the travel of the cutting oxygen stream into, through, and out of the metal.

Dross: The material (oxidized and molten metal) that is expelled from the kerf when cutting using a thermal process. It is sometimes called slag.

Ferrous metals: Metals containing iron.

Flashback: The flame burning back into the tip, torch, hose, or regulator, causing a high-pitched whistling or hissing sound.

Gouging: The process of cutting a groove into a surface.

Kerf: The gap produced by a cutting process.

Neutral flame: A flame burning with correct proportions of fuel gas and oxygen.

Oxidizing flame: A flame burning with an excess amount of oxygen.

Pierce: To penetrate through metal plate with an oxyfuel cutting torch.

Soapstone: Soft, white stone used to mark metal.

Washing: A term used to describe the process of cutting out bolts, rivets, previously welded pieces, or other projections from the metal surface.

Additional Resources

This module presents thorough resources for task training. The following resource material is suggested for further study.

ANSI Z49.1, Safety in Welding, Cutting, and Allied Processes. Miami, FL: American Welding Society.

Plasma Cutters Handbook: Choosing Plasma Cutters, Shop Safety, Basic Operation, Cutting Procedures, ANSI Z49.1, Safety in Welding, Cutting, and Allied Processes. Miami, FL: American Welding Society.

The Harris Products Group, a division of Lincoln Electric. Numerous videos are available at **www.harrisproductsgroup.com/en/Expert-Advice/videos.aspx**. Last accessed: November 30, 2014.

Uniweld Products, Inc. Numerous videos are available at **www.uniweld.com/en/uniweld-videos**. Last accessed: November 30, 2014.

Figure Credits

The Lincoln Electric Company, Cleveland, OH, USA, Module Opener, Figures 1, 2, 5, 14, 32–34, 48–50, 53, 60 (photo), 62, SA04

Topaz Publications, Inc., Figures 3, 7–9, 11, 13, 15, 16, 18, 27, 28, 29A, 30, 37–45, SA02, SA03, SA06–SA08, SA11

Courtesy of Uniweld Products, Figure 20, SA10

Victor Technologies, Figure 21

Vestil Manufacturing, Figure 29B

Koike Aronson, Inc. – Worldwide manufacturer of cutting, welding and positioning equipment, Figure 31

Courtesy of H & M Pipe Beveling Machine Company, Inc., Figures 35, 36

Zachry Industrial, Inc., Figures 46, 47, 51

Courtesy of Smith Equipment, Figure 52

© American Welding Society (AWS) *Welding Handbook* 1991, Welding Processes Volume No. 2, Edition No. 8, Miami: American Welding Society, Figure 55

Xerafy, SA01

Courtesy of Saf-T-Cart, SA09

Section Review Answer Key

Answer	Section Reference	Objective
Section One		
1. b	1.1.0	1a
2. d	1.2.2	1b
Section Two		
1. b	2.1.2	2a
2. a	2.2.3	2b
3. d	2.3.1	2c
4. a	2.4.4	2d
5. c	2.5.3	2e
Section Three		
1. b	3.1.1	3a
2. d	3.2.0	3b
3. c	3.3.1	3c
4. d	3.4.0	3d
Section Four		
1. a	4.1.1	4a
2. c	4.2.3	4b
3. d	4.3.2	4c
4. b	4.4.1	4d

NCCER CURRICULA — USER UPDATE

NCCER makes every effort to keep its textbooks up-to-date and free of technical errors. We appreciate your help in this process. If you find an error, a typographical mistake, or an inaccuracy in NCCER's curricula, please fill out this form (or a photocopy), or complete the online form at **www.nccer.org/olf**. Be sure to include the exact module ID number, page number, a detailed description, and your recommended correction. Your input will be brought to the attention of the Authoring Team. Thank you for your assistance.

Instructors – If you have an idea for improving this textbook, or have found that additional materials were necessary to teach this module effectively, please let us know so that we may present your suggestions to the Authoring Team.

NCCER Product Development and Revision
13614 Progress Blvd., Alachua, FL 32615

Email: curriculum@nccer.org
Online: www.nccer.org/olf

❏ Trainee Guide   ❏ Lesson Plans   ❏ Exam   ❏ PowerPoints   Other _____

Craft / Level: _____   Copyright Date: _____

Module ID Number / Title: _____

Section Number(s): _____

Description: _____

_____

_____

_____

_____

Recommended Correction: _____

_____

_____

_____

_____

Your Name: _____

Address: _____

_____

Email: _____   Phone: _____

28301-14

Elevated Masonry

This module presents masonry techniques and safety principles for high-rise masonry construction. It also covers bracing and shoring. Advances in construction techniques, plus the work of OSHA, have resulted in a greater number of safety procedures at the elevated job site. The mason has the responsibility to be aware of these procedures and practice them. This module covers safety rules and appropriate personal protective equipment used in elevated masonry; wall bracing for wind and backfill; the design and construction of interior and exterior elevated masonry systems; and the proper moving, storage, use, and disposal of masonry materials in an elevated work environment.

Module Ten

Trainees with successful module completions may be eligible for credentialing through the NCCER Registry. To learn more, go to **www.nccer.org** or contact us at **1.888.622.3720**. Our website has information on the latest product releases and training, as well as online versions of our *Cornerstone* magazine and Pearson's product catalog.

Your feedback is welcome. You may email your comments to **curriculum@nccer.org**, send general comments and inquiries to **info@nccer.org**, or fill in the User Update form at the back of this module.

This information is general in nature and intended for training purposes only. Actual performance of activities described in this manual requires compliance with all applicable operating, service, maintenance, and safety procedures under the direction of qualified personnel. References in this manual to patented or proprietary devices do not constitute a recommendation of their use.

Copyright © 2015 by NCCER, Alachua, FL 32615, and published by Pearson Education, Inc., New York, NY 10013. All rights reserved. Printed in the United States of America. This publication is protected by Copyright, and permission should be obtained from NCCER prior to any prohibited reproduction, storage in a retrieval system, or transmission in any form or by any means, electronic, mechanical, photocopying, recording, or likewise. To obtain permission(s) to use material from this work, please submit a written request to NCCER Product Development, 13614 Progress Blvd., Alachua, FL 32615.

From *Construction Craft Laborer, Level Two, Trainee Guide*, Third Edition. NCCER.
Copyright © 2015 by NCCER. Published by Pearson Education. All rights reserved.

28301-14
ELEVATED MASONRY

Objectives

When you have completed this module, you will be able to do the following:

1. Identify the proper personal protective equipment and safety precautions related to elevated masonry.
   a. Describe safety precautions related to an elevated work area.
   b. Discuss fall protection related to elevated work areas.
2. Describe how to properly brace a wall.
   a. Describe how to properly brace a concrete masonry wall for wind.
   b. Describe how to properly brace a wall for backfill.
3. Describe elevated masonry systems.
   a. List the construction sequence for elevated masonry systems.
   b. Describe how elevated masonry systems are designed.
   c. Identify common exterior walls used for elevated masonry systems.
   d. Identify common interior walls used for elevated masonry systems.
4. Describe how to properly handle materials at elevations.
   a. Explain safety precautions to be observed when working around cranes.
   b. Explain safety precautions to be observed when working around materials hoists.
   c. Explain safety precautions to be observed when moving and stocking materials.
   d. Explain safety precautions to be observed when working at elevated workstations.
   e. Explain how disposal chutes and waste bins are used when working from elevated workstations.

Performance Tasks

Under the supervision of your instructor, you should be able to do the following:

1. Properly brace a wall.
2. Demonstrate hand signals used for lifting materials.

Trade Terms

Controlled access zone
Cut
Guyed derrick
Lateral stress
Limited access zone
Reglet

Industry-Recognized Credentials

If you're training through an NCCER-accredited sponsor, you may be eligible for credentials from NCCER's Registry. The ID number for this module is 28301-14. Note that this module may have been used in other NCCER curricula and may apply to other level completions. Contact NCCER's Registry at 888.622.3720 or go to www.nccer.org for more information.

Code Note

Codes vary among jurisdictions. Because of the variations in code, consult the applicable code whenever regulations are in question. Referring to an incorrect set of codes can cause as much trouble as failing to reference codes altogether. Obtain, review, and familiarize yourself with your local adopted code.

Contents

Topics to be presented in this module include:

1.0.0 Elevated Masonry Safety and Personal Protective Equipment 1
  1.1.0 Observing Work-Area Safety 1
    1.1.1 Personal Protective Equipment 1
    1.1.2 Electrical Hazards 2
    1.1.3 Fire Prevention 4
    1.1.4 Clean Work Areas 4
  1.2.0 Observing Fall Protection Requirements 4
    1.2.1 Personal Fall Arrest Systems 5
    1.2.2 Falling Objects 6
    1.2.3 Personnel Lifts 8
    1.2.4 Access Zones 9
2.0.0 Concrete Masonry Wall Bracing 11
  2.1.0 Bracing for Wind 12
  2.2.0 Bracing for Backfill 13
3.0.0 Elevated Masonry Systems 16
  3.1.0 Identifying and Following Construction Sequences 16
  3.2.0 Understanding Building Design 18
  3.3.0 Constructing Exterior Walls 19
    3.3.1 Panel Walls 19
    3.3.2 Curtain Walls 20
    3.3.3 Parapet Walls 22
  3.4.0 Constructing Interior Walls 23
4.0.0 Materials Handling 26
  4.1.0 Working around Cranes 26
    4.1.1 Verbal Modes of Communication 27
    4.1.2 Nonverbal Modes of Communication 29
  4.2.0 Working Around Materials Hoists 32
  4.3.0 Moving and Stocking Materials 33
  4.4.0 Using Elevated Workstations 34
  4.5.0 Using Disposal Chutes and Waste Bins 34

Figures and Tables

Figure 1 Double insulation 2
Figure 2 Grounded three-prong plug 3
Figure 3 Wall-mounted and portable GFCIs 3
Figure 4 Personal fall arrest equipment 5
Figure 5 Lanyard with shock absorber 5
Figure 6 Rope grab and retractable lifeline 6
Figure 7 Vertical lifeline 6
Figure 8 Horizontal lifeline 6
Figure 9 Push-through eyebolt and shock absorber 7
Figure 10 Locking snap hook and carabiner 7
Figure 11 Debris netting installed on scaffold during a renovation project 8
Figure 12 Personnel lift 9
Figure 13 Personnel lift tower 9
Figure 14 Maximum height of unbraced wall 11
Figure 15 Brace-spacing requirements for a wall 12
Figure 16 Wind bracing for masonry wall 14
Figure 17 Temporary lateral bracing for a foundation wall 14
Figure 18 High-rise building under construction 16
Figure 19 Guardrails 17
Figure 20 Temporary stairway 18
Figure 21 Steel frame for high-rise building 18
Figure 22 Curtain wall under construction 19
Figure 23 Panel-wall mounting 20
Figure 24 Stone-panel wall attachment 20
Figure 25 Flashing and sealant detail 21
Figure 26 Curtain wall anchorage 21
Figure 27 Parapet 22
Figure 28 Precast coping 23
Figure 29 Through-wall flashing 24
Figure 30 Construction cranes 27
Figure 31 Noise-canceling microphone and headphones attached to a hard hat 28
Figure 32 Throat microphone 28
Figure 33 Hardwired communications system 28
Figure 34 Handheld radio 29
Figure 35 Standard hand signals 30
Figure 36 Standard hand signals (continued) 31
Figure 37 Scaffold mounted hoist 32
Figure 38 Large materials hoist 33
Figure 39 Reach-type forklift 34

Table 1 Lateral Support Requirements for Non-Reinforced Concrete Masonry Walls 13
Table 2 Maximum Wall Length- or Height-to-Thickness Ratio 24
Table 3 Maximum Wall Spans 24

SECTION ONE

1.0.0 ELEVATED MASONRY SAFETY AND PERSONAL PROTECTIVE EQUIPMENT

Objective

Identify the proper personal protective equipment and safety precautions related to elevated masonry.

a. Describe safety precautions related to an elevated work area.
b. Discuss fall protection related to elevated work areas.

Trade Terms

Controlled access zone: A designated work area in which certain types of masonry work may take place without the use of conventional fall protection systems.

Cut: A common term for a scaffold level.

Limited access zone: A restricted area alongside a masonry wall that is under construction.

Working safely on a high-rise job means that you must work smart. Be aware of what you can do to keep your work area safe. Take precautions to prevent falls and protect yourself from falling objects. Always use personnel lifts correctly and safely.

This section provides a review of appropriate personal protective equipment (PPE) for masons, along with additional information that applies to working in elevation conditions.

Fall protection is required when workers are exposed to falls from work areas with elevations that are 6 feet or higher. The types of work areas that put the worker at risk include the following:

- Scaffolds
- Ladders
- Leading edges
- Ramps or runways
- Wall or floor openings
- Roofs
- Excavations, pits, and wells
- Concrete forms
- Unprotected sides and edges

Injuries from falls happen because fall protection systems are used inappropriately or not at all. They also occur due to carelessness. It is your responsibility to learn how to set up, inspect, use, and maintain your own fall protection equipment. Not only will this keep you alive and uninjured, it could also save the lives of your co-workers.

Falls are classified into two groups: falls from an elevation and falls on the same level. Falls from an elevation can happen when you are doing work from scaffolds, work platforms, decking, concrete forms, ladders, or excavations. Falls from elevations are almost always fatal. This is not to say that falls on the same level aren't also extremely dangerous. When a worker falls on the same level, usually from tripping or slipping, head injuries often occur. Sharp edges and pointed objects such as exposed rebar could cut or stab the worker.

The following safe practices can help prevent slips and falls:

- Wear strong work boots that are in good repair.
- Watch where you step. Be sure your footing is secure.
- Install cables, extension cords, and hoses so that they will not become tripping hazards.
- Do not allow yourself to get in an awkward position. Stay in control of your movements at all times.
- Maintain clean and smooth walking and working surfaces. Fill holes, ruts, and cracks. Clean up slippery material and litter.
- Do not run on scaffolds, work platforms, decking, roofs, or other elevated work areas.

It is vital to use fall protection equipment when working at elevations. The three most common types of fall protection equipment are guardrails, personal fall arrest systems, and safety nets.

1.1.0 Observing Work-Area Safety

Working in a high-rise construction site is more dangerous than working on a single level. Trip hazards caused by poor housekeeping can be deadly in an elevated work area. Many injuries can be prevented by keeping the work area safe. Work-area safety includes following procedures for fire prevention, identifying electrical hazards, keeping pathways clean, and using appropriate personal protective equipment properly.

1.1.1 Personal Protective Equipment

Proper clothing and safety apparel is an important part of working safely. All workers exposed to overhead hazards are required to wear protective headgear. Wear your hard hat at all times. Shoes with safety toes are recommended for all construction workers. Heavy-duty work clothes

28301-14 Elevated Masonry

and gloves give protection from bruises and cuts caused by sharp objects and falling material. Fall protection equipment must be worn when you are working 6 feet or more above a lower level.

Remember to take safety precautions when dressing for masonry work. The following guidelines are recommended:

- Confine long hair in a ponytail or in your hard hat. Flying hair can obscure your view or get caught in machinery.
- Wear appropriate clothing and personal protective equipment.
  - Always wear a hard hat.
  - Wear goggles when cutting or grinding.
  - Wear a high-visibility vest.
  - Wear close-fitting clothing, including long-sleeved shirts (minimum 4-inch sleeves) to give extra protection if skin is sensitive.
  - Wear gloves when working with wet mortar.
  - Wear pants over boots to avoid getting mortar on legs or feet.
  - Keep gloves and clothing as dry as possible.
- Wear face and eye protection as required, especially if there is a risk from flying particles, debris, or other hazards such as brick dust or proprietary cleaners.
- Wear hearing protection as required.
- Wear respiratory protection as required.
- Protect any exposed skin by applying skin cream, body lotion, or petroleum jelly.
- Wear sturdy work boots or work shoes with thick soles. Never show up for work dressed in sneakers, loafers, or sport shoes.
- Wear fall protection equipment as required.

1.1.2 Electrical Hazards

The most serious danger in using electrically powered tools is that of electrocution. Electricity can also cause burns, shocks, explosions, and fires. Electrical shocks can be minor and uncomfortable, or they can be severe, causing burns or death. Even a small amount of current can cause the heart to stop pumping in rhythm. If not corrected, this condition will result in death.

Electrical shock can also cause a loss of balance, muscle control, or consciousness. This could cause the victim to fall and/or drop a tool. A fall from a ladder or scaffold can be quite serious. To prevent electrical shock, tools must provide at least one of the following types of protection:

- *Double insulated* – Double insulation is more convenient than three-wire cords. The user and tools are protected in two ways: by normal insulation on the wires inside, and by a housing that cannot conduct electricity to the operator in the event of a malfunction (*Figure 1*).
- *Powered by a low-voltage isolation transformer* – If your electrically powered tools do not have either a ground plug or double insulation, check with your supervisor to make sure that you are protected by a low-voltage isolation transformer.
- *Grounded with a three-wire cord* – Three-wire cords have two current-carrying conductors and one grounding conductor. Three-prong plugs are common on electrically powered tools (*Figure 2*). A three-prong plug should only be plugged into a three-prong, grounded receptacle. Any time an adapter is used to accommodate a two-hole receptacle, the adapter wire must be attached to a known ground. Never remove the third prong (grounding conductor) from a plug. If you are using a three-prong extension cord, make sure that it is properly grounded at its source.

Water can cause an electrical short. Prevent shock by keeping tools dry. When not in use, store power tools in a dry place. Do not use electrically powered tools in damp or wet places. Wear all appropriate personal protective equipment, such as gloves and safety footwear, when working with electrically powered tools.

The use of a ground fault circuit interrupter (GFCI) is one method used to overcome grounding and insulation problems. A GFCI is a fast-acting circuit breaker that senses small imbalances in the circuit caused by current leakage to ground. If an imbalance is detected, the GFCI interrupts the electric power within $\frac{1}{40}$ of a second. *Figure 3* shows a wall-mounted and portable GFCI.

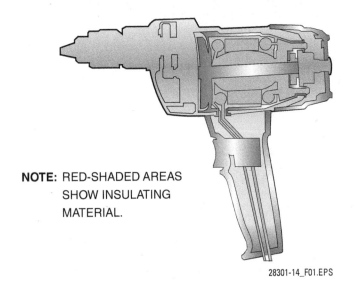

NOTE: RED-SHADED AREAS SHOW INSULATING MATERIAL.

Figure 1 Double insulation.

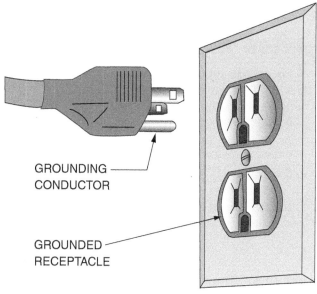

Figure 2 Grounded three-prong plug.

(A)

(B)

Figure 3 Wall-mounted and portable GFCIs.

A GFCI will not protect you from line-to-line contact hazards such as holding either two hot wires or a hot and a neutral wire in each hand. It does provide protection against the most common form of electrical shock, which is a ground fault. It also provides protection from fires, overheating, and wiring-insulation deterioration. GFCIs can be used successfully to reduce electrical hazards on construction sites. Inspect all equipment before using it. Never use electrical equipment with a damaged power cord. Use fiberglass ladders when working with electrical equipment.

Because malfunctioning electrical power tools can cause sparks, these tools can also cause fires and explosions. Make sure you know about any fire hazards in your work area, and avoid using electrically powered tools around flammable materials, fumes, and gases.

Finally, electrical cords and extension cords pose a tripping hazard. Extension cords should be brightly colored to make them more visible. Do not run cords and cables in walkways. Run them along a wall, rather than in the middle of a walkway or across a walkway. Avoid running cables and cords across elevated work areas and scaffolds. Occasionally, it may be necessary to run a cord or cable across a walkway. If this cannot be avoided, either tape the cord down and put a carpet over it, or place it in a cord runner designed to minimize the tripping hazard.

Don't Remove the Grounding Prong

An employee was climbing a metal ladder to hand an electric drill to the journeyman installer on a scaffold about 5 feet above him. When the victim reached the third rung from the bottom of the ladder, he received an electric shock that killed him. The investigation revealed that the extension cord had a missing grounding prong and that a conductor on the green grounding wire was making intermittent contact with the energized black wire, thereby energizing the entire length of the grounding wire and the drill's frame. The drill was not double insulated.

The Bottom Line: Do not disable any safety device on a power tool. A ground fault can be deadly.

Source: The Occupational Safety and Health Administration (OSHA)

1.1.3 Fire Prevention

Fire is a potential danger in all construction operations. It may present a greater hazard in high-rise construction because workers are in such vulnerable positions.

To reduce the risk of fire, protect all materials stored in the building or within 10 feet of the building with a noncombustible covering. Occasionally, combustible material must be stored within the structure. That part of the building must be fireproofed before this can happen. In either case, store combustible liquids away from other materials. Dispose of combustible waste in a secure container. Do not smoke at the work site or in the storeroom.

Make sure that at least one portable chemical fire extinguisher has been posted next to each storeroom, and that free access to fire hydrants on the street is maintained. Make sure all workers understand the fire procedures for the site. If you see that any of these items has not been taken care of, report it to your supervisor immediately.

1.1.4 Clean Work Areas

Some of the most important things that masons can do to protect themselves during elevated work are also the simplest. For example, clean work areas greatly reduce the possibility of injury. Remember to follow these guidelines:

- Keep all passageways free of materials, supplies, and other obstructions.
- Collect all scrap and waste material at the end of the day. Place it in containers or waste bins for regular removal.
- Make sure that all parts of the site are adequately lit. Replace any burnt-out bulbs.

1.2.0 Observing Fall Protection Requirements

According to OSHA, falls were the most common cause of construction worker fatalities in 2012, resulting in 278 deaths that year. That is more than three times as many fatalities as the next most-common cause. To help reduce that number, OSHA has developed construction industry standards to prevent workers from falling off, onto, or through working levels, and to protect workers from being struck by falling objects.

High-rise construction typically requires the use of fall protection systems. Areas or activities where fall protection is needed include, but are not limited to, scaffolds, excavations, hoist areas, holes, ramps, runways, and other walkways. It is required for formwork and rebar work, leading-edge work, unprotected sides and edges, and other hazardous walking/working surfaces.

Contractors must protect their workers from fall hazards and falling objects whenever a worker is 6 feet or more above a lower level, or 4 feet above open machinery. Workers must be protected from falling into dangerous equipment. Work areas must also be protected from falling objects.

In high-rise construction, workers often find it necessary to work in places requiring wire rope or slings for access, in addition to personal lifelines. These positioning slings and ropes should be inspected daily by a qualified person. According to OSHA regulations, no wire rope should be used when more than 10 percent of the total wires are frayed or broken in any running foot.

PASS Technique

A fire extinguisher can put out small fires quickly, but it will not work unless you know how to use it properly. Remember the PASS technique:

P – **P**ull the pin from the handle.
A – **A**im the nozzle at the base of the fire.
S – **S**queeze the lever, button, or handle.
S – **S**weep the extinguisher from side to side as you spray.

Note that most fire extinguishers have to be recharged after each use.

No Training + No PPE = Death

A carpenter apprentice was killed when he was struck in the head by a nail that was fired from a powder-actuated tool in another room. The tool operator was attempting to anchor a plywood form in preparation for pouring a concrete wall. When he fired the gun, the nail passed through the hollow wall and traveled 27 feet before striking the victim. The tool operator had never received training in the proper use of the tool, and none of the employees in the area were wearing personal protective equipment.

The Bottom Line: You can be injured by the actions of others. Wear your PPE as a first line of defense against injury.

Fall protection is generally provided through the use of guardrails, safety nets, and personal fall arrest systems. All systems must be tested and approved before use.

1.2.1 Personal Fall Arrest Systems

Fall arrest equipment (*Figure 4*) catches a worker after the worker has fallen. Workers are required to use a fall arrest system when there is a risk of falling more than 6 feet. It should be used when working on scaffolds, high-rise buildings, roofs, and other elevated locations. It should also be used when working near a deep hole, near a large opening in a floor, or above protruding rebar. Use a full-body harness and lanyard for fall arrest protection where vertical free-fall hazards exist.

> **NOTE**
> In the past, body belts were often used instead of a full-body harness. However, as of January 1, 1998, they were banned from such use. This is because the body belt concentrates the arresting force in the abdominal area.

A personal fall arrest system uses specialized equipment, including a body harness, lanyards, deceleration devices, lifelines, anchoring devices, and equipment connectors. The body harness goes around the legs and shoulders, with strapping across the chest and back. A D-ring on the back is used to attach a lanyard. The lanyard is then attached to a lifeline, which is anchored to a point that is capable of holding more than 5,400 pounds without failure. The line should be long enough to allow work movements but short enough to limit a fall to 6 feet or less.

Lanyards are short, flexible lines with connectors on each end (*Figure 5*). They connect the body harness to the lifeline. There are many kinds of lanyards made for different situations. All must have a minimum breaking strength of 5,000 pounds. They come in both fixed and adjustable lengths and are made of steel, rope, or nylon webbing. Most, if not all, have a shock absorber that absorbs up to 80 percent of the arresting force when a fall is being stopped.

Workers on a suspended scaffold may also use harnesses with rope grabs and retractable lifelines, shown in *Figure 6*. The lifeline is secured to a point independent of the scaffold platform. The rope grab links the lifeline to the harness. The grab has a ratchet that locks in case of a fall.

Vertical lifelines (*Figure 7*) are suspended vertically from a fixed anchor point. A fall arrest device such as a rope grab is attached to the lifeline. Vertical lifelines must have a minimum breaking strength of 5,000 pounds.

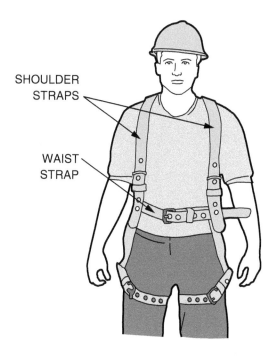

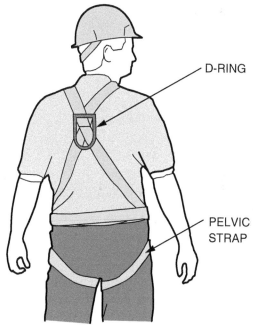

Figure 4 Personal fall arrest equipment.

Figure 5 Lanyard with shock absorber.

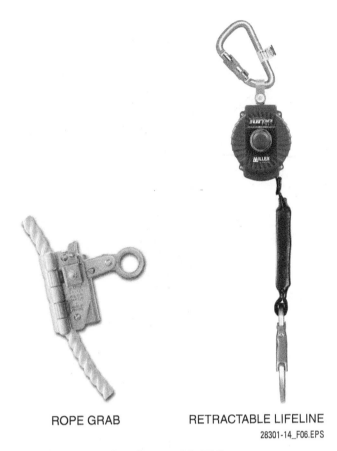

Figure 6 Rope grab and retractable lifeline.

Figure 7 Vertical lifeline.

Horizontal lifelines (*Figure 8*) are connected horizontally between two fixed anchor points. These lifelines must be designed, installed, and used under the supervision of a qualified, competent person. Horizontal lines must be able to support a minimum tensile load of 5,000 pounds per person attached to the line. Vertical and horizontal lifelines must not be used by more than one worker at the same time.

Anchor points, commonly called tie-off points, support the entire weight of the fall arrest system. The anchor point must be capable of supporting 5,000 pounds for each worker attached. Eyebolts (*Figure 9*) and overhead beams are considered anchor points to which fall arrest systems are attached.

The D-rings, buckles, snap hooks, and carabiners (*Figure 10*) that fasten and/or connect the parts of a personal fall arrest system are called connectors. There are regulations that specify how they are to be made. D-rings and snap hooks are required to have a minimum tensile strength of 5,000 pounds.

> **NOTE**
> Since January 1, 1998, only locking-type snap hooks are permitted for use in personal fall arrest systems.

Like all safety equipment, lanyards, lifelines, and safety harnesses should be carefully inspected before each use. Look for worn or frayed areas and check for metal fatigue. Do not use the equipment if you find any damage. Replace any cabling that has more than 10 percent of the total wires frayed or broken in any running foot. Only use equipment that meets or exceeds minimum OSHA standards.

1.2.2 Falling Objects

According to OSHA, falling objects killed 78 construction workers in 2012—10 percent of the total number of construction workers killed that year. In 2002, OSHA added two new rules to protect workers in high-rise construction from being struck by falling objects. First, all materials, equipment, and tools that are not being used must be secured in order to prevent them from being accidentally knocked or bumped off the platform. Second, the controlling contractor must bar other

Figure 8 Horizontal lifeline.

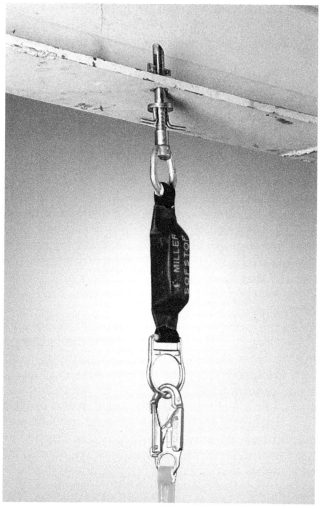

Figure 9 Push-through eyebolt and shock absorber.

Figure 10 Locking snap hook and carabiner.

construction processes below steel erection unless overhead protection is provided for the employees working below.

Debris netting, also called containment netting, is used on elevated construction sites to prevent dropped tools, materials, particulate waste such as dirt, and small items such as nails, fasteners, and screws from hitting people below. OSHA requires the installation of debris netting when a pile of materials or equipment on a scaffold is higher than the top edge of a standard toeboard. In such cases, debris netting is installed from the top rail to the deck level so that it completely encases the work area like a protective curtain (*Figure 11*).

Typical debris netting is made from a fine mesh of heavy-duty, fire-retardant polyethylene plastic, and comes in rolls in a variety of widths and lengths depending on the application. OSHA requires that debris netting maintain flexibility in extreme temperatures. Debris netting is not designed to serve as fall protection.

Keep your work area safe. Prevent injury from falling objects by following these guidelines:

- Always wear a hard hat.
- Keep openings in floors covered, secured, and properly marked.

It Really Works

A mason was moving brick on a roof. The mason was standing on a metal roof panel when his right foot slipped. In an attempt to correct himself, the mason caught his left foot in a corrugation rib. The mason then fell forward to his knees on an underlying bottom insulation sheet and broke through the layer of ceiling drywall. After a short free fall, the clutch mechanism of his retractable lanyard engaged, gently slowing the worker's descent to a complete stop.

The Bottom Line: Personal fall protection equipment can save your life.

- Do not store materials other, than masonry and mortar, within 4 feet of the working edges of a guardrail system.
- Keep the working area clear by removing excess mortar, broken or scattered masonry units, and all other material and debris on a regular basis.
- Never work or walk under loads that are being hoisted by a crane.
- Erect toeboards or guardrail systems. When guardrail systems are used to prevent material from falling from one level to another, any openings must be small enough to prevent the passage of potential falling objects.
- Erect paneling or screening from the walking/working surface or toeboard to the upper edge of the top guardrail or midrail if tools, equipment, or materials are piled higher than the top edge of the toeboard.

- Raise and lower tools and material with a rope and bucket or other lifting device. Never throw tools or material to or from a raised surface.

1.2.3 Personnel Lifts

Many contractors switch to a temporary stairway above the fourth cut, or level, of scaffolding. Continue the temporary stairway upward as the work progresses. It must be maintained in serviceable condition until at least one permanent stairway has been completed.

Stairways must be adequately lighted. If temporary stairways are required, they should be adequately braced, wide enough for two persons, and equipped with railings and toeboards. Ramps or runways used in place of stairways should also have railings. Stairways and ramps should be kept free of ice, snow, grease, mud, and other slipping hazards.

As the work level becomes higher, ladders and stairways are discarded for personnel lifts or hoists. Lifts should be plumb, securely braced, and enclosed their full height with expanded metal or wire mesh (*Figure 12*). The doors or gates should be 6 feet tall. They should lock and unlock only from the inside. The locking mechanism should operate only when the cage is stopped at the landing level. Hoisting equipment should be thoroughly inspected each day and tested whenever subject to a new use.

Personnel lifts must be equipped with guardrails, toeboards, and hand controls that operate from the ground or the platform. Only persons riding the lift should be able to lock the doors. The lifts must also have safety brakes to prevent free fall if the cabling fails.

OSHA requires that personnel lifts be given a trial lift and proof testing before they are used to lift people. This is accomplished by loading the lift to 125 percent of capacity, operating it, and inspecting the cabling. The lift capacity must be posted so that it will not be overloaded by accident.

All personnel lift towers (*Figure 13*) that are outside the structure must be enclosed their full height on the side or sides used for entrance and exit to the structure. At the lowest landing, the enclosure on the sides not used for exit or entrance to the structure should reach a height of at least 10 feet. Other sides of the tower structure adjacent to floors or scaffold platforms should be enclosed to a height of 10 feet above the level of such floors or scaffolds. Towers inside structures should be enclosed on all four sides throughout their full height.

Figure 11 Debris netting installed on scaffold during a renovation project.

Figure 12 Personnel lift.

Figure 13 Personnel lift tower.

Typically, materials hoists do not have the safety features of personnel lifts. On some jobs, however, personnel hoists do double duty because they have enclosed cabs, internal controls, and safety brakes. Be sure that materials hoists are clearly marked so that everyone knows that they do not have the safety equipment required for carrying workers.

Although using cranes to lift people was common in the past, OSHA regulations, as spelled out in 29 *CFR* 1926.550, now discourage the practice. Using a crane to lift personnel is not specifically prohibited by OSHA, but the restrictions are such that it is only permitted in special situations where no other method is suitable. When it is allowed, certain controls must be in place:

- The rope design factor is doubled.
- The lift capacity is cut in half.
- Free falling is prohibited.
- Devices are required that provide warnings and prevent the lower load block (pulley) or hook from coming into contact with the upper load block, boom point, or boom-point machinery, likely causing the failure of the rope or release of the load or hook block.
- The platform must be specifically designed for lifting personnel.
- Before the lift is used, it must be tested with a comparable weight and then inspected.
- Every intended use must undergo a trial run with weights rather than people.

1.2.4 Access Zones

To prevent injury on any project where an elevated masonry wall is to be constructed, the contractor must establish controlled access zones and limited access zones before construction can begin. This section reviews the purpose of each of these access zones, and how to establish and maintain them.

A controlled access zone, or CAZ, is a designated work area in which certain types of masonry work may take place without the use of conventional fall protection systems such as guardrails, personal fall arrest systems, or safety nets. Controlled access zones are established prior to the beginning of construction. They are created and clearly marked to limit entrance to authorized workers only. OSHA guidelines specify that if there are no guardrails, masons are the only workers allowed in the CAZ. The height of the CAZ is equal to the height of the wall plus 4 feet. If the wall is higher than 8 feet, the CAZ must remain defined until the wall is braced.

The CAZ is marked on the side of the wall without the scaffold by a rope, wire, or tape control line that runs along the entire length of the unprotected edge, and is connected to each side of the guardrail or wall. The purpose of the control line is to restrict access to workers who are not authorized to work there. OSHA also requires that supporting stanchions must be clearly flagged or marked at maximum intervals of 6 feet. The stanchion flagging must be rigged to ensure that the lowest point is not less than 39 inches and not more than 45 inches from the working surface. The control line must be able to withstand at least 200 pounds of force applied to it.

For block and brick wall construction, the control line is erected a minimum of 6 feet and a maximum of 25 feet from the unprotected edge. For precast concrete wall construction, the control line should be erected a minimum of 6 feet and a maximum of 60 feet, or half the length of the precast member, whichever is less, from the edge. If overhand bricklaying techniques are used to construct the wall, the control line must be a minimum of 10 feet and a maximum of 15 feet from the edge.

On floors and roofs where guardrail systems are not in place before the start of overhand bricklaying operations, controlled access zones should be enlarged to enclose all points of access, materials handling areas, and storage areas. On floors and roofs where guardrail systems are in place but need to be removed to allow overhand bricklaying work or leading-edge work, remove only that portion of the guardrail necessary to accomplish that day's work.

A limited access zone (LAZ) marks a restricted area alongside a masonry wall that is under construction. Like the CAZ, the LAZ is established before construction begins. The LAZ is equal to the height of the wall plus 4 feet, for the entire length of the wall on the side without scaffold. If the maximum height of the wall is 8 feet, the LAZ must remain in place until the wall has been adequately supported. If the wall is higher than 8 feet, the wall must be braced until permanently supported.

Barricades are used to mark off the LAZ. Only employees who are working on the construction of the wall are permitted access to the LAZ.

Additional Resources

Fall Protection and Scaffolding Safety: An Illustrated Guide. 2000. Grace Drennan Ganget. Government Institutes.

"Online Safety Library: Scaffold Safety." Oklahoma State University. **www.ehs.okstate.edu**

"Scaffolding." OSHA. **www.osha.gov**

1.0.0 Section Review

1. To reduce the risk of fire in a building, use a noncombustible covering to protect materials stored in the building as well as those stored within _____.

   a. 40 feet of the building
   b. 30 feet of the building
   c. 20 feet of the building
   d. 10 feet of the building

2. OSHA requires that supporting stanchions in a controlled access zone be clearly flagged or marked at maximum intervals of _____.

   a. 3 feet
   b. 6 feet
   c. 9 feet
   d. 12 feet

Section Two

2.0.0 Concrete Masonry Wall Bracing

Objective

Describe how to properly brace a wall.
a. Describe how to properly brace a concrete masonry wall for wind.
b. Describe how to properly brace a wall for backfill.

Performance Task

Properly brace a wall.

All buildings are constructed from the ground level up. Concrete masonry walls are not often designed to be freestanding, self-supporting systems. In fact, both during construction and after the structure has been completed, the stability of the structure must be maintained. During construction, and immediately after a wall section has been laid up, you must consider the need for appropriate bracing while the mortar gains its final set. You must also maintain sufficient support for intersecting walls by applying the proper anchors or ties between walls, and plan the use of anchors to tie other building materials, such as joists, to the masonry. Otherwise, the wall will collapse, potentially causing injury and death as well as damage and delays.

Wall supports for masonry walls may be in the form of vertical columns of masonry units that are directly tied into the structure itself and located at various points along the length of the wall. These vertical columns, called pilasters, increase the base area of the structure and add bonding area along their vertical distance. After the design, the mason's skill is a major factor in determining the structure's strength, durability, safety, and appearance.

In addition to the static weight of the wall, which is called the dead load, wind speed must also be taken into consideration when building a wall, and bracing provided to ensure that the wall under construction is protected from being toppled by the wind. Once the wall has been attached to bracing, wind speeds in excess of 35 miles per hour require the evacuation of the immediate work area. Leave bracing in place until the wall's final lateral support has been provided in accordance with the design.

During construction, the maximum unbraced and braced heights of a masonry wall above grade are calculated in terms of the relationship between the nominal wall thickness and the density of the masonry unit. *Figure 14* shows how to visualize the maximum height above the base or highest line of lateral support that is allowed for walls of various thickness during the construction of the wall and prior to bracing. *Figure 15* shows how to visualize the typical vertical and horizontal brace-spacing requirements for an unsupported wall.

In the past, local building codes set limits for unsupported, nonreinforced concrete walls according to the thickness and the height of the unsupported wall. Often, these regulations specified various heights and thicknesses without considering the total design features or other available technical data. Newer building codes take a number of features into consideration, including overall design, flexural stresses, shear stresses, and wind loads. Therefore, more realistic estimates of unsupported wall heights or lengths are available to the designer and mason. *Table 1* provides guidelines for lateral support requirements for nonreinforced concrete masonry walls (refer to *Figure 14*).

Lateral support may be provided by vertical supports such as intersecting walls, pilasters, or columns. Horizontal support may be given by floors, roofs, or beams. The distance between lateral supports depends on the following:

- The type of construction (reinforced, partially reinforced, and nonreinforced)
- Wall thickness
- The degree of lateral and/or vertical loading expected from outside and inside the structure

Allowing for code restrictions for the placement of supports is the job of the architect or designer. However, you must be certain that each feature is laid up with the proper technique to ensure that the structure meets the code requirements.

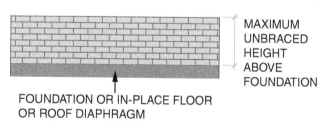

Figure 14 Maximum height of unbraced wall.

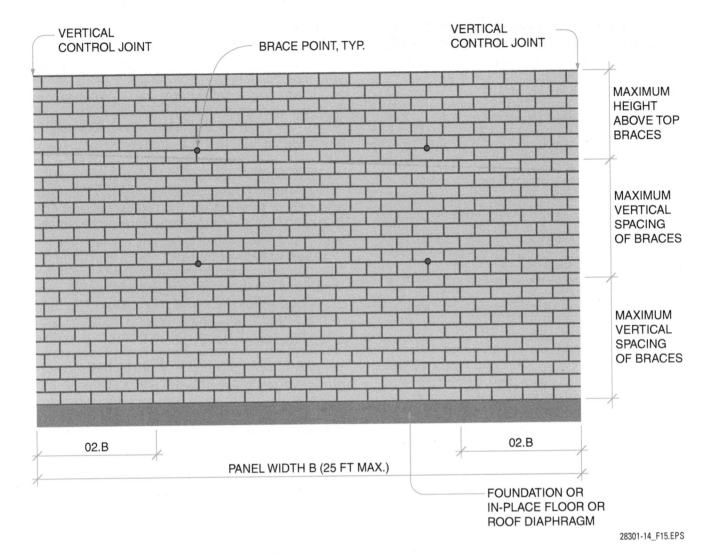

Figure 15 Brace-spacing requirements for a wall.

2.1.0 Bracing for Wind

Most local building codes and OSHA requirements insist that temporary vertical bracing or shoring be used during construction to provide adequate support for uncured wall sections. Bracing is required primarily because of the force of wind.

Wind produces lateral forces on a wall that tend to overturn the wall. These forces produce a great deal of stress near the base of a wall and must be counteracted by bracing or shoring. A newly laid wall has very little internal strength and must be supported until the mortar cures and is self-supporting. To take care of this problem, temporary bracing similar to that shown in *Figure 16* should be provided.

Generally, masonry walls over 8 feet high should be braced. The supports should be no more than 20 feet apart. The OSHA regulations for wall bracing are covered in 29 *CFR* 1926.706(b), which states: "All masonry walls over eight feet in height shall be adequately braced to prevent overturning and to prevent collapse unless the wall is adequately supported so that it will not overturn or collapse. The bracing shall remain in place until permanent supporting elements of the structure are in place."

> **NOTE**
> In some areas, the foundation must withstand frost heaving or earthquakes. Check local codes for specific requirements for foundations.

Federal OSHA requirements do not define what constitutes adequate bracing. State or local building codes may have specific rules. Always check your local building codes.

The Masonry Contractors Association of America developed two standards for wall bracing to clarify the OSHA requirements. *Standard Practice for Bracing Masonry Walls under Construction* was developed to provide a detailed definition of adequate bracing for masonry walls. The *Masonry Wallbracing Design Handbook* contains over 200 diagrams of wall bracing for specific conditions.

2.2.0 Bracing for Backfill

In the case of masonry basement or foundation walls, the lateral pressure created by backfilled earth can also damage or overturn the structure. As with wind bracing, you can use bracing techniques and common trade skills to avoid such damage.

Beyond the use of temporary bracing of such basement walls, avoid backfilling until the first-floor construction is in place. You must also avoid backfilling with wet material or using water when compacting the backfill materials. During the process of backfilling, avoid the impact stresses created by backfill materials sliding down steep slopes directly against the structure.

Temporary bracing of wood planks or steel members should be placed diagonally against the wall at different horizontal distances according to the bracing system used and the expected lateral forces. This arrangement is shown in *Figure 17*.

The braced wall section should be allowed to cure for a period of three to seven days under satisfactory weather conditions. Even more important, bracing should be used during times when construction is postponed because of weather conditions. For instance, in areas where heavy rains occur, bracing may prevent wall failure as water, mud, and backfill exert hydrostatic pressure against the wall.

Table 1 Lateral Support Requirements for Non-Reinforced Concrete Masonry Walls

Standard Building Code Table 2105.1 Lateral Support (h/t)* Ratios For Exterior Bearing And Nonbearing Walls[1]				
	Design wind pressure (psf)[2]			
Wall construction[3]	15	20	25	30
Grouted, solid, or filled-cell masonry	26	22	20	18
Hollow masonry or masonry-bonded hollow walls	23	20	18	16
Cavity walls[4]	20	18	16	15

* h = clear height or length between lateral supports
t = nominal wall thickness

1. h/t ratios required for wind pressures greater than 30 psf must be determined by an engineering analysis in accordance with the Masonry Standards Joint Committee (MSJC) *Building Code Requirements for Masonry Structures*, ACI 530/ASCE 6/TMS 402, as prescribed in Standard Building Code Section 2103.6.

2. All masonry units shall be laid in Type M, S, or N mortar unless otherwise required (see *Standard Building Code* Table 2102.9). Where Type N mortar is used and the wall spans in the vertical direction, the ratios shall be reduced by 10%.

3. These wind pressures include shape factors from *Standard Building Code* Section 1205.

4. In computing the h/t ratio for cavity walls, t shall be the sum of the nominal thicknesses of the inner and outer wythes.

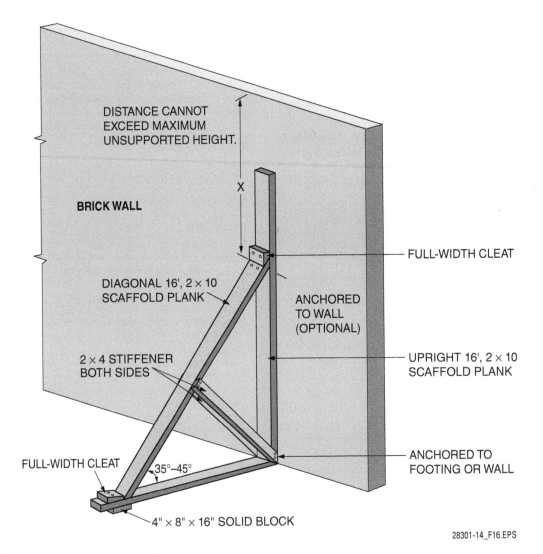

Figure 16 Wind bracing for masonry wall.

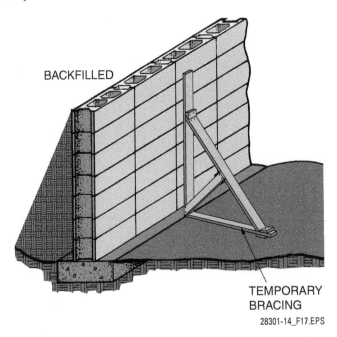

Figure 17 Temporary lateral bracing for a foundation wall.

Additional Resources

Bricklaying: Brick and Block Masonry. 1988. Brick Industry Association. Orlando, FL: Harcourt Brace & Company.

Concrete Masonry Handbook: for Architects, Engineers, Builders, Fifth Edition. 1991. W. C. Panerese, S. K. Kosmatka, and F. A. Randall, Jr. Skokie, IL: Portland Cement Association.

Masonry Wallbracing Design Handbook, Latest Edition. Rashod R. Johnson and Daniel S. Zechmeister. Algonquin, IL: Masonry Contractors Association of America.

Standard Practice for Bracing Masonry Walls under Construction. 2012. Algonquin, IL: Masonry Contractors Association of America.

2.0.0 Section Review

1. In general, masonry walls should be braced if their height exceeds _____.
    a. 8 feet
    b. 10 feet
    c. 12 feet
    d. 14 feet

2. Masonry basement and foundation walls should be braced to protect them against the lateral pressure created by _____.
    a. wind
    b. flooding
    c. adjacent construction
    d. backfilled earth

SECTION THREE

3.0.0 ELEVATED MASONRY SYSTEMS

Objective

Describe elevated masonry systems.
  a. List the construction sequence for elevated masonry systems.
  b. Describe how elevated masonry systems are designed.
  c. Identify common exterior walls used for elevated masonry systems.
  d. Identify common interior walls used for elevated masonry systems.

Trade Terms

Guyed derrick: An apparatus used for hoisting on high-rise buildings, consisting of a boom mounted on a column or mast that is held at the head by fixed-length supporting ropes or guys.

Lateral stress: Wind shear and other forces applying horizontal pressure to a wall or other structural unit.

Reglet: A narrow molding used to separate two structural elements, usually roof and wall, to divert water.

The growing popularity of high-rise masonry construction is a result of advances in masonry reinforcement, engineering, and materials. For the mason, techniques of high-rise construction require constant attention to detail and safety.

3.1.0 Identifying and Following Construction Sequences

All building construction requires substantial planning, but this is especially true with high-rise construction. The architect and engineer will plan and make design choices. The contractor will plan and be responsible for the method of construction.

There are several ways to construct a high-rise building. The most common method of high-rise construction is tier by tier. Each tier represents a vertical column height of two or three building floors. Another method involves erecting the structure in segments or bays. A bay is the distance between loadbearing supports such as beams or columns. This method is typically used on buildings whose height can be reached by ground-based lifting equipment.

Frequently, a combination of these methods is employed. The first set of horizontal or vertical segments is erected by ground-based mobile equipment. A tower crane is then erected to complete the lifting process. Most tall buildings require one or more tower cranes, as shown in *Figure 18*. The crane is raised as the tiers are finished.

For most tall buildings, the first step is setting the foundation. The second step is raising a skeleton frame of steel or concrete. Fill-in construction proceeds upward by tiers along the frame, typically in horizontal segments or bays around the building core. Masonry work may proceed by segments, panels, or floors.

The permanent structural floor does not have to be finished before construction proceeds upward. However, it must be covered over as soon as possible except for necessary openings. OSHA requires guardrails (*Figure 19*) wherever there is a danger of falling through an opening. The top guardrail is required to withstand a weight of 250 pounds and the middle rail must be able to withstand a weight of 200 pounds. The top rail should be between 38 inches and 42 inches from the bottom, with the middle rail positioned halfway between the top rail and the bottom surface. OSHA regulations also prohibit more than four uncovered floors above the highest permanently finished floor, to prevent injury from falling objects.

When construction has progressed to a height 60 feet above grade, OSHA requires that per-

Figure 18 High-rise building under construction.

manent ladders must be replaced by at least one temporary stairway (*Figure 20*). The temporary stairway must be continued upward as work progresses. It must be maintained in a serviceable condition until a permanent stairway has been completed.

An important factor affecting materials lifting, as well as the safety of all those on the job, is the loadbearing capacity of the unfinished structure. During construction, the partially completed structure must be able to support itself, any construction materials, and the added load of the lifting equipment. Structural engineers are responsible for determining the load capacity of the structure. Special bracing may be required during construction. This bracing will be removed after the structure is complete.

Call for Reinforcements

Temporary supports are often needed during high-rise construction. In this instance, shoring is in place for concrete floor slabs in a parking garage. The shoring will be removed when the concrete floor is set.

Figure 19 Guardrails.

Construction personnel must not overload the structure and must not place too much material at any given point, to avoid floor failure. Working platforms are typically designed to support 50 pounds per square foot (psf). If the anticipated load exceeds 50 psf, the contractor must take appropriate steps to ensure that the deck will carry the load.

The erection abilities of cranes and derricks are sometimes restricted by the design of the structure. A tower crane or guyed derrick operates in a circular pattern. The total area it covers depends on its boom reach. If the structure has a square or rectangular footprint, only one piece of lifting equipment may be required. However, if the building is L-shaped, additional cranes may be required to reach all points on any floor.

Because of the need to move materials upward, planning for special erection aids is an important part of constructing tall buildings. Hitches or supports for attaching hoisting lines should be added to heavy elements to be lifted. These typically include columns, beams, or panels that are not easily adapted to the use of slings. Other erection aids include special plates or angle clips. In steel frame construction, these are added in the fabricating shop to the top and bottom of welded columns. When the column is erected, these aids give it stability until the floor framing is added.

Figure 20 Temporary stairway.

3.2.0 Understanding Building Design

High-rise masonry buildings usually have frames of reinforced concrete, steel, or a combination of these materials. *Figure 21* shows the construction of a steel frame for a high-rise building. These buildings can incorporate several different types of masonry walls.

The critical element in this type of masonry design is not loadbearing so much as resistance to lateral stress. Masonry walls carry little or no loads beyond their own weight. They act as load-transfer agents and displace wind loads onto the structural frame of the building. Lateral support is critical. It is provided for exterior walls by other structural elements of the building, angle irons, anchors, and other braces. The masonry can be supported laterally by intersecting walls and floors. The roof slab also gives lateral support to masonry walls.

The *International Building Code®* bases lateral support requirements on (wind) design pressure. This is one example of code requirements. Minimum required wall thickness varies among the different codes. Lateral support can be provided by cross walls, columns, pilasters, or by buttresses where the limiting span is measured vertically.

Figure 21 Steel frame for high-rise building.

> **WARNING!** Failure to adequately provide support for a wall both during and after laying up the structure can cause wall failure, which can result in death and injury to fellow workers, property damage, construction delays, and financial losses.

Anchorage between walls and supports must be able to resist wind loads and other lateral forces acting either inward or outward. All lateral support members must have sufficient strength and stability to transfer these lateral forces to adjacent structural members or to the foundation. All of the codes contain provisions stating that specific limitations may be waived if engineering analysis is provided to justify additional height or width, or reduced thickness. Such waivers must be documented on drawings stamped by the engineer of record.

3.3.0 Constructing Exterior Walls

Solid masonry walls are typically used in foundations and residential buildings, while cavity and veneer walls are often found in high-rise construction. Two types of veneer walls are only used in high-rise construction: panel and curtain walls. Both are special veneer walls that do not support any load except their own weight. They are attached to a structural frame.

In a high-rise building, these frames are usually made of steel or concrete beams and columns. Panel and curtain walls provide weather protection, color, texture, and architectural interest to the structure (*Figure 22*). They require flashing, weepholes, and other moisture containment measures.

In addition to curtain and panel walls, parapet walls are also common in high-rise construction. The construction of these three types of structures is discussed in the following sections.

3.3.1 Panel Walls

Panel walls are exterior nonbearing walls wholly supported at each floor by a concrete slab or beam, or by steel shelf angles. Detail features of a panel-wall support using a steel shelf angle are shown in *Figure 23*. On multistory buildings, masonry veneer constructed between relieving angles and masonry infill between floors and columns of structural frames are also considered panel-wall sections.

Each panel must be made to resist lateral forces and transfer the load to adjacent structural members. Panel walls can be prefabricated masonry units, or they may be job-built. In addition to be-

Figure 22 Curtain wall under construction.

ing supported along the bottom edge, panels are also attached to the structure by anchors or clips, much like stone veneer panels. *Figure 24* shows how a stone panel is attached to the steel frame.

Since they are supported at each floor, panel walls require some type of pressure-relieving joint. This is usually located between the shelf angle and the top course of the panel below the shelf angle. The joint is usually filled with a neoprene material that allows for expansion and contraction of the panel while limiting moisture penetration.

All shelf angles must be installed with flexible joints to allow for different expansion and contraction rates of the panel wall and the support structure. The flexible anchors attached to the back of the panel permit the different expansion rates to occur without cracking the structure.

The Ingalls Building

The first reinforced-concrete high-rise office building is the 16-story Ingalls building. It was completed in 1903 in Cincinnati, Ohio. Before that the tallest concrete building was only six stories. The Ingalls building is designed to act as a monolithic unit. Each floor slab provides a rigid diaphragm to steady the building from the wind. The building stands 210 feet high and is still in use today.

28301-14 Elevated Masonry

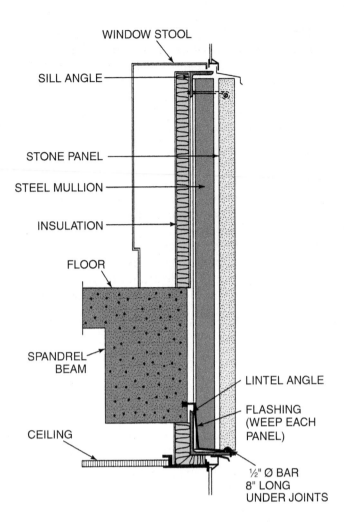

Figure 23 Panel-wall mounting.

The flashing should be brought beyond the face of the wall and turned down to form a drip. A cavity filter or mortar net may also be installed to prevent mortar from clogging the drainage area. Traditionally, a soft sealant is placed below the flashing to prevent the water from re-entering the joint. This configuration is shown in *Figure 25*. The sealant can also be placed above the flashing. This forces water to drain through the weepholes and eliminates additional drainage below the brick.

3.3.2 Curtain Walls

masonry veneer wall only used high-rise const.

Curtain walls are exterior nonbearing walls designed to span horizontally or vertically between lateral connections without intermediate support. Horizontal curtain walls span across column faces and intersecting interior walls, and are connected to them in order to transfer wind loads to the structure. Because the weight of multistory curtain walls is wholly supported by the foundation, they are built from the foundation to the roof. They

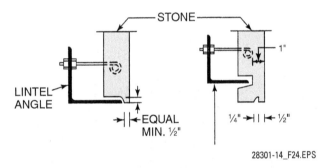

Figure 24 Stone-panel wall attachment.

are connected only at the floors and roof for lateral load transfer, without intermediate shelf angles.

Masonry curtain walls are designed by methods that rely on, or are derived from, observation or experiment. Such methods are called empirical methods. Curtain walls are also designed by engineering analysis. Regardless of the method used, masonry curtain walls are designed to span multiple structural bays. Curtain walls may be single-wythe or multiwythe design, and may incorporate reinforcing steel to increase their lateral load resistance or the distance between their lateral supports.

Curtain walls are tied to concrete or steel frames. Temperature changes will cause the masonry curtain to expand and contract at a different rate than the attached frame. Therefore, a curtain wall must be tied to the frame with flexible anchors made of galvanized steel or some other noncorrosive metal. *Figure 26* shows details of a block curtain wall anchored to a concrete frame and a steel frame. In *Figure 26A*, the anchor is embedded in the reinforced concrete beam. In *Figure 26B*, the anchor is welded to a steel beam. In both cases, the block core is grouted to hold the anchor to the block.

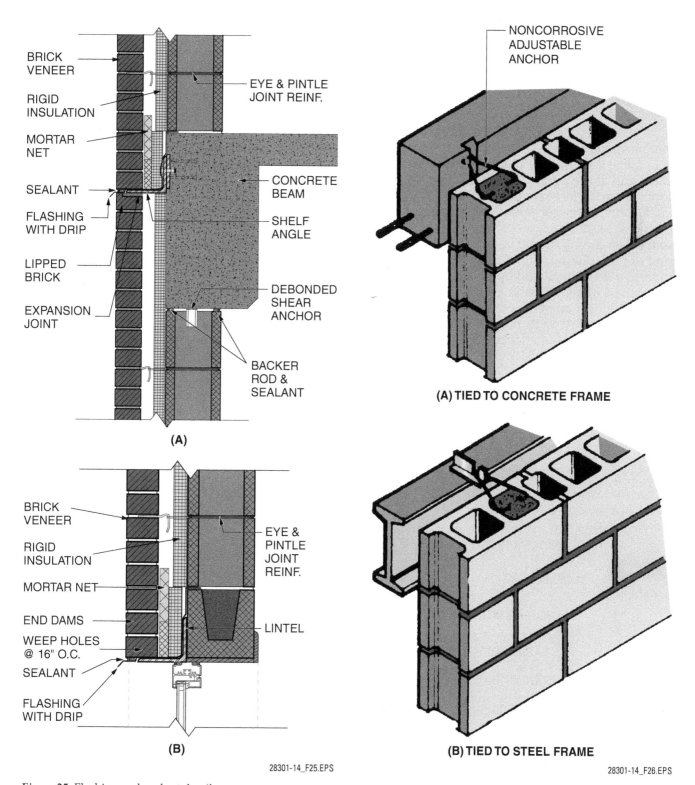

Figure 25 Flashing and sealant detail.

Figure 26 Curtain wall anchorage.

Curtain walls require flashing at the top and bottom of the walls, as well as above and below any wall openings.

3.3.3 Parapet Walls

The parapet is that part of a wall that extends above the roofline. A parapet may be added for architectural interest or it may be functional. For example, it may be used to support swing scaffolding or other equipment for window washing. Because the roof does not protect it, the parapet is exposed to the weather from both the front and the back (*Figure 27*). Wind creates increased lateral stress at the unsupported top of the parapet. Because of this exposure, it is subject to greater thermal movement.

Possible leak lines for parapets are at the coping joints and at the interface with the roofing. Metal copings are the most common because they can be made with the fewest joints. Metal copings should extend at least 2 inches below the top of the masonry. The coping should be sloped to drain to the roof of the building. Coping legs should turn out to form a drip shield and be caulked with a high-performance caulk. Flashing should be laid under the coping, and all items penetrating the flashing should be sealed with mastic. Flashing should stop inside the face shell. Flashing also provides a slip plane for differential movement.

Copings made from precast concrete, stone, and terra-cotta are commonly used on taller parapets designed to support window-washing equipment. These parapets (*Figure 28*) are structurally reinforced and anchored to the roof slab. The precast coping is flashed, sealed, and typically anchored to the wall.

Terra-cotta or other masonry coping must be carefully mortared to avoid leaking joints. The coping should overhang both sides of the wall and have integral drip notches. Because masonry coping joints are not impervious to water, through-wall flashing must be installed underneath, as shown in *Figure 29*. The coping joints should be raked out while the mortar is still plastic, then filled with elastomeric sealant. Even hairline cracks or separations at the top of the wall act as funnels for water to reach the interior of the wall.

The joint between the roof flashing and the parapet is the place on the parapet where the greatest leakage can occur. Roof flashing must be turned up onto the back face of the parapet wall and terminated above the level of the roof deck. A reglet is usually specified for this upper joint. To avoid disturbing the joint, the reglet is installed in two pieces. The mason installs the upper piece, and the roofer installs the lower piece.

In addition to the reglet over the upper edge, through-wall masonry flashing is also typically specified at the roof terminus. Where the ma-

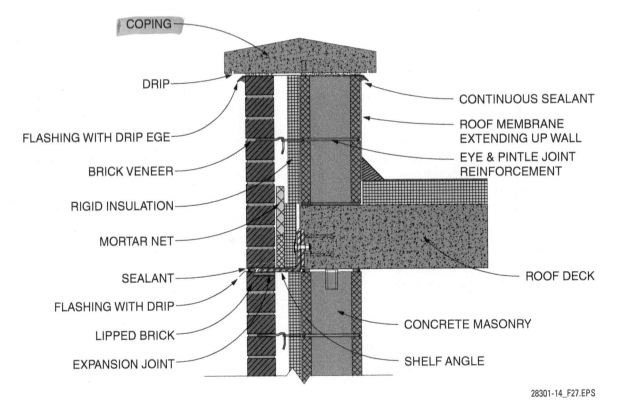

Figure 27 Parapet.

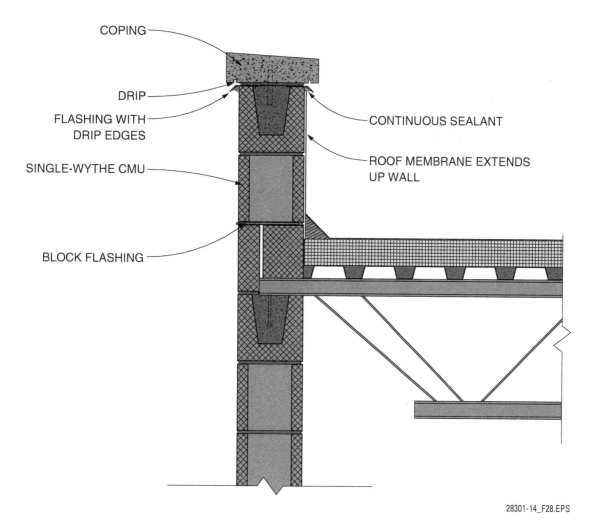

Figure 28 Precast coping.

sonry and roof slab must move independently, the specifications should call for a flexible flashing connection or a roof-edge expansion joint. The flashing provides the slip plane for differential movement between the two elements.

Block parapet walls require full mortar joints even though the wall below the roofline may have face-shell bedded joints. The flashing must be fully embedded in mortar; copings also require a full mortar bedding. The roofing material is turned up to make a flashing behind the parapet.

3.4.0 Constructing Interior Walls

Interior walls are usually partition walls. They are nonbearing walls that support only their own weight. They are used as room separators, shaft enclosures, and barriers to fire, sound, or smoke. Partition walls can be single-wythe block, tile, or brick. These materials are commonly used for stairwell and elevator-shaft enclosures, as well as room dividers. Cavity walls are used for partitions that must carry concentrated utility conduit and piping, as the cavity allows easy placement of the mechanical items.

Based on the *International Building Code®*, the ratio of height to thickness should be 36 to 1 for partition walls. *Tables 2* and *3* show the maximum wall height-to-thickness ratios. The partition must be securely anchored against lateral movement at the floor or ceiling. There is no requirement for intermediate pilasters or other lateral reinforcement inside these dimensions.

Single-wythe hollow brick, CMU, and vertical-cell tile partitions can be internally reinforced to stand over longer distances if there are no cross walls or projecting pilasters. Internal pilasters can provide the needed reinforcement to increase partition lengths. For block, a continuous vertical core can be reinforced with deformed steel bars and then grouted solid. Double-wythe cavity walls can be similarly reinforced without thickening the wall section.

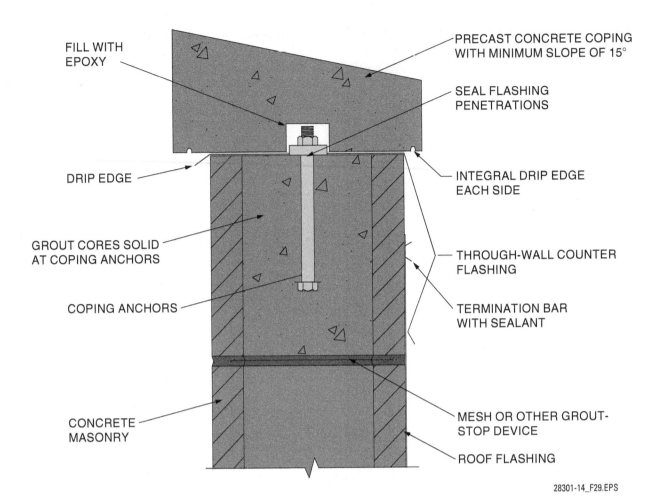

Figure 29 Through-wall flashing.

Table 2 Maximum Wall Length- or Height-to-Thickness Ratio

Construction	Maximum wall length-to thickness or height-to-thickness ratio*
Bearing walls	
Solid or solid grouted	20
	18
Nonbearing walls	
Exterior	18
Interior	36
Cantilever Walls**	
Solid	6
Hollow	4
Parapets (8" thick min.)**	3

* Ratios are determined using nominal dimensions. For multiwythe walls where wythes are bonded by masonry headers, the thickness is the nominal wall thickness. When multiwythe walls are bonded by metal wall ties, the thickness is taken as the sum of the wythe thickness.

** The ratios are maximum height-to-thickness and do not limit wall length.

Table 3 Maximum Wall Spans

Construction	Wall Thickness, (inches)			
	6	8	10	12
Bearing walls				
Solid or solid grouted	10*	13.3	16.6	20
All other	9*	12	15	18
Nonbearing walls				
Exterior	9	12	15	18
Interior	18	24	30	36
Cantilever Walls**				
Solid	3	4	5	6
Hollow	2	2.6	3.3	4
Parapets**	1.5	2	2.5	3

* 6-in. thick bearing walls are limited to one story in height.

** For these cases, spans are maximum wall heights.

Structural clay tile or glazed CMU is often used for partitioning in schools, hospitals, food-processing plants, sports facilities, airports, and public facilities. These provide a low-maintenance, high-durability surface. The partitions may be constructed of glazed materials. Double-wythe walls can provide different colors and finishes on each side. Lateral support spacing is governed by the same height-to-thickness ratio as for brick or block.

Additional Resources

"Materials Handling and Storage." OSHA. www.osha.gov

"Scaffolding." OSHA. www.osha.gov

WorkSAFE masonry safety resources. www.worksafecenter.com

3.0.0 Section Review

1. To prevent injury caused by falling objects, OSHA limits the number of uncovered floors above the highest permanently finished floor to _____.

   a. two
   b. four
   c. six
   d. eight

2. The critical element in high-rise masonry structural design is resistance to _____.

   a. lateral stress
   b. loadbearing stress
   c. torsional stress
   d. compressive stress

3. The part of a wall that extends above the roofline is called the _____.

   a. stud
   b. header
   c. parapet
   d. trimmer

4. The *International Building Code®* specifies the ratio of height to thickness for partition walls as _____.

   a. 24 to 1
   b. 36 to 1
   c. 48 to 1
   d. 60 to 1

Section Four

4.0.0 Materials Handling

[handwritten annotations: materials movement, greatest safety hazard high-rise]

Objective

Describe how to properly handle materials at elevations.

a. Explain safety precautions to be observed when working around cranes.
b. Explain safety precautions to be observed when working around materials hoists.
c. Explain safety precautions to be observed when moving and stocking materials.
d. Explain safety precautions to be observed when working at elevated workstations.
e. Explain how disposal chutes and waste bins are used when working from elevated workstations.

Performance Task

Demonstrate hand signals used for lifting materials.

In high-rise construction, materials movement poses the greatest safety hazard to workers on the job. To make sure that the job site is as safe as possible, follow these guidelines:

- Check the amount of materials ordered and the places where it will be stored.
- Plan where and when materials will be moved to elevated workstations.
- Establish clear pathways for movement of all materials.
- In advance, determine individual responsibilities for each worker as materials are lifted and transported.
- Arrange a consistent system of signals for alerting workers to materials movement.
- Be prepared to move off scaffolding when supplies are loaded onto the platforms.
- Schedule materials deliveries and crane availability with the general contractor.

Timing Is Everything

Deliveries of materials may be regulated by local laws. For example, deliveries may be prohibited during rush hours. Some communities also prohibit deliveries at night or early in the morning. Check the local ordinances when ordering materials and scheduling deliveries.

4.1.0 Working around Cranes

Cranes and derricks use large, versatile boom-arm mechanisms to lift heavy loads. A derrick has a lift arm pivoted at its base, while a crane lift arm may be moveable or fixed on its vertical axis. Both derricks and cranes operate by motorized cables with hooks for raising and lowering heavy loads of materials. The boom arms can move loads both horizontally and vertically. Several types of cranes are used on large construction projects (*Figure 30*). The most common are as follows:

- Tower cranes stand alongside or in the middle of the building under construction. They are erected on their own foundation. They have one central tower with a boom at the top.
- Mobile cranes are mounted on crawler tracks or truck beds, and move about the job site. The boom is attached to the base and extends at an angle upward and outward.
- Conventional derrick cranes stand away from the building at a distance determined by the length of the boom arm. They are mounted on a fixed base.
- Traveling cranes are placed on scaffolding that is attached to the outside of the building face and moved upwards as the work progresses.

Always be alert to any materials-handling activity going on around you. When working with lifting equipment, remember to stay out of the path of the moving load. Establish and mark materials-movement pathways before hoisting starts. The supervisor should assign each worker responsibility for particular tasks during materials movement. Masons should leave the scaffold when masonry units are being placed on elevated workstations using lifting equipment.

> **WARNING!**
> Cranes are a necessary part of large-scale construction projects, but they pose many hazards. In high-rise construction, materials movement by crane poses the greatest safety hazard to workers on the job.

In addition to being alert at all times, follow these general safety rules when working around cranes:

- Wear a hard hat and safety shoes, and eye protection as needed.
- Keep out from under loads and away from the wheels or tracks of the equipment.
- Never stand between the crane cab and the materials truck.

(A) TOWER

(B) MOBILE

Figure 30 Construction cranes.

- If you must guide a load down, use a guideline and do not get between the load and the crane.
- Before touching a load, hook, or cable, check to see that the boom is not touching or near a power line.
- If you guide a load, be alert for tipping or other movement of the crane, sling, or hook.

- When cranes pivot, so does the back of the crane. Be sure to mark off the appropriate space behind the crane to prevent workers from being hit by the back of the crane when it is pivoting.
- Unless you are assigned to such work, stay away from the crane and the materials-movement path.

Before cranes and derricks are used, they should be tested to ensure that they are capable of handling the required loads. Only certified operators should operate this equipment. The operators should recognize signals only from the designated person supervising the lift. Confused command channels can result in serious injury.

The methods and modes of communication vary widely in mobile crane operations. The method of communication refers to whether the communication is verbal (spoken) or nonverbal. The mode is what is used to facilitate the communication. This can include, for example, a bullhorn, a radio, hand signals, or flags.

4.1.1 Verbal Modes of Communication

Verbal modes of communication vary depending on the requirements of the situation. One of the most common modes of verbal communication used is a portable radio (walkie-talkie). Compact, low-power, inexpensive units enable the crane operator and signal person to communicate verbally. These units are rugged and dependable, and are widely used on construction sites and in industrial plants.

There are some disadvantages to using low-power and inexpensive equipment in an industrial setting. One disadvantage is interference. With low-power units, the frequency used to carry the signal may have many other users. The frequency can become crowded with signals from other units. Another disadvantage is high background noise. In attempting to send a signal in a high-noise area, the person sending the signal may transmit unintended noise, resulting in a garbled, unintelligible signal for the receiver. On the receiving end, the individual may not be able to hear the transmission due to a high level of background noise in the cab of the crane.

There are several solutions to the problems associated with radio use. To overcome the shortcomings associated with low-power units, more expensive units with the ability to program specific frequencies and transmit at a higher power level may be needed. Some of these more expensive units may require licensing. To overcome the background noise problem, the use of an ear-mounted noise-canceling microphone/headphone combination may be required (*Figure 31*).

Another solution is the use of an optional throat microphone (*Figure 32*). This device feeds the transmitted sound directly to the ear and picks up the voice communication from the jawbone at the ear junction. This prevents noise from entering the microphone and blocks out any background noise when listening. To avoid missed communication, the signal person's radio is usually locked in transmit so that the crane operator can tell if the unit is not transmitting. In any event, a feedback method should be established between the signal person and the crane operator so that the signal person knows the crane operator has received the signal.

Another mode of verbal communication is a hardwired system (*Figure 33*). These units overcome some of the disadvantages of radio use. When using this type of system, interference from another unit is unlikely because this system does not use a radio frequency to transmit information. As in a telephone system, occasional interference may be encountered if the wiring is not properly shielded from very strong radio transmissions. A hardwired unit is not very portable or practical when the crane is moved often. These units can also use an ear-mounted noise-canceling microphone/headphone combination to minimize the effects of background noise.

High-power handheld radios designed for use in industrial settings (*Figure 34*) are popular alternatives to communication systems that are worn on the head. They are designed to withstand dust, heat, shock, and immersion in water. They

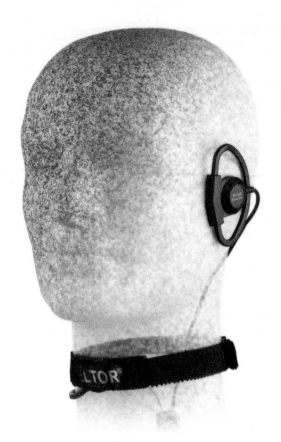

Figure 32 Throat microphone.

Figure 33 Hardwired communications system.

Figure 31 Noise-canceling microphone and headphones attached to a hard hat.

typically feature noise-suppression technology to reduce background sounds when transmitting. Their controls are designed to be used easily with gloves. Modern digital handheld radios feature displays and keypads that allow the user to send and receive text messages over the radio frequency.

4.1.2 Nonverbal Modes of Communication

Nonverbal modes of communication can vary tremendously. However, this is the most common type of communication used when performing crane operations. Several modes are available for use under this method. One mode is the use of signal flags, which may mean different-colored flags or a specific positioning of the flags to communicate the desired message. Another mode is the use of sirens, buzzers, and whistles in which the number of repetitions and duration of the sound convey the message. The disadvantage of these two modes is that there is no established meaning to any of the distinct signals unless they are pre-arranged between the sender and receiver. When sirens, buzzers, and whistles are used, background noise levels can be a problem.

The most commonly visual communication method is the set of hand signals established in American Society of Mechanical Engineers (ASME) consensus standard B30.5, *Mobile and Locomotive Cranes* (*Figures 35* and *36*). In accordance with *ASME B30.5*, crane operators are required to use standard hand signals when voice communication equipment is not used. The hand-signal chart must also be posted conspicuously at the job site. The signaler should also be versed in crane operations in order to understand and anticipate the crane's motions when signaled.

Figure 34 Handheld radio.

Crane Operator Certification

OSHA now requires crane operators to be certified. NCCER's Mobile Crane Operator Certification program is recognized by OSHA and accredited by American National Standards Institute (ANSI). It offers 13 equipment-specific certifications, including capacity maximums for each. Many states have adopted, or are in the process of adopting, the NCCER certification. In those states, certified individuals are able to apply for an operator's license in the state without further testing.

The advantage to using these standard hand signals is that they are well established and published in an industry-wide standard. This means that these hand signals are recognized by the industry as the standard hand signals to be used on all job sites. This helps ensure that there is a common core of knowledge and a universal meaning to the signals when lifting operations are being conducted. This may eliminate a significant barrier to effective communication.

Additions or modifications may be made for operations not covered by the illustrated hand signals, such as deployment of outriggers. The operator and signal person must agree upon these special signals before the crane is operated, and these signals should not be in conflict with any standard signal. If it is desired to give instructions verbally to the operator instead of by hand signals, all crane motions must be stopped before doing so.

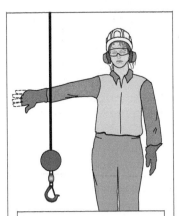
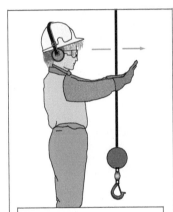

Figure 35 Standard hand signals.

Figure 36 Standard hand signals (continued).

son is also required if the crane is operating near power lines or another crane is working in close proximity.

Signal persons must be qualified by experience, and be knowledgeable in all established communication methods. They must be stationed in full view of the operator, have a full view of the load path, and understand the load's intended path of travel in order to position themselves accordingly. In addition, they must wear high-visibility gloves and/or clothing, be responsible for keeping everyone out of the operating radius of the crane, and never direct the load over anyone.

Although personnel involved in lifting operations are expected to understand these signals when they are given, it is acceptable for a signal person to give a verbal or nonverbal signal that is not part of the *ASME B30.5* standard. In cases where such nonstandardized signals are given, both the operator and the signal person must have a complete understanding of the message that is being sent.

4.2.0 Working around Materials Hoists

A materials hoist can be mounted on a scaffold (*Figure 37*), a ladder, the ground (*Figure 38*), or a truck bed. The materials hoist includes a lift platform, lift cabling, and a gasoline, diesel, or electric motor. The movement is controlled by a series of pulleys and cables. On larger hoists, the lift platform may have a cage around it.

All hoists have load limitations. The maximum rated capacity must be marked on the hoist. A typical ladder hoist can lift 400 pounds to a height of up to 40 feet. Larger hoists can lift up to 5,000 pounds up to 300 feet. Larger hoists are usually attached to the side of the building under construction, to act as temporary elevators. Check the rated capacity before using a hoist.

(A)

(B)

Figure 37 Scaffold mounted hoist.

When moving a mobile crane, audible travel signals must be given using the crane's horn:

- *Stop* – One audible signal
- *Forward* – Two audible signals
- *Reverse* – Three audible signals

There are certain requirements that mandate the presence of a signal person. When the operator of the crane cannot see the load, the landing area, or the path of motion, or cannot judge distance, a signal person is required. A signal per-

WARNING! Do not exceed a hoist's rated load. Hoist failure can cause serious injury and property damage.

Never use a hoist to transport personnel unless it has all the required safety devices. Personnel hoists have guardrails, doors, safety brakes, and hand controls in addition to the features of the materials hoist. Personnel hoists can be used for transporting materials, if needed.

When using a properly rated materials hoist to stock a workstation, follow these safety rules:

Figure 38 Large materials hoist.

- Make certain the hoist rigging is not worn, frayed, or off the pulley sheaves.
- Make certain that the load is balanced and in the middle of the hoist platform.
- Make certain that the hoist is enclosed on all sides and the gate is secured before it starts moving.
- If the hoist is operated from the ground level, make sure you understand the hand signals.
- Never ride on a materials hoist.

Oddly shaped or uncubed materials may need to be secured with safety straps or shrink wrapping to prevent them from shifting as the hoist rises. Be sure to balance the load before securing it with the straps. Check that any buckles, clamps, or ties are securely fastened before the hoist cage rises.

4.3.0 Moving and Stocking Materials

On any high-rise work site, materials handling requires special care. Masonry workstations are restocked as the job progresses, and materials for other craftworkers are moving through the job site as well. It is important to be aware of materials movement around you.

Stack masonry units carefully and safely. OSHA has guidelines regarding the stockpiling and handling of materials. Make sure you understand the following requirements before performing this work:

- Stack bagged material by stepping back the tiers and cross keying the bags at least every bags high.
- Do not store material on scaffolds or runways in excess of supplies needed for the immediate job.
- Do not stockpile palletized brick more than 7 feet in height. Taper back a loose brick stockpile when it reaches a height of 4 feet; it should be tapered back 2 inches for every foot of height above the 4-foot level.
- Taper back loose masonry units stockpiled higher than 6 feet; the stack should be tapered back one-half block per tier above the 6-foot level.
- All material stored in tiers should be stacked, racked, blocked, interlocked, or otherwise secured to prevent sliding, falling, or collapse.
- Maximum safe load limits of floors within buildings and structures should be posted in all storage areas. Do not exceed maximum safe loads.
- Keep aisles and passageways clear to provide for the free and safe movement of materials-handling equipment and workers.
- Do not place material stored inside buildings under construction within 6 feet of any hoist or inside floor opening, or within 10 feet of an exterior wall that does not extend above the top of the material stored.

Cubes of masonry and other building materials are loaded onto elevated workstations by pallet jack, forklift, crane, or derrick. *Figure 39* shows a typical reach-type forklift. Amounts and positions of elevated stockpiles must be carefully calculated so that the masons do not need to move more than the minimum amount of materials. Calculations require balancing weights across platforms and keeping within the rated loadbearing capacity of scaffold. Construction planning may call for closely set workstations; safety planning may call for stocking them with materials often instead of once daily, in order to avoid overloading the scaffold.

Masonry loads must be within the capacity of the lifting equipment and placed carefully so they do not block aisles. Any time workstations must be restocked, masons have to leave the scaffold. Lunchtime is often an appropriate time for restocking.

Even with the best planning, masons may still need to move masonry and other materials on elevated workstations. Remember to keep piles neat and vertically in line to avoid snagging clothes or

Figure 39 Reach-type forklift.

electrical cords. Keep the piles about 3 feet high so that it will not be hard to get at the brick or other material.

4.4.0 Using Elevated Workstations

Some work must be performed on an elevated workstation. Safety is the major consideration in organizing an elevated workstation, such as a personnel lift tower (refer to *Figure 13*).

Keeping materials neat and organized is particularly important on an elevated workstation, as space is tight. The danger and inconvenience from dropping items increases with height. Arrange materials and equipment with the following requirements in mind:

- Check the scaffold for proper assembly and rated loads before using it.
- Keep the work area clean and dry; water, mud, or dried mortar can create hazardous footing.
- Keep walkways and work areas free from stored materials, tools, rope, electrical cords, and trash. Store unused tools under the edges of the mortar pan.
- Stack masonry units far enough apart to leave room for mortar pans. Stacks should be no higher than 3 feet and no closer than 6 feet to any opening.
- Stack masonry directly over frame members, away from the edge of the scaffold platform, and never on the very ends of scaffold planks.

4.5.0 Using Disposal Chutes and Waste Bins

In 29 *CFR* 1926.852 of the OSHA regulations, there is a requirement for the use of waste chutes that have been set up at angles greater than 45 degrees: the chutes must be completely enclosed. On some jobs, separate chutes are used for designated materials. When dropping material down a waste chute, be careful to maintain your balance.

If internal drop holes are used inside a building framework, the area must be enclosed. The drop hole must be enclosed with a 4-foot-high barricade that surrounds the opening at a distance of 6 feet. When disposing of waste down the drop hole, do not lean over the edge.

Disposal chutes and drop holes should deliver their contents into waste bins or dumpsters. At the end of the day, fill the waste bins with dried mortar, broken masonry, and any other debris that can be removed from your workstation. Use the waste bin to keep your workstation clean.

Additional Resources

ASME B30.5, *Mobile and Locomotive Cranes*, Latest Edition. New York: ASME.

"Materials Handling and Storage." OSHA. **www.osha.gov**

"Scaffolding." OSHA. **www.osha.gov**

4.0.0 Section Review

1. The type of crane that is mounted on a fixed base situated away from the building at a distance that is determined by the length of the boom arm is called the _____.
   a. mobile crane
   b. tower crane
   c. derrick crane
   d. traveling crane

2. Materials hoists used to build a masonry wall can be mounted on each of the following *except* _____.
   a. a ladder
   b. the wall
   c. the scaffold
   d. a truck bed

3. When stacking bagged material, step the layers back and cross key the bags at least every _____.
   a. 5 bags high
   b. 10 bags high
   c. 15 bags high
   d. 20 bags high

4. When working in an elevated workstation, the maximum height of a masonry unit stack should be _____.
   a. 3 feet
   b. 4 feet
   c. 5 feet
   d. 6 feet

5. The height of the barricade that encloses an internal drop hole inside a building must be _____.
   a. 6 feet
   b. 5 feet
   c. 4 feet
   d. 3 feet

SUMMARY

High-rise construction often involves some type of masonry work. Many modern high-rise buildings have concrete or steel framework with masonry exterior walls. Exterior walls are constructed as veneer panel or curtain walls, depending on the design of the building. Masonry veneer walls do not carry any structural load, although they transmit lateral forces such as wind load to the main structure.

Personnel safety for high-rise construction requires, the practice of work-area safety rules, fire prevention, electrical hazard identification, the proper use of appropriate personal protective equipment, and maintaining a clear work area. Use of fall prevention and fall arrest equipment is an essential part of scaffold and leading-edge work.

Masonry panel walls can be supported at each floor level by shelf angles, while curtain walls are connected only at the foundation and the. Parapet walls are constructed with brick or concrete masonry units on many buildings. Panel walls, curtain walls, and parapets may require steel reinforcement and flashing.

Interior masonry walls for high-rise buildings include room partitions, stairwells, elevator shafts, and other structures. There are strict codes that specify the height-to-thickness ratios for partition walls.

Materials-handling equipment plays an important role in high-rise construction. Equipment, such as cranes and derricks, is used to lift materials and other equipment up to working floors from the ground. Materials hoists and personnel hoists are also used to move materials and personnel between levels of the building. Materials can be transported in personnel hoists.

Masons must be familiar with the basic hand signals used to direct crane and derrick operations. The crane operator should be given directions by only one designated signaler.

Review Questions

1. The most common types of fall protection equipment include guardrails, personal fall arrest systems, and _____.
   a. large air bags
   b. scaffold cages
   c. safety nets
   d. workstation enclosures

2. A ground fault circuit interrupter can break an electrical connection in as little as _____.
   a. 1/40 of a second
   b. 1/30 of a second
   c. 1/20 of a second
   d. 1/10 of a second

3. OSHA regulations require that a steel cable must be replaced if the quantity of frayed or broken wires in a running foot exceeds _____.
   a. 5 percent
   b. 10 percent
   c. 15 percent
   d. 20 percent

4. Debris netting must be installed when _____.
   a. scaffold reaches a third-story level
   b. an injury involving falling objects occurs
   c. material or equipment piles are higher than a standard toeboard
   d. masonry materials are being lifted onto scaffolds

5. Before it can be used, a personnel lift must be tested with a load equal to _____.
   a. 90 percent of its capacity
   b. 125 percent of its capacity
   c. 150 percent of its capacity
   d. 200 percent of its capacity

6. The width of a limited access zone adjacent to a wall being built is equal to _____.
   a. two-thirds the wall height
   b. the wall height
   c. the wall height plus 4 feet
   d. twice the wall height

7. Temporary bracing of a masonry wall must remain in place until _____.
   a. the final course of masonry is completed
   b. the building inspector approves its removal
   c. foundation backfilling is done
   d. permanent supporting elements of the structure are in place

8. The distance between loadbearing supports, such as beams or columns, is referred to as a _____.
   a. bay
   b. tier
   c. span
   d. panel

9. Working platforms are typically designed to support a per-square-foot load of _____.
   a. 100 pounds
   b. 75 pounds
   c. 50 pounds
   d. 25 pounds

10. Panel and curtain walls in high-rise construction require lateral bracing to _____.
    a. prevent sagging
    b. resist wind loads
    c. remain level
    d. prevent stress cracking

11. Pressure-relieving joints in masonry panel walls are usually filled with _____.
    a. butyl caulking
    b. a rubber gasket
    c. expanded foam sealant
    d. a flexible neoprene material

12. Because they have different rates of expansion, masonry curtain walls and steel or concrete building frames must be connected with _____.
    a. expansion joints
    b. flexible noncorrosive-metal anchors
    c. floating foundations
    d. sleeve connectors

28301-14 Elevated Masonry

13. Metal copings used on a parapet should extend below the top of the masonry by at least _____.
    a. 2 inches
    b. 2½ inches
    c. 3 inches
    d. 3½ inches

14. Interior nonbearing walls that support only their own weight are called _____.
    a. cubicle walls
    b. partition walls
    c. divider walls
    d. spacer walls

15. The partition lengths of internal walls can be increased if reinforcement is provided by _____.
    a. cross walls
    b. horizontal joint reinforcement
    c. internal pilasters
    d. projecting pilasters

16. A crane attached to scaffold that moves up the building face as work progresses is described as a _____.
    a. mobile crane
    b. tower crane
    c. derrick crane
    d. traveling crane

17. One of the most common modes of verbal communication between the crane operator and signal person is _____.
    a. cell phones
    b. a wired connection
    c. handheld radios
    d. bullhorns

18. Disadvantages of using radios for job-site communications include interference, crowded frequencies, and _____.
    a. equipment cost
    b. background noise
    c. battery life
    d. frequent breakdowns

19. A nonverbal means of communication sometimes used between the signal person and crane operator is _____.
    a. signs with printed commands
    b. sign language
    c. color-coded signal flags
    d. coded light signals

20. Material stored in a building under construction should be placed no closer to an inside floor opening than _____.
    a. 3 feet
    b. 6 feet
    c. 9 feet
    d. 12 feet

Trade Terms Quiz

Fill in the blank with the correct term that you learned from your study of this module.

1. A designated work area in which certain types of masonry work may take place without the use of conventional fall protection systems is called a(n) _____.

2. A(n) _____ is a narrow molding used to separate two structural elements, usually roof and wall, to divert water.

3. An apparatus used for hoisting on high-rise buildings, consisting of a boom mounted on a column or mast that is held at the head by fixed-length supporting ropes or guys, is called a(n) _____.

4. A(n) _____ is a restricted area alongside a masonry wall that is under construction.

5. Wind shear and other forces applying horizontal pressure to a wall or other structural unit causes _____.

6. _____ is a common term for a scaffold level.

Trade Terms

Controlled access zone
Cut
Guyed derrick

Lateral stress
Limited access zone
Reglet

Trade Terms Introduced in This Module

Controlled access zone: A designated work area in which certain types of masonry work may take place without the use of conventional fall protection systems.

Cut: A common term for a scaffold level.

Guyed derrick: An apparatus used for hoisting on high-rise buildings, consisting of a boom mounted on a column or mast that is held at the head by fixed-length supporting ropes or guys.

Lateral stress: Wind shear and other forces applying horizontal pressure to a wall or other structural unit.

Limited access zone: A restricted area alongside a masonry wall that is under construction.

Reglet: A narrow molding used to separate two structural elements, usually roof and wall, to divert water.

Additional Resources

This module presents thorough resources for task training. The following resource material is suggested for further study.

ASME B30.5, Mobile and Locomotive Cranes, Latest Edition. New York: ASME.
Bricklaying: Brick and Block Masonry. 1988. Brick Industry Association. Orlando, FL: Harcourt Brace & Company.
Concrete Masonry Handbook: for Architects, Engineers, Builders, Fifth Edition. 1991. W. C. Panerese, S. K. Kosmatka, and F. A. Randall, Jr. Skokie, IL: Portland Cement Association.
Fall Protection and Scaffolding Safety: An Illustrated Guide. 2000. Grace Drennan Ganget. Government Institutes.
Masonry Wallbracing Design Handbook, Latest edition. Rashod R. Johnson and Daniel S. Zechmeister. Algonquin, IL: Masonry Contractors Association of America.
"Materials Handling and Storage." OSHA. **www.osha.gov**
"Online Safety Library: Scaffold Safety." Oklahoma State University. **www.ehs.okstate.edu**
"Scaffolding." OSHA. **www.osha.gov**
Standard Practice for Bracing Masonry Walls under Construction. 2012. Algonquin, IL: Masonry Contractors Association of America.
WorkSAFE masonry safety resources. **www.worksafecenter.com**

Figure Credits

Courtesy of Dennis Neal, FMA&EF, CO01, Figure 3b, Figures 12–13, Figures 18–20, Figure 22, Figure 30a, Figure 37b, Figure 38

DBI/SALA & Protecta, Figure 5

Courtesy of Honeywell Safety Products, Figures 6–9, Figure 10a

Courtesy of PERI Formwork Systems, Inc., Figure 11, SA01

Courtesy Council for Masonry Wall Bracing, Figures 14–15

Courtesy of Associated Builders and Contractors, T01

Topaz Publications, Inc., Figure 21

Steven Fechino, Figure 23

National Concrete Masonry Association, Figure 25, Figures 27–29, T02–03

Manitowoc Cranes (The Manitowoc Company, Inc.), Figure 30b

3M Company, Figures 31–33

Product image (IC-F3001/F4001) courtesy of Icom America-Inc., Figure 34

Courtesy of Beta Max Hoist, Figure 37a

Courtesy of Skyjack, Figure 39

28301-14 Elevated Masonry

Section Review Answers

Answer	Section Reference	Objective
Section One		
1. d	1.1.3	1a
2. b	1.2.4	1b
Section Two		
1. a	2.1.0	2a
2. d	2.2.0	2b
Section Three		
1. b	3.1.0	3a
2. a	3.2.0	3b
3. c	3.3.3	3c
4. b	3.4.0	3d
Section Four		
1. c	4.1.0	4a
2. b	4.2.0	4b
3. b	4.3.0	4c
4. a	4.4.0	4d
5. c	4.5.0	4e

NCCER CURRICULA — USER UPDATE

NCCER makes every effort to keep its textbooks up-to-date and free of technical errors. We appreciate your help in this process. If you find an error, a typographical mistake, or an inaccuracy in NCCER's curricula, please fill out this form (or a photocopy), or complete the online form at **www.nccer.org/olf**. Be sure to include the exact module ID number, page number, a detailed description, and your recommended correction. Your input will be brought to the attention of the Authoring Team. Thank you for your assistance.

Instructors – If you have an idea for improving this textbook, or have found that additional materials were necessary to teach this module effectively, please let us know so that we may present your suggestions to the Authoring Team.

NCCER Product Development and Revision
13614 Progress Blvd., Alachua, FL 32615

Email: curriculum@nccer.org
Online: www.nccer.org/olf

❏ Trainee Guide ❏ Lesson Plans ❏ Exam ❏ PowerPoints Other _____

Craft / Level: Copyright Date:

Module ID Number / Title:

Section Number(s):

Description:

_____

_____

_____

_____

Recommended Correction:

_____

_____

_____

_____

Your Name:

Address:

_____

Email: Phone:

75122-13

Working from Elevations

Overview

A fall from an elevation is the most common cause of death in the construction industry. Fall protection is required on nearly every job site. This module covers the use of personal fall-protection systems. It also discusses the proper use of ladders, scaffolding, and aerial lifts.

Module Eleven

Trainees with successful module completions may be eligible for credentialing through the NCCER Registry. To learn more, go to **www.nccer.org** or contact us at **1.888.622.3720**. Our website has information on the latest product releases and training, as well as online versions of our *Cornerstone* magazine and Pearson's product catalog.

Your feedback is welcome. You may email your comments to **curriculum@nccer.org**, send general comments and inquiries to **info@nccer.org**, or fill in the User Update form at the back of this module.

This information is general in nature and intended for training purposes only. Actual performance of activities described in this manual requires compliance with all applicable operating, service, maintenance, and safety procedures under the direction of qualified personnel. References in this manual to patented or proprietary devices do not constitute a recommendation of their use.

Copyright © 2015 by NCCER, Alachua, FL 32615, and published by Pearson Education, Inc., New York, NY 10013. All rights reserved. Printed in the United States of America. This publication is protected by Copyright, and permission should be obtained from NCCER prior to any prohibited reproduction, storage in a retrieval system, or transmission in any form or by any means, electronic, mechanical, photocopying, recording, or likewise. To obtain permission(s) to use material from this work, please submit a written request to NCCER Product Development, 13614 Progress Blvd., Alachua, FL 32615.

From *Construction Craft Laborer, Level Two*, Third Edition. NCCER.
Copyright © 2015 by NCCER. Published by Pearson Education. All rights reserved.

75122-13
WORKING FROM ELEVATIONS

Objectives

When you have completed this module, you will be able to do the following:

1. Identify various types of fall-protection equipment.
   a. Explain the safety guidelines for personal fall-arrest systems.
   b. Explain the safety guidelines for other fall-protection systems.
2. Identify the safety guidelines for the use of ladders and scaffolding.
   a. State the safety requirements for various ladders.
   b. State the safety requirements for scaffolding.
3. State the guidelines for the safe operation of aerial lifts.
   a. Identify aerial lift components and operating requirements.
   b. Describe the safe operation of scissor lifts.
   c. Describe the safe operation of boom lifts.

Performance Tasks

Under the supervision of the instructor, you should be able to do the following:

1. Demonstrate how to properly inspect and don fall-protection equipment.
2. Demonstrate how to properly inspect a ladder.

Trade Terms

Aerial lift
Body harness
Connector
Duty rating
Free fall
Interlocking systems
Ladders
Lanyard
Leading edge
Lifeline
Personal fall-arrest system
Platform
Proportional control
Scaffolding
Self-retracting lanyard (SRL)
Two-point swing scaffolding

Industry-Recognized Credentials

If you are training through an NCCER-accredited sponsor, you may be eligible for credentials from NCCER's Registry. The ID number for this module is 75122-13. Note that this module may have been used in other NCCER curricula and may apply to other level completions. Contact NCCER's Registry at 888.622.3720 or go to www.nccer.org for more information.

Contents

Topics to be presented in this module include:

1.0.0 Types of Fall-Protection Equipment 1
   1.1.0 Safety Requirements for Personal Fall-Arrest Systems 2
      1.1.1 Anchor Points 4
      1.1.2 Harnesses 4
      1.1.3 Connecting Devices 7
      1.1.4 Lanyards 8
      1.1.5 Rescue After a Fall 10
   1.2.0 Safety Guidelines for Other Fall-Protection Systems 11
      1.2.1 Guardrails 11
      1.2.2 Safety Net Systems 11
      1.2.3 Safe Climbing Devices 11
2.0.0 Safety Guidelines for the Use of Ladders and Scaffolding 14
   2.1.0 Safety Requirements for Various Types of Ladders 14
      2.1.1 Straight Ladders 16
      2.1.2 Extension Ladders 18
      2.1.3 Stepladders 19
      2.1.4 Fixed Ladders 20
   2.2.0 Safety Requirements for Scaffolding 20
      2.2.1 Scaffolding Hazards 20
      2.2.2 Safety Guidelines for the Use of Scaffolding 23
3.0.0 Guidelines for the Safe Operation of Aerial Lifts 27
   3.1.0 Aerial Lift Components and Operating Requirements 27
      3.1.1 Aerial Lift Operator Qualifications 27
      3.1.2 Aerial Lift Safety Precautions 28
      3.1.3 Aerial Lift Operating Procedure 28
   3.2.0 Safe Operation of Scissor Lifts 29
   3.3.0 Safe Operation of Boom Lifts 30
      3.3.1 Maintenance 31

Figures

Figure 1   Body harness labeling 2
Figure 2   Personal fall-arrest system 3
Figure 3   Typical anchor point 4
Figure 4   Back D-ring 4
Figure 5   Full body harness 5
Figure 6   Installing the body harness 6
Figure 7   Suspension trauma strap use 7
Figure 8   Carabiner 7
Figure 9   Lanyard hook 8
Figure 10   Shock-absorbing lanyard 8
Figure 11   Non-shock absorbing lanyard 9
Figure 12   Y-configured shock absorbing lanyard 10
Figure 13   Self-retracting lanyard 10
Figure 14   Guardrails 12
Figure 15   Cable grab 12
Figure 16   Types of ladders 15
Figure 17   Proper positioning 16
Figure 18   Straight ladder 17
Figure 19   Ladder safety feet 17
Figure 20   Securing a ladder 18
Figure 21   Extension ladders 18
Figure 22   Overlap lengths for extension ladders 19
Figure 23   Typical stepladder 19
Figure 24   Ladder safety 21
Figure 25   Typical fixed ladder 22
Figure 26   Typical scaffolding 22
Figure 27   Work cage 23
Figure 28   Typical scaffolding tags 24
Figure 29   Built-up scaffolding 25
Figure 30   Swing scaffolding 25
Figure 31   Aerial lifts 28
Figure 32   Scissor lift 28
Figure 33   Operator's checklist 29
Figure 34   Self-propelled elevating work platform (scissor lift) 30
Figure 35   Articulating and straight boom lifts 30

Section One

1.0.0 Types of Fall-Protection Equipment

Objectives

Identify various types of fall-protection equipment.
 a. Explain the safety guidelines for personal fall-arrest systems.
 b. Explain the safety guidelines for other fall-protection systems.

Performance Task 1

Demonstrate how to properly inspect and don fall-protection equipment.

Trade Terms

Body harness: Straps that may be secured about the worker in a manner that will distribute the fall-arrest forces over at least the thighs, pelvis, waist, chest, and shoulders, with means for attaching it to other components of a personal fall-arrest system.

Connector: A device that is used to couple (connect) parts of the personal fall-arrest system and positioning device systems together. It may be an independent component of the system, such as a carabiner. It may also be an integral component of part of the system, such as a buckle or D-ring sewn into a body harness, or a snaphook spliced or sewn to a lanyard.

Free fall: The act of falling before a personal fall-arrest system begins to apply force to arrest the fall.

Lanyard: A flexible line of rope, wire rope, or strap that generally has a connector at each end for connecting the body harness to a deceleration device, lifeline, or anchorage.

Leading edge: The edge of a floor, roof, or formwork for a floor or other walking/working surface (such as the deck) which changes location as additional floor, roof, decking, or formwork sections are placed, formed, or constructed. A leading edge is considered to be an unprotected side and edge during periods when it is not actively and continuously under construction.

Lifeline: A component consisting of a flexible line connected vertically to an anchorage at one end (vertical lifeline), or connected horizontally to an anchorage at both ends (horizontal lifeline), and which serves as a means for connecting other components of a personal fall-arrest system to the anchorage.

Personal fall-arrest system: A system used to stop an employee in a fall from a working level. It consists of an anchorage, connectors, and a body harness, and may include a lanyard, deceleration device, lifeline, or suitable combinations of these.

Platform: A work surface elevated above lower levels. Platforms can be constructed using individual wood planks, fabricated planks, fabricated decks, and fabricated platforms.

Self-retracting lanyard (SRL): A deceleration device containing a drum-wound line that can be slowly extracted from, or retracted onto, the drum under slight tension during normal employee movement, and which, after the onset of a fall, automatically locks the drum and arrests the fall.

Falls are the leading cause of death in the construction industry. OSHA requires fall protection when workers are exposed to falls from work areas with elevations that are 6' and above. The types of work areas that put the worker at risk include the following:

- Scaffolding
- Ladders
- Leading edges
- Ramps or runways
- Wall or floor openings
- Roofs
- Excavations, pits, and wells
- Concrete forms
- Unprotected sides and edges

Falls are classified into two groups: falls from an elevation and falls on the same level. Falls from an elevation can happen when someone is doing work from scaffolding, work platforms, decking, concrete forms, ladders, or excavations. Falls from elevations are almost always fatal. This is not to say that falls on the same level aren't also extremely dangerous. When a worker falls on the same level, usually from tripping or slipping, head injuries often occur. Sharp edges and pointed objects such as exposed rebar could cut or stab the worker.

75122-13 Working from Elevations

Module Eleven 1

The following safe practices can help prevent slips and falls:

- Wear safe, strong work footwear that is in good repair.
- Watch where you step. Be sure your footing is secure.
- Do not allow yourself to get in an awkward position. Stay in control of your movements at all times.
- Maintain clean, smooth walking and working surfaces. Fill holes, ruts, and cracks. Clean up slippery material and litter.
- Install cables, extension cords, and hoses so that they will not become tripping hazards.
- Do not run on scaffolding, work platforms, decking, roofs, or other elevated work areas.

The best way to survive a fall from an elevation is to use fall-protection equipment. The two most common types of fall-protection equipment are guardrails and personal fall-arrest systems.

1.1.0 Safety Requirements for Personal Fall-Arrest Systems

Personal fall-arrest systems (PFAS) combine several pieces of equipment together into a complete fall-protection system. Many different types and styles exist, with each one designed for specific uses. In addition, the personal aspect of a PFAS must also be emphasized. One size does not fit all, and proper fit is essential to avoiding injury to the worker when a fall is arrested. *OSHA Standard 1926.502, Subpart M, Section (d)* lists specific requirements, such as the test strength and use of D-rings and snap hooks and the specifications for anchor points on the structure.

The referenced OSHA standard requires several important characteristics of a PFAS:

- It must limit the maximum arresting force imparted to the body to 1,800 pounds with a full body harness.
- The free fall distance (the distance it takes before a person comes to a stop when falling) must be limited to 6 feet and be rigged to prevent contact with anything below.
- It must bring the body to a stop within an additional 3½ feet (1.1 m).
- The system must be strong enough to withstand 5,000 pounds or twice the possible force of a body falling from a distance of 6 feet.

You will often see references to American National Standards Institute (ANSI) standards regarding the performance of PFAS products. ANSI publishes standards for PFAS equipment that provide manufacturers with guidelines for their products. Its standards cover items such as design, testing, markings, and performance. Items that bear the ANSI certification markings let the user know that a product has been built and tested to ensure its integrity. Many organizations require that PFAS equipment used by its employees or agents meets ANSI standards and is marked accordingly, without exception.

It should be obvious that any PFAS is only as good as its weakest component. They should never be used for any other task, such as lifting tools or materials to the work area. The specifications and usage instructions for components such as the full body harness (*Figure 1*) will identify the working load limit, and this limit should be strictly followed. Workers must use common sense and not overload themselves with equipment, even if the weight does fall within the limits of the PFAS specifications. Always examine fall-protection before each use and do not use it if it is worn or damaged in any way. Check with the manufacturer before mixing equipment from different manufacturers. All substitutions must be approved.

A complete PFAS (*Figure 2*) is made up of three primary components. Remember that this is a system, not a single piece of equipment. Anchor points are related to the structure, and the type and availability of anchor points help determine what other equipment should be chosen. The body harness comprises the system of belts, rings, or hooks worn by the worker. Connecting devices or connectors are used to maintain attachment

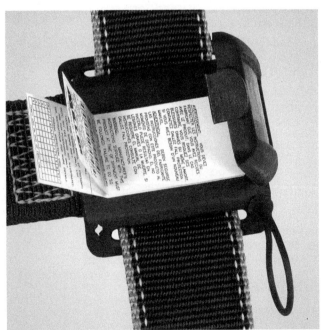

Figure 1 Body harness labeling.

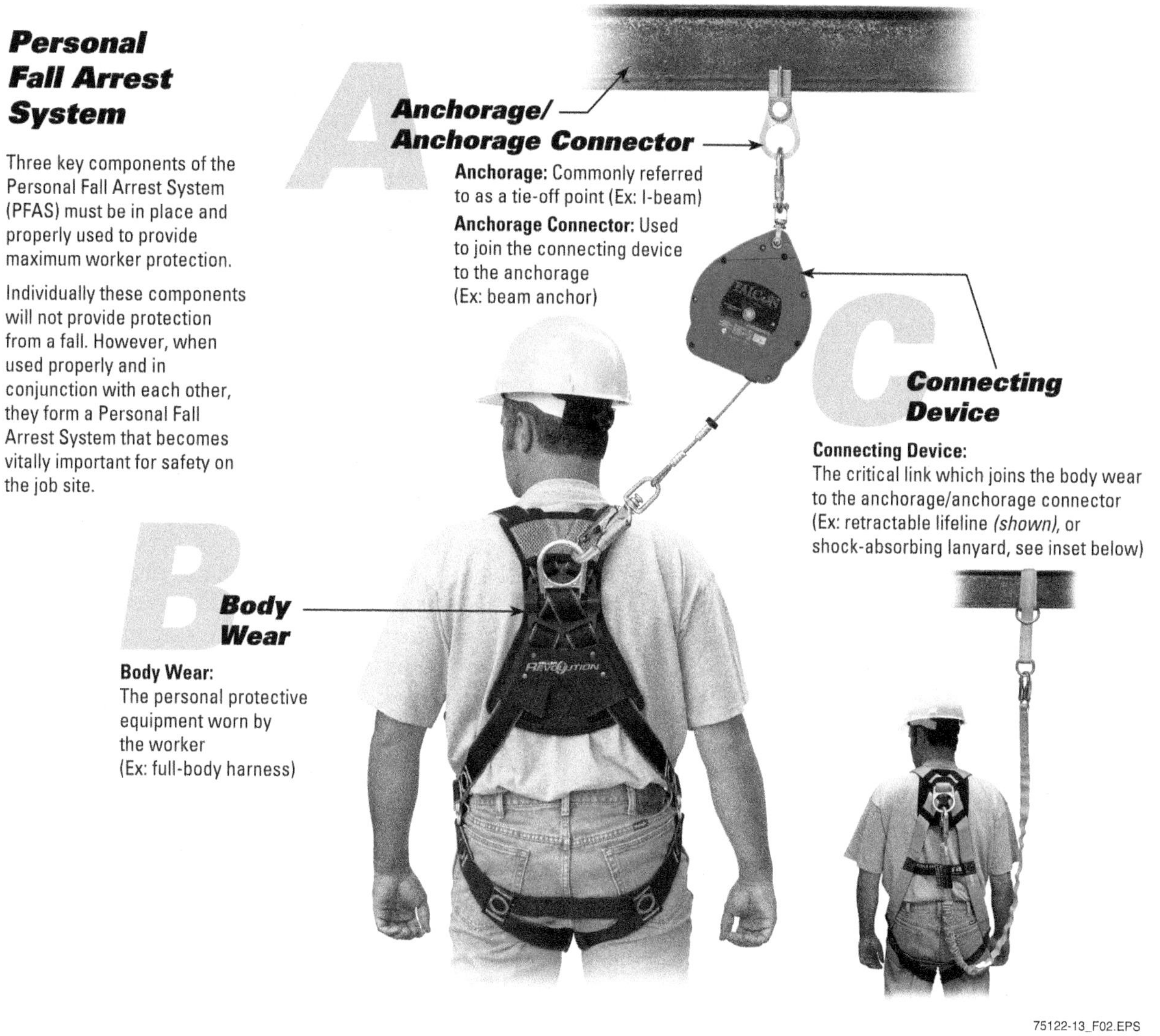

Personal Fall Arrest System

Three key components of the Personal Fall Arrest System (PFAS) must be in place and properly used to provide maximum worker protection.

Individually these components will not provide protection from a fall. However, when used properly and in conjunction with each other, they form a Personal Fall Arrest System that becomes vitally important for safety on the job site.

A - Anchorage/Anchorage Connector
Anchorage: Commonly referred to as a tie-off point (Ex: I-beam)
Anchorage Connector: Used to join the connecting device to the anchorage (Ex: beam anchor)

C - Connecting Device
Connecting Device: The critical link which joins the body wear to the anchorage/anchorage connector (Ex: retractable lifeline *(shown)*, or shock-absorbing lanyard, see inset below)

B - Body Wear
Body Wear: The personal protective equipment worn by the worker (Ex: full-body harness)

Figure 2 Personal fall-arrest system.

Case History
Obey Safety Warnings

A crew was installing the final structural steel beam (bar joist) in the roof of a new cold storage warehouse under construction. After a crane lifted the beam into place, it was not quite straight. The foreman on the job wanted to use a hammer to straighten it. He was standing on a portion of roof decking that had already been completed. To get to the beam, he reached his left foot out over an open, undecked area of the roof. He rested his left foot on the nearest joist girder. As he was preparing to strike a blow with the hammer, his foot slipped off the girder. His hands caught the bar joist, but he couldn't hold on and fell. The area where the foreman needed to work had been barricaded with wire rope safety lines on all four sides, but he removed these lines to gain access. He was not using fall-protection equipment. He was killed when he fell 38' to the floor below.

The Bottom Line: Never cross a safety line. Always wear fall-protection equipment when working on elevated surfaces.

Source: Electronic Library of Construction Occupational Safety and Health

between anchor points and the PFAS, and includes both lanyards and various pieces of hardware.

1.1.1 Anchor Points

OSHA requires that anchor points (*Figure 3*) be rated at or equal to 5,000 pounds breaking or tensile strength or twice the intended load. Two workers can connect their fall-arrest lanyard to a single anchor point. However, the anchor point must then meet a 10,000-pound standard or twice the intended load. Since the potential anchor load from a fall has doubled, so must the anchor point's rated capacity.

The ideal fall arrestance anchor point is located directly above the back D-ring (*Figure 4*), selected to minimize any swing zone hazards as well as the possible free fall distance. Swing zones are minimized when the anchor point is directly above the worker. Horizontal working situations can be far more hazardous in terms of the swing zone, as an anchor point directly above may not exist. Serious injury and damage can occur when a human body strikes an immoveable object while swinging as a pendulum. Although the PFAS may do its job by preventing the worker from falling a great distance, serious injury or death can still occur by striking an object in the swing zone.

Anchor points are often needed to secure the position of the worker, leaving the hands free to accomplish the task. Ideally, workers should select anchor points, and connect as needed, to maintain a potential fall distance of no more than two feet. Positioning connections are a factor in fall arrest, as the connection to a positioning strap will likely modify the swing zone and/or distance of the fall during an accident. However, they cannot be considered the primary anchor. Therefore, positioning anchor points are required by OSHA to be rated at a 3,000-pound strength instead of the 5,000-pound rating for primary fall arrest. Remember the positioning lanyards, connected to D-rings on the harness other than the back or front chest D-ring, are fall restraints rather than fall arresting connections. A positioning lanyard does not take the place of a fall arrestance lanyard or anchor point. If you make a connection with the intention of placing weight or stress on it in any way, consider it a positioning connection.

1.1.2 Harnesses

The full body harness (*Figure 5*) is unquestionably the center of the PFAS. It should be worn any time work is conducted 6 feet or more above the ground. A variety of D-rings are part of a full body harness assembly. The back D-ring is the only one used to connect the harness to the anchor point for primary fall arrest purposes un-

Figure 3 Typical anchor point.

ANSI

The ANSI membership is comprised of government agencies, organizations, companies, manufacturers, academic bodies, and individuals. It is a non-profit organization, not a government agency. ANSI "oversees the creation, promulgation and use of thousands of norms and guidelines that directly impact businesses in nearly every sector: from acoustical devices to construction equipment, from dairy and livestock production to energy distribution, and many more." Although its standards are not enforceable by law, it is a highly respected organization and many regulations spring from its standards.

Figure 4 Back D-ring.

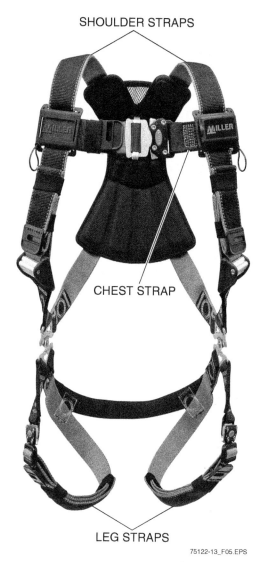

Figure 5 Full body harness.

less you are climbing a ladder. When climbing a ladder, the front chest D-ring is the likely choice. D-rings located at the hips are used for positioning and fall restraints only. D-rings mounted to shoulders are often used for rescue situations. All of them can be used for fall restraint, but the back D-ring is the primary connection for fall arrestance.

A body harness must fit correctly to ensure that it will provide proper protection. Do not place additional holes or openings in harness components under any circumstances. No field modifications to a body harness or lanyard should be attempted. Installation, maintenance, and inspection instructions are provided for every harness, and it is the responsibility of the worker to read and understand the details regarding his or her personal equipment. Harness straps are generally designed with some stretch to help absorb some of the potential force. This means good, taut installation on the body is essential so that any slack, plus stretch, does not allow the worker to fall out of the harness.

Figure 6 demonstrates the proper procedure for donning a common full body harness. The most important adjustments to be made include the chest straps, the groin straps, and the final position of the back D-ring. The related details that follow must be considered during the fitting and wearing of a full body harness:

> **NOTE**
> These guidelines are general in nature and the instructions provided for specific equipment by the manufacturer must always take precedence. It is the worker's responsibility to be intimately familiar with the duty of each and every ring and strap.

- The back D-ring location is vital to proper fall arrest. Position this ring between the shoulder blades. If it is too low, it will tend to cause the body to hang in a more horizontal position during fall arrest, increasing pressure on the diaphragm and affecting breathing. If it is positioned too high or with too much slack, the D-ring may strike the worker's head at the base of the fall, and the shoulder straps may be pulled too tightly into the neck and restrict blood flow. Remember that the impact at the base of the fall arrest can be dramatic, and that this force must be spread all around the body to prevent injury to any one portion.
- Chest straps generally form either an H pattern or an X pattern. Adjust H pattern straps to land between the bottom of the sternum and the belly button. This helps ensure the horizontal portion of the H does not contact the throat during a fall, choking the worker. Some harness designs may not allow for this adjustment, with the final position of the H being based solely on a properly sized harness.
- The position of the chest straps is also crucial for X pattern harnesses. Position the X at or just below the sternum.
- Leg straps are an integral and required part of a PFAS. Adjust the groin straps for a good, snug fit. Too much slack causes extreme discomfort in a fall, when the impact snatches them up tight and you are left suspended this way.
- A suspension trauma strap (*Figure 7*) is recommended as part of the PFAS gear. The suspension trauma strap is stored in a convenient pouch that is preconnected to the harness. This is done by either one end of the strap being permanently sewn to the harness (by the

6 Easy Steps That Could Save Your Life
How To Don A Harness

1. Hold harness by back D-ring. Shake harness to allow all straps to fall in place.

2. If chest, leg and/or waist straps are buckled, release straps and unbuckle at this time.

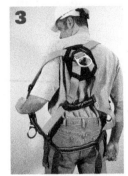

3. Slip straps over shoulders so **D-ring is located in middle of back between shoulder blades.**

4. Pull leg strap between legs and connect to opposite end. Repeat with second leg strap. If belted harness, connect waist strap after leg straps.

5. **Connect chest strap and position in midchest area.** Tighten to keep shoulder straps taut.

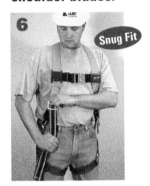

6. After all straps have been buckled, **tighten all buckles so that harness fits snug but allows full range of movement.** Pass excess strap through loop keepers.

Figure 6 Installing the body harness.

manufacturer), one end of the strap attached to the harness with a carabiner, or by choking the pouch around a harness strap or hip D-ring. The strap can then be quickly removed and used without any possibility of the user dropping it. Once connected, the strap allows the worker to stand up in the harness, relieving suspended weight and pressure from the hips and groin. This helps open the path for blood flow from the legs back to the heart, preventing blood from pooling in the lower extremities.

- A separate waist or tool belt, while not considered a necessary component of the fall-arrest system, must be fitted properly. It is best used for body positioning with the D-rings precisely located at the hip sides, rather than in the front or rear. Do not adjust it in a way that could apply pressure to the kidneys or lower back. If the waist belt is not an integral part of the harness, it must not be worn on the outside of the harness—don it first, and then add the harness over it. This is also true of any added tool belts.

Figure 7 Suspension trauma strap use.

- Saddles, like waist belts, are also not considered an integral or required part of the PFAS. They are optional and often detachable. They are generally used by workers to allow a seated, suspended position when the task may require long periods in the same location.

1.1.3 Connecting Devices

Connecting devices connect the PFAS to anchor points. They include several different types of hardware, as well as the lanyards. Lanyards are discussed in detail in the next section.

It is important to note that not all hardware will qualify as a component of a PFAS. Some hardware is to be used only for attaching tools and equipment to the worker or to structures, due to their limited specifications or testing. Hardware used as connecting devices as part of the PFAS must be drop-forged steel, and have a corrosion-resistant finish resistant to salt spray per ANSI standards. Any type of hook or carabiner (*Figure 8*) must be equipped with safety gates or keepers to prevent the hooked object from being disconnected accidentally. In most cases, these safety gates are required to be two-step, also called double action. Designs for these features vary, but those designed so that both movements required to open the gate can be done with a single hand are gen-

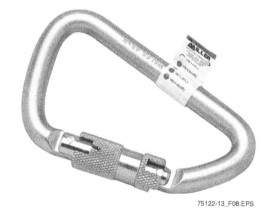

Figure 8 Carabiner.

erally better. Using both hands to manipulate a single connector can be a hazard in itself.

A carabiner is like an elongated ring, with some style of safety gate that is shaped much like a chain link. The safety gate on some may be screw-type, requiring the gate to be screwed closed. Carabiners are used to attach lines and have no sharp edges. They are tested by impact to determine their strength and are rated in newtons of force. For use in your fall-arrest system, they must be rated at a minimum of 22 kilo-newtons, or 22kN.

Double-locking snaphooks (*Figure 9*) are usually curved and have an opening to allow con-

75122-13 Working from Elevations   Module Eleven 7

nection to a line that can then be securely closed. They are usually not as consistent in appearance as carabiners, and come in somewhat different shapes. When used as part of a PFAS, the security closure should be automatic. Snaphooks are tested under direct load rather than by force applied, and thus are rated in pounds of load as the minimum breaking strength. The minimum tensile load of D-rings and snaphooks is 5,000 pounds. Snaphooks are designed to connect to D-rings primarily, not to each other. In most cases, snaphooks are already connected to a lanyard or rope to ensure the integrity of the connection and are not purchased separately.

1.1.4 Lanyards

Lanyards consist of their primary material of construction (rope, webbing, aircraft cable, etc.) with a connecting device attached to the ends. Lanyards for fall restraint or arrestance must never be field-fabricated. They are available in a great variety of lengths from as small as 1½ feet up to 30 feet, and should never be connected together to increase their length. Depending upon the use, padding may be added during fabrication or added in the field to protect against sharp edges. Never tie a knot in a lanyard or connect two lanyards together, as knots can severely reduce the load limit. Never wrap a lanyard around a structure and then choke it back into itself unless it is designed for this use. A special large D-ring is usually attached to the lanyard for this purpose. It is important to remember that whenever you are wearing a PFAS, 100 percent tie-off is required. A PFAS is absolutely useless unless you are tied off.

There are two main categories of lanyards: shock-absorbing (*Figure 10*) and non-shock absorbing (*Figure 11*). Lanyards for fall arrest should always be shock absorbing. Non-shock absorbing lanyards are used for positioning and fall restraint. You will recall that lanyards used for positioning are not considered part of the fall-arrest system, they are fall restraints and will be attached to D-rings on the harness other than the back D-ring. Since fall restraint is all about preventing a fall from happening, lanyards used for positioning should not allow a fall or movement

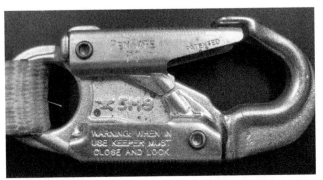

Figure 9 Lanyard hook.

Figure 10 Shock-absorbing lanyard.

Tool Lanyards

When working at elevations, a dropped tool can become a serious hazard. If the tool has moving parts, like battery-powered drills, the fall will likely destroy it.

Tool lanyards specifically designed for work in the elevated environment have been introduced by Snap-On®. Tethering tools to the tool belt or wrist can prevent injuries, save tools, prevent component damage, and prevent lost time in retrieval.

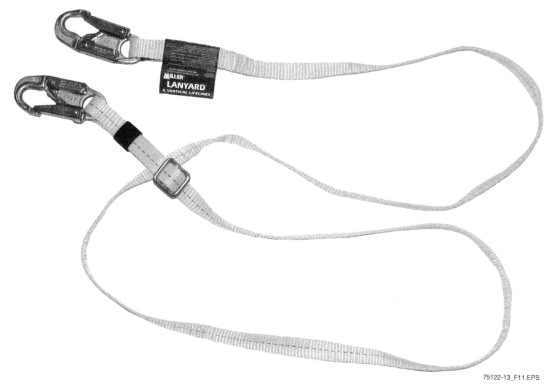

Figure 11 Non-shock absorbing lanyard.

greater than 2 feet (0.61 m). That is a relatively short distance, and it means that the positioning lanyard may have to be wrapped several times around the anchor point to get the length you need. Again, do not choke the lanyard in an effort to adjust it to the desired length. Some lanyards with special D-rings are designed for this purpose.

The line of support used for fall restraint, fall arrest, or positioning should always be toward the centerline of the body. Adjustable lanyards may have to be used to allow the correct allowable working distance. When the threat of cuts from the structure exists, use a wire or aircraft cable lanyard or a lanyard with proper padding to guard against it.

Shock-Absorbing Lanyards – Shock-absorbing, or deceleration, lanyards have shock-absorbing properties built in and are designed for fall arrest. There are also some shock-absorbing devices designed for connection to non-shock absorbing lanyards to provide the needed protection. These types of lanyards are capable of reducing the fall arrest force on the body and subsequent injuries; however, even when this force is reduced, a fall will still be painful. They should not be used for fall restraint or positioning.

A section of these lanyards is made to extend under severe stress, such as that encountered in a fall. ANSI outlines the specifics and standards of construction. The section or end that contains the energy-absorbing feature should be connected to the back D-ring rather than the anchor point. Because of their duty, deceleration lanyards should be no more than 6 feet in usable length and allow no more than an additional 3½ feet (1.06 m) of extension when the shock-absorbing feature is activated. The lanyard length must be known

What is a Newton?

A newton is a metric measurement of force combining mass, energy and time. In the meter-kilogram-second system, it is the amount of force required to accelerate a mass of one kilogram one meter per second, per second. This is expressed mathematically as $1 \text{ kg} \times \text{m/s}^2$. In the English system of measurement, it is expressed in feet-pounds-seconds. One newton equals a force of about 0.2248 pounds in the English system. It is named for Sir Isaac Newton, whose second law of motion in the late 1600s describes the change a force can produce in the movement of a body. Considered by some to be the father of modern science, he may be best known as the man who made sense out of gravity.

in order to determine the total fall. Remember that the primary anchor point should be straight above the body whenever possible to minimize the potential fall distance and swing zone. The projected fall distance must be less than the distance to any structure below that could be struck during deceleration.

The D-ring used on the PFAS for fall arrest may only be connected to one live connection at a time. This can be challenging when trying to move from one point to another, especially horizontally. The user must be able to reach back and disconnect a lanyard, while maintaining one connected at all times. There are Y-configured lanyards (*Figure 12*) used for this purpose, where a single point of attachment at the D-ring is used to accommodate two lanyards. They are also referred to as double leg or tieback lanyards.

Self-Retracting Lanyards – Self-retracting lanyards (SRLs) arrest falls as they occur by their reaction to tension. As is the case with deceleration lanyards, users must consider the operating principles as they decide on anchor points, then calculate fall distance and swing zones. Self-retracting lanyards (*Figure 13*) generally restrict free fall to 2 feet (0.61 m), then deceleration of the weight may take up to an additional 1½ feet. The total fall distance then would be 3½ feet (1.06 m), just as is the case with the deceleration lanyard by design. SRLs allow for far greater mobility for the user, both horizontally and vertically.

One very simple type of SRL with an effective length of up to roughly 10 feet (3.05 m) is used. These shorter styles are often built without true braking components and are more like deceleration lanyards. Other units allow for much greater freedom of movement for the user, and are equipped with advanced braking systems. This more sophisticated style of SRL can be found in lengths approaching 200 feet (61 m). The line

Figure 13 Self-retracting lanyard.

moves in and out slowly as the worker moves about, unless the line begins paying out at high speed. Braking operation then begins to slow and stop the fall. It is important that too much slack is not allowed to develop in the line connection to the D-ring. If slack or loose line is present, then the extra length will add to the distance of the fall before the braking mechanism begins to function.

Because of the freedom allowed by long SRLs, workers can place themselves in dangerous situations unless constant vigilance is maintained. It is all too easy to wrap the lanyard in and out of structural members as you move, or to move too far from the anchor point vertically or horizontally. Also, remember that horizontal lifelines are rated for a maximum number of connected workers; check with your supervisor or the manufacturer.

1.1.5 Rescue After a Fall

Every elevated job site should have an established rescue and retrieval plan. Planning is especially important in remote areas where help is not readily available. Before beginning work, make sure that you know what your employer's rescue plan calls for you to do in the event of a fall. Find out

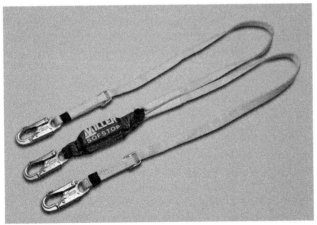

Figure 12 Y-configured shock absorbing lanyard.

what rescue equipment is available and where it is located. Learn how to use equipment for self-rescue and the rescue of others.

If a fall occurs, any employee hanging from the fall-arrest system must be rescued safely and quickly. Your employer should have previously determined the method of rescue for fall victims, which may include equipment that lets the victim perform a self-rescue, a system of rescue by co-workers, or a way to alert a trained rescue squad. If a rescue depends on calling for outside help such as the fire department or rescue squad, all the needed phone numbers must be posted in plain view at the work site. In the event a co-worker falls, follow your employer's rescue plan. Call any special rescue service needed. Communicate with and monitor the victim constantly during the rescue. After a fall, the entire fall-protection system involved in the fall must be discarded.

1.2.0 Safety Guidelines for Other Fall-Protection Systems

Other types of fall-protection systems include guardrails, safety net systems, and safe climbing assist devices. Each of these systems has its own requirements.

1.2.1 Guardrails

Guardrails are the most common type of fall protection (*Figure 14*). They protect workers by providing a barrier between the work area and the ground or lower work areas. They may be made of wood, pipe, steel, or wire rope and must be able to support 200 pounds of force applied in any direction to the top rail and 150 pounds for the mid-rail. A guardrail must be 42" high to the top rail and have a toeboard that is a minimum of 4" high. This helps to prevent the inadvertent loss of tools or material through the bottom rail. The toeboard must be securely fastened with not more than ¼" clearance above the floor level.

1.2.2 Safety Net Systems

Safety nets are used for fall protection on bridges and similar projects. They must be installed as close as possible, not more than 30', beneath the work area. There must be enough clearance under a safety net to prevent a worker who falls into it from hitting the surface below. There must also be no obstruction between the work area and the net.

Depending on the actual vertical distance between the net and the work area, the net must extend 8' to 13' beyond the edge of the work area. Mesh openings in the net must be limited to 36 square inches and 6" on the side. The border rope must have a 5,000-pound minimum breaking strength, and connections between net panels must be as strong as the nets themselves. Safety nets must be inspected at least once a week and after any event that might have damaged or weakened the system. Worn or damaged nets must be removed from service.

1.2.3 Safe Climbing Devices

There are several different systems and devices that allow a worker to climb more confidently and safely, while helping to eliminate dangerous situations. In some cases, a worker must connect to an anchor, climb, connect to another anchor, discon-

---

Case History
Communicate With Your Co-Workers

A crew was demolishing the roof of the warehouse portion of a commercial building. The work was being done at night because the coal tar on the roof would release hazardous gases if disturbed in the heat of the day. The site had adequate halogen lighting, but none of the workers on the job were using fall protection.

After the roofing material was removed, 4' × 8' sheets of plywood were exposed. Any damaged sheets needed to be replaced. A helper was to follow the workers who were replacing the plywood, and pick up the damaged sheets and put them in a chute.

One evening, a worker removed a sheet of damaged plywood, but had run out of nails to attach the replacement plywood. He walked away to get more nails. The opening where the damaged plywood had been was left unguarded. No one knew this.

The helper came along shortly after the worker left, picked up the sheet of damaged plywood, and headed for the chute. He stepped into the opening and was killed when he fell approximately 27'.

The Bottom Line: Always use the appropriate fall-protection equipment.

Source: Electronic Library of Construction Occupational Safety and Health

---

75122-13 Working from Elevations

Figure 14 Guardrails.

nect the first anchor, climb some more, etc. Safe climbing devices and systems help to eliminate the many disconnects and allow for a smoother, safer climb.

Cable grabs (*Figure 15*), also called shuttles, pucks, rope grabs, or guided fall arresters, are mounted on the cable for the worker to use as a connection point. The cable grab locks down on the cable to help restrain a fall when movement is too rapid or aggressive. In normal use, the cable grab simply rides the cable up with the user, sliding along without resistance. Its freedom of movement upward is important to ensure that it does not suddenly snag or stop.

Like other fall-protection equipment, safe climbing devices must be inspected before use and tagged out of service if they show any damage or excessive wear.

Figure 15 Cable grab.

Case History

Location is Everything

A worker was placing metal bridge decking onto the stringers of a bridge deck to be welded. After the first decking was placed down on stringers, the employee stepped onto it in order to put down the next decking. The decking was not secured in place and shifted.

Although safety nets were being used under another section of the bridge, they had not been moved forward as the crew moved to another area. The worker fell approximately 80' into the river and was killed.

The Bottom Line: Make sure that all safety equipment is in place before beginning any job.

Source: OSHA

Drop Testing Safety Nets

Safety nets should be drop-tested at the job site after the initial installation, whenever relocated, after a repair, and at least every six months if left in one place. The drop test consists of a 400-pound bag of sand of 29" to 31" in diameter that is dropped into the net from at least 42" above the highest walking/working surface at which workers are exposed to fall hazards. If the net is still intact after the bag of sand is dropped, it passed the test.

Additional Resources

Extensive information on the use of personal fall-protection equipment can be found at www.osha.gov.

1.0.0 Section Review

1. If two workers connect to the same anchor point, it must be rated for twice the intended load or _____.

    a. 3,000 pounds
    b. 5,000 pounds
    c. 7,000 pounds
    d. 10,000 pounds

2. Guardrails are the least common type of fall protection.

    a. True
    b. False

Section Two

2.0.0 Safety Guidelines for the Use of Ladders and Scaffolding

Objectives

Identify the safety guidelines for the use of ladders and scaffolding.
 a. State the safety requirements for various ladders.
 b. State the safety requirements for scaffolding.

Performance Task 2

Demonstrate how to properly inspect a ladder.

Trade Terms

Duty rating: American National Standards Institute (ANSI) rating assigned to ladders. It indicates the type of use the ladder is designed for (industrial, commercial, or household) and the maximum working load limit (weight capacity) of the ladder. The working load limit is the maximum combined weight of the user, tools, and any materials bearing down on the rungs of a ladder.

Ladder: A wood, metal, or fiberglass framework consisting of two parallel side pieces (rails) connected by rungs on which a person steps when climbing up or down. Ladders may either be of a fixed length that is permanently attached to a building or structure, or portable. Portable ladders have either fixed or adjustable lengths and are either self-supporting or not self-supporting.

Scaffolding: A temporary built-up framework or suspended platform or work area designed to support workers, materials, and equipment at elevated or otherwise inaccessible job sites.

Two-point swing scaffolding: A manual or power-operated platform supported by hangers at two points suspended from overhead supports in a way that allows it to be raised or lowered to the working position.

The most common accidents associated with ladders and scaffolding are falling, being struck by falling objects, and electrocution. In one instance, a worker died when he slipped from a fixed ladder attached to a water tower and fell 40' to the ground. He died because he was not using the proper safety equipment.

Safety is your top priority on a job. It is your responsibility to learn how to set up, use, and maintain ladders. You also need to understand the construction requirements and safety hazards associated with scaffolding systems.

2.1.0 Safety Requirements for Various Types of Ladders

Ladders are some of the most important tools on a job site. Different types of ladders should be used in different situations (*Figure 16*). Aluminum ladders are corrosion-resistant and can be used in situations where they might be exposed to the elements. They are also lightweight and can be used where they need to be frequently lifted and moved. Wooden ladders, which are heavier and sturdier than fiberglass or aluminum ladders, can be used when heavy loads must be moved up and down. Fiberglass ladders are nonconductive and also very durable, so they are useful in situations involving electrical work or where some amount of rough treatment is unavoidable. Both fiberglass and aluminum are easier to clean than wood.

Selecting the right ladder for the job at hand is important to completing a job as safely and efficiently as possible. When selecting a ladder, consider its features and how it meets the needs of the job. Always consider the highest duty rating and weight limit needed, as well as the height requirements. A ladder that is too long or too short will not allow the work surface to be reached easily, safely, or comfortably.

When selecting a ladder for a job, it is important to choose one that will extend at least 36" above the landing surface you are trying to reach. Always place the base of the ladder so that the distance between the base and the wall is one-quarter of the ladder length from the base elevation to the point where the ladder touches the wall, as shown in *Figure 17*.

Keep the following precautions in mind when setting up and using any type of ladder:

- Do not use an aluminum ladder when performing any type of electrical work or whenever there is a possibility that you might come into contact with electrical conductors.
- Place the ladder in a stable manner.
- Place the ladder so that it leaves 6" of clearance in back of the ladder and 30" of clearance in front of the ladder.
- Place the ladder so that it leans against a solid and immovable surface. Never place a ladder against a window, door, doorway, sash, loose or movable wall, or box.
- Face the ladder when climbing up or down.

(A) ALUMINUM STEPLADDER

(B) FIBERGLASS STEPLADDER

(C) FIBERGLASS EXTENSION LADDER

(D) STRAIGHT LADDER

(D) FIBERGLASS PLATFORM LADDER

(E) ROLLING WAREHOUSE LADDER

Figure 16 Types of ladders.

- Climb or descend the ladder one rung at a time. Never run up or slide down a ladder.
- Do not use ladders during high winds. If you must use a ladder in windy conditions, make sure you lash the ladder securely in order to prevent slippage.
- Check to make sure the soles of your shoes are free of oil, mud, and grease.

75122-13 Working from Elevations — Module Eleven 15

Figure 17 Proper positioning.

- Use a rope to raise and lower any tools and materials that you might need. This keeps both hands free to hold the ladder securely while climbing.
- Never rest any tools or materials on the top of a ladder.
- Move the ladder in line with the work to be done. Never lean sideways away from the ladder in order to reach the work area.
- Never stand on the top two rungs of a ladder.
- Use ladders only for short periods of elevated work. If you must work from a ladder for extended periods, use a personal fall-protection system.
- Lay the ladder on the ground when you have finished using it, unless it is anchored securely at the top and bottom where it is being used.
- Never use makeshift substitutes for ladders.
- Never use stepladders for straight-ladder work.

Each ladder is designed for a specific purpose and climbing conditions, and they generally fall into these four categories:

- Straight
- Extension
- Step
- Fixed

2.1.1 Straight Ladders

Straight ladders consist of two rails, rungs between the rails, and safety feet on the bottom of the rails (*Figure 18*). The straight ladders used in construction are generally made of wood or fiberglass. Metal ladders conduct electricity and should not be used around electrical equipment.

Ladders should always be inspected before use. Be sure to follow these guidelines when inspecting a ladder:

- Check the rails and rungs for cracks or other damage including loose rungs. If you find any damage, do not use the ladder.

Case History

Bad Weather and Unsafe Conditions Can Kill

A laborer was working on the third level of a tubular welded frame scaffolding which was covered with ice and snow. The planking on the scaffolding wasn't sturdy and the scaffolding didn't have a guardrail. There was also no access ladder for the various scaffolding levels. The worker slipped and fell approximately 20' to the pavement below. He died of a head injury.

The Bottom Line: Make sure that all scaffolding has solid planking and guardrails.

Source: OSHA

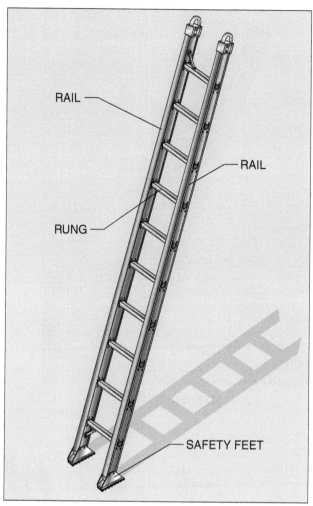

Figure 18 Straight ladder.

Figure 19 Ladder safety feet.

- Check the entire ladder for loose nails, screws, brackets, or other hardware. If you find any hardware problems, tighten the loose parts or have the ladder repaired before you use it. Ladders should only be repaired by qualified personnel.
- Make sure the feet are securely attached, that there is no damage, and that they are not worn down (*Figure 19*). Do not use a ladder if its safety feet are not in good working order.

Setup is the next step after inspecting a ladder. Use these guidelines to make sure you are setting up the ladder safely:

- Place the straight ladder at the proper angle before using it. A ladder placed at an improper angle will be unstable and could cause you to fall.
- Straight ladders should be used only on stable and level surfaces unless they are secured at both the bottom and the top to prevent any accidental movement (*Figure 20*).
- The distance between the foot of a ladder and the base of the structure it is leaning against must be one-quarter the distance between the ground and the point where the ladder touches the structure. This is also known as a 4:1 ratio.

Once you've inspected and set up the ladder, you can begin working on it. Here are some safeguards to use while working on a ladder:

- Never try to move a ladder while someone is on it. If a ladder must be placed in front of a door that opens toward the ladder, the door must be locked or blocked open so that it cannot strike the ladder.
- Never use a ladder as a work platform by placing it horizontally.
- Make sure the ladder you are about to climb or descend is properly secured.
- Make sure the ladder's feet are solidly positioned on firm, level ground.
- Always check to ensure that the top of the ladder is firmly positioned and in no danger of shifting to the right or left once you begin your climb.
- Maintain three points of contact when climbing a straight ladder.
- Always keep your body's weight in the center of the ladder between the rails.
- Never go up or down a ladder while facing away from it. Face the ladder at all times.
- Don't carry tools in your hands while you are climbing a ladder. Instead use a hand line and pull tools up once you've reached the place you will be working.

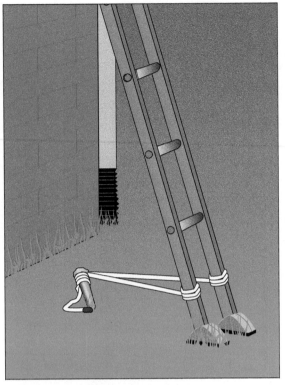

BOTTOM SECURED **TOP SECURED**

Figure 20 Securing a ladder.

> **WARNING!**
> Remember that the addition of your own weight will affect the ladder's steadiness once you mount it. It is important to test the ladder first by applying some of your weight to it without actually beginning to climb. This way, you will be sure that the ladder remains steady as you ascend.

2.1.2 Extension Ladders

An extension ladder is actually two straight ladders. They are connected so you can adjust the overlap between them and change the length of the ladder as needed (*Figure 21*).

Extension ladders are positioned and secured following the same rules as straight ladders. There are, however, some safety rules that are unique to extension ladders:

- When you adjust the length of an extension ladder, always reposition the movable section from the bottom, not the top, so you can make sure the rung locks are properly engaged after you make the adjustment.
- Make sure the section locking mechanism is fully hooked over the desired rung.
- Make sure that all ropes used for raising and lowering the extension are clear and untangled.

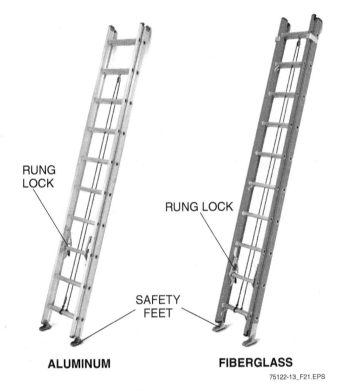

ALUMINUM **FIBERGLASS**

Figure 21 Extension ladders.

- Make sure the extension ladder overlaps between the two sections (*Figure 22*). For ladders up to 36' long, the overlap must be at least 3'.

For ladders 36' to 48' long, the overlap must be at least 4'. For ladders 48' to 60' long, the overlap must be at least 5'.

- Never stand above the highest safe standing level on a ladder. On an extension ladder, this is the fourth rung from the top. If you stand higher, you may lose your balance and fall. Some ladders have colored rungs to show where you should not stand.

2.1.3 Stepladders

Stepladders are self-supporting ladders made of two sections hinged at the top *(Figure 23)*. The section of a stepladder used for climbing consists of rails and rungs like those on straight ladders. The other section consists of rails and braces. Spreaders are the hinged arms between the sections that keep the ladder stable and prevent it from folding while in use. A stepladder may have a pail shelf to hold paint or tools.

Inspect stepladders the way you inspect straight and extension ladders. For stepladders, though, pay special attention to the hinges and spreaders to be sure they are in good repair. Also, be sure the rungs are clean. A stepladder's rungs are usually flat, so oil, grease, or dirt can easily build up on them and make them slippery.

Follow these rules when using stepladders:

- Be sure that all four feet are on a hard, even surface when you position a stepladder. If they're not, the ladder can rock from side to side or corner to corner when you climb it.
- Never stand on the top step or the top of a stepladder. Putting your weight this high will make the ladder unstable. The top of the ladder is made to support the hinges, not to be used as a step.
- Make sure the spreaders are locked in the fully open position when the ladder is in position.

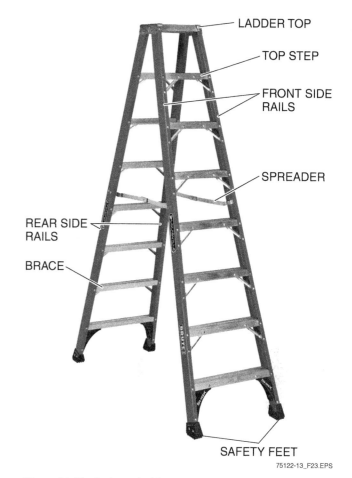

Figure 23 Typical stepladder.

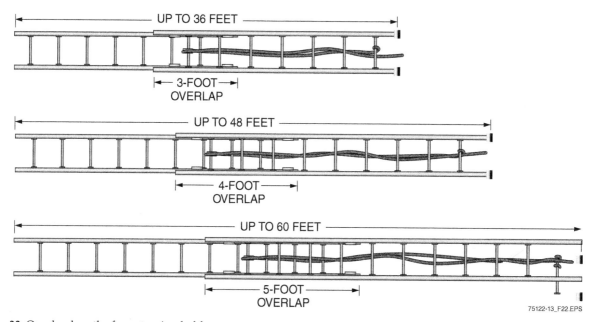

Figure 22 Overlap lengths for extension ladders.

- Never use the braces for climbing even though they may look like rungs. They are not designed to support your weight.

Figure 24 shows some common ladder safety precautions.

2.1.4 Fixed Ladders

Fixed or stationary ladders (*Figure 25*) are ladders that cannot be readily moved or carried because they are a permanent part of a building or structure. Fixed ladders must be capable of supporting at least two loads of 200 pounds (91 kilograms) each. Each step or rung must be capable of supporting a single concentrated load of at least 200 pounds applied in the middle of the step or rung.

The angle of a fixed ladder must be no greater than 90 degrees from the horizontal, as measured at the back of the ladder. The ladder must be equipped with cages, wells, ladder safety devices, or SRLs when the climb is less than 24', but the top of the ladder is a distance greater than 24' above lower levels.

Ladders where the total length of the climb equals or exceeds 24' must be equipped with one of the following:

- Ladder safety devices
- Self-retracting lifelines and rest platforms at intervals not to exceed 150' (15.2 meters)
- A cage or well and multiple ladder sections

2.2.0 Safety Requirements for Scaffolding

Scaffolding consists of elevated working platforms that support workers and materials. They are a very common sight on many construction jobs. Typical scaffolding is shown in *Figure 26*. The main part of the scaffolding is the working platform. A working platform should have a guardrail system that includes a top rail, midrail, toeboard, and screening. To be safe and effective, the top rail should be approximately 42" high, the midrail should be located halfway between the toeboard and the top rail, and the toeboard should be a minimum of 4" high.

If people will be passing or working under the scaffolding, the area between the top rail and the toeboard must be screened. Finally, the platform planks must be laid closely together. For safety purposes, the ends of the planks must overlap at least 6" and no more than 12".

2.2.1 Scaffolding Hazards

Improper or careless use of scaffolding can result in accidents, injury, or death. Those who work on scaffolding can minimize their risks by being aware of the hazards involved and following the proper safety procedures and guidelines to minimize those hazards.

The main hazards involved with the use of scaffolding are as follows:

- Falls
- Workers being struck by falling objects
- Electric shock

Falls can happen because fall protection or prevention has not been provided, is not used, or is installed or used improperly. Poorly planked scaffolding causes many falls. Working on scaffolding when conditions are dangerous such as in high winds, ice, rain, and lightning also leads to accidents. Falls also happen when scaffolding collapses because of improper construction.

Fall protection is required on any scaffolding 10' or more above a lower level. Note that this is the OSHA requirement and specific states or localities may have more stringent requirements. Fall-protection devices consist of guardrail systems, personal fall-arrest systems, and/or safety nets. A guardrail system normally serves as adequate fall protection for most scaffolding.

Case History

Think Before You Act

During the construction of a building, a masonry worker was instructed by his foreman to prepare a batch of mortar on the second level and use the stairway to carry it to the third level. This worker decided it would be quicker and easier to use the top section of an extension ladder (without safety feet) instead of the stairway. He set up the ladder by placing one end of the ladder on the wet concrete floor and leaning the other end of the ladder against the wall. He then started to climb. When he was halfway up, the ladder slipped on the wet floor, causing him to fall approximately 12' to his death.

The Bottom Line: Always follow safety instructions.

Source: The National Institute for Occupational Safety and Health (NIOSH)

DOs

- Be sure your ladder has been properly set up and is used in accordance with safety instructions and warnings.
- Wear shoes with non-slip soles.

- Keep your body centered on the ladder. Hold the ladder with one hand while working with the other. Never let your belt buckle pass beyond either ladder rail.

- Move materials with extreme caution. Be careful pushing or pulling anything while on a ladder. You may lose your balance or tip the ladder.

- Get help with a ladder that is too heavy to handle alone. If possible, have another person hold the ladder when you are working on it.

- Climb facing the ladder. Center your body between the rails. Maintain a firm grip.
- Always move one step at a time, firmly setting one foot before moving the other.

- Haul materials up on a line rather than carry them up an extension ladder.
- Use extra caution when carrying anything on a ladder.

Read ladder labels for additional information.

DON'Ts

- DON'T stand above the highest safe standing level.
- DON'T stand above the second step from the top of a stepladder and the 4th rung from the top of an extension ladder. A person standing higher may lose their balance and fall.

- DON'T climb a closed stepladder. It may slip out from under you.
- DON'T climb on the back of a stepladder. It is not designed to hold a person.

- DON'T stand or sit on a stepladder top or pail shelf. They are not designed to carry your weight.
- DON'T climb a ladder if you are not physically and mentally up to the task.

- DON'T exceed the Duty Rating, which is the maximum load capacity of the ladder. Do not permit more than one person on a single-sided stepladder or on any extension ladder.

- DON'T place the base of an extension ladder <u>too close</u> to the building as it may tip over backward.
- DON'T place the base of an extension ladder <u>too far away</u> from the building, as it may slip out at the bottom.
Please refer to the 4 to 1 Ratio Box.

- DON'T over-reach, lean to one side, or try to move a ladder while on it. You could lose your balance or tip the ladder.
Climb down and then reposition the ladder closer to your work!

4 TO 1 Ratio
Place an extension ladder at a 75-1/2° angle. The set-back ("S") needs to be 1 ft. for each 4 ft. of length ("L") to the upper support point.

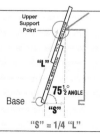

Figure 24 Ladder safety.

Figure 25 Typical fixed ladder.

Figure 26 Typical scaffolding.

Guardrail systems must extend around all open sides of the scaffolding. The side facing the work surface need not have a guardrail if it is located less than 14" away from the work surface. Any opening on a scaffolding platform must be protected by a guardrail system, including the access opening(s) and platforms that do not extend across the entire width of the scaffolding.

People who work or pass under scaffolding may be hit by falling objects. Tools, materials, debris, and scaffolding parts may fall to the surface below. Those working on scaffolding may also be injured if there are others working above them, or if the structure or workpiece extends above the work level of the scaffolding. Any worker who is exposed to the danger of falling objects is required to wear a hard hat. Depending on the situation, additional protection may be needed such as debris nets, screens or mesh, canopy structures, and toeboards. Barricades that prevent access under the scaffolding can also be used to protect workers and others.

> **WARNING!**
> If the scaffolding you are working on shifts or begins to collapse, stop what you are doing and exit the scaffolding immediately.

Because most scaffolding is made of metal, the chance of electric shock is always a hazard. Never assume that you can work around high-voltage wires just by avoiding contact. High voltages can arc through the air and cause electrocution without direct contact. When scaffolding must be erected close to power lines, the utility company must be called in to de-energize, move, and/or cover the lines with insulating protective barriers.

Case History

Watch Your Step

A maintenance employee was descending from a fixed ladder and fell approximately 5' to the floor. The employee injured his left ankle and right knee, requiring surgery. As a result, he missed several months of work.

The Bottom Line: Even a simple and repetitive task, such as climbing a fixed ladder, requires your maximum attention.

Source: OSHA

> **NOTE:** Always refer to the competent person on site if you have any questions about the safety of scaffolding.

2.2.2 Safety Guidelines for the Use of Scaffolding

Safety begins by getting training in the proper use of scaffolding. It is equally important to always use the right safety equipment, including a hard hat and personal fall-protection systems.

Never work on scaffolding, in a work cage (*Figure 27*), or on a platform if you have any of the following conditions:

- Are subject to seizures
- Become dizzy or lightheaded when working at an elevation
- Take medication that might affect your stability and/or performance
- Are under the influence of drugs and/or alcohol

When working on scaffolding, always follow these guidelines:

- Erect and use scaffolding according to the manufacturer's instructions. Scaffolding must also be erected and used in accordance with all local, state, and federal/OSHA requirements.
- An industry best practice is to attach a green, red, or yellow tag (*Figure 28*) as needed to any scaffolding that is assembled and erected to alert users of its current mechanical and/or safety status. Do not rely solely on the tag. Inspect all parts of scaffolding before each use.
- If the scaffolding shifts, exit the scaffolding immediately.

Safety Guidelines for Built-Up Scaffolding – Built-up scaffolding (*Figure 29*) is built from the ground up at a job site. Use the following guidelines when erecting and using tubular built-up scaffolding:

- Inspect all scaffolding parts before assembly.

Figure 27 Work cage.

- Never use parts that are broken, damaged, or deteriorated. Be cautious of rusted materials.
- Follow the manufacturer's recommendations for the proper methods of erecting and using scaffolding.
- Do not interchange parts from different manufacturers unless permitted by the manufacturer.
- Do not force braces or other parts to fit. Adjust the level of the scaffolding until the connections can be made easily.
- Provide adequate sills with baseplates for all scaffolding built on filled or soft ground. Compensate for uneven ground by using adjusting screws or leveling jacks.

Case History

Inspect All Materials

A crew laying bricks on the upper floor of a three-story building built a 6' platform to connect two scaffolds. The platform was correctly constructed of two 2" × 12" planks with standard guardrails. One of the planks however, was not scaffolding-grade lumber. It also had extensive dry rot in the center. When a bricklayer stepped on the plank, it broke and he fell 30' to his death.

The Bottom Line: Make sure that all planking is sound and secure. Your life depends on it.

Source: OSHA

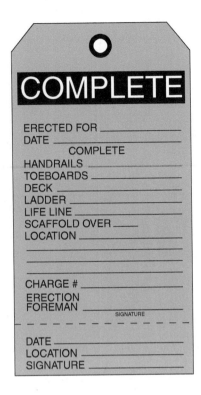 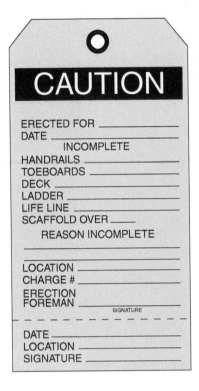 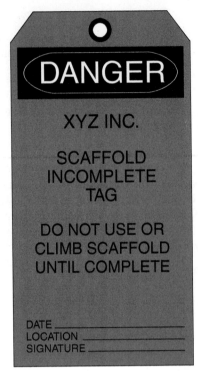

Figure 28 Typical scaffolding tags.

- Do not use boxes, concrete blocks, bricks, or other similar objects to support scaffolding.
- Keep scaffolding free of clutter and slippery material.
- Be sure scaffolding is plumb and level at all times. Follow the prescribed spacing and positioning requirements for the parts of the scaffolding. Anchor or tie-in scaffolding to the building at prescribed intervals.
- Use ladders rather than cross braces to climb the scaffolding. Position ladders with caution to prevent the scaffolding tower from tipping.
- Do not work on scaffolding that is more than 10' high without guardrails, midrails, and toeboards on open sides and ends.
- Lock or remove the casters of mobile scaffolding when it is positioned for use.
- Avoid building scaffolding near power lines.

Safety Guidelines for Swing and Other Suspended Scaffolding – Swing scaffolding (*Figure 30*) is suspended by ropes or cables in a manner that allows it to be raised or lowered as needed. Another type of suspended scaffolding is a work cage. A work cage is typically suspended with rigging devices that attach to I-beams with various sizes of clamps and rollers. It's important to follow these guidelines when erecting and using swing and other suspended scaffolding:

- Follow the manufacturer's recommendations for installation, use, and maintenance of the equipment. Before installation, inspect all parts of a structure to which rigging and tieback lines will be secured to ensure that they can support the load.

Case History

Be Prepared

Two employees were painting the exterior of a three-story building when one of the two outriggers on their two-point suspension scaffolding failed. One painter safely climbed back onto the roof while the other fell approximately 35' to his death. Neither painter was wearing an approved safety harness and lanyard attached to an independent lifeline.

The Bottom Line: Be prepared for the unexpected. When working on scaffolding, always wear approved personal protective equipment.

Source: OSHA

Figure 29 Built-up scaffolding.

Figure 30 Swing scaffolding.

- Be sure rigging devices are of the proper size and design to support the scaffolding and that they are installed properly. If counterweights are used to secure the inner end of outrigger beams, they should be fastened to the outrigger. Roofing materials and sand bags are not appropriate counterweights.
- Check that tieback lines are installed perpendicular to the face of the structure and are secured to a solid support.
- Check for power lines or electric service wires on the job site. If they pose a hazard, contact the utility company to have them temporarily de-energized.
- Observe the scaffolding's load capacity; never overload the equipment.

- Inspect all scaffolding equipment each day. Check ropes and cables thoroughly for wear, fraying, corrosion, brittleness, damage, or other conditions that may weaken them. Have them replaced as necessary by qualified personnel.
- Keep suspension ropes and cables straight and perpendicular to the platform during use. Do not affix them to anything to change the line of travel.
- Lash the scaffolding to the building or structure to prevent it from swaying.
- Stay off scaffolding during storms or high winds; watch for icy or slippery platforms.
- Guarantee safe access to the swing stage at all times.
- Use two-way radios for communication between workers on the scaffolding and on the ground.
- Do not combine two or more two-point swing scaffolding units to form one unit.
- Raise and lash the scaffolding in a safe position when not in use.

Case History

Work Independently

Two workers were sandblasting a 110'-high water tank while working on a two-point suspension scaffolding 60' to 70' above the ground. The scaffolding attachment point failed, releasing the scaffolding cables, and the scaffolding fell to the ground. Both workers died instantly because they were not tied off independently, nor was the scaffolding equipped with an independent attachment system.

The Bottom Line: Make sure that all safety equipment is in place before using any scaffolding.

Source: OSHA

75122-13 Working from Elevations

Case History

Know the Dangers of Your Job

A worker was standing on a 6"-wide plank laid between two adjacent I-beams 14' above a concrete floor. He was using a jackhammer to chip away an old concrete and brick floor from a horizontal I-beam. He lost his balance and fell to the floor below, sustaining fatal head injuries. He was not wearing any fall-protection gear.

The Bottom Line: Always use appropriate fall-arrest equipment. Do not use scaffolding unless it is constructed properly.

Source: OSHA

Additional Resources

Extensive information on the use of ladders and scaffolds can be found at www.osha.gov.

2.0.0 Section Review

1. Aluminum ladders are nonconductive and useful when performing electrical work.
   a. True
   b. False

2. The scaffolding side facing the work surface need not have a guardrail if it is located less than _____.
   a. 14" away from the work surface
   b. 16" away from the work surface
   c. 18" away from the work surface
   d. 24" away from the work surface

Section Three

3.0.0 GUIDELINES FOR THE SAFE OPERATION OF AERIAL LIFTS

Objectives

State the guidelines for the safe operation of aerial lifts.

a. Identify aerial lift components and operating requirements.
b. Describe the safe operation of scissor lifts.
c. Describe the safe operation of boom lifts.

Trade Terms

Aerial lift: A mobile work platform designed to transport and raise personnel, tools, and materials to overhead work areas.

Interlocking systems: Control systems that are set to prevent accidents by keeping one system from overriding the other.

Proportional control: A control that increases speed in proportion to the movement of the control.

Lifts are used to raise and lower workers to and from elevated job sites. There are two main types of lifts: boom lifts and scissor lifts. Both types are available in various models. Some are transported on a vehicle to a job site, where they are unloaded. Others are trailer-mounted and towed to the job site by a vehicle. Still others are permanently mounted on a vehicle. Depending on their design, aerial lifts can be used for indoor work, outdoor work, or both. *Figure 31* shows two types of commonly used aerial lifts.

Boom lifts are designed for both indoor and outdoor use. Boom lifts have a single arm that extends a work platform/enclosure capable of holding one or two workers. Some models have a jointed (articulated) arm that allows the work platform to be positioned both horizontally and vertically. Scissor lifts are typically used indoors and raise a work enclosure vertically by means of crisscrossed supports. Most models of aerial lifts are self-propelled, allowing workers to move the platform as work is performed. The power to move these lifts is provided by several means, including electric motors, gasoline, propane, or diesel engines, and hydraulic motors.

All startup and shutdown procedures must follow manufacturers' instructions.

3.1.0 Aerial Lift Components and Operating Requirements

Aerial lifts normally consist of three major assemblies: the platform, a lifting mechanism, and the base. *Figure 32* shows the components of a scissor lift.

The platform of an aerial lift is constructed of a tubular steel frame with a skid-resistant deck surface, railings, toeboard, and midrails. Entry to the platform is normally from the rear. The entry opening is closed either with a chain or a spring-returned gate with a latch. The work platform may also be equipped with a retractable extension platform. The lifting mechanism is raised and lowered either by electric motors and gears or by one or more single-acting hydraulic lift cylinders. A pump driven by either an AC or DC motor provides hydraulic power to the cylinders. The base provides a housing for the electrical and hydraulic components of the lift. These components are normally mounted in swing-out access trays. This allows easy access when performing maintenance or repairs on the unit. The base also contains the axles and wheels for moving the assembly. In the case of a self-propelled platform, electrical or hydraulic motors drive two or more of the wheels to allow movement of the lift from one location to another. Brakes are incorporated on one or more of the wheels to prevent accidental movement of the lift.

3.1.1 Aerial Lift Operator Qualifications

Only trained and authorized workers may use an aerial lift. Safe operation requires the operator to understand all limitations and warnings, operating procedures, and operator requirements for maintenance of the aerial lift. The following is a list of requirements that the operators must meet:

- Understand and be familiar with the associated operator's manual for the lift being used.
- Understand all procedures and warnings in the operator's manual and those posted on decals on the aerial lift.
- Be familiar with the employer's work rules and all related government (OSHA) safety regulations.
- Demonstrate their understanding, and operate the associated model of aerial lift during training in the presence of a qualified trainer.

Some equipment comes with interlocking systems. Most, if not all lifts have interlocking control systems to prevent an operator on the ground or in the air from causing an accident. The controls can only perform a particular action from

75122-13 Working from Elevations   Module Eleven 27

BOOM-SUPPORTED WORK PLATFORM (BOOM LIFT)

SELF-PROPELLED ELEVATING WORK PLATFORM (SCISSOR LIFT)

Figure 31 Aerial lifts.

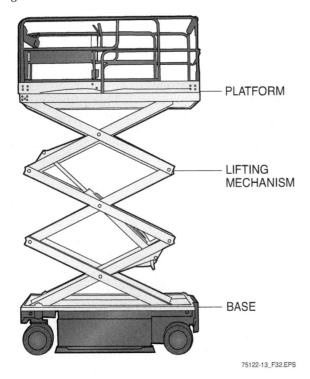

Figure 32 Scissor lift.

one station or the other. Such interlocks prevent an operator on the ground from unknowingly placing an operator on the lift at risk, or the operator in the lift from injuring the operator on the ground. Never defeat or modify such interlocks, or any other safety device.

3.1.2 Aerial Lift Safety Precautions

Safety precautions unique to aerial lifts are listed here. Remember that the other safety precautions already discussed in this module also apply. Each manufacturer provides specific safety precautions in the operator's manual that comes with the equipment. Specifically, *OSHA Standard 1926.453* defines and governs the use of aerial lifts as expressed in the following precautions:

- Avoid using the lift outdoors in stormy weather or in strong winds.
- Prevent people from walking beneath the work area of the platform.
- Use personal fall-arrest equipment (body harness and lanyard) as required for the type of lift being used. Use approved anchorage points.

- Do not use an aerial lift on uneven ground.
- Lower the lift and lock it into place before moving the equipment. Also, lower the lift, shut off the engine, set the parking brake, and remove the key before leaving it unattended.
- Stand firmly on the floor of the basket or platform. Do not lean over the guardrails of the platform, and never stand on the guardrails. Do not sit or climb on the edge of the basket or use planks, ladders, or other devices to gain additional height.

3.1.3 Aerial Lift Operating Procedure

The following procedures are for a scissor-type, self-propelled aerial lift. Remember that operating procedures vary from model to model. The operator must become familiar with all operating procedures, controls, safety features, and associated safety precautions before operating any type of aerial lift.

A typical proportional control procedure involves the operation of a lever or foot pedal to cause the aerial lift to move. The farther the control is moved, the more power is applied to the motor, and the faster the aerial lift moves.

The following tasks must be performed before operating the aerial lift:

- Carefully read and fully understand the operating procedures in the operator's manual and all warnings and instruction decals on the work platform.
- Check for obstacles around the work platform and in the path of travel, such as holes, drop-offs, debris, ditches, and soft fill. Operate the aerial lift only on firm surfaces.
- Check overhead clearances. Make sure to stay at least 10 feet away from overhead power lines.
- Make sure batteries are fully charged (if applicable).
- Make sure all guardrails are in place and locked in position.
- Perform an operator's checklist. Check all components and fluid levels, including fuel, oil, hydraulic, and coolant levels.
- Never make unauthorized modifications to the components of an aerial lift.

Figure 33 shows an example of a typical operator's checklist.

3.2.0 Safe Operation of Scissor Lifts

Scissor lifts (*Figure 34*) are powered personnel lifts. They use hydraulic rams to raise and lower the platform. Power is supplied to the hydraulic

OPERATOR'S CHECKLIST

INSPECT AND/OR TEST THE FOLLOWING DAILY OR AT BEGINNING OF EACH SHIFT

___ 1. OPERATING AND EMERGENCY CONTROLS
___ 2. SAFETY DEVICES
___ 3. PERSONNEL PROTECTIVE DEVICES
___ 4. TIRES AND WHEELS
___ 5. OUTRIGGERS (IF SO EQUIPPED) AND OTHER STRUCTURES
___ 6. AIR, HYDRAULIC, AND FUEL SYSTEM(S) FOR LEAKS
___ 7. LOOSE OR MISSING PARTS
___ 8. CABLES AND WIRING HARNESSES
___ 9. DECALS, WARNINGS, CONTROL MARKINGS, AND OPERATING MANUALS
___ 10. GUARDRAIL SYSTEM
___ 11. ENGINE OIL LEVEL (IF SO EQUIPPED)
___ 12. BATTERY FLUID LEVEL
___ 13. HYDRAULIC RESERVOIR LEVEL
___ 14. COOLANT LEVEL (IF SO EQUIPPED)
___ 15. CHECK SCISSORS FOR CRACKS, DEFORMATIONS
___ 16. CHECK FOR OPERATOR'S MANUAL

Figure 33 Operator's checklist.

system by a battery, a gasoline engine, or a propane-powered (LPG) engine. The controls for the lift are located at an operator's station on the platform itself. Some models also include an operator's station at the ground and the platform level. In addition to controlling the height of the platform, the operator is also able to move the unit to the front or rear and to steer it as it moves.

Lift heights and capacities vary by model and manufacturer. Lift heights may be as high as 41 feet and capacities may reach 2,000 pounds. Both the height and capacity of the lift can usually be found on a decal conspicuously placed on the lift. Some manufacturers offer optional manual outriggers for added stability.

No one is permitted to operate a scissor lift without training in the proper operating procedures for the particular model. Keep the following safety guidelines in mind when operating a scissor lift:

75122-13 Working from Elevations   Module Eleven   29

3.3.0 Safe Operation of Boom Lifts

Boom lifts (*Figure 35*) are work platforms mounted on the end of an arm that can be raised and lowered and may be able to extend beyond the mobile base. The boom section may consist of two or more telescoping sections. The machines usually can be driven by the operator from the platform. However, base-mounted controls are also

Figure 34 Self-propelled elevating work platform (scissor lift).

- Read and observe the warnings and cautions placed on the unit.
- Keep the vehicle and work platform at least 10 feet from electrical power lines. Scissor lifts are not insulated to protect against arcing or electrical contact.
- Check for loose or missing parts before operating the lift. Anything that is not working correctly must be taken out of service and repaired before use.
- Do not remove any parts or safety devices.
- Make certain that the immediate area is clear of all personnel and obstructions before raising or lowering the lift.
- Keep the guardrails and chains in place when operating the lift.
- Do not bridge between two scissor lifts or between a scissor lift and a building.
- Never crawl into the scissors at any point.
- Operate the vehicle only on firm, solid ground. Do not operate the vehicle near a drop-off.
- Do not impose side loads on the work platform.
- Never exceed the rated capacity of the lift.
- Do not operate the vehicle in an explosive atmosphere.
- Never climb on railings.
- Only operate per manufacturer's instructions.

(A) ARTICULATING BOOM LIFT

(B) TELESCOPIC BOOM LIFT

Figure 35 Articulating and straight boom lifts.

provided and are typically used only in an emergency. Boom lifts are generally restricted to supporting one or two workers at a time with their tools.

3.3.1 Maintenance

Maintenance of boom lifts must be done in accordance with the manufacturer's recommendations. Lubrication of the boom is a large part of the maintenance requirements. Normal engine maintenance is required on the LP- or gasoline-powered units, and recharging is required on battery-powered units. Servicing the hydraulic-powered units of boom lifts requires checking fluid levels, cleaning or replacing filters regularly, and keeping the units clean for ease in examining components for damage and detecting any leaks.

Additional Resources

Information and free e-books on aerial lifts can be found on the JLG Industries, Inc. website at www.jlg.com.

3.0.0 Section Review

1. An aerial lift should have its hydraulic reservoir level checked _____.
   a. at the start of each shift
   b. weekly
   c. monthly
   d. every three months

2. Scissor lifts can be bridged together to create a larger work platform.
   a. True
   b. False

3. The base-mounted controls on a boom lift are used _____.
   a. for maintenance purposes
   b. mainly in emergency situations
   c. for primary control
   d. for raising the boom only

SUMMARY

Falls are a very common and serious type of accident on construction sites. You have a responsibility to yourself and others to work safely and use fall-protection equipment. Personal injury and damage to equipment is less likely to happen when fall-protection equipment is used. Fall-protection equipment must be used correctly every time you work at an elevated site. It is important that everyone on a job site knows and follows the company's fall-protection procedures. If these procedures are not followed, or if protection equipment is not used, it is likely that falls will result in serious injuries or death.

Ladders and scaffolding are used in a variety of different construction jobs. The type of ladders or scaffolding will vary with each job, but the dangers won't. There is always a risk of falling and being stuck by falling objects, each of which can result in serious injury or death. It is your responsibility to be aware of and follow the safety procedures associated with ladders and scaffolding. Aerial work platforms and scissor lifts are used to raise and lower workers to and from elevated work areas.

Falls from ladders, scaffolds, lifts, or unprotected work areas can cause serious injuries or death when no fall-protection equipment or the wrong kind of equipment is used. Similar mishaps can occur when the right type of equipment is improperly used. The three types of fall protection commonly used when working at elevated levels are guardrails, personal fall-arrest systems, and safety nets. OSHA requires that all workers use fall protection to protect themselves from free falling more than 6 feet (1.8 meters) to the ground or a lower work area.

Review Questions

1. Fall protection is required when working at elevations of _____.
   a. 6' and above
   b. 8' and above
   c. 10' and above
   d. 12' and above

2. The leading cause of death in the construction industry is _____.
   a. electrocution
   b. falls
   c. vehicle accidents
   d. suffocation

3. OSHA requires that positioning anchor points be rated at _____.
   a. 1,800 pounds
   b. 2,400 pounds
   c. 3,000 pounds
   d. 5,000 pounds

4. An SRL is a(n) _____.
   a. anchor point
   b. deceleration device
   c. Y-configured lanyard
   d. saddle belt

5. Guardrails must be able to support _____.
   a. 75 pounds of force applied to the top rail
   b. 100 pounds of force applied to the top rail
   c. 150 pounds of force applied to the top rail
   d. 200 pounds of force applied to the top rail

For Questions 6 through 8, match the type of ladder to the corresponding description.

6. _____ Aluminum

7. _____ Fiberglass

8. _____ Wooden
   a. Corrosion-resistant and can be used in situations where it might be exposed to the elements
   b. Heavier than other ladders and can be used when heavy loads must be moved up and down
   c. Durable and nonconductive
   d. Illegal for use in most states

9. Straight ladders can be used on unstable surfaces if safety feet are used.
   a. True
   b. False

10. The overlap for a 50' extension ladder must be a minimum of _____.
    a. 2'
    b. 3'
    c. 4'
    d. 5'

11. It is safe to stand on the top step of a stepladder as long as you are holding onto a solid component of the building.
    a. True
    b. False

12. A fixed ladder must be able to support at least two loads of _____.
    a. 150 pounds each
    b. 200 pounds each
    c. 250 pounds each
    d. 300 pounds each

13. Electric shock is one of the hazards of working on scaffolding.
    a. True
    b. False

14. An aerial lift should have its emergency controls checked _____.
    a. daily
    b. weekly
    c. monthly
    d. every three months

15. Which of the following is true regarding scissor lifts?
    a. They are limited to a lift height of 15 feet.
    b. Their capacity can reach 2,000 pounds.
    c. The operation of scissor lifts is the same across all manufacturers.
    d. Scissor lifts are insulated against electrical contact.

Trade Terms Introduced in This Module

Aerial lift: A mobile work platform designed to transport and raise personnel, tools, and materials to overhead work areas.

Body harness: Straps that may be secured about the worker in a manner that will distribute the fall-arrest forces over at least the thighs, pelvis, waist, chest, and shoulders, with means for attaching it to other components of a personal fall-arrest system.

Connector: A device that is used to couple (connect) parts of the personal fall-arrest system and positioning device systems together. It may be an independent component of the system, such as a carabiner. It may also be an integral component of part of the system, such as a buckle or D-ring sewn into a body harness, or a snaphook spliced or sewn to a lanyard.

Duty rating: American National Standards Institute (ANSI) rating assigned to ladders. It indicates the type of use the ladder is designed for (industrial, commercial, or household) and the maximum working load limit (weight capacity) of the ladder. The working load limit is the maximum combined weight of the user, tools, and any materials bearing down on the rungs of a ladder.

Free fall: The act of falling before a personal fall-arrest system begins to apply force to arrest the fall.

Interlocking systems: Control systems that are set to prevent accidents by keeping one system from overriding the other.

Ladder: A wood, metal, or fiberglass framework consisting of two parallel side pieces (rails) connected by rungs on which a person steps when climbing up or down. Ladders may either be of a fixed length that is permanently attached to a building or structure, or portable. Portable ladders have either fixed or adjustable lengths and are either self-supporting or not self-supporting.

Lanyard: A flexible line of rope, wire rope, or strap that generally has a connector at each end for connecting the body harness to a deceleration device, lifeline, or anchorage.

Leading edge: The edge of a floor, roof, or formwork for a floor or other walking/working surface (such as the deck) which changes location as additional floor, roof, decking, or formwork sections are placed, formed, or constructed. A leading edge is considered to be an unprotected side and edge during periods when it is not actively and continuously under construction.

Lifeline: A component consisting of a flexible line connected vertically to an anchorage at one end (vertical lifeline), or connected horizontally to an anchorage at both ends (horizontal lifeline), and which serves as a means for connecting other components of a personal fall-arrest system to the anchorage.

Personal fall-arrest system: A system used to stop an employee in a fall from a working level. It consists of an anchorage, connectors, and a body harness, and may include a lanyard, deceleration device, lifeline, or suitable combinations of these.

Platform: A work surface elevated above lower levels. Platforms can be constructed using individual wood planks, fabricated planks, fabricated decks, and fabricated platforms.

Proportional control: A control that increases speed in proportion to the movement of the control.

Scaffolding: A temporary built-up framework or suspended platform or work area designed to support workers, materials, and equipment at elevated or otherwise inaccessible job sites.

Self-retracting lanyard (SRL): A deceleration device containing a drum-wound line that can be slowly extracted from, or retracted onto, the drum under slight tension during normal employee movement, and which, after the onset of a fall, automatically locks the drum and arrests the fall.

Two-point swing scaffolding: A manual or power-operated platform supported by hangers at two points suspended from overhead supports in a way that allows it to be raised or lowered to the working position.

Additional Resources

This module presents thorough resources for task training. The following resource material is suggested for further study.

Information and free e-books on aerial lifts can be found on the JLG Industries, Inc. website at **www.jlg.com**.

Extensive information on the use of ladders and scaffolds and on the use of personal fall-protection equipment can be found at **www.osha.gov**.

Figure Credist

LPR Construction, Module opener

Fall protection materials provided courtesy of Miller Fall Protection, Franklin, PA, Figures 1–6, 8, 10–13

Photo courtesy Capital Safety (DBI-SALA and PROTECTA), Figures 7, 15

Courtesy of Snap-On Industrial - Tools at Height Program, SA01

Topaz Publications, Inc., Figures 9, 25, 35B

Guardian Fall Protection, Figure 14

Courtesy of Louisville Ladders, Figures 16, 23

Werner Co., Figures 21, 24

Spider Staging, Figures 27, 30

Courtesy of JLG Industries, Inc., Figures 31, 34, 35A

Section Review Answer Key

Answer	Section Reference	Objective
Section One		
1. d	1.1.1	1a
2. b	1.2.1	1b
Section Two		
1. b	2.1.0	2a
2. a	2.2.1	2b
Section Three		
1. a	3.1.3; Figure 33	3a
2. b	3.2.0	3b
3. b	3.3.0	3c

NCCER CURRICULA — USER UPDATE

NCCER makes every effort to keep its textbooks up-to-date and free of technical errors. We appreciate your help in this process. If you find an error, a typographical mistake, or an inaccuracy in NCCER's curricula, please fill out this form (or a photocopy), or complete the online form at **www.nccer.org/olf**. Be sure to include the exact module ID number, page number, a detailed description, and your recommended correction. Your input will be brought to the attention of the Authoring Team. Thank you for your assistance.

Instructors – If you have an idea for improving this textbook, or have found that additional materials were necessary to teach this module effectively, please let us know so that we may present your suggestions to the Authoring Team.

NCCER Product Development and Revision
13614 Progress Blvd., Alachua, FL 32615

Email: curriculum@nccer.org
Online: www.nccer.org/olf

❏ Trainee Guide ❏ Lesson Plans ❏ Exam ❏ PowerPoints Other _____

Craft / Level: _____ Copyright Date: _____

Module ID Number / Title: _____

Section Number(s): _____

Description: _____

_____

_____

_____

_____

Recommended Correction: _____

_____

_____

_____

_____

Your Name: _____

Address: _____

_____

Email: _____ Phone: _____

70101-15

Your Role in the Green Environment

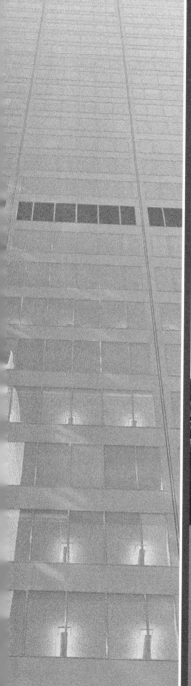

Overview

The construction industry is always changing. In this new era, the green environment is an important consideration. As a construction craft worker, you should know how daily activities at work and at home affect the green environment. With this knowledge, you can make smart choices to reduce your impact. This module explains how your daily choices make a difference. You will learn to measure your carbon footprint and reduce it. You will also learn how buildings affect the green environment and how green building rating systems work.

Module Twelve

Trainees with successful module completions may be eligible for credentialing through the NCCER Registry. To learn more, go to **www.nccer.org** or contact us at **1.888.622.3720**. Our website has information on the latest product releases and training, as well as online versions of our *Cornerstone* magazine and Pearson's product catalog.

Your feedback is welcome. You may email your comments to **curriculum@nccer.org**, send general comments and inquiries to **info@nccer.org**, or fill in the User Update form at the back of this module.

This information is general in nature and intended for training purposes only. Actual performance of activities described in this manual requires compliance with all applicable operating, service, maintenance, and safety procedures under the direction of qualified personnel. References in this manual to patented or proprietary devices do not constitute a recommendation of their use.

Copyright © 2015 by NCCER, Alachua, FL 32615, and published by Pearson Education, Inc., New York, NY 10013. All rights reserved. Printed in the United States of America. This publication is protected by Copyright, and permission should be obtained from NCCER prior to any prohibited reproduction, storage in a retrieval system, or transmission in any form or by any means, electronic, mechanical, photocopying, recording, or likewise. To obtain permission(s) to use material from this work, please submit a written request to NCCER Product Development, 13614 Progress Blvd., Alachua, FL 32615.

From *Construction Craft Laborer, Level Two, Trainee Guide*, Third Edition. NCCER.
Copyright © 2015 by NCCER. Published by Pearson Education. All rights reserved.

70101-15
YOUR ROLE IN THE GREEN ENVIRONMENT

Objectives

When you have completed this module, you will be able to do the following:

1. Select actions to improve your personal environmental impact at home and work.
   a. Describe the major challenges buildings cause directly or indirectly on the green environment.
   b. Identify choices in your personal and work life that impact the green environment.
   c. Prioritize your actions in terms of which ones matter most for the green environment.
2. Identify technologies and practices that reduce environmental impacts of a project over its life cycle.
   a. Describe the life cycle phases of a building and its impacts on the green environment.
   b. Identify green site and landscape best practices and describe their pros and cons.
   c. Identify green water and wastewater best practices and describe their pros and cons.
   d. Identify green energy best practices and describe their pros and cons.
   e. Identify green materials and waste best practices and describe their pros and cons.
   f. Identify green indoor environment best practices and describe their pros and cons.
   g. Identify green integrated strategies and describe the pros and cons of those alternatives.
3. Explain how craft workers can influence and contribute to a project's Leadership in Energy and Environmental Design (LEED) certification.
   a. Describe the LEED rating process.
   b. Identify construction activities and project features that affect a project's LEED rating.
   c. List kinds of information collected during construction to support LEED documentation.
   d. Identify common construction pitfalls that affect a project's LEED rating.

Performance Tasks

This is a knowledge-based module; there are no performance tasks.

Trade Terms

Absorptive finishes
Acidification
Aerated autoclaved concrete (AAC)
Aeration
Albedo
Alternative fuel
Aquifer depletion
Aquifers
Best Management Practices (BMPs)
Bio-based
Biodegradable
Biodiversity
Biofuel
Biomimicry
Bioswales
Blackout
Blackwater
Brownfield
Brownout
Building-integrated photovoltaics (BIPVs)
Byproducts
Carbon cycle
Carbon footprint
Carbon neutral
Carbon offsets
Carpool
Charrette
Chlorofluorocarbons (CFCs)
Cob construction
Commissioning
Compact fluorescent lamps (CFLs)
Conservation
Corporate Sustainability Report (CSR)
Deconstruction
Deforestation
Desertification
Downcycled
EarthCraft
Ecological footprint
Ecosystems
Embodied energy
Endangered species
Energy efficiency
ENERGY STAR

(continued)

Trade Terms (continued)

Environmental Product Declaration (EPD)
Equipment idling
Erosion
Flood plains
Flushout
FSC certified
Fugitive emissions
Geothermal
Global climate change
Graywater
Green Seal Certified
Greenhouse effect
Grid intertie
Halons
Hardscape
Heat island
Hybrid vehicle
Hydrochlorofluorocarbons (HCFCs)
Hydrologic cycle
Indoor air quality (IAQ)
Infiltration
Insulating concrete forms (ICFs)
Integrative design
Just-in-time delivery
Leadership in Energy and Environmental Design (LEED)
Life cycle
Life cycle assessment
Life cycle cost
Light-emitting diode (LED) lamp
Lithium ion (Li-ion)
Local materials
Location valuation
Low-emission vehicle
Maintainability
Massing
Minimum Efficiency Reporting Value (MERV)
Monofill
Multi-function materials
Nano-materials
Nickel cadmium (NiCad)
Nonrenewable
Nontoxic
Offgassing
Ozone depletion
Ozone hole
Papercrete
Passive solar design
Passive survivability
Pathogens
Payback period
Peak shaving
Persistent, bioaccumulative toxin (PBT)
Pervious concrete
Phantom loads
Phase change materials (PCMs)
Photosensor
Photovoltaics
Pollution prevention
Post-consumer
Post-industrial/pre-consumer
Preservation
Radio-frequency identification (RFID)
Rainwater harvesting
Rammed earth
Rapidly renewable
Raw materials
Recyclable
Recycled content
Recycled plastic lumber (RPL)
Recycling
Renewable
Reusable
Salvaged
Sedimentation
Sick Building Syndrome
Smart materials
Softscape
Solar
Solid waste
Solvent-based
Spoil pile
Sprawl
Stormwater runoff
Strawbale construction
Structural insulated panels (SIPs)
Sustainably harvested
Takeback
Thermal bridging
Thermal mass
Urbanization
Urea formaldehyde
Vapor-resistant
Virgin materials
Volatile organic compounds (VOCs)
Walk-off mats
Waste separation
Water efficiency
Water footprint
Water-based
Water-resistant
Waterproof
Wetlands
Xeriscaping
Zoning

Industry Recognized Credentials

If you are training through an NCCER-accredited sponsor, you may be eligible for credentials from NCCER's Registry. The ID number for this module is 70101-15. Note that this module may have been used in other NCCER curricula and may apply to other level completions. Contact NCCER's Registry at 888.622.3720 or go to **www.nccer.org** for more information.

Contents

Topics to be presented in this module include:

1.0.0 Improving Your Personal Environmental Impact 1
   1.1.0 The Nature of Change 2
      1.1.1 Changes in the Man-Made Environment 3
      1.1.2 Relationships Between Human Activities and the Green Environment 3
   1.2.0 Impact of Individual Human Activities 5
      1.2.1 The Average American Household 6
      1.2.2 The Impacts of the Products You Use 7
      1.2.3 Your Carbon Footprint and Global Climate Change 8
   1.3.0 Things You Can Do to Make a Difference 12
      1.3.1 Seeking Leverage Points 12
      1.3.2 Reducing Energy Use 14
      1.3.3 Reducing Fuel Use and Increasing Fuel Efficiency 17
      1.3.4 Rejecting, Reducing, Reusing, and Recycling Materials 18
      1.3.5 Planting Vegetation 19
      1.3.6 Finding Better Energy Sources 19

2.0.0 Best Practices for Reducing Environmental Impacts of Projects 22
   2.1.0 Facility Life Cycle 25
   2.2.0 Site and Landscape Best Practices 28
      2.2.1 Site Selection 28
      2.2.2 Building Orientation 29
      2.2.3 Site Development 29
      2.2.4 Restoring Ecosystems 31
   2.3.0 Water and Wastewater Best Practices 31
      2.3.1 Reducing Water Use 31
      2.3.2 Optimizing Water Use 32
      2.3.3 Finding Alternative Sources of Water 33
      2.3.4 Finding Alternative Sinks for Wastewater 34
   2.4.0 Energy Best Practices 35
      2.4.1 Avoiding Unneeded Energy Use 37
      2.4.2 Optimizing Energy Use 38
      2.4.3 Balancing Electrical Loads 40
      2.4.4 Finding Alternative Energy Sources 40
   2.5.0 Materials and Waste Best Practices 43
      2.5.1 Eliminating the Unnecessary Use of New Materials 44
      2.5.2 Using Materials More Efficiently 47
      2.5.3 Finding Better Sources of Materials 50
      2.5.4 Finding Better Sinks for Waste Streams 52
   2.6.0 Indoor Environment Best Practices 53
      2.6.1 Preventing Problems at the Source 54
      2.6.2 Providing Segregation and Ventilation 55
      2.6.3 Taking Advantage of Natural Forces 57
      2.6.4 Giving Users Control over Their Environment 58

Contents (continued)

- 2.7.0 Integrated Strategies ... 59
  - 2.7.1 Solving the Right Problem ... 59
  - 2.7.2 Understanding and Exploiting Relationships Between Systems ... 61
  - 2.7.3 Using Services Rather than Products ... 61
  - 2.7.4 Considering the Options ... 62
  - 2.7.5 Counting All the Costs ... 64
- 3.0.0 Contributing to a Project's LEED Certification ... 67
  - 3.1.0 LEED Green Building Rating System ... 68
    - 3.1.1 Structure of the LEED Rating System ... 69
    - 3.1.2 Types of Rating Systems and Levels of Certification ... 70
    - 3.1.3 Certification Process ... 71
  - 3.2.0 Goals of the LEED Green Building Rating System ... 71
    - 3.2.1 Selecting the Site ... 74
    - 3.2.2 Protecting and Restoring the Surrounding Environment ... 74
    - 3.2.3 Promoting Sustainable Behavior ... 77
    - 3.2.4 Conserving Resources ... 78
    - 3.2.5 Finding Better Waste Sinks ... 80
    - 3.2.6 Creating Healthy and Productive Living Environments ... 81
    - 3.2.7 Performing System Checks ... 83
    - 3.2.8 Seeking Better Methods ... 84
  - 3.3.0 LEED Documentation ... 85
  - 3.4.0 Common Pitfalls During Construction ... 86
    - 3.4.1 Poor Planning ... 86
    - 3.4.2 Poor Execution of Credit Requirements ... 86
    - 3.4.3 Poor Documentation ... 87
    - 3.4.4 Lack of Coordination with Other Trades ... 88
- Appendix A Common Acronyms ... 101
- Appendix B What it Means to be Green ... 102
- Appendix C LEED for Building Design + Construction ... 106
- Appendix D Worksheet 1: Inventory Your Household Impacts ... 108
- Appendix E Worksheet 2: Inventory Your Product Impacts ... 110
- Appendix F Worksheet 3: Determine Your Carbon Footprint ... 111

Figures

- Figure 1 Colored leaves mark the changing of seasons in many regions ... 2
- Figure 2 Satellite view of a hurricane ... 3
- Figure 3 Las Vegas has been developed in a location without much water. Its sprawl leads to increased transportation requirements ... 3
- Figure 4 The impacts of human activities on the green environment ... 5
- Figure 5 Personal choices make a difference ... 6
- Figure 6 The greenhouse effect ... 10
- Figure 7 The effects of rising sea levels ... 10

Figure 8	Santa Monica reduced its ecological footprint through recycling and energy conservation programs	12
Figure 9	Leverage points are like a carefully placed car jack	14
Figure 10	US energy consumption by source (US Department of Energy)	15
Figure 11	Energy consumption in the average US household	16
Figure 12	ENERGY STAR products meet strict requirements for energy conservation and efficiency	17
Figure 13	Interaction between a building and the green environment	27
Figure 14	Categories of best practices	28
Figure 15	Building oriented to take advantage of the sun's path	29
Figure 16	A bioswale receives stormwater runoff from a parking lot	30
Figure 17	Permeable paving options include pervious concrete and grid systems that allow water to pass into the soil	30
Figure 18	High- and low-albedo pavements	31
Figure 19	Amenities to promote alternative transportation	32
Figure 20	Waterless urinal	32
Figure 21	Components of a rainwater harvesting system	34
Figure 22	Components of a building-scale graywater system	35
Figure 23	Constructed wetland wastewater treatment system	36
Figure 24	Bio-based wastewater treatment system	36
Figure 25	High-albedo roofs	37
Figure 26	Lighting retrofits are an easy way to save energy	38
Figure 27	Using information to encourage energy conservation for peak shaving	41
Figure 28	The Green-e Energy Certified power provider logo	42
Figure 29	Small-scale wind turbine on a commercial building	43
Figure 30	Waste-based and recycled content materials	45
Figure 31	Exposed ceilings eliminate unneeded materials	46
Figure 32	Multi-function materials	47
Figure 33	Green roofs provide enclosure, stormwater management, and thermal control while cleaning the air and providing a habitat for birds and insects (A) Walls can also be vegetated (B)	48
Figure 34	Framing details for OVE wood frame construction	49
Figure 35	Carpet tile is a modular material that prevents waste	50
Figure 36	Bio-based and rapidly renewable materials	51
Figure 37	SCS Certified content label	52
Figure 38	FSC Certified wood product	52
Figure 39	Habitat for Humanity ReStores accept donations of construction materials	52
Figure 40	Look for the Green Seal Certified paint logo	54
Figure 41	VOC content can be determined from product labels	55
Figure 42	Nonabsorptive finishes such as ceramic tiles are easier to clean and do not absorb pollutants	55
Figure 43	Construction isolation measures ensure that contaminants from construction tasks do not contaminate building systems	56
Figure 44	Paint color choices can influence the user's experience of a building space	57

Figures (continued)

Figure 45	Natural ventilation system	58
Figure 46	Giving users control over their environment can increase productivity and satisfaction	59
Figure 47	Underfloor air distribution systems help conditioned air reach all parts of a space	60
Figure 48	Integrative design means that investments in one system can be offset by savings in related systems	62
Figure 49	Changes can be undertaken by professionals, laypersons, or both	63
Figure 50	Changes that require users to sacrifice comfort are unlikely to be sustained	63
Figure 51	Changes can be simple if they require no modification of infrastructure	64
Figure 52	Some green products are easy to find on the shelves of your local store	64
Figure 53	The Built Green program certified this home outside Denver, Colorado	68
Figure 54	The LEED Green Building Rating System ensures that buildings are constructed to specific environmental standards	68
Figure 55A	LEED project worksheet (1 of 2)	72
Figure 55B	LEED project worksheet (2 of 2)	73
Figure 56	Sites close to amenities reduce the need to travel by car	74
Figure 57	Projects in undeveloped areas force people to use motorized transport	74
Figure 58	Sedimentation fencing in place to manage stormwater runoff	75
Figure 59	Grass seeding and straw mulch are one way to stabilize excavated soil	75
Figure 60	Temporary fencing can be used to mark limits of disturbance	75
Figure 61	Effective tree protection fencing requires the tree to be protected as far out as the drip line	76
Figure 62	Dark pavements worsen urban heat islands	76
Figure 63	Many outdoor fixtures contribute to light pollution	76
Figure 64	Bike racks, reserved parking for carpools, and preferred parking for motorcycles and high-efficiency vehicles reward green behavior	77
Figure 65	Recycling facilities encourage users to sort their waste	78
Figure 66	An under-sink graywater system stores sink and shower water for toilet flushing	78
Figure 67	A regional material originates from within 100 miles of a job site	79
Figure 68	Bio-based materials are an important resource for green building	79
Figure 69	In some areas, mixed construction waste can be taken off site and sorted for recycling	80
Figure 70	In other areas, on-site sorting is required for recycling	81
Figure 71	On-site waste separation can earn up to two LEED credits plus one Innovation Credit	81

Figure 72	Look for levels of VOCs on product labels and buy products with low levels	82
Figure 73	Containing construction contaminants is critical to ensure indoor air quality	82
Figure 74	Low-emitting materials can help to ensure good indoor air quality	82
Figure 75	Photocopiers are a common source of indoor air pollution in office environments	83
Figure 76	Daylight and views are correlated with many positive outcomes	83
Figure 77	Building commissioning provides a chance to check that all building systems work well together	84
Figure 78	Using next-generation emissions controls for clean construction, such as the Tier IV engine on this grader, can help earn a LEED innovation Credit	85
Figure 79	Verify that products meet specifications by checking VOC levels on labels	88
Figure 80	Thermal bridging caused by a metal beam	89
Figure 81	Be careful when notching structural members to preserve structural integrity	90

Worksheet 1A: Inventory Your Household Impacts (1 of 2) 7
Worksheet 1B: Inventory Your Household Impacts (2 of 2) 8
Worksheet 2: Inventory Your Product Impacts 9
Worksheet 3: Determine Your Carbon Footprint 13

Section One

1.0.0 Improving Your Personal Environmental Impact

Objective

Select actions to improve your personal environmental impact at home and work.
  a. Describe the major challenges buildings cause directly or indirectly on the green environment.
  b. Identify choices in your personal and work life that impact the green environment.
  c. Prioritize your actions in terms of which ones matter most for the green environment.

Trade Terms

Acidification: A process that converts air pollution into acid substances, leading to acid rain. Acid rain is best known for the damage it causes to forests and lakes. Also refers to the outflow of acidic water from metal and coal mines.

Aquifer: An underground layer of water-bearing rock or soil from which groundwater can be extracted using a well.

Aquifer depletion: A situation where water is withdrawn from its underground source faster than the rate of natural recharge.

Biodiversity: A measure of the variety among organisms present in different ecosystems.

Carbon cycle: The movement of carbon between the biosphere, atmosphere, oceans, and geosphere of the Earth. It is a biogeochemical cycle. In the cycle, there are sinks, or stores, of carbon. There are also processes by which the various sinks exchange carbon.

Carbon footprint: A measure of impact human activities have on the environment. It is determined by the amount of greenhouse gases produced. It is measured in pounds or kilograms of carbon dioxide.

Carbon neutral: An action or product whose production absorbs as much carbon from the atmosphere as it produces.

Carbon offset: An agreement with another party that they will reduce their carbon production by some amount in exchange for payment.

Carpool: An arrangement in which several people travel together in one vehicle. The people take turns driving and share in the cost.

Compact fluorescent lamp (CFL): A fluorescent bulb that is designed to fit in a normal light fixture. They use less energy and last longer than incandescent bulbs.

Conservation: Using natural resources wisely and at a slower rate than normal.

Deforestation: The removal of trees without sufficient replanting.

Desertification: The creation of deserts through degradation of productive land in dry climates by human activities.

Ecological footprint: A measure of impact human activities have on the environment. It compares human consumption of natural resources with the Earth's capacity to regenerate them. Measured in hectares or acres.

Ecosystem: A combination of all plants, animals and microorganisms in an area that complement each other. These function together with all of the nonliving physical factors of the environment.

Embodied energy: The total energy required to bring a product to market. It includes raw material extraction, manufacturing, final transport, and installation.

Energy efficiency: Getting more use out of electricity already generated.

ENERGY STAR: A United States government program to promote energy efficiency. It is a joint program of the US Environmental Protection Agency (EPA) and the US Department of Energy (DOE).

Equipment idling: The operation of equipment while it is not in motion or performing work. Limiting idle times reduces air pollution and greenhouse gas emissions.

Erosion: The displacement of solids by wind, water, ice, or gravity or by living organisms. These solids include rocks and soil particles.

Geothermal: Heat that comes from within the Earth.

Global climate change: Changes in weather patterns and temperatures on a planetary scale. This may lead to a rise in sea levels, melting of polar ice caps, increased droughts, and other weather effects.

Greenhouse effect: The overall warming of a planet's surface due to retention of solar heat by its atmosphere.

Hybrid vehicle: A vehicle that uses two or more power sources for propulsion.

Life cycle: The useful life of a system, product, or building.

 70101-15 Your Role in the Green Environment

Life cycle assessment: An analytic technique to evaluate the environmental impact of a system, product, or building throughout its life cycle. This includes the extraction or harvesting of raw materials through processing, manufacture, installation, use, and ultimate disposal or recycling.

Life cycle cost: The cost of a system or a component over its entire life span.

Light-emitting diode (LED) lamp: A highly efficient, electronic light bulb using a glowing diode designed to fit a standard light fixture.

Low-emission vehicle: Vehicles that produce fewer emissions than the average vehicle. Beginning in 2001, all light vehicles sold nationally were required to meet this standard.

Ozone depletion: A slow, steady decline in the total amount of ozone in Earth's stratosphere.

Ozone hole: A large, seasonal decrease in stratospheric ozone over Earth's polar regions. The ozone hole does not go all the way through the layer.

Payback period: The amount of time it takes to break even on an investment; the period of time after which savings from an investment equals its initial cost.

Phantom loads: Electricity consumed by appliances and devices when they are switched off.

Photovoltaic: A technology that converts light directly into electricity.

Preservation: The act and advocacy of protecting the natural environment.

Raw material: A material that has been extracted directly from nature. It is in an unprocessed or minimally processed state.

Recycling: The reprocessing of old materials into new products. A goal is to prevent the waste of potentially useful materials and reduce the consumption of new materials.

Renewable: A resource that may be naturally replenished.

Solar: Energy from the sun in the form of heat and light.

Sprawl: Unplanned and inefficient development of open land.

Stormwater runoff: Unfiltered water that reaches streams, lakes, and oceans after a rainstorm by flowing across impervious surfaces.

Urbanization: Converting rural land to higher density development.

Water efficiency: Managing water use to prevent waste or overuse. Includes using less water to achieve the same benefits.

Water footprint: The volume of water used directly or indirectly to sustain something over a period of time.

Our impacts on the green environment are considerable. Resource use and global climate change are major concerns. Global climate change, also called global warming, involves changes in weather and temperature worldwide. Your daily activities have an impact on these larger problems, with your carbon footprint being a measure of your contribution to global climate change. The products you buy and the energy you consume affect the climate. It is also influenced by waste you throw away and gasoline you use. Many opportunities exist to reduce your carbon footprint, which include reducing energy and fuel use and rejecting, reducing, reusing, and recycling products. You can also plant vegetation or find better energy sources.

1.1.0 The Nature of Change

The man-made environment is changing along with the green environment. Some changes are independent of human actions. The changes in the seasons, for example, are a result of the Earth's position in relation to the sun. These changes have predictable effects on temperature and weather around the world. *Figure 1* shows colored leaves that mark the change of summer to autumn in northern climates. Seasonal changes make the setting for each building unique. Development in each region should respond to

Figure 1 Colored leaves mark the changing of seasons in many regions.

local climate, and should also reflect the demands of that climate on the built environment.

Other changes are less predictable, such as severe weather patterns. In the United States, hurricanes are common in summer and fall; snow and ice storms occur in the winter, and tornados are more common in spring and summer. In other places, cyclones or typhoons are a threat. Flooding and volcanic eruptions cause extensive damage when they occur. *Figure 2* shows a satellite view of a hurricane, also known as a cyclone or typhoon. This is one of the most severe types of weather affecting life on earth.

Severe weather also causes a challenge for the built environment. Buildings must be able to function through severe weather. The role of buildings is to preserve the safety and security of their occupants, making facilities that can survive severe storms important.

1.1.1 Changes in the Man-Made Environment

Just as the green environment changes, the built environment also changes. The role of the built environment is to help humans survive and prosper, as well as to provide shelter from the effects of weather and climate. Over time, buildings have become larger and more complex in order to meet changing human needs and desires. New technologies and practices make buildings more functional, comfortable, and efficient, while other developments result in new styles and appearances.

Expectations for built facilities have also changed. According to the National Association of Home Builders, the average house size has almost doubled in the past forty years. Houses have grown from an average of 1,400 square feet (130 square meters) in 1970 to nearly 2,700 square feet (250 square meters) in 2013. During that time, the average family size in the United States decreased from 3.14 people per household to 2.54. This means that the amount of living space per person has risen nearly 50 percent over the past 25 years.

At the same time, the development of buildings and neighborhoods has had an impact on human health and well-being. The Federal Highway Administration has found that Americans walk for fewer than 6 percent of daily trips. Much of the development in the United States is designed around the automobile. The absence of sidewalks and bicycle paths in neighborhoods discourages people from walking. Large distances between residential areas and services such as stores and schools make people more likely to drive. A University of Maryland study investigated the health effects of urban sprawl. It found that people who live in neighborhoods characterized by urban sprawl weigh about six pounds more than people living in compact neighborhoods. (Compact neighborhoods have sidewalks and shops close to residential areas.) More recent studies have confirmed the relationship between our location and our personal health and well-being. Location also has implications on the green environment. For example, many developments have been built in locations without enough water to support growth (*Figure 3*).

1.1.2 Relationships Between Human Activities and the Green Environment

Human population is growing and expectations for standards of living are increasing. This creates a greater demand for resources from the green environment. The raw materials used to build, furnish, and operate buildings are extracted

Figure 2 Satellite view of a hurricane.

Figure 3 Las Vegas has been developed in a location without much water. Its sprawl leads to increased transportation requirements.

Did You Know?

Worldwide Power Use is Growing at a Staggering Rate

According to the US Energy Information Administration, the world's total power use is expected to increase by 56 percent between 2010 and 2040. Over 85 percent of that growth will come from projects in rapidly developing countries, such as Colombia (shown here).

---

from the green environment. The fuel required to power cars, factories, and buildings is also extracted from the environment, while burning these fuels has consequences as well.

The impacts of human activities on the green environment are significant; you can see some impacts in your own backyard. There are also global impacts which have effects that may not be seen for many years (*Figure 4*). Either way, the impacts of human beings on the green environment are undeniable. Major environmental challenges include the following:

- *Global climate change* – Increasing greenhouse gases produce an overall rise in global temperatures. These gases are due in part to the burning of fossil fuels as a result of urbanization. Urbanization often increases air pollution. Temperature changes affect sea levels as ice melts at the Earth's poles, and also increase the potential for severe weather.
- *Excess wood harvesting* – The harvesting of wood resources at an unsustainable rate is known as deforestation. It leads to soil depletion, pollution of streams, and habitat loss. Deforestation also contributes to global climate change because trees reduce greenhouse gases. Preservation of forest areas can help to slow global warming.
- *Species extinction* – Habitat loss has caused thousands of species to become extinct. This is known as a loss of biodiversity. The remaining habitats are fragmented and degraded in quality. Human beings depend on diverse ecosystems to purify the air and water. These ecosystems help to stabilize climate change and provide a variety of resources from lumber to medicine. Ecosystem health is essential for human survival.
- *Decreasing water supplies* – The rise in global temperatures has caused desertification (the spread of deserts) in dryer parts of the world. In other areas, overgrazing and the overuse of groundwater has led to aquifer depletion. This is particularly true in agricultural and urban areas. Aquifers are the only reliable source of water in many parts of the country.
- *Air and water pollution* – Industrial activities and the daily activities of people contribute to the pollution of air and water. The burning of fossil fuels for transportation and energy

Figure 4 The impacts of human activities on the green environment.

produces smog, which leads to acidification (acid rain) and plant decline. The use of these fuels also contributes to ground level ozone and other forms of air pollution. Runoff from paved and deforested areas contributes to water pollution. Industrial deposits into waterways increase water temperature and contaminants. Many modern components of wastewater cannot be removed using normal treatment methods. These include such things as prescription drugs and plastic residuals, which accumulate in nature with unknown long-term consequences.

- *Soil contamination and depletion* – Contaminants from human activities often remain in the soil and eventually migrate to water sources. Sources of contamination range from leaking underground storage tanks to stormwater runoff from paved areas. Removing vegetation exposes topsoil and causes erosion losses.
- *Loss of ozone* – Release of chlorine-based gases such as refrigerants has led to ozone depletion in the upper atmosphere. Ozone molecules are necessary to shield living organisms from solar radiation. Reduced levels of ozone are thought to be responsible for the increased rate in skin cancer and cataracts in humans as well as damage to marine and terrestrial eco-systems. These chemicals also contribute to a greater-than-usual growth in the seasonal ozone hole that appears naturally in the polar regions.

None of these occur in isolation. In many cases, human activities contribute to more than one problem at a time. All of them require attention to ensure the health of the planet for future generations.

1.2.0 Impact of Individual Human Activities

Given the growing awareness of human impacts on the green environment, how do your daily activities contribute to environmental problems and their solutions? The manufacturing, transportation, use, and disposal of the products used in daily living contribute to each person's individual impacts. Your individual carbon footprint can serve as a measure of how your lifestyle contributes to global climate change. Your work on the job site and the buildings you help to create also contribute to the challenges being faced by the green environment.

1.2.1 The Average American Household

The average US household is annually responsible for the production of 4,060 pounds (1,842 kilograms) of garbage, over 46,000 gallons (174,129 liters) of wastewater, and over 96,000 pounds (43,545 kilograms) of carbon dioxide (CO_2), along with smaller amounts of sulfur dioxide (SO_2), nitrogen oxides (NO_X), and heavy metals. These impacts come from the use of products, the use of resources such as electricity, the types of waste generated, and travel (*Figure 5*).

The Consumer Electronics Association estimates that the average household spends nearly $1,200 per year on electronics, including televisions, digital cameras, and other devices. According to the US Energy Information Administration (EIA), the average household also spends about $1,400 per year to run these devices, along with all the other uses of energy in the home such as heating and lighting. In fact, according to studies conducted by Nielson Media Research, the average American household has more televisions than it does people. The US Department of Agriculture reports that American households also spend an average of more than $6,500 per year on food, much of which requires extensive use of water and fossil fuel energy to produce, process, and transport to the table.

According to the EPA, the average household in the United States spends around $520 per year on its water and sewer bill. This varies significantly based on cost of water in different parts of the country. Each person in the United States consumes 100 gallons (379 liters) per day on average and generates over 50 gallons (189 liters) of wastewater. (About half of the water consumed is for irrigation and goes back into the ground rather than the wastewater stream.) Of this wastewater, 9,000 gallons (34,069 liters) per year is used to flush away only 230 gallons (871 liters) of waste from toilets, a surprisingly wasteful use of water that has been treated to drinking water standards. Water use also requires energy for pumping, collection, treatment, and distribution. For example, letting a faucet run for five minutes uses as much energy as leaving a 60-watt light bulb on for 14 hours.

People also produce tremendous amounts of solid waste, both directly in homes and indirectly through the production of consumer goods. The average American generates 4.38 pounds (2 kilograms) of garbage per day, of which 75 percent could be recycled, but less than 35 percent typically is. That adds up to nearly 1,600 pounds (726 kilograms) per person per year. On average, each American consumes 3 gallons (11.4 liters) of gasoline per day—enough to fill up 21 bathtubs per year. This doesn't count all the trips on public transit, trains, buses, or airplanes.

All this adds up to a considerable impact on the green environment. To determine exactly how

Figure 5 Personal choices make a difference.

much impact you and your family have, complete the inventory of household impacts found in *Worksheet 1A* and *B*.

1.2.2 The Impacts of the Products You Use

Every product you use has a hidden history of harvesting, extraction, manufacturing, and transportation that cause impacts beyond what you see in the product itself. The total energy required to bring a product to market is its embodied energy. It includes everything from raw material extraction to manufacturing to final transport and installation. For example, aluminum beverage cans may be manufactured with metals mined in several different countries and then shipped on pallets made from wood harvested in another country, using fuel from yet another country. The can itself is more costly and complicated to manufacture than the beverage. Drinking the beverage takes a few minutes, throwing the can away takes a second, and yet the true impact is immense. Think of all the products you consume or use over the course of a given year. How much of an impact do you think those products have? Answer the questions in *Worksheet 2* to develop an

---

WORKSHEET 1A: INVENTORY YOUR HOUSEHOLD IMPACTS

Conducting an inventory of your household consumption, waste generation, and activities is the first step in understanding how you can reduce your impact. Answer the following questions based on your best guess. If you share a household with other people, divide the total answer for your household by the number of people who live in your home.

How many gallons of garbage do you throw away each week? An average American garbage can holds 32 gallons. Multiply by 52 to calculate the gallons of garbage you throw away per year.

Gallons of garbage per year: _____

How much electricity do you use per year? You'll find this information on your electricity bill. You can add up the total for 12 months worth of bills, or multiply a monthly average by 12. If you know how much your electricity costs per month, you can estimate the amount of energy used in kilowatt-hours by dividing your total bill by 10.

Total electricity per year in kilowatt-hours: _____

How many therms of natural gas do you use per year? Check your natural gas bill if you get one. It will tell you how many therms of gas you use per month. Your highest values will probably be during the winter heating season. If you know your monthly average, multiply by 12 to calculate your annual use.

Total therms of natural gas per year: _____

How many gallons of propane do you use each year? You can check this on your propane bill if you get one. Add up the total number of gallons per year.

Total gallons of propane per year: _____

How many gallons of fuel oil do you use per year? You can check this on your fuel oil bill if you get one. Add up the total number of gallons per year.

Total gallons of fuel oil per year: _____

On average, what is your monthly combined water and sewage bill? Check your monthly bill if you get one and add up all the amounts for a one-year period. If you don't get a water and sewage bill, you can estimate this amount as approximately $75 per person per year.

Annual cost for water and sewage: _____

How many square feet is your house or dwelling? If you don't know, draw a floor plan of your house or apartment and estimate the floor area in square feet.

Size of house (floor area) in square feet: _____

70101-14_WS01A.EPS

WORKSHEET 1B (Continued)

On average, how many miles do you drive your household vehicles per week, and what are their average fuel efficiencies? 300 miles per week per vehicle or 15,000 miles per year is about average in the United States. If you're not sure about fuel efficiency, assume your car gets 22 miles to the gallon, which is about average. If you know how many gallons of fuel you use per week, skip directly to the end of the line.

Car 1 miles per week _____ / Car 1 miles per gallon _____ = Car 1 gallons per week _____
Car 2 miles per week _____ / Car 2 miles per gallon _____ = Car 2 gallons per week _____
Car 3 miles per week _____ / Car 3 miles per gallon _____ = Car 3 gallons per week _____

Add up the gallons of gas per week for all your vehicles, and then multiply by 52 to estimate the gallons of gasoline you use per year.

Total gallons of gasoline you use per year: _____

On average, how much do you travel each year on airplanes? Estimate the number of flight segments below for each of the three distances. A round trip counts as two segments, and each segment of a multi-segment flight counts as its own flight. For example, if you fly from Baltimore to Cincinnati to Atlanta, that counts as two flight segments.

Number of short-haul flight segments (less than 700 miles or 2 hours): _____
Number of medium-haul flight segments (700 – 2,500 miles or 2 – 4 hours): _____
Number of long-haul flight segments (more than 2,500 miles or longer than 4 hours): _____

On average, how many miles do you travel on public transportation per year?

Number of miles per year on transit bus/subway: _____
Number of miles per year on intercity bus: _____
Number of miles per year on intercity train: _____

Add up the total number of miles per year on public transportation: _____

Your answers to these questions will help you calculate your own carbon footprint later in the module. Keep track of your answers in the workbook or on a separate worksheet.

---

inventory of the products you consume and begin to estimate the impacts those products have. Your answers to these questions will help you calculate your carbon footprint later in this module.

1.2.3 Your Carbon Footprint and Global Climate Change

Greenhouse gases such as CO_2 create what is known as the greenhouse effect (*Figure 6*). A certain level of greenhouse gas is essential because it prevents the loss of heat into outer space. However, if the level increases too much, it can result in global warming. The carbon cycle is the transfer of carbon (mostly in the form of CO_2) between Earth and the atmosphere. Plants absorb CO_2 and release oxygen, while people (and many of their energy-using activities, such as combustion) and animals absorb oxygen and release CO_2. The delicate balance of this cycle has shifted as people generate more and more CO_2 while depleting trees and plants that absorb CO_2 and generate oxygen.

The inventories you created are useful for estimating your own contribution to the greenhouse effect. Any changes made now can help to reduce the severity of these impacts.

For example, the rise in global temperatures is expected to cause a certain portion of polar ice caps to melt. This will cause a rise in sea levels as water from melting ice runs off into oceans. If you live in a coastal area at a low elevation, you might find that the land around you is disappearing as the ocean level rises (*Figure 7*). At a minimum, you can expect severe weather to occur more frequently—watch for stronger, more frequent

WORKSHEET 2: INVENTORY YOUR PRODUCT IMPACTS

Consider the products you buy or that are bought for you over the course of a year. Answer the following questions based on your best guess. If you share a household with other people, divide the total answer for your household by the number of people who live in your home.

Eating out: $ _____ per month × 12 = $ _____ per year

Meat, fish, & protein: $ _____ per month × 12 = $ _____ per year

Cereals & baked goods: $ _____ per month × 12 = $ _____ per year

Dairy: $ _____ per month × 12 = $ _____ per year

Fruits & vegetables: $ _____ per month × 12 = $ _____ per year

Other: $ _____ per month × 12 = $ _____ per year

On average each month, how much do you spend on the following goods and services? Multiply each amount by 12 to estimate your average annual spending in each category. If you already know how much you spend per year, you can skip the monthly amount and write the average amount per year at the end of each line.

Clothing: $ _____ per month × 12 = $ _____ per year

Furnishings & household items: $ _____ per month × 12 = $ _____ per year

Other goods: $ _____ per month × 12 = $ _____ per year

Services: $ _____ per month × 12 = $ _____ per year

hurricanes, tornados, and storms interspersed with periods of drought. There are also many other potential effects of global climate change that scientists believe might occur as temperatures rise. These include increases in forest fires due to drier conditions and possible extinction of species such as polar bears due to a loss in habitat. Many scientists believe that increases in greenhouse gases will lead to rising temperatures worldwide. Others believe that climate change may have drastically different effects in different parts of the world—some areas may get much warmer, while others may get much colder. One scientist even refers to the concept as "global weirding" since it is difficult to predict exactly how global systems will respond. While the ultimate effects of increased greenhouse gases are yet to come, scientists agree that human activities are contributing to climate change. One way to evaluate your personal impact on global climate change is through

Did You Know?

Aluminum Cans

The United States still gets three-fifths of its aluminum from virgin ore, at twenty times the energy intensity of recycled aluminum. Americans throw away enough aluminum to replace our entire commercial aircraft fleet every three months.

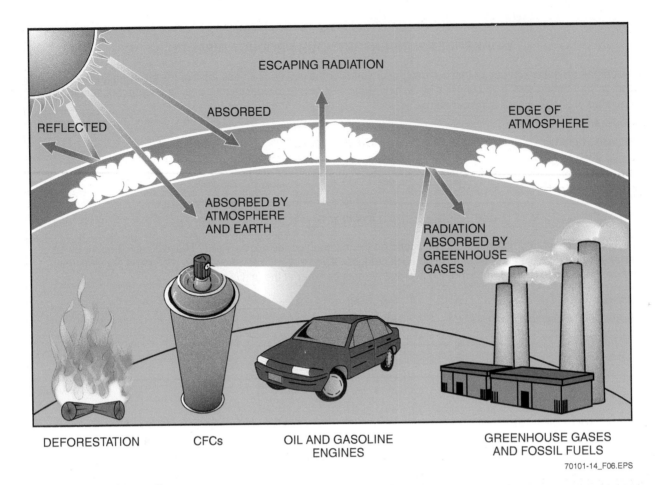

Figure 6 The greenhouse effect.

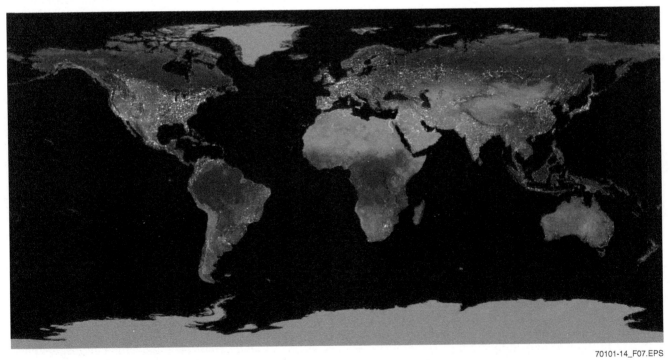

Figure 7 The effects of rising sea levels.

footprint analysis. There are three different types of footprint analysis: carbon footprint, ecological footprint, and water efficiency/water footprint. Your carbon footprint evaluates how many tons of carbon are emitted as a result of your actions. Your ecological footprint is measured in acres or hectares and represents the amount of productive land area it takes to support you, including manufacturing the products you buy, growing the food you eat, and absorbing the waste you produce. Your water footprint estimates the number of gallons or liters of water per year required to sustain your lifestyle. Each of these methods focuses on a different way to calculate the effects you personally have on the green environment. Your carbon footprint is a measure of your contribution to global climate change. Your ecological and water footprints focus on how much of the Earth's limited resources you consume. In fact, your ecological footprint is often expressed in terms of how many Earths it would take if everyone lived like you. The average American's ecological footprint is about 20 acres (8 hectares), compared to a world average of 6.7 acres (2.7 hectares).

Unfortunately, if only 12 percent of the Earth's biosphere or viable surface area is reserved for other species, there are less than 5 acres (2 hectares) available per person. With continued population growth and increase in standards of living, the amount of overshoot stands to increase even further. Right now, 20 percent of the Earth's viable ecological footprint is consumed by 2.5 percent of its population—the richest 2.5 percent. Many of the remaining population do not have enough for basic survival. Some cities are making concerted efforts to reduce their ecological footprints (*Figure 8*).

Using the information collected earlier in this module about household activities and consumption, you can estimate how many pounds of carbon per year are generated by each of these activities. For each of the sections in *Worksheet 3*, fill in the amounts from your earlier inventories and multiply by the factors shown to estimate the total pounds of carbon for each activity. Then add up the totals for each category to estimate your total carbon footprint in pounds.

Did You Know?

Surprising Impacts of Global Warming

In addition to increasing temperatures, severe weather, and rising sea levels, some surprising things that may happen because of global climate change include the following:

- Rising levels of CO_2 and warmer temperatures are prompting many plants to bloom earlier and produce more pollen, resulting in higher incidence of allergies in many parts of the United States. Poison ivy, an allergy-causing plant, has more potent itch-causing oils when grown in environments with higher CO_2 levels, and tends to thrive in warmer temperatures.
- Increased temperatures are causing changes in habitats that may be prompting animals such as squirrels, mice, and chipmunks to move to higher elevations. These changes also affect other species such as polar bears, whose habitat is threatened as polar ice caps melt away.
- Thawing of permafrost can affect the stability of surfaces and structures on top of them, causing sinkholes, damage to structures, rockslides, and mudslides. Some scientists are concerned that thawing of permafrost may also reveal bodies of animals and people formerly frozen in the ice. As these corpses are discovered, they may be a source of diseases such as smallpox to which humanity is no longer resistant.
- Due to changes in blooming schedules for many plant species and earlier arrival of spring temperatures, animals that migrate based on the number of hours of daylight may find that the food sources on which they depend are no longer available by the time they arrive in their normal migration areas.

(Source: www.livescience.com)

POISON IVY

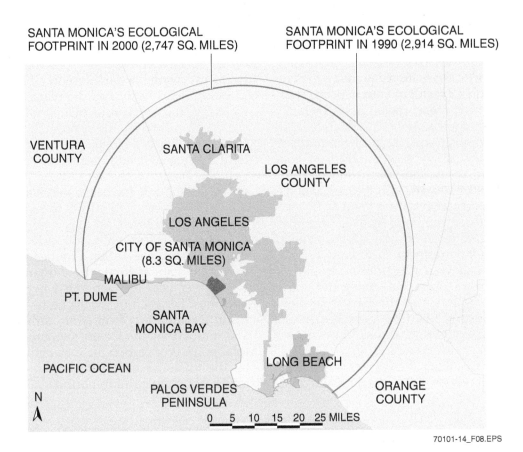

Figure 8 Santa Monica reduced its ecological footprint through recycling and energy conservation programs.

According to the World Bank, in 2010 the average American person generated about 38,800 pounds (17.6 metric tons) per person, compared to 20,060 pounds (9.1 metric tons) by the average German, 8,400 pounds (3.8 metric tons) by the average Mexican, and 660 pounds (0.3 metric tons) by the average Kenyan.

1.3.0 Things You Can Do to Make a Difference

You can do many things to reduce your personal impact on the green environment. Each action has a different level of impact. For example, according to the EPA, planting a tree seedling and allowing it to grow for 10 years can absorb about 86 pounds (0.04 metric tons) of CO_2. Avoiding the use of one gallon of gasoline saves about 19 pounds (0.009 metric tons) of CO_2, while taking an average car off the road for one year saves about 11,765 pounds (5.3 metric tons) of CO_2. You may not be able to stop driving, but you can probably stop driving as much. Small changes such as these are known as leverage points.

1.3.1 Seeking Leverage Points

Leverage points are small changes that make a big difference. To understand leverage points, think about changing a tire on your car (*Figure 9*). Placing a jack in the right location under the frame enables you to lift a heavy car off the ground. The jack and its placement are the critical factors that allow you to lift an enormous load.

GOING GREEN

Footprint Calculators

To calculate your ecological footprint, go to **www.myfootprint.org**. An online carbon footprint calculator is available at **www.carbonfootprint.com**. You can calculate your water footprint at **www.waterfootprint.org**. To see how much your own footprint equals in terms of impacts, go to **www.epa.gov/cleanenergy** and use the Greenhouse Gas Equivalencies Calculator to convert your footprint to equivalents such as tanker trucks' worth of gasoline or railcars full of coal.

WORKSHEET 3: DETERMINE YOUR CARBON FOOTPRINT

Fill in the amounts from Worksheets 1 and 2 to calculate your total carbon footprint in pounds of carbon. Which of your activities contributes the most to your carbon footprint.

Item	Quantity	Carbon Factor	Total Pounds of CO_2/Year
Gallons of garbage/year:	_____	× 2 lbs/gallon	= _____ lbs/year
Total electricity/year in kilowatt-hours:	_____	× 1.4 lbs/kWh	= _____ lbs/year
Total therms of natural gas/year:	_____	× 11.7 lbs/gallon	= _____ lbs/year
Total gallons of propane/year:	_____	× 12.7 lbs/gallon	= _____ lbs/year
Total gallons of fuel oil/year:	_____	× 22.4 lbs/gallon	= _____ lbs/year
Annual cost for water and sewage:	_____	× 8.9 lbs/dollar	= _____ lbs/year
House size (floor area) in square feet:	_____	× 2.1 lbs/sq ft	= _____ lbs/year
Gallons of gasoline/year:	_____	× 20 lbs/gallon	= _____ lbs/year
No. of short-haul flight segments/year:	_____	× 304 lbs/segment	= _____ lbs/year
No. of medium-haul flight segments/year:	_____	× 726 lbs/segment	= _____ lbs/year
No. of long-haul flight segments/year:	_____	× 2,217 lbs/segment	= _____ lbs/year
Miles per year on public transportation:	_____	× 0.5 lb/mile	= _____ lbs/year
Eating out (US dollars/year):	_____	× 0.8 lb/dollar	= _____ lbs/year
Meat, fish, & protein (US dollars/year):	_____	× 3.2 lbs/dollar	= _____ lbs/year
Cereals & baked goods (US dollars/year):	_____	× 1.6 lbs/dollar	= _____ lbs/year
Dairy (US dollars/year):	_____	× 4.2 lbs/dollar	= _____ lbs/year
Fruits & vegetables (US dollars/year):	_____	× 2.6 lbs/dollar	= _____ lbs/year
Other (US dollars/year):	_____	× 1.0 lb/dollar	= _____ lbs/year
Clothing (US dollars/year):	_____	× 1.0 lb/dollar	= _____ lbs/year
Household items (US dollars/year):	_____	× 1.0 lb/dollar	= _____ lbs/year
Other goods (US dollars/year):	_____	× 0.75 lb/dollar	= _____ lbs/year
Services (US dollars/year):	_____	× 0.4 lb/dollar	= _____ lbs/year
		Total Carbon Footprint	= _____ **lbs/year**

70101-14_WS03.EPS

When deciding what changes to make, consider how much effort is required to make the change, how easy it will be to sustain, and how much impact the change will have. Look for simple changes that you can make (such as using a jack) and the best opportunities to make them that will have the most impact (such as selecting the right place on the car to position the jack). Changes that require you to adjust how you behave on an ongoing basis are often more difficult than changes requiring a single effort.

Finally, be sure to consider all the associated costs and benefits of potential actions, not just the initial costs of purchase. Sometimes the life cycle costs of an item make it worthwhile to spend more money up front to achieve greater savings over time. Life cycle assessment involves looking at the environmental impact throughout a product's entire life cycle. For example, an energy-efficient light bulb may cost more initially but will pay for itself through reduced energy use. It may also save money through reduced maintenance costs because it lasts longer and therefore won't have to be replaced as often, and it may reduce air conditioning costs because it runs cooler. The amount of time required to save the additional money invested up front is known as the payback period.

70101-15 Your Role in the Green Environment

Module Twelve 13

Figure 9 Leverage points are like a carefully placed car jack.

1.3.2 Reducing Energy Use

One important way to reduce your carbon footprint is by reducing your overall energy use. About 87 percent of all energy used in the United States comes from burning fossil fuels, including oil, coal, natural gas, and propane (*Figure 10*). In fact, the burning of fossil fuels is the largest single contributor to carbon footprint in the United States, and it is at the root of many of the challenges faced by the green environment.

The average US household spends its energy dollar as shown in *Figure 11*. It is not surprising that most of the money goes toward heating, cooling, hot water, appliances, and lighting. What may be surprising is the large *Other* category. This category includes the growing number of battery chargers used to power cellular phones, portable tools, and other electronic devices. These chargers draw a small amount of power even when the battery is fully charged. The warmth you feel when you touch a battery charger is some of the wasted power being dissipated as heat.

Look around—how many items plugged into your walls have a clock, timer, or charger built in? All of these items draw what are called **phantom loads**. In other words, they are consuming energy even when it appears that the device itself is turned off. The only way to stop the power consumption by these devices is to unplug them completely. There are many ways to reduce your use of energy. Some of the methods recommended by **energy.gov** are listed below.

To reduce the amount of energy used to heat and cool your home, do the following:

- Choose energy-efficient furnaces or air conditioners that are the right size for your home. Check for rebates from your local utility that can help pay to replace your current systems with more efficient ones.

Going Green: Reducing your Personal Carbon Footprint

Even simple changes can make a big difference in your carbon footprint. Here are some ideas from the US Environmental Protection Agency on how much carbon you can save. To convert to metric tons, divide the number of pounds by 0.000453.

By...	You can save...
Reducing the number of car trips you take	One pound of CO_2 for every mile you don't drive
Adjusting your thermostat two degrees warmer in summer and two degrees cooler in winter	2,000 pounds of CO_2 on average per year
Unplugging chargers and turning off appliances and computers when not in use	1,000 pounds of CO_2 per year
Shortening your shower by two minutes per day	342 pounds of CO_2 per year
Updating your refrigerator to a new ENERGY STAR model	500 pounds of CO_2 per year
Using a reusable mug instead of disposable cups	135 pounds of CO_2 per year if done twice per day, every day
Recycling solid waste	One pound of CO_2 for every pound of waste recycled

U.S. ENERGY CONSUMPTION BY SOURCE

	BIOMASS RENEWABLE HEATING, ELECTRICITY, TRANSPORTATION	5.7%		PETROLEUM NONRENEWABLE TRANSPORTATION, MANUFACTURING	19.3%
	HYDROPOWER RENEWABLE ELECTRICITY	3.1%		NATURAL GAS NONRENEWABLE HEATING, MANUFACTURING, ELECTRICITY	30.5%
	GEOTHERMAL RENEWABLE HEATING, ELECTRICITY	0.3%		COAL NONRENEWABLE ELECTRICITY, MANUFACTURING	24.5%
	WIND RENEWABLE ELECTRICITY	2.0%		URANIUM NONRENEWABLE ELECTRICITY	10.1%
	SOLAR & OTHER RENEWABLE LIGHT, HEATING, ELECTRICITY	0.4%		PROPANE NONRENEWABLE MANUFACTURING, HEATING	4.2%

Figure 10 US energy consumption by source (US Department of Energy).

- Properly insulate your home's attic, walls, and slab or crawl space, including all ducts outside the conditioned space of your home.
- Contact your utility company for a free energy audit, or go online to **http://hes.lbl.gov** for a web-based energy audit tool that can help identify areas where you can improve household energy use.
- Install programmable thermostats, insulated windows, and ceiling fans to increase comfort in your home. Programmable thermostats can save up to 20 percent of your heating and cooling costs by cutting back your systems at night and during the day when you are not at home.

Think About It

Living Off the Grid

Suppose you want to build a small house in the country. The house will provide a retreat and will have all the comforts of home, but with its rural location, it is too far from existing power supplies to be connected to the grid. How will you provide power for your house and its amenities? Which electrical devices are you willing to live without in order to save energy?

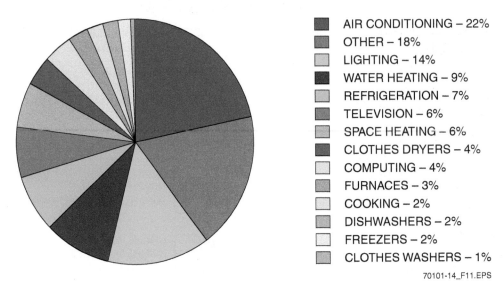

Figure 11 Energy consumption in the average US household.

To save energy used for hot water heating, do the following:

- Consider replacing your tank hot water heater with a tankless or on-demand hot water heater that only heats water as needed, or a solar hot water heater if your home design can accommodate one.
- Set your water heater to 120°F (49°C) instead of 140°F (60°C). Most appliances that require hotter water, such as dishwashers, have built-in heating coils.
- Insulate your hot water pipes and hot water tank to reduce heat loss.
- Replace faucets and showerheads with low-flow models if they are more than 10 years old. Newer showerheads can deliver a pleasant shower at 1.5 gallons (5.7 liters) per minute or less due to the introduction of air into the water stream.

To save energy used by appliances such as washers, dryers, refrigerators, and dishwashers, do the following:

- Wash your laundry in cold water. It helps your clothes last longer and fade less, and can save over $60 per year in hot water heating costs.
- Consider a front-loading washing machine. These machines save up to 25 percent of the water used for washing, along with the energy required to heat that water. They also use less detergent and remove more water from clothes during the spin cycle, reducing drying time and energy use.
- Use a clothesline instead of a dryer.
- Choose a top-freezer model of refrigerator instead of a side-by-side model. Top-freezer models are more efficient due to the placement of the compressor. Beware of built-in icemakers—they add to the energy cost and are the culprit for many repair calls.
- Keep your appliances full. Full refrigerators maintain more even temperatures and cycle on less frequently. Full washing machines and dishwashers get the maximum amount of cleaning possible for the least use of resources. Load washing machines, dryers, and dishwashers according to the manufacturer's instructions to increase effectiveness.

To save energy in lighting, household electronics, and miscellaneous areas, do the following:

- Swap your incandescent bulbs for compact fluorescent lamps (CFLs) or light-emitting diodes (LEDs). They use less energy and last longer than a standard incandescent bulb.
- Use power strips and sensors. Power strips can be used to unplug your electronic devices without disrupting cords and cables. Electric photocells and occupancy sensors turn lights off automatically when they are not needed.
- Buy ENERGY STAR appliances and devices. These products perform in the top 25 percent of their product class in terms of energy use and contribute to overall energy conservation (*Figure 12*). Listings of approved products are available online at www.energystar.gov.
- Use the sleep mode or shut down equipment such as computers when not in use. Keeping a computer and monitor running 24 hours a day consumes up to 1,100 kilowatt-hours of energy, which could cost over $100 per year depending on electricity rates.

Think About It

Life Cycle Costs and Benefits

Consider the light bulbs shown here, all equivalent to a 60-watt incandescent bulb. The 8.5 watt light-emitting diode (LED) lamp on the right costs about $10, lasts for 25,000 hours of use or more, and consumes about $24 worth of energy over its life cycle. The 60-watt incandescent lamp on the left costs about 50 cents and lasts for about 2,000 hours of use. Over 25,000 hours of use (requiring 13 bulbs), this bulb will consume about $170 worth of energy at 11.3 cents per kilowatt-hour (typical in the United States). The 15-watt compact fluorescent lamp in the center costs about $1.70, lasts for about 10,000 hours of use, and would consume about $42 worth of energy over a life cycle of 25,000 hours (not including the costs of three bulbs needed to last this long). All generate about the same amount of light. Which one is the best investment?

1.3.3 Reducing Fuel Use and Increasing Fuel Efficiency

Over half of the carbon footprint of most Americans is due to transportation choices. Some ways to reduce the overall use of fuel and increase fuel efficiency include the following:

- Drive a fuel-efficient, low-emission vehicle. Many options are available, including gasoline-electric hybrid vehicles that get over 50 miles per gallon (21.2 kilometers per liter). You may also consider driving a scooter or motorcycle. While most people can't get by with this type of vehicle alone, the fuel efficiency (up to 100 miles per gallon/42.4 kilometers per liter or more) can make scooter commuting a worthwhile investment, particularly if you live in a warmer climate.
- Carpool and/or combine trips whenever possible. The greatest amount of air pollution is released when an engine is cold. Combine errands on a single trip whenever possible. This helps vehicles run more efficiently and reduces the overall miles traveled.
- Maintain your vehicle properly. Regularly check the tire pressure and keep the engine tuned for optimal performance and fuel economy.
- Drive slower. Driving at speeds above 60 miles per hour (96.6 kilometers per hour) decreases fuel efficiency.
- Reduce equipment idling. If you will be sitting in your car for more than a few minutes, turn the engine off.
- Use mass transportation instead of individual vehicles. The fuel used per person is much lower for mass transportation than for travel in a single occupancy vehicle.
- Avoid unnecessary air travel. Airplanes are a significant source of carbon emissions, and their emissions are made worse by the fact that they occur high in the atmosphere. If possible, replace short-haul flights with travel by train or bus.

Figure 12 ENERGY STAR products meet strict requirements for energy conservation and efficiency.

Did You Know?
Energy Consumption by Appliances

According to the US Energy Information Administration, the average US household consumed 11,280 kilowatt hours of electricity in 2011. Where does all this energy go? The US Department of Energy has estimated how much electricity is required to run different appliances you may have in your home. Which appliance in your home requires the most energy to run? Considering the number of hours each appliance is used, which appliance uses the most energy overall? What is your biggest leverage point? Does it surprise you?

Appliance	Electricity Requirement
Fish aquariums	50 to 1,210 watts
Coffee maker	900 to 1,200 watts
Washer	350 to 500 watts
Dryer	1,800 to 5,000 watts
Dishwasher	1,200 to 2,400 watts
Ceiling fan	65 to 175 watts
Window fan	55 to 250 watts
House fan	240 to 750 watts
Hair dryer	1,200 to 1,875 watts
Clothes iron	1,000 to 1,800 watts
Microwave	750 to 1,100 watts
Average desktop PC	60 to 250 watts
Laptop	50 watts
Refrigerator	725 watts
Televisions	65 to 170 watts
Toaster	800 to 1,400 watts
DVD player	20 to 25 watts
Vacuum	1,000 to 1,440 watts
Water heater	4,500 to 5,500 watts

- Purchase locally produced goods and products. Many foods, particularly fruits and vegetables, are transported thousands of miles before they reach local stores. Shop at local farmers' markets or ask your grocer to stock locally produced foods to minimize the fuel used to transport goods.
- Move information, not people or things. Take advantage of the telephone and Internet as much as possible to avoid unnecessary travel. For example, rather than driving to several stores to find out which one carries a product or to compare prices, call the stores instead or research it on the Internet.

1.3.4 Rejecting, Reducing, Reusing, and Recycling Materials

The three Rs (reduce, reuse, and recycle) is a common phrase used to remember how to make greener choices. Reducing the materials used has a greater impact than reusing products or recycling them. This is true because using less of a material means less has to be produced, causing less impact from harvesting, manufacturing, and transportation. Likewise, reusing materials is better than using disposable materials that you recycle. Each reuse is one less product that has to be made. When you do use disposable goods,

Did You Know?
Energy Savings and Your Health

A common tip to save energy is to reduce the temperature on your water heater. However, sometimes this can result in unexpected problems. Researchers are learning that lowering water heater temperature can allow the growth of pathogens in water heater tanks and pipes. These pathogens may be released into the air during showers or other water uses.

For most healthy adults, these pathogens don't pose a health risk. However, for people whose immune systems are vulnerable, including young children, ill people, and the elderly, it's better to keep the tank set at a higher temperature to prevent pathogen growth. This isn't likely to be a problem with tankless water heaters. On the other hand, higher water temperatures can increase the risk of scalding, so it is important to take extra care when using hotter water, especially when there are young children in the house.

---

always try to recycle them instead of just throwing them out. Recycling is a good source of raw materials for new products. It is also an important way to save resources for the future.

The best choice of all, however, is a fourth R—reject. Be on the lookout for ways to reject the need to use raw materials in the first place. Rejecting the offer of a bag when you've only purchased one item is a good example. Most receipts can also be rejected—think twice before you say yes at the gas pump or ATM. From online statements for your bills to email instead of letters, there are many opportunities to reject the use of raw materials. Together, the four Rs—reject, reduce, reuse, and recycle—can help you remember the best way to be green in the choices you make every day.

1.3.5 Planting Vegetation

One important way to offset carbon emissions is by planting vegetation, such as trees or a garden. As mentioned earlier, plants absorb CO_2 and produce oxygen. They also provide shade, reduce stormwater runoff, promote groundwater recharge, and prevent soil erosion. There are many options for creating a beautiful and functional landscape while minimizing the impact of fertilizers, pesticides, herbicides, and emissions from lawn equipment. These include the following:

- Plant native plants in your landscape instead of alien or invasive species. Native plants are well-adapted to local climate conditions and pests, and can often be grown without irrigation, pesticides, or fertilizer.
- Group plants with similar needs together in the landscape. This allows you to focus the application of irrigation, pesticides, and fertilizer only on the areas that really need it, while avoiding over-irrigation of other parts of the landscape.
- Plant edibles as part of your landscape, or even replace your lawn with a garden that can provide food for you and your family.
- Use manual equipment instead of gasoline-powered equipment for landscape maintenance. A standard walk-behind lawnmower puts out as much emissions as 11 cars, while the average riding mower generates as much pollution as 34 cars. Hybrid mowers are now available that use small amounts of gasoline to generate electricity that powers the mower. These mowers can also be used as generators to power electrical tools and equipment in the field.

Did You Know?
The Impacts of Paper

One tree makes about 16 reams of copy paper, or 8,333 sheets. According to the USDA Forest Service, that same tree over a 50-year life span generates $31,250 worth of oxygen. It also provides $62,000 worth of air pollution control, recycles $37,500 worth of water, and controls $31,250 worth of soil erosion. Paper has a much larger impact than many people realize. The average office employee prints 10,000 pages per year, with six wasted pages per day on average. Producing each ton of paper (200,000 sheets) requires 98 tons (89 metric tons) of raw material. It requires as much energy as an average household in ten months. It also produces 19,075 gallons (72,207 liters) of wastewater. Think carefully before you print!

1.3.6 Finding Better Energy Sources

After reducing your energy consumption as much as possible, investigate alternative sources of energy. These include distributed renewable energy systems such as wind turbines, hydropower, or photovoltaics you can install on site. Other options include purchasing green power if your utility provider offers it. This generally involves paying a small surcharge per kilowatt-hour to support the development of renewable energy resources such as large-scale wind, solar, and geothermal energy. Geothermal energy is derived from heat generated within the earth.

If green power is not available in your area, you can also reduce your carbon footprint through the purchase of carbon offsets. Carbon offsets can be purchased from certification companies and represent a certain amount of carbon that your purchase helps to reduce. For example, you may elect to purchase a carbon offset to mitigate the impacts of air travel, and you may even be asked if you'd like to buy an offset when you purchase a plane ticket. The money spent on offsets is used to help fund projects that reduce the amount of carbon emissions somewhere else. Carbon offsets can be used to plant trees, increase energy efficiency, or develop renewable energy sources. For example, a company may offset the carbon generated by its production processes by helping a school replace an old heating system. Spending money to reduce carbon emissions somewhere else is one way to be carbon neutral (that is, generating no greenhouse gases at all).

Think About It

Reducing Your Carbon Footprint

Compare your carbon footprint with the list of actions in this section. Which actions can you take to reduce your carbon footprint? Which ones seem easiest to achieve? Be sure to consider what kinds of changes you will be most likely to sustain over time. Remember that changes in technology are often easier to sustain than changes in behavior.

Making Greener Choices

Even small choices have an impact. The carbon footprint of the average cheeseburger is 6.6 pounds (0.003 metric tons). This includes the energy required to grow the feed for the cattle, grow the vegetables, grow and process the grain, store and transport the components, and cook the burger.

Did You Know?

Additional Resources

Field Guide for Sustainable Construction., Department of Defense – Pentagon Renovation and Construction Office. (2004). PDF, 2.6 MB, 312 pgs. Available for download at **www.wbdg.org**.

Greening Federal Facilities, 2nd Ed. US Department of Energy Federal Energy Management Program. (2001). PDF, 2.1 MB, 211 pgs. Available for download at **www.wbdg.org**.

Natural Capitalism. Lovins, A., Hawkin, P., and Lovins, L.H. (1995). Little, Brown, & Company, Boston, MA. Available online at **www.natcap.org**.

Sustainable Buildings Technical Manual. Public Technologies, Inc./US Department of Energy. (2006). PDF, 3.1 MB, 292 pgs. Available for download at **www.greenbiz.com**.

Sustainable Buildings and Infrastructure: Paths to the Future. Pearce, A.R., Ahn, Y.H., and Hanmi Global. (2012). Routledge, London, UK.

Sustainable Construction: Green Building Design and Delivery, 3rd Ed. Kibert, C.J. (2012). Wiley, New York, NY.

The HOK Guidebook to Sustainable Design, 3rd Ed. Odell, W. and Lazarus, M.A. (2015). Wiley, New York, NY.

1.0.0 Section Review

1. Reducing your overall energy use is a way to _____.
   a. improve health and comfort
   b. save trees
   c. improve the economy
   d. reduce your carbon footprint

2. The increase in greenhouse gases is a result of _____.
   a. using biofuels
   b. burning fossil fuels
   c. using ethanol
   d. increased satellite use

3. The water used to flush toilets is commonly _____.
   a. treated to the same quality standards as drinking water
   b. nonpotable
   c. treated with antibacterial chemicals
   d. treated with special salts

Section Two

2.0.0 Best Practices for Reducing Environmental Impacts of Projects

Objective

Identify technologies and practices that reduce environmental impacts of a project over its life cycle.

a. Describe the life cycle phases of a building and its impacts on the green environment.
b. Identify green site and landscape best practices and describe their pros and cons.
c. Identify green water and wastewater best practices and describe their pros and cons.
d. Identify green energy best practices and describe their pros and cons.
e. Identify green materials and waste best practices and describe their pros and cons.
f. Identify green indoor environment best practices and describe their pros and cons.
g. Identify green integrated strategies and describe the pros and cons of those alternatives.

Trade Terms

Absorptive finish: A surface finish that will absorb dust, particles, fumes, and sound.

Aerated autoclaved concrete (AAC): A lightweight, precast building material that provides structural strength, insulation, and fire resistance. Typical products include blocks, wall panels, and floor panels.

Aeration: Mixing air into a liquid substance.

Albedo: The extent to which an object reflects light from the sun. It is a ratio with values from 0 to 1. A value of 0 is dark (low albedo). A value of 1 is light (high albedo).

Alternative fuel: Any material or substance that can be used as a fuel other than conventional fossil fuels. They produce less pollution than fossil fuels. These include biodiesel and ethanol.

Best Management Practice (BMP): A way to accomplish something with the least amount of effort to achieve the best results. This is based on repeatable procedures that have proven themselves over time for large numbers of people.

Bio-based: A material derived from living matter, either plant or animal.

Biodegradable: Organic material such as plant and animal matter and other substances originating from living organisms. These are capable of being broken down into innocuous products by the action of microorganisms.

Biofuel: A solid, liquid, or gas fuel consisting of or derived from recently dead biological material. The most common source is plants.

Biomimicry: The act of imitating nature to create new solutions modeled after natural systems.

Bioswale: An engineered depression designed to accept and channel stormwater runoff. A bioswale uses natural methods to filter stormwater, such as vegetation and soil.

Blackout: A situation where the electrical grid fails and does not provide any power.

Blackwater: Water or sewage that contains fecal matter or sources of pathogens.

Brownfield: Property that contains the presence or potential presence of a hazardous substance, pollutant, or contaminant. Cleaning up and reinvesting in these properties reduces development pressures on undeveloped, open land and improves and protects the environment.

Brownout: A situation where the electrical grid provides less power than normal, but enough for some equipment to still work.

Building-integrated photovoltaics (BIPVs): Photovoltaic systems built into other types of building materials. See *photovoltaic*.

Byproducts: Another product derived from a manufacturing process or a chemical reaction. It is not the primary product or service being produced.

Charrette: An intense meeting of project participants that can quickly generate design solutions by integrating the abilities and interests of a diverse group of people.

Chlorofluorocarbons (CFCs): A set of chemical compounds that deplete ozone. They are widely used as solvents, coolants, and propellants in aerosols. These chemicals are the main cause of ozone depletion in the stratosphere.

Cob construction: An ancient building method using hand-formed lumps of earth mixed with sand and straw.

Commissioning: A review process conducted by a third party that involves a detailed design review, testing and balancing of systems, and system turnover.

Deconstruction: Taking a building apart with the intent of salvaging reusable materials.

Downcycling: Recycling one material into a material of lesser quality. An example is the recycling of plastics into lower-grade composites.

Endangered species: A population of a species at risk of becoming extinct. A threatened species is any species that is vulnerable to extinction in the near future.

Flood plain: An area surrounding a river or body of water that regularly floods within a given period of time.

Flushout: Using fresh air in the building HVAC system to remove contaminants from the building.

FSC certified: Wood or wood products that have met the Forest Stewardship Council's tracking process for sustainable harvest.

Fugitive emissions: Pollutants released to the air other than those from stacks or vents. They are often due to equipment leaks, evaporative processes, and wind disturbances.

Graywater: A non-industrial wastewater generated from domestic processes including laundry and bathing. Graywater comprises 50 to 80 percent of residential wastewater.

Green Seal Certified: A certification of a product to indicate its environmental friendliness. Green Seal is a group that works with manufacturers, industry sectors, purchasing groups, and government at all levels to green the production and purchasing chain. Founded in 1989, Green Seal provides science-based environmental certification standards.

Grid intertie: A connection between a local source of power and the utility power grid.

Hardscape: Paved areas surrounding a project, including parking lots and sidewalks.

Hydrologic cycle: The circulation and conservation of Earth's water supply. The process has five phases: condensation, infiltration, runoff, evaporation, and precipitation.

Indoor air quality (IAQ): The content of interior air that could affect health and comfort of building occupants.

Insulating concrete form (ICF): Rigid forms that hold concrete in place during curing and remain in place afterwards. The forms serve as thermal insulation for concrete walls.

Integrative design: A collaborative design methodology that emphasizes the input of knowledge from several areas in the development of a complete design.

Infiltration: The movement of air from outside a building to inside through cracks or openings in the building envelope.

Just-in-time delivery: A material delivery strategy that reduces material inventory. The material that is delivered is used immediately.

Lithium ion (Li-ion): A type of rechargeable battery in which a lithium ion moves between the anode and cathode. They are commonly used in consumer electronics.

Local materials: Materials that come from within a certain number of miles from the project. Materials produced locally use less energy during transportation to the site. The LEED system of building certification offers points for the use of regional or local materials.

Maintainability: An indication of how difficult it is to properly maintain a building system or technology.

Massing: The three-dimensional shape of a building, including length, width, and height. Determines the amount of a building exposed to sunlight.

Minimum efficiency reporting value (MERV): A measurement scale designed in 1987 by the American Society of Heating, Refrigeration, and Air-Conditioning Engineers (ASHRAE) to rate the effectiveness of air filters. The scale is designed to represent the worst-case performance of a filter when dealing with particles in the range of 0.3 to 10 microns.

Monofill: A landfill that accepts only one type of waste, typically in bales.

Multi-function material: A material that can be used to perform more than one function in a facility.

Nano-material: A material with features smaller than a micron in at least one dimension.

Nickel cadmium (NiCad): A popular type of rechargeable battery using nickel oxide hydroxide and metallic cadmium as electrodes.

Nonrenewable: A material or energy source that cannot be replenished within a reasonable period.

Nontoxic: Substances that are not poisonous.

Offgassing: The evaporation of volatile chemicals at normal atmospheric pressure. Building materials release chemicals into the air through evaporation.

Papercrete: A fiber-cement material that uses waste paper for fiber.

Passive solar design: Designing a building to use the sun's energy for lighting, heating, and cooling with minimal additional inputs of energy.

Passive survivability: The ability of a building to continue to offer basic function and habitability after a loss of infrastructure (for example, water and power).

Pathogen: An infectious agent that causes disease or illness.

Peak shaving: An energy management strategy that reduces demand during peak times of the day and shifts it to off-peak times, such as at night.

Pervious concrete: A mixture of coarse aggregate, Portland cement, water, and little to no sand. It has a 15 to 25 percent void structure and allows 3 to 8 gallons of water per minute to pass through each square foot. Also known as permeable concrete.

Phase change material (PCM): A material that stores or releases large amounts of energy when it changes between solid, liquid, or gas forms.

Photosensor: A sensor that measures daylight and adjusts artificial lighting to save energy.

Pollution prevention: The prevention or reduction of pollution at the source.

Post-consumer: Products made out of material that has been used by the end consumer and then collected for recycling.

Post-industrial/pre-consumer: Material diverted from the waste stream during the manufacturing process.

Radio-frequency identification (RFID): An automatic identification method. It relies on storing and remotely retrieving data using devices called RFID tags.

Rainwater harvesting: The gathering and storing of rainwater.

Rammed earth: An ancient building technique similar to adobe using soil that is mostly clay and sand. The difference is that the material is compressed or tamped into place, usually with forms that create very flat vertical surfaces.

Rapidly renewable: A material that is replenished by natural processes at a rate comparable to its rate of consumption. The LEED system of building certification rewards use of rapidly renewable materials that regenerate in 10 years or less, such as bamboo, cork, wool, and straw.

Recyclable: Material that still has useful physical or chemical properties after serving its original purpose. It can be reused or remanufactured into additional products. Plastic, paper, glass, used oil, and aluminum cans are examples of recyclable materials.

Recycled content: A material containing components that would otherwise have been discarded.

Recycled plastic lumber (RPL): A wood-like product made from recovered plastic, either by itself or mixed with other materials. It can be used as a substitute for concrete, wood, and metals.

Reusable: A material that can be used again without reprocessing. This can be for its original purpose or for a new purpose.

Salvaged: A used material that is saved from destruction or waste by being reused.

Sick Building Syndrome: A variety of illnesses thought to be caused by poor indoor air. Symptoms include headaches, fatigue, and other problems that increase with continued exposure.

Smart material: Materials that have one or more properties that can be significantly changed. These changes are driven by external stimuli.

Softscape: The area surrounding a building that contains plantings, lawn, and other vegetated areas.

Solid waste:: Products and materials discarded after use in homes, businesses, restaurants, schools, industrial plants, or elsewhere.

Solvent-based: A material that consists of particles suspended or dissolved in a solvent. A solvent is any substance that will dissolve another. Solvent-based building materials typically use chemicals other than water as their solvent, including toluene and turpentine, with hazardous health effects.

Strawbale construction: A building method that uses straw as the structural element, insulation, or both. It has advantages over some conventional building systems because of its cost and availability.

Structural insulated panel (SIP): A composite building material used for exterior building envelopes. It consists of a sandwich of two layers of structural board with an insulating layer of foam in between. The board is usually oriented strand board (OSB) and the foam can be polystyrene, soy-based foam, urethane, or even compressed straw.

Sustainably harvested: A method of harvesting a material from a natural ecosystem without damaging the ability of the ecosystem to continue to produce the material indefinitely.

Takeback: A condition where manufacturers recover waste from packaging or products after use, at the end of their life cycle.

Thermal mass: A property of a material related to density that allows it to absorb heat from a heat source, and then release it slowly. Common materials used to provide thermal mass include adobe, mud, stones, or even tanks of water.

Urea formaldehyde: A transparent thermosetting resin or plastic. It is made from urea and formaldehyde heated in the presence of a mild base. Urea formaldehyde has negative effects on human health when allowed to offgas or burn.

Vapor-resistant: A material that resists the flow of water vapor.

Virgin material: A material that has not been previously used or consumed. It also has not been subjected to processing. See *raw material*.

Volatile organic compounds (VOCs): Gases that are emitted over time from certain solids or liquids. Concentrations of many VOCs are up to 10 times higher indoors than outdoors. Examples include paints and lacquers, paint strippers, cleaning supplies, pesticides, building materials, and furnishings.

Walk-off mat: Mats in entry areas that capture dirt and other particles.

Water-based: Materials that uses water as a solvent or vehicle of application.

Waterproof: A material that is impervious to or unaffected by water.

Water-resistant: A material that hinders the penetration of water.

Wetland: Lands where saturation with water is the dominant factor. This determines the way soil develops and the types of plant and animal communities living in the soil and on its surface.

Xeriscaping: Landscaping that requires little or no water for irrigation. Xeriscaping can be achieved through smart plant selection, mulching, and other tactics.

Zoning: Rules for placing similar items next to one another, as in plants within a landscape, or types of buildings within a community.

Throughout recorded history, humans have constructed buildings to protect themselves and their possessions. Buildings provide shelter from adverse climate conditions such as rain, snow, wind, and temperature extremes. They also offer privacy and security. In addition to these roles, built facilities also serve the following purposes:

- Collection, treatment, and/or storage of solid, liquid, and gaseous waste
- Provision and distribution of pure water
- Processing and distribution of agricultural products into food
- Manufacturing and distribution of various products

The impacts of buildings on the environment have not always been obvious, but the effect over time is undeniable. Buildings are responsible for over 10 percent of the world's freshwater withdrawals. They also use 25 percent of the world's wood harvest and nearly 40 percent of its material and energy flows. In the United States, 72 percent of all electricity use is related to building construction and operation. Almost a third of all new and remodeled buildings suffer from poor indoor air quality (IAQ) due to emissions from various sources, volatile organic compounds (VOCs) released from building products, and pathogens from inadequate moisture protection and ventilation. Poor indoor air results in variety of illnesses known collectively as Sick Building Syndrome, which is thought to cost more than $60 billion annually in lost productivity and medical costs nationwide.

Nearly one-quarter of all ozone-depleting chlorofluorocarbons (CFCs) are emitted by building air conditioners. The processes used to manufacture building materials also contribute. Approximately half of the CFCs produced around the world are used in buildings. This includes refrigeration and air conditioning systems and fire extinguishing systems. CFCs are also in certain insulation materials. In addition, half of the world's fossil fuel consumption is attributed to the servicing of buildings. Lighting accounts for 20 to 25 percent of the electricity used in the United States annually. Offices in the US spend 30 to 40 cents of every energy dollar for lighting. This makes it one of the most expensive and wasteful building features. Finally, the construction industry is responsible for 20 to 40 percent of the total municipal solid waste stream.

2.1.0 Facility Life Cycle

The life cycle of a built facility may range from 30 years to over 100 years. It typically runs through the following phases:

- *Planning or pre-design* – A facility's life starts with an idea during the planning or pre-design phase. This phase focuses on defining the function of the building. The physical requirements of the building are determined. Goals for the budget and construction schedule are set. Legal or regulatory constraints are identified that should be taken into account during design. The outcome of the planning phase is typically a set of requirements for the facility. These describe the functional expectations the owner has for the facility.

- *Design* – The second phase of the facility life cycle is design. This is when the facility is transformed from an idea into a set of construction drawings and specifications. A design meeting called a charrette may be used to obtain input from project stakeholders.
- *Construction* – During construction, workers follow the set of construction documents to construct a building that meets the owner's requirements. The outcome of the construction phase is a completely functional building.
- *Operation and maintenance* – After construction, the operation and maintenance phase of the life cycle begins. The building is used to meet the owner's needs. This is typically the longest phase of the life cycle. Operation is the process by which the facility performs its intended functions. Maintenance consists of all actions performed on the facility to keep it in good operating condition. It includes activities such as changing light bulbs, servicing the building systems, and cleaning the facility. It also includes minor repairs or replacement of building components that break down.
- *Rehabilitation or end of life cycle and disposal* – At some point, a facility will no longer meet the owner's requirements. A possible choice is to rehabilitate or reconstruct the facility to improve its performance. This can be even more challenging than the original construction. Another possibility is to end the life cycle of the facility through deconstruction or demolition. Deconstruction is a planned, careful disassembly of the facility to salvage building components for future use. Demolition is a more destructive process.

Built facilities affect the green environment over their life cycles. *Figure 13* shows some of the relationships between a building and the environment during its life cycle. Note the import and export of materials, energy, and waste, all of which have impacts on the green environment.

Think About It
Inventory the World Around You

Consider the built environment in which you're sitting right now. What features of the building do you think work well in terms of green performance? Which features do you think could be better? Working individually or in small teams, explore the building in which you are taking this class. Also, consider the site on which it is located. Begin by drawing a floor plan for the building. Add significant features of the site. Take an inventory of the following items and mark the location of each observation on your floor plan or site map. For each item, list why it's an example of excellence or how it could be improved.

- *Energy use* – Look around your building for technologies and features that save or waste energy. Pay special attention to the building envelope, which is the outside skin of the building including windows and doors. Examine the heating, ventilation, and cooling systems. Study the lighting in the building. What opportunities do you see to improve the energy performance of the building? What features already work well?
- *Water use* – Inspect your building for features that save or waste water. Pay special attention to faucets, fixtures, appliances, landscaping, and cooling systems that rely on water. What opportunities can you find to improve the water performance of your building? What things already work well?
- *Materials* – Now consider your building from the standpoint of materials. Examine the structure of the building, its enclosure, and interior and exterior finishes. Look at how the building is currently being used. What opportunities do you see to improve the performance of your building in terms of the materials used for construction or consumed for operations? What things are already satisfactory?
- *Indoor environment* – Look around the interior spaces of your classroom building. What about the spaces contributes to a comfortable and productive environment? What could be better? Pay special attention to lighting, acoustics, views, and indoor air quality.
- *Outdoor environment* – Step outside your building and inspect the site on which it is located. How could the building and its site be improved from the outside? In which ways is it already satisfactory? Pay special attention to the landscape (both the plants and the paved areas). Also consider the location of the building itself with regard to the people who use it. What transportation options exist now? What could be better?

When you've finished with your inventory, share your findings with the rest of the class. Have other students noticed things you didn't see?

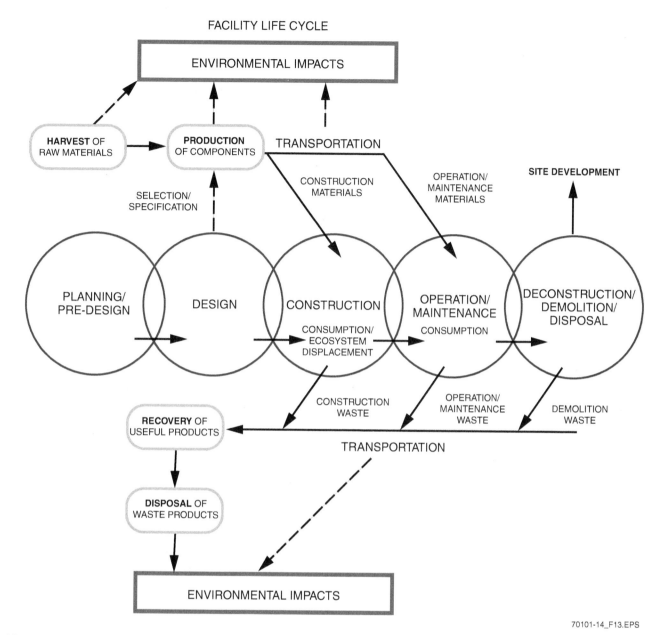

Figure 13 Interaction between a building and the green environment.

You can change your actions to improve those impacts. Some impacts are due to the materials and energy used by the facility while other impacts come from the waste streams that leave the facility. How the facility interacts with the site on which it is located creates impacts, as well as how the facility interacts with its users. The following best management practices (BMPs) represent a spectrum of options for improving the impacts of the built environment on the green environment:

- *Site and landscape* – This category involves choosing a good site for a facility. It involves the specific placement of the facility on the site and the design of landscaping. Avoiding damage to the site is important. Restoring the quality of the site after construction is also important.
- *Water and wastewater* – This category addresses optimizing the use of water. It also covers alternative water sources and wastewater sinks.
- *Energy* – This category addresses optimizing the use of energy. It also covers methods of using energy more efficiently. It includes balancing energy demands and seeking alternative sources for energy.
- *Materials* – This category includes optimizing the use of materials. It promotes using abundant, renewable, or multifunction materials. It also involves seeking alternative sources for materials.

- *Waste* – This category includes eliminating or preventing waste. It involves reusing waste within the facility system, sharing it with other systems, and storing it for future use.
- *Indoor environment* – This category includes preventing problems at the source, segregating polluters, taking advantage of natural forces, and giving users control over their environment.
- *Integrated strategies* – This category includes practices that result in multiple benefits from a single action. It includes capitalizing on construction means and methods. It also discusses making technologies do more than one thing and exploiting relationships between systems.

Each of these categories is shown in *Figure 14* and described in detail in the following sections.

2.2.0 Site and Landscape Best Practices

The first category of best practices deals with the facility site and landscape. The most important decision for a building is the selection of a site, followed by where to put the building on the site. These two decisions affect the building's performance over its life cycle. Next, the development of the site should use low-impact principles and features as well as ensure that ecosystems on the site are restored to their best quality.

2.2.1 Site Selection

The choice of site affects the building's energy use and environmental impact, and it also governs travel to and from the facility. For example, choosing a site in an already developed area provides building occupants with access to existing stores and services. It may even allow them to take advantage of walking, bicycling, or public transit to access the building instead of having to drive a car. A good location can also provide a positive contribution to an existing neighborhood. Choosing a site that is a brownfield is a possibility. A brownfield is a site that may have real or perceived environmental contamination. Cleaning up a brownfield represents a positive step for the community, and it may provide a tax credit or development incentive for the builder.

Avoid sites with valuable ecological resources or higher levels of risk from environmental damage, such as sites with wetlands or habitats of threatened or endangered species. The goal is to preserve these resources and prevent the need to mitigate or help correct any impacts. It also saves considerable expense. Choose sites that are well out of flood plains or areas where mudslides or wildfires are common. Be sure to consider the long-term risks of resource depletion as well. For example, the water supply in some areas is becoming scarce. This may mean restrictions on how facilities are developed and used in the future.

Some sites can share common resources with other sites if the facilities have different hours of operation. For example, a bank that is only open during the day may allow parking for a nightclub or restaurant at night. This eliminates the need for two separate parking lots. Choosing sites in already developed areas provides access to the existing infrastructure, which includes streets as well as water, power, and sewer lines. This can save time and money and reduce the negative effects on the site.

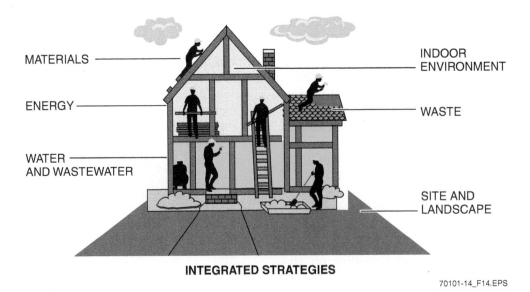

Figure 14 Categories of best practices.

2.2.2 Building Orientation

A building's relationship to the sun has a large impact on energy use. In fact, it can save 30 to 40 percent in heating and cooling costs over the life cycle of the building. *Figure 15* shows a building in relation to the sun's path in the Northern Hemisphere. The building is oriented with its long axis running from east to west, and most of the windows are on the south side of the building. In the summer, the sun passes higher in the sky and the overhangs on the building help shade the windows, which keeps the building cooler. In the winter, the sun passes lower in the sky. This allows sunlight to enter the windows under the overhangs and provides free heat and light. This is an example of passive solar design. Passive solar design takes advantage of the sun's energy to provide heat without using electrical or combustion energy. Another important idea in passive solar design is building massing, where the shape and size of the building are designed to provide access for solar energy to enter the building to provide heat, power, and daylight. Massing and orientation can also influence a building's ventilation. If oriented with openings to take advantage of prevailing winds, buildings can be naturally ventilated in some climates.

Trees can also be used to provide free cooling. Deciduous trees on the south, east, and west sides of a building can provide shade in the summer, which reduces cooling costs. In the winter, they lose their leaves and allow solar energy to reach the building. This energy provides warmth and heat gain in the winter. Trees also provide shelter from prevailing winds, which helps to reduce the heating load on the building. Preserve existing trees whenever possible. In addition to energy benefits, trees also have positive effects on water retention, air quality, and property values.

Another important consideration when placing the building on site is avoiding areas that are difficult to develop. For example, areas with steep slopes are more difficult to develop than flat areas, and require more earthwork and/or larger foundations. The same is true for areas with high water tables. After examination of the entire site, thought must be given to both initial and long-term costs of the building location.

2.2.3 Site Development

Site development involves changes made to the landscape of the site. These changes may adjust terrain, change vegetation, and add amenities for humans. Ideally, site development should avoid disturbing existing landscape, especially on sites that have not been previously developed. Contractors may set up fences to protect trees and other important landscape during construction. Damage to site ecosystems can be avoided with careful planning of the construction process. Limiting heavy equipment to areas where it is needed will also help to protect soil and avoid damage to vegetation.

Heavy equipment used for site development contributes to a project's carbon footprint. Heavy equipment should be used as efficiently as possible

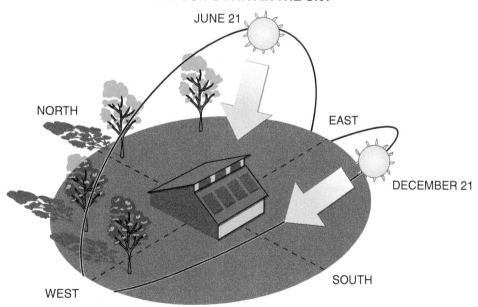

Figure 15 Building oriented to take advantage of the sun's path.

for earthwork. More efficient equipment can also be used. Heavy equipment that uses biodiesel instead of fossil fuels, including excavators and trucks can be found at some sites. Hybrid vehicles are available that combine combustion engines with electric motors and batteries. These vehicles can also reduce the carbon impacts of equipment idling.

Landscaping is divided into two main categories: softscape and hardscape. Softscape refers to the vegetated parts of a site, such as grass, trees, and bushes. Hardscape refers to areas that are paved or otherwise developed. Best practices associated with low-impact landscaping include the following:

- *Native plants* – Use native plant species. Native plants are well adapted to the climate and require little or no irrigation, fertilizer, or pesticides once established. Avoid exotic plants from other parts of the world. They often out-compete native plants and can become invasive. Kudzu is a well-known invasive species common in the southeast United States. Many common plants, such as English ivy, are also invasive and should be avoided. Visit **www.plants.usda.gov** and click on the links for noxious and invasive plants. This website also has information on native plants that may do well in your area.
- *Zoned landscaping* – Group plants with similar needs, which is called zoning. Zoned landscaping allows you to focus irrigation, fertilizer, and pesticides only on the plants that require it. It can also help reduce the risk of wildfire around buildings. This is accomplished by keeping areas near buildings clear of flammable vegetation and debris.
- *Mulching* – Use bark chips, pine straw, or other natural materials to suppress weeds and retain moisture. This reduces the need for irrigation and pesticides, and helps stabilize new plants until they become established.
- *Xeriscaping* – Xeriscaping is the selection of native plants that require no irrigation once they are established. It is common in the southwest, where xeriscape areas of rock, sand, and native plants replace green lawns that would require high volumes of water in order to survive.
- *Runoff control* – Use vegetated areas to absorb and treat stormwater runoff from paved areas. Bioswales and rain gardens are engineered basins that collect stormwater. They are similar in purpose except that rain gardens tend to have a more formal layout of plants and stones, while bioswales are planted using a more natural approach. Once collected, the water is percolated slowly back into the soil (*Figure 16*). Plants within the bioswale also help to remove contaminants from runoff. These pollutants would otherwise have to be treated at a wastewater treatment plant to avoid contaminating local water streams.
- *Permeable pavement* – Permeable pavement captures stormwater and allows it to percolate back into the soil (*Figure 17*). Options include pervious concrete (concrete that contains spaces through which water can move), stabilized soil, and grid systems. Stormwater that drains over pavement may become contaminated and require collection and treatment. This is particularly true in areas where vehicles drive or park. These contaminants include engine leaks, rubber and asbestos particles, and other residues. When concentrated in runoff, these contaminants can pollute local streams and rivers. Stormwater runoff can also cause stream damage after running across hot

Figure 16 A bioswale receives stormwater runoff from a parking lot.

Figure 17 Permeable paving options include pervious concrete and grid systems that allow water to pass into the soil.

pavement, leading to thermal pollution of local streams, which can harm aquatic plants and animals. Some permeable pavements provide basic treatment of contaminants, accomplished through plant roots or microorganisms in the pavement pores.

- *Light-colored pavement* – Pavement that is light in color helps minimize the amount of heat absorbed from the sun. Light colors have high albedo values. The term *albedo* describes the extent to which an object reflects light from the sun. It is a ratio with values from 0 to 1; a value of 0 is dark (low albedo), while a value of 1 is light (high albedo). High-albedo pavements help to minimize the urban heat island effect, which is caused in part by dark asphalt and buildings that absorb, rather than reflect, heat. Using light-colored concrete can reduce overall temperatures during summer months and lower the air conditioning requirements of surrounding buildings. *Figure 18* shows high- and low-albedo pavements.
- *Alternative transportation* – Some landscaping features promote the use of alternative transportation (*Figure 19*). These include sheltered transit stops, bike racks, and bike paths and sidewalks or pedestrian trails. Setting aside special parking for carpools or alternative fuel vehicles rewards drivers for making greener choices. One type of alternative fuel is biofuel, which is commonly made from plants.

2.2.4 Restoring Ecosystems

After the project is built, restore natural ecosystems to the extent possible by setting aside areas of the site to remain undeveloped. Native species require little or no maintenance; they provide landscape beauty and serve as a natural habitat for various creatures, including birds and butterflies. Open spaces also provide a place for building occupants to be outside and enjoy nature. Whenever possible, work with adjoining sites to create connected undeveloped areas, including local streams and waterways.

2.3.0 Water and Wastewater Best Practices

The second category of best practices involves the sources and uses of water for a facility. These practices also deal with sinks for a building's wastewater. Nearly all buildings require a source of water to meet the needs of occupants. This includes water for drinking, washing, and waste disposal.

Water may appear to be abundant, but less than 3 percent of all water on Earth is fresh water. The rest is salt water, which is mostly unusable. Of the small amount of fresh water available, 69 percent is trapped as ice and snow cover and 30 percent is stored as groundwater. Less than 1 percent is available on the surface as freshwater lakes and rivers. The hydrologic cycle continuously replenishes the freshwater supply. This process has five phases: condensation, infiltration, runoff, evaporation, and precipitation. In many areas, the rate of use exceeds the recharge rate, leading to aquifer depletion.

To conserve water, the most important action you can take is to eliminate unnecessary uses of it. Then increase the efficiency of the water you do use. After determining how to reduce the need for water as much as possible, the next steps are to look for alternative sources of water and seek alternative ways to remove and treat wastewater.

2.3.1 Reducing Water Use

If you are involved with an existing building, complete a water audit. This will help find leaks and identify opportunities for improvement.

Figure 18 High- and low-albedo pavements.

Figure 19 Amenities to promote alternative transportation.

Your local water authority may be able to provide information on completing a water audit.

Another way to eliminate water use is to install waterless toilets and urinals. Waterless urinals (*Figure 20*) take advantage of the fact that urine is a liquid, meaning it does not require water to be conveyed through wastewater pipes. Waterless urinals use a special trap that allows urine to pass through, but prevents sewer gases from escaping.

Waterless toilets are also available. For example, composting and incinerating toilets both work without the use of water. Composting toilets convert human waste into a useful product that can be used to improve the soil. Incinerating toilets use electrical energy or other fuel to convert the waste into ash. Incinerating toilets use a considerable amount of energy, which should be taken into account as a tradeoff against the water savings.

2.3.2 Optimizing Water Use

Many types of fixtures are available to reduce water use in showers and sinks. These include low-flow or aerated fixtures and alternate types of controls. Aeration is the introduction of air into the water flow. It is a common technique to reduce the actual flow of water in faucets and showerheads while maintaining the perception of high-volume flow. It works well for hand washing and showers, but can be frustrating when filling a pot with water for cooking or cleaning. Some aerated kitchen faucets allow the user to control the degree of aeration and restore full flow when needed.

Another option is foot-operated sinks. These allow precise control of flow timing while leaving hands free. More precise automated flow controls are also becoming available for sinks and toilets. The EPA's Water-Sense® program provides a labeling system to help identify low-flow fixtures and controls (see **www.epa.gov/watersense**).

Toilet technology has improved considerably over the last decade. There are now many options available to improve the efficiency of water use. In addition to conventional low-flow toilets, several manufacturers now offer dual-flush toilets. These allow a full-volume flush for solid waste and a half-volume flush for liquids. Automated dual-flush control units are also available. These units dispense either a full-volume or low-volume flush depending on how long the user is within range of the flush sensor. Manual controls are

Figure 20 Waterless urinal.

Think About It
Waterless Urinals: Balancing Water Savings with Maintenance Requirements

Waterless urinals save significant water over the life of a building. They eliminate water used to flush liquid waste down sewer pipes. However, there are tradeoffs to think about in choosing a urinal that will work well. Waterless urinals require a different maintenance routine than usual. Proper training is required for maintenance staff to keep waterless urinals working properly.

There are two main types of waterless urinals. One type relies on a special cartridge to prevent sewer gases from escaping, which can be costly and the cartridge must be periodically replaced. The second type uses an oil or gel to provide a seal that must be replenished regularly. The gel is less expensive than the cartridge. However, calcium deposits may build up in the waste pipes with this technology, and a plumber must then periodically remove these deposits. This adds to the maintenance cost of the urinal.

Choosing the best urinal for a project requires thinking about operations and maintenance. What resources will be available? What training will be provided? Choosing a urinal that does not fit with maintenance routines or expectations may result in a need to replace it in the future. It may even require changes in walls to provide water supply pipes for replacement urinals. Think carefully about the whole life cycle of a product before using it, to avoid problems later.

---

also included to allow users to control the flush level if desired.

When selecting water sources for landscaping, consider drip irrigation or irrigation controlled by moisture sensors and timers. Drip irrigation provides a slow release of water directly to the root zone of the plant. This results in less evaporation than a typical sprinkler system. Moisture sensors and timers control the operating periods of irrigation systems. This ensures that water is only dispensed when needed. It also allows watering during the early morning when water has time to reach plant roots before evaporating. With either system, be sure to properly locate and maintain all components of the system to avoid waste.

Water-efficient appliances, such as high-efficiency dishwashers and washing machines, reduce water use by up to 50 percent over traditional units. Since these units use less hot water, they also save energy and are reviewed under the EPA's ENERGY STAR program. See **www.energystar.gov** for a list of washing machines and dishwashers that meet ENERGY STAR criteria.

2.3.3 Finding Alternative Sources of Water

A third option for greening your use of water is to seek alternative sources of water. Focus on uses that do not require water to be treated to drinking water standards. Much of the water used in homes and businesses does not require this high level of water quality. Think about the extra energy and treatment required to use drinking water to flush a toilet.

One alternative source of fresh water is known as rainwater harvesting, which involves capturing and using rainwater. Rainwater harvesting has been used on a small scale since development began. Larger scale systems have been used in arid parts of the country for decades and are likely to become more common as water scarcity increases. *Figure 21* shows a rainwater harvesting system. These systems require a collection surface, typically a roof that is relatively debris-free. The surface must also be made from a chemically stable substance. Older metal roofs should not be used as they may have joints sealed with lead-based solder, which could contaminate the water. Asphalt shingles are also not recommended since they tend to shed particles as they age. Slate, synthetic, and lead-free metal roofs are all good collection surfaces.

In addition to the collection surface, rainwater systems require components to provide filtration, storage, and overflow. Most existing roof drainage systems can be retrofitted for rainwater harvesting. Depending on rainfall levels, a roof can capture tens of thousands of gallons of water per year.

Another alternative source of water is graywater. This is used water from sinks, showers, and laundry facilities. It is called graywater because it contains a small amount of contamination that can be safely treated on site. Full-building graywater systems collect and filter water for irrigation or toilet flushing (*Figure 22*). Water recovered for reuse is transmitted through purple-colored pipes instead of regular supply pipes. Graywater systems must be carefully designed to meet code requirements, as graywater supports bacterial growth while in storage. Provisions must be made to balance

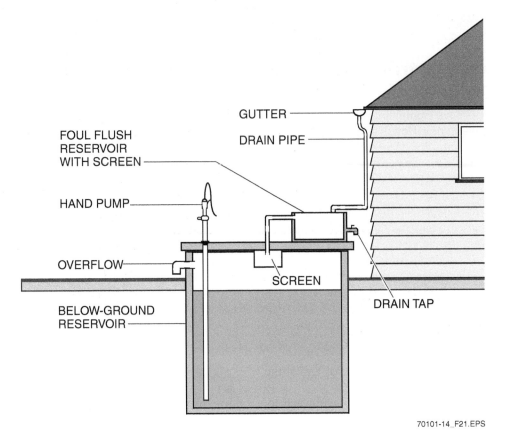

Figure 21 Components of a rainwater harvesting system.

supply with demand and allow overflow into a sewer when necessary. Graywater that remains in the system after a specified period must be disposed of rather than reused.

Small-scale graywater systems are also available. They capture water from one fixture, such as a shower, and divert it for use in another fixture, such as a toilet. Some systems fit under a bathroom sink and work well for retrofits.

Water from toilets and dishwashers is considered blackwater. This is because it contains pathogens. The pathogens are from either human waste or from animal fats associated with dishwashing. Blackwater requires separate piping and additional levels of treatment before reuse.

2.3.4 Finding Alternative Sinks for Wastewater

The final opportunity to green a building's water systems involves finding other uses or alternative treatments for wastewater. A building's wastewater stream also includes stormwater. Stormwater may be contaminated by materials from parking lots and roofs, and is also likely to be much hotter than local water streams. Alternative systems are available to treat this type of wastewater.

Graywater systems can capture water that has not been heavily contaminated. This water can be redirected for certain uses without treatment if allowed by local codes. These systems can also be the first step in a more comprehensive system. Lightly contaminated water is recycled for heavier use, and then treated on site using a separate system.

One type of alternative wastewater technology is a plant-based constructed wetland. Constructed wetlands have been used in a variety of climates and applications. *Figure 23* shows a constructed wetland for a small town with about 900 residents. This series of ponds is filled with plants that provide increasing levels of water treatment. A primary treatment step removes solids from the water stream before it enters the ponds. After treatment, the clean water is discharged to the local river.

Smaller plant-based or bio-based wastewater treatment systems can be used to treat the wastewater from a single building. One example of a bio-based system is called a Living Machine®. Living Machines® combine plants, bacteria, and other organisms into a simulated ecosystem that uses wastewater as a nutrient source. *Figure 24* shows a bio-based wastewater treatment system at the Rocky Mountain Institute in Snowmass, Colorado. This system includes hedgehogs and lizards as part of the indoor ecosystem, and it even produces bananas in winter.

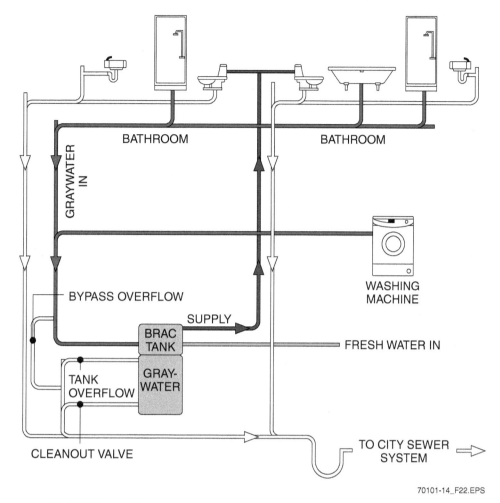

Figure 22 Components of a building-scale graywater system.

2.4.0 Energy Best Practices

The energy used in homes and buildings has considerable impacts on the green environment. As described in Section One, producing and using energy has a strong effect on the carbon footprint of our society. Many construction activities rely on carbon-intensive energy sources such as fuel for heavy equipment. The resulting buildings also consume energy over their extended life cycle. Much of that energy is still produced from non-renewable energy sources, creating greenhouse gases. Renewable energy sources represent only a small share of all energy used. Energy best practices can be used to reduce the impacts of buildings on climate change and resource depletion, including the following:

- Avoiding unnecessary energy use
- Optimizing energy use
- Balancing electrical loads
- Seeking alternative energy sources

Generators provide power where it is needed when electrical connections are not available. To do so, most generators rely on combustion of fossil fuels. This combustion results in air pollutants as well as noise pollution on the job site. Extension cords are also required to connect equipment, which can create trip hazards and unsafe conditions. However, generators allow continuous production without the need to stop and change batteries.

> **Think About It**
> ## Powering Construction Activities
>
> Advances in battery technology have made construction easier in many ways. No longer are fossil-fuel generators required for power on sites where electricity has not yet been installed. But what is the impact of batteries on the natural and human environments compared to generators? Which is greener?

Batteries allow workers to move quickly and easily around the site to get their work done. However, they must be recharged and they eventually wear out. Multiple batteries are needed to allow workers to keep working while other batteries are charging, and charging batteries at the job site may still require generators. The time it takes to change batteries can reduce productivity, although risks from long power cords are eliminated. Manufacturing batteries has high environmental impacts—they use toxic heavy metals and are energy-intensive. Their disposal can also be a problem, but they should always be properly recycled.

While some battery technologies are greener than others, not all batteries work equally well for all uses. New generators that use solar power are also becoming available. Choosing the best approach for a task requires considering tradeoffs and planning. This ensures that the best choice for health, safety, and the environment is made while not compromising productivity.

Figure 23 Constructed wetland wastewater treatment system.

Figure 24 Bio-based wastewater treatment system.

Energy-Efficient Daylighting

GOING GREEN

This classroom is lit by a carefully designed system that includes photosensor controls and daylighting louvers. It also has blinds that reduce solar transmission by 97 percent in the closed position. The louvers direct light upward into the room while minimizing heat gain in the space.

2.4.1 Avoiding Unneeded Energy Use

The best way to green a building's energy use is to reduce or eliminate the demand for energy. One approach involves setting back the thermostat and wearing a sweater in the winter or opening windows to take advantage of the breeze in the summer. Another is to include timers and occupancy sensors that can be used to control the lighting or temperature when a space is unoccupied.

Other strategies reduce demand with no changes required in occupant behavior. These include passive heating and cooling, daylighting, and heat recovery ventilators (HRVs). HRVs capture waste heat from exhaust air and use it to preheat incoming air. Some systems also equalize humidity in a similar way. Other systems take advantage of the thermal mass of building materials. Thermal mass is a property that allows a material to absorb heat and then release it slowly. For example, adobe houses have thick walls that absorb heat during the day and release it at night.

Another simple strategy is a high-albedo roof (*Figure 25*). These roof materials and coatings reduce heat gain by reflecting solar energy rather than absorbing it. As *Figure 25(B)* shows, these surfaces need not be white and shiny. New coatings are available that reduce heat gain even with darker color roofs. In many cases, the albedo of a roof can be increased without cost by choosing a different roof color during design.

(A) HIGH-ALBEDO MEMBRANE ROOF

(B) HIGH-ALBEDO METAL ROOF

Figure 25 High-albedo roofs.

> **WARNING!** Make sure to always use proper fall protection when working at height. High albedo roofs may help a building be greener, but they have also been reported to have higher glare. This can endanger workers installing and working on them. Wear proper protection from the sun, including eye protection and sunscreen, while working on high albedo roofs. Be aware of roof edges that may be hard to see. Be sure to tie off properly.

High-performance building envelopes can also eliminate unnecessary uses of energy for heating and cooling. A building's envelope is the outer shell that protects it from the elements. If you added up the area of all cracks and openings in the building envelope of a typical American house, it would equal about one square yard (nearly one square meter). That is the equivalent of leaving a window fully open year-round, even when the heater or air conditioner is on. Ways to improve a building's envelope include the following:

- Increasing insulation in wall cavities and attics
- Sealing cracks and using vapor-resistant barriers to reduce infiltration
- Installing high-efficiency windows and doors
- Using low-emissivity paint or radiant barriers in attics to reduce heat gain

All of these methods can reduce the demands on facility space conditioning equipment and increase occupant comfort. Commissioning the building envelope is also important for high-performance envelopes. Commissioning has traditionally been limited to building mechanical and electrical systems. However, having a third party check that all parts of the envelope are installed correctly and working together can be valuable to prevent future problems.

Finally, use good control systems for lighting to minimize the use of energy during hours when daylight provides enough light. Dual switching

> **Think About it**
>
> ## Choosing a Roof: Balancing Energy Savings with Durability
>
> High albedo roofs can help save significant energy over the life of a building by keeping the roof cooler. However, there are tradeoffs to think about with some types of high albedo roofs. According to the Lawrence Berkeley Laboratory, the dirt that accumulates on a high albedo roof can reduce energy savings by 20 percent or more. This means that the roof must be cleaned regularly to maintain its function. Roof cleaning requires working at height, which poses additional safety risks.
>
> Keeping roof systems cooler can help a roof last longer due to reduced stress from expansion and contraction. However, certain types of high albedo roof membranes have a much shorter service life than conventional roofs. For instance, thermoplastic polyolefin (TPO) roofs are more environmentally friendly in many ways than other roof membranes. However, their service life may be significantly less than a conventional membrane roof. Many factors must be considered in choosing a roof system. The design and details of the roof will play a large role in roof service life. Other factors include embodied energy, use of toxic materials, recyclability, and cost. Commitment to maintenance is also important.

involves wiring light switches so that combinations of lamps can be switched on or off depending on how much light is needed. This helps users to adjust the light to appropriate levels. Properly labeling the switches allows for easier use.

2.4.2 Optimizing Energy Use

There are many opportunities for optimizing energy use. One way is proper sizing of heating, ventilation, and air conditioning (HVAC) equipment. Oversized equipment operates at peak efficiency only a few days per year. A better choice is the use of two smaller units. Under normal conditions, only one unit is required and it operates at peak capacity. Both units are used together to meet peak demands that occur only a few hours per year. The energy savings offsets the initial cost of two systems. Having two units also means one can be kept running while working on the other, increasing maintainability.

Proper operation and maintenance is an often-overlooked way to save energy. Even simple maintenance, such as cleaning the lenses on lighting fixtures, can greatly increase light output and reduce the need for supplemental lighting. Continuous commissioning can be applied to existing buildings. Commissioning is a procedure to examine each building component to ensure it is operating as originally designed. It can be used to identify ways to save energy with existing equipment. Continuous commissioning typically pays back in less than two years. Operator training is also important to ensure that building equipment is used properly.

Energy audits can identify ways to retrofit existing systems with more efficient ones. A lighting retrofit has rapid payback in many buildings (*Figure 26*). Replace older magnetic ballasts with electronic ballasts in fluorescent lights. You can convert existing T-12 fixtures to T-8 when changing ballasts. Other lighting retrofits include installing solid state LED lighting fixtures or T-5 fluorescent lighting. These retrofits require replacement, not just of lamps and ballasts, but also of fixtures themselves. Maintenance requirements can be reduced due to the much longer service life of solid state lighting. Adding photosensors or occupancy sensors is another lighting retrofit, as they can shut off lights when they are not needed.

Figure 26 Lighting retrofits are an easy way to save energy.

Did You Know?

Prevention Through Design

One of the most important parts of a green building is its envelope. This system provides a boundary between outside and inside. It allows control of the climate inside the building to make occupants comfortable. It also can affect the building's energy and water performance in important ways. From high albedo roofs that reflect unwanted heat, to roof- or wall-mounted photovoltaics or solar heating systems, to vegetated roofs and walls, the building envelope requires careful design and construction.

It is particularly important to consider construction and maintenance of these innovative systems during design. Many injuries that occur on green projects are associated with falls from height while working on the building envelope. Prevention through Design (PtD) is a practice of thinking ahead during design about the ways in which people could be hurt while working on the building. The building is then designed to eliminate danger when possible, or provide ways to reduce that danger. For instance, all of these roofs require regular maintenance during operation. An access door could be added to the design to avoid the need for ladders, and tieoff points or walking surfaces may be included on the roof to make safe access easier.

PtD is especially important in green buildings, given the additional functions played by the building envelope. Not only should designers consider safety issues in the design, they should also consider maintainability in choosing building systems. Maintainability of a system refers to the difficulty associated with proper maintenance of that system. Owners must be prepared for different maintenance requirements that go along with innovative technology. Without proper maintenance, many innovative systems will not work correctly. This can cause problems during operation. Designers should discuss maintenance requirements with owners when choosing building technologies. Owners must ensure that proper training and resources are required to keep the building working properly over time.

WARNING! Fluorescent lamps contain mercury, and they must be properly disposed as hazardous waste. Old ballasts and thermostats also often contain toxic components and must be treated as hazardous waste.

WARNING! Lighting retrofits require working at height. Be sure to tie off and use appropriate safety measures.

Replacing tank-type water heaters with tankless models is another way to save energy. Tankless models cost more initially, but they eliminate standing heat loss since water is only heated as needed. This saves energy over the life cycle. Regular maintenance is important to keep both tank-type and tankless water heaters working properly.

High-performance HVAC systems also save energy. Variable speed fans, pumps, and motors enable these systems to operate at optimum levels. This makes the system cycle on and off less frequently, which increases user comfort and saves energy. Optimized distribution systems also help reduce heating, cooling, and ventilation costs. Sealing and insulating ducts is an important way to ensure that the energy used to condition air actually benefits the users of the building.

Finally, ultra high-efficiency appliances and equipment can quickly pay for themselves in energy savings. Look for ENERGY STAR appliances when making replacements or buying new equipment.

2.4.3 Balancing Electrical Loads

The electrical grid is a complex system of power plants and distribution networks. The end systems that consume the power are also complex. To keep the grid functioning, energy supply must remain balanced with demand. If this does not happen, the system becomes unstable. When demand exceeds supply, additional generators are brought online to meet the additional need. If these peak generators cannot meet demand, the system may experience brownouts, which are noticeable losses in system voltage that may damage equipment such as motors. In the worst case, the system becomes so unstable that safety mechanisms activate to take parts of the grid offline. This is known as a blackout.

Load balancing is a way to limit the loads placed on the power grid. This allows power plants for peak power generation to be smaller, and reduces the need for new power plants. The highest demand for energy is typically during the middle of the day in the hottest part of summer. At this time, industry is operating at maximum output and commercial buildings are operating air conditioning at maximum capacity. Shutting down unnecessary equipment during these periods is one way to reduce the need to bring peak generators online.

Peak shaving reduces demand during peak times of the day and shifts demand to off-peak times, such as at night. Utilities often charge lower rates during off-peak times. This provides an incentive for users to shift demands to these periods. Thermal storage systems are designed to take advantage of peak shaving. Energy is used at night to super-cool a thermal storage medium. This medium then absorbs heat to provide cooling during the day.

Energy management systems use electronic controls to balance loads within buildings during peak periods. They do this by reducing the amount of power sent to equipment that is tolerant of voltage variations, including certain types of air conditioners, pumps, fans, and motors. They also maintain a steady stream of power to equipment that requires a constant voltage input, such as computers. Depending on the local or regional climate for power production, the price for electricity during peak hours can be quite high. In many areas, an investment in energy management equipment can rapidly pay for itself in terms of reduced energy costs. Often, utility companies will finance or invest in these systems for built facilities. They may also provide technical assistance to support implementation.

Finally, information can be a useful way to balance loads and shave peak demand. Large institutions such as universities often pay for power based on time-of-day usage. During the hottest parts of the year, rates can easily triple when demand for power is highest. Some institutions send out email reminders to shut down unnecessary equipment and lighting during these periods (*Figure 27*). This saves both electrical energy and money by reducing use during high-price periods.

2.4.4 Finding Alternative Energy Sources

After reducing demand, optimizing efficiency, and balancing loads, the last step is to explore alternative sources of energy. There are two primary ways to obtain alternative energy: buying energy

---

Going Green — Reducing the Cost of the Brooklyn Bridge

The Brooklyn Bridge has gone green with LED lights, part of New York City's plan to save energy and cut greenhouse gases. The 160 LED fixtures each use only 24 watts of power instead of the 100-watt mercury bulbs formerly in place, and last three times as long as the mercury vapor lamps they replaced. This saves energy costs and reduces carbon dioxide emissions by about 24 tons each year.

WE NEED YOUR HELP!

Energy Advisory

Electricity Price Caution is in effect today for the time period from 3 pm to 7 pm.
Electricity Price Caution is issued when the electricity prices are 3 to 6 times higher than normal.

Here are some actions you can take to help reduce our electricity consumption:
1. Activate the energy saving or "sleep" mode on computers and copiers.
2. Turn off your computer monitor when you are away from your desk for more than 15 minutes.
3. Turn off lights when out of your office or cubicle.
4. Turn off lights in unused common areas such as copy rooms, break rooms, conference rooms, unoccupied rooms, and restrooms.
5. If you have control over the thermostat setting for the air conditioner, raise it by two degrees during the peak hours. Consider raising the level of the thermostat further when your facilities are unoccupied.
6. Shut off nonessential machinery, computers, and other equipment.
7. Consider reducing the number of copiers available for use during peak hours.

XYZ Power Company – Prices for Today

Hours	cents/kWh
1:00	2.1203
2:00	2.0594
3:00	2.0235
4:00	2.0263
5:00	2.0278
6:00	2.0213
7:00	2.1258
8:00	2.1257
9:00	2.1637
10:00	2.2745
11:00	2.4378
12:00	3.3367
13:00	4.9845
14:00	8.0461
15:00	11.3748
16:00	12.5938
17:00	12.4085
18:00	11.2554
19:00	8.9338
20:00	7.2654
21:00	6.3741
22:00	4.9265
23:00	3.3319
24:00	2.2695

Average price = 6.0211 per hour at end interval

Figure 27 Using information to encourage energy conservation for peak shaving.

from a green power provider, or generating power on site. Green power providers generate electricity from renewable energy sources such as wind power and solar power. They avoid nonrenewable power sources such as fossil fuels. To see if certified green power is available in your area, visit the Green-e website at **www.green-e.org**. Green-e is a nonprofit organization that certifies the renewable energy power companies sell to consumers. *Figure 28* shows the Green-e Energy Certified logo that companies can use on the certified renewable energy products they sell.

The second option is to explore the possibility of on-site renewable energy. This includes photovoltaics, wind turbines, gas-fired microturbines, and fuel cells. *Figure 29* shows a roof-mounted wind turbine providing power to a commercial building.

Did You Know?

Smart Grid

Smart grid technology allows buildings to communicate with power producers to better manage power needs. Smart grid technology also gathers information on system performance and alerts utilities to problems that require maintenance or repair. A key part of the smart grid is automation technology that allows utilities to control power-consuming equipment in a building from a remote location.

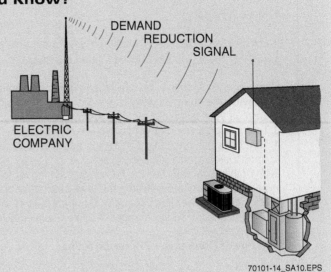

Figure 28 The Green-e Energy Certified power provider logo.

Standalone power systems require costly and high-maintenance battery storage for excess power. Electrical energy is difficult to store. The most effective way to use on-site power generation is with a grid intertie, which is a connection to the local utility grid. It allows you to route excess power back to the grid when you generate more than you can use. You can also continue to buy power from the utility when your on-site system does not meet your demands. With photovoltaic systems, peak capacity falls at the same time of day as peak demand, which is in the afternoon. In some areas, you may even be able to sell your power. If you live in a part of the country where there is not much excess generating capacity, utilities may help finance your project. They may also provide technical assistance to set up renewable energy systems. Check online at **www.dsireusa.org** to see what programs are available in your area.

> **WARNING!**
> Photovoltaic panels generate electrical current whenever they are exposed to light, whether or not they are connected to the electrical system of a building. Be aware of this risk and take proper precautions – they are still potentially dangerous even if they have been disconnected.

> **WARNING!**
> Battery banks used for power backup and storage pose both electrical and chemical hazards. Many batteries use lead-acid technology, both of which are toxic chemicals. Batteries may also produce hydrogen gas that needs to be properly vented outdoors.

With demand for energy only likely to grow in the future, you need to be aware of opportunities to manage consumption of electrical power in built facilities. Energy investments save money over the life cycle of a facility and can increase the ability of the facility to withstand fluctuations in power supply. This reduces vulnerability to natural disasters and other threats. As dependence

GOING GREEN: Rebuilding the World Trade Center

Developers of the new World Trade Center complex are committed to ensuring that it meets green goals. Each of the seven towers will be developed to meet or exceed LEED Silver standards. Sustainable design has been used in other parts of the site as well, including the WTC Transportation Hub, Memorial, and vehicle security center. Green features include water-efficient landscaping, local and recycled materials, and high-performance systems for indoor air quality and comfort. The 7 WTC Tower was the first green commercial office building in New York City, certified LEED Gold in 2006.

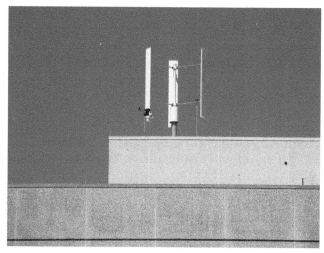

Figure 29 Small-scale wind turbine on a commercial building.

on power grows, the vulnerability of power systems will increase as well. Facilities with well-managed, minimal energy requirements will be the least vulnerable to fluctuations in power (and prices) from utility suppliers. These facilities will use on-site power generation as well as passive systems for lighting, heating, and cooling.

2.5.0 Materials and Waste Best Practices

The materials used to construct and maintain buildings are one of the biggest contributors to their impact on the green environment. As mentioned earlier, buildings are responsible for 25 percent of the world's wood harvest. Current rates of deforestation amount to an area the size of the state of Georgia each year. Buildings also account for about 40 percent of global material and energy flows. That adds up to over three billion tons each year. Each product or material used in a building has a life history that includes the following:

- Harvesting of all the raw materials required to make it
- All the byproducts generated during its production
- All the energy used to transport its components from place to place

In addition to raw materials, solid waste also contributes to a building's impact on the green environment. Buildings are responsible for up to 20 to 40 percent of the solid waste stream in some parts of the United States.

Greening your resource use starts by eliminating the unnecessary use of new materials. This includes using materials that come from waste, salvaged, or recycled sources. It also includes re-using and adapting existing buildings instead of building new ones.

The second step is to use materials more efficiently. Construction methods that reduce waste contribute to this goal. A whole new generation of multi-function materials is now available that provide structure, enclosure, insulation, and other functions. Using these products reduces the amount of raw material needed to meet building needs and save on shipping and packaging. These materials are pre-engineered to minimize waste on site.

Seeking better sources for the materials we use can help ensure an ongoing supply of products to meet future needs. Switching to abundant, renewable materials helps to preserve the limited supply of nonrenewable resources. Ensuring that materials are sustainably harvested means that the supply of those materials can continue

Going Green: Wind Farms

The US Department of Energy states that wind farms may be a major source of power as we approach the year 2030. Their analysis shows that up to 20 percent of US power needs could be handled by wind power, which would reduce pollution to the same extent as taking 140 million cars off the road.

indefinitely. Using local materials helps to reduce the impacts of transportation of raw materials and supports local economies.

Finally, finding better sinks (destinations) for waste from the building helps the green environment. Some of the ways to improve performance in this area include on-site and off-site recycling and using biodegradable or reusable packaging. Salvaging or deconstructing building systems at the end of their lives is also green. Biodegradable packaging is made of natural materials that readily break down using natural processes. Some manufacturers sponsor takeback programs that recover a product or its packaging at the end of its life cycle.

2.5.1 Eliminating the Unnecessary Use of New Materials

The first step in greening building materials is to eliminate the unnecessary use of new materials. This means finding ways to get the job done with fewer materials. It also means substituting materials or parts of materials that aren't new whenever possible. Reusing materials is a green choice. Every product that is salvaged and reused saves energy and raw materials. Recycled content prevents the need to harvest virgin materials, which helps conserve valuable resources.

Perhaps the greenest choice of all is to reuse or adapt an existing building instead of building a new one. Not only do you eliminate the need for many new materials, you also reduce site development and infrastructure costs. Be aware that existing buildings present a challenge, as certain conditions are not always as expected. Sometimes contamination from lead-based paint, asbestos, or other substances must be remediated. Some systems may need to be replaced or upgraded as well. Avoid reusing windows, older HVAC equipment and appliances, and older plumbing fixtures; they are likely to be very inefficient and may no longer meet current standards.

Pay attention to packaging when procuring raw materials. In some cases, you can specify that products be delivered with reusable packaging that can be returned to the manufacturer. In other cases, packaging is recyclable or even biodegradable. If product packaging is bio-based (that is, made from substances derived from living matter), look for opportunities to compost or chip it on site for use as a soil amendment. This also

---

Think About It

Multi-Function Materials: Structural Insulated Panels

Multi-function materials, such as structural insulated panels (SIP), are becoming more common in the industry. This is due to both time savings in the construction schedule, and energy savings during building operation. These materials replace several different building materials with a single product. What materials does a SIP replace? How does a SIP change the way a building is built? Which trades are affected, and how? What tasks are easier? What tasks are more difficult? Why?

reduces the need to purchase new materials for this purpose.

Salvaged materials can be a very green contribution to a project. If an existing building is being demolished, look for ways to reuse its materials in the new building. Often, concrete and masonry rubble can be reused for fill, subbase for pavements, or drainage. Timber in good condition can often be reused for structural purposes, or remilled for flooring or siding. Masonry units may be reused as well. Materials that cannot be reused on the new project may be worth salvaging for other projects. These may include doors, hardware, and various fixtures.

New materials with recycled content also reduce the use of virgin materials. Common materials with recycled content include steel and concrete. Recycled materials include a wide variety of finishes, structural materials, and even landscaping materials (*Figure 30*). With materials like steel and concrete, you may not even be able to tell the difference from virgin materials. Other products may include recycled content as composites. Recycled plastic lumber (RPL) is one such product. RPL combines post-consumer or post-industrial plastic waste with wood fibers or other materials. The result is an extremely durable wood substitute.

Recycling reduces the amount of waste sent to landfills and creates a source of raw materials for new products. However, not all recycling is the same. True recycling means using the waste to create the same kind of product. Many materials can only be downcycled. They are turned into something else that can never be turned back into the original product. One example is plastics recycling. Recycling plastic into clear new containers is difficult due to mixing colors and contamination during recycling. Chemical bonds can also decay in the plastic. Most recycled plastics are turned into composite products such as recycled plastic lumber, lower-grade plastic bags, or even carpet backing. Recycling plastics does reduce the amount of new materials required, but at a loss of both purity and quality. The best way to be green is to reduce the use of these products in the first place.

Recycled content can come from a variety of sources. Some recycled content is produced as a byproduct of manufacturing processes. This material is called post-industrial/pre-consumer recycled content. It can be recovered and reused without ever leaving the factory. Other recycled

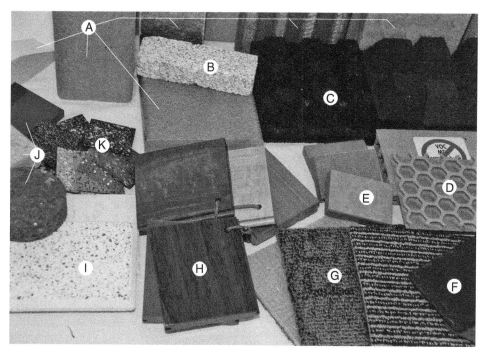

A. Recycled plastic lumber
B. Insulating concrete form material with recycled polystyrene pellets
C. Recycled rubber turf stabilizer
D. Recycled paper fiber board
E. Recycled paper fiber countertop
F. Recycled rubber shingles
G. Recycled content carpet (face fiber and backing)
H. Salvaged lumber
I. Recycled paper and glass cement tiles
J. Recycled glass countertops
K. Recycled rubber flooring

Figure 30 Waste-based and recycled content materials.

content comes from products that have been produced and used by consumers, then recovered for recycling. This material is called *post-consumer* recycled content. Post-industrial recycled content results from waste or inefficiency in manufacturing; rather than recycling this material, it would be better to improve the manufacturing process. Post-consumer content is better than post-industrial recycled content because it truly closes the material loop. Products made from post-consumer waste are valued more highly when certifying a green building than pre-consumer waste.

Pollution prevention is the careful design of products and processes to eliminate the use or waste of materials. One example is to eliminate finishes and expose structural materials instead. For example, concrete floors can be stained, polished, or textured to create beautiful surfaces that do not require additional finishes, such as wood or tile. Exposed ceilings (*Figure 31*) eliminate the need for extensive dropped ceiling systems and provide visual interest while saving costs.

In many projects, natural systems can be used to perform the functions provided by engineered systems at lower cost. One example is living machines made of plants, snails, bacteria, and other organisms. Living machines, natural drainage swales, and constructed wetlands can be used to collect and treat wastewater without requiring any input of chemicals. The result is healthy plants and purified water that can be safely used for other purposes. Some of these practices cost more than conventional systems. However, you can often save enough money to pay for the investment through savings in other systems and/or life cycle cost savings.

Figure 31 Exposed ceilings eliminate unneeded materials.

Many building strategies also take advantage of the lessons offered by the green environment. This approach is known as biomimicry—imitating nature to create new solutions modeled after natural systems. One example is African termite mounds that maintain constant internal temperatures even with large fluctuations in outdoor air temperature. Buildings have been designed to mimic these mounds using natural ventilation. The mounds use an intricate network of tunnels to control the flow of air for optimum quality, temperature, and moisture levels. These self-sufficient mounds also include a type of farming.

Think About It

Recycling vs. Downcycling

This photo shows two different products with recycled content: a carpet tile and a piece of industrial flooring. Which one do you think is recycled? Which one is downcycled?

The termites supply fungus in the mounds with chewed wood fiber, which the fungus then breaks down into a usable food source. Nature produces solutions with optimum efficiency because plants and animals that cannot compete simply do not survive. As the world heads into an age of increasingly scarce resources and more people who need them, biomimicry offers valuable lessons for buildings and technologies alike.

2.5.2 Using Materials More Efficiently

The second strategy for greening materials is to use materials more efficiently, which means getting more benefit from the materials you do use and/or using fewer materials to achieve the same result. One way to do this is to use multi-function materials. Multi-function materials do more than one thing as part of a building. They can speed up construction, save building materials, and reduce waste. Every system and technology used in a project comes with overhead, which includes extra packaging, transportation, and other costs that do not add value to the product itself. When one product serves as multiple systems, the overhead for that product is a fraction of the overhead associated with the multiple systems it replaces.

Many multi-function materials can serve as the structure of the building. These include insulating concrete forms (ICFs), structural insulated panels (SIPs), and aerated autoclaved concrete (AAC). Each of these structural systems (*Figure 32*) also serves as the building enclosure, insulation system, and mounting surface for interior and exterior finishes. By using these products, construction of one system achieves as much as three or four traditional building systems. This saves time, labor, packaging, transportation, and money.

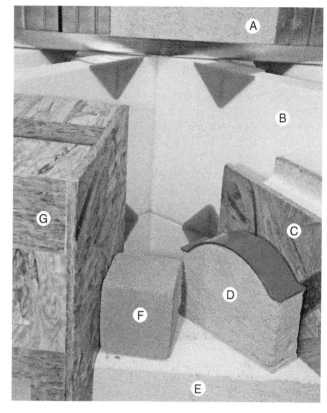

A. Embedded steel stud SIP
B. Polystyrene insulating concrete form
C. Polystyrene stress skin SIP
D. Aerated glass stress skin SIP
E. Aerated autoclaved concrete
F. Fiber-reinforced aerated concrete
G. Straw core stress skin SIP

Figure 32 Multi-function materials.

An energy-related example is building-integrated photovoltaics (BIPVs). These solar panels are built into components that play other roles in a building. Some BIPVs are used to generate

Biomimicry

Scientists have begun to look to nature for inspiration as they develop new building materials and systems. One example is a spider web. Spiders produce fibers stronger than Kevlar and spin them into self-healing structures that can resist tremendous loads. Spider webs contain many lessons for structural materials and design.

electricity when the sun shines on a window, and others are manufactured as solar shingles—they generate power while acting as the roof itself. Still others are used as shading devices for windows or parking lots.

Green roofs are another example (*Figure 33*). These vegetated roofs have energy benefits as well as benefits for roof life.

Green roofs are typically installed on top of a roof membrane. They can be of varying complexity, and help to ballast the roof itself. They also protect the membrane from solar radiation, which would otherwise cause the material to break down. Green roofs stabilize temperatures on the roof, reducing stress on the roof membrane and keeping the building cooler. They help to absorb rainfall, which reduces stormwater runoff. They are also a microhabitat for birds, insects, and plants, and they also help to clean the air. This single system, although it may cost more than a traditional roof, provides significant benefits that can make it a good investment. All benefits and costs must be considered when deciding what types of systems to use.

> **WARNING!**
> Installing and maintaining a green roof requires working at height, often by trades that are unfamiliar with fall protection. Green roofs should include design features to facilitate safe access for maintenance. Be sure to employ active fall protection when working on green roofs.

Optimal value engineered (OVE) framing for light wood frame construction is another way to use materials more efficiently. OVE framing is a green practice because it reduces the amount of wood needed. It also allows more insulation to be incorporated in the walls of the building (*Figure 34*). This reduces the amount of heat that can escape through the walls and increases the energy performance of the building.

OVE framing also saves considerable materials during construction by using wood more efficiently. This means less lumber, fewer cuts, and less waste. Key principles of OVE framing include:

- Designing buildings on 2-foot (0.6-meter) modules to make the best use of common sheet sizes
- Spacing wall studs, roof joists, and rafters up to 24" (0.6 meters) on center
- Using in-line framing where floor, wall, and roof framing members are vertically in line with one another and loads are transferred directly downward

(A)

(B)

Figure 33 (A) Green roofs provide enclosure, stormwater management, and thermal control while cleaning the air and providing a habitat for birds and insects. (B) Walls can also be vegetated.

- Using two-stud corner framing and inexpensive drywall clips or scrap lumber for drywall backing
- Eliminating headers in walls that are not loadbearing
- Using single lumber headers and top plates when appropriate

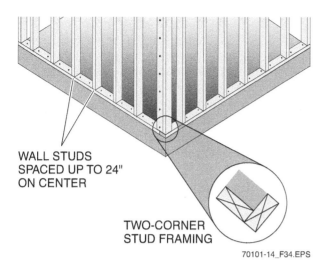

Figure 34 Framing details for OVE wood frame construction.

These techniques reduce the amount of wood used and create a better thermal envelope. They also make construction easier for crews that must install plumbing, electrical, and HVAC services in the walls. Fewer structural members mean less drilling during installation and less effort to install. All of these attributes make OVE framing a good strategy to green your building and save both first costs and life cycle costs.

Another approach is to use smart materials. Smart materials work by changing in response to environmental conditions. For example, smart windows may automatically become more opaque in response to higher levels of sunlight. Smart window blinds may automatically adjust to follow the sun's path. Some materials are under development that can change colors or other properties as well.

Phase change materials (PCMs) are one class of smart building materials. These materials are capable of storing and releasing large amounts of energy when they change between solid, liquid, or gas forms. When the material reaches a temperature at which it changes phase, it can absorb or release a large amount of heat at an almost constant temperature. Materials at their phase change temperatures can be used to absorb or release heat to cool or heat a building with far less energy than conventional cooling or heating systems.

A special class of smart materials is called nano-materials. Nano-materials use extremely small particles that add functionality to building materials. For example, nano-coatings can be used to make paint repel dirt or concrete absorb air pollution. Nano-coatings are also being used to add mold resistance or biocidal (germ killing) properties to finishes. A new generation of photovoltaic panels are being developed using carbon nano-tubes, which improve efficiency by increasing the surface area exposed to the sun.

New generations of lightweight modularized construction systems are now available that can also do more with less. Prefabricated, factory-assembled building components reduce waste. They use materials more efficiently since they are manufactured in a controlled environment. Regular production allows manufacturers to optimize the use of materials and produce the most efficient components possible.

Think About It

OVE Framing: Impact on Cost

How does an improvement to one system change the requirements for another system? OVE framing is a good practice for saving lumber and improving insulation of walls. However, using more widely spaced studs changes how other materials like drywall are used. With fewer attachment points, regular drywall may flex, causing uneven wall surfaces that are difficult to finish. Thicker drywall is generally required to do a good job with wider stud spacing. This means that while the cost of framing may decrease, the cost of drywall will increase. What other impacts might occur? How would this change the process of assembling a wall?

Modular construction components include carpet tile (*Figure 35*), raised floor systems, and demountable furniture systems. These systems allow rapid reconfiguration to meet changing user needs. They also permit replacement on a unit-by-unit basis in the case of damage. This means avoiding the need to replace entire rooms of carpet for one stained or worn area. Even regular materials can be used for modular construction, by using regular dimensions that match the standard sizes of building materials. For example, don't design decking that is 9 feet (2.7 meters) long, as decking comes in increments of 2 feet (0.6 meters). Expand your deck to match standard material sizes and avoid having to waste material.

Choose systems and materials of appropriate durability. If you know that the facility you're building is going to be remodeled every 5 to 7 years, don't use granite countertops unless you plan to recover and reuse them somewhere else. This idea is known as *design for disassembly*. Put your systems together in a way that you can take them apart at the end of their service life for possible reuse.

During the construction process, use lean construction methods to maximize efficiency. Lean construction eliminates unnecessary steps and materials, and eliminates waste by changing the way things are built. For example, just-in-time delivery of materials to the site means that products are installed right off the truck, not needing to be stored or staged. This eliminates the opportunity for materials to become damaged in storage. It also minimizes the opportunity for loss or damage as products are moved around the site.

Careful material tracking helps to avoid waste. A new method of material tracking involves the use of radio-frequency identification (RFID) tags. These tags make it easy to locate materials in crowded staging areas. If materials must be stored before installation, protect them from moisture and possible damage. Centralized cutting operations also help to increase material efficiency. Keep cutoffs in the same area where all cutting is done. This makes workers more likely to use them instead of pulling a new piece of material from the supply stack.

2.5.3 Finding Better Sources of Materials

Another way to green your use of materials is to find better sources. For example, rapidly renewable materials have become common. Rapidly renewable materials are a special class of bio-based materials that grow quickly. According to the US Green Building Council (USGBC)®, rapidly renewable materials are any material that can be sustainably harvested on less than a 10-year cycle (*Figure 36*). They may be derived from either plant or animal sources, including bamboo, cellulose fiber, and wool. Other examples are cotton insulation, corn-based carpet, blown soy insulation, agrifiber, linoleum, wheatboard, strawboard, and cork.

Other materials can be considered green because they are abundant. This includes buildings made from soil, such as rammed earth, adobe, or cob construction. Cob construction is an ancient building method using hand-formed lumps of earth mixed with sand and straw. Like cob, adobe typically includes plant fibers to help stabilize the soil. Other earth-based construction methods strengthen the soil by using small quantities of cement or lime. An innovative material in this category is papercrete, a fiber-cement that uses waste paper for fiber.

Figure 35 Carpet tile is a modular material that prevents waste.

What's Wrong with This Picture?

A. Soy-based foam insulation
B. Wool carpet
C. Corn-based carpet fiber
D. Coir/straw soil stabilization mats
E. Corn-based carpet backing
F. Coconut wood flooring
G. Wheat straw fiberboard
H. Sorghum board
I. Bamboo flooring/plywood
J. Linoleum
K. Hemp fiber fabric
L. Cork flooring
M. Cotton fiber batt insulation

Figure 36 Bio-based and rapidly renewable materials.

Strawbale construction uses both abundant and bio-based materials. The bales of straw used for this purpose are typically three to five times more densely packed than agricultural bales. Strawbale construction can be either structural or used as fill with a wood frame or other structure. Bales are stabilized using rebar or bamboo spikes. Bond beams are used to ensure a load-bearing surface along the tops of walls. Strawbale is becoming more widely used in certain parts of the United States; Arizona even has a special section of the building code for this construction method.

Proper detailing to prevent moisture problems is critical for success. Deep overhangs and moisture barriers between bales and foundations are common. Strawbale construction has a high insulating value, and has a two-hour or greater fire rating depending on the exterior and interior finish.

Recycled content materials are also a better source than virgin materials for many applications. Post-consumer recycled content is especially desirable. However, many recycled content materials are composites. This means that they are combinations of multiple different types of raw materials. The prospects for recycling composite materials are limited. However, some composite materials outperform conventional materials, requiring lower maintenance and fewer raw materials.

People often confuse the term *recyclable* with *recycled*, but there's a big difference between the two. *Recyclable* means that the product itself can be recycled after it is used; *Recycled* means that a product contains material that has been previously used, either in industry during manufacturing (post-industrial) or by consumers (post-consumer). Post-industrial recycled content is less desirable than post-consumer recycled content, as post-industrial recycled content often results from wasteful manufacturing processes that could be improved. Scientific Certification Systems is an organization that provides certification of the recycled content of materials. Look for the SCS logo on products to be sure that their recycled content has been verified (*Figure 37*).

Look for sustainably harvested materials. The Forest Stewardship Council (FSC) is an organization that evaluates wood products to determine if they are sustainably harvested. Look for FSC certified wood products (*Figure 38*). More information on the FSC certification standards is available online at **www.fsc.org**.

Figure 37 SCS Certified content label.

Figure 38 FSC Certified wood product.

Seek out local sources of materials wherever possible for your projects. Using locally produced materials reduces the energy required to transport them. This is one of the biggest impacts of building material. Often, locally harvested materials can be obtained at lesser cost than imported materials. Using local materials can save time for procurement and helps to support the local economy.

2.5.4 Finding Better Sinks for Waste Streams

The final step you can take to green your materials use is to find better uses for building waste. Depending on the type of project you are building, multiple options may be available. Consult salvage companies for projects that involve the removal or rehabilitation of existing buildings. This will ensure that all reusable materials are removed before demolition begins. Older buildings may be good candidates for deconstruction, which is the systematic disassembly of a building in the reverse order from how it was built. This process salvages timber and other materials for reuse.

Both salvaged materials from existing buildings and leftover materials from new projects can be donated to local nonprofit organizations involved with construction. Habitat for Humanity is a nonprofit organization that operates ReStores in various cities (*Figure 39*). ReStores are retail outlets that sell donated building materials. Companies that donate materials to a ReStore can take a tax deduction for their charitable contribution. At the same time, the ReStore obtains a source of materials that it can sell to raise money to build homes for needy families. If your project is likely to have usable materials left over, contact your local ReStore to ask about on-site pickups and donation opportunities. You can also shop at the ReStore yourself if you need small quantities of a certain item or would like to find salvaged or used materials for your next project.

For waste that is not suitable for donation, consider recycling as an option for disposal. Metal recycling is available throughout the United States. Organizations that recycle cardboard, drywall, ceiling tiles, carpet, concrete, masonry, and asphalt shingles are becoming more common. Consider contracting with a waste recovery company to simplify recycling. Depending on the waste streams generated, these companies deliver dumpsters at different times during the project. They may also provide worker training or housekeeping services to ensure that materials are sorted correctly. Some areas also have facilities for off-site sorting of co-mingled construction and demolition waste. These facilities take your waste at a reduced rate and sort it to recover

Figure 39 Habitat for Humanity ReStores accept donations of construction materials.

recyclable materials. Depending on market conditions and the composition of your waste stream, they may even pay you for your waste. In addition to typical building materials, batteries can also be recycled. Lithium ion (Li-ion) and nickel cadmium (NiCad) batteries are both rechargeable and recyclable.

> **WARNING!**
> On-site sorting of construction waste requires additional handling of material to be sure it is placed in the proper container. Sites with multiple waste containers can be more congested and pose hazards for workers. Additional handling of scrap and waste can also pose physical hazards such as cuts or strains. Plan site layout and logistics carefully. Use proper lifting practices and wear appropriate personal protective equipment when handling waste.

Bio-based waste is a good candidate for on-site chipping, mulching, or composting. Pallets, wood waste, cardboard, and even drywall can be chipped or mulched on site by specialty contractors. The result is a product that can be used as a soil amendment or for landscaping purposes. Some contractors also grind concrete and masonry rubble for use on site as a fill or subbase.

> **NOTE**
> Cardboard is a higher value waste that should be recycled rather than composted if possible.

> **WARNING!**
> Do not burn pressure-treated wood. Chemicals in the wood can become airborne and inhaled into the lungs. Always separate treated wood from wood being chipped for mulch. The chemicals used in pressure-treated lumber can contaminate the soil. Collect scraps and place in a dumpster reserved for pressure-treated material. Wash your hands thoroughly after handling pressure-treated material.

Certain kinds of waste, if produced in large quantities, may be good candidates for storage in a monofill. Monofills are landfills that take only one type of waste, typically in bales. Monofills are designed to store waste so that it can be easily recovered later. Most waste destined for monofills is high-value waste. Cost-effective recycling methods are now being developed for wastes in monofills. One such type of monofill is for carpet waste; recent advances in separation technology have made carpet monofills into mines for raw materials.

The world's population continues to grow. Given increasing standards of living worldwide, depletion of resources is a concern. Smart use of materials can help ensure an ongoing supply to meet human needs both now and in the future.

2.6.0 Indoor Environment Best Practices

People in developed countries spend over 90 percent of their time (on average) indoors in climate-controlled buildings. These buildings are sealed against leaks to maximize energy efficiency. At the same time, the materials used to build these structures have become largely composite and/or synthetic. From carpets to engineered wood products, much of the indoor environment now consists of products that emit chemicals over their life spans.

Together, these factors have led to a sharp increase in building-related health symptoms known as Sick Building Syndrome. Symptoms include headaches, fatigue, and other problems that increase with continued exposure. With increased concerns for energy conservation, building operators often reduce ventilation rates in unoccupied spaces or during nights and weekends. Reduced ventilation rates, along with tighter building construction and chemical emissions from building products, greatly reduce indoor air quality. In addition, many products such as drywall, carpet, and ceiling tiles serve as excellent food sources for mold. Inadequate ventilation, moisture, and food sources have led to explosions of mold growth in some construction projects. This has had disastrous results for building occupants.

People also contribute to the problem by virtue of their activities. CO_2 levels are an indicator of the effectiveness of ventilation of a room. Higher concentrations of CO_2 correlate with drowsiness, reduced productivity, or even headaches and other health effects. Activities like smoking, cooking, or using printers and photocopiers contribute pollutants to the indoor air. Shedding skin cells and the gases produced from digestion also contribute to poor indoor air quality.

Keeping occupants happy, healthy, and productive is essential for good business and a happy home life. This section explores ways to green buildings by improving indoor environmental quality. One tactic is to prevent potential problems at their source. When pollution sources cannot be avoided, take measures to isolate the source of pollution to prevent it from spreading. Use the laws of physics and psychology to create pleasant and well-functioning spaces for building occupants. Give users control over their environment to allow them to make adjustments for comfort and productivity.

2.6.1 Preventing Problems at the Source

Often, decisions made during the design of a building can have a huge impact on indoor environmental quality. For example, locating buildings upwind of major pollution sources such as power plants can reduce the need to fix air quality later. The same applies to locating buildings away from noise sources such as highways. Careful placement of air intakes away from pollution sources, such as loading docks, can also prevent problems. Be sure all vents and exhaust areas are located downwind from air intakes. Consider the features of neighboring buildings when deciding where to locate air intakes and exhausts.

The materials and processes used to construct a facility have the potential to create problems. Many modern finishes such as paints, sealants, carpets, and composites contain solvents and adhesives that release VOCs as they age. The off-gassing of VOCs is an example of fugitive emissions. These emissions come from the material itself, not from a stack or vent. That new car smell that many people like is actually caused by the materials releasing VOCs, such as urea formaldehyde. If you can smell an odor, the product that caused it is already on its way to your lungs. Select products with a low or zero VOC content. Most major paint producers now have lines of low VOC paints that perform as well and cost about the same as traditional paints. These paints are water-based rather than solvent-based. Colors may be limited for some brands and types of low VOC paints. The pigments required to achieve extremely dark colors often contain VOCs. Be sure to check the VOC content of pigments when purchasing dark colored paints. Green Seal offers an industry standard for paints to ensure that they meet the criteria for low VOCs and are lead and cadmium free (*Figure 40*). Green Seal Certified paints and sealants are listed online at **www.greenseal.org**.

> **WARNING!** Just because your project uses low VOC products does not mean you should be careless during installation. Ensure adequate ventilation when installing products that might offgas, and wear appropriate personal protective equipment.

Figure 40 Look for the Green Seal Certified paint logo.

Other products such as carpets, furnishings, adhesives, and composite wood can also be evaluated using third party standards such as the California Department of Public Health (CDPH) *Standard Method for the Testing and Evaluation of Volatile Organic Chemical Emissions from Indoor Sources Using Environmental Chambers*. Green building rating systems such as the Leadership in Energy & Environmental Design (LEED) rating system reference this and other standards to ensure that products are rigorously and consistently tested for emissions. Many products are tested in enclosed environmental test chambers. VOC emissions from these products are measured over time and reported for use in product labels (*Figure 41*).

Proper design, drainage, and landscaping can help prevent water intrusion into a building. Wherever water or moisture exists at warm temperatures with an available food source, mold or mildew is likely to grow. This can be prevented through good design, ventilation, and maintenance. Make sure the areas surrounding the building are graded properly to drain water away from the building. This helps prevent moisture problems in basements and crawl spaces. Crawl spaces should also have a waterproof barrier between the soil and the conditioned space.

Careful selection of landscape plants can also help to prevent problems. Many ornamental trees and shrubs produce a significant amount of pollen. So do species such as olive, acacia, oaks, maples, and pines. Pollen produced by these trees can aggravate allergies and can contaminate air intakes and rainwater harvesting systems. Ask

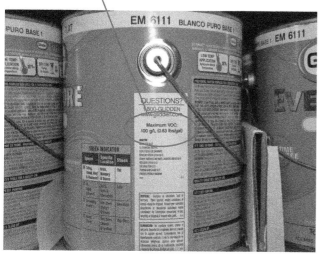

Figure 41 VOC content can be determined from product labels.

Figure 42 Nonabsorptive finishes such as ceramic tiles are easier to clean and do not absorb pollutants.

about the pollen production of landscape plants before you make your final selection.

During construction, pay attention to the sequencing of construction activities. Schedule dust-producing and dirty activities before carpets or other absorptive finishes are installed. Other absorptive finishes include wall coverings, fabrics, ceiling tiles, and many types of insulation. Ensure that materials are stored on-site in a way that keeps them safe from moisture and potential mold growth. If possible, use just-in-time delivery to minimize the amount of time materials must be stored on the site. Never install material that has become wet without testing to be sure it will not become contaminated with mold. Wait until after the building shell is complete with windows, doors, and a roof before installing any finishes or materials that can retain moisture. Be sure to allow adequate time for curing of concrete and finishes and provide adequate ventilation. Significant moisture is released during curing and will remain in the building and cause problems if it has no exit.

Select finishes that are nonabsorptive such as tile floors and painted walls (*Figure 42*). These finishes are easier to clean and are water-resistant. They also do not trap particulates and odors that can cause occupant discomfort. Choose finishes that contain biocidal (germ-killing) or mold-resistant coatings. Natural linoleum has biocidal properties, which makes it a good choice for flooring.

After the building is finished, green housekeeping practices can help maintain a healthy indoor environment. Follow manufacturer's instructions for maintaining finishes. For example, vacuum carpets often to prevent the buildup of irritants such as dust mites; read labels on cleaning chemicals, and choose nontoxic, water-based cleaners wherever possible.

Monitor building humidity and maintain relative humidity levels between 30 percent and 50 percent in all occupied spaces of the building. During summer, cooling coils may require maintenance to ensure that they are properly dehumidifying incoming air. Inspect and properly maintain all HVAC equipment; this enables you to identify and fix problems before they create unhealthy conditions.

2.6.2 Providing Segregation and Ventilation

The second step in achieving good indoor environmental quality is to segregate activities and materials that create pollution from other parts of the building. Segregation is especially important during construction activities to prevent contamination of building systems. Use separate ventilation systems for high-risk areas. Entrance control measures such as walk-off mats help block contaminants that can enter the building through entryways.

Any type of construction activity has the potential to cause future indoor air quality problems if not properly managed. Many construction activities, such as cutting operations and sanding, produce dust or airborne particulates. Other activities involve installing materials that release VOCs or moisture while they cure, including paint, sealants, adhesives, concrete, and new lumber. Even with dust collectors installed on equipment, building systems must be protected from absorbing pollutants during construction. These contaminants reduce indoor air quality and may result in damage to sensitive building systems. Proper protective measures reduce cleanup time, which speeds construction and lowers the cost (*Figure 43*).

> **NOTE**
> The Sheet Metal and Air Conditioning Contractor's National Association (SMACNA) publishes *Indoor Air Quality Guidelines for Occupied Buildings Under Construction*. This standard provides a set of best practices for contractors to preserve IAQ during construction.

> **NOTE**
> The American Society for Heating, Refrigeration, and Air Conditioning Engineers publishes standards to guide the design and operation of building mechanical systems. *ASHRAE Standard 52.2 – Method of Testing General Ventilation Air-Cleaning Devices for Removal Efficiency by Particle Size* defines the Minimum Efficiency Reporting Value (MERV) metric for air filters used in building ventilation.

At a minimum, construction isolation measures should include protection of all HVAC intakes, grills, and registers. Use plastic sheeting and duct tape or ties to seal off all possible points of entry to building ductwork or plenums. Be sure to deactivate the HVAC system before you do this, and isolate entryways to other parts of the building and stairwells or corridors outside the work areas.

Install absorptive materials after dust-producing activities are completed. Absorptive materials include carpet, wall coverings, fabrics, ceiling tiles, and many types of insulation. If this is not possible, use plastic or paper sheeting to protect absorptive surfaces.

Maintain proper ventilation during construction to ensure worker health and safety by exhausting pollutants from the workspace. If possible, maintain continuous negative pressure exhausted to outside air. Exhausting workspace air prevents dust and contaminants from migrating to other parts of the building. Ventilation should be provided by a temporary ventilation system to avoid contaminating building ductwork with pollutants. If the building's permanent ventilation system is used, filters with a **minimum efficiency reporting value (MERV)** no less than 8 should be used.

Regular housekeeping during construction keeps contaminants under control and prevents them from becoming airborne. Specialty contractors can be retained for this purpose. Alternatively, work practices can be put in place to require workers to keep their areas clean and free of debris. All construction isolation measures should be regularly inspected for leaks and tears and replaced as needed.

At the end of construction, remove all masking and sheeting used to isolate ventilation systems. Temporary high-efficiency filters may be used in the HVAC system during some construction activities and before building occupancy. High-efficiency filters trap remaining contaminants and prevent them from entering the system. Replace all filters prior to occupancy. The EPA recommends a two-week flushout period for all new buildings prior to occupancy. During flushout, ventilation systems are run continuously with full outdoor air, accelerating the offgas removal from building materials. However, it can also introduce large amounts of humidity into the building depending on the climate and season. This humidity adds to the already high humidity rates from the curing of materials inside the building itself. If a building flushout is undertaken, ensure that humidity controls are in place to handle the extra humidity load that may occur.

Research has shown that the primary sources of household pollutants are ordinary products such as paint, cleaning compounds, personal care products, and building materials. Everyday tasks such as bathing, laundering, cooking, and heating can all contribute to poor indoor air quality.

Areas of concern in commercial buildings include kitchens, office equipment rooms, housekeeping areas, and chemical mixing and storage areas. Provide separate ventilation for these areas to minimize the risk of pollutants spreading to other parts of the building. In addition, these areas may be isolated from other spaces in the

Figure 43 Construction isolation measures ensure that contaminants from construction tasks do not contaminate building systems.

building through full deck-to-deck partitions and sealed entryways, depending on the risk level of contamination. Indoor smoking areas are special cases that require careful design and construction to protect nonsmokers from exposure to second-hand smoke.

Entrance control is important to keep dirt and pollutants outside the building from getting inside. Air lock entryways and vestibules are one approach that can also save energy by keeping conditioned air inside the building and unconditioned air out. Entry areas can benefit from the use of walk-off mats and dirt collectors, particularly in areas where inclement weather may lead to snow and ice being tracked into the building. Walk-off mats are also available for use outside construction areas to help prevent contamination of other areas of the building.

2.6.3 Taking Advantage of Natural Forces

A variety of natural forces can affect the indoor environment. The laws of physical science can be used to provide natural lighting, ventilation, heating, and cooling in a space. This not only saves energy, but also can provide better indoor environmental quality for occupants.

Paint colors are one of the easiest ways to influence the experience of a space (*Figure 44*). Often, they require no additional expense. Color can influence both how you perceive spaces and how you feel about them. Dark colors can make workspaces feel cramped and depressing, and tend to make rooms look smaller and more confined. Lighter colors create a sense of openness, may improve mood, and make rooms look larger and more spacious. Dark colors make high ceilings look closer, while lighter colors make low ceilings look higher. Combinations of colors can be especially effective in creating focus areas and drawing people into a space.

Natural daylight can be used to enhance the indoor environment in many ways, but can cause problems if not managed well. Potential problems include excessive heat gain, fading, and glare. Common features to incorporate daylighting in buildings include skylights, atria, and glazing. Light shelves or louvers are one way to incorporate daylight into a space without overheating spaces along the building perimeter. Light shelves bounce sunlight onto the ceiling of a room and reflect it further back into the building space. In combination with reflective or light-colored ceilings, light shelves spread the benefits of natural light throughout a space.

> **WARNING!**
> Skylights, atria, and windows offer benefits to building occupants, but they can pose hazards for workers. Energy efficient windows with multiple panes and thermal frames are much heavier than conventional windows. They often come in large sizes in green buildings, and can result in strain injuries during installation. Use proper lifting practices. Cleaning these transparent surfaces often involves work at height. Be sure to employ proper fall protection.

Natural ventilation, when carefully designed, can also create a comfortable indoor environment with minimal energy. *Figure 45* shows an example of a natural ventilation system that would work well in an arid climate. Wind scoops are used to capture prevailing winds and direct them over a receptacle of moisture. As the air absorbs moisture, it becomes cooler and sinks down into the living space. As it mixes in the living space and becomes stale, its temperature gradually increases. The warmer stale air rises to ceiling level and is exhausted downwind with another air scoop. This type of system relies on absorption of moisture to cool the air. In climates with high humidity, other approaches are required.

Finally, geothermal heating and cooling takes advantage of the fact that the earth's temperature below the frost line stays relatively constant year round. Geothermal heating and cooling systems use pipes embedded in the ground or in nearby ponds or streams to shed heat in the summer and absorb heat in the winter. The fluid in these pipes is preheated or precooled by this contact with the earth and used to condition indoor environments. Geothermal heat pumps can also be used

Figure 44 Paint color choices can influence the user's experience of a building space.

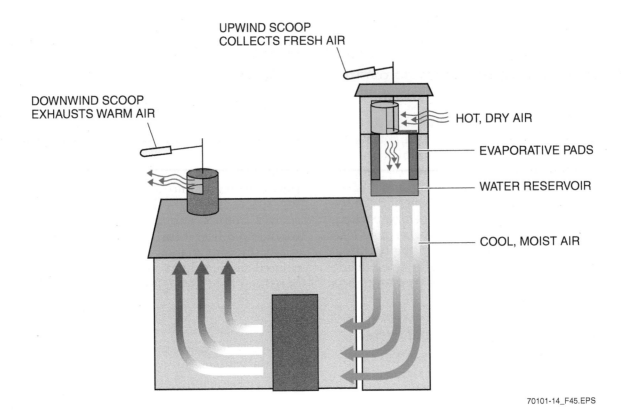

Figure 45 Natural ventilation system.

to heat hot water. They are very energy efficient, despite a relatively high initial cost.

2.6.4 Giving Users Control over Their Environment

Giving users control over their environment helps to improve indoor environmental quality for building occupants. Various methods can be used to achieve this. Operable windows, climate controls, and underfloor air distribution systems (UFADs) all provide occupants with the ability to control their individual spaces. The effects on building users can be significant. Studies have suggested that occupants who can control their spaces are happier, have greater job satisfaction, take fewer sick days, and are more productive.

Operable windows have gone in and out of vogue in architectural design over the past decades. Traditional buildings without mechanical heating and cooling had operable windows to allow users to control the indoor climate. Many contemporary buildings, however, have mechanical systems that don't work well with uncontrolled introduction of outside air. For this reason, many commercial and institutional buildings do not include operable windows. This creates a threat to passive survivability if the power to run mechanical equipment becomes unavailable. Passive survivability is the ability of a building to continue to function when the infrastructure goes down. There is growing interest in designing buildings around this idea, as many current buildings operate poorly or fail when deprived of power. In many climates, operable windows can be used in lieu of mechanical ventilation during the swing seasons (fall and spring) when outdoor temperatures are mild. This can save considerable energy. It also contributes to user satisfaction with individual workspaces.

Lighting, temperature, humidity, and ventilation controls are also becoming more sophisticated in modern buildings. Many commercial buildings rely on elaborate sensor and control systems to adjust these variables. However, each individual user of a building is different. Not all users are comfortable with the same environmental conditions. Individual controls at each workstation or office can help users adjust conditions to meet their individual needs (*Figure 46*). It can also increase their satisfaction with the space.

Underfloor air distribution systems are one way to increase the level of control users have over their individual workspaces. These systems provide the ability to control ventilation rates and perceived temperatures at each individual workspace. They also increase ventilation effectiveness and promote a uniform flow of conditioned air through the workspace (*Figure 47*).

Did You Know?
The Psychological Impacts of Color

Psychological research has shown that people react to different colors in predictable ways. Common colors and their psychological reactions are as follows:

- Yellow is highly visible and often used for safety markings. Yellow can make small rooms look larger and bring light to narrow entranceways and hallways.
- Orange is perceived to be cheerful and friendly. It is a good choice for rooms where people gather informally.
- Red stimulates the pituitary gland and raises heart rates and blood pressure. It also stimulates appetite. Red encourages action and aggressiveness. It is often used on buttons and knobs.
- Blue is a calming, soothing color. Blue works well in bedrooms.
- Green evokes feelings of relaxation and quietness. Green works well in study or work environments since it promotes concentration.
- Purples can reduce blood pressure and suppress appetite.
- Brown works well in environments where food is prepared or eaten. It also works well in general living environments. Brown is associated with comfort, reliability, and warmth.
- Gray encourages creativity. Gray interiors can be perceived as depressing unless accented with bright, clean colors.
- White is good around food and in precision work environments since it suggests sterility. Pure white can be perceived as harsh. Off-whites are better suited for many applications.
- Black is perceived to be dignified, sophisticated, and elegant. It tends to visually recede and enhances other colors used in combination with it.

These systems can be combined with modular office furniture to allow rapid reconfiguration of space. They make it easy to adjust the configuration of space while maintaining proper ventilation and space conditioning in each work area. Modular control units can be moved around to provide airflow wherever users are located. This level of flexibility and control means that users are happier in their workspace. It also means they are less likely to need supplemental space heaters or fans to be comfortable.

Figure 46 Giving users control over their environment can increase productivity and satisfaction.

2.7.0 Integrated Strategies

Many of the tactics and technologies described in the previous section offer multiple benefits for the green environment. The challenge for you is to find green solutions that meet the owner's needs for a facility. At the same time, they should not damage natural ecosystems or deplete resource bases, and should not cost more than a traditional project would cost. This section highlights strategies for green building decisions that achieve multiple benefits and affect multiple systems. Finding good integrated solutions can help you create better green buildings that cost less to build and have long-term benefits as well.

2.7.1 Solving the Right Problem

The first strategy for greening a project should be to make sure you're solving the right problem. Too often, building professionals start planning a new facility without considering other options. Possibilities include leasing, renovating an existing building, and telecommuting. In some cases, this could result in both green advantages and performance advantages at a much lower cost than building a new structure. In other cases, the most appropriate solution may still be to build new. However, considering other options early in the process can lead to a different mindset about

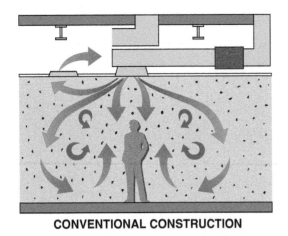
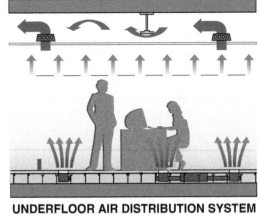

Figure 47 Underfloor air distribution systems help conditioned air reach all parts of a space.

the project that can help you be more creative on later decisions throughout the project.

The key to making sure you're solving the right problem is to focus not on solutions, but on the needs those solutions will meet. For example, how many times have you been asked at the checkout line whether you'd prefer paper or plastic bags? If you focus on these two choices as the only options available to you, you might miss other solutions. However, if you phrase the question as, "How can I transport my groceries from store to home?" you can consider other options entirely. What about reusable canvas bags, grocery delivery services, or even modular shopping carts that fold up to fit in your car? You might even decide to grow some of your own food to reduce your need for groceries. To use this principle, determine the functional necessity to be

Did You Know?

Passive Survivability

Hurricanes, tornados, floods, earthquakes, and a variety of human-created disasters or errors can cause the electricity or water supply to fail. Modern skyscrapers, such as the one shown here, typically have poor passive survivability. These buildings rely on electrical power for everything from ventilation to elevators to water pumps. Users may not even be able to open a window due to the building design. The principles of passive survivability align well with green building. Designing buildings to function in the absence of external supplies of energy, water, and materials makes them less vulnerable to external threats. Passive solar design, natural ventilation, rainwater harvesting, durable materials, and graywater reuse increase passive survivability and green a building at the same time.

Going Green: Net Zero Buildings

A goal of some high performance buildings is to be net zero. This means that over the course of a year, the building will produce as much or more of a resource as it consumes. Buildings generally try to be net zero with regard to energy or water. Energy can be produced on site through photovoltaics or other sources, and water can be obtained by rainwater harvesting or water reuse and treatment on site after it is used. The US Army is striving to make its facilities net zero, not just for energy and water but also for solid waste. This includes composting and recycling on site and accepting waste from other facilities. The amount of waste recovered from other sources is used to offset waste from the facility that cannot be recovered on site.

To successfully achieve net zero, a facility must use resources as efficiently as possible. Efficiency improvements are nearly always less expensive than on-site generation of power or water. Carefully consider what waste streams are available locally from nearby facilities when planning for net zero waste. On-site waste recovery or recycling can be expensive and take a lot of space; it may not make sense in every situation. However, sharing capacity with neighboring facilities can help make the investment worthwhile.

PHOTOVOLTAICS PRODUCING ENERGY ON A US ARMY FITNESS CENTER IN MONTEREY, CALIFORNIA. (NET ZERO ENERGY SITE)

---

met. If you find yourself defining your problem in terms of one or more solutions, chances are you need to take a step back and reassess your options.

2.7.2 Understanding and Exploiting Relationships Between Systems

Another approach is to exploit the functional relationships between systems. Building systems are related to one another, and the design of one system affects the design of another. For example, increasing the weight of a structure means that the foundation has to be upsized as well. Sometimes these relationships can be exploited to pay for investments in one system through savings in another. This is known as integrative design (*Figure 48*).

For example, consider using high-performance windows to provide daylighting. Considered in isolation, more expensive windows will raise total project cost. However, using high-performance windows will improve the energy efficiency of the building envelope. Additional daylighting will also reduce the heat load from light fixtures. Therefore, the extra costs of better windows can be offset by reducing the capacity of the building cooling system. In addition, a smaller HVAC system might mean smaller pumps, fans, and motors, reduced duct sizes, smaller plenums, and reduced floor-to-floor height, also reducing the cost of the facility. Reduced floor-to-floor height means less surface area of the building envelope, which means less material costs for the system. It also means that the overall weight of the building is reduced, meaning that foundations can be smaller and more efficient as well.

In the end, the overall increase in the total first cost of the project may be negligible if the benefits of improving one system are captured in the design of related systems. More importantly, life cycle cost savings can be even greater with these more efficiently designed systems. HVAC systems in particular will be much more efficient if they are correctly sized for the facility, allowing them to operate at maximum efficiency over the life cycle of the facility.

2.7.3 Using Services Rather than Products

A third strategy being used to green construction projects is dematerialization. Dematerialization refers to using services instead of products to meet user needs. Construction companies have successfully used this concept for years. For example, a small general contractor may rent equipment that is used infrequently, such as heavy equipment for grading.

BETTER PERFORMIING WINDOWS

REDUCED NEED FOR LIGHTING

REDUCED BUILDING ENVELOPE AREAS/FOUNDATIONS

REDUCED FLOOR-TO-FLOOR HEIGHT

LOWER COOLING LOADS/ SMALLER HVAC SYSTEM

REDUCED DUCT/PLENUM HEIGHT

Figure 48 Integrative design means that investments in one system can be offset by savings in related systems.

Rather than take ownership responsibility (and associated liability) for equipment, you pay another company to provide the benefits of that equipment to you. That company has an incentive to provide the most efficient equipment possible. The company makes its profits based on a fee per hour or unit of service provided. It also has incentive to design products that can be easily repaired, upgraded, or disassembled. This is because the company retains responsibility for the ongoing maintenance and eventual disposition of the equipment.

2.7.4 Considering the Options

When considering what actions to take on a project, it's critical to make smart choices. Every action you take comes with a cost. Recognizing those costs and considering the benefits of your actions can help you choose actions that achieve the results you intend. Think about leverage points; look for easy actions that can make a big difference.

One way to think about potential choices is to consider who has to act. Is action required on the part of building professionals, laypersons, or both? As a building professional, you might not want to change your work practices. However, your knowledge of construction means that you have a much better chance of being successful than a building user getting involved with the operation of the building (*Figure 49*).

Another issue to consider is whether users will have to change their behaviors to achieve the desired effects. Some changes are completely transparent to users, while others require significant change in habits or procedures. Changes are more likely to have the desired outcomes if they do not require people to change their behavior or comfort level. For example, asking occupants to turn down the heat to an uncomfortable level will save energy at first. However, it is a change that is unlikely to be sustained over time (*Figure 50*). When possible, choose solutions that can get the job done without requiring users to change their habits.

The degree to which the change is compatible with existing infrastructure is important. Some changes, like lighting retrofits, can be as simple as changing a light bulb (*Figure 51*). Other changes may not be compatible at all with existing buildings or may not be possible given the skills and equipment of the construction crew. Look for easy changes that can be undertaken with your existing skills and tools wherever possible.

Finally, availability is key to determine how easy a change will be to implement. Imagine if you had to special order every piece of technology you needed to get a job done. It's much easier to make a change when you know you can easily obtain the materials and systems you need to get the job done. Many building material retail stores now stock green products (*Figure 52*). Be sure to ask about these products using the factors discussed in earlier parts of this module.

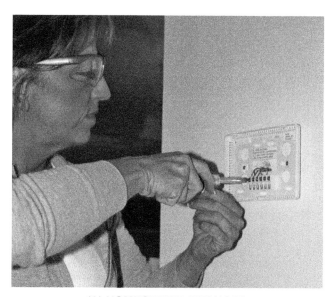

(A) HOMEOWNER CHANGES

(B) DESIGN CHANGES

Figure 49 Changes can be undertaken by professionals, laypersons, or both.

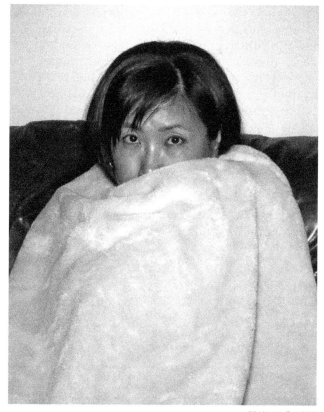

Figure 50 Changes that require users to sacrifice comfort are unlikely to be sustained.

Figure 51 Changes can be simple if they require no modification of infrastructure.

2.7.5 Counting All the Costs

Decisions about individual materials or systems are sometimes made on a first cost basis without considering cost from a life cycle perspective. This means that some green products seem more expensive than they really are. If savings in labor cost and construction schedule were taken into account, these materials could show an immediate cost advantage to owners. For example, one project manager estimated a savings of $3,000 per day due to shortening the construction schedule on a dormitory project. This dormitory was built using pre-engineered concrete panels, enabling the project to be built more quickly. In addition, the owner saved money by avoiding the hotel expenses of dorm residents, who were able to move in sooner.

Figure 52 Some green products are easy to find on the shelves of your local store.

Other factors that can make the cost of green building more reasonable include the following:

- Reduced costs of materials and waste disposal
- Reduced liability and environmental risk
- Happier occupants with increased productivity, reduced absenteeism, fewer building-related health problems, improved morale, and less employee turnover
- Reduced operational and disposal costs
- Reuse of facilities that otherwise would be disposed
- Preparedness for future regulations and requirements

Each of these benefits reflects a potential cost savings for owners. However, many of these types of costs are not typically counted as part of project costs. If these benefits can be realized, then green projects will have an economic advantage over conventional projects. Counting all the costs gives a truer picture of what green best practices will mean for the building over its whole life cycle.

Building Automation

GOING GREEN

An important trend in green building is building automation. Building automation is an integrated approach to optimize the building's use of water and energy while maintaining occupant comfort. The "nerves" of the system are a series of sensors to measure building conditions. These sensors are connected to a control system, which is the "brain" of the system. The control system uses information from the sensors to adjust different parts of the building to control environmental conditions. For example, daylight harvesting can be used to save energy by using photosensors to turn off lights when adequate daylight is available. The automation system may also activate shades or smart windows to provide shading when there is too much light or heat gain.

Building automation systems can be used to control a wide variety of interior and exterior conditions. These include lighting, temperature, humidity, and others. They may also be used to track performance of building systems and notify operators of potential problems. They also can provide useful information to building users about the effect their actions have on building performance. In this way, building automation can influence both the actions of the building systems and the actions of the building users that affect energy performance.

Did You Know?
Saving Water vs. Saving Energy

Sometimes saving one resource results in unintentionally consuming more of another resource. A recent study at Virginia Tech considered the effects of water recirculation systems, required by code in some areas. These systems prevent wasted water at the tap by keeping hot water circulating so there is no wait for hot water. At the same time, making energy at power plants also consumes water in the form of steam production. The study found that water saved by recirculation devices is less than the water used by power plants to make the energy used to run them. This example shows the need to think about all possible impacts of a choice before deciding what to do.

Additional Resources

Field Guide for Sustainable Construction, Department of Defense – Pentagon Renovation and Construction Office. (2004). PDF, 2.6 MB, 312 pgs. Available for download at www.wbdg.org.

Greening Federal Facilities, 2nd Ed. US Department of Energy Federal Energy Management Program. (2001). PDF, 2.1 MB, 211 pgs. Available for download at www.wbdg.org.

Natural Capitalism. Lovins, A., Hawkin, P., and Lovins, L.H. (1995). Little, Brown, & Company, Boston, MA. Available online at www.natcap.org.

Sustainable Buildings Technical Manual. Public Technologies, Inc./US Department of Energy. (2006). PDF, 3.1 MB, 292 pgs. Available for download at www.greenbiz.com.

Sustainable Buildings and Infrastructure: Paths to the Future. Pearce, A.R., Ahn, Y.H., and Hanmi Global. (2012). Routledge, London, UK.

Sustainable Construction: Green Building Design and Delivery, 3rd Ed. Kibert, C.J. (2012). Wiley, New York, NY.

The HOK Guidebook to Sustainable Design, 3rd Ed. Odell, W. and Lazarus, M.A. (2015). Wiley, New York, NY.

2.0.0 Section Review

1. The percentage of US buildings that suffer from poor indoor air quality is about ____.
   a. 15 percent
   b. 20 percent
   c. 33 percent
   d. 50 percent

2. The parts of a building site that are paved are called ____.
   a. softscape
   b. hardscape
   c. landscape
   d. parkscape

3. Most of the water on the planet is ____.
   a. groundwater
   b. ice
   c. salt water
   d. fresh water

4. A high-albedo roof ____.
   a. absorbs solar energy
   a. is expensive
   b. reflects solar energy
   c. reduces light pollution

5. Salvaged materials save the use of ____.
   a. raw materials
   b. labor
   c. equipment
   d. water

6. As they age, modern finishes, like paint, release ____.
   a. air
   b. VOCs
   c. carbon
   d. energy

7. Changes are easier and more likely to be successful when ____.
   a. building users are required to change their habits
   b. building occupants agree to turn down thermostats
   c. the construction crew has compatible skills and equipment
   d. suppliers special order the necessary materials

Section Three

3.0.0 Contributing to a Project's LEED Certification

Objective

Explain how craft workers can influence and contribute to a project's Leadership in Energy and Environmental Design (LEED) certification.

a. Describe the LEED rating process.
b. Identify construction activities and project features that affect a project's LEED rating.
c. List kinds of information collected during construction to support LEED documentation.
d. Identify common construction pitfalls that affect a project's LEED rating.

Trade Terms

Corporate Sustainability Report (CSR): An annual report by a company documenting its efforts to reduce negative environmental and social impacts.

EarthCraft: A residential green building program of the Greater Atlanta Home Builders Association in partnership with Southface Energy Institute.

Environmental Product Declaration (EPD): A document listing the environmental impacts caused by manufacture of a product.

Halons: Ozone-depleting compounds consisting of bromine, fluorine, and carbon. They were commonly used as fire extinguishing agents, both in built-in systems and in handheld portable fire extinguishers. Halon production in the United States ended on December 31, 1993.

Heat island: An area that is warmer than surrounding areas due to absorbing more solar radiation.

Hydrochlorofluorocarbons (HCFCs): A group of man-made compounds containing hydrogen, chlorine, fluorine, and carbon. They are used for refrigeration, aerosol propellants, foam manufacture, and air conditioning. They are broken down in the lowest part of the atmosphere and pose a smaller risk to the ozone layer than other types of refrigerants.

Leadership in Energy and Environmental Design (LEED): A point-based rating system used to evaluate the environmental performance of buildings. Developed by the US Green Building Council, LEED provides a suite of standards for environmentally sustainable construction. Founded in 1998, LEED focuses on the certification of commercial and residential buildings and the neighborhoods in which they exist.

Location valuation: A bonus under the LEED rating system for using a product produced locally. Materials from within a 100-mile radius can be counted as twice their value as a reward for using local products.

Persistent, bioaccumulative toxin (PBT): A harmful substance such as a pesticide or organic chemical that does not quickly degrade in the natural environment but does accumulate in plants and animals in the food chain.

Sedimentation: The process of depositing a solid material from a state of suspension in a fluid, usually air or water.

Spoil pile: Excavated soil that has been moved and temporarily stored during construction.

Thermal bridging: A condition created when a thermally conductive material bypasses an insulation system, allowing the rapid flow of heat from one side of a building wall to the other. Metal components, including metal studs, nails, and window frames, are common culprits.

Waste separation: Sorting waste by specific type of material and storing it in different containers to facilitate recycling.

Choosing the best green practices to use on a project can be difficult, but there are tools to help. Green building rating systems help you evaluate how buildings affect the green environment. Rating systems are particularly important to owners and purchasers of buildings. They offer valuable information about the building's overall environmental performance. Green building rating systems provide a way to help the construction market do better in terms of meeting green project goals. Most green building rating systems include performance levels that must be met for the building to be certified. They may also provide guidelines that help project teams meet or exceed those levels.

There are two major types of green building rating systems: location-specific and general. Location-specific rating systems are typically developed by local builder associations. They take into account local climate and building best practices for the area. Most location-specific rating systems apply to residential buildings. There are over 25 different local or regional residential rating systems in the United States, the first of which was established in Austin, Texas in 1991. Other local rating systems include EarthCraft homes in the southeastern United States, the Built Green rating system in Colorado (*Figure 53*), and the California Green Builder program for production builders in California.

The second type of rating system is general, and these systems can be used in multiple locations. In the United States, there are both residential and commercial general rating systems. There are also rating systems for neighborhoods and for infrastructure such as roads. Many national rating systems used in other countries are modeled after the US Green Building Council's Leadership in Energy and Environmental Design (LEED) rating system (see **www.usgbc.org**), developed in the 1990s. The LEED rating system was based on a similar system called the Building Research Establishment Environmental Assessment Method (BREEAM), which was developed earlier in the United Kingdom.

National residential rating systems include LEED for Homes and the National Association of Home Builders' National Green Building Standard (**www.nahbrc.org**). Commercial rating systems in use in the United States include the LEED rating system, Living Building Challenge (**living-future.org/lbc**), and the Green Globes rating system (**www.thegbi.org** and **www.greenglobes.com**).

The LEED Green Building Rating System is the predominant standard in the United States for rating commercial buildings (*Figure 54*). LEED is a reference standard for government agency buildings at the federal, state, and local levels. Many owners in the private sector have also adopted it. The US General Services Administration requires all of its new buildings to meet at least the Gold level of LEED certification. Other federal agencies have followed suit. Many states also require LEED certification for public buildings at various levels. Even individual cities have adopted LEED as a standard, including the cities of New York, Atlanta, Seattle, and others. By 2015, McGraw Hill Construction estimates that approximately 40 to 48 percent of all new non-residential construction projects in the United States will be green.

3.1.0 LEED Green Building Rating System

The LEED rating system applies to a wide variety of project types. It is designed to be applicable in different climates and contexts throughout the United States and worldwide. According to the

Figure 53 The Built Green program certified this home outside Denver, Colorado.

Figure 54 The LEED Green Building Rating System ensures that buildings are constructed to specific environmental standards.

> ### Did You Know?
> ## Rating Systems Around the World
> Green building rating systems were first developed in the United Kingdom in the 1990s. Since then, they have spread around the world. A recent study by the National Institutes of Occupational Safety and Health found over 40 different rating systems worldwide. Rating systems have been used on all continents except Antarctica. Certification by a rating system is required in some locations for some types of projects. However, in most places, certifying a project is voluntary. Project teams can also use rating systems as an internal tool to improve their projects without certification. Many tools are available free online.

US Green Building Council, by the end of 2013, approximately 42 percent of all square footage of buildings pursuing LEED Certification was outside the United States. LEED is characterized in terms of its structure, seven major categories of credits, and four levels of certification. A complete list of the LEED prerequisites, credits, and point values for new construction (LEED-BD+C Version 4) is provided in *Appendix A*.

> **NOTE**
> This module focuses primarily on the LEED for Building Design + Construction (LEED-BD+C) rating system, in Version 4 at the time of this writing. Other rating systems differ by project type and may vary in specific credit requirements.

3.1.1 Structure of the LEED Rating System

LEED for New Construction, now known as LEED for Building Design + Construction (LEED-BD+C), was the first of all the LEED rating systems to be developed. Each of the subsequent systems was modeled after the same structure.

The LEED rating system consists of a series of performance goals and requirements in nine primary categories:

- *Integrative Process (IP)* – This category addresses the process by which the project is planned, designed, and delivered.
- *Location and Transportation (LT)* – This category covers issues related to the location of the project site and the amenities around and near the project that affect transportation needs.
- *Sustainable Sites (SS)* – This category covers impacts to the project site during construction and the impacts the project will have on neighboring sites after completion.
- *Water Efficiency (WE)* – This category addresses water use and wastewater generation by the building during operation.
- *Energy and Atmosphere (EA)* – This category covers all aspects of the building's energy performance, energy source(s), and atmospheric impacts.
- *Materials and Resources (MR)* – This category pertains to the sources and types of materials used on the project, the amount of waste generated, and the degree to which the project makes use of existing buildings.

Seattle Focuses on Saving Power

Seattle, Washington, has written LEED certification into its City Plan. In addition to providing city staff to assist builders in creating and building green projects, they have legislated that all city-funded projects and renovations with over 5,000 square feet of occupied space must achieve a LEED Gold rating.

> **Did You Know?**
> ## LEED Minimum Program Requirements
>
> In order to be considered for LEED certification, a project must meet the LEED Minimum Program Requirements or MPRs. This means that not all types of projects are eligible to be LEED certified, although project teams can always use the LEED guidelines to create a better project. MPRs are designed to ensure that LEED is only applied in ways that make sense. Some of the projects that cannot be certified under LEED include the following:
>
> - Projects that are designed to be moved from one place to another, including boats and mobile homes
> - Projects built on artificially created land, unless the land was previously developed for another project
> - Buildings smaller than 1,000 square feet (93 square meters) of gross floor area to be certified under LEED-BD+C
>
> Moveable projects are not qualified because many LEED credits depend on the site chosen for the building. Projects on new artificially created land are not qualified because developing new land often displaces or disrupts ecosystems. Projects that are too small often do not have all the components LEED is designed to measure. Projects must also specify a reasonable site boundary to meet MPRs.

- *Indoor Environmental Quality (EQ)* – This category covers aspects of the building's indoor environment, ranging from ventilation to air quality to daylight and views.
- *Innovation in Design (ID)* – This category rewards the project for going beyond the minimum credit requirements and for using a LEED Accredited Professional.
- *Regional Priority (RP)* – This category rewards projects for addressing local environmental priorities (identified by ZIP code at www.usgbc.org).

Each category consists of a series of credits that define points that can be earned by a project. Six out of the nine categories also have prerequisites that the project must meet to be considered for certification. A project must meet all prerequisites in all categories in order to pursue certification. Projects can receive credit for features of special regional importance with Regional Priority (RP) credits. They can also receive credit for going significantly beyond basic credit requirements with Innovation in Design (ID) credits. For example, a system that pays special attention to water efficiency in a desert environment might be given an extra RP credit in addition to a WE credit.

3.1.2 Types of Rating Systems and Levels of Certification

The LEED system was initially developed to apply to new commercial construction. As the system grew in popularity, it became apparent that different types of projects would require different criteria to be properly rated. The original rating system was then customized through the development of special credits for specific project types such as hospitals and schools. These types of projects have characteristics that require special interpretation of LEED credit requirements. However, they can still use the basic LEED-BD+C structure. Separate versions of the rating system exist in the following major categories:

- LEED for Building Design and Construction (BD+C), which has special credits applicable to core and shell, schools, retail, data centers, warehouses and distribution centers, hospitality, healthcare, homes and multi-family lowrise, and multi-family midrise projects.
- LEED for Interior Design and Construction (ID+C), which addresses commercial interiors as well as special requirements of retail and hospitality projects.
- LEED for Building Operations and Maintenance (O+M), which covers existing buildings and has special credits for retail, schools, hospitality, data centers, and warehouses and distribution centers.
- LEED for Neighborhood Development (ND), which can be applied to developments in the planning phase or at the completion of the development.

Some parts of different rating systems complement each other. For instance, a LEED-BD+C project can receive up to fifteen points for choosing a project site within a development certified under LEED-ND. Likewise, a LEED-ID+C project can receive credit for being located in a building certified under LEED-BD+C.

There are 110 possible points under the LEED-BD+C rating system, and twelve prerequisites that apply to all projects. Projects must meet all prerequisites applicable to the project to pursue certification. Certification can be achieved at four different levels:

- *Certified* – 40 to 49 points
- *Silver* – 50 to 59 points
- *Gold* – 60 to 79 points
- *Platinum* – 80 points and above

Platinum is the most difficult level to reach. Currently, just over 4,000 certified projects have achieved this level of certification, out of more than 70,000 worldwide. This is only about 6 percent of all certified projects. All LEED rating systems award certification at the four listed levels.

3.1.3 Certification Process

In 2008, the United States Green Building Council established the Green Building Certification Institute (GBCI) to administer LEED project certifications and professional credentials and certificates. The process of certifying a building under LEED has several steps. These steps are guided by the LEED Online documentation system. Certification occurs in parallel with an integrative process that involves participation of many stakeholders. The first step in the plan is to undertake discovery and set a direction for the project. Based on the type of project, the appropriate rating system is chosen. Then, Minimum Program Requirements are verified. The team will review project information and hold a goal-setting workshop including the owner, design team, and construction team. This should all occur before design begins. A formal LEED project scope will be defined, including a project boundary.

Formal certification begins by registering the project with GBCI to declare intent to pursue certification. This allows the project team to access GBCI databases as well as set up an online workspace to manage project documentation. Registering the project requires paying a fee to GBCI that is the same for all projects.

After the project has been registered, the next step is to identify LEED credits that can be pursued. This is based on the project goals, and typically involves a meeting to review the LEED checklist (*Figure 55A* and *B*). The process of documentation to prove compliance with LEED credit requirements continues throughout construction. It may also extend into the first year of occupancy depending on the credits pursued.

The project team can submit documentation to GBCI for review at two points in time. The first time is at the end of the design process, and the second is after construction is complete. The project team may also opt to submit everything at once when the project is complete. All documentation is compiled by the project team using LEED Online. Submittal of documentation for review involves paying a review fee based on project size and the rating system used. This prompts GBCI to conduct the review. If the team submits a design review, GBCI will evaluate points under the rating system that can be measured at the end of the design process. It will not evaluate credits that require documentation during the construction phase. Design phase review is useful for the project team to get an idea of how many points they are likely to obtain in the project. It provides a basis for deciding how many additional points will need to be earned during construction to meet the project goals.

At the conclusion of the project, final documentation is assembled online. A fee is paid, and the package is reviewed by GBCI. Upon review of the full project documentation, GBCI may elect to request additional clarification as part of a point audit. GBCI makes a final determination as to which points should be awarded and determines a level of certification for the project. The project team has the right to appeal any credits declined in the review. This requires paying an additional fee and providing additional documentation. The ruling of GBCI following appeals is final.

3.2.0 Goals of the LEED Green Building Rating System

The basic concepts of the LEED rating system can be captured in the form of eight simple goals:

1. Choose a good site for your project.
2. Protect and restore the surrounding environment.
3. Promote sustainable behavior.
4. Conserve resources.
5. Find better waste sinks.
6. Create healthy and productive living environments.
7. Check systems to be sure they work right.
8. Look for better ways to do things.

> **NOTE**
> The details of LEED credit requirements are beyond the scope of this module. If you're interested in learning more about specific LEED credit requirements, you can view LEED standards for any of the current rating systems at **www.usgbc.org**.

Credit	Integrative Process	1
Location and Transportation		**16**
Credit	LEED for Neighborhood Development Location	16
Credit	Sensitive Land Protection	1
Credit	High Priority Site	2
Credit	Surrounding Density and Diverse Uses	5
Credit	Access to Quality Transit	5
Credit	Bicycle Facilities	1
Credit	Reduced Parking Footprint	1
Credit	Green Vehicles	1
Sustainable Sites		**10**
Prereq	Construction Activity Pollution Prevention	Required
Credit	Site Assessment	1
Credit	Site Development - Protect or Restore Habitat	2
Credit	Open Space	1
Credit	Rainwater Management	3
Credit	Heat Island Reduction	2
Credit	Light Pollution Reduction	1
Water Efficiency		**11**
Prereq	Outdoor Water Use Reduction	Required
Prereq	Indoor Water Use Reduction	Required
Prereq	Building-Level Water Metering	Required
Credit	Outdoor Water Use Reduction	2
Credit	Indoor Water Use Reduction	6
Credit	Cooling Tower Water Use	2
Credit	Water Metering	1
Energy and Atmosphere		**33**
Prereq	Fundamental Commissioning and Verification	Required
Prereq	Minimum Energy Performance	Required
Prereq	Building-Level Energy Metering	Required
Prereq	Fundamental Refrigerant Management	Required
Credit	Enhanced Commissioning	6
Credit	Optimize Energy Performance	18
Credit	Advanced Energy Metering	1
Credit	Demand Response	2
Credit	Renewable Energy Production	3
Credit	Enhanced Refrigerant Management	1
Credit	Green Power and Carbon Offsets	2

Figure 55A LEED project worksheet (1 of 2).

Materials and Resources		13
Prereq	Storage and Collection of Recyclables	Required
Prereq	Construction and Demolition Waste Management Planning	Required
Credit	Building Life-Cycle Impact Reduction	5
Credit	Building Product Disclosure and Optimization - Environmental Product Declarations	2
Credit	Building Product Disclosure and Optimization - Sourcing of Raw Materials	2
Credit	Building Product Disclosure and Optimization - Material Ingredients	2
Credit	Construction and Demolition Waste Management	2

Indoor Environmental Quality		16
Prereq	Minimum Indoor Air Quality Performance	Required
Prereq	Environmental Tobacco Smoke Control	Required
Credit	Enhanced Indoor Air Quality Strategies	2
Credit	Low-Emitting Materials	3
Credit	Construction Indoor Air Quality Management Plan	1
Credit	Indoor Air Quality Assessment	2
Credit	Thermal Comfort	1
Credit	Interior Lighting	2
Credit	Daylight	3
Credit	Quality Views	1
Credit	Acoustic Performance	1

Innovation		6
Credit	Innovation	5
Credit	LEED Accredited Professional	1

Regional Priority		4
Credit	Regional Priority: Specific Credit	1
Credit	Regional Priority: Specific Credit	1
Credit	Regional Priority: Specific Credit	1
Credit	Regional Priority: Specific Credit	1

TOTAL Possible Points:	**110**

Figure 55B LEED project worksheet (2 of 2).

3.2.1 Selecting the Site

The first goal of the LEED rating system is to help you choose a good site for your project. One way to get points is to choose a project site in a neighborhood certified under LEED-ND (LT Credit: LEED for Neighborhood Development Location). LEED rewards projects that avoid sensitive land, which includes wetlands, habitats of threatened or endangered species, or flood plains (LT Credit: Sensitive Land Protection). The rating system also awards credits for choosing a site in areas that are already developed to a certain density (LT Credit: Surrounding Density and Diverse Uses), as shown in *Figure 56*. These sites are preferred over undeveloped sites that can require the use of cars to get to amenities such as stores, parks, banks, and others, as shown in *Figure 57*. Locating near transit can also obtain a point under LT Credit: Access to Quality Transit. LEED also encourages projects to locate on sites that are brownfields or located in priority development areas such as historic districts (SS Credit: High-Priority Site). Brownfields are sites that have real or perceived environmental contamination, such as old gas stations, industrial sites, and dry cleaners.

3.2.2 Protecting and Restoring the Surrounding Environment

The second major goal of the LEED system is avoiding or repairing damage to the site, building, surroundings, and world at large during and following construction. Toward this end, project teams can get credit for assessing the ecological resources available on site (SS Credit: Site Assessment). They also set credit for designing

Figure 57 Projects in undeveloped areas force people to use motorized transport.

projects to preserve or restore site ecology (SS Credit: Site Development – Protect or Restore Habitat and SS Credit: Open Space).

The construction team plays a critical role in achieving the credits and prerequisite under this goal. A prerequisite for all LEED projects is to meet all requirements of the 2003 US EPA Construction General Permit, which includes measures to protect soil and water streams, and to mitigate dust and noise (SS Prerequisite: Construction Activity Pollution Prevention). All projects over one acre in size must comply with these requirements under federal law. This prerequisite requires projects to

Figure 56 Sites close to amenities reduce the need to travel by car.

Did You Know?

Accreditation vs. Certification

A common mistake when discussing LEED projects is to confuse accreditation and certification. Accreditation is a process that is applied to building professionals who manage the process of LEED certification for projects. Becoming a LEED Accredited Professional requires you to pass the LEED Accreditation Exam (more information is available at **www.gbci.org**, the website of the Green Building Certification Institute that manages and administers the LEED exam). Certification, on the other hand, applies to buildings. Buildings are certified under the LEED rating system and are awarded certification at the Certified, Silver, Gold, or Platinum levels. Keeping these terms straight will help your credibility with respect to the LEED rating system. Remember, people are accredited and buildings are certified.

use erosion controls to prevent stormwater runoff, sedimentation, and dust from leaving the site (*Figure 58*). Another consideration is managing excavated soil (*Figure 59*). The EPA General Permit Requirements include stabilizing spoil piles.

LEED encourages builders to limit site disturbance during construction to a minimal area (SS Credit: Site Development – Protect or Restore Habitat), using techniques such as fencing shown in *Figure 60*. LEED also encourages ecological site restoration at the end of construction. You can help meet this requirement by being aware of how far you can go outside your work boundary. If you are working on a LEED project, ask your supervisor to mark the boundaries and be sure you stay within them.

Contractors should install protective fencing when working near trees. Place the fencing at or beyond the drip line of the tree's outermost branches to protect the root system (*Figure 61*). Notify your supervisor if you see unprotected trees.

Since parking lots are significant sources of heat and pollutants in the built environment, LEED rewards projects that minimize the area of the site used for parking (LT Credit: Reduced Parking Footprint). Parking is also an important consideration in SS Credit: Rainwater Management and SS Credit: Heat Island Reduction. These credits focus on reducing specific negative impacts of pavements used for parking. Treating and/or retaining rainwater on site using other means is also encouraged (SS Credit: Rainwater Management). This helps prevent contamination of local waterways by runoff from parking lots and roofs.

Figure 58 Sedimentation fencing in place to manage stormwater runoff.

FENCING TO LIMIT DISTURBED AREA

Figure 59 Grass seeding and straw mulch are one way to stabilize excavated soil.

TREE PROTECTION FENCING

Figure 60 Temporary fencing can be used to mark limits of disturbance.

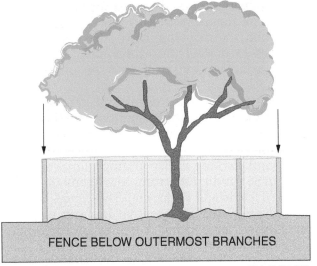

Figure 61 Effective tree protection fencing requires the tree to be protected as far out as the drip line.

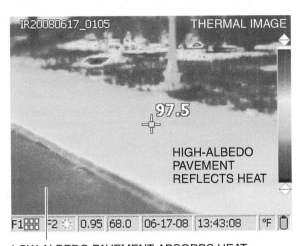

Figure 62 Dark pavements worsen urban heat islands.

Projects that use light-colored roofs are also recognized for reducing urban heat islands (SS Credit: Heat Island Reduction). Urban heat islands are developed areas where ambient temperatures are higher than surrounding areas. They are caused by dark-colored pavements and buildings that absorb solar energy (*Figure 62*).

LEED also acknowledges projects that avoid using unnecessary levels of lighting (SS Credit: Light Pollution Reduction) due to unshaded light fixtures, such as the one shown in *Figure 63*. Light pollution disturbs endangered species such as loggerhead turtles. It is suspected of disrupting sleep and mating cycles in animals, and may have adverse effects on people.

To protect the environment outside the job site, projects are encouraged to reduce or eliminate the use of ozone-depleting chemicals. This includes chemicals in air conditioning and refrigeration equipment and propellants in insulation (EA Prerequisite/Credit: Fundamental/Enhanced Refrigerant Management). Chlorofluorocarbons (CFCs), hydrochlorofluorocarbons (HCFCs), and Halons all deplete atmospheric ozone. Damaging atmospheric ozone has negative effects on plants, animals, and people. Producing energy from renewable sources instead of fossil fuel is also encouraged under LEED. This helps reduce the amount of carbon released into the atmosphere contributing to the greenhouse effect. EA Credit: Renewable Energy Production and EA Credit: Green Power and Carbon Offsets both encourage renewable energy. Renewable energy can be produced on-site or can be purchased from off-site utilities.

Figure 63 Many outdoor fixtures contribute to light pollution.

Several LEED credits recognize the impacts that building materials have on the natural environment. LEED rewards projects that use materials where manufacturers openly disclose these impacts (MR Credit: Building Product Disclosure and Optimization – Environmental Product Declarations). Project teams can receive additional points if the products they choose are environmentally preferable. One way to do this is if the raw materials used to make those products have been developed responsibly (MR Credit: Building Product Disclosure and Optimization – Sourcing of Raw Materials). Another way is to reuse buildings or building materials to prevent the need to dispose of them and make new materials (MR Credit: Building Life-Cycle Impact Reduction). Project teams can also use techniques such as life cycle assessment to measure impacts of their choices in building design.

For Materials and Resources credits, it is very important to track products used in construction throughout the construction process. Be sure to ask your supervisor what criteria are important for different building products. Be sure to verify any substitutions before you make them. Keep track of documentation for products you use, since it will be needed to document the certification. For instance, MR Credit: Building Product Disclosure and Optimization – Material Ingredients requires information from manufacturers about the ingredients used to make materials. MR Credit: Building Product Disclosure and Optimization – Sourcing of Raw Materials requires information about how those ingredients were extracted from the natural environment and what effects were caused.

3.2.3 Promoting Sustainable Behavior

The next major goal of the LEED rating system encourages projects to include amenities that promote sustainable behavior. This includes providing amenities that encourage people to use low-impact transportation such as transit, bicycling, green vehicles, or carpooling (LT Credits: Access to Quality Transit; Bicycle Facilities; Green Vehicles), as shown in *Figure 64*. LT Credit: Surrounding Density and Diverse Uses provides the opportunity to walk or bicycle to nearby amenities instead of driving. SS Credit: Open Space rewards projects that provide green space, which may encourage building occupants to spend time being active outdoors.

Water and energy performance credits require the building to include efficient fixtures that save water and energy during its life cycle (WE Prerequisite/Credit: Indoor Water Use Reduction

Figure 64 Bike racks, reserved parking for carpools, and preferred parking for motorcycles and high-efficiency vehicles reward green behavior.

and EA Prerequisite/Credit: Minimum Energy Performance; Optimize Energy Performance).

Projects that include water or energy meters can also help building occupants behave more sustainably. WE Prerequisite: Building-Level Water Metering and EA Prerequisite: Building-Level Energy Metering require all LEED projects to have at least one meter for the whole building's water and energy use. WE Credit: Water Metering and EA Credit: Advanced Energy Metering offer additional points for measuring water or energy use for individual systems or devices.

Recycling facilities (*Figure 65*) encourage occupants to sort their solid waste and reduce the amount of trash that goes to landfills (MR Prerequisite: Storage and Collection of Recyclables). Having facilities to collect and store recyclables is required of all LEED certified buildings.

Providing smoking facilities as required by EQ Prerequisite: Environmental Tobacco Smoke (ETS) Control helps protect nonsmokers from environmental tobacco smoke. When adequate smoking facilities are not provided, smokers tend to huddle near the door, which may cause discomfort for those entering and leaving the building. Many hospitals, schools, and government buildings prohibit smoking within a certain distance of building entrances.

3.2.4 Conserving Resources

The fourth goal encompassed by LEED is to conserve resources as much as possible throughout the project. This includes using environmentally preferable materials wherever possible. Environmentally preferable materials include products whose manufacturers have used best practices for reducing impacts during manufacture, or whose products are extracted responsibly (MR Credit: Building Product Disclosure and Optimization – Sourcing of Raw Materials). Products made from raw materials such as FSC Certified wood, rapidly renewable materials, or with recycled content can be counted in this credit. LEED also rewards manufacturers who make information about their product available to help designers make better choices (MR Credit: Building Product Disclosure and Optimization – Environmental Product Declaration and MR Credit: Building Product Disclosure and Optimization – Material Ingredients). This includes information about a product's environmental impacts and ingredients. Finally, this goal also includes using less material, or reusing existing materials instead of new materials (MR Credit: Building Life-Cycle Impact Reduction).

Buildings are also rewarded by LEED for using graywater or rainwater instead of potable water for landscaping (WE Prerequisite/Credit: Outdoor Water Use Reduction). LEED also rewards indoor uses of graywater such as toilet flushing (WE Prerequisite/Credit: Indoor Water Use Reduction), such as the under-sink system shown in *Figure 66*. Conserving water or finding alternate sources of water for cooling towers is rewarded under MR Credit: Cooling Tower Water Use.

Finding better sources for electrical energy is encouraged via installation of on-site power generation (EA Credit: Renewable Energy Production). This helps a building be more energy independent from outside fuel sources such as foreign oil. You can also get credit for contracting for green power from a third party provider or for purchasing carbon offsets (EA Credit: Green Power and Carbon Offsets). Including capabilities for peak shaving and controlling a building's energy demand is also rewarded (EA Credit: Demand Response). Using natural daylighting (EA Credit: Daylighting) also conserves energy that would be used to artificially light a building.

Not only is the type of material important for a green building, but also where the components come from. This helps reduce impacts associated

Figure 65 Recycling facilities encourage users to sort their waste.

Figure 66 An under-sink graywater system stores sink and shower water for toilet flushing.

with transportation of materials. Several LEED MR credits reward the use of regional materials to reduce transportation impacts. *Figure 67* shows an example of the 100-mile radius that defines a regional material according to LEED. Products whose extraction, manufacture, and purchase occurs within 100 miles of the project site can be counted double in LEED calculations.

As a craft worker, you may be required to procure green materials and systems to meet the project specifications (*Figure 68*). When procuring these materials, carefully review the specifications to be sure you understand all product requirements. Keep cut sheets or other documentation to support compliance. These materials will help when preparing the project documentation.

Better sources for construction materials that are encouraged under LEED include the following:

- Salvaged or reused materials
- Materials with recycled content, especially post-consumer recycled content
- Materials from locations near the project site
- Rapidly renewable bio-based materials
- Wood products that are sustainably harvested
- Materials from a manufacturer who will take back or recycle the product at the end of its life

A. Paper fiber panel core
B. Laminated strand lumber (LSL)
C. Engineered wood I-joist
D. Abundant American hardwoods
E. Laminated veneer lumber (LVL)
F. Adobe block with straw
G. Straw-core SIP with oriented strand board (OSB) skin
H. Parallel strand lumber (PSL)

Figure 68 Bio-based materials are an important resource for green building.

Figure 67 A regional material originates from within 100 miles of a job site.

Manufacturers must provide corporate sustainability reports (CSR), environmental product declarations (EPD), or other objective documentation to prove their claims about their products. A corporate sustainability report provides information about a company's programs and achievements each year to benefit the environment and society. An environmental product declaration is an objective report that documents the environmental impacts of a product and allows it to be compared to other products. Some of the impacts in an EPD include global warming potential, ozone depletion, depletion of nonrenewable resources, and acidification of water sources. Products whose EPDs are produced by the manufacturer instead of a third party receive only half value in LEED calculations. Third party-verified EPDs are more objective, since the third party has no motive to rate the product better than it actually is.

3.2.5 Finding Better Waste Sinks

Minimize or prevent waste wherever possible using the best practices discussed earlier in this module. During design and construction, the focus of LEED is on material use and waste. LEED offers credit for designing buildings to use fewer materials under MR Credit: Building Life-Cycle Impact Reduction. Healthcare projects also can get special credits for designing spaces for flexibility (MR Credit: Design for Flexibility). This is because healthcare facilities change often to accommodate new practices. Each change can produce a lot of solid waste unless spaces are designed to allow easy changes.

With waste that is produced, be sure you understand the project policy for waste separation and recycling. All LEED projects must develop and implement a construction and demolition waste management plan (MR Prerequisite: Construction and Demolition Waste Planning). A final report is also required to show what waste was produced and what happened to it. Projects that divert a significant amount of construction and demolition waste from landfills or incineration can get additional points (MR Credit: Construction and Demolition Waste Management). Construction projects can meet this credit requirement by sending mixed waste to an off-site separation facility (*Figure 69*) or through on-site separation of construction and demolition waste for recycling (*Figure 70*).

Projects can also achieve credits for reducing the amount of waste generated during construction to less than 2.5 pounds of construction waste per square foot (12.2 kilograms of waste per square meter). This goal can be achieved in various ways discussed in Section Two, including the following:

- Prefabrication of components
- Modular construction
- Designs that use standard-size materials

If on-site recycling separation is part of the project, be sure to put waste in the correct dumpster (*Figure 71*). If you don't see a place to recycle, ask for one. Do not contaminate dumpsters with other waste. In particular, keep treated wood

Figure 69 In some areas, mixed construction waste can be taken off site and sorted for recycling.

Did You Know?
Pre-Consumer vs. Post-Consumer Recycled Content

Not all recycled content is the same. Some waste is produced as a byproduct of manufacturing, then recovered for recycling. This is called pre-consumer recycled content because it is recycled before it enters the market to be used by consumers. It is also sometimes called post-industrial recycled content. Often, the manufacturing process could be made more efficient to reduce or eliminate this waste. Post-consumer recycled content is the result of a consumer actively recycling waste after a product has been used. One example is recycled plastic bottles being made into carpet. LEED rewards both types of recycled content, but pre-consumer content can only be counted half as valuable as post-consumer content in LEED calculations.

Figure 70 In other areas, on-site sorting is required for recycling.

Figure 71 On-site waste separation can earn up to two LEED credits plus one Innovation Credit.

separate from other materials. Also, keep anything with food on it out of cardboard dumpsters. Cleaner dumpster loads are more likely to be accepted for recycling.

If you have excess materials or salvaged goods on a project, ask your supervisor to arrange a place for temporary storage. Consider donating unused materials to a Habitat for Humanity ReStore. If there is a ReStore near you, take the time to visit it; you may find products that you want to buy.

LEED also includes credits to encourage finding better waste sinks during building operation. SS Credit: Rainwater Harvesting encourages behavior to prevent stormwater from being treated as waste. Various Water Efficiency credits also reduce the amount of wastewater that is produced by the building. MR Prerequisite: Storage and Collection of Recyclables provides user amenities to divert solid waste from being sent to landfills.

3.2.6 Creating Healthy and Productive Living Environments

The sixth major goal of the LEED rating system is to create healthy and productive living environments for building occupants. LEED rewards projects for encouraging human-powered transport. This can be achieved by locating within walking distance from other amenities (LT Credit: Surrounding Density and Diverse Uses). Projects can also provide bicycle facilities (LT Credit: Bicycle Facilities) and outdoor open space (SS Credit: Open Space) to be used by occupants.

At the building scale, LEED strives to make buildings healthy and productive by the following actions:

- Avoiding toxic ingredients when choosing building materials (MR Credit: Building Product Disclosure and Optimization – Material Ingredients).
- Achieving minimum levels of indoor air quality (EQ Prerequisite: Minimum Indoor Air Quality Performance).
- Keeping nonsmokers away from environmental tobacco smoke (EQ Prerequisite: Environmental Tobacco Smoke [ETS] Control)
- Making sure spaces receive adequate ventilation (EQ Credit: Enhanced Indoor Air Quality Strategies).
- Making sure air quality is good before allowing occupants to move in (EQ Credit: Indoor Air Quality Assessment).
- Avoiding the use of materials that emit high levels of VOCs (EQ Credit: Low-Emitting Materials). *Figure 72* shows a product label indicating VOC content.
- Separating areas of the building that contain polluting activities from other areas (EQ Credit: Enhanced Indoor Air Quality Strategies).
- Designing building conditioning systems to provide thermal comfort (EQ Credit: Thermal Comfort).
- Providing users with the ability to control conditions in their own spaces (EQ Credit: Thermal Comfort and EQ Credit: Interior Lighting).
- Providing users with access to natural daylight and views (EQ Credit: Daylight and EQ Credit: Quality Views).
- Enhancing well-being and productivity through controlling acoustic conditions (EQ Credit: Acoustic Performance).

Consider whether your activities will produce airborne dust or offgas VOCs. Take measures to protect the building from these activities (*Figure 73*).

To create buildings that are free of problems for future occupants, focus on work inside the building. Using low-emitting materials can help to ensure good indoor air quality (*Figure 74*).

LEED encourages separate ventilation for chemical mixing and storage areas and copy rooms (*Figure 75*). These areas can produce air pollution that is unhealthy for building occupants. Copiers and printers produce ozone during operation, which can cause respiratory irritation.

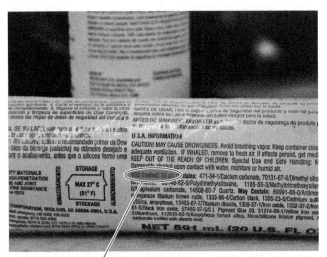

Figure 72 Look for levels of VOCs on product labels and buy products with low levels.

If you work inside the building, make sure you understand how your activities affect indoor air quality. LEED provides a credit for creating and following an indoor air quality management plan during construction. This is based on SMACNA's *Indoor Air Quality Guidelines for Occupied Buildings Under Construction* and other measures, including the following:

- Prohibiting the use of tobacco products on site during construction. This includes smoking and smokeless tobacco inside the building. It also forbids use of tobacco products within 25 feet (7.5 meters) of the building's entrances.
- Protecting absorptive materials from moisture damage before, during, and after installation.
- Keeping contaminants out of all HVAC equipment. This includes not operating building air handling equipment during construction unless it is properly protected with filters. It also includes sealing ductwork during construction.
- Avoiding toxic materials where possible.
- Isolating areas of work to prevent contamination of other spaces.
- Scheduling construction activities to finish dirt-producing activities before installing absorptive materials.
- Maintaining good job site housekeeping.

Managing moisture during construction is key for the health of building occupants. Prevent moisture damage by storing materials in a protected staging area. Never install wet materials. Installing wet or damp materials can create mold problems in the building before it is even complete. If materials do get wet, they may need to be replaced.

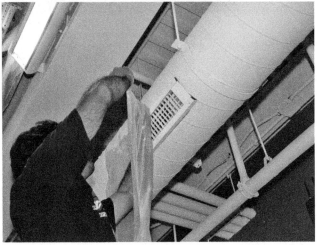

Figure 73 Containing construction contaminants is critical to ensure indoor air quality.

Figure 74 Low-emitting materials can help to ensure good indoor air quality.

Figure 75 Photocopiers are a common source of indoor air pollution in office environments.

Other building products can also be unhealthy. Healthcare projects receive special consideration under LEED for this goal. Persistent, bioaccumulative, and toxic (PBT) materials should be avoided. These materials degrade very slowly and can accumulate in body tissues. They can cause problems with the central nervous system. Special actions that apply to Healthcare projects include the following:

- Reducing the use of mercury-containing products and devices, and preventing mercury release (MR Prerequisite/Credit: PBT Source Reduction - Mercury).
- Reducing the use of lead, cadmium, and copper (MR Credit: PBT Source Reduction – Lead, Cadmium, and Copper).
- Choosing furnishings that don't contain toxic materials (MR Credit: Furniture and Medical Furnishings).

Mercury is a common material used in fluorescent lamps, mercury vapor high-intensity discharge lamps, and some thermostats and switches. These products should be carefully handled if they are used in the project, and avoided if possible. Lead is often found in solder and flux for plumbing connections. It may also be used in roofing and flashing, wiring, and paints. Cadmium is used in paints. Copper can be released into the water supply when copper pipes corrode. All joints in copper piping should be either mechanically crimped or properly soldered to prevent corrosion.

Furnishings may also contain toxic materials, and LEED encourages projects to use furnishings that do not contain them. Toxins in furniture can come from stain and non-stick treatments, antimicrobial treatments, heavy metals, and urea formaldehyde. Furnishings also can offgas VOCs, which may be released in large quantities into a new building when furnishings are installed. Consider unpacking furnishings off-site to allow them to offgas before moving them into the building.

LEED credits to promote daylighting, acoustic performance, and quality views are included because of growing evidence that these factors provide many benefits to occupants (*Figure 76*). Exposure to natural daylight has been correlated with effects ranging from better test scores to improved dental health. Being able to look out the window has also been linked with increased productivity in offices and faster healing in hospitals. LEED encourages the use of daylighting to increase human comfort as well as reduce the need to rely on electric lighting. Better acoustics enhances learning, and has been made a prerequisite in LEED for Schools projects.

3.2.7 Performing System Checks

LEED puts special emphasis on checking the design, construction, and initial operation of a building to be sure it meets the design intent. This requires a process called commissioning (EA Prerequisite: Fundamental Commissioning and Verification). Commissioning can also be expanded to cover more systems at a more detailed level (EA Credit: Enhanced Commissioning).

Figure 76 Daylight and views are correlated with many positive outcomes.

Commissioning involves a third party in design review, testing and balancing of systems, and system turnover (*Figure 77*).

At the prerequisite level, mechanical, electrical, plumbing, and renewable energy systems must be commissioned to obtain LEED certification. Commissioning agents will evaluate these systems for energy and water consumption, indoor environmental quality, and durability. They will also review the building's exterior enclosure design to ensure it has been properly designed and constructed. Projects can receive additional points for expanding commissioning activities to include in-depth design review, training, and post-construction verification.

LEED also rewards projects that include the ability to measure the performance of systems. This includes water consumption (WE Prerequisite: Building-Level Water Metering and WE Credit: Water Metering) and energy consumption (EA Prerequisite: Building-Level Energy Metering and EA Credit: Advanced Energy Metering). These systems may also be able to adjust the building's mechanical and electrical systems to optimize performance. Another example is using CO_2 monitors to control building ventilation (EQ Credit: Enhanced Indoor Air Quality Strategies). The building itself should also be designed to keep occupants comfortable (EQ Credit: Thermal Comfort and EQ Credit: Acoustic Performance). LEED also encourages post-occupancy evaluation after the building has been occupied. The purpose of this survey is to ensure that building users are still comfortable after the building has been broken in.

Figure 77 Building commissioning provides a chance to check that all building systems work well together.

3.2.8 Seeking Better Methods

An important part of green building is improving the process by which projects are delivered. LEED requires an integrative planning and design process that involves stakeholders outside the normal process (IP Prerequisite: Integrative Project Planning and Design). Designers from multiple disciplines along with owners, constructors, and operators participate throughout the planning and design phases. The combination of perspectives, along with an open mind for new ideas, can lead to design breakthroughs. It also helps to ensure that what is designed can actually be built and operated properly. Teams that go beyond the basic process can earn additional points (IP Credit: Integrative Process). This

---

High-Performance Building Envelopes

Going Green

Rinker Hall on the University of Florida's Gainesville campus earned LEED Gold certification for its energy-efficient design. The architect was challenged to design a building that would fit in with the other brick buildings on campus, but not trap heat. The compromise was a freestanding masonry shade wall that serves as a second skin and helps match the design of the building to the site. The high-performance building envelope reflects heat and includes an air-infiltration barrier, a thermal break, and high-performance glass.

includes analyses to identify interrelationships among building systems.

LEED also rewards projects for finding new ways to make projects green that are not currently included in the rating system (ID Credit: Innovation). Project teams can propose new credits for future versions of LEED based on best practices they have achieved. Or, they can pursue a credit from the LEED Pilot Credit Library. The Pilot Credit Library contains credits still under development that have been field-tested by at least one project. However, these credits are not yet a part of the official rating system standard. One example of a current pilot credit is Clean Construction. This credit reduces negative impacts from equipment idling and exhaust from construction vehicles and equipment (*Figure 78*). Projects can also receive innovation credits for greatly exceeding regular credit thresholds in areas including water efficiency, energy performance, materials use, and waste diversion. These are called exemplary performance credits. Finally, projects can be awarded a point for involving a LEED Accredited Professional (LEED AP) as part of the project team (ID Credit: LEED Accredited Professional).

Be aware of the project's goals for LEED certification and speak up if you have ideas for improvement. On the job, various Innovation Credits such as Clean Construction may impact you. Another possible impact may happen if your project is pursuing Innovation Credits for education. You may see students on the job site who are observing or recording your work as part of LEED documentation of the project. Or you may be asked to participate in special training yourself. Feel free to ask questions. If you see students observing, it is important to show them your typical work practices so they can accurately capture what happens on the job site.

3.3.0 LEED Documentation

Managing documentation for a LEED project can be challenging. The LEED AP handling the documentation of your project will be grateful if your materials and activities are well documented. Be sure you understand the types of information you will be required to provide, and ask questions if needed.

Projects pursuing Materials and Resources credits require information about each product used in the project. This varies based on what specific credits are being pursued. All calculations are based on the material cost of each product, so the actual cost of each product must be provided or estimated. This does not include labor or other installation costs.

For recycled content, you need the percent by weight of both post-consumer and pre-consumer recycled content for all parts of the material. As you might imagine, this can get quite complicated for a product such as carpeting, which includes several different materials. Product documentation may also require information about VOC content, Forest Stewardship Council chain of custody, or other information.

Another challenging factor to calculate is a product's location valuation. This calculation requires that you know the origin of the material and its components. A material may be counted double for many Materials and Resources credits if it originated and was processed and sold within a 100-mile radius of the site.

For Construction and Demolition Waste Recycling, documentation is needed about the total weight or volume of waste for the project, along with the weight or volume of material diverted for recycling. Your project team will choose either weight or volume as the basis for calculations. After the choice has been made, all calculations must use that basis. You should retain waste hauler receipts showing the destination and weight or volume of each different material claimed for this credit. Receipts are also required for products donated to charity. Materials reused on site or reused on other projects require a recorded estimated weight or volume. Photographs of recycling containers are also required.

Photographs are an important part of LEED documentation. They are required as proof of what was done during construction, since many measures will not be visible after the fact.

Figure 78 Using next-generation emissions controls for clean construction, such as the Tier IV engine on this grader, can help earn a LEED innovation Credit.

There is a limited window of opportunity to take required photos. It is critical that photos be taken at the right times for LEED credits being pursued. Construction indoor air quality measures such as HVAC protection are one type of photograph required. Another type is photographs of recycling containers for construction and demolition waste management. Stormwater protection measures such as sedimentation fencing must also be photographed. Be sure to ask your supervisor what photographs you are required to take for LEED documentation. Show the date the picture was taken in the photograph if possible, and take photos on a regular basis.

3.4.0 Common Pitfalls During Construction — Design errors

With all the ways you can contribute to a LEED project, there are also plenty of pitfalls to avoid. Four of the most serious pitfalls include poor planning, poor execution of credit requirements, poor documentation, and lack of coordination with other trades.

3.4.1 Poor Planning

Some LEED pitfalls occur before the project starts. Contractors must be familiar with all LEED requirements for the project, and they must make all subcontractors aware of those requirements when planning the job. Often, LEED requirements are included as a separate section in the specifications. However, many subcontractors only read the specs pertaining to their trade. This means they can miss LEED requirements that would affect their bid. Contractors must ensure all subcontractors understand LEED requirements when bidding on a project. Ideally, LEED requirements will be included in the specifications in each section where they apply. Including contractors in integrative design can help to avoid these pitfalls.

It is also important to keep LEED goals in mind when planning project tasks. The sequence of tasks in the schedule is important. Absorptive materials should be properly protected until it is safe to install them. Concrete and other materials should be cured before moisture-sensitive products are installed. Dirt-producing activities should be completed before finishes are installed.

Following a site plan is also important, as construction activities can be damaging to the natural environment. Limits of disturbance should be marked off before clearing begins. LEED projects may require extra dumpsters to account for recycling. Dumpsters should be located to minimize traffic. Material storage and staging is also important; just-in-time delivery can help avoid disturbance to the site.

Make sure that all products used in the project meet LEED requirements. For instance, it is easy to remember that paint should be low- or zero-VOC. However, fire caulking and duct sealants must also be low-VOC to meet air quality limits. Subcontractors should plan to be sure these materials are available. They may need to allow extra time for delivery for some products.

Coordination with the owner and other stakeholders is key. Air quality testing can be disturbed if the owner moves in before testing is done. The owner's furniture may offgas VOCs into the space, leading to bad test results. Housekeeping at the end of a project can also cause problems. If toxic chemicals are used to clean the building, air quality can suffer. This happens even though the materials in the building are low VOC.

Roles in managing LEED requirements are important. Someone must be assigned to verify all products that come on site to be sure they comply. Someone should review all product substitution requests to make sure new products meet LEED goals. Someone must be in charge of collecting documentation. This includes FSC chain of custody information, photographs, and waste hauling receipts. That person should be asked if documentation is complete before final payments are made.

3.4.2 Poor Execution of Credit Requirements

Even if you know exactly what is necessary to achieve a LEED credit, it is not always easy to meet those requirements. Some credits can be difficult to achieve. For example, EQ Credit: Construction Indoor Air Quality Management Plan requires you to protect the indoor air quality

---

Did You Know?

When in Doubt, Document

If you ever have questions about whether or not you should document something, always err on the side of caution. Gathering documentation after the fact is nearly impossible. Develop a documentation plan with your team and create a system for collecting and storing receipts, bills of lading, cut sheets, and other documentation relevant for LEED credits. Coordinate with other members of the project team to make sure you know what to keep and what to do with it.

of the building. This requires various measures to isolate your work area and keep it clean. Failure to maintain these measures, even if you set them up correctly, can cause the project's goals not to be met.

A project can fail its indoor air quality test for many reasons. This means it would not be awarded the associated LEED credit. Common reasons projects fail IAQ tests include the following:

- Not controlling products brought on site for use in the building
- Allowing high VOC products to be used for fire caulking, duct sealing, and other applications not specifically mentioned by LEED
- Allowing owners to occupy the building before IAQ testing, including moving furniture into the building
- Using toxic materials for final cleaning of the project. These include toxic chemicals and strippers, window washing chemicals, and spot removers, especially spot cleaners for vinyl tile

It is important to make sure everyone involved in the project knows about LEED goals and what must be done to meet them. Even the cleaners brought in at the end of the project are an important part of the project team.

Another potential problem area is SS Prerequisite: Construction Activity Pollution Prevention. As a prerequisite, these requirements must be met. If they are not, the project's LEED certification can be denied. One example of a pollution prevention activity is sedimentation and erosion control. This includes measures like sedimentation fencing and protection around stormwater drains. If you are responsible for maintaining these controls, be sure you understand exactly what needs to be done. Check these systems regularly and repair them if they become damaged. Even if they are not directly your responsibility, inform your supervisor of any problems so they can be corrected.

Similarly, be sure you understand project goals for SS Credit: Site Development – Protect or Restore Habitat. Often areas of the site will be selected to keep from damage during construction. Protection fences may be put up to remind equipment operators to avoid these areas. Ask your supervisor if you are unclear on what can and cannot happen in those areas. Bring problems to the attention of your supervisor or crew leader.

MR Credit: Construction and Demolition Waste Management is also a common credit where things go wrong. Be sure you understand the project's recycling goals and how they will be met. Ask questions in weekly or daily toolbox meetings, especially if you notice that others are not recycling properly. If you see contamination in any of the recycling bins, mention it to your supervisor. Talk to your crewmates to be sure they understand the importance of correct separation for recycling.

Buying green materials for a project can also be a challenge. The specifications for a green project have been carefully developed to meet project green goals. They should be followed whenever you buy materials. Ask vendors for documentation to support the products you buy. Make sure you allow enough time to purchase any unusual materials.

3.4.3 Poor Documentation

The third major pitfall that can impede your project's LEED certification is poor documentation. The amount of information needed to certify a project is extensive; you must know the properties of the products you use, and you also must report their cost. Therefore, keeping track of information as it occurs can help you successfully provide the information the project team needs.

One credit that requires careful documentation is MR Credit: Construction and Demolition Waste Management. For this credit, you must prove where all of the waste for your project has gone. This is necessary to calculate the percent of waste that was diverted from landfills. To do this, you must save waste hauling receipts and receipts from recycling or salvage companies. Set up a process for tracking this information.

What's Wrong with This Picture?

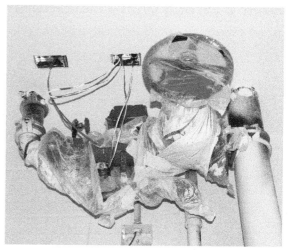

Another common error in waste documentation is to vary the units used to measure waste. You can choose whether to measure waste by weight (pounds or tons; kilograms) or volume (cubic yards or cubic meters). However, you must choose one way or the other, then use only that basis for all documentation. Be sure to ask about what units will be used by your waste hauling provider. Make them aware of the project's LEED goals.

Photos are also required for many LEED credits. Often, these photos must be taken at a specific time to capture a construction activity. Make sure those times are noted in the schedule. Someone must have responsibility to take photos at the right time. All photos should be properly stored and labeled. Photos are required for EQ Credit: Construction Indoor Air Quality Management Plan, SS Prerequisite: Construction Activity Pollution Prevention, and MR: Construction and Demolition Waste Management.

Another credit requiring product information is EQ Credit: Low-Emitting Materials. This credit requires you to show that carpets, paints, sealants, adhesives, composite wood, and other products meet VOC emission requirements. The easiest way to do this is to ensure that each product meets requirements before you bring it on the job site. After they are on site, keep an inventory of products you have used and their labeled levels of VOCs. For carpets and paints that are certified via a label, be sure to note the label when checking for compliance (*Figure 79*).

The same approach applies for Materials and Resources Credits, which deal with various green materials. For each product used, retain a cut sheet or other data showing relevant properties

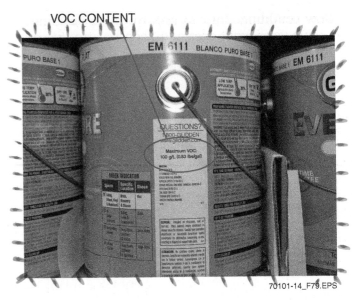

Figure 79 Verify that products meet specifications by checking VOC levels on labels.

for each credit. Keep track of the costs of each product so that the LEED AP can track it as part of LEED calculations.

One product that requires special attention is FSC certified wood. To be counted for the credit, the complete chain of custody for that product must be documented, from the initial harvest through to the final supplier. If the FSC chain of custody documentation is not available, it may be possible to ask the supplier to provide additional documentation to fix the problem. However, this must be done as soon as the product is received. If you wait until the end of the project to deal with this problem, you may lose the credit because you can no longer get documentation.

It is important to track documentation as you go, so errors or missing information can be addressed immediately. Do not wait to organize documentation until the end of the project. The person responsible for LEED documentation should be consulted before all final payments are approved. A final review by USGBC should be complete before final payment, ensuring that contractors address any remaining questions for certification.

3.4.4 Lack of Coordination with Other Trades

The fourth major pitfall to achieving LEED certification is a lack of coordination with other trades. Poor coordination may put LEED certification at risk, resulting in poor energy performance and other problems.

Energy performance typically suffers the worst from poor coordination. Cracks and gaps between materials installed by different trades can

What's Wrong with This Picture?

allow conditioned air to leak from the building, causing significant energy loss or moisture problems. Careful construction follow-up can seal these leaks, though this is often overlooked.

Coordination problems can also contribute to thermal bridging. Thermal bridging occurs where materials of higher conductivity (like metal and wood) pass completely through an exterior wall without a thermal break (*Figure 80*). These materials act as thermal pipelines. They can result in cold spots in the wall or ceiling surface where moisture can condense. Over time, these moist areas serve as a breeding ground for mold.

Lack of coordination can also affect structural integrity. Conflicts inevitably occur when multiple trades must fit their systems into the same space. In wood frame construction, improper notching can occur when plumbing or electrical trades install piping or conduit through wood members. The *International Building Code* limits the depth and location of cuts in all structural members (*Figure 81*). This is to ensure that structural integrity is preserved.

All field modifications should be checked to be sure they do not impact the building's integrity. Careful coordination of the trades in these areas prevents the need for rework. Rework can be difficult and expensive after finishes are complete. Coordination can also reduce the risk of failure.

Another common problem is disturbing insulation in walls. The energy performance of insulation is significantly reduced when it is compressed or when gaps are left in wall cavities. It may also be compromised if the insulation gets wet. Check all insulation before interior finishes are installed, allowing it to be fixed while the walls are still open.

During the project, speak up if you see anything that might compromise the project's LEED goals. Make your supervisor aware if you see problems with stormwater management, dust containment, or any other protective measures. If you see others doing things incorrectly, talk with them to find out why. Encourage them to follow the LEED requirements to meet the project goals.

Everyone involved in the project should be made aware of LEED goals. They should know not only the goals that affect them, but all the goals in general. Sometimes trades are familiar with one aspect of green building but not others. Mistakes can be made that can cause the project to lose points. For instance, HVAC workers are familiar with energy performance: they know to seal ductwork to enhance the energy performance of their systems. However, they may not realize that duct sealant contains VOCs. They

Figure 80 Thermal bridging caused by a metal beam.

may use materials that offgas and cause the project to fail its IAQ test. All stakeholders need to understand how their work influences the project's green goals, both directly and indirectly.

Projects seeking LEED certification can perform better than conventional buildings. However, achieving LEED certification requires careful attention during construction. To successfully reach LEED goals, make sure LEED requirements are in the specs and that the project plan meets the requirements. Be sure that what gets done on the project matches the project plan. Together, these actions will help a project be successfully certified.

What's Wrong with This Picture?

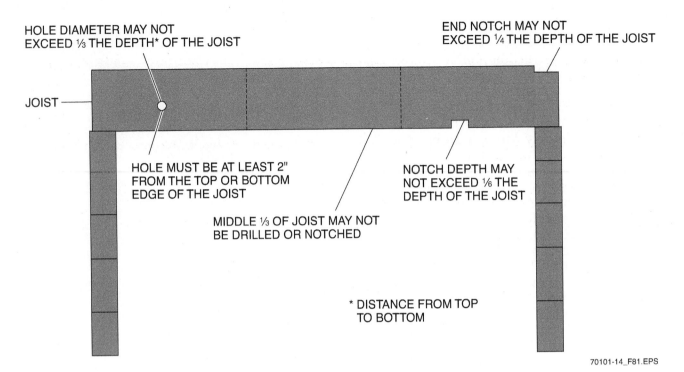

Figure 81 Be careful when notching structural members to preserve structural integrity.

Think About It
Your Responsibilities

Perhaps the most important question you can ask is whether your project is seeking a LEED certification or not. If the project is seeking certification, the next step is to ask what your role will be in getting the certification. If you are procuring materials, what information is needed and who should receive it? What specifications and requirements do the materials need to meet? If you are working outside the building or inside in a protected area, what do you need to do to protect the work area? How should waste be managed? Are there any other special requirements that will be your responsibility? Do you see any opportunities for improvement?

Additional Resources

Field Guide for Sustainable Construction. Department of Defense – Pentagon Renovation and Construction Office. (2004). PDF, 2.6 MB, 312 pgs. Available for download at **www.wbdg.org**.

Greening Federal Facilities, 2nd Ed. US Department of Energy Federal Energy Management Program. (2001). PDF, 2.1 MB, 211 pgs. Available for download at **www.wbdg.org**.

Natural Capitalism. Lovins, A., Hawkin, P., and Lovins, L.H. (1995). Little, Brown, & Company, Boston, MA. Available online at **www.natcap.org**.

Sustainable Buildings Technical Manual. Public Technologies, Inc./US Department of Energy. (2006). PDF, 3.1 MB, 292 pgs. Available for download at **www.greenbiz.com**.

Sustainable Buildings and Infrastructure: Paths to the Future. Pearce, A.R., Ahn, Y.H., and Hanmi Global. (2012). Routledge, London, UK.

Sustainable Construction: Green Building Design and Delivery, 3rd Ed. Kibert, C.J. (2012). Wiley, New York, NY.

The HOK Guidebook to Sustainable Design, 3rd Ed. Odell, W. and Lazarus, M.A. (2015). Wiley, New York, NY.

3.0.0 Section Review

1. The LEED rating system was developed by the ____.
   a. Lean Building Institute
   b. US Green Building Council
   c. Sustainable Built Environment Initiative
   d. United Kingdom

2. Erosion controls used to prevent stormwater runoff during construction are rewarded by ____.
   a. LT Credit: Sensitive Land Protection
   b. SS Prerequisite: Construction Activity Pollution Prevention
   c. SS Credit: Rainwater Management
   d. LT Credit: High Priority Site

3. Which credit requires proof of disposal, such as waste hauling receipts?
   a. MR Prerequisite: Storage and Collection of Recyclables.
   b. MR Prerequisite: Construction and Demolition Waste Management Planning.
   c. MR Credit: Building Life-Cycle Impact Reduction.
   d. MR Credit: Construction and Demolition Waste Management.

4. A common pitfall to LEED certification during construction is ____.
   a. lack of coordination with other trades
   b. using diesel engines on the job site
   c. not following directions
   d. using nonrenewable materials

Summary

As the green environment changes, the construction industry must also change. Past practices have had many negative impacts on the green environment. Today, there are many things that can be done to change for the better.

You can estimate your individual impact by calculating your carbon footprint. Your carbon footprint measures how much carbon your activities emit to the atmosphere. Some of the ways to reduce your carbon footprint include reducing product and energy use, or recycling waste. You can also plant vegetation, find better sources of energy, or buy carbon offsets to help other people improve. As you consider your options, be sure to look for leverage points: small actions you can take to make a big difference.

As a craft worker, there are six major areas where you can improve a construction project's impact on the green environment. These include the site and landscape, water and wastewater, energy, materials and waste, indoor environment, and integrated strategies. Within each of these areas, you can have the most impact by eliminating unnecessary uses of resources. You can also find more efficient ways to use resources. Look for better sources for resources, and better sinks for waste. Seek out integrated strategies where one action has multiple benefits. Again, look for leverage points, where small actions can make a big difference.

The LEED Green Building Rating System provides a useful measuring stick for industry improvement. LEED has nine categories of credits (Integrative Design, Location and Transportation, Sustainable Sites, Water Efficiency, Energy and Atmosphere, Materials and Resources, Indoor Environmental Quality, Innovation in Design, and Regional Priority). Projects can be certified at four levels: Certified, Silver, Gold, and Platinum. Platinum is the most difficult rating level. LEED projects must meet certain basic requirements called prerequisites, and then can choose from points in the nine categories to achieve a rating level.

As a craft worker, you have many opportunities to contribute to a project's LEED certification. How you manage the project site is important. Pay attention to stormwater management and control systems. Be aware of how you protect building systems from potential indoor environmental contamination. Be careful to buy materials that meet specifications. Look for better ways to dispose of waste streams. Remember, your actions can make a real difference.

Review Questions

1. A natural change in the green environment is ____.
   a. deforestation
   b. urban sprawl
   c. changes in the seasons
   d. aquifer depletion

2. The overuse of groundwater is leading to ____.
   a. loss of biodiversity
   b. global climate change
   c. aquifer depletion
   d. deeper coal mines

3. One of the contributors to global warming is ____.
   a. aquifer depletion
   b. deforestation
   c. rising sea levels
   d. recycling

4. Airplanes are among the largest contributors to ____.
   a. helium emissions
   b. nitrogen emissions
   c. carbon emissions
   d. oxygen emissions

5. The longest phase of the building life cycle is typically ____.
   a. planning
   b. design
   c. construction
   d. operation and maintenance

6. A way to minimize the urban heat island effect is to ____.
   a. pave with asphalt
   b. choose light-colored pavements
   c. choose dark-colored pavements
   d. coat paved areas with a sealer

7. The first step in greening water use is to ____.
   a. identify alternative toilets
   b. eliminate unnecessary uses
   c. install low-flow toilets
   d. install a water meter

8. A common technique to reduce the actual flow of water in faucets is ____.
   a. turning off the water
   b. aeration
   c. heating
   d. cooling

9. A green alternative source of water is ____.
   a. groundwater
   b. rainwater harvesting
   c. melting ice
   d. salt water

10. Lighting retrofits to a building have a ____.
    a. slow payback period
    b. rapid payback period
    c. large payback period
    d. negative payback period

11. Green power generates energy from ____.
    a. nonrenewable resources
    b. coal
    c. renewable resources
    d. natural gas

12. Material that is recycled from manufacturing processes is termed ____.
    a. post-consumer waste
    b. consumer waste
    c. post-industrial waste
    d. pre-industrial waste

13. An example of a multi-function material is ____.
    a. structural insulated panels
    b. hardwood flooring
    c. concrete
    d. steel framing

14. A smart material changes in response to ____.
    a. funding
    b. the user's needs
    c. environmental conditions
    d. a schedule

15. Rapidly renewable materials are ____ .
    a. recycled
    b. bio-based
    c. unsustainable
    d. expensive

16. Construction isolation measures to prevent building contamination should protect ____ .
    a. storage areas
    b. HVAC intakes
    c. water supplies
    d. construction equipment

17. Passive survivability is the ability for a building to function when ____ .
    a. the infrastructure goes down
    b. there is a labor strike
    c. there is a shortage of supplies
    d. the air is contaminated

18. The LEED rating system that is for the individual tenant spaces in commercial buildings is ____ .
    a. LEED-EB
    b. LEED-NC
    c. LEED-CI
    d. LEED-CS

19. A project team can submit documentation for LEED credits ____ .
    a. only at the beginning of construction
    b. only at the end of construction
    c. only at the end of design
    d. both at the end of construction and design

20. A prerequisite for all LEED projects is to meet all the requirements of the ____ .
    a. Lean Construction Institute
    b. EPA General Construction Permit
    c. LEED requirements
    d. local building code

Trade Terms Introduced in This Module

Absorptive finish: A surface finish that will absorb dust, particles, fumes, and sound.

Acidification: A process that converts air pollution into acid substances, leading to acid rain. Acid rain is best known for the damage it causes to forests and lakes. Also refers to the outflow of acidic water from metal and coal mines.

Aerated autoclaved concrete (AAC): A lightweight, precast building material that provides structural strength, insulation, and fire resistance. Typical products include blocks, wall panels, and floor panels.

Aeration: Mixing air into a liquid substance.

Albedo: The extent to which an object reflects light from the sun. It is a ratio with values from 0 to 1. A value of 0 is dark (low albedo). A value of 1 is light (high albedo).

Alternative fuel: Any material or substance that can be used as a fuel other than conventional fossil fuels. They produce less pollution than fossil fuels. These include biodiesel and ethanol.

Aquifer: An underground layer of water-bearing rock or soil from which groundwater can be extracted using a well.

Aquifer depletion: A situation where water is withdrawn from its underground source faster than the rate of natural recharge.

Best Management Practice (BMP): A way to accomplish something with the least amount of effort to achieve the best results. This is based on repeatable procedures that have proven themselves over time for large numbers of people.

Bio-based: A material derived from living matter, either plant or animal.

Biodegradable: Organic material such as plant and animal matter and other substances originating from living organisms. These are capable of being broken down into innocuous products by the action of microorganisms.

Biodiversity: A measure of the variety among organisms present in different ecosystems.

Biofuel: A solid, liquid, or gas fuel consisting of or derived from recently dead biological material. The most common source is plants.

Biomimicry: The act of imitating nature to create new solutions modeled after natural systems.

Bioswale: An engineered depression designed to accept and channel stormwater runoff. A bioswale uses natural methods to filter stormwater, such as vegetation and soil.

Blackout: A situation where the electrical grid fails and does not provide any power.

Blackwater: Water or sewage that contains fecal matter or sources of pathogens.

Brownfield: Property that contains the presence or potential presence of a hazardous substance, pollutant, or contaminant. Cleaning up and reinvesting in these properties reduces development pressures on undeveloped, open land and improves and protects the environment.

Brownout: A situation where the electrical grid provides less power than normal, but enough for some equipment to still work.

Building-integrated photovoltaics (BIPVs): Photovoltaic systems built into other types of building materials. See *photovoltaic*.

Byproducts: Another product derived from a manufacturing process or a chemical reaction. It is not the primary product or service being produced.

Carbon cycle: The movement of carbon between the biosphere, atmosphere, oceans, and geosphere of the Earth. It is a biogeochemical cycle. In the cycle, there are sinks, or stores, of carbon. There are also processes by which the various sinks exchange carbon.

Carbon footprint: A measure of impact human activities have on the environment. It is determined by the amount of greenhouse gases produced. It is measured in pounds or kilograms of carbon dioxide.

Carbon neutral: An action or product whose production absorbs as much carbon from the atmosphere as it produces.

Carbon offset: An agreement with another party that they will reduce their carbon production by some amount in exchange for payment.

Carpool: An arrangement in which several people travel together in one vehicle. The people take turns driving and share in the cost.

Charrette:: An intense meeting of project participants that can quickly generate design solutions by integrating the abilities and interests of a diverse group of people.

Chlorofluorocarbons (CFCs): A set of chemical compounds that deplete ozone. They are widely used as solvents, coolants, and propellants in aerosols. These chemicals are the main cause of ozone depletion in the stratosphere.

Cob construction: An ancient building method using hand-formed lumps of earth mixed with sand and straw.

Commissioning: A review process conducted by a third party that involves a detailed design review, testing and balancing of systems, and system turnover.

Compact fluorescent lamp (CFL): A fluorescent bulb that is designed to fit in a normal light fixture. They use less energy and last longer than incandescent bulbs.

Conservation: Using natural resources wisely and at a slower rate than normal.

Corporate Sustainability Report (CSR): An annual report by a company documenting its efforts to reduce negative environmental and social impacts.

Deconstruction: Taking a building apart with the intent of salvaging reusable materials.

Deforestation: The removal of trees without sufficient replanting.

Desertification: The creation of deserts through degradation of productive land in dry climates by human activities.

Downcycling: Recycling one material into a material of lesser quality. An example is the recycling of plastics into lower-grade composites.

EarthCraft: A residential green building program of the Greater Atlanta Home Builders Association in partnership with Southface Energy Institute.

Ecological footprint: A measure of impact human activities have on the environment. It compares human consumption of natural resources with the Earth's capacity to regenerate them. Measured in hectares or acres.

Ecosystem: A combination of all plants, animals and microorganisms in an area that complement each other. These function together with all of the nonliving physical factors of the environment.

Embodied energy: The total energy required to bring a product to market. It includes raw material extraction, manufacturing, final transport, and installation.

Endangered species: A population of a species at risk of becoming extinct. A threatened species is any species that is vulnerable to extinction in the near future.

Energy efficiency: Getting more use out of electricity already generated.

ENERGY STAR: A United States government program to promote energy efficiency. It is a joint program of the US Environmental Protection Agency and the US Department of Energy.

Environmental Product Declaration (EPD): A document listing the environmental impacts caused by manufacture of a product.

Equipment idling: The operation of equipment while it is not in motion or performing work. Limiting idle times reduces air pollution and greenhouse gas emissions.

Erosion: The displacement of solids by wind, water, ice, or gravity or by living organisms. These solids include rocks and soil particles.

Flood plain: An area surrounding a river or body of water that regularly floods within a given period of time.

Flushout: Using fresh air in the building HVAC system to remove contaminants from the building.

FSC Certified: Wood or wood products that have met the Forest Stewardship Council's tracking process for sustainable harvest.

Fugitive emissions: Pollutants released to the air other than those from stacks or vents. They are often due to equipment leaks, evaporative processes, and wind disturbances.

Geothermal: Heat that comes from within the Earth.

Global climate change: Changes in weather patterns and temperatures on a planetary scale. This may lead to a rise in sea levels, melting of polar ice caps, increased droughts, and other weather effects.

Graywater: A nonindustrial wastewater generated from domestic processes. These include laundry and bathing. Graywater comprises 50 to 80 percent of residential wastewater.

Greenhouse Effect: The overall warming of a planet's surface due to retention of solar heat by its atmosphere.

Green Seal Certified: A certification of a product to indicate its environmental friendliness. Green Seal is a group that works with manufacturers, industry sectors, purchasing groups, and government at all levels to green the production and purchasing chain. Founded in 1989, Green Seal provides science-based environmental certification standards.

Grid intertie: A connection between a local source of power and the utility power grid.

Halons: Ozone-depleting compounds consisting of bromine, fluorine, and carbon. They were commonly used as fire extinguishing agents, both in built-in systems and in handheld portable fire extinguishers. Halon production in the US ended on December 31, 1993.

Hardscape: Paved areas surrounding a project, including parking lots and sidewalks.

Heat island: An area that is warmer than surrounding areas due to absorbing more solar radiation.

Hybrid vehicle: A vehicle that uses two or more power sources for propulsion.

Hydrochlorofluorocarbons (HCFCs): A group of man-made compounds containing hydrogen, chlorine, fluorine, and carbon. They are used for refrigeration, aerosol propellants, foam manufacture, and air conditioning. They are broken down in the lowest part of the atmosphere and pose a smaller risk to the ozone layer than other types of refrigerants.

Hydrologic cycle: The circulation and conservation of Earth's water supply. The process has five phases: condensation, infiltration, runoff, evaporation, and precipitation.

Indoor air quality (IAQ): The content of interior air that could affect health and comfort of building occupants.

Infiltration: The movement of air from outside a building to inside through cracks or openings in the building envelope.

Insulating concrete form (ICF): Rigid forms that hold concrete in place during curing and remain in place afterwards. The forms serve as thermal insulation for concrete walls.

Integrative design: A collaborative design methodology that emphasizes the input of knowledge from several areas in the development of a complete design.

Just-in-time delivery: A material delivery strategy that reduces material inventory. The material that is delivered is used immediately.

Leadership in Energy and Environmental Design (LEED):: A point-based rating system used to evaluate the environmental performance of buildings. Developed by the US Green Building Council, LEED provides a suite of standards for environmentally sustainable construction. Founded in 1998, LEED focuses on the certification of commercial and residential buildings and the neighborhoods in which they exist.

Life cycle: The useful life of a system, product, or building.

Life cycle assessment: An analytic technique to evaluate the environmental impact of a system, product, or building throughout its life cycle. This includes the extraction or harvesting of raw materials through processing, manufacture, installation, use, and ultimate disposal or recycling.

Life cycle cost: The cost of a system or a component over its entire life span.

Light-emitting diode (LED) lamp: A highly efficient, electronic light bulb using a glowing diode designed to fit a standard light fixture.

Lithium ion (Li-ion): A type of rechargeable battery in which a lithium ion moves between the anode and cathode. They are commonly used in consumer electronics.

Local materials: Materials that come from within a certain number of miles from the project. Materials produced locally use less energy during transportation to the site. The LEED system of building certification offers points for the use of regional or local materials.

Location valuation: A bonus under the LEED rating system for using a product produced locally. Materials from within a 100-mile radius can be counted as twice their value as a reward for using local products.

Low-emission vehicle:: Vehicles that produce fewer emissions than the average vehicle. Beginning in 2001, all light vehicles sold nationally were required to meet this standard.

Maintainability: An indication of how difficult it is to properly maintain a building system or technology.

Massing: The three-dimensional shape of a building, including length, width, and height. Determines the amount of a building exposed to sunlight.

Minimum efficiency reporting value (MERV): A measurement scale designed in 1987 by the American Society of Heating, Refrigeration, and Air-Conditioning Engineers (ASHRAE) to rate the effectiveness of air filters. The scale is designed to represent the worst-case performance of a filter when dealing with particles in the range of 0.3 to 10 microns.

Monofill: A landfill that accepts only one type of waste, typically in bales.

Multi-function material: A material that can be used to perform more than one function in a facility.

Nano-material: A material with features smaller than a micron in at least one dimension.

Nickel cadmium (NiCad): A popular type of rechargeable battery using nickel oxide hydroxide and metallic cadmium as electrodes.

Nonrenewable: A material or energy source that cannot be replenished within a reasonable period.

Nontoxic: Substances that are not poisonous.

Offgassing: The evaporation of volatile chemicals at normal atmospheric pressure. Building materials release chemicals into the air through evaporation.

Ozone depletion: A slow, steady decline in the total amount of ozone in Earth's stratosphere.

Ozone hole: A large, seasonal decrease in stratospheric ozone over Earth's polar regions. The ozone hole does not go all the way through the layer.

Papercrete: A fiber-cement material that uses waste paper for fiber.

Passive solar design: Designing a building to use the sun's energy for lighting, heating, and cooling with minimal additional inputs of energy.

Passive survivability: The ability of a building to continue to offer basic function and habitability after a loss of infrastructure (i.e., water and power).

Pathogen: An infectious agent that causes disease or illness.

Payback period: The amount of time it takes to break even on an investment; the period of time after which savings from an investment equals its initial cost.

Peak shaving: An energy management strategy that reduces demand during peak times of the day and shifts it to off-peak times, such as at night.

Persistent, bioaccumulative toxin (PBT): A harmful substance such as a pesticide or organic chemical that does not quickly degrade in the natural environment but does accumulate in plants and animals in the food chain.

Pervious concrete: A mixture of coarse aggregate, Portland cement, water, and little to no sand. It has a 15 to 25 percent void structure and allows 3 to 8 gallons of water per minute to pass through each square foot. Also known as permeable concrete.

Phantom loads: Electricity consumed by appliances and devices when they are switched off.

Phase change material (PCM): A material that stores or releases large amounts of energy when it changes between solid, liquid, or gas forms.

Photosensor: A sensor that measures daylight and adjusts artificial lighting to save energy.

Photovoltaic: A technology that converts light directly into electricity.

Pollution prevention: The prevention or reduction of pollution at the source.

Post-consumer: Products made out of material that has been used by the end consumer and then collected for recycling.

Post-industrial/pre-consumer: Material diverted from the waste stream during the manufacturing process.

Preservation: The act and advocacy of protecting the natural environment.

Radio-frequency identification (RFID): An automatic identification method. It relies on storing and remotely retrieving data using devices called RFID tags.

Rainwater harvesting: The gathering and storing of rainwater.

Rammed earth: An ancient building technique similar to adobe using soil that is mostly clay and sand. The difference is that the material is compressed or tamped into place, usually with forms that create very flat vertical surfaces.

Rapidly renewable: A material that is replenished by natural processes at a rate comparable to its rate of consumption. The LEED system of building certification rewards use of rapidly renewable materials that regenerate in 10 years or less, such as bamboo, cork, wool, and straw.

Raw material: A material that has been extracted directly from nature. It is in an unprocessed or minimally processed state.

Recyclable: Material that still has useful physical or chemical properties after serving its original purpose. It can be reused or remanufactured into additional products. Plastic, paper, glass, used oil, and aluminum cans are examples of recyclable materials.

Recycled content: A material containing components that would otherwise have been discarded.

Recycled plastic lumber (RPL): A wood-like product made from recovered plastic, either by itself or mixed with other materials. It can be used as a substitute for concrete, wood, and metals.

Recycling: The reprocessing of old materials into new products. A goal is to prevent the waste of potentially useful materials and reduce the consumption of new materials.

Renewable: A resource that may be naturally replenished.

Reusable: A material that can be used again without reprocessing. This can be for its original purpose or for a new purpose.

Salvaged: A used material that is saved from destruction or waste by being reused.

Sedimentation: The process of depositing a solid material from a state of suspension in a fluid, usually air or water.

Sick Building Syndrome: A variety of illnesses thought to be caused by poor indoor air. Symptoms include headaches, fatigue, and other problems that increase with continued exposure.

Smart material: Materials that have one or more properties that can be significantly changed. These changes are driven by external stimuli.

Softscape: The area surrounding a building that contains plantings, lawn, and other vegetated areas.

Solar: Energy from the sun in the form of heat and light.

Solid waste: Products and materials discarded after use in homes, businesses, restaurants, schools, industrial plants, or elsewhere.

Solvent-based: A material that consists of particles suspended or dissolved in a solvent. A solvent is any substance that will dissolve another. Solvent-based building materials typically use chemicals other than water as their solvent, including toluene and turpentine, with hazardous health effects.

Spoil pile: Excavated soil that has been moved and temporarily stored during construction.

Sprawl: Unplanned and inefficient development of open land.

Stormwater runoff: Unfiltered water that reaches streams, lakes, and oceans after a rainstorm by flowing across impervious surfaces.

Strawbale construction: A building method that uses straw as the structural element, insulation, or both. It has advantages over some conventional building systems because of its cost and availability.

Structural insulated panel (SIP): A composite building material used for exterior building envelopes. It consists of a sandwich of two layers of structural board with an insulating layer of foam in between. The board is usually oriented strand board (OSB) and the foam can be polystyrene, soy-based foam, urethane, or even compressed straw.

Sustainably harvested:: A method of harvesting a material from a natural ecosystem without damaging the ability of the ecosystem to continue to produce the material indefinitely.

Takeback:: A condition where manufacturers recover waste from packaging or products after use, at the end of their life cycle.

Thermal bridging:: A condition created when a thermally conductive material bypasses an insulation system, allowing the rapid flow of heat from one side of a building wall to the other. Metal components, including metal studs, nails, and window frames, are common culprits.

Thermal mass:: A property of a material related to density that allows it to absorb heat from a heat source, and then release it slowly. Common materials used to provide thermal mass include adobe, mud, stones, or even tanks of water.

Urbanization:: Converting rural land to higher density development.

Urea formaldehyde:: A transparent thermosetting resin or plastic. It is made from urea and formaldehyde heated in the presence of a mild base. Urea formaldehyde has negative effects on human health when allowed to offgas or burn.

Vapor-resistant: A material that resists the flow of water vapor.

Virgin material: A material that has not been previously used or consumed. It also has not been subjected to processing. See also *raw material*.

Volatile organic compounds (VOCs): Gases that are emitted over time from certain solids or liquids. Concentrations of many VOCs are up to 10 times higher indoors than outdoors. Examples include paints and lacquers, paint strippers, cleaning supplies, pesticides, building materials, and furnishings.

Walk-off mat: Mats in entry areas that capture dirt and other particles.

Waste separation: Sorting waste by specific type of material and storing it in different containers to facilitate recycling.

Water efficiency: Managing water use to prevent waste or overuse. Includes using less water to achieve the same benefits.

Water footprint: The volume of water used directly or indirectly to sustain something over a period of time.

Water-based: Materials that uses water as a solvent or vehicle of application.

Water-resistant: A material that hinders the penetration of water.

Waterproof: A material that is impervious to or unaffected by water.

Wetland: Lands where saturation with water is the dominant factor. This determines the way soil develops and the types of plant and animal communities living in the soil and on its surface.

Xeriscaping: Landscaping that requires little or no water for irrigation. Xeriscaping can be achieved through smart plant selection, mulching, and other tactics.

Zoning: Rules for placing similar items next to one another, as in plants within a landscape, or types of buildings within a community.

Appendix A

Common Acronyms

AAC	Aerated autoclaved concrete	LED	Light-emitting diode
ACH	Air changes per hour	Li-ion	Lithium ion
BIPV	Building-integrated photovoltaic	Low-E	Low-emissivity
BMP	Best management practice	LT	Location and Transportation (LEED credit category)
BTU	British thermal unit		
CFC	Chlorofluorocarbon	MCS	Multiple chemical sensitivity
CFL	Compact fluorescent lamp	MDF	Medium-density fiberboard
CO	Carbon monoxide	MERV	Minimum efficiency reporting value
CO_2	Carbon dioxide	MPR	Minimum Program Requirements
CSR	Corporate Sustainability Report	MR	Materials and Resources (LEED credit category)
EA	Energy and Atmosphere (LEED credit category)		
		N_2O	Nitrous oxide
EEM	Energy efficient mortgage	NACH	Natural air changes per hour
EMF	Electromagnetic field	NAHB	National Association of Home Builders
EPA	Environmental Protection Agency	NFRC	National Fenestration Rating Council
EPD	Environmental Product Declaration	NiCad	Nickel cadmium
EQ	Indoor Environmental Quality (LEED credit category)	OPC	Off-peak cooling
		OSB	Oriented strand board
ERV	Energy recovery ventilator	OVE	Optimum value engineering
FSC	Forest Stewardship Council	PB	Particleboard
GBCI	Green Building Certification Institute	PBT	Persistent bioaccumulative toxin
GHG	Greenhouse gas	PCM	Phase change material
HCFC	Hydrochlorofluorocarbon	PF	Phenyl formaldehyde
HERS	Home energy rating system	PtD	Prevention through design
HID	High intensity discharge	PV	Photovoltaic
HVFA	High volume flyash	PVC	Polyvinyl chloride
IAQ	Indoor air quality	REC	Renewable Energy Credit
ICF	Insulating concrete form	RFID	Radio-frequency identification
ID	Innovation in Design (LEED credit category)	RP	Regional Priority (LEED credit category)
IDP	Integrative design process	RPL	Recycled plastic lumber
ISO	International Standards Organization	R-value	Resistance to heat flow
LCA	Life cycle assessment	SEER	Seasonal Energy Efficiency Ratio
LEED	Leadership in Energy and Environmental Design	SFI	Sustainable Forest Initiative
		SHGC	Solar heat gain coefficient
LEED-AP	Leadership in Energy and Environmental Design Accredited Professional	SIP	Structural insulated panel
		SMACNA	Sheet Metal and Air Conditioning Contractors' National Association
LEED-BD+C	Leadership in Energy and Environmental Design for Building Design and Construction and Major Renovation		
		SS	Sustainable Sites (LEED credit category)
LEED-ID+C	Leadership in Energy and Environmental Design for Interior Design and Construction	TPO	Thermoplastic polyolefin
		UF	Urea formaldehyde
LEED-O+M	Leadership in Energy and Environmental Design for Operations and Maintenance of Existing Buildings	USGBC	United States Green Building Council
		UV	Ultraviolet
		U-value	Resistance to heat loss (also known as U-factor)
LEED-H	Leadership in Energy and Environmental Design for Homes	VOC	Volatile organic compound
LEED-ND	Leadership in Energy and Environmental Design for Neighborhood Development	WE	Water Efficiency (LEED credit category)

Appendix B

WHAT IT MEANS TO BE GREEN

What makes a building green? Green buildings are designed to use resources efficiently. This includes energy, water, and building materials. Green buildings must also provide a healthy, comfortable, and safe environment for the occupants. They should be constructed in a way that minimizes their impact on the environment. The qualities that make a building green depend on the systems, methods, equipment, and materials used in its construction.

Craft professionals make a large impact on the green environment simply by how they use materials. Do you know that the building industry generates over thirty million tons of construction waste each year that ends up in landfills? Yet almost eighty percent of that waste could be recycled. Think about the waste that you generate. Can that copper wire be recycled? What about the cast iron or PVC? Is the waste on your project recycled? What happens to excess materials at the end of the project? Are they recovered, or are they thrown away?

Carpenters, electricians, plumbers, and HVAC technicians all play significant roles in the greening of the built environment. As a craft worker, you must be aware of the things you do every day that help to make a building green, and be willing to learn new methods and work with new technologies.

This section provides an overview of what it means to be a green craft professional. Whatever your trade, there are specific methods that can be applied towards building more energy-efficient commercial and residential structures with less environmental impact.

WHAT DOES IT MEAN TO BE A GREEN CARPENTER?

As a carpenter, you contribute to the efficiency of a building in many ways. For example, the method of framing you use can improve the energy efficiency of a building. The materials you install on the exterior can improve the durability of the building. You can conserve resources by reducing the amount of material used to build a structure. You can also recycle waste materials. Your daily activities on a job site affect the efficiency of a building.

The emphasis on efficiency has led to changes in materials and construction methods. One of these methods is Optimum Value Engineering (OVE), a resource-efficient method of framing. This method allows for greater energy efficiency and reduction in construction costs. It uses a single top plate, eliminates headers in interior walls, and uses wider stud spacing. This method consumes less wood and creates more space for insulation in exterior walls. All of this directly improves the efficiency of the structure. Learning and using the principles of OVE is a direct way that carpenters help the green environment.

The green movement has also been marked by the use of building components for various purposes. A roofing system is an example. The roof may be designed to do more than protect the occupants from the elements. It may be connected to a rainwater collection system, or it may be a green or vegetated roof. These non-traditional systems require the skills of carpenters to construct major components.

Over the last few decades, carpenters have started to use new greener building materials. An example is the supplementing of traditional dimensional lumber used in framing with engineered lumber and I-joists. More recently, multi-function building materials have been introduced. These include structural insulated panels and insulated concrete forms, each providing structural support and insulation. These materials will change the way carpenters frame structures and build foundations. Other green building materials include insulated foam sheathing and recycled flooring. These innovations require carpenters to successfully use them as part of the green strategy.

You should also be aware of the types of materials being used on a job site. If a project is seeking to be green, be sure to look for the Forest Stewardship Council (FSC) certification label on lumber. If the lumber on the job site does not have an FSC seal, alert your supervisor. Another way you can be green-minded is to read labels. Check the recycled content of materials you are using. Also check the labels on adhesives to determine if they are low VOC or non-toxic. These steps promote green building.

WHAT DOES IT MEAN TO BE A GREEN ELECTRICIAN?

As an electrician, you will contribute significantly to the success of an energy-efficient building. Both residential and commercial buildings consume large amounts of energy. This energy is used in lighting, heating, cooling, and operating the building. Building owners are looking for ways to reduce their energy bills through conservation and efficiency. The equipment you install can directly assist with this. It can also produce energy more responsibly. You can help by carefully following specifications when installing fixtures, and installing only the minimum specified. This reduces energy use and can directly reduce the amount of greenhouse gases generated from coal-fired power plants. Your daily activities on a job site affect the efficiency of a building. They also impact the environment.

The emphasis on energy efficiency has led to innovations in energy conservation in buildings. This includes energy management systems. These systems have advanced controls that automatically adjust lighting during the day. They also include environmental controls that manage the temperature and humidity of a building. Electronic sensors and meters are also used to monitor energy use and building conditions. These systems work together to ensure the comfort of the occupants. Some buildings have controls that can be used directly by occupants. Learning how to install and trouble-shoot these systems is a direct way that electricians help the green environment.

The green movement also supports the use of alternative energy systems. Alternative energy systems include sources such as solar, wind, and biomass. Solar photovoltaic panels are an example of an alternative energy system. The installation of these panels requires skilled electricians who know how to wire them correctly. Alternative energy sources need proper power distribution to work well. These non-traditional systems require the skills of electricians to correctly install and maintain major components.

Electricians have adapted to new greener building technologies and methods. An example is the reduction of incandescent lighting fixtures. Electricians have been installing fluorescent lighting and replacing the ballasts in these fixtures for decades. More recently, solid state lighting including light-emitting diodes is becoming popular. This is due to their energy efficiency and durability.

Light pollution reduction is also a critical area for green building. The correct installation and location of outdoor lighting fixtures can help limit light pollution. Each of these innovations requires electricians to successfully use them as part of the green strategy. Remember, careful and proper disposal of ballasts and fluorescent lamps is important, as these contain hazardous materials.

Methods are just as important. Using only the amount of wire needed in a building is a way to use resources efficiently. Another example is to be sure not to disturb the insulation in the wall when replacing wiring or installing new electrical boxes. This will cause gaps in the insulation and reduce the efficiency of the building. Taking these steps on the job site will help to maintain the energy efficiency of a building.

70101-15 Your Role in the Green Environment

WHAT DOES IT MEAN TO BE A GREEN HVAC TECHNICIAN?

As an HVAC technician, your work has a large impact on the comfort and health of building occupants. Your work also impacts the ability of a building to be green. Heating, cooling, and ventilation consume the largest amount of energy in a building. This energy consumption increases the burning of fossil fuel and greenhouse gas emissions. HVAC systems designed, installed, commissioned, and maintained properly will reduce these impacts. This is where HVAC technicians can make the greatest impact on the green environment.

A challenge facing green HVAC technicians is the correct sizing of air ducts. Another challenge is determining the conditioning capacity required in a facility. The rule of thumb based on square footage is not good enough for modern green facilities. Correct sizing is critical for energy efficiency and the comfort of the occupants. You will begin to see a shift in the size and types of equipment and ductwork you install.

Over the past decade, there has been a shift in the types of equipment used for heating and air conditioning. These include high efficiency heat pumps and air conditioning units with higher Seasonal Energy Efficiency Ratios (SEER). If a project is seeking to be green, then be sure to look for the appropriate ratings on equipment. If the air conditioning unit does not have the appropriate SEER rating, alert your supervisor. Another way to be green-minded is to be sure you are installing enough returns to keep a facility balanced. If you know there are not enough returns, tell your supervisor. Keep in mind the new greener building codes. In today's building environment, HVAC technicians are responsible for ensuring the appropriate equipment is installed.

Conserving resources and efficiently using materials go hand-in-hand for HVAC systems. How and where ductwork is installed can save energy and materials. An example is compact air distribution systems. These tend to keep ductwork runs as short as possible to minimize the amount of ductwork needed. Another example is to lay out ductwork so air can be discharged from inside walls or ceilings. This may also reduce the amount of ductwork that runs through unconditioned space. If it must be run through unconditioned space for practical purposes, be sure to insulate the ducts, as this improves energy efficiency.

Quality workmanship is critical to maintain healthy indoor air quality. No other trade has a more direct impact on air quality than HVAC. Sealing ductwork is extremely important. Improperly sealed return air ducts draw in humidity that may promote mold growth. Also, leaking ducts cause air handlers to work harder and use more energy. Unsealed return ducts may also draw dust or harmful gases such as carbon monoxide into conditioned space. This may place the health of the occupants at risk. Unconnected ductwork can cause pressure differences in a building. This difference may cause conditioned air to move through the building envelope. This will decrease the efficiency of any facility and increase energy use. As an HVAC technician in the green environment, you will help ensure occupant health and safety and the energy efficiency of a building.

HVAC technicians also play a critical role in protecting the green environment. Refrigerants used in air conditioning equipment can damage the ozone layer if allowed to leak. Antifreeze and other chemicals leaked to the soil can be harmful to the water supply. As an HVAC technician in the green environment, you will help maintain the health of the surrounding community and world at large.

In addition to improved construction methods, new technologies are being used to help improve the efficiency of buildings. An example is Energy Recovery Ventilation (ERV). During the heating season, this system recovers heat from indoor air exhausts and then transfers this heat to fresh incoming air. In the cooling season, it transfers heat from incoming fresh air to the exhaust. This reduces the amount of energy required to condition fresh air. These new systems will be part of greener buildings, and installation will require your experience.

Energy-efficient HVAC systems will require involvement of contractors early in the design phase of a building. This will help to ensure that owners understand how important the location of equipment and ductwork is for energy efficiency. They will also need to know the benefits of high efficiency heat pumps and air conditioners. During the building's life, they will need the expertise of HVAC technicians to maintain and recommission systems to ensure efficient operation. As a technician, you will install the equipment, ductwork, and returns indicated in the design. The days of on-site location of ducts is soon to be a thing of the past in the green workforce.

WHAT DOES IT MEAN TO BE A GREEN PLUMBER?

As a plumber, your work will affect the efficient use of water in a building. By installing and maintaining water-efficient systems, plumbers help reduce energy costs. The EPA estimates that three to four percent of all energy consumed in the United States is associated with treating drinking water and wastewater. Water-efficient systems directly help reduce energy use and greenhouse gas emissions. Gas emissions in homes and businesses can occur if water heaters and other combustion fixtures are not installed properly. Your work will directly impact the health and safety of the occupants.

The emphasis on efficiency has led to new innovations in water and energy conservation. The use of tankless hot water heaters is an example of a way to save energy in a building. Solar hot water heating and drain water heat recovery systems are examples, too. Many building owners are installing low-flow or dual-flush toilets and waterless urinals to reduce water use. These fixtures and systems require a plumber who is knowledgeable about their installation and maintenance.

Green technologies are shaping the future of water conservation efforts. Graywater systems are becoming more popular as new building codes allow their use. Plumbers will be responsible for installing the piping network for these systems. Rainwater harvesting systems are another example. Proper sizing of pipes, storage tanks, and collection systems may be your responsibility. On-site wastewater treatment systems are becoming more common as an alternative to traditional sewage connections. These systems require special piping and on-site treatment vessels. The effective design of plumbing systems for green roofs will also be an important role for green plumbers. These non-traditional plumbing systems require the skills of plumbers to correctly install and maintain major components.

Plumbers have adapted to new greener building technologies. An example is the increase in low-flow fixtures being installed. Sensor-operated fixtures save water and reduce the transmission of germs between users. These technologies require plumbers to successfully install and maintain them as part of the green strategy. They also require careful system design to be sure water isn't kept too long in pipes. This increases water age and can cause water quality problems for building users.

The greener methods are important too. Use your resources wisely by running the shortest length of pipe required. Consider the size of the cuts you make in the floors and walls when installing fixtures. Your cuts can create gaps that allow heating and cooling energy to be wasted. Be sure to make holes the appropriate size and always seal them. Taking these steps on a job site will help to maintain the energy efficiency of a building.

Appendix C

LEED for Building Design + Construction, Version 4.0
Registered Project Checklist

Location and Transportation		**16**
Credit	LEED for Neighborhood Development Location	16
Credit	Sensitive Land Protection	1
Credit	High Priority Site	2
Credit	Surrounding Density and Diverse Uses	5
Credit	Access to Quality Transit	5
Credit	Bicycle Facilities	1
Credit	Reduced Parking Footprint	1
Credit	Green Vehicles	1
Sustainable Sites		**10**
Prereq	Construction Activity Pollution Prevention	Required
Credit	Site Assessment	1
Credit	Site Development - Protect or Restore Habitat	2
Credit	Open Space	1
Credit	Rainwater Management	3
Credit	Heat Island Reduction	2
Credit	Light Pollution Reduction	1
Water Efficiency		**11**
Prereq	Outdoor Water Use Reduction	Required
Prereq	Indoor Water Use Reduction	Required
Prereq	Building-Level Water Metering	Required
Credit	Outdoor Water Use Reduction	2
Credit	Indoor Water Use Reduction	6
Credit	Cooling Tower Water Use	2
Credit	Water Metering	1
Energy and Atmosphere		**33**
Prereq	Fundamental Commissioning and Verification	Required
Prereq	Minimum Energy Performance	Required
Prereq	Building-Level Energy Metering	Required
Prereq	Fundamental Refrigerant Management	Required
Credit	Enhanced Commissioning	6
Credit	Optimize Energy Performance	18
Credit	Advanced Energy Metering	1
Credit	Demand Response	2
Credit	Renewable Energy Production	3
Credit	Enhanced Refrigerant Management	1
Credit	Green Power and Carbon Offsets	2

	Materials and Resources	13
Prereq	Storage and Collection of Recyclables	Required
Prereq	Construction and Demolition Waste Management Planning	Required
Credit	Building Life-Cycle Impact Reduction	5
Credit	Building Product Disclosure and Optimization - Environmental Product Declarations	2
Credit	Building Product Disclosure and Optimization - Sourcing of Raw Materials	2
Credit	Building Product Disclosure and Optimization - Material Ingredients	2
Credit	Construction and Demolition Waste Management	2

	Indoor Environmental Quality	16
Prereq	Minimum Indoor Air Quality Performance	Required
Prereq	Environmental Tobacco Smoke Control	Required
Credit	Enhanced Indoor Air Quality Strategies	2
Credit	Low-Emitting Materials	3
Credit	Construction Indoor Air Quality Management Plan	1
Credit	Indoor Air Quality Assessment	2
Credit	Thermal Comfort	1
Credit	Interior Lighting	2
Credit	Daylight	3
Credit	Quality Views	1
Credit	Acoustic Performance	1

	Innovation	6
Credit	Innovation	5
Credit	LEED Accredited Professional	1

	Regional Priority	4
Credit	Regional Priority: Specific Credit	1
Credit	Regional Priority: Specific Credit	1
Credit	Regional Priority: Specific Credit	1
Credit	Regional Priority: Specific Credit	1

TOTAL Possible Points:		**110**

Version published on 6 June 2014

Appendix D

WORKSHEET 1: INVENTORY YOUR HOUSEHOLD IMPACTS

WORKSHEET 1: INVENTORY YOUR HOUSEHOLD IMPACTS

Conducting an inventory of your household consumption, waste generation, and activities is the first step in understanding how you can reduce your impact. Answer the following questions based on your best guess. If you share a household with other people, divide the total answer for your household by the number of people who live in your home.

How many gallons of garbage do you throw away each week? Find out the size of your trash bin in liters. Multiply by 52 to calculate the liters of garbage you throw away per year.

Liters of garbage per year: _____

How much electricity do you use per year? You'll find this information on your electricity bill. You can add up the total for 12 months worth of bills, or multiply a monthly average by 12. Most energy bills tell you usage in kilowatt hours. If your bill uses a different unit, you can look up a unit converter at www.converterunits.com.

Total electricity per year in kilowatt-hours: _____

How much natural gas do you use per year? Check your natural gas bill if you get one. It will tell you how many therms of gas you use per month. Your highest values will probably be during the winter heating season. If you know your monthly average, multiply by 12 to calculate your annual use. In the United States, the unit for natural gas is the therm. If your bill uses a different unit, you can look up a unit converter at www.converterunits.com.

Total therms of natural gas per year: _____

How many liters of propane do you use each year? You can check this on your propane bill if you get one. Add up the total number of liters per year.

Total liters of propane per year: _____

How many liters of fuel oil do you use per year? You can check this on your fuel oil bill if you get one. Add up the total number of liters per year.

Total liters of fuel oil per year: _____

On average, what is your monthly combined water and sewage bill? Check your monthly bill if you get one and add up all the amounts for a one-year period. If you don't get a water and sewage bill, you can estimate this amount as approximately $75 USD per person per year. You can look up a currency converter to convert your currency to US dollars at www.xe.com.

Annual cost in US dollars for water and sewage: _____

How many square meters is your house or dwelling? If you don't know, draw a floor plan of your house or apartment and estimate the floor area in square meters.

Size of house (floor area) in square meters: _____

70101-14_A01A.EPS

WORKSHEET 1 (Continued)

On average, how many kilometers do you drive your household vehicles per week, and what are their average fuel efficiencies? 483 kilometers per week per vehicle or 24,140 kilometers per year is about average in the United States. If you're not sure about fuel efficiency, assume your car gets 9.4 kilometers per liter, which is about average. If you know how many liters of fuel you use per week, skip directly to the end of the line.

Car 1 kilometers per week _____ */ Car 1 kilometers per liter* _____ *= Car 1 liters per week* _____

Car 2 kilometers per week _____ */ Car 2 kilometers per liter* _____ *= Car 2 liters per week* _____

Car 3 kilometers per week _____ */ Car 3 kilometers per liter* _____ *= Car 3 liters per week* _____

Add up the liters of gas per week for all your vehicles, and then multiply by 52 to estimate the liters of gasoline you use per year.

Total liters of gasoline you use per year: _____

On average, how much do you travel each year on airplanes? Estimate the number of flight segments below for each of the three distances. A round trip counts as two segments, and each segment of a multi-segment flight counts as its own flight. For example, if you fly from Baltimore to Cincinnati to Atlanta, that counts as two flight segments.

Number of short-haul flight segments (less than 1,126 kilometers or 2 hours): _____

Number of medium-haul flight segments (1,126 – 4,023 kilometers or 2 – 4 hours): _____

Number of long-haul flight segments (more than 4,023 kilometers or longer than 4 hours): _____

On average, how many kilometers do you travel on public transportation per year?

Number of kilometers per year on transit bus/subway: _____

Number of kilometers per year on intercity bus: _____

Number of kilometers per year on intercity train: _____

Add up the total number of *kilometers* per year on public transportation: _____

Your answers to these questions will help you calculate your own carbon footprint later in the module. Keep track of your answers in the workbook or on a separate worksheet.

70101-14_A01B.EPS

Appendix E

WORKSHEET 2: INVENTORY YOUR PRODUCT IMPACTS

WORKSHEET 2: INVENTORY YOUR PRODUCT IMPACTS

Consider the products you buy or that are bought for you over the course of a year. Answer the following questions based on your best guess. If you share a household with other people, divide the total answer for your household by the number of people who live in your home. You can use your own currency to estimate how much you spend on each item. You will need to convert to US dollars before you calculate your carbon footprint. Find your currency's US dollar equivalent at www.xe.com.

1 of your currency = $ _____ US dollars. Write this amount in the third space in each line below to convert your currency to US dollars.

Eating out: $ _____ per month × 12 = _____ per year × _____ $US currency = $ _____ per year
Meat, fish, & protein: $ _____ per month × 12 = _____ per year × _____ $US currency = $ _____ per year
Cereals & baked goods: $ _____ per month × 12 = _____ per year × _____ $US currency = $ _____ per year
Dairy: $ _____ per month × 12 = _____ per year × _____ $US currency = $ _____ per year
Fruits & vegetables: $ _____ per month × 12 = _____ per year × _____ $US currency = $ _____ per year
Other: $ _____ per month × 12 = _____ per year × _____ $US currency = $ _____ per year

On average each month, how much do you spend on the following goods and services? Multiply each amount by 12 to estimate your average annual spending in each category. If you already know how much you spend per year, you can skip the monthly amount and write the average amount per year at the end of each line. Don't forget to convert to US dollars at the end.

Clothing: $ _____ per month × 12 = _____ per year × _____ $US currency = $ _____ per year
Furnishings &
   household items: $ _____ per month × 12 = _____ per year × _____ $US currency = $ _____ per year
Other goods: $ _____ per month × 12 = _____ per year × _____ $US currency = $ _____ per year
Services: $ _____ per month × 12 = _____ per year × _____ $US currency = $ _____ per year

70101-14_A02.EPS

Appendix F

WORKSHEET 3: DETERMINE YOUR CARBON FOOTPRINT

WORKSHEET 3: DETERMINE YOUR CARBON FOOTPRINT

Fill in the amounts from Worksheets 1 and 2 to calculate your total carbon footprint in kilograms of carbon. Which of your activities contributes the most to your carbon footprint?

Item	Quantity	Carbon Factor	Total Kilograms of CO_2/Year
Liters of garbage/year:	_____	× 0.24 kg per liter	= _____ kg per year
Total electricity/year in kilowatt-hours:	_____	× 0.64 kg per kWh	= _____ kg per year
Total therms of natural gas/year:	_____	× 1.40 kg per liter	= _____ kg per year
Total liters of propane/year:	_____	× 1.52 kg per liter	= _____ kg per year
Total liters of fuel oil/year:	_____	× 2.68 kg per liter	= _____ kg per year
Annual cost (USD) for water and sewage:	_____	× 4.04 kg per dollar	= _____ kg per year
House size (floor area) in square meters:	_____	× 10.25 sq m	= _____ kg per year
Liters of gasoline/year:	_____	× 2.40 kg per liter	= _____ kg per year
No. of short-haul flight segments/year:	_____	× 137.89 kg per segment	= _____ kg per year
No. of medium-haul flight segments/year:	_____	× 329.31 kg per segment	= _____ kg per year
No. of long-haul flight segments/year:	_____	× 1,005.61 kg per segment	= _____ kg per year
Kilometers per year on public transportation:	_____	× 0.14 kg per km	= _____ kg per year
Eating out (US dollars/year):	_____	× 0.36 kg per dollar	= _____ kg per year
Meat, fish, & protein (US dollars/year):	_____	× 1.45 kg per dollar	= _____ kg per year
Cereals & baked goods (US dollars/year):	_____	× 0.73 kg per dollar	= _____ kg per year
Dairy (US dollars/year):	_____	× 1.91 kg per dollar	= _____ kg per year
Fruits & vegetables (US dollars/year):	_____	× 1.18 kg per dollar	= _____ kg per year
Other (US dollars/year):	_____	× 0.45 kg per dollar	= _____ kg per year
Clothing (US dollars/year):	_____	× 0.45 kg per dollar	= _____ kg per year
Household items (US dollars/year):	_____	× 0.45 kg per dollar	= _____ kg per year
Other goods (US dollars/year):	_____	× 0.34 kg per dollar	= _____ kg per year
Services (US dollars/year):	_____	× 0.18 kg per dollar	= _____ kg per year

Total Carbon Footprint: = _____ kg per year

70101-14_A03.EPS

Additional Resources

This module presents thorough resources for task training. The following resource material is suggested for further study.

Field Guide for Sustainable Construction. Department of Defense – Pentagon Renovation and Construction Office. (2004). PDF, 2.6 MB, 312 pgs. Available for download at **www.wbdg.org**.

Greening Federal Facilities, 2nd Ed. US Department of Energy Federal Energy Management Program. (2001). PDF, 2.1 MB, 211 pgs. Available for download at **www.wbdg.org**.

Natural Capitalism. Lovins, A., Hawkin, P., and Lovins, L.H. (1995). Little, Brown, & Company, Boston, MA. Available online at **www.natcap.org**.

Sustainable Buildings Technical Manual. Public Technologies, Inc./U.S. Department of Energy. (2006). PDF, 3.1 MB, 292 pgs. Available for download at **www.greenbiz.com**.

Sustainable Buildings and Infrastructure: Paths to the Future. Pearce, A.R., Ahn, Y.H., and Hanmi Global. (2012). Routledge, London, UK.

Sustainable Construction: Green Building Design and Delivery, 3rd Ed. Kibert, C.J. (2012). Wiley, New York, NY.

The HOK Guidebook to Sustainable Design, 3rd Ed. Odell, W. and Lazarus, M.A. (2015). Wiley, New York, NY.

The following list is a compilation of websites referenced in this module:

American Society of Heating, Refrigerating, and Air-Conditioning Engineers: **www.ashrae.org**

Arid Solutions Inc.: **www.aridsolutionsinc.com**

Carbon Footprint: **www.carbonfootprint.com**

Database of State Incentives for Renewables & Efficiency: **www.dsireusa.org**

Energy Star: **www.energystar.gov**

Forest Stewardship Council: **www.fsc.org**

Green Building Certification Institute: **www.gbci.org**

Green Building Initiative: **www.thegbi.org**

Green Globes: **www.greenglobes.com**

Green Seal: **www.greenseal.org**

Green-e: **www.green-e.org**

Habitat for Humanity: **www.habitat.org**

International Initiative for a Sustainable Built Environment: **www.iisbe.org**

Living Building Challenge: **www.living-future.org/lbc**

NAHB Research Center: **www.nahbrc.org**

Natural Capitalism: **www.natcap.org**

Office of the Federal Environmental Executive: **www.ofee.gov**

Refining Process: **www.myfootprint.org**

Sheet Metal and Air Conditioning Contractors' National Association: **www.smacna.org**

Smart Communities Network: **www.smartcommunities.ncat.org**

The Carpet and Rug Institute: **www.carpet-rug.org**

The PLANTS Database: **www.plants.usda.gov**

Unit Conversion: **www.convertunits.com**

U.S. Environmental Protection Agency: Clean Energy **www.epa.gov/cleanenergy**

U.S. Green Building Council: **www.usgbc.org**

WaterSense: **www.epa.gov/watersense**

Whole Building Design Guide: **www.wbdg.org**

Figure Credits

NOAA, *Figure 1*

NASA/NOAA GOES Project, Figure 2

Library of Congress, Prints & Photographs Division, photograph by Carol M. Highsmith (Reproduction Number, LC-DIG-highsm-13380), Figure 3

US Environmental Protection Agency, Figure 4A

US Geological Survey/photo by Robert Kamilli, Figure 4B

Deere & Company, Figures 4C, 78

US Geological Survey, Figures 4D, 67

Steve Hillebrand/USFWS, Figure 4E

USDA Forest Service/Frank Koch, Figure 4F

Henry Rose, Figure 5, SA06

Dr. Christine Fiori/Virginia Tech, Figures 6, 45

NASA, Figure 7

The City of Santa Monica, Figure 8

Sushil Shenoy/Virginia Tech, Figures 9, 16–18, 26, 29, 30-32, 35-44, 46, 51, 52, 56-61, 63, 65, 68, 72-77, 79, SA03, SA14, SA23, SA24

US Energy Information Administration (Sept 2014), *Figure 10*

Dr. Annie Pearce/Virginia Tech, Figures 11, 24, 25B, *27*, *48*, 49B, 53, *54*, *62*, *69–71* SA01, SA05, SA13, SA16, SA17

US Department of Energy Office of Energy Efficiency and Renewable Energy, Figure 15

Jamie Carroll/NCCER, Figures 19, 64

Waterless Co., Figure 20

Brac Systems, Figure 22

Tim Davis/NCCER, Figure 25A, SA07, SA22

Courtesy of the Center for Resource Solutions (**www.green-e.org**), Figure 28

Courtesy of USDA_NRCS, Figure 33A

Elizabeth Westfall, Figure 33B

Image provided by Tate Access Floors, Inc., Figure 47

Tim Dean, Figure 49A, SA25

Topaz Publications, Inc., *Figure 50*, SA02

© US Green Building Council, Figure 55, Appendix A

Arid Solutions Inc.com, Figure 66

Ren Solutions, Figure 80

USDA Forest Service, SA04

US Army photo by Todd Plain, SA08

@esperng (Foap.com), SA09

Pratik Doshi/Virginia Tech, SA11, SA18

Joshua Winchell/USFWS, SA12

NPS photo by Ed Austin/Herb Jones, SA15

US Army photo by Capt. Michael N. Meyer, SA19

Library of Congress, Prints & Photographs Division, photograph by Carol M. Highsmith (Reproduction Number, LC-DIG-highsm-13019), SA20

Library of Congress, Prints & Photographs Division, photograph by Carol M. Highsmith (Reproduction Number, LC-DIG-highsm-12111), SA21

Section Review Answer Key

Answer	Section Reference	Objective
Section One		
1. d	1.0.0	1a
2. b	1.1.2	1b
3. a	1.2.1	1c
Section Two		
1. c	2.0.0	2a
2. b	2.2.3	2b
3. c	2.3.0	2c
4. c	2.4.1	2d
5. a	2.5.1	2e
6. b	2.6.1	2f
7. c	2.7.4	2g
Section Three		
1. b	3.0.0	3a
2. b	3.2.2	3b
3. d	3.3.0	3c
4. a	3.4.4	3d

NCCER CURRICULA — USER UPDATE

NCCER makes every effort to keep its textbooks up-to-date and free of technical errors. We appreciate your help in this process. If you find an error, a typographical mistake, or an inaccuracy in NCCER's curricula, please fill out this form (or a photocopy), or complete the online form at **www.nccer.org/olf**. Be sure to include the exact module ID number, page number, a detailed description, and your recommended correction. Your input will be brought to the attention of the Authoring Team. Thank you for your assistance.

Instructors – If you have an idea for improving this textbook, or have found that additional materials were necessary to teach this module effectively, please let us know so that we may present your suggestions to the Authoring Team.

NCCER Product Development and Revision
13614 Progress Blvd., Alachua, FL 32615

Email: curriculum@nccer.org
Online: www.nccer.org/olf

❏ Trainee Guide ❏ Lesson Plans ❏ Exam ❏ PowerPoints Other _____

Craft / Level: _____ Copyright Date: _____

Module ID Number / Title: _____

Section Number(s): _____

Description: _____

_____

_____

_____

_____

Recommended Correction: _____

_____

_____

_____

_____

Your Name: _____

Address: _____

_____

Email: _____ Phone: _____

Glossary

Absorptive finish: A surface finish that will absorb dust, particles, fumes, and sound.

Abutment: The supporting substructure at each end of a bridge.

Acidification: A process that converts air pollution into acid substances, leading to acid rain. Acid rain is best known for the damage it causes to forests and lakes. Also refers to the outflow of acidic water from metal and coal mines.

Aerated autoclaved concrete (AAC): A lightweight, precast building material that provides structural strength, insulation, and fire resistance. Typical products include blocks, wall panels, and floor panels.

Aeration: Mixing air into a liquid substance.

Aerial lift: A mobile work platform designed to transport and raise personnel, tools, and materials to overhead work areas.

Albedo: The extent to which an object reflects light from the sun. It is a ratio with values from 0 to 1. A value of 0 is dark (low albedo). A value of 1 is light (high albedo).

Alternative fuel: Any material or substance that can be used as a fuel other than conventional fossil fuels. They produce less pollution than fossil fuels. These include biodiesel and ethanol.

Anti-two-blocking devices: Devices that provide warnings and prevent two-blocking from occurring. Two-blocking occurs when the lower load block or hook comes into contact with the upper load block, boom point, or boom point machinery. The likely result is failure of the rope or release of the load or hook block.

Aquifer: An underground layer of water-bearing rock or soil from which groundwater can be extracted using a well.

Aquifer depletion: A situation where water is withdrawn from its underground source faster than the rate of natural recharge.

Arc blast: An explosion similar to the detonation of dynamite that occurs during an arc flash incident.

Arc fault: A high-energy discharge between two or more conductors.

Arc fault circuit interrupter (AFCI): A device intended to provide protection from the effects of arc faults by recognizing characteristics unique to arcing and by functioning to de-energize the circuit when an arc fault is detected.

Arc flash boundary: An approach limit at a distance from exposed energized electrical conductors or circuit parts within which a person could receive a second-degree burn if an electrical arc flash were to occur.

Architectural concrete: Concrete that serves as the architectural finish material.

Assured equipment grounding conductor program: A detailed plan specifying an employer's required equipment inspections and tests and a schedule for conducting those inspections and tests.

Backfire: A loud snap or pop as a torch flame is extinguished.

Band: Reinforcing steel in columns that is wrapped around the vertical bars to counteract compression forces.

Bar list: A bill of materials for a job site that shows all bar quantities, sizes, lengths, grades, placement areas, and bending dimensions to be used.

Baseplate: Bottom component on a shore, typically a 6"-square steel plate with nail holes for fastening to a mudsill.

Batten: Strip of lumber laid flat against the form panel to provide reinforcement.

Beam: A horizontal structural member.

Bent: A self-supporting frame having at least two legs and placed at right angles to the length of the structure it supports, such as the columns and cap supporting the spans of a bridge.

Best Management Practice (BMP): A way to accomplish something with the least amount of effort to achieve the best results. This is based on repeatable procedures that have proven themselves over time for large numbers of people.

Bio-based: A material derived from living matter, either plant or animal.

Biodegradable: Organic material such as plant and animal matter and other substances originating from living organisms. These are capable of being broken down into innocuous products by the action of microorganisms.

Biodiversity: A measure of the variety among organisms present in different ecosystems.

Biofuel: A solid, liquid, or gas fuel consisting of or derived from recently dead biological material. The most common source is plants.

Biomimicry: The act of imitating nature to create new solutions modeled after natural systems.

Bioswale: An engineered depression designed to accept and channel stormwater runoff. A bioswale uses natural methods to filter stormwater, such as vegetation and soil.

Blackout: A situation where the electrical grid fails and does not provide any power.

Blackwater: Water or sewage that contains fecal matter or sources of pathogens.

Body harness: Straps that may be secured about the worker in a manner that will distribute the fall-arrest forces over at least the thighs, pelvis, waist, chest, and shoulders, with means for attaching it to other components of a personal fall-arrest system.

Bolted fault: A short circuit or electrical contact between two conductors at different potentials, in which the impedance or resistance between the conductors is essentially zero.

Boom lift: Aerial lift with a single arm that extends a work platform/enclosure capable of holding one or two workers; some models may have an articulated joint.

Brace: A diagonal supporting member used to reinforce a form against the weight of the concrete.

Bracing collar: Metal strap used to brace a round column form; it is attached near the top of the form and extends to the ground.

Breakdown voltage: The voltage at which an insulator has a breakdown and ceases to act as a resistor.

Brownfield: Property that contains the presence or potential presence of a hazardous substance, pollutant, or contaminant. Cleaning up and reinvesting in these properties reduces development pressures on undeveloped, open land and improves and protects the environment.

Brownout: A situation where the electrical grid provides less power than normal, but enough for some equipment to still work.

Buck: A frame placed inside a concrete form to provide an opening for a window or door.

Building-integrated photovoltaics (BIPVs): Photovoltaic systems built into other types of building materials. See *photovoltaic*.

Bulkhead: Form component fastened vertically inside a form to stop the flow of concrete at a certain location.

Bundle of bars: A bundle consisting of one size, length, or mark (bent) of bar, with the following exceptions: very small quantities may be bundled together for convenience, and groups of varying bar lengths or marks that will be placed adjacent to one another may be bundled together.

Byproducts: Another product derived from a manufacturing process or a chemical reaction. It is not the primary product or service being produced.

Caissons: Piers usually extending through water or soft soil to solid earth or rock; also refers to cast-in-place, drilled-hole piles.

Capacity: The total amount of weight capable of being lifted. This includes personnel, tools and materials, and/or equipment.

Capital: A flared section at the top of a concrete column.

Carbon cycle: The movement of carbon between the biosphere, atmosphere, oceans, and geosphere of the Earth. It is a biogeochemical cycle. In the cycle, there are sinks, or stores, of carbon. There are also processes by which the various sinks exchange carbon.

Carbon footprint: A measure of impact human activities have on the environment. It is determined by the amount of greenhouse gases produced. It is measured in pounds or kilograms of carbon dioxide.

Carbon neutral: An action or product whose production absorbs as much carbon from the atmosphere as it produces.

Carbon offset: An agreement with another party that they will reduce their carbon production by some amount in exchange for payment.

Carburizing flame: A flame burning with an excess amount of fuel; also called a reducing flame.

Carpool: An arrangement in which several people travel together in one vehicle. The people take turns driving and share in the cost.

Center of gravity: The point around which all of an object's weight is evenly distributed.

Charrette:: An intense meeting of project participants that can quickly generate design solutions by integrating the abilities and interests of a diverse group of people.

Chlorofluorocarbons (CFCs): A set of chemical compounds that deplete ozone. They are widely used as solvents, coolants, and propellants in aerosols. These chemicals are the main cause of ozone depletion in the stratosphere.

Circuit breaker: A device designed to protect circuits from overloads and to enable them to be reset after tripping.

Climbing form: A form used to construct vertical walls in successive pours. It is raised to a new level for each pour.

Cob construction: An ancient building method using hand-formed lumps of earth mixed with sand and straw.

Column: A post or vertical structural member supporting a floor beam, girder, or other horizontal member and carrying a primarily vertical load.

Column spirals: Columns in which the vertical bars are enclosed within a spiral that functions like a column tie.

Column ties: Bars that are bent into square, rectangular, U-shaped, circular, or other shapes for the purpose of holding vertical column bars laterally in place and that prevent buckling of the vertical bars under compression load.

Combined center of gravity: When the weight of two items is combined, the center of gravity shifts to one point for both items.

Commissioning: A review process conducted by a third party that involves a detailed design review, testing and balancing of systems, and system turnover.

Compact fluorescent lamp (CFL): A fluorescent bulb that is designed to fit in a normal light fixture. They use less energy and last longer than incandescent bulbs.

Compactor: A machine used to compact soil to prevent the settling of the soil after construction.

Compressor: A motor-driven machine used to supply compressed air for pneumatic tools.

Concrete cover: The distance from the face of the concrete to the reinforcing steel; also referred to as fireproofing, clearance, or concrete protection.

Connector: A device that is used to couple (connect) parts of the personal fall-arrest system and positioning device systems together. It may be an independent component of the system, such as a carabiner. It may also be an integral component of part of the system, such as a buckle or D-ring sewn into a body harness, or a snaphook spliced or sewn to a lanyard.

Connector: An employee who, working with hoisting equipment, is placing and connecting structural members and/or components.

Conservation: Using natural resources wisely and at a slower rate than normal.

Contact splice: A means of connecting reinforcing bars by lapping in direct contact.

Continuous beam: A beam that extends over three or more supports (including end supports).

Controlled access zone: A designated work area in which certain types of masonry work may take place without the use of conventional fall protection systems.

Controlled Decking Zone (CDZ): An area in which certain work (for example, initial installation and placement of metal decking) may take place without the use of guardrail systems, personal fall arrest systems, fall restraint systems, or safety net systems and where access to the zone is controlled.

Corporate Sustainability Report (CSR): An annual report by a company documenting its efforts to reduce negative environmental and social impacts.

Corrugated: Material formed with parallel ridges or grooves.

Crab steering: A steering mode where all wheels may move in the same direction, allowing the machine to move sideways on a diagonal; also known as oblique steering.

CRSI: Concrete Reinforcing Steel Institute.

Cut: A common term for a scaffold level.

Dead load: The actual weight of the deck itself.

Deconstruction: Taking a building apart with the intent of salvaging reusable materials.

Deforestation: The removal of trees without sufficient replanting.

Desertification: The creation of deserts through degradation of productive land in dry climates by human activities.

Double-curtain wall: A concrete wall that contains a layer of reinforcement at each face.

Dowel: A bar connecting two separately cast sections of concrete. A bar extending from one concrete section into another is said to be doweled into the adjoining section.

Downcycling: Recycling one material into a material of lesser quality. An example is the recycling of plastics into lower-grade composites.

Drag lines: The lines on the edge of the material that result from the travel of the cutting oxygen stream into, through, and out of the metal.

Drive/steer controller: One-handed joystick lever control used to control the speed and steering of an aerial lift.

Drop-head: Shoring hardware that allows formwork to be released and dropped a few inches prior to stripping the forms.

Dross: The material (oxidized and molten metal) that is expelled from the kerf when cutting using a thermal process. It is sometimes called slag.

Duty rating: American National Standards Institute (ANSI) rating assigned to ladders. It indicates the type of use the ladder is designed for (industrial, commercial, or household) and the maximum working load limit (weight capacity) of the ladder. The working load limit is the maximum combined weight of the user, tools, and any materials bearing down on the rungs of a ladder.

EarthCraft: A residential green building program of the Greater Atlanta Home Builders Association in partnership with Southface Energy Institute.

Ecological footprint: A measure of impact human activities have on the environment. It compares human consumption of natural resources with the Earth's capacity to regenerate them. Measured in hectares or acres.

Ecosystem: A combination of all plants, animals and microorganisms in an area that complement each other. These function together with all of the nonliving physical factors of the environment.

Embodied energy: The total energy required to bring a product to market. It includes raw material extraction, manufacturing, final transport, and installation.

Endangered species: A population of a species at risk of becoming extinct. A threatened species is any species that is vulnerable to extinction in the near future.

Energy efficiency: Getting more use out of electricity already generated.

Energy source: Any source of electrical, mechanical, hydraulic, pneumatic, chemical, thermal, or other energy.

ENERGY STAR: A United States government program to promote energy efficiency. It is a joint program of the US Environmental Protection Agency and the US Department of Energy.

Energy-isolating device: Any mechanical device that physically prevents the transmission or release of energy. These include, but are not limited to, manually operated electrical circuit breakers, disconnect switches, line valves, and blocks.

Environmental Product Declaration (EPD): A document listing the environmental impacts caused by manufacture of a product.

Equipment grounding conductor: A wire that connects metal enclosures and containers to ground.

Equipment idling: The operation of equipment while it is not in motion or performing work. Limiting idle times reduces air pollution and greenhouse gas emissions.

Equipment maintenance: The care, cleaning, inspection, and proper use of machinery and equipment.

Equipment operator: A person skilled in operating certain equipment.

Erosion: The displacement of solids by wind, water, ice, or gravity or by living organisms. These solids include rocks and soil particles.

Far face: The face farthest from the viewer (as of a wall); may be the outside or inside face, depending on whether one is inside looking out or outside looking in.

Ferrous metals: Metals containing iron.

Fiberglass-reinforced plastic (FRP): A durable material used for column forms and as an overlay on plywood panels.

Fibrillation: Very rapid, irregular contractions of the muscle fibers of the heart that result in the heartbeat and pulse going out of rhythm with each other.

Flashback: The flame burning back into the tip, torch, hose, or regulator, causing a high-pitched whistling or hissing sound.

Flat slab: A concrete slab reinforced in two or more directions, with drop panels but generally without beams, and with or without column capitals.

Flight: Continuous series of steps from one floor to another or from a floor to a landing.

Flood plain: An area surrounding a river or body of water that regularly floods within a given period of time.

Flushout: Using fresh air in the building HVAC system to remove contaminants from the building.

Forklift: A machine designed to facilitate the movement of bulk items around the job site.

Form liner: Plastic or wood liners used on the inside of wall forms to add a design or special feature to the concrete.

Four-wheel steering: A steering mode where the front and rear wheels may move in opposite directions, allowing for very tight turns; also known as independent steering or circle steering.

Free fall: The act of falling before a personal fall-arrest system begins to apply force to arrest the fall.

FSC Certified: Wood or wood products that have met the Forest Stewardship Council's tracking process for sustainable harvest.

Fugitive emissions: Pollutants released to the air other than those from stacks or vents. They are often due to equipment leaks, evaporative processes, and wind disturbances.

Fulcrum: A point or structure on which a lever sits and pivots.

Generator: A machine designed to generate electricity.

Geothermal: Heat that comes from within the Earth.

Girder: The principal beam supporting other beams.

Global climate change: Changes in weather patterns and temperatures on a planetary scale. This may lead to a rise in sea levels, melting of polar ice caps, increased droughts, and other weather effects.

Gouging: The process of cutting a groove into a surface.

Graywater: A nonindustrial wastewater generated from domestic processes. These include laundry and bathing. Graywater comprises 50 to 80 percent of residential wastewater.

Green Seal Certified: A certification of a product to indicate its environmental friendliness. Green Seal is a group that works with manufacturers, industry sectors, purchasing groups, and government at all levels to green the production and purchasing chain. Founded in 1989, Green Seal provides science-based environmental certification standards.

Greenhouse Effect: The overall warming of a planet's surface due to retention of solar heat by its atmosphere.

Grid intertie: A connection between a local source of power and the utility power grid.

Ground fault circuit interrupter (GFCI): A fast-acting circuit breaker that senses small imbalances in the circuit caused by current leakage to ground and, in a fraction of a second, shuts off the electricity.

Grounding: The process of directly connecting an electrical circuit to a known ground to provide a zero-voltage reference level for the equipment or system.

Guyed derrick: An apparatus used for hoisting on high-rise buildings, consisting of a boom mounted on a column or mast that is held at the head by fixed-length supporting ropes or guys.

Halons: Ozone-depleting compounds consisting of bromine, fluorine, and carbon. They were commonly used as fire extinguishing agents, both in built-in systems and in handheld portable fire extinguishers. Halon production in the US ended on December 31, 1993.

Hardscape: Paved areas surrounding a project, including parking lots and sidewalks.

Haunch: A structure provided on all bridges with steel girders or prestressed concrete I-beams. They provide a means of final adjustment of the deck-slab elevation to match the design roadway profile and cross slope. The haunch allows this adjustment without having the top flange of the girder project into the structural deck.

Heat island: An area that is warmer than surrounding areas due to absorbing more solar radiation.

Hickey bar: A hand tool with a side-opening jaw used in developing leverage for making in-place bends on bars or pipes.

Hook: A 180-degree (semicircular) or 90-degree turn at the free end of a bar to provide anchorage in concrete. For stirrups and column ties only, turns of either 90 degrees or 135 degrees are used.

Hybrid vehicle: A vehicle that uses two or more power sources for propulsion.

Hydraulic: Tools and equipment that are powered or moved by liquid under pressure.

Hydrochlorofluorocarbons (HCFCs): A group of man-made compounds containing hydrogen, chlorine, fluorine, and carbon. They are used for refrigeration, aerosol propellants, foam manufacture, and air conditioning. They are broken down in the lowest part of the atmosphere and pose a smaller risk to the ozone layer than other types of refrigerants.

Hydrologic cycle: The circulation and conservation of Earth's water supply. The process has five phases: condensation, infiltration, runoff, evaporation, and precipitation.

Indoor air quality (IAQ): The content of interior air that could affect health and comfort of building occupants.

Infiltration: The movement of air from outside a building to inside through cracks or openings in the building envelope.

Inserts: Devices that are positioned in concrete to receive a bolt or screw to support shelf angles, machinery, etc.

Insulating concrete form (ICF): Concrete form system in which concrete is cast between two expanded polystyrene (EPS) foam panels; rigid forms that hold concrete in place during curing and remain in place afterwards. The forms serve as thermal insulation for concrete walls.

Insulation: The practice of placing nonconductive material such as plastic around the conductor to prevent current from passing through it.

Integrative design: A collaborative design methodology that emphasizes the input of knowledge from several areas in the development of a complete design.

Interlocking systems: Control systems that are set to prevent accidents by keeping one system from overriding the other.

Just-in-time delivery: A material delivery strategy that reduces material inventory. The material that is delivered is used immediately.

Kerf: The gap produced by a cutting process.

Ladder: A wood, metal, or fiberglass framework consisting of two parallel side pieces (rails) connected by rungs on which a person steps when climbing up or down. Ladders may either be of a fixed length that is permanently attached to a building or structure, or portable. Portable ladders have either fixed or adjustable lengths and are either self-supporting or not self-supporting.

Landing: Horizontal slab or platform in a stairway to break the run of stairs.

Lanyard: A flexible line of rope, wire rope, or strap that generally has a connector at each end for connecting the body harness to a deceleration device, lifeline, or anchorage.

Lapped splice: The joining of two reinforcing bars by lapping them side by side, or the length of overlap of two bars; similarly, the side and end overlap of sheets or rolls of welded-wire fabric reinforcement.

Lateral stress: Wind shear and other forces applying horizontal pressure to a wall or other structural unit.

Leadership in Energy and Environmental Design (LEED): A point-based rating system used to evaluate the environmental performance of buildings. Developed by the US Green Building Council, LEED provides a suite of standards for environmentally sustainable construction. Founded in 1998, LEED focuses on the certification of commercial and residential buildings and the neighborhoods in which they exist.

Leading edge: The edge of a floor, roof, or formwork for a floor or other walking/working surface (such as the deck) which changes location as additional floor, roof, decking, or formwork sections are placed, formed, or constructed. A leading edge is considered to be an unprotected side and edge during periods when it is not actively and continuously under construction.

Life cycle: The useful life of a system, product, or building.

Life cycle assessment: An analytic technique to evaluate the environmental impact of a system, product, or building throughout its life cycle. This includes the extraction or harvesting of raw materials through processing, manufacture, installation, use, and ultimate disposal or recycling.

Life cycle cost: The cost of a system or a component over its entire life span.

Lifeline: A component consisting of a flexible line connected vertically to an anchorage at one end (vertical lifeline), or connected horizontally to an anchorage at both ends (horizontal lifeline), and which serves as a means for connecting other components of a personal fall-arrest system to the anchorage.

Lifting eye: Load-lifting device attached to a heavy form panel; used to rig the panel with a crane or other lifting equipment.

Light-emitting diode (LED) lamp: A highly efficient, electronic light bulb using a glowing diode designed to fit a standard light fixture.

Limited access zone: A restricted area alongside a masonry wall that is under construction.

Lithium ion (Li-ion): A type of rechargeable battery in which a lithium ion moves between the anode and cathode. They are commonly used in consumer electronics.

Live load: The carrying capacity of the deck including the dead load of the deck.

Local materials: Materials that come from within a certain number of miles from the project. Materials produced locally use less energy during transportation to the site. The LEED system of building certification offers points for the use of regional or local materials.

Location valuation: A bonus under the LEED rating system for using a product produced locally. Materials from within a 100-mile radius can be counted as twice their value as a reward for using local products.

Lockout: The placement of a lockout device on an energy-isolating device, in accordance with an established procedure, ensuring that the energy-isolating device and the equipment being controlled cannot be operated until the lockout device is removed.

Lockout device: Any device that uses positive means such as a lock to hold an energy-isolating device in a safe position, thereby preventing the energizing of machinery or equipment.

Low-emission vehicle:: Vehicles that produce fewer emissions than the average vehicle. Beginning in 2001, all light vehicles sold nationally were required to meet this standard.

Maintainability: An indication of how difficult it is to properly maintain a building system or technology.

Massing: The three-dimensional shape of a building, including length, width, and height. Determines the amount of a building exposed to sunlight.

Minimum efficiency reporting value (MERV): A measurement scale designed in 1987 by the American Society of Heating, Refrigeration, and Air-Conditioning Engineers (ASHRAE) to rate the effectiveness of air filters. The scale is designed to represent the worst-case performance of a filter when dealing with particles in the range of 0.3 to 10 microns.

Momentum: A physical force that causes an object in motion to stay in motion.

Monofill: A landfill that accepts only one type of waste, typically in bales.

Monolithically: Used to describe concrete that is placed in forms continuously and without construction joints.

Mudsill: Hardwood planks or square pads used to support baseplates and prevent shores from sinking into the soil and causing the floor to collapse.

Multi-function material: A material that can be used to perform more than one function in a facility.

Nano-material: A material with features smaller than a micron in at least one dimension.

Near face: The face nearest the viewer, which may be inside or outside, depending on whether one is inside looking out or outside looking in.

Neutral flame: A flame burning with correct proportions of fuel gas and oxygen.

Nickel cadmium (NiCad): A popular type of rechargeable battery using nickel oxide hydroxide and metallic cadmium as electrodes.

Nonrenewable: A material or energy source that cannot be replenished within a reasonable period.

Nontoxic: Substances that are not poisonous.

Oblique steering: A steering mode where all wheels may move in the same direction, allowing the machine to move sideways on a diagonal; also known as crab steering.

Offgassing: The evaporation of volatile chemicals at normal atmospheric pressure. Building materials release chemicals into the air through evaporation.

Operator's compartment: The portion of a forklift where the operator is positioned to control the forklift.

Outrigger: An extension that projects from the main body of a crane to add stability and support.

Oxidizing flame: A flame burning with an excess amount of oxygen.

Ozone depletion: A slow, steady decline in the total amount of ozone in Earth's stratosphere.

Ozone hole: A large, seasonal decrease in stratospheric ozone over Earth's polar regions. The ozone hole does not go all the way through the layer.

Papercrete: A fiber-cement material that uses waste paper for fiber.

Passive solar design: Designing a building to use the sun's energy for lighting, heating, and cooling with minimal additional inputs of energy.

Passive survivability: The ability of a building to continue to offer basic function and habitability after a loss of infrastructure (i.e., water and power).

Pathogen: An infectious agent that causes disease or illness.

Payback period: The amount of time it takes to break even on an investment; the period of time after which savings from an investment equals its initial cost.

Peak shaving: An energy management strategy that reduces demand during peak times of the day and shifts it to off-peak times, such as at night.

Persistent, bioaccumulative toxin (PBT): A harmful substance such as a pesticide or organic chemical that does not quickly degrade in the natural environment but does accumulate in plants and animals in the food chain.

Personal fall-arrest system: A system used to stop an employee in a fall from a working level. It consists of an anchorage, connectors, and a body harness, and may include a lanyard, deceleration device, lifeline, or suitable combinations of these.

Pervious concrete: A mixture of coarse aggregate, Portland cement, water, and little to no sand. It has a 15 to 25 percent void structure and allows 3 to 8 gallons of water per minute to pass through each square foot. Also known as permeable concrete.

Phantom loads: Electricity consumed by appliances and devices when they are switched off.

Phase change material (PCM): A material that stores or releases large amounts of energy when it changes between solid, liquid, or gas forms.

Phosphatized: Used to describe metal that has been treated with a phosphoric acid to prepare it for a finish coat. In composite floor decking, this unpainted surface will contact the concrete.

Photosensor: A sensor that measures daylight and adjusts artificial lighting to save energy.

Photovoltaic: A technology that converts light directly into electricity.

Pierce: To penetrate through metal plate with an oxyfuel cutting torch.

Pile cap: A structural member placed on the tops of piles and used to distribute loads from the structure to the piles.

Pitch: The center-to-center spacing between the turns of a spiral.

Placing drawings: Detailed drawings that give the bar size, location, spacing, and all other information required to place the reinforcing steel.

Platform: A work surface elevated above lower levels. Platforms can be constructed using individual wood planks, fabricated planks, fabricated decks, and fabricated platforms.

Plyform®: An APA performance-rated panel used for concrete forms.

Pollution prevention: The prevention or reduction of pollution at the source.

Post-consumer: Products made out of material that has been used by the end consumer and then collected for recycling.

Post-industrial/pre-consumer: Material diverted from the waste stream during the manufacturing process.

Power hop: Action in heavy equipment that uses pneumatic tires to create a bouncing motion between the fore and aft axles. Once started, the oscillation back and forth usually continues until the operator either stops or slows down significantly to change the dynamics.

Powered industrial trucks: An OSHA term for several types of light equipment that include forklifts. See *forklift*.

Preservation: The act and advocacy of protecting the natural environment.

Profile: The appearance of a floor deck viewed cross-sectionally.

Proportional control: A control that increases speed in proportion to the movement of the control.

Radio-frequency identification (RFID): An automatic identification method. It relies on storing and remotely retrieving data using devices called RFID tags.

Rainwater harvesting: The gathering and storing of rainwater.

Rammed earth: An ancient building technique similar to adobe using soil that is mostly clay and sand. The difference is that the material is compressed or tamped into place, usually with forms that create very flat vertical surfaces.

Rapidly renewable: A material that is replenished by natural processes at a rate comparable to its rate of consumption. The LEED system of building certification rewards use of rapidly renewable materials that regenerate in 10 years or less, such as bamboo, cork, wool, and straw.

Raw material: A material that has been extracted directly from nature. It is in an unprocessed or minimally processed state.

Rebar horses: Wood or metal supports that are used in groups of two or more to hold main reinforcing in a convenient position for placing ties while prefabricating column, beam, or pile cages.

Recyclable: Material that still has useful physical or chemical properties after serving its original purpose. It can be reused or remanufactured into additional products. Plastic, paper, glass, used oil, and aluminum cans are examples of recyclable materials.

Recycled content: A material containing components that would otherwise have been discarded.

Recycled plastic lumber (RPL): A wood-like product made from recovered plastic, either by itself or mixed with other materials. It can be used as a substitute for concrete, wood, and metals.

Recycling: The reprocessing of old materials into new products. A goal is to prevent the waste of potentially useful materials and reduce the consumption of new materials.

Reglet: A narrow molding used to separate two structural elements, usually roof and wall, to divert water.

Reinforced concrete: Concrete that has been placed around some type of steel reinforcement material. After the concrete cures, the reinforcement provides greater tensile and shear strength for the concrete. Almost all concrete is reinforced in some manner.

Renewable: A resource that may be naturally replenished.

Reshoring: Process performed for multistory construction when shoring equipment is removed from a partially cured slab. As shores for forms are removed, the elevated deck is reshored by placing shores against the bottom of the slab.

Retaining wall: A wall that has been reinforced to hold or retain soil, water, grain, coal, or sand.

Reusable: A material that can be used again without reprocessing. This can be for its original purpose or for a new purpose.

Rustication line: An indentation made in a concrete panel by attaching a thin strip of tapered wood or plastic to the form.

Safety interlock: A safety mechanism used to prevent incorrect control operation.

Salvaged: A used material that is saved from destruction or waste by being reused.

Scaffolding: A temporary built-up framework or suspended platform or work area designed to support workers, materials, and equipment at elevated or otherwise inaccessible job sites.

Schedule: A table on placing drawings that lists the size, shape, and number of bars each way, and the mark number of the bars if they are bent.

Scissor lift: Aerial lift used to raise a work enclosure vertically by means of crisscrossed supports.

Sedimentation: The process of depositing a solid material from a state of suspension in a fluid, usually air or water.

Self-retracting lanyard (SRL): A deceleration device containing a drum-wound line that can be slowly extracted from, or retracted onto, the drum under slight tension during normal employee movement, and which, after the onset of a fall, automatically locks the drum and arrests the fall.

Sheathing: Plywood, planks, or sheet metal that make up the surface of a form.

Shiplap: Boards that have been rabbeted on both edges so that they overlap when they are placed together.

Shock hazard: A dangerous condition associated with the possible release of energy caused by contact or approach to energized electrical conductors or circuit parts.

Shoring: Construction of a temporary support system to carry the dead load of fresh cast-in-place concrete for horizontal decks and slabs, including the dead load of the reinforcing steel and forms as well as the live loads that occur during construction.

Sick Building Syndrome: A variety of illnesses thought to be caused by poor indoor air. Symptoms include headaches, fatigue, and other problems that increase with continued exposure.

Simple beam: A beam supported at each end (two points) and not continuous.

Single-curtain wall: A concrete wall that contains a single layer of vertical or horizontal reinforcing bars in the center of the wall.

Skid-steer loader: A small, highly maneuverable machine equipped with a bucket for moving stone, dirt, and bulk items around a job site.

Sleeve: A tube that encloses a bar, dowel, anchor bolt, or similar item.

Slip form: A form that is moved continuously while the concrete is being placed.

Smart material: Materials that have one or more properties that can be significantly changed. These changes are driven by external stimuli.

Soapstone: Soft, white stone used to mark metal.

Softscape: The area surrounding a building that contains plantings, lawn, and other vegetated areas.

Solar: Energy from the sun in the form of heat and light.

Solid waste: Products and materials discarded after use in homes, businesses, restaurants, schools, industrial plants, or elsewhere.

Solvent-based: A material that consists of particles suspended or dissolved in a solvent. A solvent is any substance that will dissolve another. Solvent-based building materials typically use chemicals other than water as their solvent, including toluene and turpentine, with hazardous health effects.

Span: The horizontal distance between supports of a member such as a beam, girder, slab, or joist; also, the distance between the piers or abutments of a bridge.

Spoil pile: Excavated soil that has been moved and temporarily stored during construction.

Sprawl: Unplanned and inefficient development of open land.

Spreader: A wood or metal device used to hold the sides of a form apart.

Spud wrench: Tool to align holes in adjoining form panels; it has a long, tapered steel handle with an open-end wrench on one end.

Staggered splices: Splices in bars that are not made at the same point.

Stirrups: Reinforcing bars used in beams for shear reinforcement; typically bent into a U shape or box shape and placed perpendicular to the longitudinal steel; rebar bent into a loop; used to provide shear reinforcement for concrete.

Stormwater runoff: Unfiltered water that reaches streams, lakes, and oceans after a rainstorm by flowing across impervious surfaces.

Strawbale construction: A building method that uses straw as the structural element, insulation, or both. It has advantages over some conventional building systems because of its cost and availability.

Stringer: Timber placed at the tops of shores to support joists and deck form panels.

Strips: Bands of reinforcing bars in flat-slab or flat-plate construction. The column strip is a quarter-panel wide on each side of the column center line and runs from column to column. The middle strip is half a panel in width, filling in between column strips, and runs parallel to the column strips.

Strongback: An upright supporting member attached to the back of a form to stiffen or reinforce it, especially around door and window openings.

Structural insulated panel (SIP): A composite building material used for exterior building envelopes. It consists of a sandwich of two layers of structural board with an insulating layer of foam in between. The board is usually oriented strand board (OSB) and the foam can be polystyrene, soy-based foam, urethane, or even compressed straw.

Support bars: Bars that rest upon individual high chairs or bar chairs to support top bars in slabs or joists, respectively. They are usually #4 bars and may replace a like number of temperature bars in slabs when properly lap spliced; also used longitudinally in beams to provide support for the tops of stirrups. Also called raiser bars.

Sustainably harvested: A method of harvesting a material from a natural ecosystem without damaging the ability of the ecosystem to continue to produce the material indefinitely.

Tagout: The placement of a tagout device on an energy-isolating device, in accordance with an established procedure, to indicate that the energy-isolating device and the equipment being controlled may not be operated until the tagout device is removed.

Tagout device: Any prominent warning device, such as a tag and a means of attachment, that can be securely fastened to an energy-isolating device in accordance with an established procedure. The tag indicates that the machine or equipment to which it is attached is not to be operated until the tagout device is removed in accordance with the energy-control procedure.

Takeback: A condition where manufacturers recover waste from packaging or products after use, at the end of their life cycle.

Telehandler: A type of powered industrial truck characterized by a boom with several extendable sections known as a telescoping boom; another name for a shooting boom forklift.

Temperature bars: Bars distributed throughout the concrete to minimize cracks due to temperature changes and concrete shrinkage.

Template: A device used to locate and hold dowels, to lay out bolt holes and inserts, etc.

Thermal bridging: A condition created when a thermally conductive material bypasses an insulation system, allowing the rapid flow of heat from one side of a building wall to the other. Metal components, including metal studs, nails, and window frames, are common culprits.

Thermal mass: A property of a material related to density that allows it to absorb heat from a heat source, and then release it slowly. Common materials used to provide thermal mass include adobe, mud, stones, or even tanks of water.

Tie: A reinforcing bar bent into a box shape and used to hold longitudinal bars together in columns and beams. Also known as stirrup ties.

Tie wire: Wire (generally #16, #15, or #14 gauge) used to secure rebar intersections for the purpose of holding them in place until concreting is completed.

Tines: A prong of an implement such as a fork. For forklifts, tines are often called forks.

Twist-lock outlet: A type of electrical connector.

Two-point swing scaffolding: A manual or power-operated platform supported by hangers at two points suspended from overhead supports in a way that allows it to be raised or lowered to the working position.

Urbanization: Converting rural land to higher density development.

Urea formaldehyde: A transparent thermosetting resin or plastic. It is made from urea and formaldehyde heated in the presence of a mild base. Urea formaldehyde has negative effects on human health when allowed to offgas or burn.

Vapor-resistant: A material that resists the flow of water vapor.

Virgin material: A material that has not been previously used or consumed. It also has not been subjected to processing. See also *raw material*.

Volatile organic compounds (VOCs): Gases that are emitted over time from certain solids or liquids. Concentrations of many VOCs are up to 10 times higher indoors than outdoors. Examples include paints and lacquers, paint strippers, cleaning supplies, pesticides, building materials, and furnishings.

Waler: Horizontal wood or metal members installed on the outside of form walls to strengthen the sheathing and stiffen the walls.

Walk-off mat: Mats in entry areas that capture dirt and other particles.

Washing: A term used to describe the process of cutting out bolts, rivets, previously welded pieces, or other projections from the metal surface.

Waste separation: Sorting waste by specific type of material and storing it in different containers to facilitate recycling.

Water efficiency: Managing water use to prevent waste or overuse. Includes using less water to achieve the same benefits.

Water footprint: The volume of water used directly or indirectly to sustain something over a period of time.

Water-based: Materials that uses water as a solvent or vehicle of application.

Water-resistant: A material that hinders the penetration of water.

Waterproof: A material that is impervious to or unaffected by water.

Weephole: A drainage opening in a wall.

Wetland: Lands where saturation with water is the dominant factor. This determines the way soil develops and the types of plant and animal communities living in the soil and on its surface.

Whip check: A cable attached to the ends of an air supply hose and a tool, or to the ends of two hoses that are connected to each other, to prevent the hose from whipping uncontrollably when charged with compressed air.

Xeriscaping: Landscaping that requires little or no water for irrigation. Xeriscaping can be achieved through smart plant selection, mulching, and other tactics.

Yoke assembly: Clamping and support device used in slip forms.

Zoning: Rules for placing similar items next to one another, as in plants within a landscape, or types of buildings within a community.

Index

A
A. *See* Amp (A)
AAC. *See* Aerated autoclaved concrete (AAC)
Absorptive finish, (70101):22, 55, 95
Abutment, (27304):1, 5
Accelerator lever, (27406):30
Access zones
   controlled, (28301):1, 9–10, 40
   limited, (28301):1, 10, 40
Accidents. *See also* Safety
   causes of, (28301):1
   deaths from, (28301):4, 7
   falls, (28301):1, 4
   scaffolding and, (75122):14
   steel erection, prevention, (75110):1–2
Acetylene cylinders, (29102):4–5, 9–13
Acetylene flow rates, (29102):22
Acetylene for oxyfuel cutting, (29102):9–10, 12–13
ACI. *See* American Concrete Institute (ACI)
Acidification, (70101):1, 5, 95
Acronyms, common, (70101):101
AEM equipment safety manuals. *See* Association of
   Equipment Manufacturers (AEM), equipment safety
   manuals
Aerated autoclaved concrete (AAC), (70101):22, 47, 95
Aeration, (70101):22, 32, 95
Aerial lifts
   boom lift, (27406):4
   components, (27406):5–6, (75122):27
   defined, (27406):4, 41, (75122):27, 34
   function, (27406):4
   ground controls, (27406):6, 7
   hydraulic systems controls, (27406):6–7
   inspections and maintenance, (27406):8–10
   interlocking control systems, (75122):27–28, 29, 34
   operating procedure, (27406):7–8, (75122):29
   operator qualifications, (27406):6, (75122):27–28
   operator responsibilities, (27406):8–10
   platform controls, (27406):6, 7
   scissor lifts
      function, (27406):4
      ground controls, (27406):6
      platform controls, (27406):6
      safety, (75122):29–30
   transporting, (27406):4, 6–7
   types of, (75122):27
Aerial lift safety
   boom lifts operation and maintenance, (75122):30–31
   inspections and maintenance, (27406):8–10
   OSHA *Standard 29 CFR 1926.453*, (75122):28–29
   precautions, (27406):7
   pre-operation checks, (75122):29
   scissor lifts operation, (75122):29–30
AFCI. *See* Arc fault circuit interrupter (AFCI)
Agriculture Department, U.S., (70101):6
Air compressors
   function, (27406):20
   hoses, connecting, (27406):21–22

inspections, pre-operational, (27406):22
   operating procedures, (27406):22–23
   operator's maintenance responsibility, (27406):23
   safety precautions, (27406):20–21
   sizing, (27406):23
   tow-behind, (27406):20
   types of, (27406):20
Air conditioners, (70101):25
Air-filter meter, (27406):16
Air hoses, connecting, (27406):21–22
Air pollution, (70101):4–5, 19, 48
Air-pressure gauge, (27406):20
Albedo, (70101):22, 31, 95
All-metal panels, (27308):4, 6
All-plastic bar supports, (27304):22
Alternator not charging, (27406):15
Aluminum cans, (70101):9
Aluminum ladders, (75122):15
Aluminum post shores, (27309):38–39
Ambient-light sensor, (27406):12
American Concrete Institute (ACI)
   bar bend designations, (27304):13
   rebar fabrication guidelines, (27304):48
   rebar placement guidelines, (27304):35–36
   standards
      315, *Details and Detailing of Concrete Reinforcement*,
         (27304):13
      318, lap splices, (27304):34
      318-95, *Building Code Requirements for Structural
         Concrete*, (27304):10
American households
   family size, decrease in, (70101):3
   televisions, number of, (70101):6
   weight of Americans, effect of sprawl on, (70101):3
American households environmental impacts
   average production of
      carbon dioxide, (70101):6
      garbage, (70101):6
      heavy metals, (70101):6
      nitrogen oxides, (70101):6
      sulfur dioxide, (70101):6
      wastewater, (70101):6
   average spending on
      electronics, (70101):6
      energy, (70101):6
      food, (70101):6
      water and sewage, (70101):6
   house size, (70101):3
   inventory worksheet, (70101):7–8, 108–109
American National Standards Institute (ANSI)
   ladder duty ratings, (75122):14, 34
   membership, (75122):4
   Mobile Crane Operator Certification accreditation,
      (28301):29
   purpose, (75122):4

American National Standards Institute (ANSI) *(continued)*
   standards
      lanyard construction, (75122):9
      PFAS equipment, (75122):2, 7
      Z49.1, Safety in Welding Cutting, and Allied Processes, (29102):2, 3
American Society for Testing and Materials International (ASTM) standards
   A82, plain welded-wire fabric reinforcement, (27304):26
   A495, deformed welded-wire fabric reinforcement, (27304):26
   A497, deformed welded-wire fabric reinforcement production, (27304):26
   A615, Standard Specification for Deformed and Plain Carbon-Steel Bars for Concrete Reinforcement, (27304):10, 12
   A706, Standard Specification for Low-Alloy Steel Deformed Bars and Plain Bars for Concrete Reinforcement, (27304):10
   A955M, stainless steel reinforcing bars, (27304):12
   A970, headed reinforcing bars, (27304):11
   A996, Standard Specification for Rail-Steel and Axle-Steel Deformed Bars for Concrete Reinforcement, (27304):10
   Standard Metric and Inch-Pound Reinforcing Bars, (27304):10
   steel reinforcement, (27304):1
American Society of Mechanical Engineers (ASME)
   B30.5, Mobile and Locomotive Cranes, (28301):29
   forklift defined by, (22206):1
American Welding Society (AWS)
   EG2.0, Guide for the Training and Qualification of Welding Personnel: Entry-Level Welder, (29102):57
      Module 8-Thermal Cutting Processes, (29102):57
   F4.1, Safe Practices for the Preparation of Containers and Piping for Welding and Cutting, (29102):3
   School Excelling through National Skills Standards Education (SENSE) program, (29102):57
Ammeter gauge, (27406):15
Amp (A), (75121):1
Anchor points, personal fall arrest systems, (28301):6
Anchors
   curtain walls, (28301):21
   function, (28301):19
   panel walls, (28301):19
ANSI. *See* American National Standards Institute (ANSI)
Anti-two blocking devices, (75123):1, 6, 26
Appliances
   energy consumption by specific, (70101):18
   ENERGY STAR, (70101):14, 39
   reducing energy use with, (70101):15–16, 17
Aquifer, (70101):1, 4, 95
Aquifer depletion, (70101):1, 4, 95
Arc blast, (75121):1, 13, 25
Arc burns, (75121):4
Arc fault, (75121):1, 5, 25
Arc fault circuit interrupter (AFCI), (75121):1, 9–10, 25
Arc fault current, (75121):5
Arc flash, (75121):5, 13–14
Arc flash boundary, (75121):1, 12, 13, 25
Arch bridge, (27304):6
Arches, reinforced structural concrete, (27304):5
Architectural concrete, (27308):1, 5, 44
Architectural finishes. *See* Finishes
ASME. *See* American Society of Mechanical Engineers (ASME)
Association of Equipment Manufacturers (AEM), equipment safety manuals, (27406):33
Assured equipment grounding conductor program, (75121):1, 9, 10, 25
ASTM standards. *See* American Society for Testing and Materials International (ASTM) standards

At-risk work areas, (28301):1
Audible signals, (28301):29, 32
Automobiles
   alternative fuel vehicles, (70101):31
   built environment and, (70101):3
   carbon footprint, (70101):14
   carpools, (70101):1, 17, 95
   fuel efficiency, (70101):17–18, 31
Auxiliary light switches, (27406):16
AWS. *See* American Welding Society (AWS)
Axles, prestart inspection checks, (22206):25

B

Backfill, bracing for, (28301):13–14
Backfire, (29102):7, 13, 61
Backhoe
   all-wheel drive/steering, (27406):30, 32
   controls, (27406):30, 32
   function, (27406):30
   inspections, pre-operational, (27406):34
   operating guidelines, (27406):34
   operator qualifications, (27406):30
   operator's maintenance responsibility, (27406):35
   safety precautions, (27406):33
   stabilizer controls, (27406):32
   two-wheel drive/steering, (27406):30
Backhoe bucket, (27406):32, 34
Backhoe loader bucket, (27406):32, 34
Backshoring, (27309):34
Bale handler, (22206):17, 18
Ballasting tires, (22206):25–26
Band, (27304):32, 40, 53
Bar list
   defined, (27304):9, 13, 53
   labels, (27304):19, 21
   rebar, (27304):13, 15–16, 18, 20
   sample, (27304):16, 18, 20
Barrel-shell roof, (27304):5
Barricades, fall protection requirements, (28301):10
Bar supports
   accessories, (27304):24, 25
   all-plastic, (27304):22
   coatings, (27304):19
   identification of, (27304):24
   precast concrete blocks, (27304):22, 24
   purpose, (27304):19
   steel standee, (27304):22, 25
   steel-wire, (27304):19, 22, 23
Baseplate, (27309):1, 49
Batten, (27308):20, 24, 44
Batteries, (70101):35–36, 42, 53
Battery banks, (70101):42
Battery charger connection, (27406):16
Battery-powered platform trucks, (22206):2
Battery safety, (27406):2–3, (70101):42
Battery-voltage warning indicator, (27406):12
B-B Plyform®, (27309):22
BD+C. *See* LEED for Building Design + Construction (BD+C)
Beam, (27304):1, 53
Beam-and-girder forms, (27309):13
Beam and slab forms, one-way and two-way
   manufactured, (27309):17, 18
   site-built, (27309):13–17, 18
Beam bridge, (27304):5–6
Beam cage, (27304):3, 4
Beam forms, (27309):13, 15–16

Beam pockets, framing, (27309):16
Beams
   overhead, fall arrest and, (28301):7
   rebar placement in, (27304):42, 45–48
   types of, (27309):13, 14
Beam table flying decks, (27309):27–31
Bending bars, (27304):16
Bending jib, (27304):29
Bending rebar, (27304):17, 19, 29–31
Bent, (27304):1, 5, 53
Best management practices (BMPs)
   biomimicry, (70101):46–47
   building location, (70101):54
   building orientation, (70101):29
   categories, (70101):27–28
   costs, (70101):46, 49, 52, 54, 63
   defined, (70101):22, 95
   energy
      alternative sources for, (70101):40–43
      avoiding unnecessary use, (70101):37–38
      balancing electrical loads, (70101):40
      batteries, (70101):35–36
      category overview, (70101):27
      generators, (70101):35
      optimizing, (70101):38–39
   indoor air quality (IAQ)
      construction materials impacts, (70101):54
      construction sequencing, (70101):55
      finishes, (70101):55
      flushout period, (70101):56
      furnishings, (70101):54
      HVAC systems, (70101):57–58
      landscaping impacts, (70101):54–55
      maintenance, (70101):55
      preventing problems at the source, (70101):54–55
      segregation, (70101):55–57
      using natural forces, (70101):57–58
   indoor environment
      building design impact on, (70101):54
      category overview, (70101):28
      color, use of, (70101):57, 59
      daylighting, (70101):57
      housekeeping, (70101):55, 56
      humidity levels, (70101):55, 56, 57
      user controls, (70101):58–59
   indoor environmental quality (EQ), (70101):28
   integrated strategies
      category overview, (70101):28
      consider options, (70101):62–63
      counting costs, (70101):64
      integrative design, (70101):61, 62
      solving the right problem, (70101):59–61
      system relationships, (70101):61
      use services not products, (70101):61–62
   landscaping, (70101):27, 29–31, 33, 54
   materials
      category overview, (70101):27
      efficient use of, (70101):43, 47–50
      finding better sources for, (70101):43–44, 50–52
      new, unnecessary use of, (70101):43, 44–47
      smart, (70101):49
      tracking, (70101):50
   site development, (70101):29–31
   site selection, (70101):27, 28
   water
      alternative sources for, (70101):33–34
      category overview, (70101):27
      for landscaping, (70101):33
      optimizing use, (70101):32–33
      reducing use, (70101):31–32
      wastewater sinks, (70101):34
Bevel cutting, (29102):48–50, 52
Bio-based, (70101):22, 44, 95
Bio-based waste, (70101):53
Biodegradable, (70101):22, 44, 95
Biodiversity, (70101):1, 4, 95
Biofuel, (70101):22, 31, 95
Biomass energy, (70101):15
Biomimicry, (70101):22, 46–47, 95
Bioswale, (70101):22, 30, 95
BIPVs. *See* Building-integrated photovoltaics (BIPVs)
Birds vs. squirrels on power lines, (75121):7
Blackout, (70101):22, 40, 95
Blackwater, (70101):22, 34, 95
Blockouts, (27309):30
Blowing tips, (29102):22, 23
BMPs. *See* Best management practices (BMPs)
Body harness
   defined, (75122):1, 34
   highlighted in text, (75122):2
   personal fall-arrest system (PFAS)
      chest straps, (75122):5
      D-rings, (75122):4–5, 8–10
      illustrated, (75122):3
      leg/groin straps, (75122):5
      OSHA standards, (75122):2
      suspension trauma strap, (75122):5–6, 7
      waist/tool belt, (75122):6–7
Bolted fault, (75121):1, 5, 25
Bolted fault current, (75121):5
Bonded posttensioning, (27304):6–7
Boom attachment controls, (22206):11–13
Boom extensions, (22206):18–21, 41, 45
Boom lift, (27406):4, 41, (75122):27, 28, 30–31
Braces
   defined, (27308):1, 44
   function, (27308):3–4
   materials, (27308):3
   wall-form, (27308):3–4
Bracing collar, (27308):20, 44
Bracing concrete masonry walls
   for backfill, (28301):13–14
   building codes, (28301):11, 12
   lateral supports, (28301):11, 13
   safety, (28301):11
   spacing requirements, (28301):11, 12
   standards, (28301):12–13
   vertical columns, (28301):11
   for wind, (28301):11, 12–13, 14
Brakes
   parking brakes
      controls, (22206):13
      control switch, (22206):13
      indicator lights, (27406):12
      manual release, (27406):7
   prestart inspection checks, (22206):25
   service brakes, (22206):13
Breakdown voltage, (75121):1, 4, 25
Breaker, (27406):6
BREEAM. *See* Building Research Establishment Environmental Assessment Method (BREEAM)
Bridge decks, (27309):41–43
Bridges, (27304):5–6
Brooklyn Bridge LED retrofit, (70101):40
Brownfield, (70101):22, 28, 74, 95
Brownout, (70101):22, 40, 95

Buck, (27308):10, 13, 44
Buckets
   backhoe bucket, (27406):32, 34
   backhoe loader bucket, (27406):32, 34
   clamshell buckets, (22206):46–47
   rough-terrain forklift attachments, (22206):22, 45–46
   skid-steer loader buckets, (27406):13
Building automation, (70101):65
Building codes, concrete masonry wall bracing, (28301):11, 12
Building envelope, high-performance, (70101):37, 39, 49, 84
Building-integrated photovoltaics (BIPVs), (70101):22, 47, 95
Building Research Establishment Environmental Assessment Method (BREEAM), (70101):68
Buildings
   locating, (70101):54
   orienting, (70101):29
Built environment. *See also* Construction projects
   automobiles and the, (70101):3
   compact neighborhoods, (70101):3
   environmental impacts, (70101):25, 26–28
   expectations for, (70101):3
   facility life cycle, (70101):25–28
   function, (70101):3, 25
   global energy use and, (70101):4
   indoor air quality (IAQ), (70101):25
   weather impacts on the, (70101):3
   weight of Americans, effect of sprawl on, (70101):3
Built-up scaffolding safety, (75122):23–24
Bulkhead, (27308):17, 44
Bundle of bars, (27304):9, 17, 19, 53
Burns
   electrical, (75121):4–5, 13
   thermal, (75121):1, 5
Byproducts, (70101):22, 43, 95

C
Cab guard, (22206):8
Cable grabs, (75122):12
Caissons
   in bridges, (27304):5
   defined, (27304):1, 53
Caldwell fixed angle telescoping boom Model FB-30, (22206):18–19
California Department of Public Health (CDPH) *Standard Method for the Testing and Evaluation of Volatile Organic Chemical Emissions from Indoor Sources Using Environmental Chambers*, (70101):54
Cam-operated twist lock, (27406):21–22
Cantilevered beams, (27304):2, 4
Cantilevered stairs, (27308):35
Cantilever walls
   maximum length- or height-to-thickness, by type, (28301):24
   maximum spans, by type, (28301):24
   retaining walls, (27304):3, 39
Capacity, (75110):11
Capitals, (27309):4, 49
Caps, cotton, (29102):2
Cap tie, (27304):46
Carabiners, (28301):7, (75122):7
Carbon cycle, (70101):1, 8, 95
Carbon dioxide, (70101):6, 53
Carbon dioxide monitors, (70101):84
Carbon footprint
   Americans vs. others, (70101):12
   calculating, (70101):5–6, 12, 111
   defined, (70101):1, 2, 95
   global climate change and, (70101):8–12
   of heavy equipment use, (70101):29
   per cheeseburger, (70101):20
   reducing, (70101):14
Carbon neutral, (70101):1, 20, 95
Carbon offset, (70101):1, 20, 95
Carburizing flame, (29102):32, 40, 61
Cardboard, (70101):53
Carpenters, green, (70101):102–103
Carpools, (70101):1, 17, 95
Cast-in-place construction, (27304):11
Cavity walls, (28301):23
CAZ. *See* Controlled access zone (CAZ)
CDPH. *See* California Department of Public Health (CDPH)
CDZ. *See* Controlled Decking Zone (CDZ)
Center of gravity (CG)
   combined, (75123):15, 18, 26
   defined, (75123):15, 26
   forklift operation safety and, (75123):17–18
CFCs. *See* Chlorofluorocarbons (CFCs)
CFL. *See* Compact fluorescent lamp (CFL)
CG. *See* Center of gravity (CG)
Change, environmental, (70101):2–3
Charette, (70101):22, 26, 95
Check valves, (29102):15–16, 34
Chemical hazards, battery banks, (70101):42
Chlorofluorocarbons (CFCs), (70101):22, 25, 96
Circuit breaker, (27406):4, 41
Circuit interrupters
   arc fault circuit interrupter (AFCI), (75121):1, 9–10, 25
   ground fault circuit interrupter (GFCI), (28301):2–3, (75121):1, 6, 8–9, 25
Clamps for column forms, (27308):24–26
Clamshell buckets, (22206):46–47
Clamshell cylinder valve protection caps, (29102):11
Cleanouts, (27304):41
Climbing devices fall protection, (75122):11–12
Climbing forms, (27308):27, 29–30, 44
Clips, panel wall, (28301):19
Closed ties, (27304):46
Clothing appropriate for safety, (75110):2–3
Clutch control lever, (27406):24
Coal energy, (70101):15
Cob construction, (70101):22, 50, 96
Color, using indoors, (70101):57, 59
Column forms
   clamps, (27308):24–26
   sections, (27308):20
   supports, (27304):42
   types of
      fiber, (27308):20, 21, 22, 23
      fiberglass, (27308):21, 23
      job-built, (27308):24–25
      oval, (27308):20
      plastic, (27308):22
      plywood, (27308):21
      rectangular, (27308):24
      round, (27308):20, 22
      square, (27308):23, 24
      steel, (27308):20–21, 23–24
Column-mounted tables, (27309):31–33
Columns
   defined, (27304):1, 53
   rebar placement in, (27304):3, 39–44
   vertical, (28301):11
Column spirals, (27304):1, 4, 53
Column ties, (27304):1, 4–5, 40–44, 53
Combination torch, (29102):19, 21

Combined center of gravity, (75123):15, 18, 26
Commissioning, (70101):22, 37, 83–84, 96
Communication
   crane operations, (28301):27, 29–31, 32
   in heavy equipment operations, (75123):7, 9, 10–11
   nonverbal modes of
      audible signals, (28301):29, 32
      hand signals, (28301):29–31, (75123):10–11
      signal flags, (28301):29
   verbal modes of, (28301):27–29
Communication equipment
   background noise, (28301):27–28
   hardwired systems, (28301):28
   high-power, (28301):28–29
   low-power, (28301):27
   microphones
      ear-mounted noise-cancelling, (28301):28
      throat, (28301):28
   radios
      handheld, (28301):28–29
      portable, (28301):27
   walkie-talkies, (28301):27
Compact fluorescent lamp (CFL), (70101):1, 16, 96
Compaction equipment
   flat-plate, (27406):24–25
   function, (27406):23
   inspections, pre-operational, (27406):25
   operating procedures, (27406):25
   operator qualifications, (27406):24
   operator's maintenance responsibility, (27406):25–26
   safety precautions, (27406):25
   soil conditions used for, (27406):23
   types of, (27406):23–24
Compact neighborhoods, (70101):3
Compactor, (27406):4, 41
Composite decking safety, (27309):19
Composite-slab deck forms, (27309):17
Composite slabs, (27309):6–7, 30–31
Compressive strength of concrete, (27304):1
Compressor, (27406):4, 41
Concrete. *See also Specific types of*
   applications, (27304):1
   importance of, (27304):1
   ingredients comprising, (27304):1
   strength of, (27304):1
Concrete bond, (27304):2
Concrete cover, (27304):1, 2, 53
Concrete hoppers, (22206):22, 23
Concrete masonry wall bracing
   for backfill, (28301):13–14
   building codes, (28301):11, 12
   lateral supports, (28301):11, 13
   safety, (28301):11
   spacing requirements, (28301):11, 12
   standards, (28301):12–13
   vertical columns, (28301):11
   for wind, (28301):11, 12–13, 14
Concrete masonry walls, maximum heights, (28301):11
Concrete Reinforcing Steel Institute (CRSI)
   bar bend designations, (27304):13
   defined, (27304):9, 53
Connectors/connecting devices
   defined, (75110):11, (75122):1, 34
   personal fall-arrest system (PFAS), (75122):2–4, 7–8
Conservation, (70101):1, 16, 96
Construction drawings, design section, (27308):37
Construction equipment operation safety. *See also Safety*
   AEM equipment safety manuals, (27406):33
   aerial lifts, (27406):6–7
   air compressors, (27406):20–21
   backhoes, (27406):33
   batteries, (27406):2–3
   compaction equipment, (27406):25
   forklifts, (27406):28
   fueling, (27406):2
   generators, (27406):16–17
   hydraulic systems, (27406):2
   qualified operators
      aerial lifts, (27406):6
      air compressors, (27406):30
      forklifts, (27406):26
      skid-steer loader, (27406):11
   skid-steer loader, (27406):12–13, 14
   transporting equipment, (27406):1
Construction projects. *See also* Built environment
   building rating systems, (70101):67–68. *see also* LEED Rating System
   carbon footprint, (70101):29
   design for disassembly, (70101):50
   energy conservation in, (70101):15–16
   green practices, choosing, (70101):67
   isolation measures, (70101):55–57, 81
   optimal value engineered framing, (70101):48–49
   sequencing in, (70101):55
Construction safety, (70101):39
Construction waste
   amount generated per year, (70101):102
   best management practices (BMPs), (70101):52–53
   donating, (70101):52, 81
   environmental impacts, (70101):25
   LEED documentation requirements, (70101):85–86, 87
   LEED performance goals and requirements for, (70101):80–81
   net zero buildings, (70101):61
   recycling
      best management practices, (70101):52–53
      LEED documentation requirements, (70101):85
      LEED Green Building Certification, (70101):87
      LEED Green Building Rating System, (70101):78
      packaging, (70101):44
   separating, (70101):67, 80, 100
   storage, (70101):52–53, 78, 81
   waste streams
      finding better sinks for, (70101):44, 52–53, 80–81
      impact of, (70101):43
Consumer electronics, average spending on, (70101):6
Consumer Electronics Association, (70101):6
Contact splice, (27304):32, 34–35, 53
Containers, explosion hazards, (75123):5
Containment netting, (28301):7
Continuity testers, (75121):10, 11
Continuous beam, (27304):1, 2, 3, 53
Continuous footings, (27304):4
Control joysticks, (27406):12, 13
Controlled access zone (CAZ), (28301):1, 9–10, 40
Controlled Decking Zone (CDZ), (75110):3–4, 11
Control points, ACI specifications, (27304):35
Coolant temperature indicator, (27406):12
Coping, (28301):22, 23
Coping joints, (28301):23
Corporate Sustainability Report (CSR), (70101):67, 80, 96
Corrugated, (27309):4, 6, 49
Corrugated metal deck panels, (27309):20
Corrugated steel decking, (27309):6–7

Costs
　of best management practices, (70101):46, 49, 52, 54, 63–64
　LEED projects, (70101):81, 86–87
Cotton cap, (29102):2
Couplers, (22206):44–45
Coupling splices, (27304):35
Crab steering, (22206):1, 3, 53
Cracking
　causes of, (27304):2, 6
　controlling, rebar for, (27304):2, 6
Crane operations
　audible signals, (28301):29, 32
　hand signals, (28301):29–31
　high-rise building construction, (28301):17
　nonstandardized signals, (28301):32
　safety guidelines, (28301):26–27
　signal persons, mandated, (28301):32
　used as personnel lifts, (28301):9
Crane operator certification, (28301):29
Cranes, types of
　derrick cranes, (28301):26
　mobile cranes, (28301):26
　tower cranes, (28301):26
　traveling cranes, (28301):26
Crane safety
　basics, (75110):6–7, (75123):20
　power line hazards, (75110):8, (75123):20–22
　site hazards and restrictions, (75123):22–23
　weather hazards, (75123):22
CRSI. See Concrete Reinforcing Steel Institute (CRSI)
CSR. See Corporate Sustainability Report (CSR)
Cube forks, (22206):17, 18
Culvert forms, (27309):43
Cup-type strikers, (29102):25
Curing, (27304):2. See also Finishing
Current
　high- vs. low-voltage, (75121):7–8
　measures of, (75121):1
　typical effects, by current value, (75121):3
Curtain walls, (27304):5, (28301):20
Curved wall forms
　kerfs, (27308):10, 15
　patented, (27308):10, 15
　wood construction, (27308):10, 15
Cut, meaning in elevated masonry, (28301):1, 40
Cut level, stairways above the fourth, (28301):8
Cutting guides, (29102):29
Cutting tips
　flue cutting tips, (29102):22, 23
　riser cutting tips, (29102):22, 23
　rivet cutting tips, (29102):21, 22
Cylinder cart, (29102):24, 26
Cylinders
　explosion prevention, (29102):10
　fuse plugs, (29102):10
　handling and storage safety, (29102):4–5
　markings on, (29102):4, 7–8, 10, 12
　regulations for, (29102):7, 10
　transporting and securing, (29102):32
Cylinder valve protection caps
　clamshell, (29102):11
　function, (29102):5
　high-pressure, (29102):11
　oxygen cylinders, (29102):9
Cylinder valves
　acetylene cylinders, (29102):9–10
　cracking, (29102):31–32
　opening, (29102):36
　oxygen cylinders, (29102):7, 9
　safety plugs, (29102):9

D

Daylighting, (70101):57, 78, 81, 83
Dead load, (27309):4, 49
Death and injury
　electricity related, (75121):1, 3
　falling objects, (28301):7
　falls, (28301):4
Debris inspection check, (22206):24–25
Debris netting, (28301):7
Decks, elevated
　flying decks, (27309):26–33, 36
　formwork
　　composite slabs, (27309):6–7
　　one-way joist slabs, (27309):5–6
　　one-way solid slabs, (27309):4
　　posttensioned concrete slabs, (27309):7
　　posttensioned slabs-on-grade, (27309):8
　　two-way flat plate slabs, (27309):5, 7
　　two-way flat slabs, (27309):4–5
　　two-way joist slabs, (27309):6
　framing, proprietary multicomponent aluminum system, (27309):23–24
　hand-set
　　grading, (27309):36
　　multicomponent, (27309):23–24
　　panelized, (27309):24–25
　outriggers, (27309):25–26
　shoring
　　aluminum posts, (27309):38–39
　　backshoring, (27309):34
　　drop-heads, (27309):38
　　horizontal, adjustable, (27309):38
　　installation, (27309):34–35
　　manufactured, (27309):36–40
　　mudsills, (27309):34
　　preshoring, (27309):34
　　wood, adjustable, (27309):35–36
　surfaces, (27309):22–23
Deconstruction, (70101):22, 26, 52, 96
Deforestation, (70101):1, 4, 43, 96
Deformed reinforcing bars, (27304):11
Deformed welded-wire fabric reinforcement production, (27304):26–27
Dematerialization, (70101):61
Derricks, (28301):17, 26
Desertification, (70101):1, 4, 96
Design drawings, (27308):37
Design for disassembly, (70101):50
Differential lock control, (22206):13–14
Differential lock pedal, (27406):30
Disconnect switches, (22206):9–10
Display select button, (27406):12
Disposal chutes, (28301):34
Dome pans, (27309):6, 10–11
Double-curtain wall, (27304):32, 38, 53
Double-locking snaphooks, (75122):7–8
Doublewythe cavity walls, (28301):23
Dowel cage, (27304):40, 42
Dowels
　applications, (27304):11
　defined, (27304):9, 36, 53
　placement, (27304):40, 43
Downcycled, (70101):45
Downcycling, (70101):23, 46, 96
Drag lines, (29102):46, 61

D-rings, personal fall-arrest system (PFAS), (28301):6, (75122):4–5, 8–10
Drive-speed selector, (27406):6
Drive/steer controller, (27406):4, 6, 41
Drop-head, (27309):22, 24, 38, 49
Drop holes, (28301):34
Drop-panel formwork, (27309):17
Dross, (29102):1, 61
Dust mitigation, (70101):35, 74, 75
Duty rating, (75122):14, 34

E

EA. *See* LEED energy and atmosphere (EA)
Ear-mounted noise-cancelling microphones, (28301):28
Earplugs, (29102):2
EarthCraft, (70101):67, 68, 96
Earth-supported stairs, (27308):34–35
Ecological footprint, (70101):1, 11, 96
Ecosystem, (70101):1, 4, 96
Ecosystem restoration, (70101):31, 74–77
Edge forms, (27309):30
EIA. *See* Energy Information Administration (EIA)
Electrical accessory socket, (27406):12, 13
Electrical faults, (75121):5
Electrical hazards. *See also* Electrical shock
    arc blast, (75121):1, 25
    arc flash, (75121):1, 5, 12, 13–14, 25
    arc flash blast, (75121):13
    battery banks, (70101):42
    burns, (75121):1, 4–5, 13
    death and injury statistics, (75121):1
    electrocution, (28301):2
    elevated masonry work, (28301):2–3
    equipment failure, (75121):5
    extension/power cords, (28301):2, 3, (75121):5–6
    grounding problems, (28301):2–3
    high-voltage, (75121):6–7
    housings and enclosures, (75121):8
    insulation problems, (28301):2
    overhead power lines, (75110):8, (75121):6–7, (75123):20–22
    power tools, (28301):2–3
    shorts, (28301):2
    solar energy, (70101):42
Electrically safe work condition, (75121):10
Electrical safety
    assured equipment grounding conductor program, (75121):1, 9, 10, 25
    boundaries
        approach boundaries, (75121):11–12
        arc flash boundary, (75121):1, 12, 13, 25
        hazard boundaries, (75121):10–15
        shock protection boundaries, (75121):13–14
    circuit interrupters
        arc fault circuit interrupter (AFCI), (75121):1, 9–10, 25
        ground fault circuit interrupter (GFCI), (28301):2–3, (75121):1, 6, 8–9, 25
    grounding
        defined, (75121):25
        equipment grounding conductor, (75121):1, 8, 25
        grounding conductor, (75121):6
        insulation for, (75121):8
        three-prong grounded receptacle, (75121):6–7
    high-voltage equipment, (75121):9
    over-voltage installation categories, (75121):11
    qualified persons and, (75121):1, 11, 13, 29
    tool design for, (28301):2
    training programs, (75121):13

Electrical shock
    basics, (75121):6
    circuit interrupters to prevent, (75121):1, 6, 8–9, 25
    effects on the body, (75121):1–4
    emergency response, (75121):5
    extension cords, (28301):2
    power cords and, (75121):5–6
    on scaffolding, (75122):22
    severity factors, (75121):4
    shock protection boundaries, (75121):13–14
Electrical shock hazard analysis, (75121):11–13
Electricians, green, (70101):103
Electricity
    construction impact on, (70101):25
    current, measures of, (75121):1
    generation and transmission, (75121):1–2
    green power providers, (70101):41
    for lighting, (70101):25
    load balancing, (70101):40
    peak shaving, (70101):24, 40, 78, 98
    power line voltage, (75121):6–7
Electrocution, (28301):2, (75121):8
Electronics, average spending by American households, (70101):6
Elevated loads, pickup and place, (22206):42, 44
Elevated masonry safety
    cranes used as personnel lifts, (28301):9
    debris netting, (28301):7
    electrical hazards, (28301):2–3
    falling objects, (28301):7
    fall protection
        access zones, (28301):1, 9–10, 40
        barricades, (28301):10
        contractor responsibilities, (28301):4, 6–7
        falling objects, (28301):6–8
        OSHA requirements, (28301):4–5, 7, 8, 9
        personal fall arrest systems, (28301):5–6
        personnel lifts, (28301):8–9
        positioning slings and ropes, (28301):4
        at-risk work areas, (28301):1
    fire prevention, (28301):3, 4
    guardrails, (28301):16, 17
    lighting for, (28301):7
    materials hoists, (28301):9
    OSHA standards and regulations, (28301):4–5, 7, 8, 9, 16, 17
    permanent flooring, (28301):16
    personal protective equipment, (28301):1–2, 5–6
    personnel lift towers, (28301):8
    railings, (28301):7
    ramps, (28301):7
    stairways, (28301):8
    stairways replacing ladders, (28301):16–17, 18
    work-area, (28301):1–4, 7–8
Elevated masonry systems
    bracing, (28301):17
    building design, (28301):18–19
    construction sequence
        bays, (28301):16
        fill-ins, (28301):16
        foundations, (28301):16
        permanent flooring, (28301):16
        segments, (28301):16
        skeleton frames, (28301):16
        tier-by-tier, (28301):16
    curtain walls, (28301):20–22
    disposal chutes, (28301):34
    drop holes, (28301):34

Elevated masonry systems (*continued*)
    exterior wall construction
        curtain walls, (28301):20–22
        panel walls, (28301):19–20
        parapet walls, (28301):22–23
    IBC® code requirements, (28301):18
    interior wall construction, (28301):23, 25
    lateral supports, (28301):18–19
    lifting equipment
        cranes, (28301):17
        derricks, (28301):17
    loadbearing capacity, (28301):17
    materials handling and storage, (28301):4, 33–34
    panel walls, (28301):19–20
    parapet walls, (28301):22–23
    reinforcements, (28301):17–19
    temporary supports, (28301):17
    veneer walls, (28301):19
    waste bins, (28301):34
    working platforms, (28301):17
Elevated workstations, (28301):34
Embodied energy, (70101):1, 7, 96
Emergency lowering valve, (27406):6
Emergency removal of lockout/tagout device, (75121):22
Emergency responders
    electrical shock, (75121):5
    falls, rescue and retrieval plans, (75122):10–11
Emergency stop button, (27406):6
Emergency stop switch, (27406):15
Endangered species, (70101):23, 28, 74, 96
End-bearing splices, (27304):35
Energy advisories, (70101):40–41
Energy audits, (70101):38
Energy conservation
    avoiding unnecessary use, (70101):37–38
    batteries, (70101):35–36, 42
    best management practices, (70101):35–43
    building orientation, (70101):29
    carbon footprint reduction, (70101):14
    construction for, (70101):15–16
    fuel alternatives, (70101):22, 31, 95
    generators, (70101):35
    green power, (70101):41
    grid interties for, (70101):23, 42, 97
    lighting and, (70101):16–17, 37–38, 40
    load balancing, (70101):40
    methods of, (70101):14–17
    net zero buildings, (70101):61
    peak shaving, (70101):24, 40, 78, 98
    smart grid technology, (70101):42
Energy consumption
    alternative sources for, (70101):40–43, 78
    average spending on, (70101):6
    built environment impact on, (70101):25, 26, 27
    fossil fuels use, (70101):6, 15, 17–18, 25
    global increases in, (70101):4
    hot water heating, (70101):16
    LEED performance goals and requirements for, (70101):78
    by lighting, (70101):25
    by source, US, (70101):15
Energy efficiency
    defined, (70101):1, 20, 96
    LEED Green Building Rating System, (70101):77
    operation and maintenance for, (70101):38
Energy Information Administration (EIA), (70101):6
Energy-isolating device, (75121):16, 25
Energy source, (75121):16, 25
ENERGY STAR, (70101):1, 16, 96

ENERGY STAR appliances, (70101):14, 39
Engine air-filter restriction indicator, (27406):12
Engine bed, (27406):25
Engine coolant temperature, (27406):15
Engine diagnostic display panel, (27406):20
Engine guard, (27406):25
Engine oil-pressure warning indicator, (27406):12
Engine overspeed indicator, (27406):15
Engine power switch, (27406):20
Engine start switch, (22206):10
Engine tachometer gauge, (27406):15
Environmental Product Declaration (EPD), (70101):67, 80, 96
Environmental Protection Agency (EPA), (70101):6
Environmental Protection Agency (EPA) Construction
    General Permit, (70101):74–77
Environmental Protection Agency (EPA) Water-Sense®
    program, (70101):32
Environmental protections, spill containment and cleanup, (22206):32–33
EPA. *See* Environmental Protection Agency (EPA)
EPD. *See* Environmental Product Declaration (EPD)
EQ. *See* Indoor environmental quality (EQ)
Equipment grounding conductor, (75121):1, 8, 25
Equipment idling, (70101):1, 17, 96
Equipment operator, defined, (75123):1, 26
Ergonomics, (75123):9
Erosion, (70101):1, 5, 96
Erosion controls, (70101):19, 75
Exothermic oxygen lances, (29102):29–30
Explosion prevention, (29102):10
Extension cords, (27406):19, (75121):5–6
Extension cord safety, (28301):2, 3
Extension ladders, (75122):18–19
Exterior walls
    curtain walls, (28301):20–22
    maximum length- or height-to-thickness, by type, (28301):24
    maximum spans, by type, (28301):24
    panel walls, (28301):19–20
    parapet walls, (28301):22–23
Eyebolts, personal fall arrest systems, (28301):6, 7

F

Face shield, (29102):2
Fall arrest systems, (28301):5–6
Fall hazards, (27308):13, (27309):1
Falling/flying objects, (28301):6–8, (75110):4–5, (75122):22
Fall protection. *See also* Personal fall-arrest system (PFAS)
    access zones, (28301):9–10
    barricades, (28301):10
    contractor responsibilities, (28301):4, 6–7
    falling objects, (28301):6–8
    guardrails, (75122):11
    OSHA standards, (28301):4–5, 7, 8, 9, (75122):1, 20
    personnel lifts, (28301):8–9
    positioning slings and ropes, (28301):4
    at-risk work areas, (28301):1
    safe climbing devices, (75122):11–12
    safety net systems, (75122):11, 13
Fall protection devices, (75122):20–21
Fall protection equipment
    common types of, (28301):1
    OSHA requirements, (27309):1
Falls
    causes of, (28301):1
    classes of, (28301):1, (75122):1
    death statistics, (28301):4
    preventing, (75122):2

rescue after, (75122):10–11
from slipping and tripping, (75122):1
surviving, (75122):2
Family size, decrease in, (70101):3
Far face, (27304):32, 37, 53
Farming operations, (22206):17
Federal Highway Administration, (70101):3
Fencing, (70101):75, 76, 87
Ferrous metals, (29102):1, 61
Fiber column forms, (27308):20, 21, 22, 23
Fiberglass column forms, (27308):21, 23
Fiberglass ladders, (75122):14
Fiberglass-reinforced plastic (FRP), (27308):20, 44
Fibrillation, (75121):1, 3, 4, 25
50A circuit breakers, (27406):16
Figure-eight tie, (27304):33
Finishes
  indoor air quality and, (70101):55, 81
  patterned, (27308):39
  smooth, (27308):37
  textured, (27308):37, 38
Finishing, (27308):37. *See also* Curing
Fire, fuel sources for, (75123):5
Fire extinguishers, PASS technique for using, (28301):4
Fire extinguishing systems, (70101):25
Fire hydrants, (28301):4
Fire prevention, (28301):3, 4
Fittings, oxyfuel cutting equipment, (29102):16
Fixed ladders, (75122):20, 22
Fixed-mast, rough terrain forklift, (27406):26, 27
Fixed-mast forklift, (22206):1, 2, 3–6
Flashback, (29102):7, 13, 24, 61
Flashback arrestors, (29102):15–17, 34
Flashing
  curtain walls, (28301):22
  panel walls, (28301):20, 21
  parapet walls, (28301):22–23
  through-wall, (28301):23–24
Flatbed trucks, unloading, (22206):44
Flat-plate compaction equipment, (27406):24–25
Flat slab, (27304):32, 48, 53
Flat-slab/flat-plate forms, (27309):17, 19
Flight (of steps), (27308):31, 44
Flood plain, (70101):23, 28, 74, 96
Floor decks, structural concrete
  composite slabs, (27309):6–7
  one-way joist slabs, (27309):5–6
  one-way solid slabs, (27309):4
  posttensioned concrete slabs, (27309):7
  posttensioned slabs-on-grade, (27309):8
  two-way flat plate slabs, (27309):5, 7
  two-way flat slabs, (27309):4–5
  two-way joist slabs, (27309):6
Flue cutting tips, (29102):22, 23
Fluorescent lamps, (70101):39, 83
Flushout, (70101):23, 56, 96
Flying decks
  applications, (27309):26
  beam tables, (27309):27–31
  blockouts, (27309):30
  column-mounted tables, (27309):31–33
  elevated-slab edge forms, (27309):30
  embedments, (27309):30
  grading, (27309):36
  jointing, (27309):30–31
  safety hazards, (27309):26–27
  truss tables, (27309):27–31

Food
  average spending by American households, (70101):6
  gardening for, (70101):19
Footings, rebar placement in, (27304):3, 35–37, 43
Footwear, protective, (75110):3
Forest fires, (70101):9
Forest Stewardship Council (FSC), (70101):51
Forest Stewardship Council (FSC) Certified, (70101):23, 52, 96
Forest Stewardship Council (FSC) Certified wood products, (70101):51, 78, 88
Fork attachments, (22206):17–18
Forklift
  applications, (75123):15
  assemblies, (27406):26
  controls, (27406):26
  defined, (22206):1, (27406):4, 41
  four-wheel drive, (27406):26–27
  function, (27406):26
  limitations, (75123):16
  operation
    parking, (27406):30
    picking up and placing loads, (22206):41–42, 43, 44, (27406):28, 29
    placing elevated loads, (27406):29
    rigging loads, (27406):29
    traveling with a load, (27406):29
    traveling with long loads, (27406):29
  safety
    basic precautions, (27406):28
    driving and, (75123):15
    load handling, (75123):17–18
    operator responsibility, (75123):15–16
  side-shift devices, (27406):28
  two-wheel drive, (27406):26–27
  types of
    fixed-mast, rough terrain, (27406):26, 27
    telescoping-boom, (27406):27
    typical, (27406):26
Forklift operators
  maintenance responsibility, (27406):29–30, 31
  qualifications, (27406):26
  safety responsibilities, (75123):15–16
  training, (75123):15
Fork trucks, classifications, (22206):1
Form jacks, (27308):29
Form liner, (27308):10, 44
Formwork. *See also* Horizontal formwork; Vertical formwork
  advances in, (27308):1
  custom-made, (27308):21
  failure
    avoiding, (27308):2
    common causes of, (27308):25
  historically, (27308):1
  preparing for, (27308):1–2
Fossil fuels use, (70101):6, 15, 17–18, 25
Foundations, rebar placement in, (27304):35–37, 43
Four-wheel-drive forklift, (22206):3, 7
Four-wheel steering, (22206):3, 53
Frame level indicator, (22206):16–17
Frame-leveling controls, (22206):38
Frame shores, (27309):36, 37
Free fall, (75122):1, 2, 34
Freestanding stairs, (27308):36
Free-wheeling valve, (27406):6
Fresh water percent of all water, (70101):31

Freshwater withdrawals, (70101):25
Friction lighters, (29102):23–24
FRP. *See* Fiberglass-reinforced plastic (FRP)
FSC. *See* Forest Stewardship Council (FSC)
Fuel alternatives, (70101):22, 31, 95
Fuel cylinder wrench, (29102):38
Fuel gas cylinders, (29102):5
Fuel gas cylinder valve leak repair, (29102):38
Fuel gas hoses, (29102):18
Fuel gas regulators, (29102):14–15
Fuel gauge, (22206):14–15, (27406):12, 15
Fueling safety, (27406):2
Fuel sources for fires, (75123):5
Fuel/water separator check, (22206):26
Fugitive emissions, (70101):23, 54, 96
Fulcrum, (22206):30, 39, 53
Furnishings, indoor air quality and, (70101):83
Fuse plugs, (29102):10

G

Gang forms, (27308):7–8, 10, 13
Garbage produced per day, per person, (70101):6
Gardening, (70101):19
Gas cylinders RFID tags, (29102):10
Gas distribution manifolds, (29102):17–18
Gasoline, explosive force of, (75123):13
Gasoline vapors, (75123):13
GBCI. *See* Green Building Certification Institute (GBCI) (USGBC)
General Services Administration, U.S., (70101):68
Generator output-connection lugs, (27406):16
Generators
    defined, (27406):4, 41
    extension cords used with, (27406):19
    function, (27406):14, (70101):35
    grounding requirements, (27406):16, 17
    inspections, pre-operational, (27406):18
    operating procedures, (27406):17–18
    operator's maintenance responsibility, (27406):18–19
    safety precautions, (27406):16–17
    tow-behind
        components, (27406):14
        engine control panel, (27406):14, 17
        generator control panel, (27406):15–17
Geothermal, (70101):1, 20, 96
Geothermal energy, (70101):15, 57–58
GFCI. *See* Ground fault circuit interrupter (GFCI)
Girder forms, (27309):13, 15
Girders
    defined, (27304):1, 2, 53
    rebar placement in, (27304):42, 45–48
Global climate change, (70101):1, 2, 4, 8–12, 96
Gloves, (28301):2, (75110):3
Gouging, (29102):7, 50–51, 61
Gouging tips, (29102):21, 22, 23
Graywater, (70101):23, 33, 36, 78, 96
Green Building Certification Institute (GBCI) (USGBC), (70101):71
Green buildings
    craftworker contributions, (70101):102–105
    elements of, (70101):102
    Prevention through Design (PtD), (70101):39
    University of Florida, (70101):84
    World Trade Center, (70101):42
Green cities
    Austin, TX, (70101):68
    Brooklyn Bridge LED lights, (70101):40
    LEED standard used in, (70101):68
    Santa Monica, (70101):12
    Seattle, (70101):69
Green-e Certified energy providers, (70101):41, 42
Green environment
    environmental challenges, (70101):4–5
    human activities relation to the, (70101):3–5
    seasonal changes, (70101):2–3
    weather patterns, (70101):3
Greenhouse effect
    defined, (70101):1, 8, 96
    LEED Green Building Rating System, (70101):76
    overview, (70101):10
Green power, (70101):78
Green Seal Certified, (70101):23, 54, 97
Grid intertie, (70101):23, 42, 97
Ground fault circuit interrupter (GFCI), (28301):2–3, (75121):1, 6, 8–9, 25
Grounding
    assured equipment grounding conductor program, (75121):1, 9, 10, 25
    defined, (75121):25
    insulation for, (75121):8
    three-prong grounded receptacle, (75121):6–7
Grounding conductor, (75121):1, 6, 8, 25
Grounding generators, (27406):16, 17
Guardrails, (28301):16, 17, (75122):11, 20–21
Guyed derrick, (28301):16, 17, 40

H

Habitat for Humanity ReStores, (70101):52, 81
Habitat restoration, (70101):87
Habitats, avoiding sensitive, (70101):78
Halons, (70101):67, 76, 97
Handheld radios, (28301):28–29
Hand-set decks
    grading, (27309):36
    multicomponent, (27309):23–24
    panelized, (27309):24–25
Hand signals, (22206):33–34, (28301):29–31, (75123):10–11
Hand trucks, (22206):1
Hard hats, (75110):3
Hardscape
    albedo values, (70101):30–31
    defined, (70101):23, 30, 97
    LEED Green Building Rating System, (70101):75
    runoff, (70101):5, 30–31
Haunch, (27309):41, 42, 49
Hazardous materials, (75110):4–5
Hazardous waste, (70101):39, 83
HCFCs. *See* Hydrochlorofluorocarbons (HCFCs)
HDO resins. *See* High-density overlay (HDO) resins
Headed reinforcing bars (HRBs), (27304):11
Headgear, (28301):1, (75110):3
Heads, (27309):6
Health and well-being. *See also* Indoor air quality (IAQ)
    effects of urban sprawl on, (70101):3
    energy savings and, (70101):19
    LEED Green Building Rating System goals, (70101):81–83
Healthcare projects, (70101):80, 83
Health hazards. *See also* Electrical shock
    chemical, (70101):42
    falling/flying objects, (28301):6–8, (75110):4–5, (75122):22
    falls
        causes of, (28301):1
        classes of, (75122):1
        classifications of, (28301):1
        death statistics, (28301):4
        preventing, (75122):2

rescue after, (75122):10–11
scaffolding, (75122):20
from slipping and tripping, (75122):1
surviving, (75122):2
hazardous waste, (70101):39, 83
hydraulic fluid, (27406):2
impalement, (27304):19
indoor air quality (IAQ), (70101):24, 25, 53, 99
noise-induced hearing loss, (75121):13
OSHA identified, (75110):4
respiratory-related, (29102):3, (75110):6
soils, (75123):5
vision-related, (29102):2, (75110):6
Hearing protection, (29102):2
Heart, (75121):4
Heating, ventilation, and air conditioning (HVAC) systems.
　See also Ventilation
best management practices (BMPs), (70101):57–58
geothermal, (70101):57–58
LEED documentation, (70101):86
LEED Green Building Rating System goals, (70101):81
optimizing energy use, (70101):38, 39
Heating, ventilation, and air conditioning (HVAC)
　technicians, green, (70101):104
Heat islands, (70101):67, 75, 76, 97
Heat recovery ventilators (HRVs), (70101):37
Heavy bending bars, (27304):16
Heavy equipment, types of, (75123):2
Heavy equipment operation safety
communication in, (75123):7, 9, 10–11
daily checklist, (75123):6–7, 8
fueling, (75123):12
maintenance, (75123):9, 12
moving equipment, (75123):7
Heavy equipment operators, training and certification, (22206):30–31
Heavy metals, (70101):6
Heavy steel-framed panels, (27308):6
Helmets, tinting requirements, (29102):2
HEPA filters. See High-efficiency particulate arresting (HEPA) filters
Hickey bar, (27304):9, 13, 30, 53
High-density overlay (HDO) resins, (27309):22
High-efficiency particulate arresting (HEPA) filters, (29102):3
High engine-temperature indicator, (27406):15
High-speed travel range indicator, (27406):12
High-top safety shoes, (29102):2
High-voltage electrical hazards, (75121):6–7, 9
Hoists, (28301):9, 32–33
Hooks
forklift attachments, (22206):18–21
reinforcing bars, (27304):9, 13, 16, 53
Horizontal curtain walls, (28301):21
Horizontal elevated slab forms
beam pockets, framing, (27309):16
composite-slab deck forms, (27309):17
flat-slab/flat-plate, (27309):17, 19
I-joist pan, (27309):11–12
noncomposite floor slabs, (27309):20
one- and two-way beam
　manufactured, (27309):17, 18
　site-built, (27309):13–17, 18
pan forms, (27309):10–11
shoring, (27309):9, 35
supporting, (27309):9–10
vertical stiffeners, (27309):13, 16

Horizontal formwork
bridge decks, (27309):41–43
culverts, (27309):43
safety, (27309):1–2
shoring standards, (27309):1–2
tunnels, (27309):44
Horizontal lifelines, personal fall arrest systems, (28301):6
Horn button, (27406):6
Horses, rebar, (27304):32, 41, 53
Hoses
air hoses, connecting, (27406):21–22
color coding, (29102):18
oxyfuel cutting
attaching to the torch, (29102):35
color coding, (29102):18
connecting to regulators, (29102):34–35
function, (29102):18
leak testing, (29102):39
maintenance, (29102):18
Hot water heating, (70101):16, 19, 39
Hour meter, (22206):16, (27406):15
Household impacts inventory worksheet, (70101):7–8, 108–109
Housekeeping practices, (70101):55–56, 81
House sizes, increases in, (70101):3
HRBs. See Headed reinforcing bars (HRBs)
HRVs. See Heat recovery ventilators (HRVs)
Human body
breakdown voltage, (75121):1, 4, 25
electrical shock, typical effects of, (75121):1–4
the heart, (75121):4
the muscles, (75121):4
skin characteristics, (75121):3–4
Humidity levels in indoor environments, (70101):55, 56, 57
Hurricanes, (70101):3
HVAC. See Heating, ventilation, and air conditioning (HVAC) systems
Hybrid vehicle, (70101):1, 17, 97
Hydraulic, (75123):1, 6, 26
Hydraulic jacks, (27304):6, (27308):29
Hydraulic oil-filter warning indicator, (27406):12
Hydraulic oil-temperature warning indicator, (27406):12
Hydraulic systems safety, (27406):2
Hydrochlorofluorocarbons (HCFCs), (70101):67, 76, 97
Hydrologic cycle, (70101):23, 31, 97
Hydropower, (70101):15

I

IAQ. See Indoor air quality (IAQ)
ICF. See Insulating concrete form (ICF)
ID. See Innovation in design (ID)
ID+C. See LEED for Interior Design and Construction (ID+C)
Idling equipment, (70101):1, 17, 96
IEC. See International Electrotechnical Commission (IEC)
Ignition switch, (27406):12, 13, 15
I-joist pan forms, (27309):11–12
Indoor air quality (IAQ)
best management practices, (70101):54–58, 60
construction materials impacts, (70101):54–55
defined, (70101):23, 97
finishes and, (70101):55, 81
flushout period, (70101):56
furnishings and, (70101):54
HVAC systems and, (70101):57–58, 81, 86
landscaping impacts, (70101):54–55
LEED documentation requirements, (70101):87
LEED Green Building Rating System goals, (70101):78, 81–83

Indoor air quality (IAQ) (*continued*)
  maintenance for, (70101):55
  poor, causes of, (70101):25
  segregation, (70101):55–57
  sequencing for, (70101):55
  Sick Building Syndrome, (70101):24, 25, 53, 99
  smoking facilities, (70101):78, 81, 82
  using natural forces, (70101):57–58
  ventilation and, (70101):55–57, 58, 60
Indoor environmental quality (EQ)
  best management practices (BMPs), (70101):28, 52–59
  built environment impact on, (70101):26, 27
  category overview, (70101):28
  climate-controlled buildings, (70101):53
  color and, (70101):57, 59
  daylighting, (70101):57, 78, 81, 83
  landscaping impacts, (70101):48, 54–55
  LEED performance goals and requirements for, (70101):70
  problems, preventing, (70101):54–55
  user controls, (70101):58–59, 81
  using natural forces, (70101):57–58
Industrial Truck Association (ITA), (22206):1
Inert gas cylinders, (29102):5
Infiltration, (70101):23, 37, 97
Ingalls building, (28301):19
Injector torch, (29102):19
Innovation in design (ID), (70101):70
Inserts, (27304):9, 53
Inside corners, (27304):3
Insulating concrete form (ICF), (27308):1, 38, 44, (70101):23, 47, 97
Insulation, (75121):1, 4, 8, 25
Integrative design, (70101):23, 61, 97
Integrative process (IP), (70101):69
Interior support beams, (27304):2
Interior wall construction
  elevated masonry systems, (28301):23, 25
  maximum length- or height-to-thickness, by type, (28301):24
  maximum spans, by type, (28301):24
Interlocking systems, (75122):27–28, 29, 34
Interlock systems safety, (27406):1
International Building Code®
  height to thickness ratio for partition walls, (28301):23, 24
  ladder support requirements, (28301):18
International Electrotechnical Commission (IEC), (75121):11
IP. *See* Integrative process (IP)
I-pattern loader maneuvering, (22206):46, 47
ITA. *See* Industrial Truck Association (ITA)

J

JHA. *See* Job hazard analysis (JHA)
JLGTrak rough-terrain telehandlers, (22206):6
Job hazard analysis (JHA), (27308):13, (27309):1
Joints
  block parapet walls, (28301):23
  coping joints, (28301):23
  pressure-relieving, (28301):19
Joist bands, (27309):5
Joist floors, (27304):5
Joists, rebar placement in, (27304):46–47
Joysticks, rough-terrain forklifts, (22206):11–13
Jump forms, (27308):29–30
Just-in-time delivery, (70101):23, 50, 55, 97

K

Kerfs, (27308):10, 15, (29102):7, 24, 61
Key switch, (27406):6

Kickers, (27309):13, 16, 30

L

Labeling. *See also* Markings
  arc flash equipment, (75121):5
  lockout/tagout, (75121):17, (75123):12
  scaffolding, (75122):24
Ladder hoists, (28301):32–33
Ladders
  aluminum, (75122):14, 15
  defined, (75122):14, 34
  duty rating, (75122):14, 34
  fiberglass, (75122):14
  hoists mounted on, (28301):18
  OSHA requirements, (28301):16–17
  positioning, (75122):16
  replaced by stairs, (28301):16–17
  selecting, (75122):14
  straight, (75122):16–18
  uses for, by type, (75122):14
  wooden, (75122):14
Ladder safety
  basics, (75122):14–16, 21
  extension ladders, (75122):18–19
  fixed ladders, (75122):20
  stepladders, (75122):19–20
  straight ladders, (75122):17–18
Landfills, (70101):23, 53, 98
Landings, (27304):3, (27308):31, 44
Landscaping
  best management practices, (70101):27, 29–31, 33
  ecosystem restoration, (70101):31
  green roofs, (70101):48
  indoor environment impacts, (70101):29, 54–55
  LEED Green Building Rating System, (70101):74–77
  LEED performance goals and requirements for, (70101):78
  water conservation systems, (70101):78
Lanyards
  defined, (75122):1, 34
  highlighted in text, (75122):4
  personal fall arrest systems, (28301):5
  self-retracting (SRL), (75122):1, 10, 34
  shock- and non-shock absorbing, (75122):8–10
Lapped splice, (27304):32, 34, 53
Laser levels, (27309):36, (75110):6
Lateral stresses, (28301):16, 18, 22, 40
LAZ. *See* Limited access zone (LAZ)
Leadership in Energy and Environmental Design (LEED), (70101):67, 97
Leading edge, (75122):1, 34
Leaks, testing and repair
  fuel gas cylinder valve, (29102):38
  oxyfuel cutting equipment
    fuel gas cylinder valve, (29102):38
    hoses, (29102):39
    initial and periodic, (29102):38–39
    leak points, (29102):37–38, 39
    regulators, (29102):39
    torches, (29102):39–40, 41
  rough-terrain forklift prestart inspection, (22206):24–25
Lean construction methods, (70101):50
Leaning Tower of Pisa, (27406):27
LED. *See* Light-emitting diode (LED) lamps
Ledgers, (27309):13, 16
LEED (Leadership in Energy and Environmental Design), (70101):67–68, 68, 97
  standards, (70101):42, 54, 58–69, 71
LEED accreditation, (70101):74

LEED Accredited Professional (LEED AP), (70101):85
LEED Certification
   accreditation vs., (70101):74
   documentation requirements, (70101):85–88
   levels of, (70101):70–71
   outside the US, (70101):69
   potential problems
      lack of coordination with other trades, (70101):88–89
      poor documentation, (70101):87–88
      poor execution of credit requirements, (70101):86–87
      poor planning, (70101):86
   process, (70101):71–73
LEED energy and atmosphere (EA), (70101):69
LEED for Building Design + Construction (BD+C), (70101):69, 70, 106–107
LEED for Building Operations and Maintenance (O+M), (70101):70
LEED for Homes, (70101):68
LEED for Interior Design and Construction (ID+C), (70101):70
LEED for Neighborhood Development (ND), (70101):70, 74
LEED for New Construction, (70101):69
LEED for Schools, (70101):83
LEED Gold standard, (70101):68, 69, 71
LEED location and transportation (LT), (70101):69
LEED materials and resources (MR), (70101):69
LEED Minimum Program Requirements (MPRs), (70101):70
LEED Online documentation system, (70101):71
LEED Pilot Credit Library, (70101):85
LEED Platinum standard, (70101):71
LEED Rating System
   basis for, (70101):68
   defined, (70101):67
   function, (70101):68
   goals
      by category, (70101):69–70
      costs, (70101):86–87
      healthy and productive living environments, (70101):81–83
      promoting sustainable behavior, (70101):77–78
      protecting and restoring the surrounding environment, (70101):74–77
      resource conservation, (70101):78–80
      site selection, (70101):74
   requirements by category, (70101):69–70
   structure, (70101):69–70
   types of, (70101):70–71
LEED Silver standard, (70101):42, 71
Leveling instruments
   frame level indicator, (22206):16–17
   frame-leveling controls, (22206):38
   laser levels, (27309):36, (75110):6
Leveling operations, (22206):37–38
Leverage points, (70101):12–14
Life cycle, (70101):1, 13, 25–28, 97
Life cycle assessment, (70101):2, 13, 77, 97
Life cycle cost, (70101):2, 13, 97
Lifelines, personal fall arrest systems, (28301):5–6, (75122):1, 10, 34
Lift-arm float indicator, (27406):12
Lift attachment controls, (22206):11–13
Lifting devices, stone masonry, (28301):9, 32–33
Lifting eye, (27308):17, 44
Lifting hooks, (22206):18–21, 45
Lifts, aerial. *See* Aerial lifts
Light bending bars, (27304):16
Light-emitting diode (LED) lamps
   Brooklyn Bridge retrofit, (70101):40
   defined, (70101):2, 16, 97
Lighting
   controls, (70101):37–38
   daylighting, (70101):57, 78, 81, 83
   energy used for, (70101):16–17, 25
   health hazards, (70101):39, 83
   retrofits, (70101):38, 40
   for safety, (28301):7
Lightning, (75123):22
Light pollution, (70101):76
Lights and beacons, (22206):13, 14
Lights and beacons control switches, (22206):13, 14
Li-ion. *See* Lithium ion (Li-ion)
Limited access zone (LAZ), (28301):1, 10, 40
Limited approach shock protection boundary, (75121):13
Liquefied fuel gases for oxyfuel cutting, (29102):13–14
Liquefied fuel gas torch tips, (29102):21, 22
Liquefied petroleum (LP) gas, (29102):13
Liquid propane (LP) gas, (75123):5
Lithium ion (Li-ion), (70101):23, 53, 97
Live load, (27309):4, 49
Living Machine®, (70101):34
Loaders, hydraulic system components, (22206):7–8
Loading operations, flatbed trucks, (22206):44
Loads, forklift operations
   elevated loads, (27406):29
   picking up and placing, (22206):41–42, 43, 44, (27406):28, 29
   rigging loads, (27406):29
   traveling with a load, (27406):29
   traveling with long loads, (27406):29
Local materials, (70101):23, 44, 97
Location valuation, (70101):67, 85, 97
Lockout, defined, (75121):16, 25
Lockout devices, (75121):16, 17, 25
Lockout/tagout devices, (75121):17–18, (75123):12
Lockout/tagout emergency removal, (75121):22
Lockout/tagout procedure, (75121):19–21
Longitudinal stability indicator, (22206):16–17
Long pans, (27309):11
Low- and high-speed lift-enable buttons, (27406):6
Low-emission vehicles, (70101):2, 17, 97
Low engine-oil pressure indicator, (27406):15
Low VOC paint, (70101):54
LP gas. *See* Liquefied petroleum (LP) gas; Liquid propane (LP) gas
LT. *See* LEED location and transportation (LT)

M

mA. *See* Milliamp (mA)
Main circuit breaker, (27406):16
Maintainability, (70101):23, 38, 97
Maintenance
   best management practices, (70101):55
   boom lifts, (75122):30–31
   of equipment, defined, (75123):1, 26
   for heavy equipment safety, (75123):9, 12
Man-made environment, (70101):2–5
Markings. *See also* Labeling
   color coding hoses, (29102):18
   cylinders, (29102):4, 7–8, 10, 12, 44
   soapstone markers, (29102):25, 26
Masonry Contractors Association of America standards
   Masonry Wallbracing Design Handbook, (28301):13
   Standard Practice for Bracing Masonry Walls Under Construction, (28301):12–13
Masonry curtain walls, (28301):20
Massing, (70101):23, 29, 97

Materials. *See also* Construction waste
  best management practices, (70101):27, 43–50
  efficient use of, (70101):43, 47–50, 78
  finding better sources for, (70101):43–44, 50–52, 79
  LEED documentation requirements, (70101):85, 87
  LEED Green Building Rating System, (70101):77, 78–82
  manufacturing, (70101):25
  new, unnecessary use of, (70101):43, 44–47, 78
  regional use of, (70101):79, 85
  smart, (70101):49
  storage, (70101):52–53, 55, 78, 81
  tracking, (70101):50
Materials handling
  forklifts
    picking up and placing loads, (22206):41–42, 43, 44, (27406):28, 29
    placing elevated loads, (27406):29
    rigging loads, (27406):29
    traveling with a load, (27406):29
    traveling with long loads, (27406):29
  guidelines, (28301):26
  hoists, (28301):9, 32–33
  moving materials, (28301):33–34
  safety, (28301):26–27, 33–34
  steel erection, (75110):7
  stockpiling, (28301):33–34
  storage, (28301):4
MDO resins. *See* Medium-density overlay (MDO) resins
Mechanical-coupling splices, (27304):34
Mechanical guides, oxyfuel cutting, (29102):25–27
Medium-density overlay (MDO) resins, (27309):22
Menu buttons, (27406):6
Mercury vapor discharge lamps, (70101):83
MERV. *See* Minimum efficiency reporting value (MERV)
Metal copings, parapet walls, (28301):22
Metal decking, (27309):20
Metal washing tips, (29102):22, 23
Microhabitats, (70101):48
Microphones
  ear-mounted noise-cancelling, (28301):28
  throat, (28301):28
Milliamp (mA), (75121):1
Minimum efficiency reporting value (MERV), (70101):23, 56, 98
Mirror adjustment, (22206):14
Mixed waste, (70101):80
MMFX$_2$® rebar, (27304):13
Mobile cranes, (28301):26
Modular materials and systems, (70101):50, 59
Moisture control, (70101):55, 81
Momentum, (75123):1, 4, 26
Monofill, (70101):23, 53, 98
Monolithically, (27309):4, 5, 49
Motor-driven oxyfuel cutting equipment, (29102):27–29
Motorized equipment hazards, (75123):1
MR. *See* LEED materials and resources (MR)
MRPs. *See* LEED Minimum Program Requirements (MPRs)
Mudsills, (27309):1, 34, 49
Multi-function material, (70101):23, 43, 47–48, 98
Multiwythe curtain walls, (28301):20
Muscles, (75121):4

N

Nail-head tie, (27304):33, 39
Nano-material, (70101):23, 49, 98
Narrow pans, (27309):10–11
National Association of Home Builders, (70101):3
National Association of Home Builders' National Green Building Standard, (70101):68
National Electrical Code® (*NEC*®)
  AFCI requirement, (75121):10
  generator grounding requirements, (27406):16, 17
  Section 250.20, (27406):17
  Section 250.34, (27406):17
National Fire Protection Association (NFPA) standards, *70E*®, *Electrical Safety in the Workplace*, (75121):10–15
Natural gas, (29102):13, (70101):15
ND. *See* LEED for Neighborhood Development (ND)
Near face, (27304):32, 37, 53
NEC®. *See National Electrical Code*® (*NEC*®)
Net zero buildings, (70101):61
Neutral flame, (29102):32, 40, 61
NiCad. *See* Nickel cadmium (NiCad)
Nickel cadmium (NiCad), (70101):23, 53, 98
Nitrogen oxides, (70101):6
Noise-induced hearing loss, (75121):13
Noise pollution, (70101):35, 74, 81
Noncomposite floor slabs, (27309):20
Nonrenewable, (70101):23, 41, 98
Nonrenewable energy, (70101):15. *See also specific types of*
Non-shock absorbing lanyards, (75122):8–9
Nonstandardized signals, (28301):32
Nontoxic, (70101):23, 55, 98
Nonverbal communication
  audible signals, (28301):29, 32
  crane operations, (28301):29–32
  hand signals, (28301):29–31, (75123):10–11
  nonstandardized signals, (28301):32
  signal flags, (28301):29

O

O+M. *See* LEED for Building Operations and Maintenance (O+M)
Oblique steering, (22206):1, 3, 53
Occupational Safety and Health Administration (OSHA)
  forklift, term used for and definition of, (22206):1
  major hazards identified by, (75110):4
  manifold system safety precautions, (29102):18
  Standard 29 CFR
    926.550, cranes used as personnel lifts, (28301):9
    1910.16(a)(2)(x), (22206):30
    1910.16(b)(2)(xiv), (22206):30
    1910.178 (l), (22206):30
    1915.120, (22206):30
    1917.1(a)(2)(xiv), (22206):30
    1918.1(b)(10), (22206):30
    1925.701(b), reinforcing steel, guarding of protruding, (27304):16
    1925.703, safety regulations for cast-in-place concrete and formwork, (27308):3
    1926, Subpart Q, Concrete and Masonry Construction, (27308):3
    1926.453, aerial lifts, (75122):28–29
    1926.453, safe use of aerial lifts, (27406):7
    1926.502, Subpart M, Section (d), fall-arrest systems, (75122):2, 4
    1926.602 (d), (22206):33
    1926.703, shoring and reshoring, (27309):1–2
    1926.706(b), concrete masonry wall bracing, (28301):12
    1926.760, (75110):3–4
    1926.852, waste chutes, (28301):34
  training requirements
    crane operator certification, (28301):29
    fall protection training, (75110):4

forklift operator training, (75123):15
powered industrial truck operators, (22206):30
Occupational Safety and Health Administration (OSHA) regulations
  assured equipment grounding conductor program, (75121):10
  debris netting, (28301):7
  electrical shock prevention, (75121):9
  fall-protection, (28301):4–5, 7, 8, 9, (75110):3–4, (75122):1, 20
  high-rise building construction
    guardrails, (28301):16
    stairways replacing ladders, (28301):16–17, 18
  maintenance checklist, (22206):54–56
  OSHA Directive CPL 02-01-028, Powered Industrial Truck Operations, (22206):30
  personal fall-arrest system (PFAS), (27309):1, (75122):2, 4
  ROPS on heavy equipment, (22206):33
  seat belts on heavy equipment, (22206):33
  stockpiling and handling materials, (28301):33–34
  working at elevations, (27308):13
Offgassing, (70101):23, 54, 98
One-Call Notification System, (27406):10
One-piece closed ties, (27304):46
One-piece hand cutting torch, (29102):18–19, 20
120V/240V twist-lock outlets, (27406):16
120V GFCI duplex loads, (27406):16
One-way beam and slab forms
  manufactured, (27309):17, 18
  site-built, (27309):13–17, 18
One-way joist slab, (27309):5–6
One-way joist slab pans, (27309):10, 11
One-way reinforced slab, (27304):47–48
One-way solid slabs, (27309):4
Operating handle, (27406):24
Operator cab
  cab guards, (22206):8
  environmental controls, (22206):8
  mirror adjustments, (22206):14
  purposes, (22206):8
  Rollover Protective Structure (ROPS), (22206):8
  seat adjustment, (22206):10, 14
  seat belts, (22206):33
  steering wheel adjustment, (22206):10
Operator's compartment, (75123):15, 26
Optical pattern-tracing machine, (29102):27
Optimal value engineered (OVE) framing, (70101):48–49
OSHA. *See* Occupational Safety and Health Administration (OSHA)
Outrigger controls, (22206):38
Outriggers, (22206):8, (27309):25–26, (27406):7, (75123):20, 26
Oval column forms, (27308):20
OVE. *See* Optimal value engineered (OVE) framing
Overhang beams, (27304):2
Overhead power lines
  birds vs. squirrels on, (75121):7
  crane operation safety, (75110):8, (75123):20–22
  electrical hazards, (75121):6–7
Over-voltage installation categories, (75121):11
Oxidizing flame, (29102):32, 40, 41, 61
Oxyacetylene flames, (29102):41
Oxyfuel cutting
  bad cuts, causes of, (29102):46–47
  bevel cutting, (29102):48–50, 52
  good cuts, recognizing, (29102):46–47
  gouging, (29102):50–51
  metals of, (29102):1
  procedure
    lay out the cut, (29102):46
    prepare the metal, (29102):46
  process, (29102):1
  safety practices
    cutting/welding containers, (29102):3–4
    cylinder handling and storage, (29102):4–5
    explosion prevention, (29102):3
    fire prevention, (29102):3
    personal protective equipment (PPE), (29102):1–3
    ventilation, (29102):4
  of steel
    piercing a plate, (29102):48, 49
    thick steel, (29102):48
    thin steel, (29102):47–48
  straight-line cutting, (29102):51
  washing, (29102):50
Oxyfuel cutting equipment
  cylinder cart, (29102):24, 26
  cylinders
    exchanging, (29102):44
    explosion prevention, (29102):10
    fuse plugs, (29102):10
    handling and storage safety, (29102):4–5
    marking and tagging, (29102):4, 7–8, 10, 12, 44
    regulations for, (29102):7, 10
    transporting and securing, (29102):32
  cylinder valve, (29102):7, 9–10, 31–32, 36, 38
  cylinder valve protection caps, (29102):5, 9, 11
  disassembling, (29102):43
  exothermic oxygen lances, (29102):29–30
  fittings, (29102):16
  flashback arrestors, (29102):34
  friction lighters, (29102):23–24
  fuel gas cylinder valve, leak repair, (29102):38
  gas distribution manifords, (29102):17–18
  hoses
    attaching to the torch, (29102):35
    color coding, (29102):18
    connecting to regulators, (29102):34–35
    function, (29102):18
    leak testing, (29102):39
    maintenance, (29102):18
    setting up, (29102):35
  leaks, testing and repair
    fuel gas cylinder valve, (29102):38
    hoses, (29102):39
    initial and periodic, (29102):38–39
    leak points, (29102):37–38, 39
    regulators, (29102):39
    torches, (29102):39–40, 41
  mechanical guides, (29102):25–27
  motor-driven, (29102):27–29
  Performance Accreditation Tasks (PATs)
    cutting a shape, (29102):59–60
    set up, igniting, adjusting, shut down, (29102):58
  portable, (29102):37
  regulator adjusting screws, (29102):35
  regulator pressure-adjusting screws, (29102):14
  regulators
    check valves, (29102):15–16
    damage preventions, (29102):15
    flashback arrestors, (29102):15–17
    fuel gas, (29102):14–15
    function, (29102):14–15
    gauges, (29102):14
    leak testing, (29102):39
    oxygen, (29102):14–15

Oxyfuel cutting equipment
   regulators (*continued*)
      setting up, (29102):33–34
      single-stage, (29102):15
      two-stage, (29102):15
   setting up
      attaching hoses to the torch, (29102):35
      attaching regulators, (29102):33–34
      closing torch valves, (29102):35
      connecting cutting attachments, (29102):35
      connecting hoses to regulators, (29102):34–35
      cracking cylinder valves, (29102):31–32
      installing cutting tips, (29102):35
      installing flashback arrestors/check valves, (29102):34
      loosening regulator adjusting screws, (29102):35
      opening cylinder valves, (29102):36
      purging the torch, (29102):36–37
      regulators, (29102):33–34
      setting working pressures, (29102):36–37
      torches, (29102):35
      transporting and securing cylinders, (29102):32
   soapstone markers, (29102):25, 26
   tip cleaners, (29102):24
   tip drills, (29102):24
   torches
      check valves, (29102):15–17
      combination torch, (29102):19, 21
      flashback arrestors, (29102):15–17
      leak testing, (29102):39–40, 41
      one-piece hand cutting torch, (29102):18–19, 20
      setting up, (29102):35
   torch tips, (29102):19, 20, 21–24, 35
   torch wrench, (29102):16–17
   track burners
      bevel cutting, (29102):52
      straight-line cutting, (29102):51
      torch adjustment, (29102):51
   universal torch wrench, (29102):16–17
Oxyfuel cutting gases
   acetylene, (29102):9–10, 12–13
   liquefied fuel gases, (29102):13–14
   oxygen, (29102):7–9, 13
Oxyfuel torch flames
   adjusting, (29102):42–43
   carburizing, (29102):40
   controlling
      backfires, (29102):42
      flashbacks, (29102):42
   neutral, (29102):40
   oxidizing, (29102):41
   oxyacetylene, (29102):41
   oxypropane, (29102):41
   temperatures, (29102):13
Oxygen
   flame temperatures, (29102):13
   hoses, (29102):18
   for oxyfuel cutting, (29102):7–9, 13
   regulators, (29102):14–15
Oxygen cylinders, (29102):5, 7
Oxygen cylinder valves, (29102):9
Oxypropane flames, (29102):41
Ozone-depleting chemicals, (70101):76, 82
Ozone depletion, (70101):2, 5, 98
Ozone hole, (70101):2, 5, 98

P

Packaging, recycling, (70101):44
Paint, impact on indoor air quality, (70101):54

Panel form systems
   all-metal, (27308):4, 6
   heavy steel-framed, (27308):5, 6
   plastic, (27308):5
   plywood
      metal frames and, (27308):5
      unframed, (27308):4
Panelized deck systems
   hand-set, (27309):24–25
   site-built, (27309):25
Panel walls, elevated masonry systems, (28301):19–20
Pan forms, (27309):10–12
Paper, impacts of, (70101):19
Papercrete, (70101):23, 50, 98
Parapet, (28301):22
Parapet walls
   elevated masonry systems, (28301):22–23
      maximum length- or height-to-thickness, by type, (28301):24
      maximum spans, by type, (28301):24
Parking areas, (70101):75
Parking brakes
   controls, (22206):13
   control switch, (22206):13
   indicator lights, (27406):12
   manual release, (27406):7
Partition walls, (28301):23, 24
Passive solar design, (70101):23, 29, 98
Passive survivability, (70101):24, 58, 60, 98
Pathogens, (70101):19, 24, 98
PATs, oxyfuel cutting. *See* Performance Accreditation Tasks (PATs), oxyfuel cutting
Pattern-tracing machines, (29102):27
Payback period, (70101):2, 13, 98
PBT. *See* Persistent, bioaccumulative toxin (PBT)
PCM. *See* Phase change material (PCM)
Peak shaving, (70101):24, 40, 78, 98
Performance Accreditation Tasks (PATs), oxyfuel cutting
   cutting a shape, (29102):59–60
   set up, igniting, adjusting, shut down, (29102):58
Persistent, bioaccumulative toxin (PBT), (70101):67, 83, 98
Personal environmental impacts
   calculating, (70101):9, 13, 110, 111
   carbon footprint, (70101):5–6, 8–9, 11–12
   fuel use, (70101):6, 12–18
   garbage, production and recycling, (70101):6
   product impacts, (70101):7–8
   reducing, (70101):14–18, 20
   wastewater produced per day, (70101):6
   water consumption per day, (70101):6
Personal fall-arrest system (PFAS). *See also* Fall-protection
   anchor points, (75122):2–4
   ANSI standards, (75122):2
   body harness
      chest straps, (75122):5
      D-rings, (75122):4–5, 8–10
      fit, (75122):4–5
      illustrated, (28301):5, (75122):3
      leg/groin straps, (75122):5
      procedure for donning, (75122):6
      snap hooks, (28301):6, 7, (75122):7–8
      suspension trauma strap, (75122):5–6, 7
      waist/tool belt, (75122):6–7
   connecting devices, (75122):2–4, 7–8
   defined, (75122):1, 34
   illustrated, (75122):3
   lanyards
      defined, (75122):1, 34

highlighted in text, (75122):4
    self-retracting (SRL), (75122):1, 34
    shock- and non-shock absorbing, (75122):8–10
lifelines, (28301):5–6, (75122):1, 10, 34
OSHA standards
    anchor points, (75122):2, 4
    body harness, (75122):2
    maximum arresting force, (75122):2
    working at elevations, (27308):13
working load limits, (75122):2
Personal protective equipment (PPE). *See also* Safety
    arc hazards, (75121):5, 13–14
    elevated masonry work, (28301):1–2, 5–6
    job-site safety, (75123):5
    oxyfuel cutting, (29102):1–3
    for welders, (75110):6
Personnel lifts
    fall protection requirements, (28301):8–9
    hoists used as, (28301):32–33
    OSHA standards, (28301):8
Personnel lift towers, (28301):8
Personnel platform, (22206):21–22
Pervious concrete, (70101):24, 30, 98
PFAS. *See* Personal fall-arrest system (PFAS)
Phantom loads, (70101):2, 14, 98
Phase change material (PCM), (70101):24, 49, 98
Phosphatized, (27309):9, 17, 49
Photosensor, (70101):24, 38, 98
Photovoltaic, (70101):2, 20, 98
Photovoltaic panels, (70101):42, 49
Pierce, (29102):27, 48, 61
Pilasters, (28301):11, 23
Pile caps, (27304):32, 36, 53
Pitch, (27304):9, 17, 53
Placing drawings, (27304):9, 35–36, 53
Plain reinforcing bars, (27304):11
Plain welded-wire fabric reinforcement
    common styles, (27304):27
    designations, (27304):25
    roll form, (27304):25, 26
    sizes, (27304):26
    standards, (27304):25
Planting vegetation, (70101):19, 48
Plastic, recycling, (70101):45
Plastic column forms, (27308):22
Plastic panels, (27308):5
Plate cutting machine, (29102):27
Plates, (27308):5
Platform
    defined, (75122):1, 34
    elevated masonry systems, (28301):17
    lockout switch, (22206):22
    personnel, (22206):21–22
Platform controls
    aerial lifts, (27406):6, 7
    scissor lifts, (27406):6
Platform trucks, (22206):2
Platform-up/platform-down buttons, (27406):6
Plumbers, green, (70101):105
Plyform®, (27309):22, 49
Plywood column forms, (27308):21
Plywood deck surfaces, (27309):22–23
Plywood panels
    and metal frames, (27308):5
    steel-framed, (27308):6
    unframed, (27308):4
Pneumatic jack, (27308):29
Poke-outs, (27309):25–26

Pollution
    air pollution, (70101):4–5, 19, 48
    light pollution, (70101):76
Pollution prevention, (70101):24, 46, 87, 98
Population growth, (70101):11
Portable radios, (28301):27
Portable track cutting machines, (29102):27
Positive-pressure torch, (29102):19–20
Post-consumer, (70101):24, 45, 98
Post-consumer recycled content, (70101):46, 51
Post-industrial/pre-consumer, (70101):24, 45, 98
Posttensioned concrete, (27304):6–8
Posttensioned slabs, (27309):7
Posttensioned slabs-on-grade, (27309):8
Posttensioning, function of, (27309):7
Posttensioning tendons, (27309):7
Power cords, (75121):5–6
Power distribution system, (75121):1–2
Powered industrial trucks, (22206):1, 53, (27406):4, 26, 41
Power hop, (22206):24, 25–26, 53
Power lines
    energized, (75123):21–22
    overhead, (75110):8, (75121):6–7, 7, (75123):20–22
Power line safety
    basics, (75123):21–22
    crane operations, (75110):8, (75123):20–22
Power tools
    electrically-powered, hazards of, (28301):2–3
    safety basics, (28301):2–3, (75121):8, 9
    three-prong plugs on, (28301):2, 3
PPE. *See* Personal protective equipment (PPE)
Precast concrete block, (27304):22, 42
Precast concrete block bar supports, (27304):24
Preheat indicator, (27406):12
Preservation, (70101):2, 4, 98
Preshoring, (27309):34
Pressure-relieving joints, (28301):19
Pressure selector switch, (27406):20
Prevention through Design (PtD), (70101):39
Products, personal use impacts, (70101):7–8
Products, personal use impacts inventory worksheet, (70101):9, 110
Profile, (27309):9, 19, 49
Prohibited approach shock protection boundary, (75121):13–14
Propane, (29102):13, (70101):15
P.R.O.P.E.R. lockout/tagout procedure, (75121):21
Proportional control, (27406):4, 8, 41, (75122):27, 34
Propylene fuel gas, (29102):13
PtD. *See* Prevention through Design (PtD)
Pulley cover, (27406):25

Q
Qualified person, (75121):13

R
Radial bending, (27304):17, 19
Radio-frequency identification (RFID), (70101):24, 50, 98
Radio-frequency identification (RFID) tags, gas cylinders, (29102):10
Radios
    handheld, (28301):28–29
    portable, (28301):27
Railings, (28301):7, 16, 17
Rainwater harvesting, (70101):24, 33, 34, 98
Rainwater management, (70101):75
Rammed earth, (70101):24, 50, 98
Ramps, (28301):7

Rapidly renewable, (70101):24, 50, 98
Raw material, (70101):2, 3, 99
Rebar. *See* Reinforcing bar (rebar)
Rectangular column forms, (27308):24
Recyclable, (70101):24, 44, 51, 99
Recycled content, (70101):24, 44–46, 99
Recycled plastic lumber (RPL), (70101):24, 45, 99
Recycled vs. recyclable, (70101):51
Recycling
   for carbon footprint reduction, (70101):14
   construction waste
      best management practices, (70101):52–53
      LEED documentation requirements, (70101):85
      LEED Green Building Certification, (70101):87
      LEED Green Building Rating System, (70101):78
   defined, (70101):2, 99
   for energy reduction, (70101):18–19
   garbage, (70101):6
   packaging, (70101):44
   plastic, (70101):24, 45, 99
   timber, (70101):45
Regional priority (RP), (70101):70
Reglet, (28301):16, 22, 40
Regulator adjusting screws, (29102):35
Regulator pressure-adjusting screws, (29102):14
Regulators, oxyfuel cutting
   attaching hoses to, (29102):34–35
   check valves, (29102):15–16
   damage prevention, (29102):15
   flashback arrestors, (29102):15–17
   fuel gas, (29102):14–15
   function, (29102):14–15
   gauges, (29102):14
   leak testing, (29102):39
   oxygen, (29102):14–15
   single-stage, (29102):15
   two-stage, (29102):15
Reinforced concrete
   defined, (27304):1, 54
   economic value, (27304):3
   materials used, (27304):1
   production value, (27304):4
Reinforced concrete construction
   arches, (27304):5
   bridges, (27304):5–6
   buildings, (27304):5
   concrete cover, (27304):9–10
   curtain walls, (27304):5
   joist floors, (27304):5
   posttensioning in, (27304):6–8
   roofs, (27304):5
   standards, (27304):9
Reinforcement
   bracing concrete masonry walls
      for backfill, (28301):13–14
      building codes, (28301):11, 12
      lateral supports, (28301):11, 13
      safety, (28301):11
      spacing requirements, (28301):11, 12
      standards, (28301):12–13
      vertical columns, (28301):11
      for wind, (28301):11, 12–13, 14
   high-rise building construction, (28301):17
Reinforcing bar (rebar)
   applications, (27304):2
   bar list, (27304):13, 15–16, 18, 20
   bar list labels, (27304):19, 21
   bend designations, (27304):13
   benders, (27304):29
   bending, (27304):17, 19, 29–31
   bending bars, (27304):16
   bends, standardized, (27304):13, 14
   bundling, (27304):9, 17, 19, 53
   cages, (27304):40
   classifications
      heavy bending bars, (27304):16
      light bending bars, (27304):16
      special bending bars, (27304):16
      straight bars, (27304):16
   coatings, (27304):10
   cracking and, (27304):2, 6
   cutters, (27304):29, 30
   cutting, (27304):29, 30
   fabrication, (27304):11–13, 47–48
   fabrication details
      hooks, (27304):15, 16
      radial bending, (27304):17, 19
      spirals, (27304):17
   fabrication tolerances, (27304):17
   grade markings, (27304):13
   grades of, (27304):10, 11
   grouping, (27304):16
   hooks, (27304):13, 15, 16
   horses, (27304):32, 41, 53
   impalement hazard, (27304):19
   manufacturing, (27304):12
   markings, (27304):11, 12, 13, 22
   mats, (27304):36, 38
   placement
      ACI guidelines, (27304):35
      in beams, (27304):2–3, 42, 45–48
      in cantilevered retaining walls, (27304):39
      in columns, (27304):4, 39–42, 43–44
      in footings and foundations, (27304):4, 35–37, 43
      in girders, (27304):42, 45–48
      in pile caps, (27304):36
      placing drawings indicating, (27304):9–10
      in walls, (27304):37–39
   plain, (27304):11
   purpose, (27304):2
   radial bending, (27304):17, 19
   resistance of forces by, (27304):2
   safety, (27304):10, 19
   splicing, (27304):34–35, 37
   stainless steel, (27304):12
   standard configuration for, (27304):11
   standards, (27304):10–11
   standard sizes, identification numbers, (27304):22
   supports
      accessories, (27304):24, 25
      all-plastic, (27304):22
      coatings, (27304):19
      identification of, (27304):24
      precast concrete blocks, (27304):22, 24
      purpose, (27304):19
      steel standee, (27304):22
      steel-wire, (27304):19, 22, 23
   tagging, (27304):17
   tools for working with, (27304):13, 34, 38
   tying, (27304):32–34, 38
   uncoated, (27304):11
Reinforcing materials, (27304):1
Rejecting for energy reduction, (70101):18–19
Relief valves, (29102):10
Renewable, (70101):2, 20, 99

Renewable energy. *See also specific types of*
   consumption by source of, (70101):15
     green power providers, (70101):41
     LEED performance goals and requirements for, (70101):78
   on-site, (70101):41
Rescue and retrieval plans, (75122):10–11
Reshoring, (27309):1, 49
Resource conservation, LEED Ratings, (70101):78–80
Respiratory protection, (29102):3, (75110):6
ReStores, (70101):52, 81
Restricted approach shock protection boundary, (75121):13
Retaining walls, (27304):1, 2, 3, 54
Reusable, (70101):24, 44, 99
Reusing, (70101):18–19, 44
RFID. *See* Radio-frequency identification (RFID)
Ribbed slabs, (27309):5–6
Ribs, (27308):29
Rigging operations, (75110):5
Rigid-frame bridge, (27304):6
Riser cutting tips, (29102):22, 23
Rivet blowing tips, (29102):22, 23
Rivet cutting tips, (29102):21, 22
Rocky Mountain Institute, (70101):34
Rollover Protective Structure (ROPS), (22206):8
Roof decks
   composite slab, (27309):6–7
   one-way joist slab, (27309):5–6
   one-way solid slabs, (27309):4
   posttensioned slabs, (27309):7
   posttensioned slabs-on-grade, (27309):8
   two-way flat plate slabs, (27309):5, 7
   two-way flat slabs, (27309):4–5
   two-way joist slabs, (27309):6
Roofs
   high-albedo, (70101):37
   vegetated, (70101):48
Rope grabs, personal fall arrest systems, (28301):5–6
Ropes, positioning for fall protection, (28301):4
ROPS. *See* Rollover Protective Structure (ROPS)
Rotating-beam laser levels, (27309):36
Rotating telehandlers, (22206):4
Rough-terrain forklift
   attachments
     bale handler, (22206):17, 18
     boom extensions, (22206):18–21, 41, 45
     buckets, (22206):22, 45–46
     clamshell buckets, (22206):46–47
     concrete hoppers, (22206):22, 23
     couplers, (22206):44–45
     cube forks, (22206):17, 18
     forks, (22206):17–18
     lifting hooks, (22206):20–21, 41, 45
     personnel platform, (22206):21–22
   categories
     fixed-mast, (22206):1, 2, 3–6
     telehandlers, (22206):1, 4, 6, 8, 53
   maintenance
     battery check, (22206):27
     boom wear pads, (22206):28
     engine coolant check, (22206):27–28
     50-hour (weekly) checks, (22206):26–27, 55
     long interval, (22206):29, 55–56
     lubrication, (22206):28
     preventive, (22206):24
     record keeping, (22206):29
     service intervals, (22206):24
     10-hour checks, (22206):25, 26, 55
     typical requirements example, (22206):54–56

   transmission, (22206):7, 11
Rough-terrain forklift chassis components
   axles, (22206):1, 3, 7
   engine area, (22206):7
   fixed-mast forklifts, (22206):3–6
   loader hydraulic system components, (22206):7–8
   operator cab
     environmental controls, (22206):8
     mirror adjustment, (22206):14
     purposes, (22206):8
     seat adjustment, (22206):10, 14
     steering wheel adjustment, (22206):10
     typical, (22206):8
   outriggers, (22206):8
   steering, (22206):3
   transmission, (22206):7
Rough-terrain forklift controls
   boom attachment, (22206):11–13
   differential lock, (22206):13–14
   F-N-R transmission, (22206):11, 12
   frame-leveling, (22206):38
   joysticks, (22206):11–13
   lift attachment, (22206):11–13
   mirror adjustment, (22206):14
   for operator comfort, (22206):8, 10, 14
   outriggers, (22206):38
   for safety
     night use, (22206):14
     platform lockout switch, (22206):22
   steering wheel, (22206):11
   steering wheel adjustment, (22206):10
   switches
     disconnect, (22206):9–10
     engine start, (22206):10
     lights and beacons, (22206):13, 14
     parking brake control, (22206):13
     platform lockout, (22206):22
     service brake control, (22206):13
     stabilizers, (22206):13
     steering mode, (22206):13
     transmission neutralizer, (22206):13
Rough-terrain forklift instrumentation
   frame level indicator, (22206):16–17
   fuel level gauge, (22206):14–15
   indicator lights, (22206):16
   longitudinal stability indicator, (22206):16–17
   service hour meter, (22206):16
   speedometer, (22206):16
   tachometer, (22206):16
   temperature gauges
     engine coolant, (22206):15
     hydraulic oil, (22206):15–16
     transmission oil, (22206):15
   warning lights, (22206):16
Rough-terrain forklift operations
   F-N-R transmission controls, (22206):11, 12
   load charts/maximum capacity, (22206):37–41
   maneuvering
     backward, (22206):37
     forward, (22206):36–37
     I-pattern, (22206):46, 47
     steering, (22206):11, 37
     turning, (22206):37
     Y-pattern, (22206):46, 47
   nighttime, (22206):14
   prestart inspection checks
     attachments, (22206):25
     axles, (22206):25

Rough-terrain forklift operations
    prestart inspection checks (*continued*)
        brakes, (22206):25
        cooling system check, (22206):24–25
        debris, (22206):24–25
        differentials, (22206):25
        drive belts, (22206):25
        electrical devices, (22206):25
        engine compartment, (22206):25
        fuel service, (22206):26
        fuel/water separator, (22206):26
        hydraulic system, (22206):25
        leaks, (22206):24–25
        lights, windows, mirrors, (22206):25
        OSHA checklist, (22206):57–58
        ROPS, (22206):25
        safety decals, (22206):25
        tires, (22206):25–26
    prestart preparations, (22206):35
    safety
        co-workers, (22206):31–32
        equipment safety, (22206):32
        operator, (22206):31
        the public, (22206):31–32
        spill containment and cleanup, (22206):32–33
        spotters and hand signals, (22206):33
    shutdown procedure, (22206):36
    stability triangle, (22206):39, 41
    startup procedures, (22206):35
    warm-up procedure, (22206):36
    work activities
        farming, (22206):17
        leveling, (22206):37–38
        lifting, (22206):37–41
        loading, (22206):41–42, 45–47
        loads, picking up and placing, (22206):41–42, 43, 44
        traveling with a load, (22206):42–43
        unloading, (22206):43–44
        in unstable soils, (22206):48
        waste handling, (22206):17
Rough-terrain forklift operators, training and certification, (22206):30–31
Round column forms, (27308):20, 22
RP. *See* Regional priority (RP)
RPL. *See* Recycled plastic lumber (RPL)
Rupture disks, (29102):10
Rustication line, (27308):37, 44

S

Saddle tie, (27304):33
Safe climbing devices, (75122):11–12
Safety. *See also* Electrical safety
    basics, (75122):23, (75123):2, 4
    batteries, (27406):2–3
    clothing, appropriate, (75110):2–3
    composite decking, (27309):19
    concrete masonry wall bracing, (28301):11
    construction equipment operations
        AEM equipment safety manuals, (27406):33
        aerial lifts, (27406):6–7
        air compressors, (27406):20–21, 30
        backhoes, (27406):33
        batteries, (27406):2–3
        compaction equipment, (27406):25
        forklifts, (27406):26, 28
        fueling, (27406):2
        generators, (27406):16–17
        hydraulic systems, (27406):2
        interlock systems, (27406):1
        skid-steer loader, (27406):11, 12–13, 14
        transporting equipment, (27406):1
    cutting/welding containers, (29102):3–4
    cylinder handling and storage, (29102):4–5
    explosion prevention, (29102):3
    extension cords, (28301):2, 3
    falling objects, (28301):7
    fire prevention, (28301):3, 4, (29102):3
    flying decks, (27309):26–27
    forklift
        basic precautions, (27406):28
        driving and, (75123):15
        load handling, (75123):17–18
        operator responsibility, (75123):15–16
    formwork
        horizontal, (27309):1–2
        vertical, (27308):3
    hand signals, (22206):33–34
    hazardous waste, (70101):83
    heavy equipment operation
        communication in, (75123):7, 9, 10–11
        daily checklist, (75123):6–7, 8
        fueling, (75123):12
        maintenance, (75123):9, 12
        moving equipment, (75123):7
    interlocking systems, (27406):1
    job-site, (75123):2, 4
    ladders
        basics, (75122):14–16, 21
        extension ladders, (75122):18–19
        fixed ladders, (75122):20
        stepladders, (75122):19–20
        straight ladders, (75122):17–18
    lifting devices, (28301):33
    lighting for, (28301):7
    manifold systems, (29102):18
    nighttime operations, (22206):14
    oxyfuel cutting
        cutting/welding containers, (29102):3–4
        cylinder handling and storage, (29102):4–5
        explosion prevention, (29102):3
        fire prevention, (29102):3
        personal protective equipment (PPE), (29102):1–3
        ventilation, (29102):4
    power lines
        basics, (75123):21–22
        crane operations, (75110):8, (75123):20–22
    power tools, (28301):2–3, (75121):8, 9
    reinforcing bar (rebar), (27304):10, 19
    reinforcing steel, (27304):10
    rough-terrain forklift operations
        co-workers, (22206):31–32
        equipment safety, (22206):32
        operator, (22206):31
        the public, (22206):31–32
        spill containment and cleanup, (22206):32–33
        spotters and hand signals, (22206):33
    scaffolding
        basics, (75122):23
        built-up scaffolding, (75122):23–24
        fall-protection devices, (75122):20–21
        suspended scaffolding, (75122):24–25
        swing scaffolding, (75122):24–25
    skid-steer loader, (27406):12–13, 14
    spill containment and cleanup, (22206):32–33
    spotters, (22206):33

steel erection
    accident prevention guidelines, (75110):1–2
    clothing, appropriate, (75110):2–3
    controlled decking zones (CDZs), (75110):3–4
    cranes, working with, (75110):6–7, 8
    falling/flying objects, (75110):4–5
    fall protection, (75110):3
    hazardous materials, (75110):4–5
    health hazards, (75110):4
    lifting and material handling, (75110):7
    rigging operations, (75110):5
    tools, (75110):5–6
    warning signs, (75110):4
    welding, (75110):6
ventilation, (29102):4
wall form construction, (27308):13
welding, (75110):6
Safety bar, (27406):7, 12, 13
Safety boundaries
    approach boundaries, (75121):11–12
    arc flash boundary, (75121):1, 12, 13, 25
    crane operations, (75123):21
    hazard boundaries, (75121):10–15
    power line clearances, (75123):21
    shock protection boundaries, (75121):13–14
Safety-circuit bypass, (27406):15
Safety glasses, (29102):2
Safety interlock, (27406):1, 41
Safety nets, (28301):1
Safety net systems, (75122):11, 13
Safety shoes, (29102):2
Salt water percent of all water, (70101):31
Salvaged, (70101):24, 43, 99
Salvaged buildings and materials, (70101):44–45
Santa Monica, (70101):12
SAR. *See* Supplied-air respirators (SAR)
Scaffolding
    accidents associated with, (75122):14
    defined, (75122):14, 34
    hazards
        electric shock, (75122):22
        falling/flying objects, (75122):22
        falls, (75122):20
    labeling, (75122):24
    types of
        built-up scaffolding, (75122):23–24
        suspended scaffolding, (75122):24–25
        swing scaffolding, (75122):24–25
        two-point swing scaffolding, (75122):14, 25, 34
    typical, (75122):22
Scaffolding safety
    basics, (75122):23
    built-up scaffolding, (75122):23–24
    fall-protection devices, (75122):20–21
        guardrails, (75122):20–21
        OSHA requirements, (75122):20
    suspended scaffolding, (75122):24–25
    swing scaffolding, (75122):24–25
Schedule, (27304):32, 36, 54
School Excelling through National Skills Standards Education (SENSE) program (AWS), (29102):57
Scissor lifts, (27406):4–6, 41, (75122):27, 28, 29–30
Scissor lift tires, (27406):4
Screw jacks, (27308):29
Sealant
    coping joints, (28301):22
    panel wall flashing, (28301):21
Sea levels, effects of rising, (70101):8, 10

Seat adjustment, (22206):10
Seat belts, (22206):33, (27406):12
Sedimentation, (70101):67, 75, 99
Segregation
    best management practices (BMPs), (70101):55–57
    LEED Green Building Rating System goals, (70101):81
Seismic stirrup/tie, (27304):15
Self-retracting lanyard (SRL), (75122):1, 10, 34
SENSE. *See* School Excelling through National Skills Standards Education (SENSE) program (AWS)
Sequencing, (70101):55
Service brake control, (22206):13
Service hour meter, (22206):16
Shade number guide, (29102):2
Sheathing, (27308):1, 5, 44
Sheet metal cutting tip, (29102):21, 22
Shelf angles, (28301):19
Shell roofs, (27304):5
Shiplap, (27308):37, 44
Shock absorbers, personal fall arrest systems, (28301):7, (75122):9–10
Shock hazard. *See* Electrical shock
Shock protection boundaries, (75121):13–14
Shoes, protective, (28301):1
Shooting-boom forklifts, (22206):6
Shoring
    concrete floors, (27309):9
    defined, (27309):1, 49
    elevated decks
        adjustable wood, (27309):35–36
        backshoring, (27309):34
        installation, (27309):34–35
        manufactured, (27309):36–40
        mudsills, (27309):34
        preshoring, (27309):34
    horizontal elevated slab forms, (27309):35
    manufactured
        aluminum post, (27309):38–39
        frame shores, (27309):36, 37
        steel post, (27309):36–38
    wood, adjustable, (27309):38
Shorts, causes of, (28301):2
Showers, (70101):14
Shrinkage, (27304):2
Sick Building Syndrome, (70101):24, 25, 99
Signage
    voltage hazards, (75121):9
    warning signs, (75110):4
Signal flags, (28301):29
Signal persons, mandated, (28301):32
Signals, nonverbal
    audible, (28301):29, 32
    hand, (28301):29–31
    nonstandardized, (28301):32
Simple beam, (27304):1, 2, 54
Single-curtain wall, (27304):32, 38, 54
Single-stage regulators, (29102):15
Single-wythe cavity walls, (28301):23
Single-wythe curtain walls, (28301):20
Sinks and sink fixtures, (70101):32
SIP. *See* Structural insulated panel (SIP)
Site development, (70101):29–31, 87
Site selection, (70101):27, 28, 74–77
Skid-steer loader
    assemblies, (27406):11
    buckets, (27406):13
    components, (27406):10
    controls, (27406):12–13

Skid-steer loader (*continued*)
    defined, (27406):4, 41
    function, (27406):10
    operating procedures, (27406):13–14
    operator qualifications, (27406):11
    operator's maintenance responsibility, (27406):14, 15
    safety, (27406):12–13, 14
Skin, human, (75121):3–4
SkyTrak rough-terrain telehandlers, (22206):6, 8
Slab-on-grade, posttensioned, (27309):8
Slab pan forms, (27309):10–12
Slab reinforcement, (27304):47–48
Sleeve, (27304):9, 54
Slings and ropes, positioning for fall protection, (28301):4
Slip form, (27308):1, 44
Slipping and tripping. *See* Walking and working surface hazards
SMACNA, *Indoor Air Quality Guidelines for Occupied Building Under Construction*, (70101):82
Smart grid technology, (70101):42
Smart material, (70101):24, 49, 99
Smoking facilities, (70101):78, 81, 82
Snap hooks, personal fall arrest systems, (28301):6, 7, (75122):7–8
Snap ties, (27304):32–33
Soapstone, (29102):1, 4, 61
Soapstone markers, (29102):25, 26
Society of Automotive Engineers, (22206):24
Softscape, (70101):24, 30, 99
Soil, operations in unstable, (22206):48
Soil contamination and depletion, (70101):5
Soil erosion control, (70101):19
Solar, (70101):2, 5, 99
Solar energy
    building-integrated photovoltaics (BIPVs), (70101):47–48
    electrical hazards, (70101):42
    energy consumption percentage using, (70101):15
    passive design, (70101):23, 29, 98
    roofs, high-albedo, (70101):37
Solid waste, (70101):24, 25, 99. *See also* Construction waste
Solvent-based, (70101):24, 54, 99
Spaced-lap splice, (27304):34–35
Span, (27304):1, 2, 54
Spandrel-beam form, (27309):16
Species extinction, (70101):4, 9
Speedometer, (22206):16
Spill containment and cleanup, (22206):32–33
Spirals, (27304):17, 42
Splices, types of
    contact, (27304):32, 34–35, 53
    coupling, (27304):35
    end-bearing, (27304):35
    lapped, (27304):32, 34, 53
    mechanical-coupling, (27304):34
    spaced-lap, (27304):34–35
    staggered, (27304):32, 40, 54
    welded, (27304):34
Splicing reinforcing steel, (27304):34–35, 37
Spoil pile, (70101):67, 75, 99
Sprawl, (70101):2, 3, 99
Spreaders
    defined, (27308):1, 44
    function, (27308):3
    wall-form, (27308):3
Spread footings, (27304):3
Spud wrench, (27308):17, 44
Square column forms, (27308):23, 24
Square pans, (27309):10

Squirt booms, (22206):6
SRL. *See* Self-retracting lanyard (SRL)
SS. *See* Sustainable sites (SS)
SSRBs. *See* Stainless steel reinforcing bars (SSRBs)
Stabilizer controls, (22206):13
Staggered splices, (27304):32, 40, 54
Stainless steel reinforcing bars (SSRBs), (27304):12
Stair forms, (27308):31–36
Stairways, (27304):3, 5, (28301):8, 16–17, 18
Stakes, (27308):3–4
Start switch, (27406):15
Static electricity, (75123):13
Steel, oxyfuel cutting
    piercing a plate, (29102):48
    thick steel, (29102):48
    thin steel, (29102):47–48
Steel-beam forms, (27309):17
Steel column forms, (27308):20–21, 23–24
Steel deck-form material, (27309):17
Steel erection, applications, (75110):1
Steel erection safety
    accident prevention guidelines, (75110):1–2
    clothing, appropriate, (75110):2–3
    controlled decking zones (CDZs), (75110):3–4
    cranes, working with, (75110):6–7, 8
    falling/flying objects, (75110):4–5
    fall protection, (75110):3
    hazardous materials, (75110):4–5
    health hazards, (75110):4
    lifting and material handling, (75110):7
    rigging operations, (75110):5
    tools, (75110):5–6
    warning signs, (75110):4
    welding, (75110):6
Steel-framed panels, (27308):5
Steel post shores, (27309):36–38
Steel standee bar supports, (27304):22, 25
Steel-wire bar supports, (27304):19, 22, 23
Steering, types of
    crab steering, (22206):1, 3, 53
    four-wheel steering, (22206):3, 53
    oblique steering, (22206):1, 3, 53
Steering mode control switch, (22206):13
Steering wheel, (22206):11
Steering wheel adjustment, (22206):10
Stepladders, (75122):19–20
Stirrup
    defined, (27304):1, 54, (27309):4, 49
    function, (27304):2
    in joists, (27309):5
    placement, (27304):3
Stirrup hooks, (27304):15
Stone masonry lifting devices, (28301):9, 32–33
Stormwater runoff
    controls, (70101):30–31, 48
    defined, (70101):2, 99
    LEED documentation, (70101):86
    LEED Green Building Rating System, (70101):75
    water pollution from, (70101):5
Straight bars, (27304):16
Straight ladders, (75122):16–18
Straight-line cutting, (29102):51
Straight-line cutting guide, (29102):29
Strawbale construction, (70101):24, 51, 99
Strength
    compressive, (27304):1
    of concrete, (27304):1
    tensile, (27304):2

Stringers, (27309):1, 2, 49
Strips, (27304):32, 47, 54
Strongback
   defined, (27308):1, 44
   function, (27308):2, 3, 4
   wall-form, (27308):3, 4
Structural deck, (27309):17
Structural insulated panel (SIP), (70101):24, 44, 47, 99
Sulfur dioxide, (70101):6
Supplied-air respirators (SAR), (29102):3
Support bars, (27304):32, 45, 54
Suspended scaffolding, (27308):29, (75122):24–25
Suspended stairs, (27308):33–34
Sustainable behavior, promoting, (70101):77
Sustainable sites (SS), (70101):69
Sustainably harvested, (70101):24, 43, 50, 99
Swing scaffolding, (75122):24–25
Switches
   disconnect, (22206):9–10
   engine start, (22206):10
   lights and beacons, (22206):13, 14
   parking brake control, (22206):13
   platform lockout, (22206):22
   service brake control, (22206):13
   stabilizers, (22206):13
   steering mode control, (22206):13
   transmission neutralizer, (22206):13
Synthetic-fiber reinforcement, (27304):28

T

Tachometer, (22206):16
Tagging reinforcing bars, (27304):17
Tagout, (75121):16, 25
Tagout device, (75121):16, 18, 25. *See also* Lockout/tagout devices
Takeback, (70101):24, 44, 99
Tankless water heaters, (70101):39
Telehandlers
   applications, (22206):6
   boom, (22206):8
   defined, (22206):1, 53
   example, (22206):2
   outriggers, (22206):8
   rotating type, (22206):4
   squirt boom, (22206):6
Telescoping-boom forklift, (22206):1, 6–7, (27406):27
Televisions, number of, (70101):6
Temperature bars, (27304):32, 47, 54
Temperature gauges
   engine coolant, (22206):15
   hydraulic oil, (22206):15–16
   transmission oil, (22206):15
Temperature steel, (27304):2
Template, (27304):32, 40, 54
Tendons, (27304):6, 7
Tensile strength, (27304):2
Termite mounds, (70101):46–47
Thermal bridging, (70101):67, 89, 99
Thermal burns, (75121):1, 5
Thermal comfort, (70101):81, 84
Thermal mass, (70101):24, 37, 99
Thermostats, (70101):14, 15, 37, 39
Three-prong grounded receptacle, (75121):6–7
Three-wheeled forklifts, (22206):3
Throat microphones, (28301):28
Throttle knob, (27406):12, 13
Throttle lever, (27406):24
Throttle pedal, (27406):12, 13

Through-wall flashing, (28301):23
Thumb rocker switch, (27406):6
Tie-off points, personal fall arrest systems, (28301):6
Ties
   defined, (27304):9, 54
   function, (27308):3
   light bending bars category, (27304):16
   wall-form, (27308):3
Tie wire, (27304):9, 10, 54
Timber. *See also* Trees
   deforestation, (70101):1, 4, 96
   efficient use of in framing, (70101):48–49
   percent of wood harvest used in construction, (70101):25, 43
   reusing, (70101):45
Tines, (22206):1, 53
Tip cleaners, (29102):24
Tip drills, (29102):24
Tires
   inspection checks, (22206):25–26
   liquid ballasting, (22206):25–26
   power hop, (22206):24, 25–26, 53
Tobacco smoke, (70101):78, 81, 82
Toilets, (70101):6, 32–33
Torches and attachments, oxyfuel cutting
   attaching hoses to, (29102):35
   check valves, (29102):15–17
   closing valves, (29102):35
   combination torch, (29102):19, 21
   flashback arrestors, (29102):15–17
   igniting the torch, (29102):42
   leak testing, (29102):39–40, 41
   one-piece hand cutting torch, (29102):18–19, 20
   purging the torch, (29102):36–37
   shutting off the torch, (29102):43
Torch tips, (29102):19, 20, 21–24, 35
Tow-behind air compressors, (27406):20
Tow-behind generators
   components, (27406):14
   engine control panel, (27406):14, 17
   generator control panel, (27406):15–17
Tower cranes, (28301):26
Track burners
   bevel cutting, (29102):52
   operating, (29102):27–29, 51
   straight-line cutting, (29102):51
Training and certification
   fall protection, (75110):4
   forklift operators, (75123):15
   powered industrial truck operators, (22206):30
   rough-terrain forklift operators, (22206):30–31
Transmission neutralizer switch, (22206):13
Transmission oil temperature gauges, (22206):15
Transportation
   alternative, landscape features promoting, (70101):31, 32
   low-impact, (70101):69, 77
Transportation accessibility, (70101):74
Traveling cranes, (28301):26
Trees. *See also* Timber
   copy paper produced per, (70101):19
   indoor environment, impacts on, (70101):29, 54–55
   protective fencing for, (70101):75, 76
Tripods, (27309):37
True recycling, (70101):45
Truss table flying decks, (27309):27–31
Tunnel forms, (27309):44
20A circuit breakers, (27406):16
Twisted-wire strand, (27304):7

Twist-lock, cam-operated, (27406):21–22
Twist-lock outlet, (27406):4, 16, 41
Two-point swing scaffolding, (75122):14, 25, 34
Two-stage regulators, (29102):15
Two-way beam and slab forms
   manufactured, (27309):17, 18
   site-built, (27309):13–17, 18
Two-way flat plate slabs, (27309):5, 7
Two-way flat slabs, (27309):4–5
Two-way joist slab pans, (27309):10
Two-way joist slabs, (27309):6
Two-way reinforced slab, (27304):47–48
Two-wheel-drive forklift, (22206):1, 3, 7

U

UFADs. *See* Underfloor air distribution systems (UFADs)
Unbonded posttensioning, (27304):7–8
Underfloor air distribution systems (UFADs), (70101):58, 60
Underground utilities, locating, (27406):10
Universal torch wrench, (29102):16–17
Unstable soil, operations in, (22206):48
Uranium, (70101):15
Urbanization, (70101):2, 4, 99
Urban sprawl, (70101):3
Urea formaldehyde, (70101):25, 54, 100
Urinals, waterless, (70101):32
User control of indoor environments, (70101):58–59, 81
USGBC®. *See* US Green Building Council (USGBC)®
US Green Building Council (USGBC)®, (70101):50, 71, 88. *See also* Leadership in Energy and Environmental Design (LEED)

V

Vapor-resistant, (70101):25, 37, 100
Veneer walls, elevated masonry systems, (28301):19
Ventilation, (29102):4. *See also* Heating, ventilation, and air conditioning (HVAC) systems
   of indoor environments, best management practices, (70101):55–58, 60
   LEED Green Building Rating System goals, (70101):82, 84
   natural systems, (70101):57
   reduced rates of, (70101):53
   underfloor air distribution systems (UFADs), (70101):58
Verbal modes of communication, (28301):27–29
Vertical formwork
   architectural forms, (27308):37
   climbing forms, (27308):29–30
   custom-made, (27308):21
   erecting, (27308):17–18
   failure, avoiding, (27308):2
   finishes
      patterned finishes, (27308):39
      smooth finishes, (27308):37
      textured finishes, (27308):37, 38
   gang forms, (27308):7–8, 10
   insulating concrete (ICF), (27308):38
   maintenance, (27308):18
   safety, (27308):3
   stair forms, (27308):31–36
   stretching, avoiding, (27308):2
Vertical lifelines, personal fall arrest systems, (28301):5, 6
Vertical panel form systems
   all-metal, (27308):4, 6
   heavy steel-framed, (27308):5, 6
   light-weight, (27308):10, 11
   plastic, (27308):5
   plywood
      metal frames and, (27308):5
      unframed, (27308):4
Vertical slip-forms
   assembly, (27308):28
   climbing forms vs., (27308):29
   components
      form jacks, (27308):29
      form panels, (27308):27–28
      ribs, (27308):29
      suspended scaffold, (27308):29
      walers, (27308):29
      working deck, (27308):29
      yoke assembly, (27308):29
   process, (27308):27
Vestil Manufacturing Corporation extension boom, (22206):20, 21
Vibratory plate, (27406):25
Virgin material, (70101):25, 44–45, 100
VOC paint. *See* Volatile organic compound (VOC) paint
VOCs. *See* Volatile organic compounds (VOCs)
Volatile organic compound (VOC) paint, (70101):54
Volatile organic compounds (VOCs)
   defined, (70101):25, 100
   LEED Green Building Rating System goals, (70101):81, 82
   LEED performance goals and requirements for, (70101):87

W

Waffle slabs, (27309):6
Walers
   curved, (27308):16
   defined, (27308):1, 44
   function, (27308):3, 4
   vertical slip-form construction, (27308):29
   wall-form, (27308):1, 2, 3, 4
Walkie-talkies, (28301):27
Walking, amount done by Americans, (70101):3
Walking and working surface hazards
   extension/power cords, (75121):5–6
   falls resulting, (75122):1
Walk-off mat, (70101):25, 55, 100
Wall-form construction
   parts and accessories
      braces, (27308):3–4
      plates, (27308):5
      spreaders, (27308):3
      stakes, (27308):3–4
      strongback, (27308):3, 4
      ties, (27308):3
      walers, (27308):1, 2, 3, 4
   rebar placement in, (27304):37–39
   safety precautions, (27308):12
   systems, (27308):1–2
Wall-form systems, patented
   curved wall forms, (27308):10, 15
   framing wall openings, (27308):13, 16
   hardware, (27308):10, 14, 15
   heavy-duty, (27308):10, 14
   inspection windows in, (27308):13
Wall openings, (27308):10, 13, 16
Walls
   maximum length- or height-to-thickness, by type, (28301):24
   maximum spans, by type, (28301):24
Warm-up controls, (27406):20
Warning lights, (22206):16

Washing
    defined, (29102):7, 61
    oxyfuel cutting, (29102):50
Washing tips, (29102):21, 22, 23
Waste. *See also* Construction waste
    bio-based, (70101):53
    hazardous, (70101):39, 83
Waste bins, (28301):34
Waste heat recovery, (70101):37
Waste separation, (70101):67, 80, 100
Waste sinks, (70101):44, 52–53, 80–81
Wastewater
    best management practices (BMPs), (70101):34
    built environment impact on, (70101):27
    produced per day, per person, (70101):6
Wastewater sinks, (70101):34
Wastewater treatment systems, (70101):5, 34
Water
    alternative sources for, (70101):33–34, 78
    category overview, (70101):27
    decrease in available, (70101):4
    fresh water percent of all water, (70101):31
    net zero buildings, (70101):61
Water-based, (70101):25, 54, 100
Water conservation
    best management practices (BMPs), (70101):31–33
    graywater systems, (70101):23, 33, 36, 78, 96
    LEED Green Building Rating System goals, (70101):84
    LEED performance goals and requirements for, (70101):78
    outdoor systems, (70101):24, 33, 34, 78, 98
    rainwater harvesting, (70101):24, 33, 34, 98
Water consumption
    aquifer depletion, (70101):1, 4, 95
    built environment impact on, (70101):25, 26, 27
    freshwater withdrawals, (70101):25
    LEED Green Building Rating System, (70101):77–78
    LEED Green Building Rating System goals, (70101):84
    per day, per person, (70101):6
    reducing energy use for, (70101):16
Water efficiency
    defined, (70101):2, 11, 100
    LEED Green Building Rating System, (70101):77
    LEED performance goals and requirements for, (70101):69, 81
Water footprint, (70101):2, 11, 100
Water penetration, parapet walls, (28301):22
Water pollution, (70101):4–5
Waterproof, (70101):25, 54, 100
Water recirculation systems, (70101):63
Water-resistant, (70101):25, 100
Water-Sense® program (EPA), (70101):32
Water streams, protecting, (70101):74
Weather hazards, crane safety, (75123):22
Weather patterns, (70101):3, 9

Weepholes, (27304):32, 39, 54, (28301):20
Weight of Americans, effect of sprawl on, (70101):3
Welded splices, (27304):34
Welded-wire fabric reinforcement
    applications, (27304):24
    deformed-wire
        designations, (27304):26
        sizes, (27304):26, 27
    designations, (27304):26
    forms available, (27304):24–25
    plain-wire
        common styles, (27304):27
        designations, (27304):25
        roll form, (27304):25, 26
        sizes, (27304):26
        standards, (27304):25
    standard sizes, (27304):25
Welding safety, (75110):6
Wetlands
    defined, (70101):25, 28, 100
    LEED Green Building Rating System, (70101):74
    plant-based constructed, (70101):34
Whip check, (27406):4, 21, 41
Wide-module joist slabs, (27309):6
Wide pans, (27309):10–11
Wind
    bracing for, (28301):11, 12–13, 14, 19
    lateral stresses, (28301):22
Wind energy, (70101):15
Wind farms, (70101):43
Wind loads
    anchorage and, (28301):19
    transferring, (28301):20
Windows and window blinds, (70101):49, 58, 83
Wind scoops, (70101):57
Wooden ladders, (75122):14
Wood shoring, (27309):35–36, 38
Work-area safety, (28301):1–4, 7–8
Work-lights switch, (27406):12, 13
World Trade Center, (70101):42
Wrap-and-saddle tie, (27304):33
Wrap-and-snap tie, (27304):33

X
Xeriscaping, (70101):25, 30, 100

Y
Yoke assembly, (27308):27, 29, 44
Y-pattern loader maneuvering, (22206):46, 47

Z
Zoning, (70101):25, 30, 100